AF829923

Body Oxygen Homeostasis and Mitochondrial Function

Avraham Mayevsky • Efrat Barbiro-Michaely
Michael Tolmasov • Mira M. Mandelbaum-Livnat

Body Oxygen Homeostasis and Mitochondrial Function

From Animal Models to Clinical Applications

Springer

Avraham Mayevsky
Faculty of Life Sciences
Bar-Ilan University
Ramat-Gan, Israel

Efrat Barbiro-Michaely
Faculty of Life Sciences
Bar-Ilan University
Ramat-Gan, Israel

Michael Tolmasov
Faculty of Life Sciences
Bar-Ilan University
Ramat-Gan, Israel

Mira M. Mandelbaum-Livnat
Faculty of Life Sciences
Bar-Ilan University
Ramat-Gan, Israel

ISBN 978-3-031-94112-2 ISBN 978-3-031-94113-9 (eBook)
https://doi.org/10.1007/978-3-031-94113-9

© The Editor(s) (if applicable) and The Author(s), under exclusive license to Springer Nature Switzerland AG 2025

This work is subject to copyright. All rights are solely and exclusively licensed by the Publisher, whether the whole or part of the material is concerned, specifically the rights of translation, reprinting, reuse of illustrations, recitation, broadcasting, reproduction on microfilms or in any other physical way, and transmission or information storage and retrieval, electronic adaptation, computer software, or by similar or dissimilar methodology now known or hereafter developed.
The use of general descriptive names, registered names, trademarks, service marks, etc. in this publication does not imply, even in the absence of a specific statement, that such names are exempt from the relevant protective laws and regulations and therefore free for general use.
The publisher, the authors and the editors are safe to assume that the advice and information in this book are believed to be true and accurate at the date of publication. Neither the publisher nor the authors or the editors give a warranty, expressed or implied, with respect to the material contained herein or for any errors or omissions that may have been made. The publisher remains neutral with regard to jurisdictional claims in published maps and institutional affiliations.

This Springer imprint is published by the registered company Springer Nature Switzerland AG
The registered company address is: Gewerbestrasse 11, 6330 Cham, Switzerland

If disposing of this product, please recycle the paper.

Preface by Avraham Mayevsky

This book tells the real story and achievements of my research activities since my postdoctoral training with Prof. Britton Chance at the University of Pennsylvania, Philadelphia, during the years 1972–1974. In this book, I am integrating the knowledge accumulated in my laboratory, related to the function of the intracellular organelle—the mitochondrion—under in vivo conditions.

Normal mitochondrial function is an essential factor in all tissues and organs in the body. Mitochondrial activity provides all cells in the body the energy source (ATP) in order to enable normal functions. Mitochondrial dysfunction will create a major pathological state in various organs due to a decrease in an energy source (ATP). Much pathology in the human body is related to dysfunction of the mitochondria and decrease in ATP production.

One of the options to evaluate mitochondrial function in vivo is to measure the autofluorescence of the Nicotine-amide Adenine Dinucleotide (NADH) molecule bound to the enzymatic system in the respiratory chain embedded in the mitochondria. This approach to monitor the mitochondria in vivo was started in the early 1950s by the late Prof. Britton Chance from the University of Pennsylvania, Philadelphia. More than 800 papers were published in this area by various investigators around the world.

Most of the studies involved experimental animals, using various protocols, mimicking the pathological state in patients. In the last years, the same methodology was applied to monitor patients under various pathological states.

During the years, I realized that many young graduate students or other scientists, that try to enter this specific field of research, have difficulties to find the appropriate source of information in order to perform a well-controlled study using the monitoring of NADH fluorescence in vivo. For this reason, I decided to integrate all the relevant information in a set of books that are available to all new comers to this specific field in research of mitochondrial function in vivo. This is the first time that this subject is described in detail citing almost all of the papers published in this field.

The first general book was published in 2015 by Springer. In the last 3 years (2022–2024) I published three new specific books also by Springer. The current new book is summarizing the subject of Body Oxygen Homeostasis in experimental animals as well as in patients.

The first chapter is a sort of general introduction to the significance of mitochondrial function in health and disease as well as its part in tissue energy metabolism. Chapter 2 describes the basic concept of brain energy metabolism and its monitoring in the brain and in other organs in the body. The metabolism of oxygen in the brain and in less vital organs is presented in this chapter. In Chaps. 3 and 4 the comparison between most vital organ (the Brain) and less vital organ (Small intestine) is presented. Chapter 5 describes the effects of hemorrhage and resuscitation on the most vital and less vital organs in the rat model. In Chaps. 6 and 7 the transfer of the developed technology from experimental animals to clinical use is presented. In the last chapter, Chap. 8, a short discussion and future perspectives are presented.

Ramat-Gan, Israel Avraham Mayevsky

Preface by Michael Tolmasov

My life has never followed a straight line—and I never wanted it to.

I found myself drawn to the human brain long before I fully understood its complexity. My early years in a Moscow medical school were filled with ambition, but like many who immigrate, I had to leave one dream behind to pursue another in a new country, in a new language, with no guarantees—only determination.

Israel became my home, and science became my anchor.

For years, I have been monitoring mitochondria, blood flow, and neuronal responses in living systems. Through every mouse model, every preclinical trial, and every tissue slice under the microscope, I searched not only for answers but for meaning. The brain, with all its beauty and fragility, taught me not only how humans survive but how we decide, feel, and heal.

I have worn many hats in my journey—scientist, consultant, entrepreneur, mentor. I led complex animal trials at institutions like the Weizmann Institute and Bar-Ilan University. I contributed to early-stage pharmaceutical research and worked on life-saving technologies in cancer and neuroscience. I took part in the development of monitoring systems that provided early detection of various critical conditions. Each project, each publication, was another chapter in an ongoing dialogue with the unknown.

But behind the technical reports and scientific posters was a deeper truth: science is not just about data. It is about perseverance. It is about having the humility to learn and the courage to innovate.

Today, as a founder and CEO of a fintech company, I bridge the gap between brain science and financial behavior. What began as academic curiosity about dopamine has become a bold fintech vision. Because science, when applied with imagination, can change lives in unexpected places—from hospitals to boardrooms.

If life has taught me anything, it is that the brain is never done learning. And neither are we.

Ramat-Gan, Israel — Michael Tolmasov

Preface by Mira M. Mandelbaum-Livnat

My academic path started at Bar-Ilan University, Ramat Gan, Israel, where I earned my MSc and PhD in Biotechnology under the supervision of Professor Avraham Mayevsky. My thesis focused on the influence of hemorrhage and hemorrhagic shock on the vitality of brain and small intestinal tissue in rats.

My academic foundations enabled me to continue my journey in the field of life sciences, expanding my professional expertise. Throughout the years, I was involved in clinical trials in the field of pulmonology at Meir Medical Center in Kfar Saba, and oncology in Rabin Medical Center, Petah Tikva. I also served as the Director of the Research Center for Nerve Reconstruction at Tel Aviv Sourasky Medical Center, developing an innovative, proprietary biodegradable hydrogel designed to promote nerve regeneration for treating nerve injuries in the peripheral and central nervous systems. During this position I build scientific collaborations with study groups and industries in Israel and worldwide.

Building on my extensive experience in research and clinical project management, I have taken on the role of Vice President (VP) at Forum MD, a company specializing in medical expert opinions across all legal fields related to malpractice and personal injuries.

This position allows me to integrate my scientific background with expertise I gained along my professional path. My path was paved thanks to my advisor, Professor Avraham Mayevsky, who laid the foundations for my professional career and instilled in me the importance of critical thinking. I am truly privileged to be part of this publication proudly initiated by Professor Avraham Mayevsky, which will surely contribute to a deeper understanding of body oxygen homeostasis and mitochondrial function—subjects to which he had dedicated his academic career.

Ramat-Gan, Israel — Mira M. Mandelbaum-Livnat

Acknowledgments

Acknowledgments by Avraham Mayevsky

This book is dedicated to my dearest high school mathematics teacher Dr. Shimshon Novick from Kefar Haroeh in Israel, for his vision, patient, and support during a critical period of my life. Without his help and encouragement, my academic achievements and writing this book were impossible to be accomplished.

Also, I dedicate this book to my dearest wife, Zipora and four children, Amotz, Sarit, Hagit, and Shalev that understood my devotion to the unusual career and enabled me to fulfill all of my obligations and academic activities.

My deep appreciation is given to my parents, Shabtai and Rachel Mayevsky, for their devotion and support throughout my life span.

The significant contribution of Dr. Eliahu Pewzner and Dr. Assaf Deutsch to the development of the monitoring devices is very much appreciated.

My appreciation is given to, the late Prof. Britton Chance, an outstanding scientist, teacher, and colleague for almost four decades of fruitful collaboration.

During the 60 years of research activities at Bar Ilan University I trained many MSc and PhD students. I want to thank all of them for their confidence and the efforts that they did during the experimental work as well as in preparation and publication of the manuscripts.

Special appreciation is given to a few of my graduate students that joined my research team after their graduation. Their assistance in training new under graduate and graduate students is appreciated very much. Nili Zarchin, MSc, joined my laboratory in the early 1980s. Judith Sonn, PhD, started her activities in my laboratory in the early 1990s. Efrat Barbiro-Michaeli, PhD, started her activities in the mid-1990s and collaborated with me until my nomination as Professor Emeritus at Bar Ilan University more than 10 years ago.

Prof. Gennady Rogatsky immigrated to Israel in the early 1990s and was active in my laboratory until his retirement. He was involved in research activities and helped me to establish the literature classification for this book. Special appreciation is given to his activities.

I want to thank Avrille Goldreich, from the Faculty of Life Sciences at Bar Ian University, for her many years of help and support in the preparation of my manuscripts and especially this book.

Acknowledgments by Michael Tolmasov

None of this would have been possible without the unwavering support and extraordinary influence of those who believed in me from the beginning.

My parents, Rachel and Izhak Tolmasov, whose strength, sacrifice, and relentless belief in education formed the very foundation upon which my journey was built. They gave me not only the tools to pursue my dreams but the courage to do so against all odds. Their love was the quiet force behind every discovery, every challenge overcome, and every milestone achieved.

To my PhD advisor, Professor Avraham Mayevsky, I owe more than I can express. His visionary guidance during my doctoral studies transformed how I think, how I observe, and how I create. He taught me that science is not just a method—it is a mindset. A discipline of precision, a practice of humility, and a gateway to innovation.

And above all, I thank the heart of my life—my wife, Julia Brainin Tolmasov. Her love, brilliance, and unwavering belief in me have been the light through every dark tunnel, the calm through every storm. She is not only my partner but my truest inspiration, grounding me with grace and walking beside me with courage through every chapter.

To my beloved children, Maya and Yonatan—you are my legacy and my future. Your wonder, laughter, and love remind me daily why we strive to make the world a better place. You have given my work purpose, my life color, and my dreams wings.

Acknowledgments by Mira M. Mandelbaum-Livnat

I dedicate this work to my dear parents, Zvi and Shoshana Mandelbaum, whose love, sacrifices, and enduring support have shaped me into the person I have grown to become. I am profoundly grateful for their unwavering devotion and guidance throughout my life.

My deepest love and appreciation to my husband, Dr. Amir Livnat, for being a life companion and for his endless encouragement, patience, and belief in me. To my children, Uri, Yonatan, and Shira, I thank you for being my constant source of joy and happiness. Your curiosity provides a constant inspiration for me.

I would also like to express my heartfelt gratitude to my siblings, Hana, Yehudit, and Yechezkel and their families, for their love and support, and being a significant part of my life.

I extend my deepest gratitude to my advisor, Prof. Avraham Mayevsky, for his invaluable guidance and mentoring throughout my MSc and PhD thesis. His expertise, dedication, and encouragement have been instrumental in shaping my academic journey. It was also through Prof. Mayevsky's laboratory that I met my husband, Amir, who was also his MSc and PhD student. I am forever grateful to Prof. Mayevsky not only for his academic guidance but also for being the catalyst for the meeting that changed my life.

Additionally, I would like to thank all my colleagues in the lab over the years for their companion, collaboration, and support, which enriched my research experience and made this journey even more memorable.

Competing Interests The authors have no competing interests to declare that are relevant to the content of this manuscript.

Contents

Part I Introduction

Part I
Introduction

Chapter 1
General Introduction and Historical Background

1.1 Introduction to Oxygen in the Human Body

The functional capacity of any tissue is related to its ability to perform its work. It is possible to assess this ability through the knowledge of changes in the oxygen balance, i.e., the ratio of oxygen supply to oxygen demand in the tissue.

Figure 1.1 shows the gradient of oxygen between room air through the lungs, the large arteries, and small arterioles to the intracellular space and finally the mitochondrion. Since 90–95% of the total body consumed oxygen is utilized in the mitochondria, it is important to understand its gradient from the atmospheric air (the source) to the site of utilization in the mitochondria. In Fig. 1.1, the decreased level of oxygen is illustrated in the main sites between air and the mitochondria. In insert A, the association/dissociation curve of hemoglobin (Hb/HbO_2) is presented. Under normal conditions, the blood collected from the various organs of the body is reoxygenated in the alveoli of the lungs that are surrounded by blood capillaries. The oxygenated arterial blood ($HbO_2 = 97–98\%$ saturation) reached the left ventricle of the heart and distributed to all organs of the body via the arterial macro circulation part of the blood system. The systemic oxygenated blood is also shown in insert A Part 1. It is important to note that the release or download of oxygen from oxygenated hemoglobin is possible only by a passive diffusion process dependent on the oxygen gradient across the blood vessels and the extracellular and intracellular spaces. Therefore, most of the oxygen will be provided to the cellular compartment of the various organs via the microcirculation and not the macrocirculation as seen in the lower right part of Fig. 1.2. The level of HbO_2 saturation in the microcirculation of the various organs is dependent on the blood flow to the organ and the rate of oxygen consumption by the specific tissue. For example, in skeletal muscles at the resting state, the saturation (around 80%) is much higher (Fig. 1.1A3) as compared to the resting brain (40–50%) seen in point 2 in the insert A. The reason is that the "resting" brain consumes much more oxygen as compared to the muscle and the download of oxygen from the HbO_2 is much higher.

© The Author(s), under exclusive license to Springer Nature Switzerland AG 2025

A. Mayevsky et al., *Body Oxygen Homeostasis and Mitochondrial Function*,
https://doi.org/10.1007/978-3-031-94113-9_1

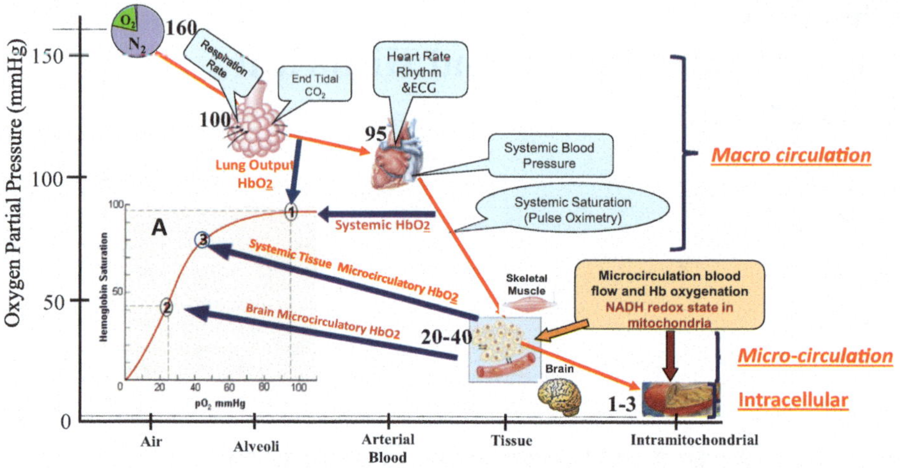

Fig. 1.1 Oxygen gradient from air to the mitochondria. Clinical monitoring of patients include the various parameters along the oxygen gradient. (Mayevsky and Barbiro-Michaely 2013b; Mayevsky et al. 2013) [Modified]

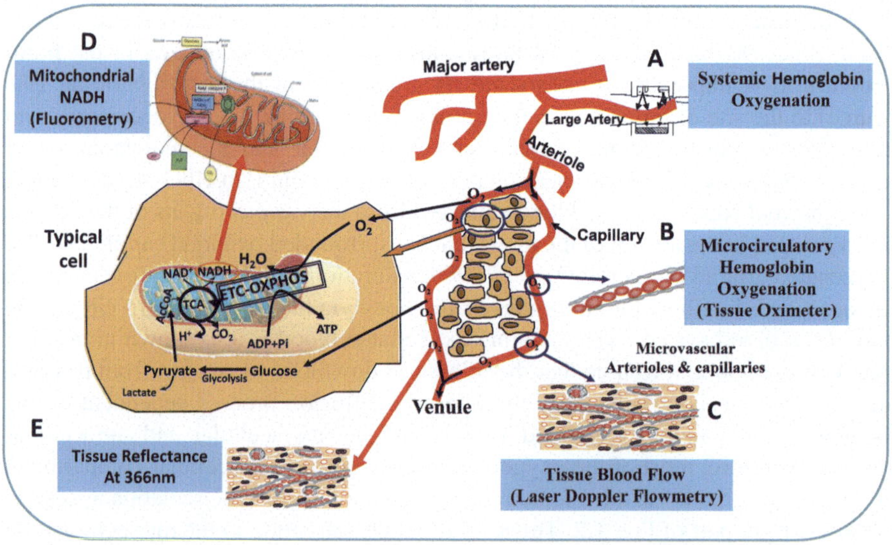

Fig. 1.2 The available techniques for real-time monitoring of energy metabolism at the tissue level. Part (**a**) shows the coupling between the macrocirculation monitored by the Pulse oximeter and the microcirculation. (**b**–**f**) monitoring of cellular and intracellular compartments (see text for details). (Mayevsky and Barbiro-Michaely 2013b; Mayevsky et al. 2013) [Modified]

The main approaches to monitor oxygen metabolism under *in vivo* conditions and in real time are to measure oxygen levels in the tissue, and the second way is to measure hemoglobin saturation. In this scheme, the various points of patients' monitoring are presented. As can be seen, the last usual parameter, in the oxygen gradient, is the pulse oximeter that measures the saturation of hemoglobin in the relatively large arteries. As of today, monitoring of the microcirculation and especially mitochondrial function is not a standard approach and clinical tool.

Figure 1.2d presents the involvement of the mitochondria in cellular and tissue energy metabolism. Substrates and O_2 are supplied and regulated by the blood in the microcirculation, namely from the very small arterioles and the capillary bed. The main function of the mitochondria is to convert the potential energy stored in various substrates and their metabolites (e.g., glucose) into adenosine triphosphate (ATP). The inner membrane of the mitochondria contains five complexes of integral membrane proteins, including NADH dehydrogenase (Complex 1). Three of those proteins are involved in the respiratory chain activity. The main function of the respiratory chain is to gradually transfer electrons from NADH and $FADH_2$ (originating from the Krebs cycle) to oxygen (O_2). With the addition of protons (H^+), water molecules (H_2O) are generated in Complex 4. NADH is a substrate or a coenzyme for the enzymatic activity of dehydrogenases that form part of the respiratory chain and reside in the inner membrane of the mitochondria. Further details on the biochemical properties of NADH can be found in publications by Chance and his collaborators.

The formation of ATP depends on the sufficiency of substrate (i.e., glucose) and oxygen supply to the tissue by the blood flow and hemoglobin oxygenation in the microcirculation (Fig. 1.2) as well as on the integrity of mitochondrial activity. Without a sufficient supply of ATP, cells cannot function properly and can, ultimately, die. Since most of the energy consumed by tissues is dependent upon the availability of oxygen, under normal mitochondrial function, there is high coupling between oxygen supply and energy production. When mitochondrial activity is damaged, due to various reasons, this coupling is disrupted. The electron transfer (oxidation/reduction) down the respiratory chain results in the production of ATP. Concomitantly, with the electron transport, the respiratory chain components switch between reduced and oxidized states, each of which has different spectroscopic properties. Hydrolysis of the pyrophosphate bonds provides the energy necessary for the cell's work. In order to assess the energy demand, it is necessary to measure different organ-specific parameters. For example, in the brain, the energy demand can be evaluated by measuring the extracellular levels of K^+ that reflect the activity of the major ATP consumer—Na^+/K^+ATPase (Mayevsky and Chance 1982; Mayevsky 1984). In the heart, most of the energy is consumed by muscle contraction activity. On the other hand, the energy supply mechanism is the same for all tissues: oxygenated blood reaching the capillary bed releases O_2 that diffuses into the cells. Therefore, it is possible to evaluate tissue energy supply by monitoring the following four parameters in all tissues: tissue blood flow (TBF), tissue blood volume (TBV), tissue oxyhemoglobin (HbO_2), and mitochondrial NADH redox state (See Fig. 1.2).

The production of available energy (ATP) depends on pO_2 (partial oxygen pressure) in the various compartments of the tissue. Information regarding pO_2 in the tissue therefore is helpful for the evaluation of tissue metabolic activity. Oxygen levels can be measured by oxygen electrodes; however, the information collected is an average of the compartments around the electrode. In the past, the sensitivity and accuracy of oxygen electrodes in the range of 1 mmHg (intracellular level) were not sufficient for the evaluation of mitochondrial function. As of today, the newly developed techniques could provide more accurate values of oxygen. Other recently published studies have suggested that intracellular oxygen levels are in a higher range (Springett and Swartz 2007; Wilson 2008; Pittman 2011; Harms et al. 2012; Schober and Schwarte 2012).

The need for an intracellular pO_2 indicator, as a physiological and biochemical parameter of living tissue, has emerged more than 60 years ago. Mitochondria are the intracellular organelles that consume most of the oxygen. Therefore, the redox state of electron carriers in isolated mitochondria *in vitro* as well as *in vivo* as a function of oxygen concentration has been extensively studied. Chance et al. concluded that "For a system at equilibrium, NADH is at the extreme low potential end of the chain, and this may be the oxygen indicator of choice in mitochondria and tissue as well" (1973). Lubbers in 1995 concluded that "the most important intrinsic luminescence indicator is NADH, an enzyme of which the reaction is connected with tissue respiration and energy metabolism" (1995).

As of today, the ability to measure tissue energy metabolism at the microcirculation and cellular level is not available for clinical applications, although a device was developed.

1.2 The Discovery of Oxygen

In 2022, J.B. West reviewed the history of atmospheric oxygen as follows: "Many of us think a lot about oxygen. This includes how the normal body handles oxygen in health, but particularly how this is complicated by lung disease. Few of us are aware that as human inhabitants of the earth, we have a unique privilege. This is that, as air breathers, we and most other animals on Earth are the only living creatures in the known universe that have an unlimited supply of oxygen. This situation came about through one of the greatest miracles of nature, that is, photosynthesis, the ability to release oxygen from water using the energy of sunlight. One consequence of this was that the first atmospheric oxygen came from the metabolism of microorganisms, the cyanobacteria, that used photosynthesis, but for which oxygen was an unwanted by-product. In fact, the oxygen had to be discarded for the organisms to thrive. When a major increase of oxygen concentration in the atmosphere occurred some 2 billion years ago, and the partial pressure of oxygen in the air rose to perhaps 200 mmHg, this Great Oxidation Event, as it was called, was a death sentence for the large population of anaerobic animals for whom oxygen was toxic. Today much of the oxygen in the atmosphere comes from photosynthesis in microorganisms, including

Main Contributors to Oxygen Discovery	Discovery year
1. Ibn al Nafis (1213-1288).	1250
2. Leonardo da Vinci (1452-1519).	1515
3. Michael Servetus (1511-1553).	1552
4. Michael Sendivogius (1566- 1636).	1603
5. John Mayow (1641-1679).	1668
6. Carl Wilhelm Scheele (1742-1786).	1771
7. Joseph Priestley (1733-1804).	1774
8. Antoine Laurent Lavoisier (1743-1794).	1775
9. Henry Cavendish (1731-1810).	1783

Fig. 1.3 The list and pictures of the nine scientists that were involved in and contributed to the discovery of oxygen. (Severinghaus 2016a, b) [Modified]

the cyanobacteria, and the recently discovered *Prochlorococcus*, which discard this unwanted by-product. The result is that the PO_2 in our atmosphere at sea level remains nearly constant at about 150 mmHg, although the factors responsible for this are not understood."

The discovery of oxygen is a process that started more than 750 years ago (about 1250) and finished after around 500 years later in 1780. There are at least nine names that are mentioned as related to this discovery and are presented in Fig. 1.3. The first five names contribute indirect and minor part to the discovery process while the last four scientists added the main information to the process of oxygen discovery within a 10-year period.

1.2.1 *Ibn al-Nafis*

The first one is Ibn al-Nafis, an Arab physician that was active in the thirteenth century (1213–1288). He was the first scientist that described the pulmonary circulation (Haddad and Khairallah 1936; Loukas et al. 2008; West 2008). West (2008) summarized the discovery as follows: "Ibn al-Nafis (1213–1288) was an Arab physician who made several important contributions to the early knowledge of the pulmonary circulation. He was the first person to challenge the long-held contention of the Galen School that blood could pass through the cardiac interventricular septum, and in keeping with this he believed that all the blood that reached the left ventricle passed through the lung. He also stated that there must be small communications or pores (manafidh in Arabic) between the pulmonary artery and vein, a prediction that preceded by 400 years the discovery of the pulmonary capillaries by

Marcello Malpighi. Ibn al-Nafis and another eminent physiologist of the period, Avicenna (ca. 980–1037), belong to the long period between the enormously influential school of Galen in the second century and the European Scientific Renaissance in the sixteenth century. This is an epoch often given little attention by physiologists but is known to some historians as the Islamic Golden Age. Its importance is briefly discussed here."

1.2.2 Leonardo da Vinci

Leonardo da Vinci, the second scientist that was cited as a discoverer of oxygen in 1515 (Reti 1952; Mitzner and Wagner 1992; West 2017). As mentioned by West in 2017 regarding Leonardo (West 2017), "Leonardo da Vinci (1452–1519) enjoys a reputation as one of the most talented people of all time in the history of science and the arts. However, little attention has been given to his contributions to physiology. One of his main interests was engineering, and he was fascinated by structural problems and the flow patterns of liquids. He also produced a large number of ingenious designs for warfare and a variety of highly original flying machines. But of particular interest to us are his contributions to bioengineering and how he used his knowledge of basic physical principles to throw light on physiological function. For example, he produced new insights into the mechanics of breathing, including the action of the ribs and diaphragm. He was the first person to understand the different roles of the internal and external intercostal muscles. He had novel ideas about the airways including the mode of airflow in them. He also worked on the cardiovascular system and had a special interest in the pulmonary circulation. But, interestingly, he was not able to completely divorce his views from those of Galen, in that although he could not see pores in the interventricular septum of the heart, one of his drawings included them. Leonardo was a talented anatomist who made many striking drawings of the human body. Finally, his reputation with many people is based on his paintings including the Mona Lisa that apparently attracts more viewers than any other painting in the world."

1.2.3 Michael Servetus

Michael Servetus was the third person involved in the discovery of oxygen via the description of the pulmonary circulation in 1552. Stefanadis et al. (2009) summarized his discovery as follows: "Michael Servetus was the first doctor ever to challenge and scientifically argue against the theories of Galen, which predominated for 14 centuries in medical schools worldwide. Even though he was relatively correct in scientific terms, Servetus was punished because of his boldness in challenging Galen's theories and was condemned to death by the Holy Inquisition. Yet, by publicly challenging Galen's and Hippocrates' predominant and unquestionable lessons on medicine for the first time, Servetus opened the door for other doctors to challenge and correct those theories and subsequently to bring about a new view of human anatomy and physiology. This article underlines the contribution of Servetus to the description of the pulmonary circulation."

1.2.4 Michael Sendivogius (1566–1636)

The fourth sage was a famous sixteenth-century Polish noble physician and alchemist, Michael Sendivogius. In 1604, he published that air contains what he named the "secret food of life." He had realized that this "food" was the same gas emitted by saltpetre (potassium nitrate) when heated. He called it part of the "saltnitre of the earth." Sendivogius' writings were frequently copied and read for over a century. However, no one grasped the chemical importance of his discovery, not even physicists Boyle and Newton at Oxford, who read his work 60 years later. Sendivogius' "food of life" story was only recently rediscovered. He discovered that air is not a single substance and contains a life-giving substance—later called oxygen—170 years before Scheele's discovery of the element.

1.2.5 John Mayow (1641–1679)

John Mayow was the fifth scientist that was involved in the discovery of oxygen. He was a chemist, physician, and physiologist who is remembered today for conducting early research into respiration and the nature of air. Mayow worked in a field that is sometimes called pneumatic chemistry. As a chemist and physiologist, he discovered, about a 100 years before Joseph Priestley and Antoine-Laurent Lavoisier, and identified spiritus nitroaereus (oxygen) as a distinct atmospheric entity. He also found out that animals take up the same gas in their blood during respiration and that the air exhaled by animals had a less amount of spiritus nitro-aereus gas than the one present in the open air.

Mayow's first experiment (as seen in Fig. 1.4) involved placing a candle in an inverted jar and placing the jar in a tub of water. The purpose of the water was to seal the vessel. To equalize the initial air pressure on both sides of the jar, Mayow used a straw but quickly withdrew the straw once the pressures were equalized. As the candle burned, the water in the jar was drawn slightly upward until the candle was extinguished (because it had consumed the oxygen, not because the water had extinguished it). Mayow correctly explained that the air inside the jar had lost some of its "elasticity" and was no longer able to supply a force to the water equal to the atmospheric pressure. He suggested that this occurred because the air had been "deprived of nitro aerial and elastic particles" by the burning of the flame "so that the air is not able to resist the pressure of the atmosphere as before."

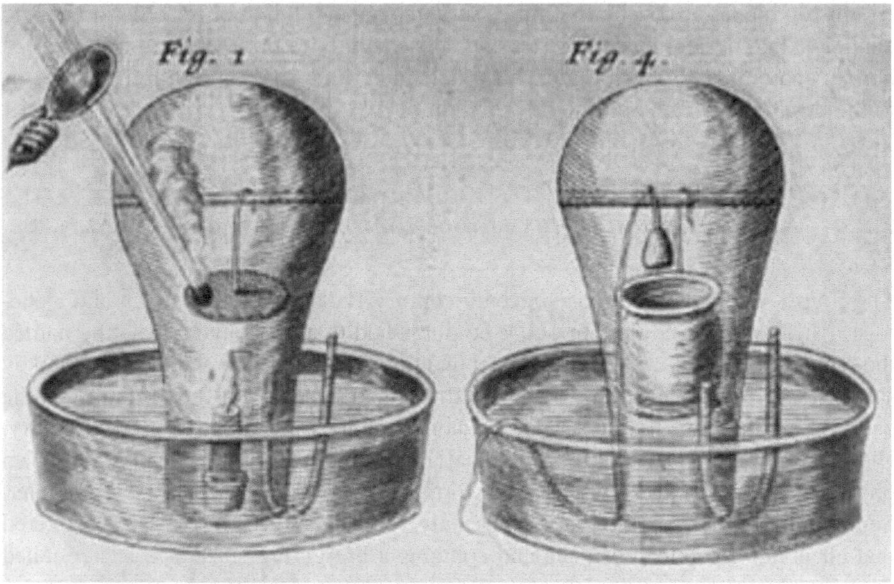

Fig. 1.4 From John Mayow experiments related to the discovery of oxygen. (By John Mayow, 1674. Public domain …www.researchgate.net)

Despite the critical need for oxygen and its significant use in body energy metabolism, the final steps of oxygen discovery in the atmosphere happened about 250 years ago (1771–1775). Until about 250 years ago, the story of oxygen discovery was described and discussed in terms of oxygen as a chemical element and the most important discovery of chemistry and science (Anon 1970; Cassebaum and Schufle 1975; Smeaton 1992; Marshall and Marshall 2005; Palmer 2006; West 2014a).

1.2.6 Carl Wilhelm Scheele (1742–1786)

The sixth discoverer of oxygen, Carl Wilhelm Scheele, was born in German-speaking Pomerania. Trained in Sweden, he became a very clever apothecary and chemist. In 1768, he published new studies of metal chemistry. In 1770, he became the director of Locke Pharmacy in Uppsala. He worked at the university there with the famous chemist Torbern Bergman, who helped him publish his work in Latin in the journal Nova Acta. In 1774, Lavoisier sent his new chemistry textbook to Bergman, including a copy for Scheele, whose published work he much admired. Scheele promptly wrote Lavoisier to thank him for the book. He explained how, in 1771, he had generated a strange new gas by heating certain metallic earths. He had named it fire air because it greatly brightened a candle flame and supported life in mice. Because neither he nor Bergman could relate it to the phlogiston theory, Scheele delayed publishing. He asked Lavoisier to repeat the experiment and then help him explain it.

The activity of Scheele was summarized by West (2014a) as follows "Carl Wilhelm Scheele (1742–1786) has an important place in the history of the discovery of respiratory gases because he was undoubtedly the first person to prepare oxygen and describe some of its properties. Despite this, his contributions have often been overshadowed by those of Joseph Priestley and Antoine Lavoisier, who also played critical roles in preparing the gas and understanding its nature. Sadly, Scheele was slow to publish his discovery and therefore Priestley is rightly recognized as the first person to report the preparation of oxygen. This being said, the thinking of both Scheele and Priestley was dominated by the phlogiston theory, and it was left to Lavoisier to elucidate the true nature of oxygen. In addition to his work on oxygen, Scheele was enormously productive in other areas of chemistry. Arguably, he discovered seven new elements and many other compounds. However, he kept a low profile during his life as a pharmacist, and he did not have strong links with contemporary prestigious institutions such as the Royal Society in England or the French Académie des Sciences. He was elected to the Royal Swedish Academy of Science but only attended one meeting. Partly as a result, he remains a somewhat nebulous figure despite the critical contribution he made to the history of respiratory gases and his extensive research in other areas of chemistry. His death at the age of 43 may have been hastened by his habit of tasting the chemicals that he worked on."

1.2.7 Joseph Priestley (1733–1804)

Oxygen's seventh discoverer was Unitarian Minister Joseph Priestley. Two churches fired him as too radical, but he soon became famous while teaching grammar and science at Warrington Academy near Manchester, helping to make it a leading school for dissenters—in fact, the "cradle of Unitarianism," as one scholar called it. On August 1, 1774, Priestley, by heating red mercury calc, made a gas that caused a glowing splinter to burst into flame and supported life in a mouse in a bottle. He named it dephlogisticated air. In October 1774, in Paris, he described his method of making it to Lavoisier, who never acknowledged his help.

West in (2014b) summarized his activities as follows: "Joseph Priestley (1733–1804) was the first person to report the discovery of oxygen and describe some of its extraordinary properties. As such, he merits a special place in the history of respiratory physiology. In addition, his descriptions in elegant eighteenth-century English were particularly arresting, and rereading them never fails to give a special pleasure. The gas was actually first prepared by Scheele (1742–1786) but his report was delayed. Lavoisier (1743–1794) repeated Priestley's initial experiment and went on to describe the true nature of oxygen that had eluded Priestley, who never abandoned the erroneous phlogiston theory. In addition to oxygen, Priestley isolated and characterized seven other gases. However, most of his writings were in theology because he was a conscientious clergyman all his life. Priestley was a product of the Enlightenment and argued that all beliefs should be able to stand the scientific scrutiny of experimental investigations. As a result, his extreme liberal views were severely criticized by the established Church of England. In addition, he was a supporter of both the French and American Revolutions. Ultimately, his political and religious attitudes provoked a riot during which his home and his scientific equipment were destroyed. He therefore emigrated to America in 1794, where his friends included Thomas Jefferson and Benjamin Franklin. He settled in Northumberland, Pennsylvania, although his scientific work never recovered from his forced departure. But the descriptions of his experiments with oxygen will always remain a high point in the history of respiratory physiology."

1.2.8 Antoine Laurent Lavoisier (1743–1794)

Lavoisier had become the world's most brilliant chemist by the age of 30. In the spring of 1775, he began his studies of Priestley's newly published gas. He named it principe oxigene but continued to refer to it as "vital air." However, after 8 years of extensively examining its chemistry, he was still unable to prove whether vital air was a new element or a compound. And he doubted but still could not disprove the phlogiston theory. By 1783, he was stalled by these dilemmas. In his Elements of Chemistry, Lavoisier falsely wrote that "This species of air was discovered almost at the same time by Mr. Priestley, Mr. Scheele, and myself," both a bold lie and proof that he had read Scheele's 1774 letter (Fig. 1.5b), using his method of making fire air while failing to acknowledge Scheele.

West in (2013) discussed and summarized the collaboration between Antoine and his wife Marie-Anne Lavoisier in the discovery of oxygen as well as the first measurements of human oxygen consumption as follows: "ANTOINE-LAURENT LAVOISIER (1743–1794) enjoys a reputation as one of the most eminent scientists of the late eighteenth century (Fig. 1.5a). His wife, Marie-Anne Pierrette Paulze Lavoisier (1758–1836), collaborated in many of his studies, but her contributions have received relatively little attention. Antoine Lavoisier's interests ranged over a wide area of chemistry. One of his major advances was to clarify the concept of an element as a substance that could not be further broken down by chemical analysis. His extensive classification of known elements was described in his book Traité élémentaire de chimie (Elementary Treatise on Chemistry) published in 1789." "In the life sciences, Lavoisier was the first person to recognize the true nature of oxygen. This had been isolated by Joseph Priestley (1733–1804) and Carl Wilhelm Scheele (1742–1786), but its role in respiration was not understood because of confusion caused by the erroneous phlogiston theory. Antoine with the help of his wife demolished this. Lavoisier also clarified the similarities between respiration and combustion, and he was responsible for the first studies of human oxygen consumption under various conditions." "Lavoisier was the first person to clearly state the role of oxygen, carbon dioxide, and nitrogen in respiration. Here he built on the previous

Fig. 1.5 Antoine Laurent Lavoisier (1747–1794) with his wife Marie-Anne (1759–1836). On the far left is the drawing board used by Marie-Anne. On the far right is a glass jar containing water and resting on a porcelain dish. This jar was presumably used for collecting gas. Next to it is a narrow tube partly filled with mercury. This is an eudiometer for measuring the oxygen concentration of gases. To the left of that is a mercury receiver for collecting gas. On the floor, there is a large glass vessel that was used for experiments to make water from hydrogen and oxygen. Next to it is a hydrometer for measuring the specific gravity of fluids. This magnificent portrait by Jacques Louis David (1788) is in the Metropolitan Museum of Art, New York. Reproduced by permission. The Metropolitan Museum of Art, Purchase, Mr. and Mrs. Charles Wrightsman Gift, in honor of Everett Fahy, 1977 (1977.10). (Image© The Metropolitan Museum of Art [Modified])

work of other investigators, particularly Priestley. This English Nonconformist minister carried out an experiment in August 1774, on heating some red mercuric oxide, he found that a remarkable gas was produced. He stated "but what surprised me more than I can yet well express was that a candle burned in this air with a remarkably vigorous flame… and a piece of red-hot wood sparkled in it" (Priestley 1775). Furthermore, he showed that a mouse was able to survive longer in this gas than in ordinary air, and he famously surmised that it might be useful for people with disease. Thus, there was no doubt that Priestley had produced oxygen. However, unfortunately, he did not understand its nature. Priestley was a follower of the phlogiston theory that had been promoted by Georg Ernst Stahl (1659–1734) and that stated that all combustible materials are composed of ash (calx) and phlogiston (Greek for inflammable), and that during burning, phlogiston escaped leaving the dephlogisticated ash behind. This is in fact the reverse of what happens during combustion when oxygen combines with a combustible material, but the theory was enormously influential in the mid-eighteenth century. Priestley is often credited with "discovering" oxygen just as Columbus "discovered" America but neither Priestley nor Columbus correctly identified what they found. Carl Wilhelm Scheele

(1742–1786) in Sweden had independently produced oxygen even before Priestley, although the publication of his findings was delayed (Scheele 1777). Scheele called the gas "fire-air" but again he was influenced by the phlogiston theory and did not recognize its true nature. Priestley visited Paris in October 1774 with his patron Lord Shelburne and they had dinner with Lavoisier and some other chemists. At that time, Priestley described the experiments that he had carried out with mercuric oxide; naturally, these evoked great interest. As a result, Lavoisier repeated the experiments and was eventually able to understand the chemical processes. For example, in 1775, in what became known as the Easter Memoir, Lavoisier stated that "the substance which combines with metals during calcination, thereby increasing their weight, is nothing else than the pure portion of the air which surrounds us and which we breathe." This was the coup de grâce to the phlogiston theory. As we shall see later, Marie-Anne Lavoisier played an important part in this advance because it was she who translated the English texts of Priestley and another proponent of phlogiston, Richard Kirwan (1733–1812), so that Antoine could read them.

1.2.9 Henry Cavendish (1731–1810)

In 1766, Cavendish discovered and published how to make an inflammable gas (H_2) by putting iron filings in strong acid. In the late 1770s, he and Priestley noted that burning inflammable air caused dew to form on glass walls. In 2014, J.B. West published a paper summarizing the activities of Cavendish as follows: "Henry Cavendish (1731–1810) was an outstanding chemist and physicist. Although he was not a major figure in the history of respiratory physiology, he made important discoveries concerning hydrogen, carbon dioxide, atmospheric air, and water. Hydrogen had been prepared earlier by Boyle but its properties had not been recognized; Cavendish described these in detail, including the density of the gas. Carbon dioxide had also previously been studied by Black, but Cavendish clarified its properties and measured its density. He was the first person to accurately analyze atmospheric air and reported an oxygen concentration very close to the currently accepted value. When he removed all the oxygen and nitrogen from an air sample, he found that there was a residual portion of about 0.8% that he could not characterize. Later, this was shown to be argon. He produced large amounts of water by burning hydrogen in oxygen and recognized that these were its only constituents. Cavendish also worked on electricity and heat. However, his main contribution outside chemistry was an audacious experiment to measure the density of the earth, which he referred to as 'weighing the world'. This involved determining the gravitational attraction between lead spheres in a specially constructed building. Although this was a simple experiment in principle, there were numerous complexities that he overcame with meticulous attention to experimental details. His result was very close to the modern accepted value. The Cavendish Experiment, as it is called, assures his place in the history of science."

In many published papers and Internet sites, the credit of oxygen discovery was given either to the American Joseph Priestley (Partington 1962; Miller 1987; Williams 2003) or to the German-Swedish Carl Wilhelm Scheele (Partington 1962; Smeaton 1963; Taylor 2019). The available information indicates that Scheele was the first one to discover oxygen in 1771–1772, but he published it officially a few years later in 1777. He claimed that in 1774, he sent a letter to the famous French chemist Antoine de Lavoisier, regarding his discovery, but Lavoisier claimed that he never received the letter. Priestley discovered oxygen and in 1774 he notified his findings to Antoine de Lavoisier who studied oxygen chemistry for a number of years. Priestley actually published his discovery in 1775. Wilkinson published a paper in (2004) describing the significant contribution of Lavoisier, Scheele, and Priestley to the understanding of respiratory physiology in the eighteenth century. Here are his conclusions: "There appears to be a great desire in history to define who was the first to do something. In reality, it makes no great difference to today's work who was first to define oxygen in the eighteenth century. What can be seen from the above is the ability of these three great men to set new paradigms and work to prove their theories. Some were more accurate than others, but it is the ability to take that lateral step that sets these characters apart from their peers. It could be considered that if Scheele first isolated oxygen then Priestley defined its properties and Lavoisier showed it was an element but in fact they all added a significant set of pieces to the physiological jigsaw puzzle of which oxygen was just a part."

1.2.10 John Severinghaus (1922–2021)

In 1996, John Severinghaus (Fig. 1.6) became interested in the story of the discovery of oxygen and published his first paper on this subject (2002). In the introduction to the cited paper, Severinghaus wrote the next paragraph: "AMERICANS are taught that the English Unitarian minister Joseph Priestley discovered oxygen in 1774. Scandinavians are taught that the Swedish apothecary Carl Wilhelm Scheele generated oxygen in Uppsala in 1771–1772, several years before Priestley. Scheele claimed that he wrote Lavoisier, describing the experiments in September 1774. However, Lavoisier denied seeing or receiving his letter. Among Nordic historians, Lavoisier has never been forgiven for this rebuff. Some historians have doubted whether such a letter was ever sent or received, because Scheele first published it in 1777. Priestley brought the news of his discovery to Lavoisier in October and in publications overturning the conventional Phlogiston Theory, causing a revolution in chemistry."

One year later, Severinghaus published an extended paper on this topic (2003) and came to the conclusion that Mary Ann Lavoisier received the original letter of Scheele and hid it until 219 years later the letter was found (Fig. 1.5b). Later, Severinghaus combined the discovery of oxygen as an element, in the eighteenth century with the description of the pulmonary circulation in the thirteenth century (2016a, b). This subject was discussed in other papers describing the Islamic Golden Age in the thirteenth century (Haddad and Khairallah 1936; West 2008).

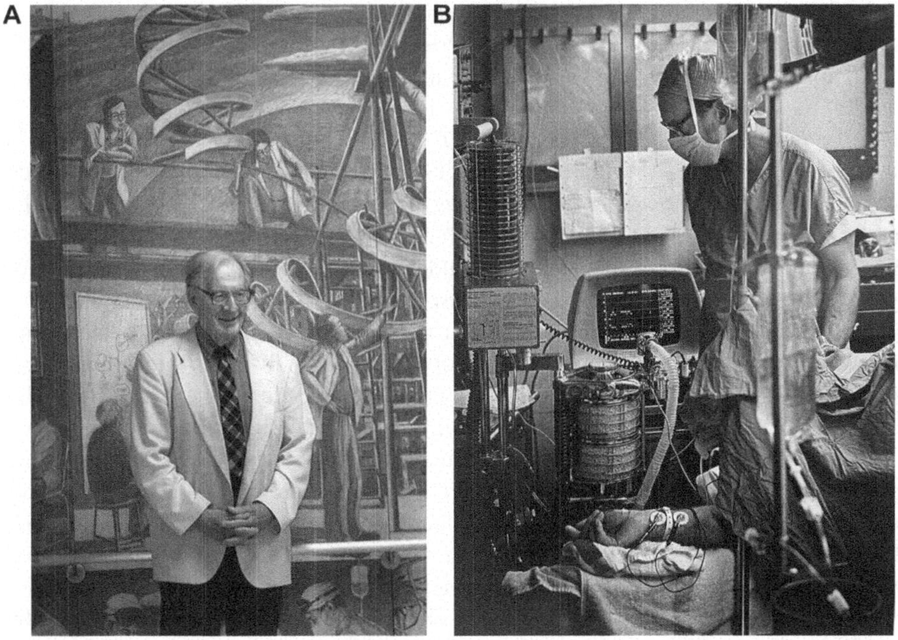

Fig. 1.6 (**a**) The "Master Tinker" and the blood–Gas analyzer. (**b**) John Severinghaus near a patient that was monitored in the hospital of the University of California, San Francisco, California, USA. (From: John W. Severinghaus, M.D., 1922 to 2021, Anesthesiology. 2021; 135(4): 555–557.) [Modified]

1.3 The Discovery of Mitochondria

The understanding of mitochondrial function has been a challenge for various investigators since its discovery more than 150 years ago. In 1981, Ernster and Schatz (1981) reviewed the history of mitochondrial structure and function studies. There is no real single answer regarding who discovered mitochondria. The process of discovery and identification was a gradual one, and four scientists involved in the discovery of the mitochondria are presented in the next few figures.

The Swiss anatomist and histologist Albert von Kölliker (1856) shown in Fig. 1.7 described in 1856 what he called "granules" in the sarcoplasm of striated muscle (Fig. 1.7). According to A. L. Lehninger, "Kölliker was among the first to notice the arrangement of granules in the sarcoplasm of striated muscle over a period of years beginning around 1850. These granules were later called sarcosomes by Retzius (Fig. 1.8) in 1890. These sarcosomes have come to be known as the mitochondria—the powerhouses of the cell. In the words of Lehninger, "Kölliker should also be credited with the first separation of mitochondria from cell structure. In 1888, he teased these granules from insect muscle, in which they are very profuse, found them to swell in water, and showed them to possess a membrane." As mentioned, Retzius in 1890 called the granules Sarcosomes, which later on were identified as the

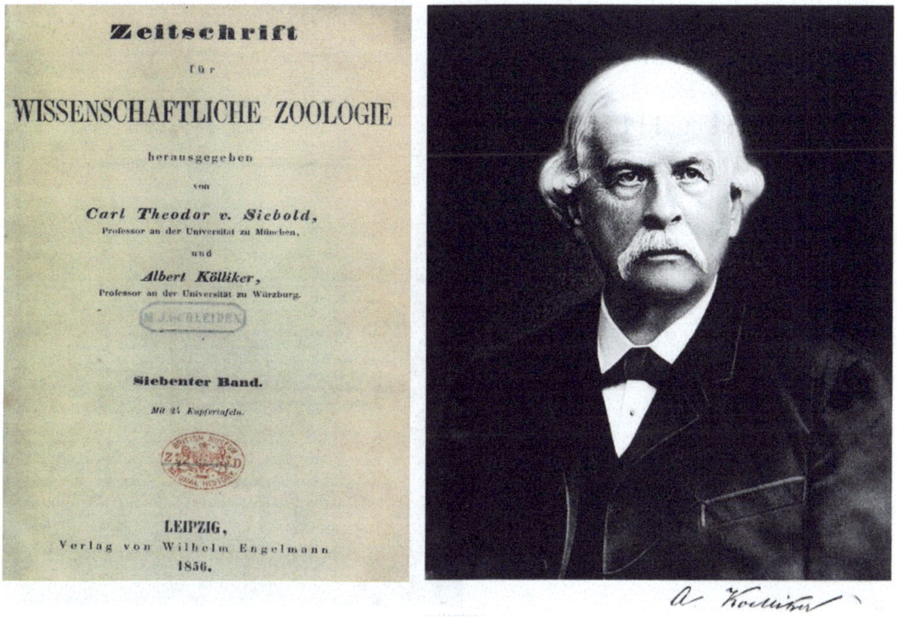

Fig. 1.7 Albert von Kölliker (**a**) contributed initially to the discovery of the mitochondrion (**b**) [Modified]

mitochondria (1890). Retzius published more than 300 scientific works in anatomy, embryology (Fig. 1.8), eugenics, craniometry, zoology, and botany. He gave his name to the 60 mm-diameter Retzius cells in the central nervous system of the leech (*Hirudo medicinalis*). During his time at the Karolinska Institute, he made important contributions to anatomical descriptions of the muscles of the eardrum, the bones of the middle ear, and the Eustachian tube.

The discovery of mitochondria in general came in 1890 when Richard Altmann (Fig. 1.9a), a cytologist, identified the organelles and dubbed them "bioblasts" as presented in his publication seen in Fig. 1.9b (1890). The fourth scientist involved in the discovery of the mitochondria was Carl Benda (Fig. 1.10), that in 1898, coined the term *mitochondria* from Greek thread, mitos, and granule, chondros (1898).

1.4 History of Mitochondrial Function Monitoring Systems

In 2006, I (Avraham Mayevsky) asked Prof Britton Chance to write a short statement regarding the history of monitoring of NADH redox state. Here is Prof. Chance's response:

> We will start by bringing a citation from Chance et al. in 1973. "The accumulation of evidence since the pioneer work of Otto Warburg an 'Atmungsferment' (1949) and David Keilin on cytochromes (1925, 1966) as the keystones of cellular oxygen utilization led us to

Fig. 1.8 Gustav Retzius and his publication that called the granules "sarcosomes" [Modified]

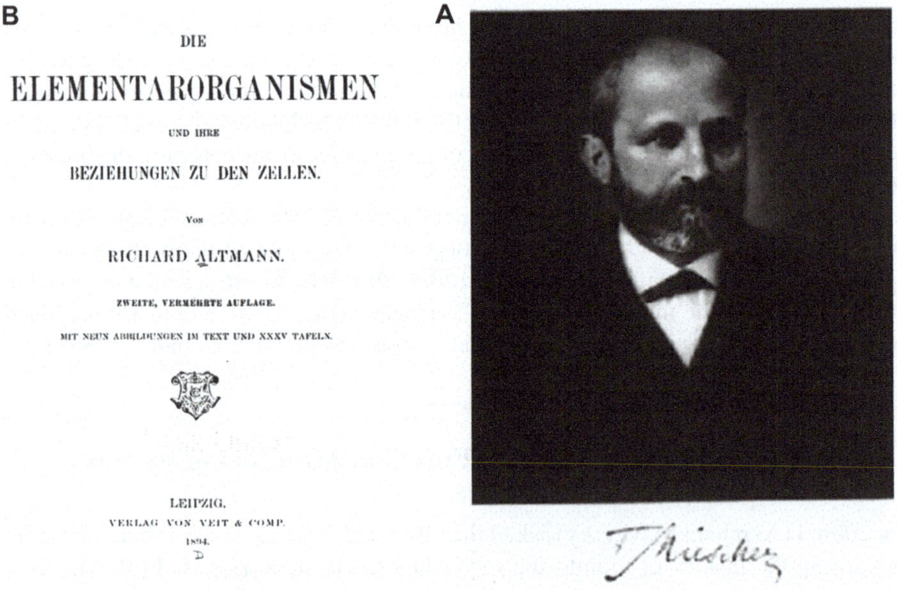

Fig. 1.9 (**a**) Richard Altman that gave the name "bioblasts" that later on identified as the mitochondria. (**b**) The book published by Altmann (1890) [Modified]

B

Carl Benda (30 December 1857 Berlin – 24 May 1932 Turin) was one of the first microbiologists to use a microscope in studying the internal structure of cells. In an 1898 experiment using crystal violet as a specific stain, Benda first became aware of the existence of hundreds of these tiny bodies in the cytoplasm of eukaryotic cells and assumed that they reinforced the cell structure. Because of their tendency to form long chains, he coined the name mitochondria ("thread granules"). These bodies had first been noted in 1857 by the physiologist and pioneer of the light microscope, Albert von Kölliker, and were later termed "bioblasts" by Richard Altmann in 1886.

A

Fig. 1.10 Prof. Carl Benda (**a**) that coined the term mitochondria in 1898 (**b**) [Modified]

the study of the redox states of electron carriers in isolated mitochondria as a function of oxygen concentration and to develop techniques to measure the states of anoxia and normoxia in living tissues."

Although Warburg isolated the reduced di-phosphopyridine nucleotide (DPNH) and the reduced tri-phosphopyridine nucleotide (TPNH), he was apparently reluctant to imagine that those molecules were part of the respiratory chain. He also studied their optical properties and identified the 340 nm absorption band as being characteristic of the reduced form. The idea that these might also belong to Keilin's respiratory chain (1925) was not considered and, in fact, was to some extent abhorrent to Warburg to imagine that he should support Keilin's views.

Thus, the observation made spectroscopically in Chance's difference spectrum of the respiratory chain showing prominent peaks of what was soon to be called NADH and a trough due to flavoproteins and finding that they were respectively over 40 and over 10 times greater concentration than Keilin's cytochrome c was in a sense an overwhelming blow to Keilin. In my (Britton Chance) close contact with him, he never wished to discuss these two components as new and overwhelmingly present members of the "respiratory chain" (taking note that the Keilin and Hartree preparation (1945) was carried out in such a way that no trace of NADH survived the rigorous sand grinding and acid extraction). In fact, personal contact with Keilin subsequent to him being awarded the Copley Medal, evoked no response whatsoever of these novel components, which certainly must have been prominent in his mind. Nor in any conversation with him did he discuss Otto Warburg's discovery of these components.

The blue autofluorescence of cells and tissues was recognized from the first studies of its fluorescence microscopy and was termed "autofluorescence." There was no attempt to characterize it or to do what was obvious to later researchers, which was to see whether oxygen or nitrogen affects the fluorescence, a study which Keilin relied upon to identify his cytochromes (the transition from trivalent to divalent caused a significant change in the electron configuration of the cytochromes). Since the

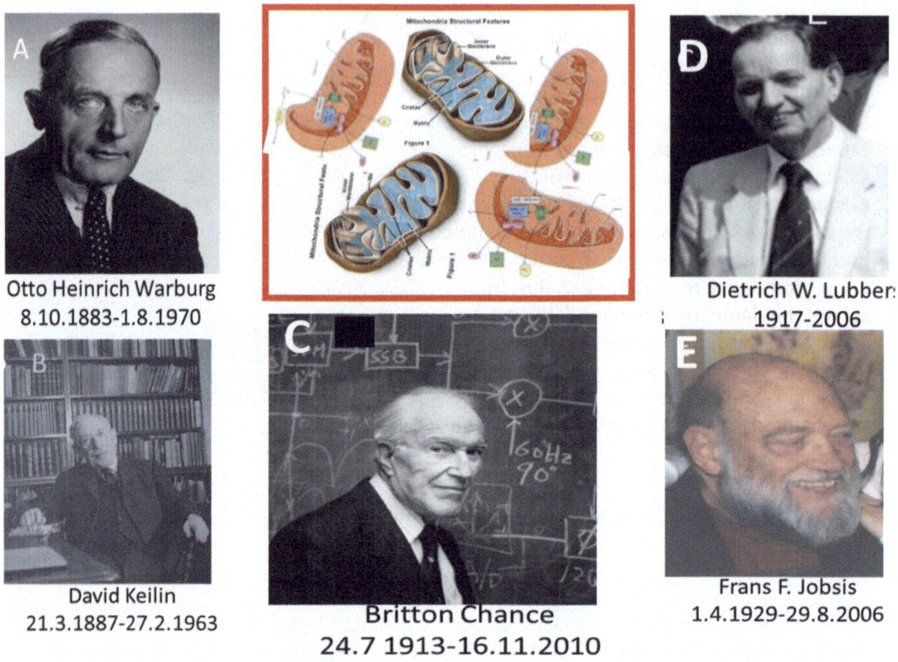

Fig. 1.11 The five scientists that affected our understanding of mitochondrial function and NADH monitoring in vitro and in vivo situations [Modified]

oxidized form—NAD^+— is not absorbing light in this range, it was possible to evaluate the redox state of the mitochondrion by monitoring the UV absorbance or blue fluorescence of NADH.

Lubbers (1995) concluded that "the most important intrinsic luminescence indicator is NADH, an enzyme of which the reaction is connected with tissue respiration and energy metabolism."

Figure 1.11 presents the pictures of the five cited scientists that significantly affected the development of the theoretical and experimental technology for the monitoring of mitochondrial NADH function *in vitro* and *in vivo*.

The discovery of the optical properties of reduced Nicotine Amide Adenine Dinucleotide (NADH) (earlier names: DPNH [diphosphopyridine nucleotide] or PN [pyridine nucleotide]) has led to very intensive research since the early 1950s. The reduced form of this molecule, NADH, shown in Fig. 1.12a, b (Kohen et al. 1969), absorbs light at 320–380 nm (Fig. 1.12c) and emits fluorescent light at 420–480 nm range (Fig. 1.12d) (Mayevsky and Barbiro-Michaely 2013a).

As the oxidized form NAD^+ does not absorb light in this range, it was possible to evaluate the redox state of the mitochondria by monitoring the UV absorbance or Blue fluorescence of NADH.

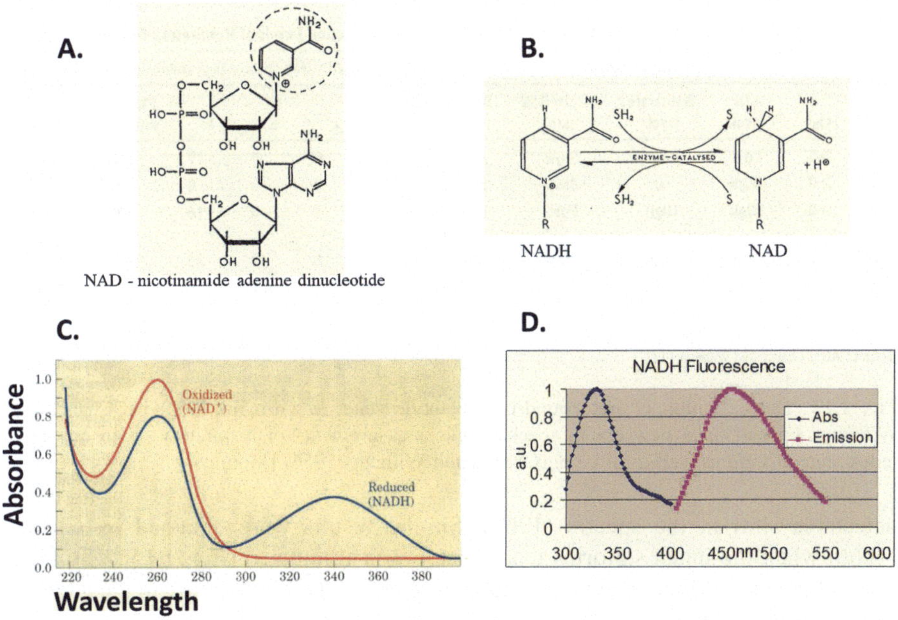

Fig. 1.12 (**a**) The structures of NAD^+. The nicotinamide group (broken ring) is the "functional" part of both molecules, i.e., the portion of the molecules where oxidation and reduction take place and (**b**) The transition between oxidized and reduced NADH. © Springer Science, reprinted with permission (Kohen et al. 1969). (**c**) Absorption Spectra of NAD^+ and NADH. (**d**) Excitation and emission spectra of NADH. (© Reprinted with kind permission of Springer Science + Business Media (Mayevsky and Barbiro-Michaely 2013a, b))

The foundations for the use of NADH fluorescence as a marker of mitochondrial activity were presented in detail by Chance and Williams (1955a, b) for isolated mitochondria *in vitro* (see Fig. 1.13). The "resting" state of the mitochondria *in vitro* was defined as "State 4," where NADH was 99% in the reduced form, and ADP was the rate-limiting substance. If ADP is added to a suspension of mitochondria, ATP synthesis will be stimulated, oxygen consumption will increase, and the rate-limiting factor will be the activity of the respiratory chain. During this State 3, or the "active state," the NADH redox state will decrease or become more oxidized (about 50%). When the "resting" mitochondria are deprived of oxygen, the activity of the mitochondria will stop and NADH will reach its maximum redox state (State 5).

The first detailed study using fluorescence spectrophotometry of NADH in intact Baker's yeast cells and Algae cells was published in 1957 by Duysens and Amesz (1957) as seen in Fig. 1.14.

The use of NADH fluorometry for *in vivo* studies started at the end of the 1950s and has been presented in various publications. In a summative study, Chance compared spectra of NADH measured from rat kidney cortex to those of rat brain cortex (Fig. 1.15a, b) and found that the main effect of anoxic transition, both in the kidney and the brain, was a large increase in the fluorescence intensity with no

Metabolic States of Mitochondria and Associated Oxidation-Reduction Levels of Respiratory Enzymes

		Characteristics				Steady state percentage reduction of components				
State	$[O_2]$	ADP level	Substrate level	Respiration rate	Rate-limiting substance	a	c	b	Flavo-protein[a]	$NADH_1$
1	>0	Low	Low	Slow	ADP	0	7	17	21	~90
2	>0	High	~0	Slow	Substrate	0	0	0	0	0
3	>0	High	High	Fast	Respiratory chain	4	6	16	20	53
4	>0	Low	High	Slow	ADP	0	14	35	40	99
5	0	High	High	0	Oxygen	~100	~100	~100	~100	~100

Source: After Chance and Williams [13].

Fig. 1.13 The definition of mitochondrial metabolic state, in vitro, in 1955, by Chance and Williams, opened up a new era in spectroscopic measurements of respiratory chain enzyme's redox state in vitro as well as in vivo. (Chance and Williams 1956) [Modified]

detectable shift in the spectra (1967). Similar results were obtained by Chance (1966) when the intact sartorius toad muscle was stimulated (Fig. 1.15c). Stimulation of the muscle (State 4 to State 3 transitions) led only to a decrease in the intensity of the fluorescence emission spectrum.

In 1975, Sundt and Anderson (1975a, b) applied *in vivo* fluorometry to study the brain of the squirrel monkey (Fig. 1.15d), and recorded a greatly increased intensity in the dead brain as opposed to normal brain, with intermediate values in the ischemic brain. At the same time, Harbig et al. showed a clear increase in the NADH spectrum between normoxic and anoxic conditions in cat brain seen in Fig. 1.16a (1976).

Corderio et al. compared the excitation spectra in porcine myocutaneous flap with those of NADH in solution shown in Fig. 1.16b (1995). Exposing the flap to 6 hours' ischemia dramatically increased the intensity of 450 nm light. Furthermore, the present authors compared two sources of excitation light: the usual mercury arc and laser excitation at 324 nm (Mayevsky 1984). The main problem in applying such spectra, under *in vivo* conditions, is the effects of hemodynamic changes under anoxia. To overcome this problem, we induced repetitive cycles of anoxia that suppress the autoregulatory compensation mechanisms (Mayevsky 1984).

Figure 1.16 (Part C) shows six scans obtained from the normoxic (A1, A2) and the anoxic brain (B1, B2) using a Hg arc, and comparative scans using Hg (C1) and the laser source (C2) in the dead brain. The results indicate that the location of the emission spectrum peak is identical in normoxia and anoxia, and is close when the laser is used. We also used the laser source in several *in vivo* studies and found that the light intensity did not harm the brain during several hours of measurement. Under anoxia, a clear increase in NADH fluorescence spectra was recorded.

The first time that NADH was monitored *in vivo* in more than one site simultaneously was reported by Chance et al in 1962. They used 2 units of the same type of fluorometer placed on two organs in the same rat. Two organs were exposed in the

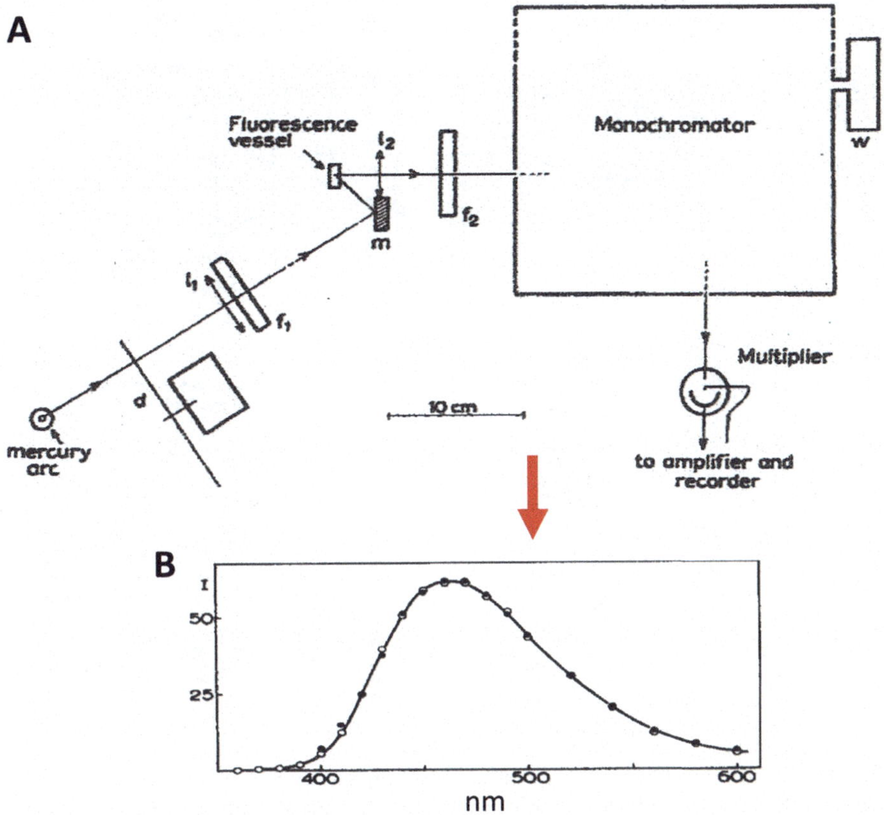

Fig. 1.14 (**a**) Apparatus for measurement of "fluorescence spectra." By means of a lens l_1 and a mirror in the mercury arc was imaged in a vessel containing the fluorescing solution or cell suspension. The fluorescing image was focused by lens l_2 on the entrance slit of the monochromator. Filter set l_1 isolated the spectral region of the exciting light, filter set l_2 absorbed the exciting radiation and transmitted light of longer wavelengths. The wavelength drum was driven by a synchronous motor. A sectioned disc *d*, mounted on another synchronous motor, modulated the incident light. The a.c. component of the current, caused by the light falling on the iP 21 multiplier, was amplified, rectified, and fed into the recorder. (**b**) Fluorescence spectra of DPNH. The open circles are for excitation by wavelength 313 mμ and the black ones for excitation by 366 mμ. The spectrum for 313 mμ has been multiplied by a certain factor to make its maximum of equal height as the maximum of the spectrum excited by 366 mμ. The spectra appear to be identical; the maximum is at about 462 mμ. (Duysens and Amesz 1957) [Modified]

same rat, namely the brain and the kidney. In this study, they exposed the animal to various perturbations including anoxia and adrenaline injection. Figure 1.17a shows the probes located above the brain (right side) and the kidney (left side). Typical responses of the kidney and the brain to anoxia are presented in part B of Fig. 1.17. The results were presented on one recorder therefore the quantitative comparison between the responses was possible.

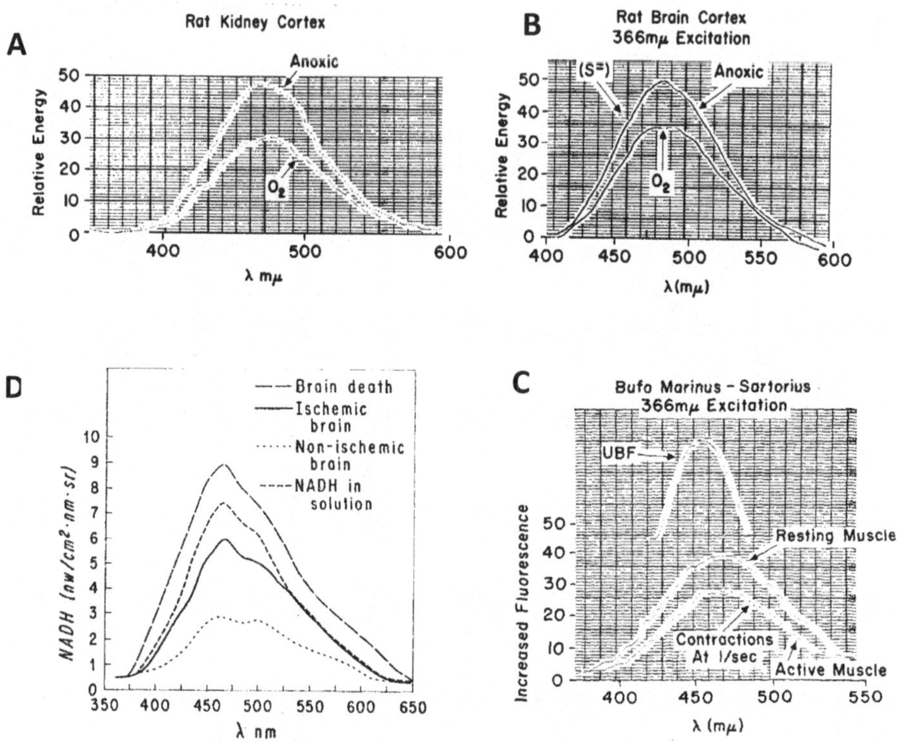

Fig. 1.15 (**a** + **b**) The emission spectra of rat kidney (**a**) and rat brain cortex (**b**), under aerobic conditions (lower trace) and under anoxic conditions (upper trace) (Chance 1967). (**c**) Emission spectra for the resting and stimulated sartorius muscle of the toad. Note that the spectrofluorometer is calibrated in a separate experiment by the emission of 4-methylumbelliferone shown at the top of the chart (UBF) (Chance 1966). (**d**) Emission spectral fluorescence curves for NADH in solution, in normal brain, in ischemic cortex, and in cortex 5 min after death. (Sundt and Anderson 1975a) [Modified]

The responses of the brain to hypoxia is presented in Fig. 1.18a. According to the definition of Chance and Williams (1955a, 1956), a shift toward State 5 involves an increase in NADH proportional to a decrease in O_2 supply. On those days, the fluorescence was measured and displayed without correction for hemodynamic artifacts, which was developed later in time. A clear increase in NADH fluorescence was recorded under the deprivation of oxygen.

In the early 1960s, Chance (1963) tested the effect of irreversible ischemia on brain NADH using the decapitation model as seen in Fig. 1.18b. Here is the citation from the paper by Chance.

"One would expect that breathing nitrogen would lead to a very rapid transition from aerobiosis to anaerobiosis since it has already been pointed out in this symposium that breathing nitrogen would serve to deplete the blood of oxygen. As shown in Fig. 1.18b, about 8 s is required between the start of DPN reduction and the

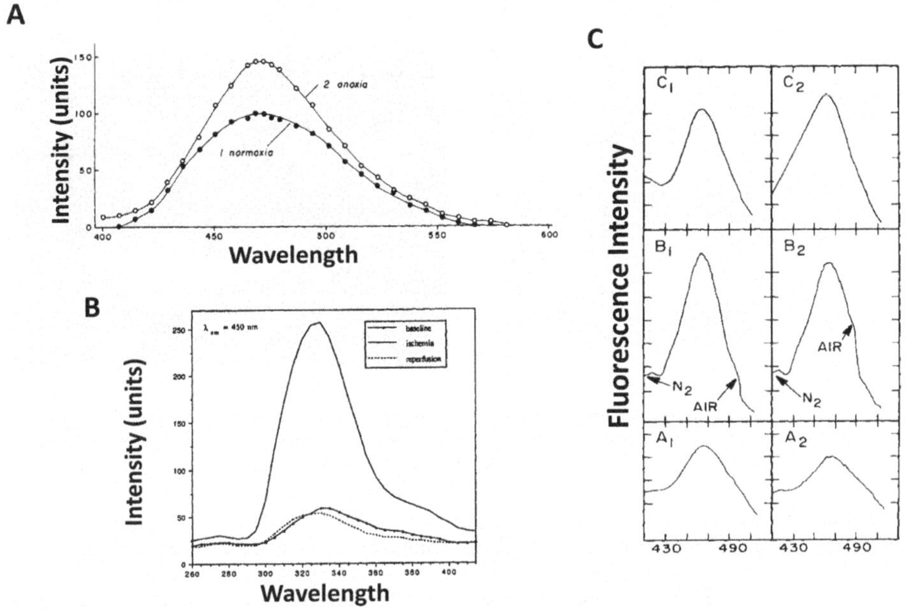

Fig. 1.16 (**a**) Pyridine nucleotide fluorescence emission spectra from cat brain cortex *in vivo* in (1) normoxia and (2) nitrogen anoxia. Both spectra peak at 469–473 nm. The spectra are not corrected for photocell sensitivity (Harbig et al. 1976). (**b**) Representative excitation spectra from an experimental myocutaneous flap at baseline, after 6 hours of ischemia, and after 4 h of reperfusion. Fluorescence is recorded at a wavelength of 450 nm, and the excitation energy ranges from 260 to 430 nm. Excitation maximum of all spectra is 330–340 nm (Cordeiro et al. 1995). (**c**) Emission spectra of the brain under excitation of 366 nm light (A1, B2, B1, B2, and C1) or laser 324 nm light (C2). C1 and C2 were measured from a dead brain. (Mayevsky 1984) [Modified]

attainment of half-maximal reduction. In an attempt to observe the time for DPN reduction in ischemia, we have employed a decapitation technique with the mouse; the optical system is arranged so that the slight mechanical artifact occurring on decapitation would not disturb the fluorescence excitation."

In order to enable the monitoring of NADH fluorescence in unanesthetized animals or other *in vivo* preparations, a flexible means was needed to connect the fluorometer with the tested organ, for example, the brain. This was achieved in 1972, when UV transmitting quartz fibers became available (Schott Jena Glass, Germany). We have used the light-guide-based fluorometer for *in vivo* monitoring of the brain (Chance et al. 1973; Mayevsky and Chance 1973) subjected to anoxia or cortical spreading depression. The historical development of light-guide-based fluorometery-reflectometry is shown in Fig. 1.19. The original device functioned on the time sharing principle (Fig. 1.19a), where four filters were placed in front of a

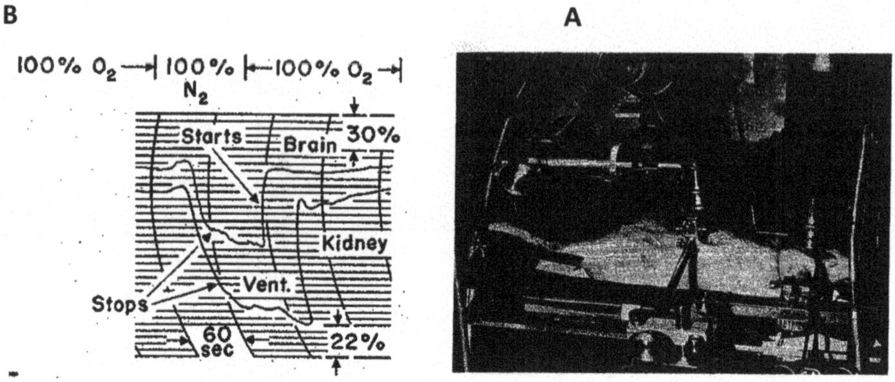

Fig. 1.17 (**a**) Experimental arrangement for simultaneous microfluorometry of brain and kidney cortex in the rat. Two microfluorometers are focused on the exposed surfaces of the brain and kidney. By means of a tracheal cannula, the oxidation–reduction level of the intracellular pyridine nucleotide can be altered and the corresponding fluorescence changes can be recorded by the two microfluorometers. For the kidney fluorometer, the water-cooled lamp housing attached to the Leitz "Ultrapak" illumination system is shown. Light choppers and photomultipliers are not shown (Chance et al. 1962). (**b**) Simultaneous recordings of fluorescence changes in rat brain and rat kidney in a cycle of anoxia. Fluorescence increases are indicated in a downward direction. (Chance et al. 1963) [Modified]

2 arms light guide. Filters 1 and 3 enabled the measurement of NADH fluorescence, while filters 2 and 4 were used to measure tissue reflectance at the excitation wavelength. The reflectance trace was used to correct the NADH signal for hemodynamic artifacts, and to indicate changes in the blood volume of the sampled tissue.

In this original system, only one photomultiplier tube was used for the detection of the two signals. Figure 1.19b presents one of the first *in vivo* brain monitoring time sharing setups, the time-sharing approach (AC mode) was replaced by splitting the light emitted from the tissue into two unequal fractions for the measurement of fluorescence and reflectance signals. This model, named the DC type fluorometer, had originally a three-way light guide which was later replaced by a two arms light-guide probe (Fig. 1.19c). In all the three configurations, the reflectance signal was used for the correction of the fluorescence signal.

Our group developed and used the model shown in Fig. 1.19c in the late 1970s. This model is still being used in our laboratory to monitor the brain, heart, liver, kidney, as also in multisite or multi-organ monitoring. The responses to anoxia using different diameter of the fiber optic bundle are shown in Fig. 1.19d.

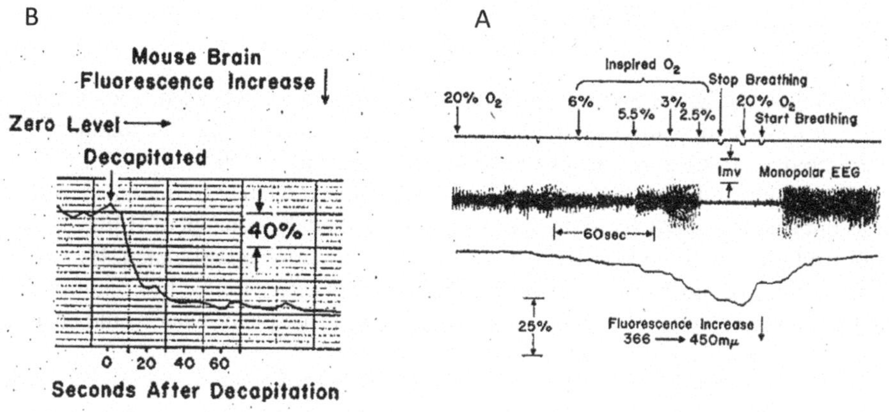

Fig. 1.18 (**a**) Correlation between EEG and pyridine nucleotide reduction in the slow transition from aerobiosis to anoxia. The top trace includes the measurements of inspired oxygen concentration according to the Pauling oximeter (Chance and Schoener 1962). (**b**) Record of the kinetics of fluorescence changes of the mouse brain cortex in ischemia caused by decapitation of a mouse. (Chance et al. 1963) [Modified]

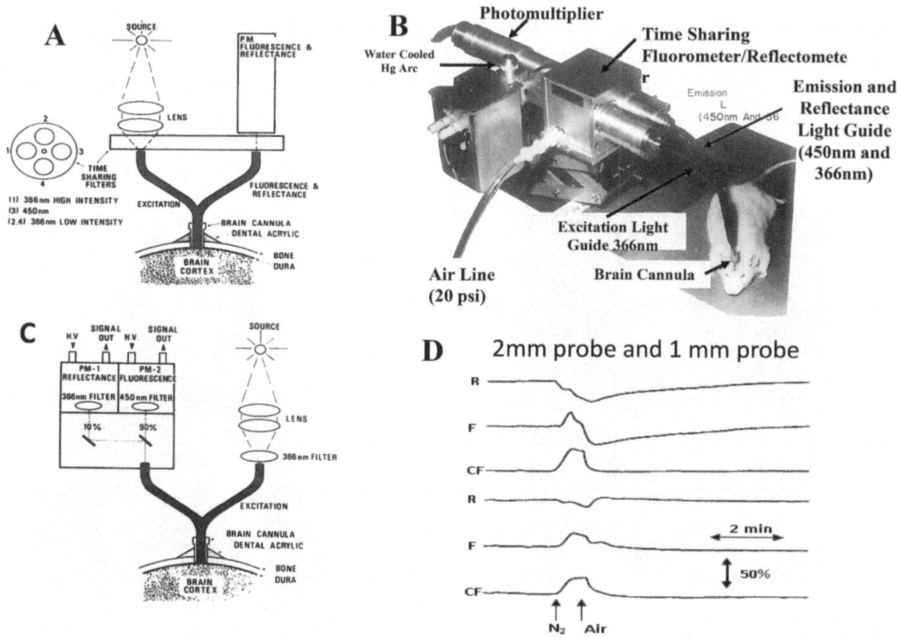

Fig. 1.19 The structure of the time sharing (**a**) and standard DC fluorometer/reflectometer (Mayevsky 1984) (**c**) connected to the brain of a small animal (Mayevsky 1976). (**b**) *Ex* Excitation, *Em* Emission optical fibers for the monitoring of NADH redox state. *H.V.* high voltage, *PM* photomultiplier (Mayevsky and Chance 1976). (**d**) The effect of the monitored tissue volume (the diameter of the fiber optic probe) was tested under anoxia as shown in D upper 3 traces (2 mm diameter) and D lower 3 traces (1 mm diameter). (Mayevsky 1984)

1.5 Homeostasis of Oxygen

According to the Royal Society of Chemistry, definition for "Oxygen Homeostasis" could be described as follows: "A homeostatic process involved in the maintenance of an internal steady state of oxygen within an organism or cell."

The two physiologists presented in Fig. 1.20, namely, Claude Bernard and Walter Bradford Cannon, were the first scientists that described the concept of Homeostasis. Milieu intérieur was the key concept with which Bernard was associated. He wrote, "The stability of the internal environment [the *milieu intérieur*] is the condition for the free and independent life." This is the underlying principle of what would later be called homeostasis, a term coined by Walter Cannon.

Figure 1.21 demonstrates the relationship between oxygen supply and oxygen demand at the tissue level that could be evaluated by monitoring the redox state of the mitochondria. It is important to note that the mechanisms that regulate the oxygen supply to any tissue are common to all organs and tissues in the body. The process of using the oxygen or the energy in the various organs is different in the various organs and tissues, as seen in the right side of Fig. 1.21. Figure 1.22 demonstrates the components of the brain oxygen balance typical of the brain. As seen, the supply of oxygen is dependent on the microcirculatory blood flow and volume, oxygenation of hemoglobin, and partial pressure of oxygen. The oxygen demand in the brain is shown on the right side of the figure and is used mainly for the electrical activities in the neurons and glia cells, as well as for synaptic activities and ionic homeostasis.

Fig. 1.20 The two physiologists (Claude Bernard and Walter Bradford Cannon) were the first scientists that described the concept of Homeostasis [Modified]

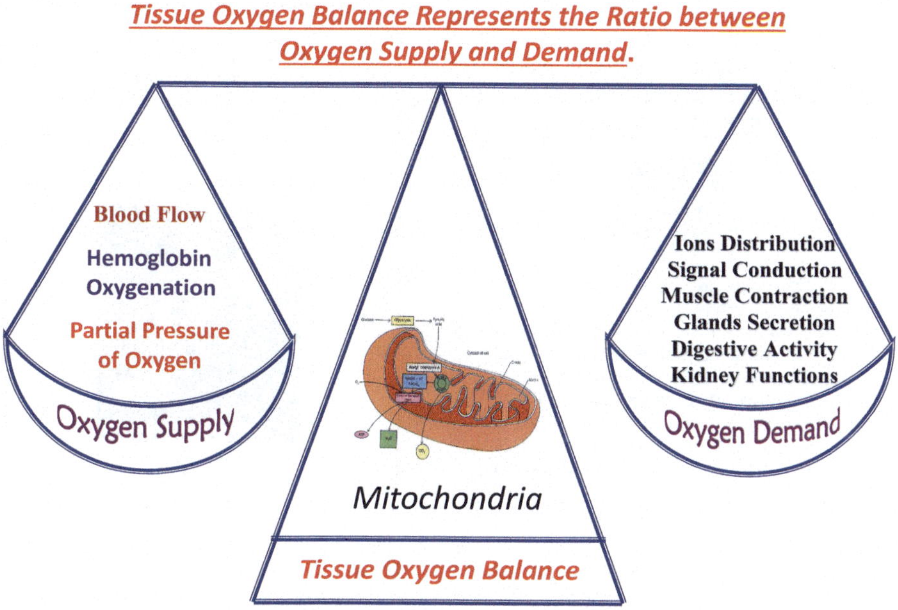

Fig. 1.21 Schematic presentation of tissue energy balance evaluated by energy supply and demand. Energy supply is measured by tissue blood flow (TBF), hemoglobin Oxygenation, and partial pressure of Oxygen, which is similar in all tissues. Energy demand varies between the different tissues and may include: Ionic Homeostasis, Signal Conduction, Muscle contraction, Glandular Secretion, and G-I tract and kidney function. Mitochondrial NADH redox state serves as an indicator for tissue energy or oxygen balance (Mayevsky 2015)

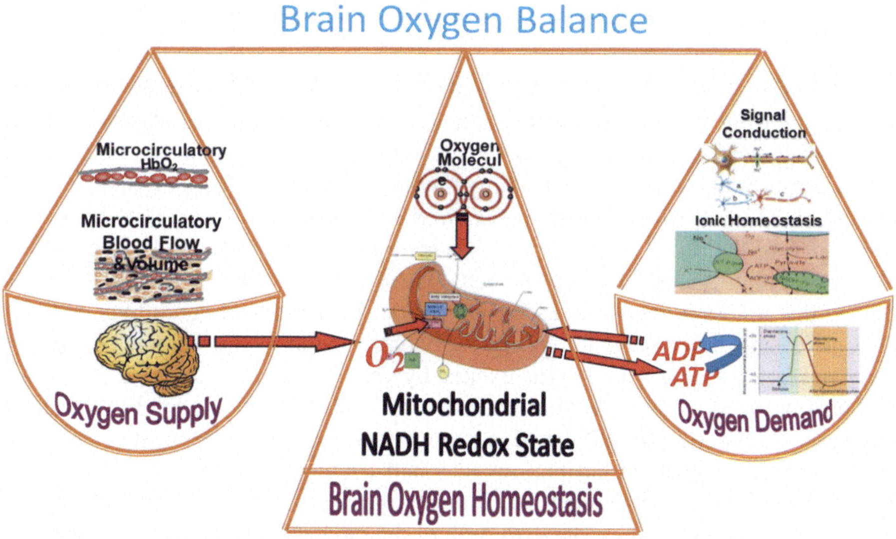

Fig. 1.22 Presentation of Brain Oxygen Balance or Brain Oxygen Homeostasis evaluated by the ratio between brain oxygen supply and brain oxygen demand. (Mayevsky 2013)

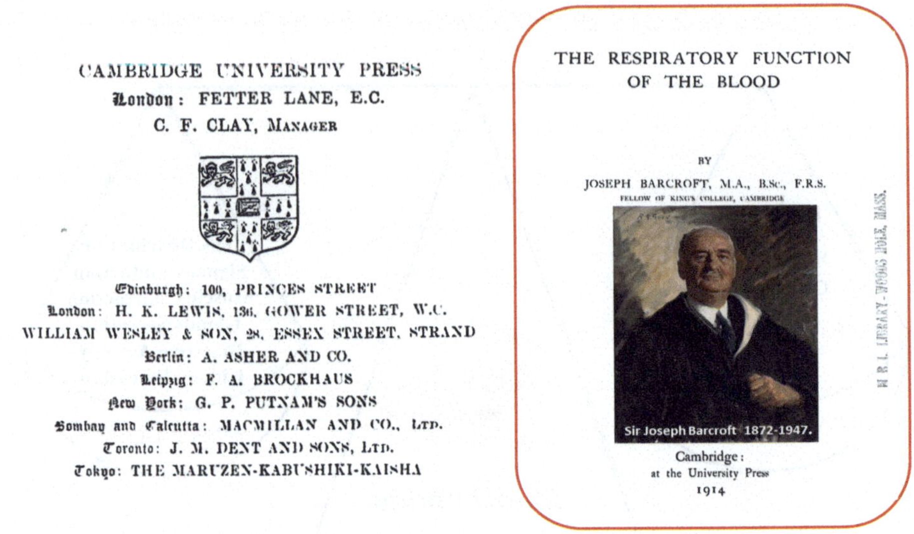
CAMBRIDGE UNIVERSITY PRESS
London: FETTER LANE, E.C.
C. F. CLAY, Manager

Edinburgh: 100, PRINCES STREET
London: H. K. LEWIS, 136, GOWER STREET, W.C.
WILLIAM WESLEY & SON, 28, ESSEX STREET, STRAND
Berlin: A. ASHER AND CO.
Leipzig: F. A. BROCKHAUS
New York: G. P. PUTNAM'S SONS
Bombay and Calcutta: MACMILLAN AND CO., Ltd.
Toronto: J. M. DENT AND SONS, Ltd.
Tokyo: THE MARUZEN-KABUSHIKI-KAISHA

THE RESPIRATORY FUNCTION OF THE BLOOD

BY

JOSEPH BARCROFT, M.A., B.Sc., F.R.S.
FELLOW OF KING'S COLLEGE, CAMBRIDGE

Cambridge:
at the University Press
1914

M B L LIBRARY-WOODS HOLE, MASS.

Fig. 1.23 The title page of the book published by J. Barcroft in 1914 [Modified]

This concept—oxygen homeostasis in the brain was investigated at the end of the nineteenth century and was published by Roy and Sherrington in 1890. They studied the effect of various factors on the regulation of blood flow to the brain and came to the following conclusion:

> We conclude then, that the chemical products of cerebral metabolism contained in the lymph which bathes the walls of the arterioles of the brain can cause variations of the caliber of the cerebral vessels: that in this reaction the brain possesses an intrinsic mechanism by which its vascular supply can be varied locally in correspondence with local variations of functional activity.

The interpretation of their conclusion is that, in the brain, changes in oxygen consumption due to changes in functional activity are due to variations in blood vessel diameter. The next step in the development of the concept of Oxygen Homeostasis was made by the leading English physiologist Joseph Barcroft that published the book *The Respiratory Function of Blood* in 1914 as seen in Fig. 1.23. In Chap. 6 of the book, seen in Fig. 1.24, he discussed the subject of "The call for Oxygen by the tissues." He proved that "There is no instance in which it can be proved that an organ increases its activity, under physiological conditions, without also increasing its demand for oxygen". Barcroft came to the second conclusion (seen in Fig. 1.25) that "in no organ excited by any form of stimulus can it be shown that positive work is done without the blood supply having to respond to a call for oxygen." Longo (2016) published a review entitled "Sir Joseph Barcroft: one Victorian physiologist's contributions to a half century of discovery." "Barcroft's scientific career may be divided into four major phases of inquiry: that of oxygen (O_2) consumption in tissues, the definition of the oxyhemoglobin saturation (HbO_2) curve, establishing the limits of human tolerance to high altitude, and pioneering the exploration of various aspects of oxygenation of the fetus in utero."

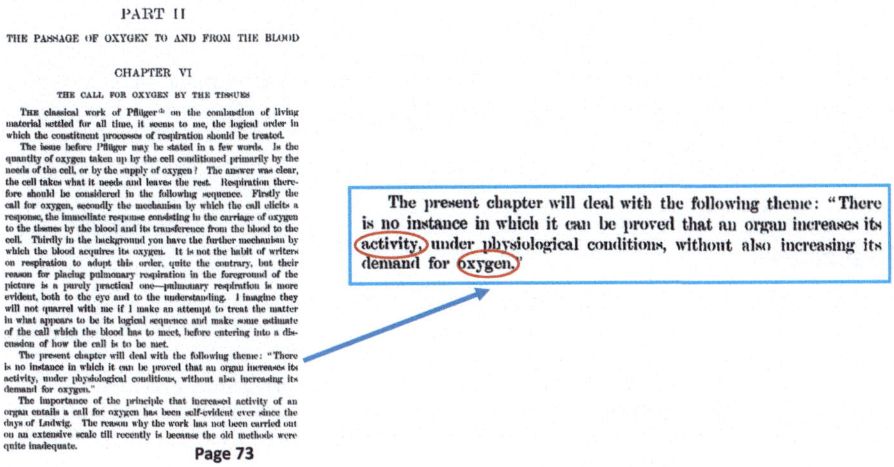

PART II

THE PASSAGE OF OXYGEN TO AND FROM THE BLOOD

CHAPTER VI

THE CALL FOR OXYGEN BY THE TISSUES

THE classical work of Pflüger[1] on the combustion of living material settled for all time, it seems to me, the logical order in which the constituent processes of respiration should be treated.

The issue before Pflüger may be stated in a few words. Is the quantity of oxygen taken up by the cell conditioned primarily by the needs of the cell, or by the supply of oxygen? The answer was clear, the cell takes what it needs and leaves the rest. Respiration therefore should be considered in the following sequence. Firstly the call for oxygen, secondly the mechanism by which the call elicits a response, the immediate response consisting in the carriage of oxygen to the tissues by the blood and its transference from the blood to the cell. Thirdly in the background you have the further mechanism by which the blood acquires its oxygen. It is not the habit of writers on respiration to adopt this order, quite the contrary, but their reason for placing pulmonary respiration in the foreground of the picture is a purely practical one—pulmonary respiration is more evident, both to the eye and to the understanding. I imagine they will not quarrel with me if I make an attempt to treat the matter in what appears to be its logical sequence and make some estimate of the call which the blood has to meet, before entering into a discussion of how the call is to be met.

The present chapter will deal with the following theme: "There is no instance in which it can be proved that an organ increases its activity, under physiological conditions, without also increasing its demand for oxygen."

The importance of the principle that increased activity of an organ entails a call for oxygen has been self-evident ever since the days of Ludwig. The reason why the work has not been carried out on an extensive scale till recently is because the old methods were quite inadequate.

Fig. 1.24 Page 73 copied from the book of Barcroft explaining the relationship between the activity of a tissue and the demand for more oxygen (1914) [Modified]

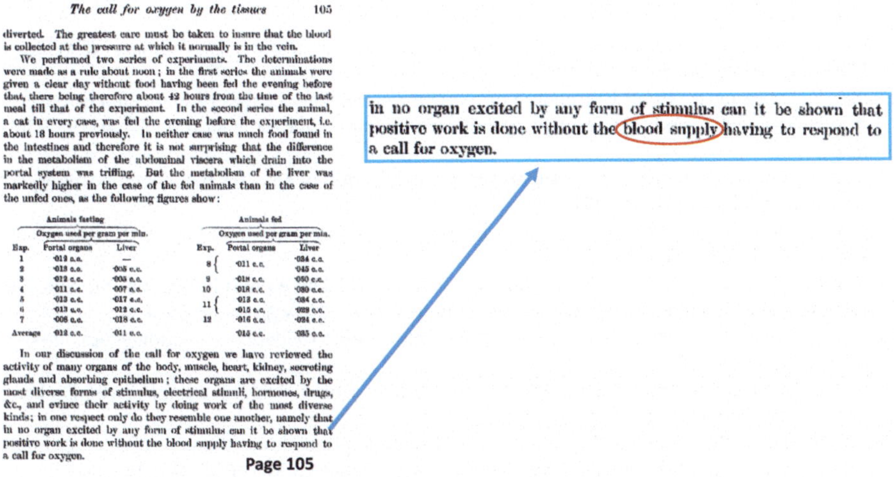

The call for oxygen by the tissues 105

diverted. The greatest care must be taken to insure that the blood is collected at the pressure at which it normally is in the vein.

We performed two series of experiments. The determinations were made as a rule about noon; in the first series the animals were given a clear day without food having been fed the evening before that, there being therefore about 42 hours from the time of the last meal till that of the experiment. In the second series the animal, a cat in every case, was fed the evening before the experiment, i.e. about 18 hours previously. In neither case was much food found in the intestines and therefore it is not surprising that the difference in the metabolism of the abdominal viscera which drain into the portal system was trifling. But the metabolism of the liver was markedly higher in the case of the fed animals than in the case of the unfed ones, as the following figures show:

Animals fasting		
	Oxygen used per gram per min.	
Exp.	Portal organs	Liver
1	·012 c.c.	—
2	·013 c.c.	·005 c.c.
3	·012 c.c.	·005 c.c.
4	·011 c.c.	·007 c.c.
5	·013 c.c.	·017 c.c.
6	·013 c.c.	·012 c.c.
7	·008 c.c.	·018 c.c.
Average	·012 c.c.	·011 c.c.

Animals fed		
	Oxygen used per gram per min.	
Exp.	Portal organs	Liver
8	·011 c.c.	·034 c.c.
		·045 c.c.
9	·018 c.c.	·050 c.c.
10	·018 c.c.	·030 c.c.
11	·013 c.c.	·034 c.c.
	·015 c.c.	·029 c.c.
12	·016 c.c.	·024 c.c.
	·015 c.c.	·035 c.c.

In our discussion of the call for oxygen we have reviewed the activity of many organs of the body, muscle, heart, kidney, secreting glands and absorbing epithelium; these organs are excited by the most diverse forms of stimulus, electrical stimuli, hormones, drugs, &c., and evince their activity by doing work of the most diverse kinds; in one respect only do they resemble one another, namely that in no organ excited by any form of stimulus can it be shown that positive work is done without the blood supply having to respond to a call for oxygen.

Fig. 1.25 Page 105 copied from the book of Barcroft describing the increase in blood supply to an active tissue that consumes more oxygen (1914) [Modified]

It is important to remember that Oxygen is a drug that may need to be optimized in the level given to patients thus avoiding under as well as overdosing (Branson and Robinson 2011; Martin and Grocott 2013).

As seen in Figs. 1.21 and 1.22, the mitochondrial NADH redox state represents the tissue oxygen balance. The supply of oxygen is dependent upon microcirculatory blood flow (TBF), blood volume (TBV), and the level of oxygen bound to the hemoglobin (HbO_2) in the small blood vessels, namely, in the microcirculation. The level of oxyhemoglobin in the microcirculation is affected by two factors, namely, oxygen consumption by the mitochondria and the microcirculatory blood flow. The demand for oxygen is affected by the specific activities taking place in each organ. The intracellular level of mitochondrial NADH (the reduced form) is a parameter related to oxygen balance.

References

Altmann R (1890) Die Elementarorganismen Und Ihre Beziehungen Zu Den Zellen. Veit & comp, Leipzig

Anon (1970) Carl Wilhelm Scheele (1742–1786) Swedish apothecary. JAMA 212(13):2258–2259

Barcroft J (1914) The respiratory function of blood. Cambridge University Press, Cambridge

Benda C (1898) Ueber die Spermatogenese der Vertebraten und höherer Evertebraten, II. Theil: Die Histiogenese der Spermien. Arch Anat Physiol 73:393–398

Branson RD, Robinson BR (2011) Oxygen: when is more the enemy of good? Intensive Care Med 37(1):1–3

Cassebaum H, Schufle JA (1975) Scheele's priority for the discovery of oxygen. J Chem Educ 52: 442

Chance B (1966) The identification and control of metabolic states. Genootschap ter Bevordering van Natuur-, Genees-, en Heelkunde te Amsterdam:5–37

Chance B (1967) Biochemical studies of transitions from rest to activity. The Williams & Wilkins Company, Baltimore

Chance B, Schoener B (1962) Correlation of oxidation-reduction changes of intracellular reduced pyridine nucleotide and changes in electro-encephalogram of the rat in anoxia. Nature 195:956–958

Chance B, Williams GR (1955a) A method for the localization of sites for oxidative phosphorylation. Nature 176(4475):250–254

Chance B, Williams GR (1955b) Respiratory enzymes in oxidative phosphorylation (III- the steady state). J Biol Chem 217:409–427

Chance B, Williams GR (1956) The respiratory chain and oxidative phosphorylation. In: Nord FF (ed) Advances in enzymology. Interscience Publisher, Inc, New York, pp 65–134

Chance B, Cohen P, Jobsis F, Schoener B (1962) Intracellular oxidation-reduction states *in vivo*. Science 137:499–508

Chance B, Schoener B, Schindler F (1963) The intracellular oxidation-reduction state, F. Dickens and E. Neil. Pergamon Press, London, pp 367–392

Chance B, Oshino N, Sugano T, Mayevsky A (1973) Basic principles of tissue oxygen determination from mitochondrial signals. Oxyg Transp Tissue Adv Exp Med Biol 37A:277–292

Cordeiro PG, Kirschner RE, Hu Q-Y, Chiao JJC, Savage H, Alfano RR, Hoffman LA, Hidalgo DA (1995) Ultraviolet excitation fluorescence spectroscopy: a noninvasive method for the measurement of redox changes in ischemic myocutaneous flaps. Plast Reconstr Surg 96:673–680

Duysens LNM, Amesz J (1957) Fluorescence spectrophotometry of reduced phosphopyridine nucleotide in intact cells in the near-ultraviolet and visible region. Biochim Biophys Acta 24: 19–26

Ernster L, Schatz G (1981) Mitochondria: a historical review. J Cell Biol 91:227s–255s

Haddad SI, Khairallah AA (1936) A forgotten chapter in the history of the circulation of the blood. Ann Surg 104(1):1–8

Harbig K, Chance B, Kovach AGB, Reivich M (1976) *In vivo* measurement of pyridine nucleotide fluorescence from cat brain cortex. J Appl Physiol 41(4):480–488

Harms FA, Bodmer SIA, Raat NJH, Stolker RJ, Mik EG (2012) Validation of the protoporphyrin IX-triplet state lifetime technique for mitochondrial oxygen measurements in the skin. Opt Lett 37:2625–2627

Keilin D (1925) On cytochrome, a respiratory pigment, common to animals, yeast, and higher plants. Proc R Soc Lond B Biol Sci 98:312–339

Keilin D (1966) The history of cell respiration and cytochrome. Cambridge University Press, Cambridge

Keilin D, Hartree EF (1945) Purification and properties of cytochrome C. Biochem J 39:289

Kohen E, Kohen C, Thorell B (1969) Use of microfluorimetry to study the metabolism of intact cells. Biomed Eng:554–565

Kolliker A (1856) Einige Bemerkungen uber die endigungen der Hautnerven un deu bau der Muskelzellen. z Wiss Zool 8:311–325

Longo LD (2016) Sir Joseph Barcroft: one victorian physiologist's contributions to a half century of discovery. J Physiol 594(5):1113–1125

Loukas M, Lam R, Tubbs RS, Shoja MM, Apaydin N (2008) Ibn al-Nafis (1210–1288): the first description of the pulmonary circulation. Am Surg 74(5):440–442

Lubbers DW (1995) Optical sensors for clinical monitoring. Acta Anaesth Scand Suppl 39(104): 37–54

Marshall JL, Marshall VR (2005) Rediscovery of the Elements: Carl Wilhelm Scheele, University of North Texas Libraries, UNT Digital Library. Spring 2005. Crediting UNT College of Arts and Sciences, Indianapolis

Martin DS, Grocott MP (2013) III. Oxygen therapy in anaesthesia: the yin and yang of O_2. Br J Anaesth 111(6):867–871

Mayevsky A (1976) Brain energy metabolism of the conscious rat exposed to various physiological and pathological situations. Brain Res 113:327–338

Mayevsky A (1984) Brain NADH redox state monitored *in vivo* by fiber optic surface fluorometry. Brain Res Rev 7:49–68

Mayevsky A (2013) Shedding light on life: optical assessment of mitochondrial function and tissue vitality in biology and medicine. J Clin Monit Comput 27:1–34

Mayevsky A (2015) Mitochondrial function *in vivo* evaluated by NADH fluorescence. Springer International, Cham

Mayevsky A, Barbiro-Michaely E (2013a) Shedding light on mitochondrial function by real time monitoring of NADH fluorescence: I. Basic methodology and animal studies. J Clin Monit Comp 27:1–34

Mayevsky A, Barbiro-Michaely E (2013b) Shedding light on mitochondrial function by real time monitoring of NADH fluorescence: II: human studies. J Clin Monit Comp 27:125–145

Mayevsky A, Chance B (1973) A new long-term method for the measurement of NADH fluorescence in intact rat brain with implanted cannula. Adv Exp Med Biol 37A:239–244

Mayevsky A, Chance B (1976) The effect of decapitation on the oxidation-reduction state of NADH and ECoG in the brain of the awake rat. Adv Exp Med Biol 75:307–312

Mayevsky A, Chance B (1982) Intracellular oxidation-reduction state measured *in situ* by a multicannel fiber-optic surface fluorometer. Science 217:537–540

Mayevsky A, Sonn J, Barbiro-Michaely E (2013) Physiological mapping of brain functions *in vivo*: surface monitoring of hemodynamic metabolic ionic and electrical activities in real-time. J Neurosci Neuroeng 2:150–177

Miller FA (1987) Joseph Priestley, preeminent amateur chemist. J Chem Educ:64

Mitzner W, Wagner E (1992) On the purported discovery of the bronchial circulation by Leonardo da Vinci. J Appl Physiol (1985) 73(3):1196–1201

Palmer W (2006) Carl Wilhelm Scheele (1742–1786): the life of a great chemist. IPS-USA-2006 internet conference on advances in the internet, processing, systems and interdisciplinary research. Palo Alto, California, USA

Partington JR (1962) The discovery of oxygen. J Chem Educ 39:123–125

Pittman RN (2011) Oxygen gradients in the microcirculation. Acta Physiol (Oxford, England) 202: 311–322
Priestley J (1775) Experiments and observations on different kinds of air. Johnson, London
Reti I (1952) Leonardo da Vinci's experiments on combustion. J Chem Edu 29(12):590–596
Retzius G (1890) Muskelfibrflle und Sarcoplasman. Biol Untersuch Stockholm, NF 1:51–88
Roy CS, Sherrington CS (1890) On the regulation of the blood-supply of the brain. J Physiol 11(1–2):85–158. 117
Scheele CW (1777) Chemische Abhandlung von der Luft und dem Feuer (Chemical studies on air and fire). Verlegt von Magn. Swederus Buchhandler, zu finden bey, Uppsala/Leipzig
Schober P, Schwarte LA (2012) From system to organ to cell: oxygenation and perfusion measurement in anesthesia and critical care. J Clin Moni Comput 26:255–265
Severinghaus JW (2002) Priestley, the furious free thinker of the enlightenment, and Scheele, the taciturn apothecary of Uppsala. Acta Anaesthesiol Scand 46(1):2–9
Severinghaus JW (2003) Fire-air and dephlogistication. Revisionisms of oxygen's discovery. Adv Exp Med Biol 543:7–19
Severinghaus JW (2016a) Eight sages over five centuries share oxygen's discovery. Adv Physiol Educ 40(3):370–376
Severinghaus JW (2016b) The most important discovery of science. In: Elwell CE, Leung TS, Harrison DK (eds) Oxygen transport to tissue XXXVII, Advances in experimental medicine and biology, vol 876. Springer, New York, pp 1–16
Smeaton WA (1963) New light on Lavoisier: the research of the last ten years. Hist Sci 2:51–69
Smeaton WA (1992) Carl Wilhelm Scheele (1742–1786): provincial Swedish pharmacist and world-famous chemist. Endeavour 16:128–131
Springett R, Swartz HM (2007) Measurements of oxygen *in vivo*: overview and perspectives on methods to measure oxygen within cells and tissues. Antiox Redox Sig 9:1295–1301
Stefanadis C, Karamanou M, Androutsos G (2009) Michael Servetus (1511–1553) and the discovery of pulmonary circulation. Hell J Cardiol 50(5):373–378
Sundt TM Jr, Anderson RE (1975a) Reduced nicotinamide adenine dinucleotide fluorescence and cortical blood flow in ischemic and nonischemic squirrel monkey cortex. I. Animal preparation, instrumentation, and validity of model. Stroke 6:270–278
Sundt TM Jr, Anderson RE (1975b) Reduced nicotinamide adenine dinucleotide fluorescence and cortical blood flow in ischemic and nonischemic squirrel monkey cortex. 2. Effects of alterations in arterial carbon dioxide tension, blood pressure, and blood volume. Stroke 6(3):279–283
Taylor G (2019) The chemical works of Carl Wilhelm Scheele. Ambix 66:90–92
Warburg O (1949) Heavy metal prosthetic groups and enzyme action. T. A. Lawson. Clarendon Press, Oxford
West JB (2008) Ibn al-Nafis, the pulmonary circulation, and the Islamic Golden age. J Appl Physiol 105(6):1877–1880
West JB (2013) The collaboration of Antoine and Marie-Anne Lavoisier and the first measurements of human oxygen consumption. Am J Physiol Lung Cell Mol Physiol 305(11):L775–L785
West JB (2014a) Carl Wilhelm Scheele, the discoverer of oxygen, and a very productive chemist. Am J Physiol Lung Cell Mol Physiol 307(11):L811–L816
West JB (2014b) Joseph Priestley, oxygen, and the enlightenment. Am J Physiol Lung Cell Mol Physiol 306(2):L111–L119
West JB (2017) Leonardo da Vinci: engineer, bioengineer, anatomist, and artist. Am J Physiol Lung Cell Mol Physiol 312(3):L392–L397
Wilkinson DJ (2004) Resuscitation great. The contributions of Lavoisier, Scheele and Priestley to the early understanding of respiratory physiology in the eighteenth century. Resuscitation 61(3): 249–255
Williams KR (2003) The discovery of oxygen and other Priestley matters. J Chem Educ 80:1129–1131
Wilson DF (2008) Quantifying the role of oxygen pressure in tissue function. Am J Physiol Heart Circ Physiol 294:H11–H13

Chapter 2
Body Organs' Energy Metabolism

2.1 Brain Energy Metabolism

Evaluation of brain physiological functions, in real time, could be done using two methodological approaches.

1. The first approach that could be classified according to the *invasiveness* of the technologies used is shown in Fig. 2.1.
2. The second approach presented in Fig. 2.2 is based on the *orientation* of the monitoring, namely, System-oriented or Organ-oriented monitoring.

The lists of techniques that are presented in Fig. 2.1 are the major ones used, and additional new methodologies may not be included or may appear in another category as well. The noninvasive group includes most of the techniques that are used in patients on a daily basis. In this approach, the sensors may touch the skin but will not penetrate the skull. This group could be subdivided into the mapping/imaging option or the local measurement approach. In the second group, the sensors may be located epidural (below the bone) or subdural touching the pia mater. In the third approach, the sensor or the probe is introduced into the tissue itself creating a new microenvironment and some damage. The number of parameters that are monitored in each option is one or two, as shown in Fig. 2.1. We introduce here the term "physiological mapping"as presented in the minimally invasive group named the Multi Parametric Approach or Multi Parametric Assembly (MPA). The aim of the MPA is to provide real-time data describing the relationship between hemodynamic, metabolic, ionic, and electrical activities in the cerebral cortex.

Monitoring of patient in real time could be divided into two possible categories as shown in Fig. 2.2. In part A, when a specific organ is monitored, the information accumulated represents the events occurring in the organ itself. For example, monitoring the intracranial pressure (ICP) or the Electroencephalography (EEG) in neurosurgical patients will provide information on the events developed in the brain itself. The information is dependent initially upon local pathological events that

© The Author(s), under exclusive license to Springer Nature Switzerland AG 2025

A. Mayevsky et al., *Body Oxygen Homeostasis and Mitochondrial Function*,
https://doi.org/10.1007/978-3-031-94113-9_2

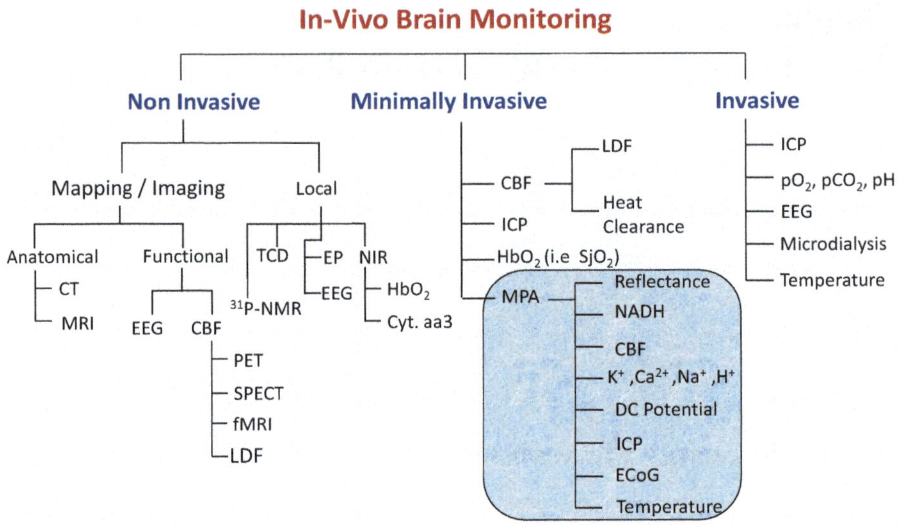

Fig. 2.1 Schematic presentation of the various techniques available for patients or experimental animal monitoring. The techniques are classified according to the invasiveness of the monitoring. (Mayevsky et al. 2013)

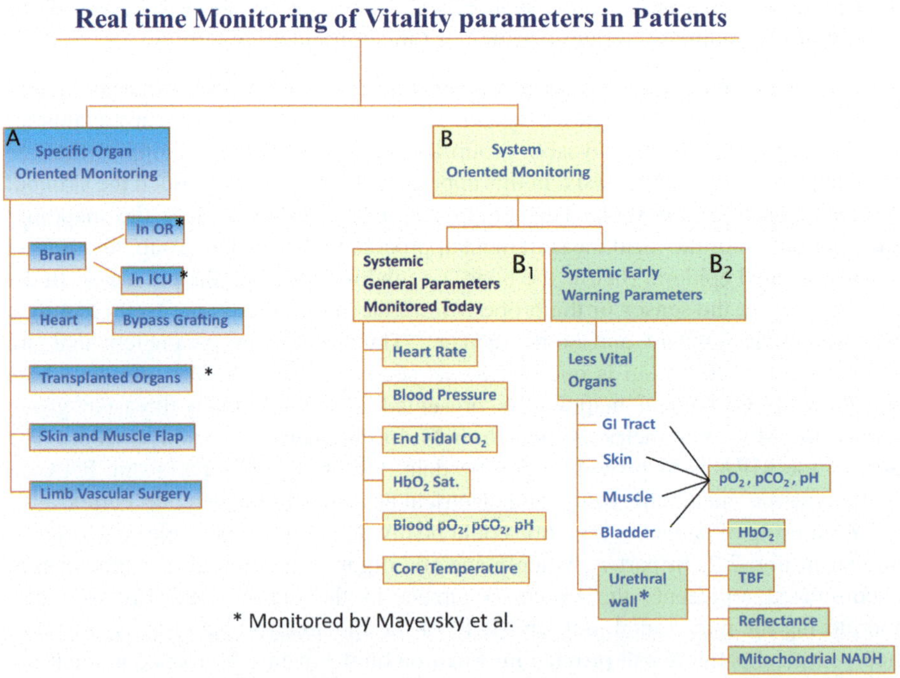

Fig. 2.2 Patients monitoring takes place at two levels. **A**—Monitoring of the function of specific organs. **B**—Monitoring of systemic parameters. **B1**—Systemic general parameters which are monitored today and **B2**—systemic early warning parameters which could be monitored by the clinical devices developed by Mayevsky et al. © Reprinted with kind permission of Springer Science + Business Media. (Mayevsky and Barbiro-Michaely 2013b)

occurred in the brain. Monitoring the Electrocardiography (ECG) represents the electrical activity of the heart, and most of the changes are correlated to local and not systemic events initially.

The second approach in monitoring patients is shown in Fig. 2.2b. This approach is divided into two subgroups labeled as B1 and B2. The parameters shown in B1 are measured routinely from the cardiovascular and respiratory systems. It provides information on systemic events such as changes in gas exchange in the respiratory system or cardiac output affecting the perfusion of various organs in the body. The second group of parameters, shown in part B2, is measured at the tissue level and the accumulated results represent the oxygen balance of the entire body.

Normal brain mitochondrial function is a precondition for the performance of all other brain functions; therefore, a short introduction on brain energy metabolism is presented here. The aim of this chapter is to demonstrate the historical development of the technology used in monitoring the brain and other organ functions under various pathophysiological conditions. We present the stages of the development and typical results obtained in our laboratory. It is important to note that in our monitoring system, all the probes were placed on the surface of the brain and never penetrated the tissue itself. Most of the references cited in this chapter were published by our group.

The functional capacity of the brain is related to its ability to perform its work. It is possible to assess this ability through the knowledge of changes in the oxygen balance, i.e., the ratio between oxygen supply to oxygen demand in the brain. Healthy brain cells perform various types of activities, as presented on the right side of Fig. 1.22 (energy or oxygen demand). The energy is derived through several complex enzyme systems, in which oxygen is the ultimate electron acceptor. The electron transfers down the respiratory chain resulting in the production of ATP. Concomitantly with the electron transport, the respiratory chain components switch between reduced and oxidized states, each of which has different spectroscopic properties. The formation of the pyrophosphate bonds depends on the sufficiency of sugar and oxygen functions, whose inadequacy can, ultimately, lead to death. Since most of the energy consumed by tissues is dependent on the availability of oxygen, the terms "energy" and "oxygen" are used here synonymously. In a normal healthy brain, the ratio or balance between oxygen supply and oxygen demand is positive and reflects the brain cell's functional capacity to do work. That is, the supply mechanism of blood flow and the oxygenated blood circulation is able to provide the spectroscopic properties of the respiratory chain components that are unique to their redox status and used as internal markers of the state of oxidative phosphorylation.

In excitable tissues, such as brain or muscle tissue, as well as in other cells, the activity of Na-K-ATPase is very sensitive to alterations in ionic homeostasis. An increase in extracellular potassium ion concentration, K^+, will stimulate pumping activity in order to bring the extracellular K^+ back to normal levels, i.e., 3 mM range. The activation of Na-K-ATPase will increase the hydrolysis of ATP, and thus the mitochondria will phosphorylate the ADP molecules that are released. The accelerated activity of the mitochondria will be accompanied by a more oxidized state, more

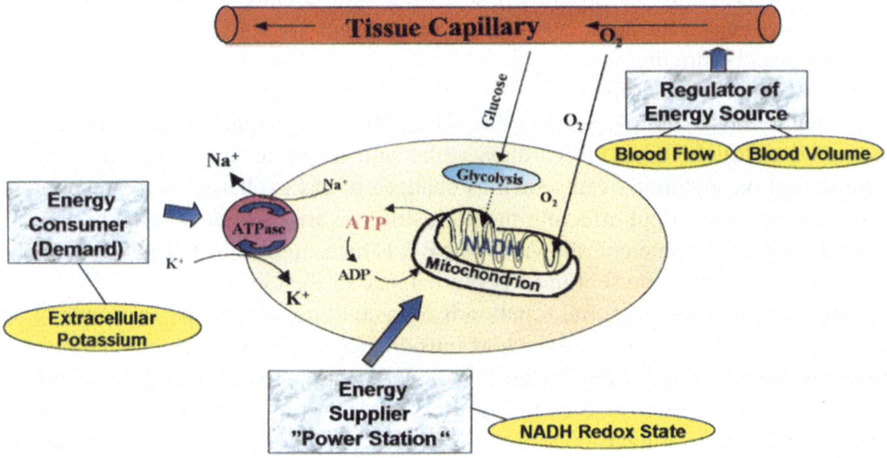

Fig. 2.3 The basic concept of tissue energy balance demonstrated by energy demand, energy supply, and the regulator

oxygen delivery to the cells, blood flow, and blood supply. This coupling between energy consumption and energy production is maintained as long as the O_2 supply is well regulated, as seen in Fig. 2.3.

Under conditions where the oxygen supply or delivery is limited, e.g., after a stroke or heart attack, the energy supplier, i.e., the mitochondria, will not be able to produce the amount of ATP needed. As a result, energy-demanding processes will be restricted. The net effect of the imbalance between energy demand and supply will be manifested by a decrease in the tissue's ability to do work. This can lead to the development of various pathological states. The left side of Fig. 1.22 presents the various parameters that could be monitored in addition to mitochondrial function as a representative of oxygen supply. In order to evaluate brain oxygen balance, it is necessary to measure parameters that represent the oxygen supply and demand at the same time. Mitochondrial NADH redox state represents the supply as well as the balance. Therefore, it is necessary to measure more parameters in addition to mitochondrial NADH. As can be seen in Fig. 2.4, our approach aims to monitor, in real time, a small volume of the cerebral cortex containing all the tissue elements that are parts of a typical functioning brain. We are interested in the microenvironment of the brain containing neurons, glia, synapses, and the microcirculatory elements (small arterioles and capillaries). The various parameters and the technology developed are presented in Fig. 2.4. During the development process, we pursued the goal of being minimally invasive in terms of the penetration into the cortical tissue itself. It was obvious that the various probes could not monitor the same volume of tissue due to the size of each probe used. Therefore, we attempted to minimize the diameter of the various probes located in the MPA that had a 5–6 mm contact area with the cerebral cortex. In most of the perturbations used, such as global ischemia, anoxia, hypoxia, or hemorrhage, most of the areas in the cortex will

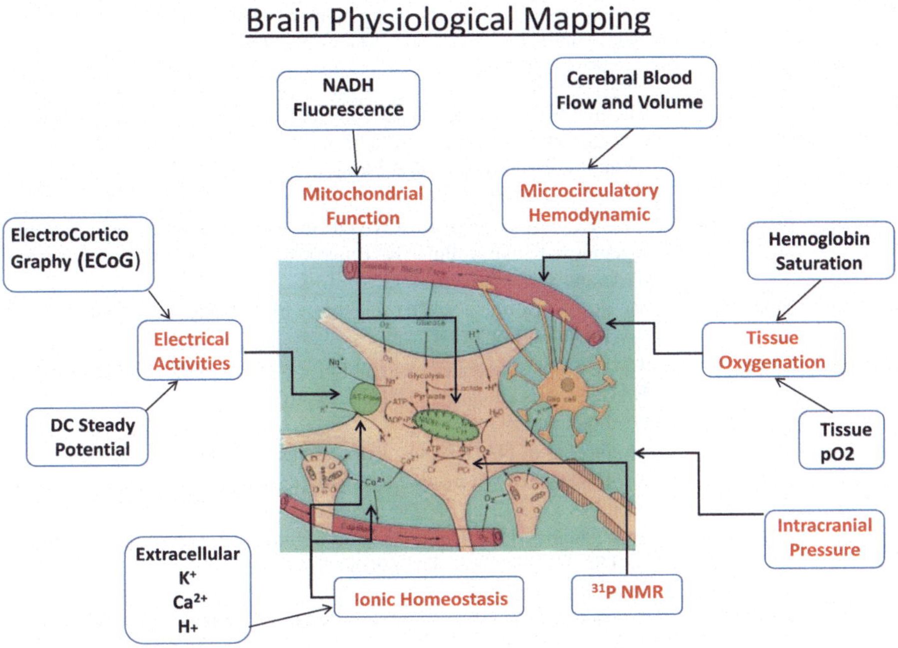

Fig. 2.4 Schematic presentation of the concept "Brain Physiological Mapping." All the presented techniques were developed and used in our laboratory. (Mayevsky et al. 2013)

respond in the same way. We have tested this concept by the development of the multisite monitoring of NADH and other parameters as was presented (Mayevsky and Chance 1982; Mayevsky and Breuer 1992; Mayevsky and Barbiro-Michaely 2013a). As can be seen in Table 2.1, the initial step in the development of the MPA was the establishment of the fiber optic-based NADH monitoring system in 1972 when the first UV transmitting optical fibers appeared. It was a continuation of the long-term usage of old devices for NADH monitoring in vivo, where the animal was located in an optic-based rigid device. The connection of the brain to the fluorometer via optical fibers enabled us to monitor, for the first time, the brain of unanesthetized animals. The initial data on the use of this technology appeared in two papers (Chance et al. 1973; Mayevsky and Chance 1973a).

The list in Table 2.1 is organized in chronological order. All details of the technological aspects and animal preparation appear in the original relevant publications; therefore, a short description of the technology relevant to each parameter will appear in the initial part of the methods section. In the results section, typical responses to various types of perturbations will be presented together with the technology used in the specific study. Our approach was to develop a new upgraded version of the monitoring system and present initial preliminary results. The next step was to run a large well-designed study on a few groups of animals, and the data was quantified and analyzed for its statistical significance.

Table 2.1 Milestones in the Development of Brain Multiparametric Monitoring of NADH Fluorescence and other Physiological Parameters In Vivo by Mayevsky et al.

Protocol	Year	Discovery/activity	Author(s)
1	1973	The first fiber optic-based fluorometer-reflectometer used in the brain of an unanesthetized animal. Monitoring of NADH and electrocortical activity (ECoG)	Chance et al. (1973), Mayevsky and Chance (1973b)
2	1974 + 1977	Simultaneous monitoring of NADH in vivo together with extracellular K^+ (microelectrode and surface electrode) and ECoG	Mayevsky et al. (1974, 1980a) and Crowe et al. (1977)
3	1980	Monitoring of brain NADH together with tissue pO_2 and ECoG	Mayevsky et al. (1980b)
4	1982	The first multiparametric assembly for NADH, extracellular K^+, H^+, DC steady potential, and ECoG	Friedli et al. (1982)
5	1983	Monitoring of NADH, pO_2, extracellular K^+, DC, and ECoG inside hyperbaric chamber	Mayevsky (1983)
6	1990–1992	Simultaneous real-time monitoring of brain NADH, HbO_2, ECoG, DC potential, extracellular K^+, and Ca^{2+}	Mayevsky et al. (1990, 1992)
7	1995	Simultaneous monitoring of brain NADH, CBF, ECoG, DC potential, extracellular K^+ and Ca^{2+}, and H^+	Mayevsky et al. (1995)
8	1996	Multiparametric monitoring of neurosurgical patients	Mayevsky et al. (1996)
9	1997	Monitoring of brain NADH, CBF, DC potential, extracellular K^+, Ca^{2+} together with high energy phosphates by ^{31}P NMR spectroscopy	Mayevsky et al. (1997)
10	2000	Monitoring of the mechanism of CSD propagation	Sonn and Mayevsky (2000)
11	2001	Multiparameteric monitoring under ICP elevation	Barbiro-Michaely and Mayevsky (2001)
12	2003	Multiparametric monitoring of rats under traumatic brain injury	Rogatsky et al. (2003)

2.1.1 Methodology Used

NADH Monitoring

NADH can be measured by utilizing its absorption spectrum in the UV range, as well as by the blue fluorescence spectrum under UV illumination. All details regarding the monitoring of NADH appear in Chap. 1.

Local Cerebral Blood Flow (CBF)

To measure in real time the cerebral blood flow (CBF) from the same cortical area as the MPA location, we used the Laser Doppler flow (LDF) technique (Dirnagl et al.

1989; Haberl et al. 1989; Wadhwani et al. 1990). The LDF measures relative flow changes, and readings have been shown to correlate with the relative changes in CBF measured by the two other quantitative approaches (for a review see Wadhwani et al. 1990). The principle of the LDF is to utilize the Doppler shift, namely the frequency change that light undergoes when reflected by moving red blood cells. A beam of low-power or diode laser light is transmitted by an optical fiber to the tissue. After the multiple scattering of the light, another optical fiber picks up the reflected light which is recorded by a photo detector. The run signal is analyzed by a complicated algorithm developed by the manufacturers, and the results are presented in percentage of a full scale (0–100%), thereby providing arbitrary relative flow values. The change in the total back-scattered light is an indirect measure of the blood volume in the sampled tissue. To quantify and normalize CBF values, we defined the reading value after death as 0 CBF. The 100% value was defined as the percent CBF read on the LDF scale during the control period.

Oxygen electrodes

The electrodes were constructed inside 1.6 mm (OD) polyethylene tubing (PE-160). Platinum wire (25 mm) was sealed in glass by a flame, and an insulated lead wire was attached to the other end of the platinum wire. Two assemblies, along with Teflon-coated 250-micron silver wire (for use as a reference), were pulled into the PE tubing so that the glass-sealed ends of the platinum were flush with the end of the PE tubing. The Teflon on the end of the silver wire was removed, and this bare section could extend beyond the edge of the PE tubing. The electrode's zero response was tested in saline bubbled with N_2, and its linearity was tested in saline bubbled with different mixtures of O_2 and N_2. A cellulose diacetate membrane was put on the tip by dipping the electrode in a 5% cellulose diacetate solution. The calibration and calculation of pO_2 in the brain are only a relative term, since the absolute value of pO_2 is very sensitive to the electrode location on the brain.

Ions-Selective Electrodes and DC Potential

To monitor the extracellular levels of K^+, Ca^{2+}, and H^+, we used specially designed mini electrodes made by World Precision Instruments (WPI; Sarasota, FL). A flexible tubing made of polyvinyl chloride was sealed at one end with a membrane sensitive to a specific ion. The tube was filled with the appropriate solution and connected to an electrode holder with a salt bridge between the membrane and an Ag-AgCl pellet located inside the holder. The interface between the polyvinyl chloride tubing and the holder was glued with epoxy, and such electrodes were usable for a few weeks. The sensitivity of the electrodes to the specific ion was close to the Nernstian value, namely, 50–60 mV/decade for K^+ and H^+ or 25–30 mV/decade for Ca^{2+}.

DC potential was measured concentrically around the ion-selective electrodes. Each electrode had a saline bath around its perimeter, and an Ag-AgCl electrode (WPI) was connected to it.

Reference Electrode

An Ag-AgCl electrode connected via a saline bridge to the neck area of the animal was used as the reference electrode. Polyethylene tubing stuffed with a cotton string that expanded when wet was inserted into the electrode holder (WPI) and glued with 5-min epoxy.

Electrocorticography-ECoG

Spontaneous electrical activity of the brain surface was measured by two polished stainless steel rods or silver wire inserted into the MPA.

Temperature Measurements

Brain temperature was measured with a thermistor probe) Yellow Springs Instruments) located inside the MPA and connected to a thermometer.

Data Collection and Analysis

Relative CBF. The LDF provided arbitrary units (blood flow and volume) that were calibrated in relative terms compared with the normoxic brain that served as the control (0–100% range).

NADH redox state. Measurements of NADH provided a method for evaluating the mitochondrial redox state. The normoxic NADH level is considered to be 100%, and the changes due to the treatment given were calculated in comparison with the normoxic level. Mayevsky (1976) has shown in the past that the NADH redox state correlates well with the brain's functional activity. The origin of the NADH signal in our system is mainly mitochondrial (for details see Mayevsky and Rogatsky 2007) whereas the cytoplasmic NADH contribution to the signal is negligible. Nevertheless, because the NADH is not measured in a purified form, we prefer to label the trace in the records as corrected fluorescence (CF).

Multiparametric Calculation. The MPA provides analog signals of electrical activities and NADH redox states as well as extracellular concentrations of K^+, H^+, and Ca^{2+}. The mV values measured by the electrodes were transformed to mM values by using the standard calibration technique (Friedli et al. 1982) and the principles of the Nernst equation. All calculations were performed via special software programs.

Real-time Data Acquisition. Because of the large number of parameters measured and the need for the logarithmic transformation of the data from the K^+, Ca^{2+}, and H^+, electrodes, it was necessary to use a computerized system for data acquisition and storage. Furthermore, to record all the analog signals (at least 16 channels), it was more appropriate to use a computer display instead of paper for continuously collecting information as presented in Fig. 2.5.

Animal Preparation for Monitoring

All experimental protocols were approved by the institutional animal care committee under the instruction of the National Institutes of Health. The experimental procedures were detailed previously (Mayevsky and Chance 1982; Mayevsky

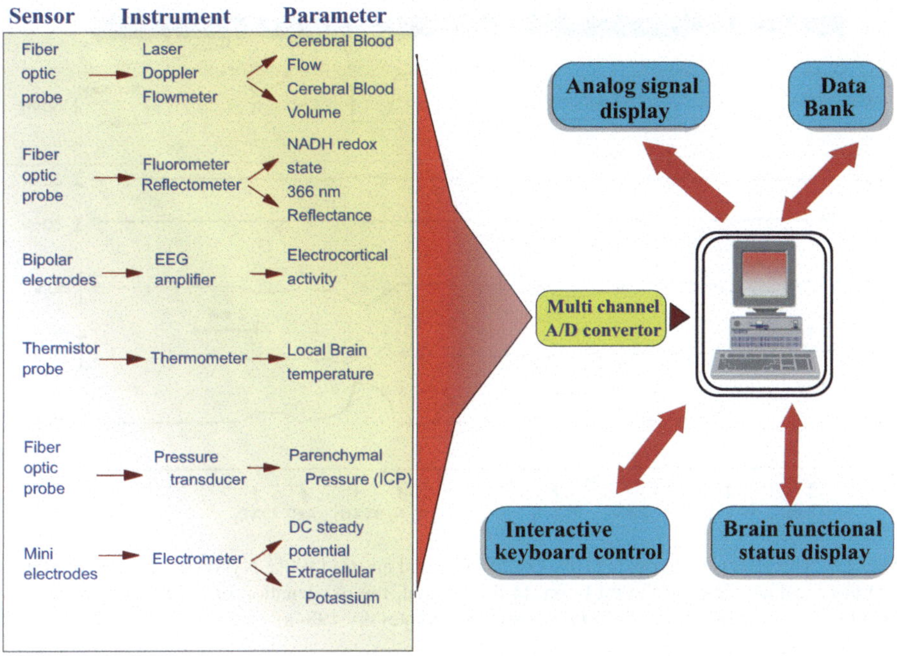

Fig. 2.5 Schematic presentation of the multiparametric monitoring system applied in our studies

1984a, b). In order to demonstrate the performance of the various monitoring devices, we used male rats and male Mongolian gerbils. The preparation of the animals for monitoring is presented in Chap. 1.

2.1.2 Typical Results of Brain Physiological Mapping

The effects of complete deprivation of oxygen (anoxia) on the brain were detected when the animal was exposed to hypoxia, as shown in Fig. 2.6 (Mayevsky 1984a). The rat was exposed to 10%, 5%, or 100% N_2. The decrease in oxygen supplied to the rat resulted in a gradual decrease in brain pO_2, as well as in an increase in NADH. The ECoG showed a clear response only to 100% N_2. This response corresponded to a slight increase in extracellular K^+.

Figure 2.7 shows the use of a multiparametric system that includes a Laser Doppler Flowmeter to record microcirculatory blood flow (Mayevsky et al. 1992). As seen, two sets of responses to Cortical Spreading Depression (CSD) were recorded after 0.1 M KCl application to the surface of the brain 2 mm anterior to the MPA. A clear correlation between the oxidation of NADH (CF) and a large increase in relative CBF can be seen during the six cycles recorded. The changes in relative CBF are calculated in relation to the control values before the waves were

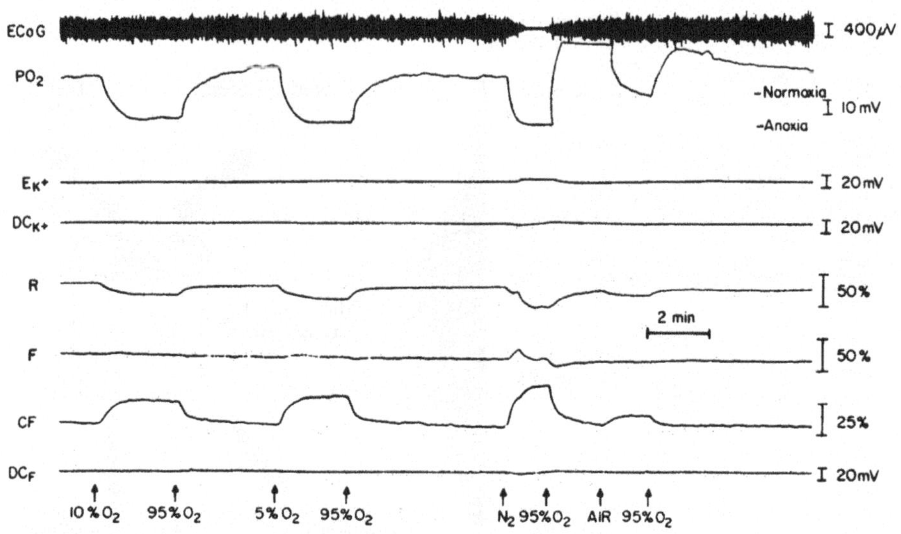

Fig. 2.6 The effects of hypoxia (10% O_2, 5% O_2) and anoxia (100% N_2) on the metabolic, ionic, and electrical activities of the rat brain. In this animal, the oxygenation of the brain responses to breathing air as compared to 95% O_2 are shown. (Mayevsky 1984a)

initiated, and in this animal, it was 4–5 times higher. The cycles of CSD were not so clear on the volume trace, and we suspect it was a reading problem, since in other studies a clear connection between flow and volume was found using the same Laser Doppler Flowmeter.

2.1.3 Effects of Elevated Intracranial Pressure (ICP) on Brain Functions

2.1.3.1 Introduction

Raised intracranial pressure (ICP) develops in most cases involving severe head injury and is a major cause of morbidity and mortality in such cases. Intracranial monitoring of various physiological and pathophysiological events has a crucial effect on patient treatment as well as short- and long-term outcome. Since ICP monitoring is of clinical importance, many experimental models have been developed for the investigation of intracranial hypertension's effects on brain function. Cerebral blood volume (CBV), cerebrospinal fluid (CSF) pressure, and brain tissue volume directly influence the ICP; thus, alteration of one of these parameters is the primary approach employed in such experiments.

An established method for ICP modification is the "balloon" method, in which a balloon is inserted into either the subdural (Johnston et al. 1973) or extradural space (Shalit and Cotev 1974; Fitch et al. 1977; Nitta et al. 1990; Daley et al. 1995), and

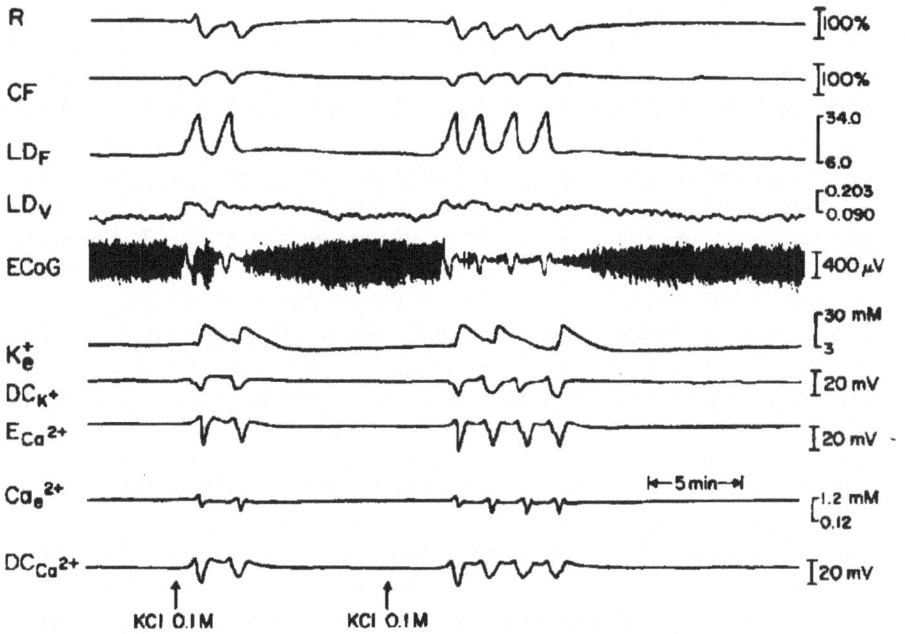

Fig. 2.7 Metabolic, hemodynamic, ionic, and electrical responses to cortical spreading depression measured in rat brain. The MPA used in this study contained two different bundles of fibers for monitoring the relative CBF and NADH redox state. R, CF—reflectance and corrected fluorescence; LDF, LDV—Laser Doppler flow and volume monitored from another optic fiber located in the MPA; ECoG—electrocorticogram; K^+, ECa^{2+}, Ca_e^{2+}—corrected potassium, uncorrected, and e a e corrected calcium ion concentrations, respectively; DCK^+, $DCCa^{2+}$—DC steady potential around the K^+ and Ca^{2+} electrodes (Mayevsky et al. 1992)

inflated with fluid (Johnston et al. 1973; Daley et al. 1995) or gas (Shalit and Cotev 1974; Fitch et al. 1977). Manipulating the volume of fluid or gas leads to ICP elevation to a desired level. A second method for ICP elevation involves the continuous infusion of artificial CSF into the ventricular space of the brain (Malkinson et al. 1985; Tsai and Wang 1988) or the rapid injection of mock CSF volumes into the cerebral ventricle (Malkinson et al. 1985). The rate of infusion or injection of fluid into the brain has a major influence on the experiment (Johnston et al. 1973). For example, a high rate of infusion or a large volume injection leads to more pronounced alterations in ICP level (Botel and Brinker 1994). Since the density of brain tissue differs from one lobe to another, the damage caused by ICP elevation also depends on the location or site in which ICP is manipulated. For instance, ICP elevation in the temporal lobe is more dangerous than comparable changes in the occipital or parietal lobe (Marmarou and Tabaddor 1993).

Another method utilized for ICP alteration is the cisterna magna infusion technique (Botel and Brinker 1994, Hauerberg and Juhler 1994). This method involves the insertion of a catheter through the atlantooccipital membrane or through the occipital bone; the catheter is connected to a bottle or an infusion bag filled with mock CSF or saline, and ICP is manipulated by elevating or decreasing the CSF bottle height. This model enables manipulation of the volume of the CSF system.

When an ICP hypertension model is selected, one must also take into consideration its similarity to the clinical situation. For example, brain swelling, which is one of the major factors in generating the ICP increase in patients following head injury or an acute expanding intracranial mass, is absent in the "balloon" model, where brain water tissue content is intact. However, methods that create local ICP elevation can simulate clinical cases of brain herniation (Shalit and Cotev 1974). Likewise, the rapid volume injection of fluid into one of the ventricles has a local influence. However, its rapid change does not resemble cases of brain edema or brain herniation. Nevertheless, it simulates cases of pharmacological agent administration into the lateral cerebral ventricle (Malkinson et al. 1985). Infusion of fluid into the lateral cerebral ventricle can mimic local brain swelling but not global brain swelling injury. In the current study, we utilized a cisterna magna infusion method for ICP elevation for several reasons. First, the fluid added is distributed through the different brain sections, avoiding a local disturbance. Second, using this model for ICP elevation enables the placement of the MPA device on the surface of the brain, allowing constant monitoring in real time. Third, penetration of the cisterna magna space is less risky than penetration of the brain ventricles. Thus, it reduces the risk of infection or other forms of brain tissue damage. Fourth, this technique enables better manipulation of the ICP level desired, since it is possible to obtain CSF without force when large volumes of mock CSF enter the area and ICP reaches elevated levels due to the high CSF pressure within the cisterna magna (Bouman and Van Wimersma Greidanus 1979). Lastly, as intracranial hypertension is associated with acute brain injury, in which a regional ICP elevation is obtained, and since we sought to isolate the ICP parameter from other parameters associated with acute brain injury, we chose this infusion model for our experiments.

Previous studies evaluating ICP have included the measurement of ICP, CSF pressure, the dynamics of the CSF drainage, CBF and/or vascular resistance, systemic blood pressure, and measurements of phosphocreatine, lactate, and ATP in the brain (Siesjo and Zwetnow 1970; Johnston et al. 1973; Shalit and Cotev 1974; Sutton et al. 1987; Andrews et al. 1988; Morrow et al. 1990; Nitta et al. 1990; Botel and Brinker 1994; Yamane et al. 1994; Nawashiro et al. 1995). To monitor many of these parameters we have developed the multiprobe assembly (MPA) device (Mayevsky and Chance 1982; Mayevsky et al. 1995; Meilin et al. 1996) which is located on the rat brain surface and affixed to the skull. Through this approach, we could monitor in real-time brain energetic state (CBF and mitochondrial NADH redox state), the ionic homeostasis (including K^+, Ca^{2+}, and H^+ extracellular levels), the electrical changes (ECoG and DC potential), and systemic changes. The provision of such detailed information from the same area of the brain should help us understand the complexity of events involved in ICP elevation. The purpose of the current study was to introduce a new approach for estimating brain functional state, by the MPA device, following ICP elevation induced by the developed cisterna magna infusion model.

MPA was developed (Friedli et al. 1982; Mayevsky et al. 1996) to enable its use in experimental animals as well as in human studies. All the probes sample the tissue in the zone of contact and the sensors do not penetrate the brain. In this study, we use a variety of probes that provide information on the various critical physiological and biochemical activities taking place in the brain. Our approach enabled us to monitor, in real time, the relationship between O_2 supply and demand in the brain reflected in mitochondrial function. Figure 2.4 demonstrates our concept of the multimodel monitoring of brain function. The O_2 or energy demand processes are represented by the Na^+ K^+ ATPase activity, which may consume at least 50% of the total energy used by the brain (Erecinska and Silver 1989). In order to evaluate the pump activity in vivo, we monitored the level of extracellular potassium, which is very sensitive to the pump activity. Normal activity of the pump requires a continuous supply of ATP produced by the mitochondria, which serves as the main source of intracellular energy supply. To evaluate in real time, the in vivo mitochondrial function, we monitored the redox state of the NADH by the surface fluorescence approach (Mayevsky and Chance 1982; Mayevsky 1984a) using the fiber optic probe. The regulation of O_2 supply to the mitochondria is accomplished by changes in microcirculatory blood flow. To monitor CBF in real time, the most accepted approach employs Laser Doppler Flowmetry (LDF) (Stern et al. 1977), which provides relative values that correlate with absolute flow values (Dirnagl et al. 1989; Haberl et al. 1989; Wadhwani et al. 1990; Sioutos et al. 1996). The combination of these three monitored parameters shown in Fig. 2.4 enabled us to detect transient changes in brain O_2 balance, which occurred in real time. The simultaneous monitoring of CBF and NADH via an optical approach, that could not be calibrated in absolute values also enables us to avoid possible artifacts (Mayevsky et al. 1998). In addition to the three parameters shown in Fig. 2.4, we also employed other parameters as indicators of brain function (Fig. 2.5).

The monitoring system described in Figs. 2.4 and 2.5 was employed in three studies reported by our laboratory (Barbiro-Michaely and Mayevsky 2001, 2003; Barbiro-Michaely et al. 2005). The results obtained in the first paper (Barbiro-Michaely and Mayevsky 2001) appears also in the second paper (Barbiro-Michaely and Mayevsky 2003) therefore only the results published in the last two papers (Barbiro-Michaely and Mayevsky 2003; Barbiro-Michaely et al. 2005) will be presented here.

2.1.3.2 First Study: Elevated ICP and the Cushing Response

Introduction

Intracranial hypertension occurs in most cases of acute head injury and is one of the major causes of death in those patients (Pickard and Czosnyka 1993). Acute hypertension is commonly considered to accompany cases in which high ICP levels are present (Shalit and Cotev 1974; Schrader et al. 1985; Nichols et al. 1996). This phenomenon is accompanied by a decrease in both heart rate and respiration. Since

Kocher and Cushing's observation in 1901, this phenomenon was called the "Cushing Response" (Fitch et al. 1977). The role and mechanism of the "Cushing Response" is not absolutely clear. According to several studies, the increase in blood pressure, observed in the "Cushing Response" may prevent secondary ischemic damage to the brain. Ungersbock (1995) showed the development of "Cushing Response" in rabbits 5.5 min after the cerebral perfusion pressure (CPP) decreased to 0. In other studies, "Cushing Response" developed when the ICP reached levels close to the arterial blood pressure level (Kearney et al. 1993) or when a decline in CBF was observed (Siesjo and Zwetnow 1970). Successful resuscitation following severe traumatic brain injury requires rapid evaluation of ICP, the autoregulation mechanism, and cerebral metabolism (Nichols et al. 1996). It has been hypothesized that the measurement of systemic blood pressure and ICP waves can be used to differentiate between intact and impaired autoregulation. The autoregulation system is responsible for keeping the cerebral blood supply intact. Two major factors are involved in this mechanism: the CPP and cerebrovascular resistance (Johnston et al. 1973). Yet the exact mechanism in which it developed is still not fully understood. The elevation of ICP to levels in which the CPP reaches the lower limit of autoregulation, leads to changes in cerebral microcirculation. Using a video microscope placed on the surface of the cerebral cortex, the Hudetz study (Hudetz et al. 1995) showed a reduction in cerebral blood velocity and flux during ICP elevation; however, within the autoregulation range, these parameters remained steady. Since the measurement of CBF is also an important factor in autoregulation evaluation, many scientists have studied the influence of ICP on CBF (Hauerberg and Juhler 1994). According to their work, it seems that when ICP is raised, the CBF is maintained at a relatively constant level through a wide range of both ICP and CPP. Beyond these levels, the CBF becomes progressively more directly dependent on the CPP. Although this phenomenon can be seen following ICP elevation or CPP reduction, a similar variability exists regarding the lower limits of autoregulation. For example, Hudetz (1995) showed a reduction in cerebral blood velocity when the CPP decreased below 60–70 mmHg in rats. Another study (Siesjo and Zwetnow 1970) showed a reduction in CBF when the CPP was in the range of 30–50 mmHg. Obviously, different animals have different ranges of autoregulation. For example, in baboons, the lower limit of autoregulation is below 40 mmHg (Nitta et al. 1990) while in cats, the lower limit of autoregulation is considered to be 30–50 mmHg CPP. However, a remarkable decrease in CBF was seen at CPP levels of 72–110 mmHg (Sutton et al. 1987). In a sheep study, CBF measured by the Microsphere technique fell when CPP was below 50 mmHg (Koehler et al. 1989).

Miller and his colleagues (1972, 1973) found different values for the lower limit of autoregulation under different methods of CPP reduction. The threshold level of perfusion pressure at which the CBF began to fall was 40 mmHg with raised ICP versus 60 mmHg after hypotension was induced by hemorrhage (Miller et al. 1972, 1973). Elevated ICP is an important pathophysiological feature of head injury, post-ischemic reperfusion, and in a variety of other diseases. Reduction in CBF causes an imbalance between the oxygen supply and oxygen demand. These events are considered to be the key element in the mechanism of injury developed due to intracranial hypertension. Yet, despite the importance of these mechanisms, there is still very limited information concerning the relationship of ICP to brain tissue oxygenation (Bissonnette et al. 1991). In a previous study, performed on rabbits, cerebrocortical NADH fluorescence was measured using a fiber optic fluororeflectometer probe, under ICP-hypertension conditions (Tranquart et al. 1994). Changes in lactate, phosphocreatine, ATP, NADH, and other metabolites were observed following ICP (Bissonnette et al. 1991). The ATP level, measured in frozen brain samples, after an experimentally induced decrease in CPP to 30–35 mmHg, showed a decrease, whereas an increase in the lactate level began at higher CPP values (Siesjo and Zwetnow 1970). According to Tranquart (1994), severe and extended cerebral anaerobic glycolysis would occur only when the ICP is close to the mean arterial pressure (MAP), and then, at that point, it would induce stroke and death. The aim of this study was to compare the effects of two levels of elevated ICP induced by the infusion of CSF into the cisterna magna. Group A, in which the ICP was elevated to 50–60 mmHg for a period of 13–15 min, was compared to Group B, in which the ICP was elevated to 20 mmHg for a period of 30 min. In addition, we tested the conditions under which the "Cushing Response" developed. The hemodynamic, metabolic, ionic, and electrical activities of the brain were monitored continuously throughout the ICP elevation period and the recovery period. In this study, brain tissue metabolism and function were assessed by the measurement of CBF, mitochondrial NADH redox state, and the level of extracellular K^+ and H^+. It is believed that the information collected by the MPA, in addition to the systemic information, can help to understand some of the responses associated with various levels of intracranial hypertension, including the development of the "Cushing Response."

Methods

The MPA monitoring brain functions during ICP elevation was performed by the MPA developed in our laboratory (Mayevsky 1978; Friedli et al. 1982; Mayevsky et al. 1985, 1995). The MPA (Fig. 2.8a) includes a bundle of optical fibers for mitochondrial redox state measurement (NADH Fluorometry) and CBF using LDF. ICP monitoring was done by a Camino transducer (Camino Laboratories, San Diego, CA, USA). Two mini-electrodes were used for the measurement of extracellular levels of K^+ and H^+, each surrounded by one DC steady potential electrode. A brain temperature thermistor and two ECoG electrodes were also located inside the MPA. The principle of NADH monitoring from the surface of the brain is that excitation light (366 nm) is passed from the fluorometer to the brain via a bundle of optical fibers made of quartz. The emitted light (450 nm), together with the reflected light at the excitation wavelength (366 nm), is transferred to the fluorometer via another bundle of fibers (Mayevsky 1984a). In order to measure the CBF in real time from the same cortical area, we applied the Laser Doppler Flowmeter technique (Perimed model PF 2B, Perimed, Sweden) (Inao et al. 1993; Bederson et al. 1995; Vinas et al. 1995). Extracellular ion levels were monitored using ion-selective membranes

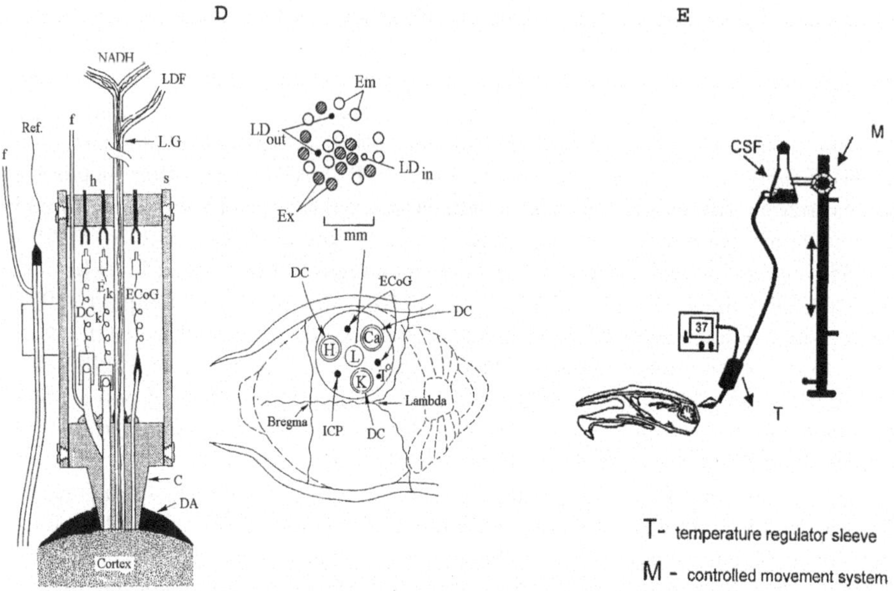

Fig. 2.8 Schematic presentation of the Multiprobe Assembly (MPA) used in this study. (**a**) The MPA includes optic fibers for: NADH redox state fluorometry (*Em* Emission, *Ex* Excitation), Laser Doppler Flowmetery ($LD_{out/in}$), and ICP monitoring. The mini-electrodes for the measurement of the extracellular levels of K^+, Ca^{2+}, H^+ each were surrounded by DC electrode, brain temperature thermistor (T), and two ECoG electrodes. (**b**) The model for intracranial pressure elevation: a controlled movement system (M) enables the elevation and reduction of a container full with mock CSF which is connected to the cisterna magna space. The mock CSF is being wormed to a temperature of 37 °C by a temperature regulator sleeve (T)

connected to an Ag/AgCl mini-electrode (WPI, Sarasota, FL, USA). The DC steady potential was measured by highly stable Ag/AgCl electrodes connected via a salt bridge located around the K^+ and H^+ electrodes. A reference electrode was located below the skin of the animal's neck. All the signals during each experiment were transferred to a multichannel computerized data acquisition and recording system (Labview A/D software, National Instruments Co., TX, USA.), as well as a Grass polygraph paper recorder for further analysis.

Animal Preparation and the Experimental Protocol All experimental protocols were approved by the Institutional Animal Care Committee under the instructions of the National Institute of Health. Wistar male rats weighing 250–300 g were anesthetized by an IP injection of 0.3 ml/100 g Equithesin (each ml contains: pentobarbital 9.72 mg; chloral hydrate 42.51 mg; magnesium sulfate 21.25 mg; propylene glycol 44.34% w/v; alcohol 11.5%; and water). The femoral artery was cannulated for blood pressure monitoring. In Group B, the trachea was also cannulated to enable artificial ventilation. Later, the rat was placed in a head holder, and a midline incision of the skin was made in the dorsal neck area in order to permit an easy approach to the cisterna magna. The atlanto-occipital membrane was punctured using a 21-gauge needle; the appearance of clear CSF indicated a good entrance. The catheter was fixed in the cisterna magna space using cyanoacrylate adhesive (Fig. 2.8b). After the midline incision of the skull, an appropriate 6 mm hole was drilled in the parietal bone, the dura mater was removed gently and the MPA was placed on the brain surface using a micromanipulator. Four stainless steel screws were used to connect the probe to the skull by dental acrylic cement. To induce ICP elevation, the cisterna magna catheter was connected to a bottle filled with mock CSF, which was heated to 37 °C. Raising and lowering the bottle, using a small step motor system, changed the ICP level in the brain. After connecting the MPA to the cortex of the rat brain, as shown in Fig. 2.8b, and a period of 30 min for stabilization, the ICP was elevated in two different protocols. Group A: The ICP was elevated to 50–60 mmHg for a period of 13–15 min in a group of seven rats. The rats were not artificially ventilated because previous experiments showed that respiratory problems start only 20 min following ICP elevation. Group B: The ICP was elevated to 20 mmHg for 30 min for a group of 12 artificially ventilated rats. In both groups, monitoring proceeded for 2 h after ICP reduction.

Statistical Analysis The Student's t-test was used to examine the effects of ICP elevation on the physiological parameters measured from the rat brain. A value of $p < 0.05$ was considered to be statistically significant. Correlation tests were performed in order to determine whether a correlation exists between the parameters monitored by the MPA. All values are presented as Mean ± S.E.

Results

The results of two typical experiments are presented as analog signals in Figs. 2.9 and 2.10. In Fig. 2.9, a typical experiment of Group A is presented. The ICP was elevated to 50–60 mmHg, and as ICP reached its maximal value, CBF decreased to 20% of the normoxic level, followed by a two-step increase in the NADH level. A rapid increase of up to 50% was followed by a partial recovery and a second progressive increase of up to 20% compared to the normoxic level. A biphasic increase is also seen in the reflectance trace, preceded by a progressive decrease. However, the NADH signal remained high until the ICP was reduced. The deterioration of the metabolic state of the brain was associated with an increase in the level of extracellular K^+ of up to 30 mM, but without a clear change in pH in this specific animal. The blood pressure remained stable during the entire elevated ICP period. The ionic changes took place several minutes after the ICP reached its maximal level. A depression in the electrical activity was seen approximately 1 min after the elevation of the ICP. During the recovery period, ICP recovered to its baseline level. A slight hyperemic response was seen in the CBF trace, and NADH became oxidized and returned to the original values. The extracellular ion levels returned to their baseline level. A progressive recovery in K^+ levels started 1 min after ICP reduction. In Fig. 2.10, the results of a typical experiment in Group B are presented.

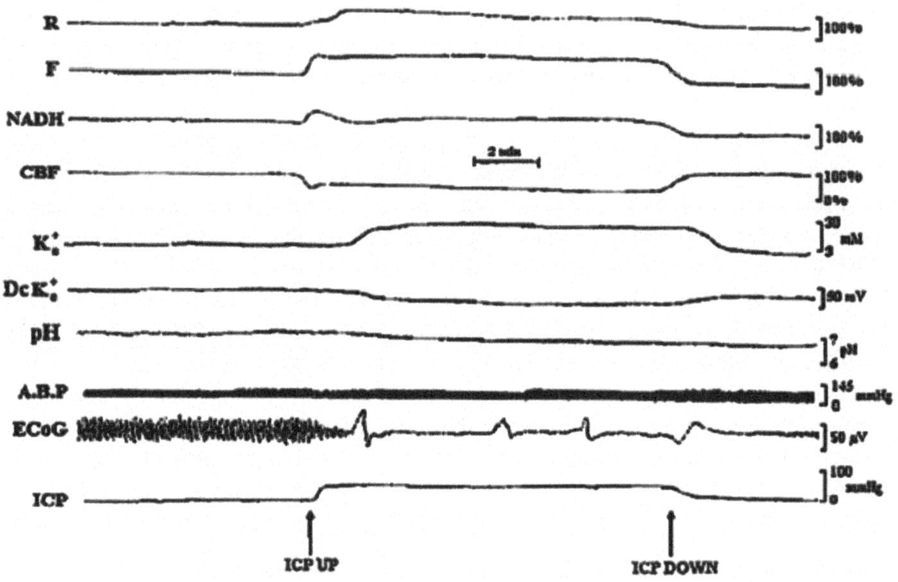

Fig. 2.9 Typical analog records showing the results of ICP elevation (up to 53 mmHg for 13 min) in a rat model presented in Fig. 2.8

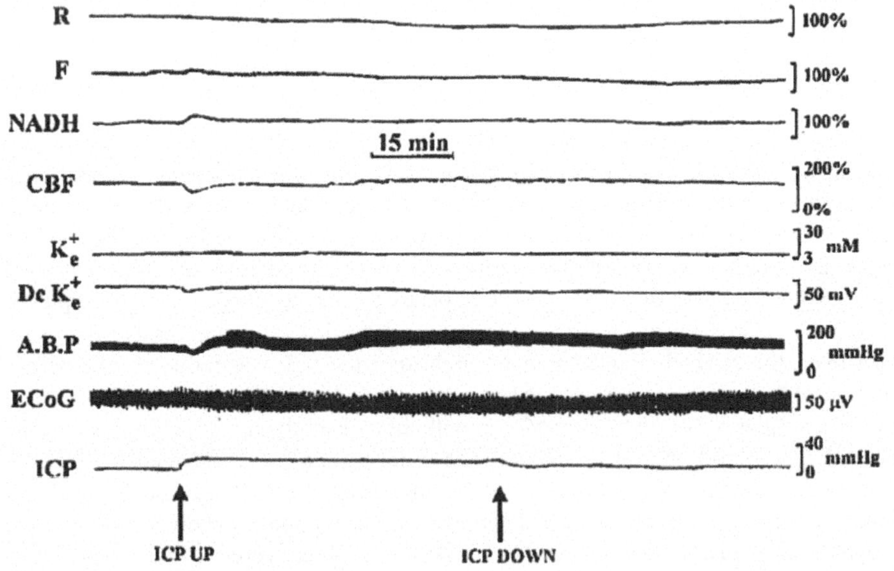

Fig. 2.10 Typical analog records showing the results of ICP elevation (up to 20 mmHg for 30 min) in a rat model presented in Fig. 2.8

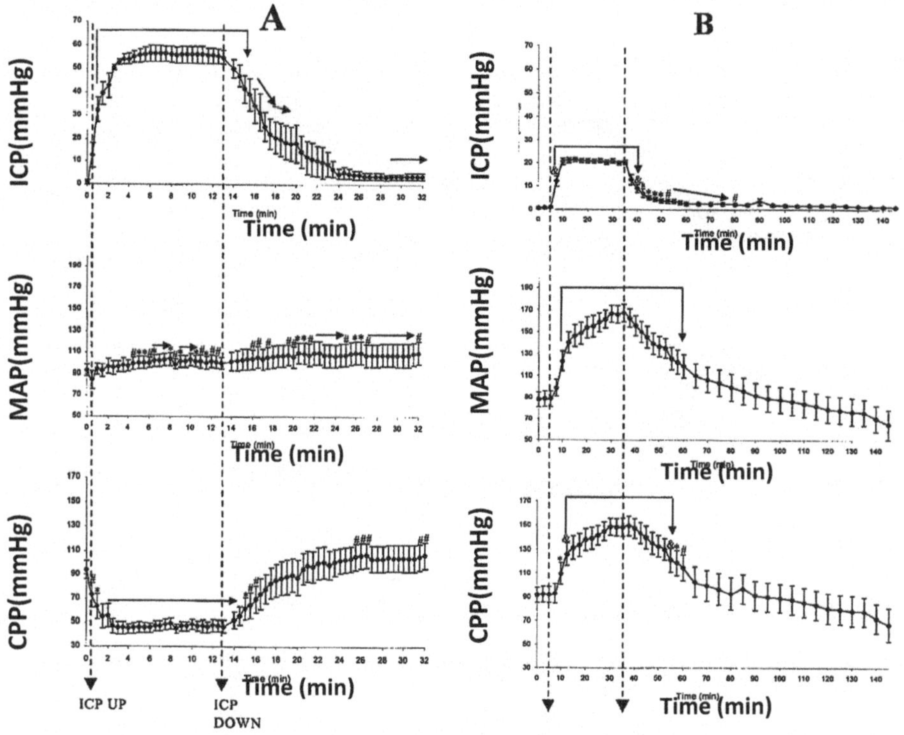

Fig. 2.11 The Mean ± S.E. levels of ICP, MAP, and CPP, recorded during ICP elevation and recovery in Group A ($n = 7$) and in Group B ($n = 12$)

As the ICP level increased, the CBF showed a transient decrease, followed by a rapid recovery and a minor hyperemia. The NADH level was transiently increased by 28%. Extracellular K^+ did not change significantly. The blood pressure decreased slightly and then increased extensively, ultimately reaching a mean blood pressure of 146 mmHg. The electrical activity remained stable throughout the entire experimental monitoring period. While recovery was associated with a progressive decrease in ICP, the CBF remained slightly above the baseline level and the systemic arterial blood pressure began to decrease (Fig. 2.11).

Quantitative analysis of the results was done, and the mean ± S.E. level of the various parameters monitored by the MPA in the two groups of rats is presented in Figs. 2.12, 2.13, and 2.14. Each value was examined for its statistical significance as compared to the baseline level. Figure 2.12 shows the changes in MAP, ICP, and the calculated CPP recorded during the ICP elevation period and throughout the 30-minute recovery period of Group A and 2 h of Group B. As seen, the ICP increased in Group A to the level of 50–60 mmHg, followed by a two-step decrease during the recovery period. We noted a fast decrease in the ICP level in the first 2 min of the recovery period, followed by a slower phase where it decreased to

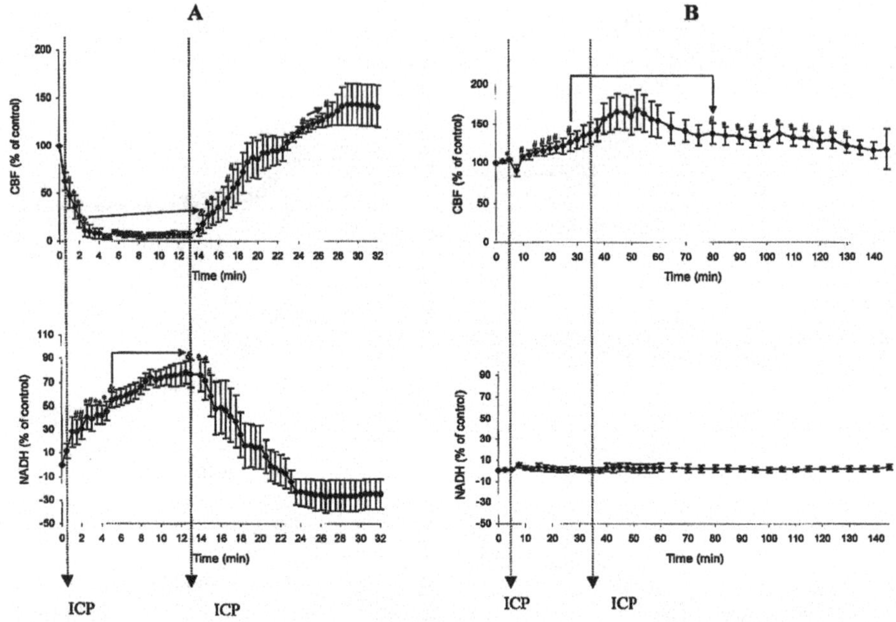

Fig. 2.12 The Mean ± S.E. levels of CBF and NADH recorded during ICP elevation and recovery in Group A ($n = 7$) and in Group B ($n = 12$)

41 mmHg. In Group B, the ICP increased to 20.3 mmHg. The recovery period also showed two-step kinetics. Within the first 5 min, the ICP decreased rapidly and, over the next 40 min, the ICP decreased gradually, reaching the basal level. The systemic mean arterial blood pressure (MABP) in Group A showed a significant increase ($p < 0.05$) 4.5 min after ICP elevation. However, this increase did not stay significantly high through the entire high ICP period (there are samples in which the MAP is not significantly high). The increase in MAP in Group B began 10 min following ICP elevation and persisted for 35 min after ICP decrease. Accordingly, the calculated CPP in Group A showed a significant decrease within 30 s after ICP elevation, followed by a gradual but full recovery, whereas the CPP in Group B showed a significant increase within 10 min following ICP elevation, which persisted throughout the first 30 min of the recovery period. Here, also recovery was completed. The changes in the hemodynamic parameters led to a clear response of the microcirculatory blood flow and the intramitochondrial NADH redox state, as shown in Fig. 2.12. In Group A, the CBF decreased significantly within 30 s and reached a very low level (close to 0) 2.5 min after ICP elevation ($P < 0.001$). The CBF returned to baseline within 3.5 min during the recovery period, followed by a hyperemic level of 132 ± 10% and then a return to the control level 14 min after ICP reduction. Yet, the CBF in Group B showed a slight but insignificant decrease as the ICP started to increase, which subsequently led to a significant ($P < 0.001$) increase in the hyperemic levels to 169 ± 25%. This increase in CBF continued for 95 min

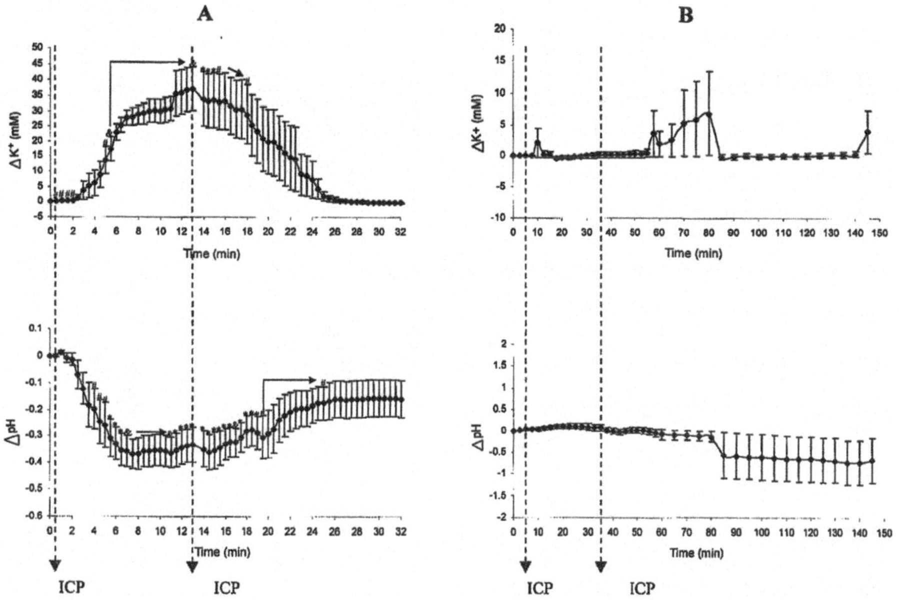

Fig. 2.13 The Mean ± S.E. levels of extracellular K^+ and H^+ during ICP elevation and recovery in Group A ($n = 7$) and in Group B ($n = 12$)

during the recovery period. The CBF decrease in Group A was associated with a significant increase in NADH level 1 min after ICP elevation ($p < 0.05$), while the hyperemic response of the CBF was recorded concomitantly with NADH oxidation (a nonsignificant decrease of 25 ± 12%). In Group B, NADH levels did not change significantly. The extracellular level of K^+ was affected by the elevated ICP via the decrease in CBF and by the inhibition of mitochondrial activity, as seen in Fig. 2.13. A progressive and significant increase in the K^+ level started 30 s following ICP elevation in Group A, reaching the level of 40 ± 17 mM. The K^+ level remained significantly high for 5.5 min within the recovery period. No significant change was seen in the K^+ extracellular level in Group B. A significant decrease in pH level in Group A was seen 4 min after ICP elevation, and it recovered within 12 min during the recovery phase. In Group B, no significant change in pH level was seen.

Correlation analyses were performed in each group of rats in order to assess the relationship between the different monitored parameters, as introduced in Figs. 2.14, 2.15, and 2.16. In each group, the correlation was calculated separately during the ICP elevation period, as well as during the recovery period. Figure 2.14 shows the correlation between the ICP and the hemodynamic parameters. It can be seen that in both groups the MAP correlates with the ICP during both experimental phases, namely, the ICP elevation and the recovery periods ($p < 0.01$). Microcirculation blood flow was also highly correlated with the increase in ICP ($p < 0.01$) in Group A in both phases. In Group B, the correlation between CBF and ICP has a lower

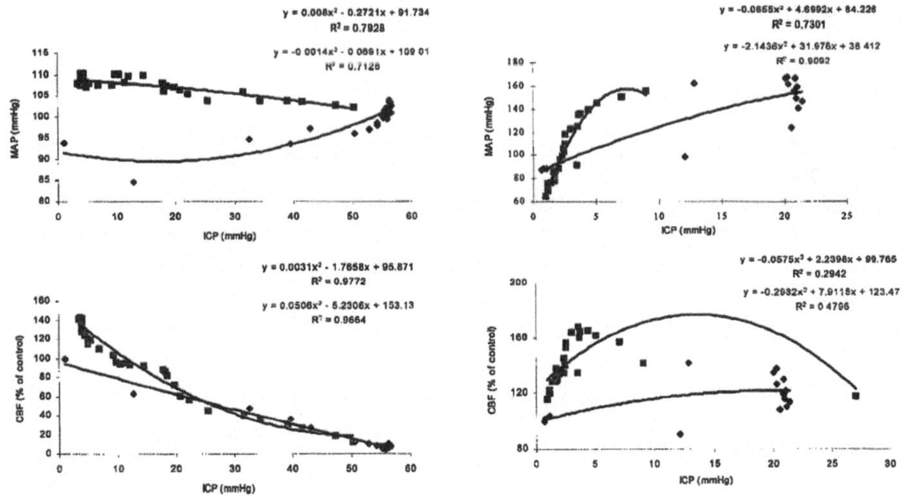

Fig. 2.14 The relationship between MAP and CBF to ICP during ICP elevation (■) and during ICP recovery (♦) in Group A ($n = 7$) and in Group B ($n = 12$). The equation and regression values for each correlation test are presented

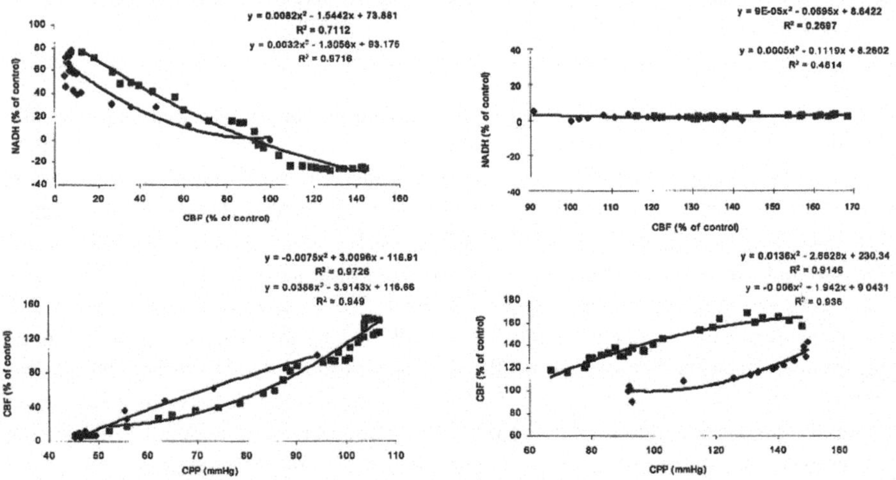

Fig. 2.15 The relationship between NADH and CBF, CBF and CPP during ICP elevation (■) and during ICP recovery (♦) in Group A ($n = 7$) and in Group B ($n = 12$). The equation and regression values for each correlation test are presented

correlation coefficient under the ICP elevation period ($p < 0.05$) as opposed to its value in the recovery period ($p < 0.01$). When the correlation between the CPP and the CBF was examined (Fig. 2.15), it was found that a highly significant correlation ($p < 0.01$) exists in both groups and in the two stages of the experiments. The relationship between the microcirculatory blood flow and the energetic state of the

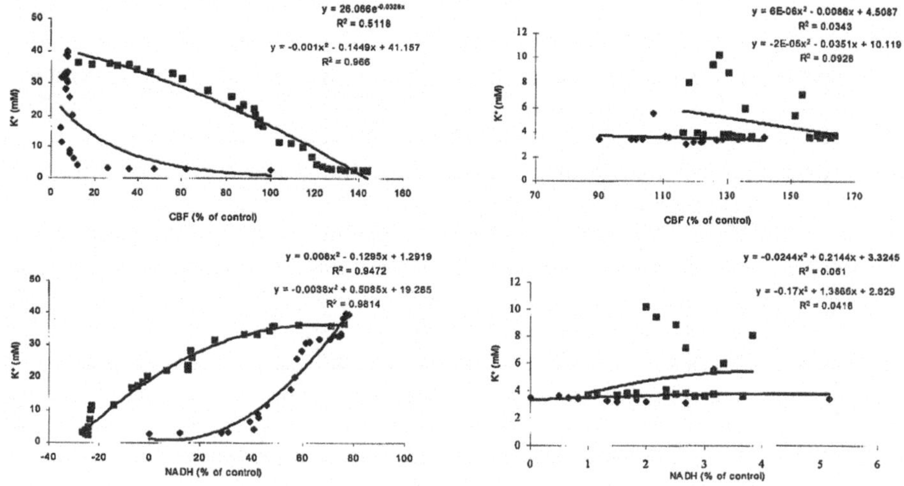

Fig. 2.16 The relationship between extracellular K^+ level and CBF/NADH during ICP elevation (■) and during ICP recovery (♦) in Group A ($n = 7$) and in Group B ($n = 12$). The equation and regression values for each correlation test are presented

brain tissue (NADH) is also presented in Fig. 2.15. As shown, a high correlation exists between CBF and the intra mitochondrial NADH redox state in Group A ($p < 0.01$). The relationship between the extracellular K^+ level and the CBF or NADH is presented in Fig. 2.16. In Group A, a positive exponential correlation exists between the extracellular K^+ level and the CBF during the ICP elevation period and a negative polynomic correlation exists during the recovery period. The correlation between the extracellular K^+ level and the mitochondrial NADH level within the ICP elevation period in Group A is described by a positive polynomic regression line, while in the recovery period a negative polynomic regression line describes the correlation between this parameter. However, in Group B no significant correlation was found between either CBF or NADH and the extracellular K^+ level.

Discussion

Elevated ICP as a result of head injury, brain tumor, or hematoma causes brain-tissue damage, neurological dysfunction, and may finally lead to death (Nitta et al. 1990). Pressure autoregulation mechanisms, which include changes in pia artery diameter, are responsible for maintaining an adequate CBF throughout different CPPs in order to prevent such a negative outcome. Our study supports previous studies showing that ICP elevation can lead to different paths of events depending on the method used, the period of intracranial hypertension, and the degree of ICP elevation.

As shown in both groups of rats (Fig. 2.11), the ICP during the recovery period, following ICP elevation, is characterized by a two-step process. The first is a rapid decrease phase, which occurs due to mock CSF leaving the cisterna space, and the second step involves a slower decrease in the ICP level, in which the physiological state of the brain tissue has a major effect. The systemic vascular response to increased ICP showed a significant increase in arterial blood pressure following ICP elevation in Group B. This phenomenon was described in previous studies as the "Cushing Response." Although considerable literature has been accumulated on the "Cushing Response," there is no unanimous view among scientists regarding the mechanism, effect, and value of this response (Fitch et al. 1977; Schrader et al. 1985; Tsai and Wang 1988; Ungersbock et al. 1995). There are two major explanations for the "Cushing Response." According to one of them, this increase in blood pressure is a compensatory mechanism intended to keep blood pressure above ICP, thereby maintaining adequate perfusion of the vital centers in the brain (Plets 1989). In due course, it was suggested that ischemia in certain parts of the brain stem is responsible for the hemodynamic changes observed in intracranial hypertension, such as the "Cushing Response," which are mediated by the secretion of catecholamines (Ogilvy and DuBois 1987; Kocsis et al. 1991; van Loon et al. 1993). On the other hand, Pasztor et al. (1986) claimed that the "Cushing Response" developed during ICP elevation is an outcome of sympathetic over-activity, caused due to the mechanical distortion of the medulla, and not a result of ischemia or hypoxia involved in intracranial hypertension situations. The importance of the "Cushing Response" plays a role also during parturition in lambs. As seen in Harris study (1989), monitoring of the peripheral organ blood flow, as ICP approached baseline MAP, showed that fetal lambs were capable of sustaining cerebral perfusion by initiating profound visceral vasoconstriction without curtailing placental blood flow. Nevertheless, the second approach claims that the "Cushing Response" signals the breakdown of the brain's mechanisms. This is based on the fact that a prolonged increase in blood pressure endangers the brain, especially when the ICP is high and the brain is already fully compressed (Shalit and Cotev 1974). Ebel et al. (1996) showed that the "Cushing Response" after the induction of subarachnoid hemorrhage, led to a deterioration of the cerebral perfusion. Fitch et al. (1977) concluded that the "Cushing Response" is a very late phenomenon and is indeed a preterminal event. Accordingly, it is the last of six indices that evolve, with a certain sequence, from the start of intracranial elevation. Other signs of a breakdown that are mentioned are bradycardia, arrhythmia, papillary constriction, unilateral papillary dilation, an increase in pulse pressure, and finally an increase in MAP. However, according to Cushing himself, who observed this phenomenon in 1901, systemic hypertension would result only in cases in which it would be advantageous to the animal, and our results support Cushing's assumption. Accordingly, since ICP hypertension in Group A was too radical for autoregulation mechanisms to overcome, the "Cushing Response" was not observed. Therefore, the CPP decreased below the lower limit of autoregulation, which led to a cascade of events including a decrease in CBF, an increase in intramitochondrial NADH, an increase in extracellular K^+, and a decrease in pH level. Similar results were presented in Marshman's study (1997), under a subarachnoid hemorrhage model, where hypertension was seen only with gradual delta ICPs such as ICP of 30 mmHg, but not at severe ICP levels, such as 50 mmHg. Although a progressive increase in arterial blood pressure is seen in Group A, this hypertension is not similar to the "Cushing Response" since it is not drastic and rapid enough. Yet, in Group B when the ICP increased up to a level found within the autoregulation limits, the "Cushing Response" that developed led to the increase in CPP. An increase in systolic blood pressure without the appearance of the "Cushing Response" was already described in a rabbit model where the ICP was raised to 46 mmHg, was left high for 10 min, and then further increased to even higher levels in which the CPP became 0 mmHg. In this study, the mean blood pressure increased progressively under the first stage of ICP elevation, probably due to autoregulation processes. However, the "Cushing Response," including bradycardia and arrhythmia, occurred with a constant time lag of 5.5 ± 0.7 s after the loss of CPP (Ungersbock et al. 1995). The "Cushing Response" could also develop under variant levels of CPP, as indicated from previous studies. In a study using the cisterna magna infusion model and the epidural balloon model in dogs, the "Cushing Response" developed under a very low CPP value. Still, in this model, the dogs were not artificially ventilated, the experiment included a step of respiratory arrest and only regional CBF was monitored (Schrader et al. 1985). Harris' study showed that "Cushing Response" normally acts to mitigate an ischemic insult when the increase in ICP approaches MAP values (Harris et al. 1998). In addition, Shalit and Cotev (1974) showed in cats that the "Cushing Response" developed only when the ICP reaches the level of systolic blood pressure and concluded that the relationships between ICP, BP, CBF, and brain metabolic activity (as reflected in the EEG) differ under various experimental models. However, in this study the increase in BP was accompanied by a progressive decrease in BP and a decrease in rCBF, they concluded that this phenomenon is not a favorable compensatory mechanism designed to maintain brain function as Cushing postulated; more likely, it is a sign of the breakdown of brain mechanisms (Shalit and Cotev 1974). A similar conclusion was obtained in another study where the subarachnoid hemorrhage technique was used for ICP elevation. In this study, the "Cushing Response" during the hemorrhage could not improve the cerebral circulation, but in contrast caused a further increase of ICP. Thus, it was concluded that the "Cushing Response" during a subarachnoid hemorrhage should be regarded as a deleterious rather than a beneficial mechanism (Brinker et al. 1992). Previous studies have shown various levels of CPP as related to the lower limit of autoregulation in the rat model. For example, Hudetz (1995) found that the lower limit of autoregulation in a rat model is 60 mmHg. Miller et al. (1972, 1973) showed that the threshold levels of perfusion pressure at which the CBF began to fall was 40 mmHg with a raised ICP versus 60 mmHg after blood pressure reduction via hemorrhage. Hauerberg and Juhler (1994) found a significant shift in the lower limit of autoregulation toward lower CPP levels during severe intracranial hypertension. Under normal physiological circumstances, a moderate change in CPP between the limits of 50 and 130 mmHg produces little or no change in the CBF of human or other mammalian subjects (1987). Our results support the findings of Hudetz et al. (1995). Accordingly, under the model in which the ICP was elevated to

50–60 mmHg, the CPP reached a value beneath the lower limits of autoregulation thus causing the reduction in CBF (Fig. 2.12). When autoregulation is lost, CBF becomes directly dependent on CPP (Johnston et al. 1973). A formula, presented in a recent study, for evaluating cerebral autoregulation in the rat claimed that a high correlation coefficient between the CPP and CBF, within normal perfusion pressure levels, indicates the loss of autoregulation, while a low correlation coefficient points to an impaired cerebral autoregulation (Botel and Brinker 1994). However, in our study, as seen in Fig. 2.15, a high, significant correlation exists between CPP and CBF in both groups, although pressure autoregulation was intact in Group B and impaired in Group A. Consequently, it may be that using only the CPP and CBF values for pressure autoregulation assessing is not always enough; extended parameters are needed, such as the parameters monitored by the MPA. Stepinska et al. (1995) showed that the use of the transcranial Doppler sonography (TCD) technique in cats, for monitoring the middle cerebral artery flow, can be misleading; therefore, other parameters should be used in similar pathophysiological conditions in order to achieve improved evaluation of the cerebral state. In this study the use of the MPA has the advantage of providing simultaneously monitored hemodynamic, energetic, ionic, and electrical information on the cerebral cortex. During the recovery period, hyperemia was seen in both groups of rats. This phenomenon was described and explained in previous studies (Tsai and Wang 1988; Kawamura et al. 1990; Nawashiro et al. 1995; Ungersbock et al. 1995). An active vasomotor response during the reperfusion phase due to an increase in CO_2 level may explain this phenomenon in Group A where the rats breathe spontaneously and are not respirated (Tsai and Wang 1988; Kawamura et al. 1990). These cardiovascular responses to ICP elevation may contain b-adrenergic mechanisms since propranolol, a b-adrenergic blocker, reduced this response (Tsai and Wang 1988). Furthermore, in Group A, the high concentration of extracellular K^+ and H^+ levels can cause vascular dilatation. Moreover, the disturbed ionic homeostasis increased the energy demand, and this can result in an increase in the CBF (Miller et al. 1972). In Group B, the hyperemia seen is probably due to the increase in MAP, as was also observed in an ischemic heart model (Osbakken and Mayevsky 1996). Impaired energy metabolism was observed under high ICP levels of 50–60 mmHg, as expressed by an increase in the intramitochondrial NADH level, and elevated K^+ and H^+ extracellular levels. Nevertheless, Sutton found a large variability in the CPP levels, which caused a reduction in brain energy metabolism (Sutton et al. 1987). This study showed a significant correlation between the MAP versus ICP, CBF versus ICP, and CBF versus CPP. Yet, it would not be correct to classify a single quantitative relationship between either of these parameters. According to Johnston et al. (Johnston et al. 1973) response pattern of cerebral circulation to ICP hypertension depends on the model used for intracranial hypertension. Moreover, different techniques for ICP monitoring, as well as the duration and severity of the treatment, can increase the variability among the results of different studies. A correlation between the metabolic state of the tissue and the CBF exists in both groups, due to the activation of metabolic autoregulation mechanisms. Disruption in the microcirculatory blood flow in the brain led to impairment of the oxygen supply and accumulation of intramitochondrial NADH. The metabolic state of the brain, evaluated by the extracellular K^+ level, showed a high correlation with the CBF and NADH level ($p < 0.05$) in Group A. Nevertheless, there was no correlation between these parameters in Group B. Moreover, almost all the correlation coefficients' values of Group A are higher than those of Group B. One explanation for this phenomenon is the fact that when the ICP reached very high levels (such as 50–60 mmHg), autoregulation mechanisms could not improve the brain's condition. Hence, the CPP decreased, the CBF decreased, the mitochondrial energy redox state was impaired, and the ionic homeostasis was damaged, thereby causing the extracellular K^+ and H^+ levels to increase. However, when the ICP was raised to 20 mmHg, pressure autoregulation mechanisms responded and improved the brain's hemodynamic, energetic, and metabolic state. Thus, although the ICP increased, the CPP and the CBF also increased, the NADH level remained stable, and the extracellular K^+ and H^+ levels did not change.

Conclusion

According to our results, an ICP of 50–60 mmHg in a rat model reduces CPP to a level which is below the lower limit of autoregulation. However, a full and complete recovery is achieved following such a treatment for a short period of time. Whereas, ICP levels around 20 mmHg, obtained for an even longer period of time, promote autoregulation mechanisms that ensure a steady, hemodynamic, energetic, and metabolic state of the brain. It is also clear that the "Cushing Response" is a beneficial event that ensures an appropriate CBF level when the cerebral blood supply is impaired by elevated ICP. In addition, our results show that the measurement of CBF and CPP is not sufficient for an autoregulation evaluation, and more information is needed. Here, the MPA used in this study may have the advantage of continuous and simultaneous multiparameter monitoring of the cerebral cortex.

2.1.3.3 Second Study: Elevated ICP and Mannitol Administration

Introduction

Intracranial hypertension develops in most cases of traumatic brain injury (TBI). Several ways to treat intracranial hypertension exist including: hyperventilation (Lang and Chesnut 1995; Geraci and Geraci 1996a, b), CSF drainage (Chesnut and Marshall 1993; Pickard and Czosnyka 1993), high dose barbiturate therapy (Go 1991; Bullock and Povlishock 1996b), sedation and paralysis drug therapy (Bullock and Povlishock 1996a, b), raising head position (Williams and Razumovsky 1993), and the use of osmotic and diuretic drugs (Wilkinson and Rosenfeld 1983; McManus and Strange 1993; Pickard and Czosnyka 1993). Hypertonic dehydration of the brain by the administration of osmotic agents, either alone or in combination with "loop" diuretics, has been an accepted treatment for increased

ICP for decades (McManus and Strange 1993). In the last years, mannitol administration has become a cornerstone in the management of increased ICP in patients with head injury (Bullock and Povlishock 1996c). The aim of the present study was to evaluate mannitol influence on the brain physiological state following increased ICP, monitored by the multiparametric assembly (MPA). We chose the cisterna magna infusion technique for ICP elevation, since it enables the isolation of the ICP parameter, and, hence, the direct influence of elevated ICP on the metabolic parameters can be examined. Moreover, since the cisterna magna is located at the posterior area of the head, placing the MPA on the brain surface (at the parietal section) simultaneously with ICP elevation, it enabled real-time monitoring of brain metabolism during the entire elevated ICP period, as well as the recovery phase. According to previous studies mannitol has two probable distinct effects on the brain. The first one is an immediate plasma-expanding effect that reduces the hematocrit and blood viscosity, and increases CBF. Mannitol also increases cerebral oxygen delivery by altering the deformability potential of red blood cells, improving flow characteristics (Fitch et al. 1977; Muizelaar et al. 1983; Muizelaar et al. 1984; Bullock and Povlishock 1996c; Kirkpatrick et al. 1996). These effects are best accomplished by bolus administration of mannitol. The second effect is osmotic dehydration of the brain parenchyma (Wise and Chater 1962). However, scientists have different views regarding the mechanism of action of mannitol in the brain. Recent studies showed that mannitol had no effect on brain water content (Takagi et al. 1983; Constantini et al. 1988; Vinas et al. 1995), while other studies showed that mannitol administration led to a decrease in the water content of the brain (Shalit and Cotev 1974; Jame et al. 1980). The decrease in ICP after mannitol infusion is due to hemodynamic changes, as well as a small reduction in CSF volume (Donato et al. 1994). As indicated earlier, mannitol affects CBF, CBV, ICP, and CPP. Previous experimental studies in the field of elevated ICP monitored mainly ICP and systemic arterial blood pressure (SABP). However, in order to obtain more accurate information, monitoring of other parameters such as CBF and cerebral metabolic activity (mitochondrial function and extracellular ions' level) may have a significant advantage. In the present study, we used the MPA to evaluate the effect of mannitol on brain hemodynamic, metabolic, ionic, and electrical activities following ICP elevation.

Methods

Multiparametric Monitoring System

Monitoring of brain functions was performed by the MPA developed in our laboratory (Mayevsky et al. 1985, 1995; Meilin et al. 1996). The MPA (Fig. 2.8a) includes a mixed bundle of optical fibers for NADH redox state fluorometry and CBF using LDF. The same holder contained a probe for ICP monitoring, two mini-electrodes for the measurement of the extracellular levels of K^+ and H^+, each surrounded by a DC electrode, a brain temperature thermistor, and two ECoG electrodes (Barbiro-Michaely and Mayevsky 2001). The principle of NADH monitoring from the surface of the brain is that excitation light (366 nm) passes from the fluorometer, which contains appropriate filters, to the brain via a bundle of optic fibers made of quartz. The emitted light (450 nm), together with the reflected light at the excitation wavelength (366 nm), is transferred to the fluorometer via another bundle of optical fibers. In order to measure the CBF in real time from the same cortical area, we applied the LDF technique (Bederson et al. 1995; Vinas et al. 1995). Extracellular K^+ and H^+ ions level was monitored using selective mini-electrodes attached to Ag/AgCl holder. The DC steady potential was measured by highly stable Ag/AgCl electrodes connected via a salt bridge located around the K^+ and H^+ electrodes. A reference electrode was located below the skin of the animal's neck.

Animal Preparation

All experimental protocols were approved by the Institutional Animal Care Committee under the instructions of the National Institute of Health. Wistar male rats, weighing 250–300 g, were anesthetized by IP injection of 0.3 ml/100 g Equithesin (each milliliter contains 9.72 mg pentobarbital, 42.51 mg chloral hydrate, 21.25 mg magnesium sulfate, 44.34% w/v propylene glycol, alcohol 11.5%, and water). The trachea was cannulated to enable artificial ventilation. Later, the rat was fixed in a head holder and a midline incision in the skin was made up to the neck area in order to permit easy approach to the cisterna magna. The atlanto-occipital membrane was punctured using a 21-gauge needle; the appearance of clear CSF indicated good entrance. The needle was fixed in the cisterna magna space using cyanoacrylate adhesive (Fig. 2.8b). A 6-mm diameter hole was drilled in the parietal bone, the dura mater was removed gently, and the MPA was placed on the brain surface using a micromanipulator. Four stainless steel screws were used for holding the probe attached to the skull by dental acrylic cement. In addition, the femoral artery was cannulated for blood pressure monitoring using UTR disposable pressure transducer (Biomatrix, Israel) and the vein was cannulated for drug administration. ICP elevation was induced by elevating a container filled with mock CSF (37 °C), which was connected to the catheter inserted into the cisterna magna space. All monitored signals during the entire experimental period were transferred to a multichannel computerized equisition and recording system (LabView A/D software, National Instruments Co., USA), in addition to a Grass polygraph for further analysis.

Experimental Protocols

Group A—ICP (30 m) + M: ICP was elevated to the level of 30 mmHg for a period of 30 min, in which mannitol 20% (1 ml) was injected IV, 15 min post ICP elevation (n-10). **Group B**—ICP (30 m): ICP was elevated to the level of 30 mmHg for a period of 30 min (n58). In addition, two control groups were tested: N mannitol was injected in normal rats; N ICP was elevated to 30 mmHg for 30 min. A bolus of saline was IV injected 15 min before ICP reduction. All groups were monitored during the elevated ICP period, as well as following 2 h of recovery.

Statistical Analysis

The Student's t-test was used to examine the effects of ICP elevation on the various physiological parameters measured from the rat brain. Each experiment was divided into two parts, the elevated ICP period and the recovery period. Therefore, the changes developed in each parameter were compared statistically with the initial value recorded during each period of the experiment. A value of $p < 0.05$ was considered to be significant. All values are presented as Mean ± S.E. For NADH calculations, the basal level of NADH was considered to be 0% and a range of 500 mV as 100%. CBF level at death was considered as 0%, and the range between basal CBF level and the level monitored at death was considered as 100%. For MAP and CPP, the values presented are the change in mmHg during the recovery phase.

Results

In order to evaluate the effects of mannitol or saline administration on normal brain tissue, two control groups were injected with mannitol or saline solution and monitored for a period of 2.5 h. The results showed no significant changes in all parameters monitored. A typical experiment in which mannitol was injected 15 min post-ICP elevation is presented in Fig. 2.17. As seen, the increase in ICP was associated with a minor decrease in CBF, followed by a progressive increase in CBF to the level of 120% of its basal level. No significant changes were observed in the reflectance (Ref) and NADH parameters. ICP elevation caused an increase in MAP from 103 to 158 mmHg. No significant changes in K^+ and pH extracellular levels were observed. Mannitol administration 15 min after ICP elevation caused a slight decrease in ICP and caused a further increase in MAP to the level of 212 mmHg. When infusion of mock CSF to the cisterna magna space stopped (ICP down), ICP decreased to its basal level. CBF also returned to its basal level, and NADH was gradually oxidized to the level of 212%. In order to evaluate mannitol influence in the cisterna magna infusion model, all parameters were tested during the recovery phase. Our results showed a 50% survival rate under ICP elevation to 30 mmHg for 30 min, whereas mannitol administration dramatically increased the survival rate of rats to 100%.

Fig. 2.18 presents the Mean ± S.E. values of each parameter during the recovery phase monitored in the two experimental groups. The rats that survived 2 h of recovery are presented separately from the nonsurviving rats. As seen, the kinetics of ICP reduction was equal in all groups; however, ICP levels 40–100 min post-ICP reduction were significantly lower in group A compared with both subgroups B (s) and B (n.s) ($p < 0.05$). Mannitol prevented the dramatic decrease in MAP and CPP observed in the rats from group B (n.s). In addition, when mannitol was injected, the decrease in CBF and the increase in mitochondrial NADH levels observed in group B (n.s) were abolished. The pH level in group A, 50–110 min within the recovery phase, was significantly lower than in group B ($p < 0.01$).

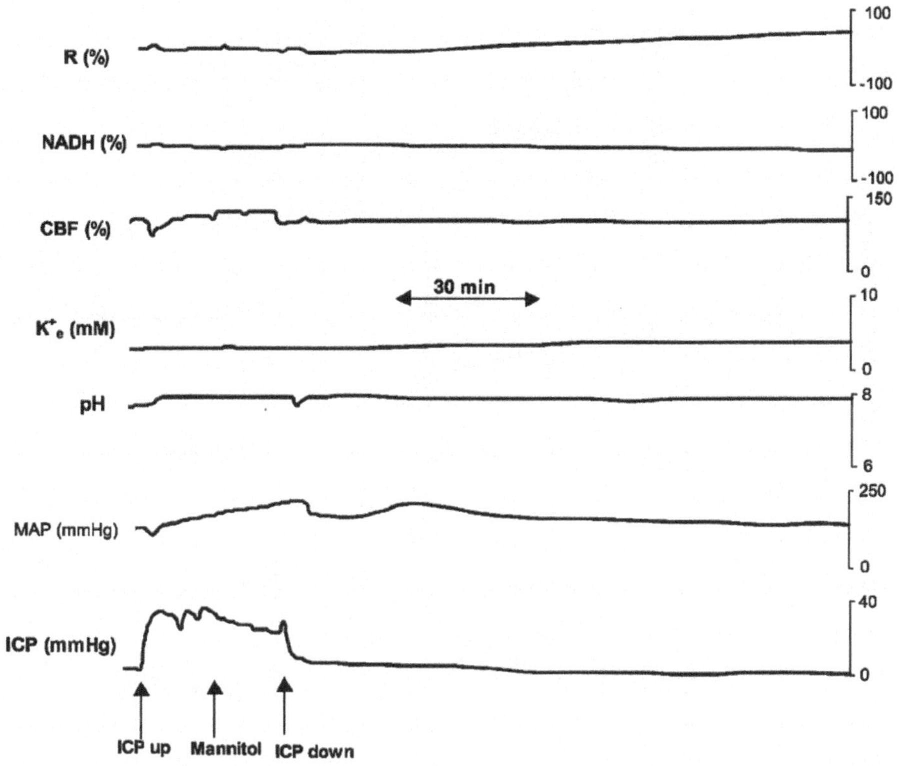

Fig. 2.17 Analog results of a typical experiment in which ICP was elevated up to 30 mmHg and mannitol was IV injected 15 min post-ICP elevation. *R* reflectance, *NADH* corrected NADH, *CBF* cerebral blood flow, K^+_e, extracellular net potassium, *pH* surface extracellular H^+ level, *MAP* mean arterial blood pressure, *ICP* intracranial pressure

Discussion

Elevated ICP is the final common pathway for many intracranial problems and has a profound influence on the final outcome (Pickard and Czosnyka 1993). Mannitol, a 6-carbon hexahydric alcohol, is one of the most popular osmotic agents in use under intracranial hypertension (Vinas et al. 1995). The purpose of mannitol administration in the present study, was to evaluate its influence on brain vitality assed using the multiparametric system, as well as its effect on the final outcome in an intracranial hypertension rat model. Our hypotheses were that since mannitol has a beneficial effect in various pathological conditions involved with intracranial hypertension, it would improve brain vitality in the cisterna infusion model, used in the present study. Moreover, the use of the multiparametric approach will contribute to the understanding of mannitol effects on the hemodynamic, metabolic, and ionic state of the brain.

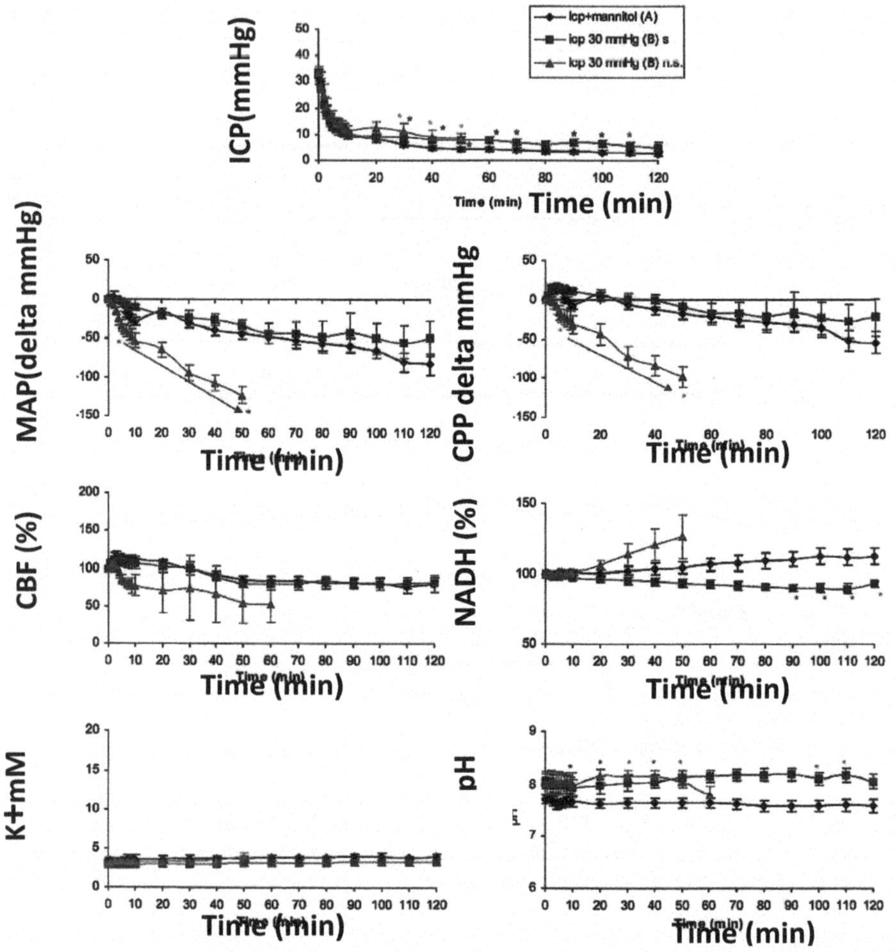

Fig. 2.18 The Mean ± S.E. levels of ICP, CBF, NADH, MAP, CPP, K^+ and pH recorded during the recovery period in groups A and B. The significance for each parameter was calculated according to the unpaired Student's t-test. $*P < 0.05$

Our results showed that mannitol prevented the mortality of 50% of the rats in the increased ICP group. This clearly indicates its beneficial effect in improving the final outcome following ICP hypertension in a rat model. In addition, as seen in Fig. 2.18, the minimum levels of ICP monitored in the mannitol-treated rats were significantly lower than ICP levels monitored in the group of rats that was not injected with mannitol. This effect of mannitol in reducing ICP was previously reported in experimental, as well as clinical studies (Donato et al. 1994; Vinas et al. 1995). In our model, the effect of mannitol in reducing ICP level initiated 45 min post-injection and lasted for 100 min. However, the experimental model in use has to be carefully considered when the total decrease in ICP level, as well as the time schedule of mannitol influence, is examined. Moreover, the rate and amount of ICP decrease following mannitol injection are dependent on the initial ICP or CPP level, brain tissue volume, and intact BBB activity (Rosner and Daughton 1990). Mannitol increases water delivery from the brain tissue into the circulation, leading to a decrease in blood viscosity and an increase in peripheral blood vessel resistance (Rosner and Daughton 1990; Pickard and Czosnyka 1993; Donato et al. 1994; Rosner 1995; Bullock and Povlishock 1996c). In the present study, mannitol improved MAP level, preventing its decline in Group B (n.s). The importance of high systemic blood pressure or CPP in head injury patients is not completely clear. As has been previously published, a pharmacological treatment intended to increase arterial blood pressure improved the final outcome of head-injured patients (Bullock and Povlishock 1996b). However, a recent study showed that the most powerful predictor of neurological worsening was the presence of high ICP (or 5 20 mmHg) and not a high CPP level (above 60 mmHg) (Juul et al. 2000). Nevertheless, it seems that in our experimental model, high MAP had an advantage, since hypotension rats did not survive the entire recovery phase, whereas hypertension rats survived 2 h of recovery post-ICP elevation. When mannitol was injected into normal rats, it had no influence on brain function, probably because mannitol has no effect when ICP is low and CPP is high. At this stage, cerebral arteries are at maximal vasoconstriction, and no additional decrease in CBV or ICP can occur (Rosner and Coley 1987; Rosner and Daughton 1990). However, at high ICP and low CPP levels, cerebral arteries are in a position at which further vasoconstriction can occur, leading to a decrease in CBV and, consequently, a decrease in ICP level (Rosner and Coley 1987). Our results showed that mannitol injection stabilized CBF levels, preventing the trend of decrease in CBF observed in the group of rats that did not survive the whole recovery period. The effect of mannitol on CBF was previously observed. Fortune et al. found that intravenous mannitol has a greater effect in improving CBF than in reducing ICP (Fortune et al. 1995). Another study showed that early use of mannitol did not increase the microcirculatory blood flow, but stabilized it, preventing further decrease (Vinas et al. 1995). Consequently, the increase in NADH was blocked, and its level stayed stable. The intact energetic balance may result from the ability of mannitol to improve the deformability of red blood cells, enabling better oxygen delivery to brain tissue (Fortune et al. 1995). As for the ionic changes following mannitol injection, an increase in pH level was observed during the recovery phase in Group B(s) compared with Group A (Fig. 2.18). One possible explanation is that a decrease in NADH/NAD$^+$ ratio (NADH oxidation) seen in Group B(s) can induce an increase in pH level.

Conclusion

In conclusion, our results showed that mannitol injection in the cisterna infusion model in rats decreased ICP levels compared with rats that were not injected. In addition, mannitol prevented the decrease of systemic blood pressure, CPP level, and CBF, leading to a dramatic decrease in the mortality rate of rats post-elevated ICP.

2.2 Metabolism in Most Vital vs. Less Vital Organs

2.2.1 First Study—Multiorgan Monitoring of Hemodynamic and Mitochondrial Responses to Anoxia and Cardiac Arrest in the Rat

2.2.1.1 Introduction

All tissues in the body are dependent upon the continuous supply of energy to perform their various functions. Glucose and oxygen are brought to the tissue through the bloodstream to fulfill this demand. The basic need for glucose as an energy substrate and oxygen for respiration is common to all tissues. However, the different organs in the body have various levels of demand for these substances. For this reason, the body distributes its resources differently to different organs (Ganong 1991).

The transfer of electrons through the respiratory chain is involved with the ionization of mitochondrial NADH, driven by glycolysis. The energy elicited from this electron transfer is used to maintain an intact cellular homeostasis. A decrease in the supply of blood, as well as oxygen, may affect the balance between O_2 demand and supply. As a result, the accumulation of reduced NADH in the tissue will occur (van Holde and Mathews 1990). Under normal conditions, the flow of blood through the body is regulated in order to maintain a steady supply of blood to various parts of the body under varying cardiovascular conditions. However, each organ has its own internal autoregulation mechanism to maintain required perfusion rates regardless of the circulatory state of the entire organism (Hinojosa-Laborde et al. 1991). As mentioned above, each of the organs in the body has unique hemodynamic and metabolic characteristics. Consequently, the response of each organ to the lack of O_2 induced by respiration of pure N_2 or cardiac arrest is expected to be different. As of today, only in one study the brain, liver, kidney, and testis were continuously monitored simultaneously (Mayevsky and Chance 1982).

In this study, surface fluorescence was used as an indicator of mitochondrial NADH under hypoxia and anoxia in those organs. In both models, an increase in NADH level was observed in each of the organs monitored. In 1998, a single combined probe for LDF and NADH fluorometry was developed and used for the measurement of tissue viability (Mayevsky et al. 1998). In other studies, these two techniques were used separately to monitor various organs. During cardiac arrest, the heart muscle ceases to contract, MABP drops to zero, and blood ceases to flow through the body. As a result, ischemia is induced in all organs. Ischemia in the brain causes a decrease in CBF and an increase in NADH level. Ischemia of the liver was shown to cause an increase in NADH (Chance and Schoener 1965) and a drop in LDF (Arvidsson et al. 1988). Occlusion of the renal artery led to a rise in renal NADH (Thorniley et al. 1995; Zurovsky and Gispaan 1995). In the present study, we monitored continuously the rat brain, liver, kidney, and testis, simultaneously. The significance of such a model is to investigate the effect of various pathophysiological conditions or different drugs on the hemodynamic and metabolic responses in different organs. In addition, such a model can reveal the order in which different organs respond to the same pathophysiological event.

2.2.1.2 Methods

Animal Preparation

Wistar male rats (250–300 g) were anesthetized by 0.3 ml/100 g Equithesin solution (each ml contains: pentobarbital 9.72 mg; chloral hydrate 42.51 mg; magnesium sulfate 21.25 mg; propylene glycol 44.34% wi/v; alcohol 11.5%; and water). During the entire experiment, the rats were maintained anesthetized by the addition of 0.1 ml Equithesin every half an hour. In addition, a heating pad was placed under the rat to maintain body temperature at 37 °C. A midline incision was made in the skin, exposing the skull. Two holes were drilled in the skull. A 3.5 mm hole was drilled in the left parietal bone for the fixation of a cannula in which the probe was inserted, and a second small hole, in which a screw was inserted for better fixation to the skull. The cannula was then fixed to the skull using acrylic cement. Then the rat was turned over on its back for further operation. A hole in the experimental table, beneath the head, allowed the insertion of the probe into the brain cannula. The femoral artery and vein were exposed and cannulated for blood pressure monitoring, via a UTR-disposable pressure transducer (Biometrix), and KCL injection to the heart, respectively. For the exposure of the kidney and liver, an abdominal midline section below the rib cage was created. The central lobe of the liver was exposed. Additionally, the left kidney was isolated from the juxtaposed spleen and intestine by placing a clamp around it. Then the right testis was exposed, and the probes were placed on each organ. All probes, except for the brain, were held in place with micromanipulators during the entire experiment. Parafilm was placed around the tip of these probes and glued to the tissue using cyanoacrylate adhesive. Parafilm was also used for the prevention of hydration. A black cloth was placed over the Parafilm to avoid room light from entering the monitoring site and causing monitoring artifacts.

Multisite-Multi Parametric Monitoring System (MSMP)

Monitoring of the brain, liver, kidney, and testis in the rat was performed using the Multisite-Multi Parametric monitoring system (MSMP) (Manor et al. 1995). This monitoring system, shown in Fig. 2.19 includes four bundles of optical fibers for the NADH redox state measurement, by surface fluorometry, as well as for tissue blood flow (TBF), using LDF. Each probe was placed on the surface of the organ monitored seen in Fig. 2.20. The principle of NADH monitoring from the surface of the brain is that excitation light (366 nm) passes from the fluorometer to the organ via a bundle of optic fibers made of quartz. The emitted light (450 nm), together with the reflected light, at the excitation wavelength (366 nm), was transferred to the fluorometer via another bundle of fibers. Monitoring of the TBF was accomplished

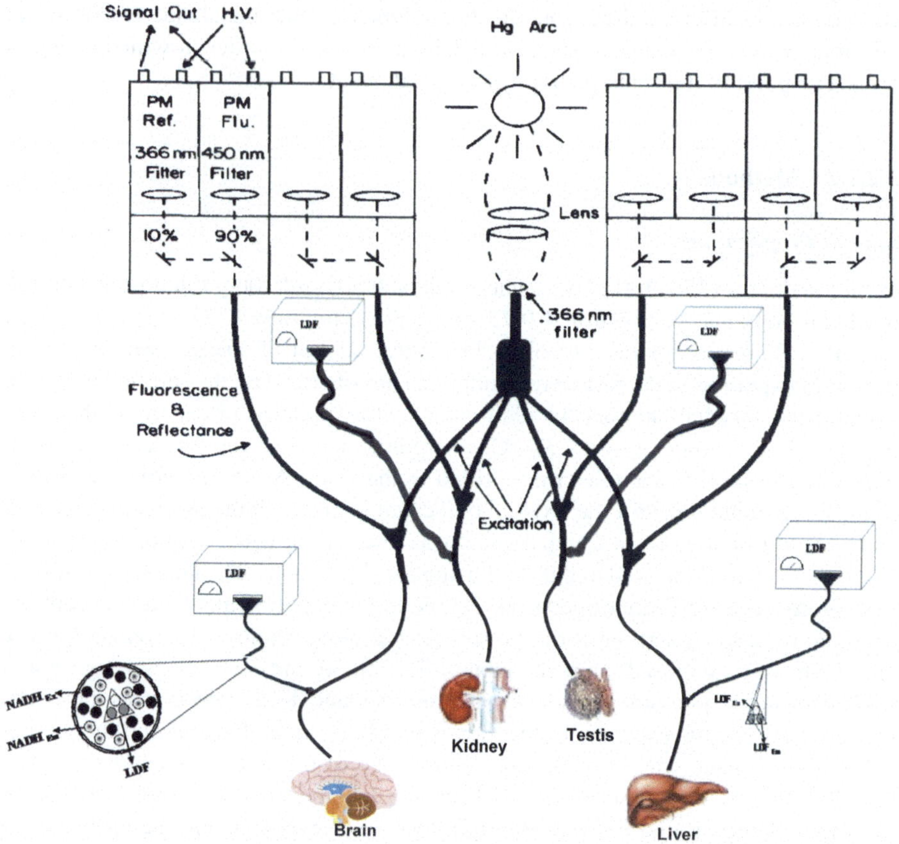

Fig. 2.19 Fluorometric monitoring of NADH and Tissue Blood Flow in four organs simultaneously. In this demonstration, we monitored the brain, kidney, liver, and testis in a rat model. (Mayevsky and Barbiro-Michaely 2013a)

using the Perimed® Periflux PF 2B. This device works on the physical principle of the Doppler shift (Stern et al. 1977; Dirnagl et al. 1989). All the signals monitored during the experiment were transferred to a multichannel computerized data acquisition and recording system (Labview A/D software, National Instruments Co., USA.) in addition to a Grass polygraph paper recorder for further analysis.

Experimental Protocol

One half-hour after surgery, short anoxia was induced by inhalation of pure N_2 until respiratory arrest. The rats were immediately resuscitated. Three hours after the induction of anoxia, 1 ml of KCL 1 M was injected IV to induce cardiac arrest.

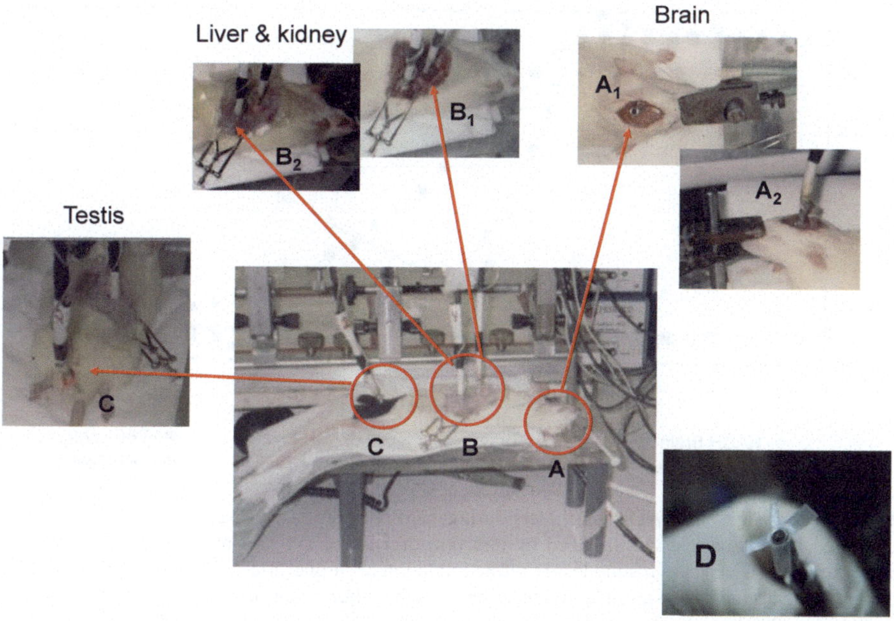

Fig. 2.20 Setup for the monitoring of four organs in the same animal. **Parts A, A1,** and **A2** show the prepared brain. Parts **B, B1,** and **B2** show the connection to the liver and the kidney. Part **C** shows the probe connected to a testis. Part **D** shows a photo of the optical probe ready for the connection to one of the visceral organs. (Mayevsky and Barbiro-Michaely 2013a)

Statistical Analysis

The ANOVA and Duncan's multiple range test were used to detect significant differences between the four organs. A value of $p < 0.05$ was considered to be significant. Mean ± S.E. values were calculated for each parameter in a group of 18 rats.

2.2.1.3 Results

The rate of change in both NADH and TBF during cardiac arrest, induced by injection of 1 ml of 1 M KCL, is presented in Fig. 2.21. The initial basal level of 0% NADH and 100% TBF were set before the injection of KCL. Approximately 3 s after KCL injection, an increase in NADH was observed in all four organs. This increase in NADH was associated with a fall in TBF to its lowest level. The rate of fall in TBF in the testis was significantly slower as compared to all other organs ($P < 0.001$). Figure 2.22 presents the maximum levels of NADH during anoxia from four organs. As seen, three significantly different groups were observed. The brain and liver showed NADH levels of 49 ± 11% and 52 ± 10.5%, respectively. A second group was the level of NADH in the kidney, 66 ± 33%. The third group was the

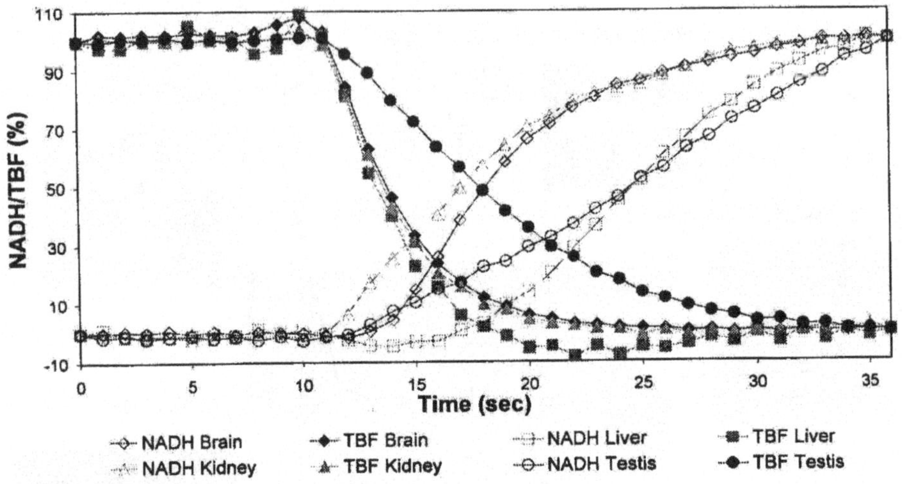

Fig. 2.21 Changes in NADH and TBF during cardiac arrest. One experiment out of nine is displayed. Cardiac arrest was induced by injection of 1 ml of IM KCI at time 10 s. 0% NADH and 100% TBF were set as resting levels before induction of cardiac arrest. The rate of fall in TBF in the testis was found to be significantly different from the other organs using ANOVA ($P < 0.001$). (Kraut et al. 2003)

Fig. 2.22 Maximum NADH values during anoxia ($n = 18$). Values are expressed in percentage, where 0% is set as NADH level before anoxia induction. Three significantly different groups (A, B, and C) were found using ANOVA and Duncan's multiple range test ($P < 0.01$). (Kraut et al. 2003)

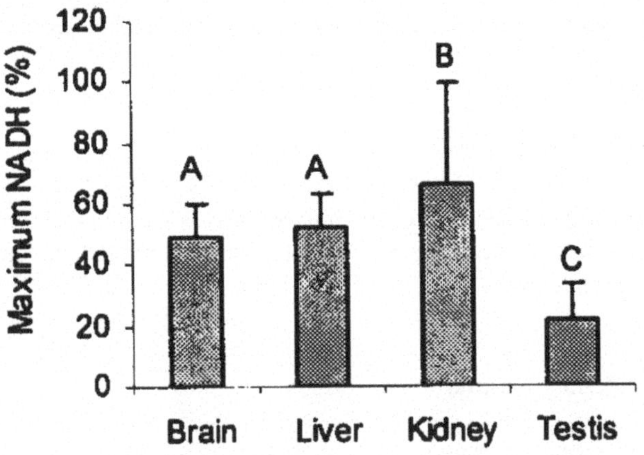

NADH level measured in the testis, which was the lowest, reaching values of 22 ± 12%. An increase in NADH levels indicates an imbalance between oxygen demand and supply. As NADH reached its maximum values, the lowest level of the TBF is usually observed.

Figure 2.23 shows the levels of minimum TBF in the four organs during anoxia (100% N_2). The use of ANOVA and Duncan's multiple range test exposed two significantly different groups regarding the lowest level of TBF. In the brain, TBF decreased to 66 ± 28% of the basal level (100%). TBF in the liver, kidney, and testis

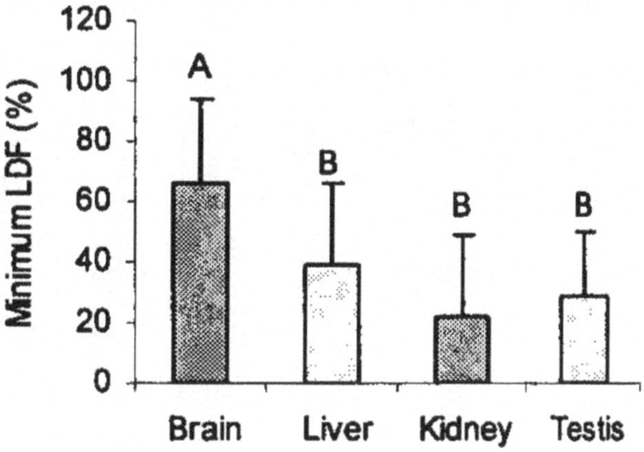

Fig. 2.23 Minimum LDF values during anoxia ($n = 18$) Values are expressed in percentage, where 100% is set as LDF level before anoxia induction and 0% is set as the level at death. Two significantly different groups (A and B) were found using ANOVA and Duncan's multiple range test ($P < 0.01$). (Kraut et al. 2003)

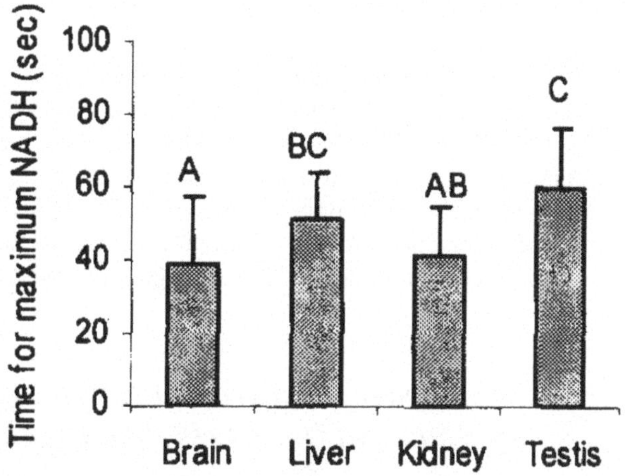

Fig. 2.24 Time to Maximum NADH Values During Anoxia ($n = 18$). The times from induction of anoxia until maximum NADH levels are shown. Three significantly different groups (A, C, and BC/AB) were found using ANOVA and Duncan's multiple range test ($P < 0.01$). (Kraut et al. 2003)

was 39 ± 27, 22 ± 27, and 29 ± 21%, respectively. The time needed for the NADH to reach its maximum level showed three significant groups (Fig. 2.24). In the brain, 38 ± 18.5 s passed until the maximum level of NADH was recorded. However, in the testis, a longer period of 60 ± 16 s was needed to retain maximum NADH values. Maximum NADH levels were observed in the liver and kidney, 52 ± 12.5 and 42 ± 12.5 s after the induction of anoxia (Fig. 2.25). The reduction in TBF during the anoxic episode showed two significantly different groups. In the brain, 33 ± 20 s for minimum TBF were needed, while in the kidney and testis, 46 ± 12 and 51 ± 15 s, respectively, were needed for the TBF to get to its minimum values.

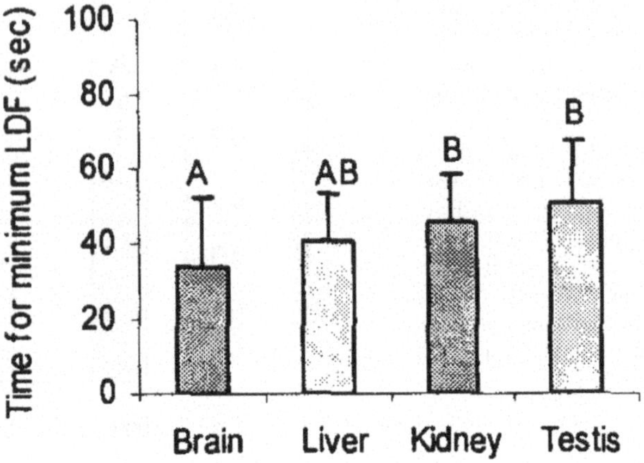

Fig. 2.25 Time to minimum TBF values during anoxia ($n = 18$). The times from induction of anoxia until minimum TBF levels are shown. Two significantly different groups (A and B) were found using ANOVA and Duncan's multiple range test ($P < 0.01$). (Kraut et al. 2003)

2.2.1.4 Discussion

As seen, the brain, liver, kidney, and testis have distinctively different responses to anoxia. The brain was affected by anoxia more rapidly than the testis, probably due to its massive consumption of O_2. Moreover, the brain apparently has the best mechanisms for autoregulation of blood flow, and thus, the decrease in CBF during anoxia was the smallest one. When cardiac arrest was induced, the testis had the most delayed reaction probably due to the fact that the testis requires less blood than the brain, liver, and kidney, which are highly perfused organs.

2.2.2 Second Study: Differential Effects of Norepinephrine on Brain and Other Less Vital Organs

2.2.2.1 Introduction

Pathophysiological events develop in many emergency clinical situations, such as hypoxemia, hypotension, sepsis, and cardiac arrest, and are mostly involved with the deterioration of body metabolic state, negative oxygen balance, and energy failure. As a compensatory mechanism, the sympathetic pathways are activated, via the secretion of epinephrine and norepinephrine (NE), leading to blood flow redistribution in the body in such a manner that blood flow to the most vital organs (brain, heart, and adrenal medulla) will increase, whereas blood flow to "less vital" organs (GI tract, skin, and kidney) will decrease. This is the reason why NE injection serves as a conventional animal model for the investigation of various critical situations in patients.

It has been stated often that the autonomic nervous system integrates diverse organ systems by redistributing the cardiac output (CO) in response to the requirements of the whole animal rather than to the needs of local tissues (Goldman 1966).

The brain, which is considered the most vital organ, is highly affected by NE. The central adrenergic neurons have several functions in the brain: they enhance cerebral vascular tone by action on alpha-receptor sites (Berntman et al. 1978; Edvinsson et al. 1978) and influence the metabolic rate of brain tissue by acting on Beta-receptors sites on the cell membrane (Kogure et al. 1979). Additionally, NE modulates the response of the cerebral cortex to an increase in brain metabolic demand. NE may mediate its effect by potentiating Na^+/K^+-ATPase or through its effects on vascular reactivity or both (LaManna et al. 1981). Studies in which a depletion of brain NE levels was induced showed a disturbance in cerebral microvascular tone and an increase in cerebral blood vessel vulnerability to hypertension (Kobayashi et al. 1991).

Since catecholamines influence the CBF, they also have an influence on the energetic state of the brain. As was observed, NE administration caused the oxidation of NADH (Dora and Kovach 1983) and a change in the rate of oxidative metabolism activity (LaManna et al. 1976). The relationship between oxygen demand and supply is not affected by NE depletion when energy demand is low.

The kidney which is a "less vital" organ (as compared to the brain) poses also its own autoregulation mechanisms, which is independent of circulating humoral or neurogenic factors but governed by intrinsic renal mechanisms (Dworkin 2000). Several studies, using different techniques, have clearly revealed the presence of adrenergic fibers on tubular and vascular elements of the kidney (Barajas and Wang 1975; Johns 1993). The activation of Alpha1-adrenergic receptors in the afferent and efferent arterioles induces arteriolar constriction leading to RBF and GFR decrease, although not at the same degree (Stern et al. 1977; Vandar 1991; Sharma and Gulati 1994; Yang et al. 1995; Parekh et al. 1996). As for the metabolic activity of the renal tissue, here, the aerobic respiratory activity is the main source for energy, producing 95% of the ATP in the kidney (Dworkin 2000). In 1962, Chance et al. (1962) found that NADH is the most sensitive component, in the respiratory chain, to tissue oxygen levels, since then several studies used the fluorometric monitoring technique to evaluate renal energetic state (Zurovsky and Sonn 1992; Mayevsky et al. 2001b).

It was demonstrated that NE increases microvasculature resistance and decreases vascular volume in a dose-dependent manner in the liver (McCuskey 1994; Wheatley and Almond 1997; Rothe and Maass-Moreno 1998). The first measurements of NADH in the liver using the fluorometric technique were made by Chance, who monitored NADH in liver slices from anesthetized rats (Chance and Schoener 1963). Following studies measured NADH fluorescence in rat hepatocyte suspensions and on the surface of perfused rat liver, using a fiber optic probe (Chance et al. 1965; Obi-Tabot et al. 1993; Thorniley et al. 1995). The changes in fluorescence intensity monitored in the perfused liver were explained by a change in the amount of reduced pyridine nucleotides in the tissue (Wakita et al. 1995).

While in most of the organs blood flow is proportional to the metabolic demands of the tissue, in the testis blood flow is low in relation to its rate of oxygen uptake (Gunn and Gould 2002). Several mechanisms of autoregulation exist in the testis, including the neurological mechanism mediated via adrenergic stimulation, which affects small blood vessels (adrenergic nerve terminals were observed in the epididymal duct tissue) (Pholpramool and Triphrom 1984; Schlegel and Chang 1997).

As previously indicated, catecholamines could induce a rapid and dose-dependent decrease in testicular blood flow (Free and Jaffe 1972; Damber and Janson 1978b) via an increase in testis vascular resistance (Damber and Janson 1978a; Damber et al. 1983).

The aim of the present study was to develop and use the multiorgan (including vital and "less vital" organs) monitoring approach in real time following NE administration in a rat model.

By using NE injection in a rat, together with the multisite multiparametric (MSMP) monitoring system, we were able to evaluate the hemodynamic and metabolic state of four different organs simultaneously. Our monitoring approach also enabled us to estimate the differences in the responses of the brain, a vital organ and "less vital" organs such as the kidney, liver, and testis. Consequently, the monitoring of "less vital" organs may indirectly provide information concerning the state of vital organs.

2.2.2.2 Methods

Animal Preparation

All experimental protocols were approved by the Institutional Animal Care Committee under the instructions of the National Institute of Health.

Wistar male rats (250–300 g) were anesthetized by 0.3 ml/100 g Equithesin IP injection (each ml contains: pentobarbital 9.72 mg; chloral hydrate 42.51 mg; magnesium sulfate 21.25 mg; propylene glycol 44.34% w/v; alcohol 11.5%; and water). During the entire experiment, the rats were maintained anesthetized by the addition of 0.1 ml Equithesin every half an hour. In addition, a heating pad was placed under the rat to maintain body temperature at 37 °C. At the end of the experiments, the rats were sacrificed by inhalation of pure N_2.

A midline incision was made in the skin, exposing the skull. Two holes were drilled in the skull. A 3.5 mm hole was drilled in the left parietal bone for the fixation of a cannula in which the monitoring probe was inserted, and a second small hole, in which a screw was inserted for better fixation of the light guide holder to the skull. The cannula was then fixed to the skull using dental acrylic cement. Then the rat was turned over on its back for further operation. A hole in the experimental table, beneath the head, allowed the insertion of the probe into the brain cannula. The femoral artery and vein were exposed and cannulated for blood pressure monitoring, via an UTR-disposable pressure transducer (Biometrix, Israel) and NE (A7256 Sigma) injection, respectively. For the exposure of the kidney and liver, an abdominal midline section below the rib cage was created. The central lobe of the liver was exposed. Additionally, the left kidney was isolated from the juxtaposed spleen and intestine. Then the right testis was exposed and the probes were placed on each organ. In the liver, the probe was placed on the central lobe in its flat area. Another probe was placed at the center of the left kidney where a flat surface exists. All probes, except for the brain, were held in place with micromanipulators during the entire experiment. Parafilm was placed around the tip of these probes and glued to the tissue using cyanoacrylate adhesive (Mayevsky et al. 1995). Parafilm was also used for the prevention of dehydration. A black cloth was placed over the parafilm to avoid room light from entering the monitoring sites and causing artifacts.

The Multisite Multiparametric Monitoring System (MSMP)

Monitoring of the brain, kidney, liver, and testis in the rat was performed using the multisite-multiparametric (MSMP) monitoring system presented in Fig. 2.19. Figure 2.26 presents the principle of one channel of the monitoring system that includes four bundles of optical fibers for the NADH redox state measurement, by surface fluorometry, as well as for TBF, using LDF. Each probe was placed on the surface of the organ monitored as presented in Fig. 2.20. The principle of NADH monitoring from the surface of the brain is that excitation light (366 nm) passes from the fluorometer to the organ via a bundle of optic fibers made of quartz. The emitted light (450 nm), together with the reflected light, at the excitation wavelength

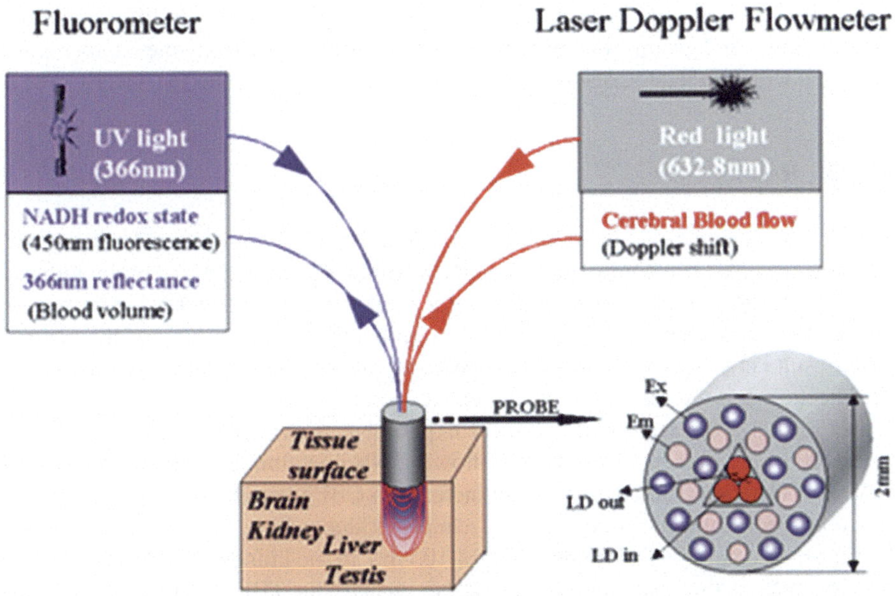

Fig. 2.26 The Multisite-Multi Parametric Monitoring System (MSMP). A schematic presentation of the one out of four probes and its way of connection to the photomultiplier (for NADH monitoring) and to the Laser Doppler Flowmeter (for TBF monitoring). (Kraut et al. 2004)

(366 nm), was transferred to the fluorometer (containing appropriate filters) via another bundle of fibers (Mayevsky and Zarchin 1987; Mayevsky et al. 1995, 2001a). Monitoring of the TBF was accomplished using the Perimed® Periflux PF 2B. This device works on the physical principle of the Doppler shift (Stern et al. 1977). All the signals monitored during the experiment were transferred at a rate of 100 samples per second to a multichannel computerized data acquisition and recording system (Labview A/D software, National Instruments Co., USA) for further analysis.

Experimental Protocol

One half-hour after surgery, short anoxia (30 s) was induced by inhalation of pure N_2 for the assessment of organ vitality, followed by a period of 30 min for stabilization. Later on, a volume of 1 ml of NE (5 μg/100 g) was injected IV during 1 min and monitoring proceeded for a period of 2.5 h. In the control group, a volume of 1 ml saline was IV injected.

Statistical Analysis

Mean ± S.E. values were calculated for each parameter in a group of nine rats. For NADH calculations, the basal level of NADH was considered to be 0% and a range of 500 mV as 100%. As follows, TBF level in death was considered as 0% and the range between basal TBF level and the level monitored in death was considered as 100%. The ANOVA and Duncan's multiple range tests were used to detect significant differences between the four organs. A value of $P < 0.05$ was considered to be significant. Significantly different groups were presented with different letters (A, B, C, etc.), and groups that were not significantly different from one another were presented with two letters, one of each group that they were similar to (AB, BC, etc.).

2.2.2.3 Results

In order to ensure reliable monitoring of four organs simultaneously using the MSMP device, we run control experiments in which monitoring of the brain, kidney, liver, and testis was performed following a 1 ml saline injection (IV). The results showed that monitoring is stable, with no significant changes in any of the parameters (results not shown). In the second step, we monitored these four organs under NE injection and the results are as indicated.

In Fig. 2.27, the results of a typical experiment in which NE was injected IV (5 μg/100 g) are presented. As seen, following NE injection, a dramatic increase in MAP was observed. Consequently, an increase in CBF (from 85 mV up to 332 mV) was observed while the other organs, namely, kidney, liver, and testis, showed a significant decrease in TBF (a decrease of 103 mV in the kidney, 57 mV in the liver, and 150 mV in the testis). Following these changes in TBF, NADH in the brain decreased by 6 mV (became oxidized), while in the rest of the organs an increase in NADH levels was observed (31 mV, 8.4 mV, and 141 mV in the kidney, liver, and testis, respectively) due to the decrease in O_2 supply.

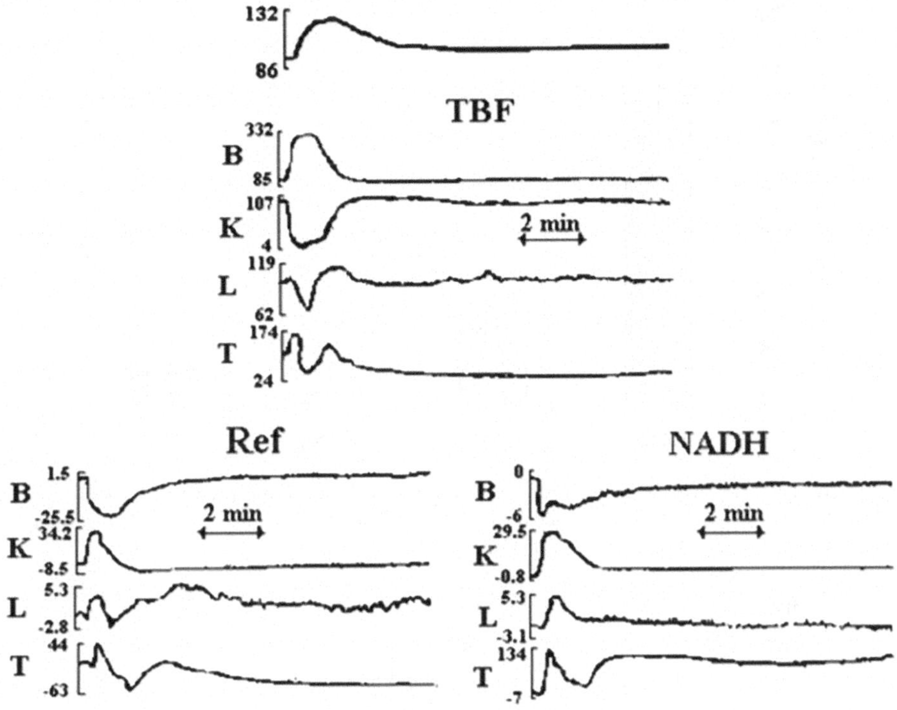

Fig. 2.27 A typical experiment in which 5 μg/100 g IV of Norepinephrine (NE) was injected. The four organs monitored were: *b* brain, *K* kidney, *L* liver, and *T* testis. The parameters shown include: *MAP* Mean Arterial Pressure, *TBF* Tissue Blood Flow, *Ref* reflectance, and *NADH* corrected fluorescence. (Kraut et al. 2004).

As for the Ref, while the brain showed a decrease in Ref (increase in blood volume) of approximately 25 mV, the other organs showed an increase in the Ref signal (decrease in blood volume) of about 30 mV in the kidney, 5 mV in the liver, and 44 mV in the testis.

The average ± S.E changes in all parameters monitored simultaneously in the four organs are presented in Figs. 2.28 and 2.29. As seen in Fig. 2.28, TBF in the brain following NE injection showed a major increase in TBF to the level of 270 ± 47% ($p < 0.05$). However, in the other organs, kidney, liver, and testis, a decrease in TBF to the levels of 10 ± 3%, 76 ± 6%, and 28 ± 9%, respectively, was seen ($p < 0.05$ for all organs). In the testis, however, a rapid phase of hyperemia to the levels of 200 ± 34% was monitored prior to the major decrease in TBF (as demonstrated in Fig. 2.27). This increase in TBF developed 19 ± 3 s after NE injection (average results not shown).

The changes in the Ref and NADH levels are presented in Fig. 2.29. As seen, the increase in NADH level is higher in the kidney and testis, 83 ± 6% and 43 ± 15%, respectively. In the brain, no significant increase in NADH was measured 4 ± 2%,

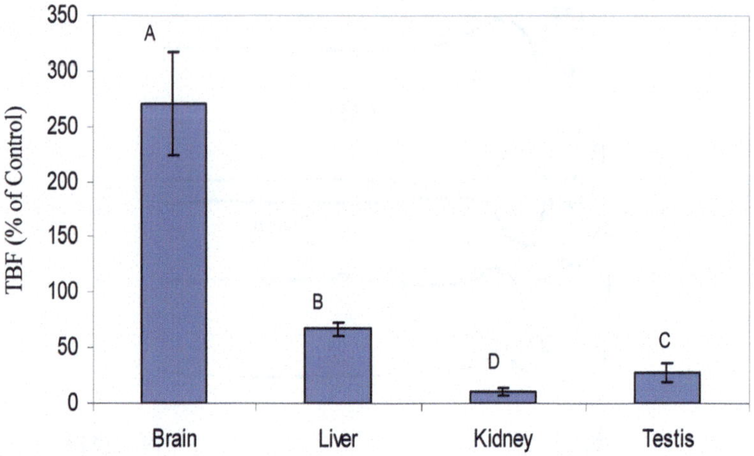

Fig. 2.28 The effect of NE (5 μg/100 g I.V.) on Tissue Blood Flow (TBF) monitored in four different organs (brain, kidney, liver, and testis, abbreviations are as in Fig. 2.15). Results presented are mean ± S.E ($n = 9$). (Kraut et al. 2004)

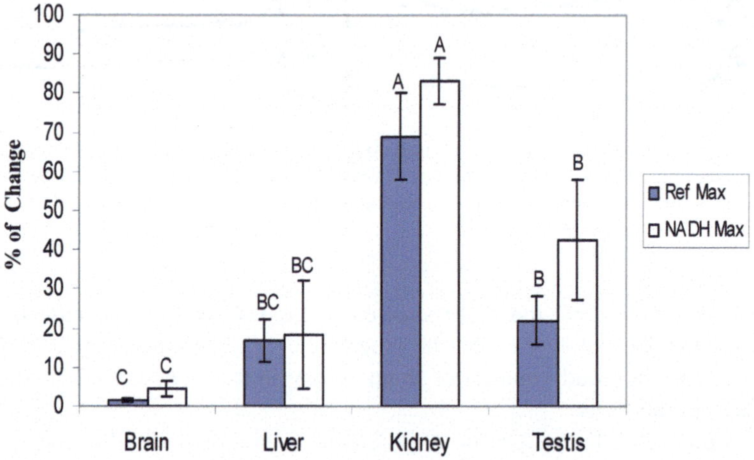

Fig. 2.29 The effect of NE (5 μg/100 g IV) on the reflectance (Ref) and NADH levels monitored in four different organs (brain, kidney, liver, and testis, abbreviations are as in Fig. 2.15). Results presented are mean ± S.E ($n = 9$). (Chance and Legallias 1959)

whereas in the liver, the increase in NADH, 18 ± 14%, was significantly different ($p < 0.05$) than the NADH level monitored in the kidney but not different from NADH level monitored in the testis and the brain. The results of Waller-Duncan T-test of variable showed three significantly different levels of NADH (A, B, and C) measured in the kidney, testis, and brain, respectively. The changes in NADH level in the kidney, liver, and testis were inversely correlated with the changes in the TBF, namely, the maximal decrease in TBF, which was observed in the kidney, was

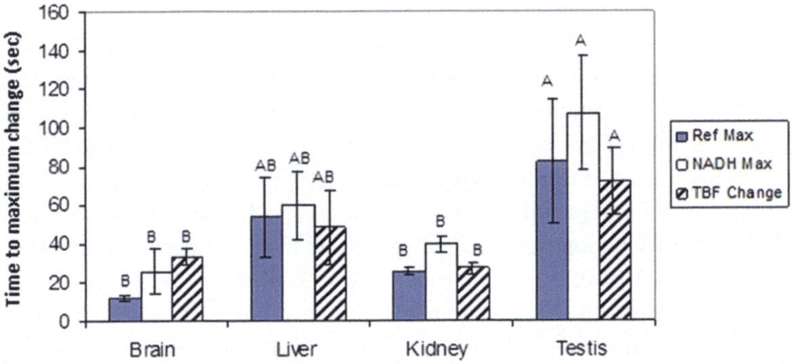

Fig. 2.30 The time needed for the reflectance—Ref, NADH, and TBF to reach their maximal change in four different organs (brain, kidney, liver, and testis, abbreviations are as in Fig. 2.15). Results presented are mean ± S.E ($n = 9$) (Chance and Legallias 1959)

associated with a maximal increase in NADH levels. As for the testis and liver, the increase in NADH level was proportional to the decrease observed in TBF level. However, in the brain, the increase of TBF was associated with stable NADH levels.

Regarding the changes in the Ref levels following NE injection (Fig. 2.29). It was observed that the increases in Ref in the kidney were higher as opposed to the rest of the organs (69 ± 11%). The maximum Ref levels in the kidney, testis, and brain showed three significantly different groups (A, B, and C, respectively). While the Ref level in the liver (17 ± 5.5%) was significantly different from the Ref level monitored in the kidney ($p < 0.05$), but not different from the level monitored in the testis and brain.

Figure 2.30 presents the time required for the different parameters to reach their maximum change (mean ± S.E.). TBF in the brain reached its maximum level within 33 ± 4 s. The decrease in TBF was the first to develop in the kidney (27 ± 3 s) and the liver and testis reached their minimum levels within 48 ± 19 and 72 ± 17 s, respectively (Fig. 2.30).

Following the changes in TBF levels, a change in Ref and NADH levels was observed. The time needed for NADH to reach its maximum level was shorter in the brain and kidney (26 ± 12 and 39 ± 4 s, respectively). The liver was the next organ to respond (59 ± 17 s), and the testis was the last organ in which NADH increased (107 ± 29 s). As for the kinetics of change in the Ref, the organs responded in the following order: first the brain (12 ± 32 s) followed by the kidney (26 ± 20 s), and finally the liver and testis (54 ± 2 and 82 ± 2 s).

The results of Waller-Duncan T-test of the variable revealed two significantly different groups regarding the kinetics of TBF, NADH, and Ref changes following NE administration. The brain and kidney are in one group and the testis on the other group, while the liver stands between these two significantly different groups.

2.2.2.4 Discussion

In cases like hypoxemia, hypotension, sepsis, dysoxia, and cardiac arrest, the metabolic state of the body will be deteriorated and energy failure will develop. Consequently, a central protection mechanism will be activated toward the redistribution of blood flow to the most vital organs like brain, heart, and adrenal gland, while the "less vital" and peripheral organs and tissues like the skin, muscles, and GI tract will undergo vasoconstriction and a decrease in blood flow and oxygen supply will occur (Barbara 1994).

The aim of the present study was to use the simultaneous multiple organ monitoring system in real time in vital and "less vital" organs following NE injection. It is assumed that knowing the kinetics of changes in the hemodynamic as well as the metabolic state of various organs can be used as a detection tool for oxygen deficiency, in different pathophysiological situations. It is for the first time that microcirculatory blood flow and mitochondrial function (NADH redox state) were monitored simultaneously in vital (the brain) and various "less vital" organs. The changes in TBF were evaluated using the LDF. Cortical studies, in the kidney using the LDF demonstrated that it is a reliable technique and could be used as an indicator for the directional changes in cortical blood flow, as well as whole kidney blood flow, compared to the electromagnetic flow probe and the radioactive microspheres method (Stern et al. 1979; Lu et al. 1993; Chiu et al. 1996). It was also demonstrated that the LDF provide a good measure of whole organ blood flow in the perfused rat liver (Wheatley and Almond 1997), as well as in the rat testis (Damber et al. 1983; Shepherd et al. 1983; Arvidsson et al. 1988), recording rapid changes in blood flow (Knobil and Neill 1994).

In the present study, an increase of 40 mmHg in arterial blood pressure developed immediately post NE injection (Goldman 1966; Holmes et al. 1980; Onesti et al. 1989). The increase in blood pressure was followed by a significant increase in CBF and a decrease in TBF of the three "less vital" organs monitored. As previously indicated, epinephrine improves blood flow to the brain probably because of its alpha-adrenergic activity (Edvinsson et al. 1978; Ballet et al. 1988; Sharma and Gulati 1994). Several investigators showed that the destruction of the central noradrenergic system produces an increase in resting CBF and CO_2 reactivity. However, others have found no effects of this procedure on these parameters (Parekh and Zou 1996).

In the present study, NE injection also caused the reduction of blood flow in the liver, probably via the activation of Beta-adrenergic receptors, leading to hepatic microvasculature vasoconstriction (Inokuchi and Malik 1984; Rothe and Maass-Moreno 1998). The decrease in renal blood flow was also documented (Stern et al. 1979; Conger et al. 1991; Nobes et al. 1991). It is important to clarify that all of the changes in TBF and NADH parameters presented in the present chapter were a direct result of NE injection since the results of the control group (saline injection) showed no effect on TBF and NADH in the brain as well as the other organs.

The results of the present study show a high correlation between TBF (O_2 supply) monitored by the LDF technique and the energetic state of the liver, kidney, and testis (mitochondrial NADH redox state). However, in the brain, although TBF increased, NADH remained stable. This is probably since the normoxic brain is well oxidized in normal perfusion levels, and thus, an increase in TBF has no further beneficial effect on mitochondrial NADH levels. Our study showed a high correlation between TBF and the Ref parameter. This is explained due to the fact that Ref is mostly influenced by the absorption properties of the tissue, which is mainly affected by tissue blood volume (TBV). Hence, a reduction in TBF caused a decrease in TBV, leading to an increase in the Ref. Although these four organs have different optical characters (the brain and testis are white tissues and the liver and kidney are red tissues), the comparison between TBF and NADH between the organs was possible since both of these monitoring techniques give relative results meaning each organ was calibrated against its own values in time.

The brain, a vital organ, showed an increase in TBF with no damage to its mitochondrial function and energetic state, while in the kidney, liver, and testis, which are defined as "less vital" organs (relatively to the brain), TBF decreased and NADH increased. Our results showed that the brain and kidneys are the first organs to respond to NE injection, probably due to their own autoregulation mechanisms. The testis showed the latest response in TBF as well as in NADH and Ref parameters. These results may imply that the testis is less important in stress conditions. However, in the liver, TBF and NADH parameters were changed in a time course, which was found to be in between the rate of changes in the brain and kidney, as opposed to the testis, probably due to its role in glucose metabolism, which increases in stress situations.

2.2.2.5 Conclusion

In conclusion, our approach of MSMP could be a practical tool to evaluate organ vitality under various pathophysiological situations. Moreover, the use of the MSMP in "less vital" organs may serve as a tool for the evaluation of the severity of oxygen deficiency in the whole body.

References

Andrews BT, Levy M, McIntosh TK, Pitts LH (1988) An epidural intracranial pressure monitor for experimental use in the rat. Neurol Res 10(2):123–126

Arvidsson D, Svensson H, Haglund U (1988) Laser-Doppler flowmetry for estimating liver blood flow. Am J Phys 254(4 Pt 1):G471–G476

Ballet F, Chretien Y, Rey C, Poupon R (1988) Differential response of normal and cirrhotic liver to vasoactive agents. A study in the isolated perfused rat liver. J Pharmacol Exp Ther 244(1): 283–289

Barajas L, Wang P (1975) Demonstration of acetylcholinesterase in the adrenergic nerves of the renal glomerular arterioles. J Ultrastruct Res 53(2):244–253
Barbara A (1994) Shock. In: Schwartz SI (ed) Principles of surgery. McGraw-Hill, New York, pp 101–122
Barbiro-Michaely E, Mayevsky A (2001) Multiparametric monitoring of brain under elevated intracranial pressure in a rat model. J Neurotrauma 18:711–725
Barbiro-Michaely E, Mayevsky A (2003) Effects of elevated ICP on brain function: can the multiparametric monitoring system can detect the "Cushing response"? Neurol Res 25:42–52
Barbiro-Michaely E, Mayevsky A, Knoller N, Hadani M (2005) In vivo multiparametric monitoring of brain functions under intracranial hypertension following mannitol administration. Neurol Res 27(1):88–93
Bederson JB, Germano IM, Guarino L (1995) Cortical blood flow and cerebral perfusion pressure in a new noncraniotomy model of subarachnoid hemorrhage in the rat. Stroke 26(6):1086–1091; discussion 1091-1082
Berntman L, Dahlgren N, Siesjo BK (1978) Influence of intravenously administered catecholamines on cerebral oxygen consumption and blood flow in the rat. Acta Physiol Scand 104(1): 101–108
Bissonnette B, Bickler PE, Gregory GA, Severinghaus JW (1991) Intracranial pressure and brain redox balance in rabbits. Can J Anaesth 38(5):654–659
Botel C, Brinker T (1994) Measurement of the dynamics of the cerebrospinal fluid system in the rat. J Exp Anim Sci 36(2–3):78–83
Bouman HJ, Van Wimersma Greidanus TB (1979) A rapid and simple cannulation technique for repeated sampling of cerebrospinal fluid in freely moving rats. Brain Res Bull 4(4):575–577
Brinker T, Seifert V, Dietz H (1992) Cerebral blood flow and intracranial pressure during experimental subarachnoid haemorrhage. Acta Neurochir 115(1–2):47–52
Bullock MR, Povlishock JT (1996a) Indications for intracranial pressure monitoring. Brain trauma foundation. J Neurotrauma 13(11):667–679
Bullock MR, Povlishock JT (1996b) The use of barbiturates in the control of intracranial hypertension. Brain trauma foundation. J Neurotrauma 13(11):711–714
Bullock MR, Povlishock JT (1996c) The use of mannitol in severe head injury. J Neurotrauma 13: 705–709
Chance B, Legallias V (1959) Differential microfluorimeter for the localization of reduced pyridine nucleotide in living cells. Rev Sci Instr 30(8):732–735
Chance B, Schoener B (1963) Control of oxidation-reduction state of NADH in the liver of anesthetized rats. Symp Regul Enzyme Act Synth Norm Neoplast Tissues Proc:169–185
Chance B, Schoener B (1965) A correlation of absorption and fluorescence changes in ischemia of the rat liver, *in vivo*. Biochem Z 341:340–345
Chance B, Cohen P, Jobsis F, Schoener B (1962) Localized fluorometry of oxidation-reduction states of intracellular pyridine nucleotide in brain and kidney cortex of the anesthetized rat. Science 136:325
Chance B, Schoener B, Krejci K, Russmann W, Wesemann W, Schnitger H, Bucher T (1965) Kinetics of fluorescence and metabolite changes in rat liver during a cycle of ischemia. Biochem Z 341:325–333
Chance B, Oshino N, Sugano T, Mayevsky A (1973) Basic principles of tissue oxygen determination from mitochondrial signals. Oxyg Transp Tissue Adv Exp Med Biol 37A:277–292
Chesnut RM, Marshall LF (1993) Management of severehead injury. In: Ropper AH (ed) Neurological and neurosurgical intensive care. Raven Press, New York, pp 203–246
Chiu AW, Chang LS, Birkett DH, Babayan RK (1996) A porcine model for renal hemodynamic study during laparoscopy. J Surg Res 60(1):61–68
Conger JD, Robinette JB, Hammond WS (1991) Differences in vascular reactivity in models of ischemic acute renal failure. Kidney Int 39(6):1087–1097

Constantini S, Cotev S, Rappaport ZH, Pomeranz S, Shalit MN (1988) Intracranial pressure monitoring after elective intracranial surgery. A retrospective study of 514 consecutive patients. J Neurosurg 69(4):540–544

Crowe W, Mayevsky A, Mela L (1977) Application of a solid membrane ion selective electrode to *in vivo* measurements. Am J Phys 233:C56–C60

Daley ML, Pasupathy H, Griffith M, Robertson JT, Leffler CW (1995) Detection of loss of cerebral vascular tone by correlation of arterial and intracranial pressure signals. IEEE Trans Biomed Eng 42(4):420–424

Damber JE, Janson PO (1978a) The effects of LH, adrenaline and noradrenaline on testicular blood flow and plasma testosterone concentrations in anaesthetized rats. Acta Endocrinol 88(2): 390–396

Damber JE, Janson PO (1978b) Testicular blood flow and testosterone concentration in spermatic venous blood of anaesthetized rats. J Reprod Fertil 52(2):265–269

Damber JE, Lindahl O, Selstam G, Tenland T (1983) Rhythmical oscillations in rat testicular microcirculation as recorded by laser Doppler flowmetry. Acta Physiol Scand 118(2):117–123

Dirnagl U, Kaplan B, Jacewicz M, Pulsinelli W (1989) Continuous measurement of cerebral cortical blood flow by laser-Doppler flowmetry in a rat stroke model. J CBF Metab 9:589–596

Donato T, Shapira Y, Artru A, Powers K (1994) Effect of mannitol on cerebrospinal fluid dynamics and brain tissue edema. Anesth Analg 78(1):58–66

Dora E, Kovach AGB (1983) Effect of topically administered epinephrine, norepinephrine, and acetylcholine on cerebrocortical circulation and the NAD/NADH redox state. J CBF Metab 3: 161–169

Dworkin LD (2000) The renal circulation. In: Brenner BM (ed) Brenner & Rector's: the kidney. Elsevier, Philadelphia, pp 277–318

Ebel H, Rust DS, Leschinger A, Ehresmann N, Kranz A, Hoffmann O, Boker DK (1996) Vasomotion, regional cerebral blood flow and intracranial pressure after induced subarachnoid haemorrhage in rats. Zentralbl Neurochir 57(3):150–155

Edvinsson L, Hardebo JE, MacKenzie ET, Owman C (1978) Effect of exogenous noradrenaline on local cerebral blood flow after osmotic opening of the blood-brain barrier in the rat. J Physiol 274:149–156

Erecinska M, Silver IA (1989) ATP and brain function. J Cereb Blood Flow Metab 9(1):2–19

Fitch W, McDowall DG, Keaney NP, Pickerodt VW (1977) Systemic vascular responses to increased intracranial pressure. 2. The 'Cushing' response in the presence of intracranial space-occupying lesions: systemic and cerebral haemodynamic studies in the dog and the baboon. J Neurol Neurosurg Psychiatry 40(9):843–852

Fortune JB, Feustel PJ, Graca L, Hasselbarth J, Kuehler DH (1995) Effect of hyperventilation, mannitol, and ventriculostomy drainage on cerebral blood flow after head injury. J Trauma 39(6):1091–1097; discussion 1097-1099

Free MJ, Jaffe RA (1972) Dynamics of circulation in the testis of the conscious rat. Am J Phys 223(1):241–248

Friedli CM, Sclarsky DS, Mayevsky A (1982) Multiprobe monitoring of ionic, metabolic, and electrical activities in the awake brain. Am J Phys 243(3):R462–R469

Ganong WF (1991) Circulation through special regions. Appleton and Lange medical book. In: Review of medical physiology. McGraw-Hill Medical, New York, pp 562–582

Geraci E, Geraci T (1996a) A look at recent hyperventilation studies: outcomes and recommendations for early use in the head-injured pateint. J Neurosci Nurs 28:222–233

Geraci EB, Geraci TA (1996b) Hyperventilation and head injury: controversies and concerns. J Neurosci Nurs 28:381–387

Go KG (1991) Intracranial pressure. Cerebral pathophysiology. Elsevier, Amsterdam

Goldman H (1966) Catecholamine-induced redistribution of blood flow in the unanesthetized rat. Am J Phys 210(6):1419–1423

Gunn SA, Gould TC (2002) Vasculature of the testes and adnexa. In: Handbook of physiology and endocrinology, pp 117–142

Haberl RL, Heizer ML, Marmarou A, Ellis EF (1989) Laser-Doppler assessment of brain microcirculation: effect of systemic alterations. Am J Physiol Heart Circ Physiol 256:H1247–H1254
Harris AP, Koehler RC, Gleason CA, Jones MD Jr, Traystman RJ (1989) Cerebral and peripheral circulatory responses to intracranial hypertension in fetal sheep. Circ Res 64(5):991–1000
Harris AP, Helou S, Traystman RJ, Jones MD Jr, Koehler RC (1998) Efficacy of the cushing response in maintaining cerebral blood flow in premature and near-term fetal sheep. Pediatr Res 43(1):50–56
Hauerberg J, Juhler M (1994) Cerebral blood flow autoregulation in acute intracranial hypertension. J Cereb Blood Flow Metab 14(3):519–525
Hinojosa-Laborde C, Frohlich BH, Cowley AWJ (1991) Contribution of regional vascular responses to whole body autoregulation in conscious areflexic rats. Hyoertension 17:1078–1084
Holmes HR, Babbs CF, Voorhees WD, Tacker WA Jr, de Garavilla B (1980) Influence of adrenergic drugs upon vital organ perfusion during CPR. Crit Care Med 8(3):137–140
Hudetz AG, Feher G, Weigle CG, Knuese DE, Kampine JP (1995) Video microscopy of cerebrocortical capillary flow: response to hypotension and intracranial hypertension. Am J Phys 268(6 Pt 2):H2202–H2210
Inao S, Kuchiwaki H, Sugita K (1993) Microcirculation in experimental brain oedema assessed by laser-Doppler flowmetry. Neurol Res 15(4):264–268
Inokuchi K, Malik KU (1984) Inhibition by bradykinin of renal adrenergic effects in anesthetized rats. Am J Phys 246(4 Pt 2):F387–F394
Jame HE, Harbough RD, Marshall LF, Shapiro HM (1980) The response of ultiple therapeutics in experiental vasogenic edema. In: Shulman K (ed) ICP IV. Springer-Verlag, Berlin, pp 272–276
Johns EJ (1993) Effect of BRL 38227 on the adrenergic regulation of the kidney vasculature of the rat. Eur J Pharmacol 230(1):47–51
Johnston IH, Rowan JO, Harper AM, Jennett WB (1973) Raised intracranial pressure and cerebral blood flow. 2. Supratentorial and infratentorial mass lesions in primates. J Neurol Neurosurg Psychiatry 36(2):161–170
Juul N, Morris GF, Marshall SB, Marshall LF (2000) Intracranial hypertension and cerebral perfusion pressure: influence on neurological deterioration and outcome in severe head injury. The executive Committee of the International Selfotel Trial. J Neurosurg 92(1):1–6
Kawamura S, Yasui N, Shirasawa M, Fukasawa H (1990) Effects of hyperbaric oxygenation on experimental acute cerebral ischemia. J Hyper Med 5:111–123
Kearney ML, Backofen JE, Koehler RC, Jones MD Jr, Traystman RJ (1993) Effect of hypoxemia on the cardiovascular response to intracranial hypertension in postnatal lambs. Am J Phys 265(5 Pt 2):H1557–H1563
Kirkpatrick PJ, Smielewski P, Piechnik S, Pickard JD, Czosnyka M (1996) Early effects of mannitol in patients with head injuries assessed using bedside multimodality monitoring. Neurosurgery 39(4):714–720; discussion 720-711
Knobil E, Neill JD (1994) Male reproduction tract. In: Knobil E, Neill JD, Greenwald GS, Market CL, Donald W (eds) The physiology of reproduction. Raven Press, New York, pp 757–800
Kobayashi H, Hayashi M, Kawano H, Handa Y, Kabuto M, Ide H (1991) Effect of chemical sympathectomy on cerebral blood flow in rats. J Neurosurg 75(6):906–910
Kocsis B, Fedina L, Pasztor E (1991) Effect of preexisting brain ischemia on sympathetic nerve response to intracranial hypertension. J Appl Physiol (1985) 70(5):2181–2187
Koehler RC, Backofen JE, McPherson RW, Jones MD Jr, Rogers MC, Traystman RJ (1989) Cerebral blood flow and evoked potentials during Cushing response in sheep. Am J Phys 256(3 Pt 2):H779–H788
Kogure K, Scheinberg P, Kishikawa H, Utsunomiya Y, Busto R (1979) Adrenergic control of cerebral blood flow and energy metabolism in the rat. Stroke 10(2):179–184
Kraut A, Barbiro-Michaely E, Zurovsky Y, Mayevsky A (2003) Multiorgan monitoring of hemodynamic and mitochondrial responses to anoxia and cardiac arrest in the rat. Adv Exp Med Biol 510:299–304

Kraut A, Barbiro-Michaely E, Mayevsky A (2004) Differential effects of norepinephrine on brain and other less vital organs detected by a multisite multiparametric monitoring system. Med Sci Monit 10(7):BR215–BR220

LaManna JC, Sylvia AL, Martel D, Rosenthal M (1976) Fluorometric monitoring of the effects of adrenergic agents on oxidative metabolism in intact cerebral cortex. Neuropharmacology 15: 17–24

LaManna JC, Harik SI, Light AI, Rosenthal M (1981) Norepinephrine depletion alters cerebral oxidative metabolism in the 'active' state. Brain Res 204:87–101

Lang EW, Chesnut RM (1995) Intracranial pressure and cerebral perfusion pressure in severe head injury. New Horiz 3(3):400–409

Lu S, Mattson DL, Roman RJ, Becker CG, Cowley AW Jr (1993) Assessment of changes in intrarenal blood flow in conscious rats using laser-Doppler flowmetry. Am J Phys 264(6 Pt 2): F956–F962

Malkinson TJ, Cooper KE, Veale WL (1985) Induced changes in intracranial pressure in the anesthetized rat and rabbit. Brain Res Bull 15(3):321–328

Manor T, Meilin S, Mayevsky A (1995) A new "multiprobe-multisite" system for in vivo monitoring of traumatic brain injury in the rat. In: 4th annual meeting of Israel Society for Neuroscience, Eilat, Israel

Marmarou A, Tabaddor K (1993) Intracranial pressure: physiology and pathophysiology. In: Cooper PR (ed) Head Injury, New York, Williams and Williams, pp 203–225

Marshman LA (1997) Cushing's 'variant' response (acute hypotension) after subarachnoid hemorrhage. Association with moderate intracranial tensions and subacute cardiovascular collapse. Stroke 28(7):1445–1450

Mayevsky A (1976) Brain energy metabolism of the conscious rat exposed to various physiological and pathological situations. Brain Res 113:327–338

Mayevsky A (1978) The responses of an awake brain to HPO under increased CO_2 concentration. Adv Exp Med Biol 92:735–740

Mayevsky A (1983) Multiparameter monitoring of the awake brain under hyperbaric oxygenation. J Appl Physiol 54(3):740–748

Mayevsky A (1984a) Brain NADH redox state monitored *in vivo* by fiber optic surface fluorometry. Brain Res Rev 7:49–68

Mayevsky A (1984b) Brain oxygen toxicity. Invited review. In: 8th symposium on underwater physiology, Bethesda, Maryland, Undersea Medical Society Inc.

Mayevsky A, Barbiro-Michaely E (2013a) Shedding light on mitochondrial function by real time monitoring of NADH fluorescence: I. Basic methodology and animal studies. J Clin Monit Comp 27:1–34

Mayevsky A, Barbiro-Michaely E (2013b) Shedding light on mitochondrial function by real time monitoring of NADH fluorescence: II: human studies. J Clin Monit Comp 27:125–145

Mayevsky A, Breuer Z (1992) Brain vasculature and mitochondrial responses to ischemia in gerbils. I. Basic anatomical patterns and biochemical correlates. Brain Res 58:242–250

Mayevsky A, Chance B (1973a) A new long-term method for the measurement of NADH fluorescence in intact rat brain with implanted cannula. Bicher HI, Bruley DF (eds). New York, Plenum Press vol 37A, pp 239–244

Mayevsky A, Chance B (1973b) A new long-term method for the measurement of NADH fluorescence in intact rat brain with implanted cannula. Adv Exp Med Biol 37A:239–244

Mayevsky A, Chance B (1982) Intracellular oxidation-reduction state measured *in situ* by a multicannel fiber-optic surface fluorometer. Science 217:537–540

Mayevsky A, Rogatsky G (2007) Mitochondrial function *in vivo* evaluated by NADH fluorescence: from animal models to human studies. Am J Physiol Cell Physiol 292:C615–C640

Mayevsky A, Zarchin N (1987) Metabolic ionic and electrical activities during and after incomplete or complete cerebral ischemia in the Mongolian gerbil. Adv Exp Med Biol 215:265–273

Mayevsky A, Zeuthen T, Chance B (1974) Measurements of extracellular potassium, ECoG and pyridine nucleotide levels during cortical spreading depression in rats. Brain Res 76:347–349

Mayevsky A, Crowe W, Mela L (1980a) The interrelation between brain oxidative metabolism and extracellular potassium in the unanesthetized gerbil. Neurol Res 1:213–226

Mayevsky A, Lebourdais S, Chance B (1980b) The interrelation between brain PO_2 and NADH oxidation- reduction state in the gerbil. J Neurosci Res 5:173–182

Mayevsky A, Friedli CM, Reivich M (1985) Metabolic, ionic and electrical responses of the gerbil brain to ischemia. Am J Phys 248:R99–R107

Mayevsky A, Frank KH, Nioka S, Kessler M, Chance B (1990) Oxygen supply and brain function *in vivo*: a multiparametric monitoring approach in the Mongolian gerbil. In: Piper J, Goldstick TK, Meyer M (eds) Oxygen transport to tissue XII. Plenum Publication, pp 303–313

Mayevsky A, Frank K, Muck M, Nioka S, Kessler M, Chance B (1992) Multiparametric evaluation of brain functions in the Mongolian gerbil *in vivo*. J Basic Clin Physiol Pharmacol 3:323–342

Mayevsky A, Meilin S, Rogatsky GG, Zarchin N, Sonn J (1995) Multiparametric monitoring of the awake brain exposed to carbon monoxide. J Appl Physiol 78:1188–1196

Mayevsky A, Doron A, Manor T, Meilin S, Zarchin N, Ouaknine GE (1996) Cortical spreading depression recorded from the human brain using a multiparametric monitoring system. Brain Res 740:268–274

Mayevsky A, Nioka S, Wang DJ, Chance B (1997) The functioning gerbil brain *in vivo*. Correlation between 31P NMR spectroscopy and the multiparametric monitoring approach. Adv Exp Med Biol 411:41–53

Mayevsky A, Meilin S, Manor T, Ornstein E, Zarchin N, Sonn J (1998) Multiparametric monitoring of brain oxygen balance under experimental and clinical conditions. Neurol Res 20:S76–S80

Mayevsky A, Kraut A, Manor T, Sonn J, Zurovsky Y (2001a) Optical monitoring of tissue viability using reflected spectroscopy in vivo. Proceeding of SPIE 4241:409–417

Mayevsky A, Nakache R, Luger-Hamer M, Amran D, Sonn J (2001b) Assessment of transplanted kidney vitality by a multiparametric monitoring system. Transplant Proc 33(6):2933–2934

Mayevsky A, Sonn J, Barbiro-Michaely E (2013) Physiological mapping of brain functions *in vivo*: surface monitoring of hemodynamic metabolic ionic and electrical activities in real-time. J Neurosci Neuroeng 2:150–177

McCuskey RS (1994) The hepatic microvascular system. In: Arias IM, Boyer JM, Fausta N et al (eds) The liver: biology and pharmacology. Raven Press, Ltd, New York, pp 1089–1106

McManus ML, Strange K (1993) Acute volume regulation of brain cells in response to hypertonic challenge. Anesthesiology 78(6):1132–1137

Meilin S, Rogatsky GG, Thom SR, Zarchin N, Guggenheimer-Furman E, Mayevsky A (1996) Effects of carbon monoxide exposure on the brain may be mediated by nitric oxide. J Appl Physiol 81:1078–1083

Miller JD, Stanek A, Langfitt TW (1972) Concepts of cerebral perfusion pressure and vascular compression during intracranial hypertension. Prog Brain Res 35:411–432

Miller JD, Stanek AE, Langfitt TW (1973) Cerebral blood flow regulation during experimental brain compression. J Neurosurg 39(2):186–196

Morrow BA, Starcevic VP, Keil LC, Seve WB (1990) Intracranial hypertension after cerebroventricular infusions in conscious rats. Am J Phys 258(5 Pt 2):R1170–R1176

Muizelaar JP, Wei EP, Kontos HA, Becker DP (1983) Mannitol causes compensatory cerebral vasoconstriction and vasodilation in response to blood viscosity changes. J Neurosurg 59(5): 822–828

Muizelaar JP, Lutz HA 3rd, Becker DP (1984) Effect of mannitol on ICP and CBF and correlation with pressure autoregulation in severely head-injured patients. J Neurosurg 61(4):700–706

Nawashiro H, Shima K, Chigasaki H (1995) Immediate cerebrovascular responses to closed head injury in the rat. J Neurotrauma 12(2):189–197

Nichols JS, Beel JA, Munro LG (1996) Detection of impaired cerebral autoregulation using spectral analysis of intracranial pressure waves. J Neurotrauma 13(8):439–456

Nitta M, Tsutsui T, Ueda Y, Ladds A, Symon L (1990) The effects of an extradural expanding lesion on regional intracranial pressure, blood flow, somatosensory conduction and brain herniation: an experimental study in baboons. Acta Neurochir 104(1–2):30–37

Nobes MS, Harris PJ, Yamada H, Mendelsohn FA (1991) Effects of angiotensin on renal cortical and papillary blood flows measured by laser-Doppler flowmetry. Am J Phys 261(6 Pt 2):F998–F1006

Obi-Tabot ET, Hanrahan LM, Cachecho R, Beer ER, Hopkins SR, Chan JCK, Shapiro JM, LaMorte WW (1993) Changes in hepatocyte NADH fluorescence during prolonged hypoxia. J Surg Res 55:575–580

Ogilvy CS, DuBois AB (1987) Effect of increased intracranial pressure on blood pressure, heart rate, respiration and catecholamine levels in neonatal and adult rabbits. Biol Neonate 52(6): 327–336

Onesti ST, Strauss RC, Mayol B, Solomon RA (1989) The effects of norepinephrine depletion on cerebral blood flow in the rat. Brain Res 477(1–2):378–381

Osbakken M, Mayevsky A (1996) Multiparameter monitoring and analysis of *in vivo* ischemic and hypoxic heart. J Basic Clin Physiol Pharmacol 7:97–113

Parekh N, Zou AP (1996) Role of prostaglandins in renal medullary circulation: response to different vasoconstrictors. Am J Phys 271(3 Pt 2):F653–F658

Parekh N, Dobrowolski L, Zou AP, Steinhausen M (1996) Nitric oxide modulates angiotensin II- and norepinephrine-dependent vasoconstriction in rat kidney. Am J Phys 270(3):R630–R635

Pasztor E, Fedina L, Kocsis B, Berta Z (1986) Activity of peripheral sympathetic efferent nerves in experimental subarachnoid haemorrhage. Part I: observations at the time of intracranial hypertension. Acta Neurochir 79(2–4):125–131

Pholpramool C, Triphrom N (1984) Effects of cholinergic and adrenergic drugs on intraluminal pressures and contractility of the rat testis and epididymis in vivo. J Reprod Fertil 71(1): 181–188

Pickard JD, Czosnyka M (1993) Management of raised intracranial pressure. J Neurol Neurosurg Psychiatry 56(8):845–858

Plets C (1989) Arterial hypertension in neurosurgical emergencies. Am J Cardiol 63(6):40C–42C

Rogatsky GG, Sonn J, Kamenir Y, Zarchin N, Mayevsky A (2003) Relationship between intracranial pressure and cortical spreading depression following fluid percussion brain injury in rats. J Neurotrauma 20:1315–1325

Rosner MJ (1995) Introduction to cerebral perfusion pressure management. Neurosurg Clin N Am 6(4):761–773

Rosner MJ, Coley I (1987) Cerebral perfusion pressure: a hemodynamic mechanism of mannitol and the postmannitol hemogram. Neurosurgery 21(2):147–156

Rosner MJ, Daughton S (1990) Cerebral perfusion pressure management in head injury. J Trauma 30(8):933–940; discussion 940-931

Rothe CF, Maass-Moreno R (1998) Hepatic venular resistance responses to norepinephrine, isoproterenol, adenosine, histamine, and ACh in rabbits. Am J Phys 274(3):H777–H785

Schlegel PN, Chang TSK (1997) Physiology of male reproduction: the testis, epididymis and ducts deferens. In: Walsh PC (ed) Campbell's urology. Saunders, pp 1254–1286

Schrader H, Zwetnow NN, Morkrid L (1985) Regional cerebral blood flow and CSF pressures during Cushing response induced by a supratentorial expanding mass. Acta Neurol Scand 71(6): 453–463

Shalit MN, Cotev S (1974) Interrelationship between blood pressure and regional cerebral blood flow in experimental intracranial hypertension. J Neurosurg 40(5):594–602

Sharma AC, Gulati A (1994) Effect of diaspirin cross-linked hemoglobin and norepinephrine on systemic hemodynamics and regional circulation in rats. J Lab Clin Med 123(2):299–308

Shepherd AP, Riedel GL, Ward WF (1983) Laser Doppler measurements of blood flow within the intestinal wall and on the surface of the liver. In: Koo A, Lam SK, Smaje LH (eds) Microcirculation of the Alimentary Tract. World Scientific Publishing Co., Inc, pp 115–129

Siesjo BK, Zwetnow NN (1970) Effects of increased cerebrospinal fluid pressure upon adenine nucleotides and upon lactate and pyruvate in rat brain tissue. Acta Neurol Scand 46(2):187–202

Sioutos PJ, Orozco JA, C. L.P (1996) Regional cerebral blood flow techniques. In: Narayan RK, Wilberger JE, Povlishock JT (eds) Neurotrauma. McGraw-Hill, New York, pp 503–517

Sonn J, Mayevsky A (2000) Effects of brain oxygenation on metabolic, hemodynamic, ionic and electrical responses to spreading depression in the rat. Brain Res 882(1–2):212–216

Stepinska G, Czernicki Z, Berdyga J, Jurkiewiz J (1995) Transcranial Doppler sonography in experimental Cushing response. Acta Neurochir 133(1–2):80–82

Stern MD, Lappe DL, Bowen PD, Chimosky JE, Holloway GA Jr, Keiser HR, Bowman RL (1977) Continuous measurement of tissue blood flow by laser-Doppler spectroscopy. Am J Phys 232(4):H441–H448

Stern MD, Bowen PD, Parma R, Osgood RW, Bowman RL, Stein JH (1979) Measurement of renal cortical and medullary blood flow by laser-Doppler spectroscopy in the rat. Am J Phys 236(1): F80–F87

Sutton LN, McLaughlin AC, Kemp W, Schnall MD, Cho BK, Langfitt TW, Chance B (1987) Effects of increased ICP on brain phosphocreatine and lactate determined by simultaneous ^{1}H and ^{31}P NMR spectroscopy. J Neurosurg 67(3):381–386

Takagi H, Saltoh T, Kitahara T (1983) The mechanism of ICP reducing effect of mannitol. In: Ishli VS (ed) Intracranial pressure. Springer-verlag, Berlin, pp 729–733

Thorniley MS, Simpkin S, Fuller B, Jenabzadeh MZ, Green CJ (1995) Monitoring of surface mitochondrial NADH levels as an indication of ischemia during liver isograft transplantation. Hepatology 21:1602–1609

Tranquart F, de Bray JM, Berson M, Akoka S, Bodard S, Pourcelot L (1994) Concurrent changes in intracranial pressure, cerebral blood flow velocity, and brain energy metabolism in rabbits with acute intracranial hypertension. Childs Nerv Syst 10(5):285–292

Tsai ML, Wang BH (1988) Effects of propranolol on the cardiovascular responses to intracranial pressure elevation in rats. Chin J Physiol 31(2):113–124

Ungersbock K, Tenckhoff D, Heimann A, Wagner W, Kempski OS (1995) Transcranial Doppler and cortical microcirculation at increased intracranial pressure and during the Cushing response: an experimental study on rabbits. Neurosurgery 36(1):147–156; discussion 156-147

van Holde KE, Mathews DK (1990) Biochemistry. The Benjamin/Cummings Publishing Company, Redwood

van Loon J, Shivalkar B, Plets C, Goffin J, Tjandra-Maga TB, Flameng W (1993) Catecholamine response to a gradual increase of intracranial pressure. J Neurosurg 79(5):705–709

Vandar AJ (1991) Control of renal hemdynamics. In: Vandar AJ (ed) Renal physiology. McGraw-Hill, New York, pp 68–82

Vinas FC, Dujovny M, Hodgkinson D (1995) Early hemodynamic changes at the microcirculatory level and effects of mannitol following focal cryogenic injury. Neurol Res 17(6):465–468

Wadhwani KC, Rapoport SI, Shepherd AP, Oberg PA (1990) Blood flow in the central and peripheral nervous systems. In: Shephrd AP, Oberg PA (eds) Laser doppler blood flowmetry, vol 107. Kluwer Academic Publication, Boston, pp 265–304

Wakita M, Nishimura G, Tamura M (1995) Some characteristics of the fluorescence lifetime of reduced pyridine nucleotides in isolated mitochondria, isolated hepatocytes, and perfused rat liver in situ. J Biochem 118(6):1151–1160

Wheatley AM, Almond NE (1997) Effect of hepatic nerve stimulation and norepinephrine on the laser Doppler flux signal from the surface of the perfused rat liver. Int J Microcirc Clin Exp 17(1):48–54

Wilkinson HA, Rosenfeld SR (1983) Furosemide and mannitol in the treatment of acute experimental intracranial hypertension. Neurosurgery 12(4):405–410

Williams MA, Razumovsky A (1993) Cerebrospinal fluid circulation, cerebral edema, and intracranial pressure. Curr Opin Neurol 6(6):847–853

Wise BL, Chater N (1962) The value of hypertonic mannitol solution in decreasing brain mass and lowering cerebro-spinal-fluid pressure. J Neurosurg 19:1038–1043

Yamane K, Shima T, Okada Y, Takeda T, Uozumi T (1994) Acute brain swelling in cerebral embolization model of rats. I: epidural pressure monitoring. Surg Neurol 41(6):477–481

Yang S, Silldorff EP, Pallone TL (1995) Effect of norepinephrine and acetylcholine on outer medullary descending vasa recta. Am J Phys 269(2 Pt 2):H710–H716

Zurovsky Y, Gispaan I (1995) Antioxidants attenuate endotoxin-induced acute renal failure in rats. Am J Kidney Dis 25(1):51–57

Zurovsky Y, Sonn J (1992) Fiber optic surface fluorometry-reflectometry technique in the renal physiology of rats. J Basic Clin Physiol Pharmacol 3(4):343–358

Part II
Detailed Animal Studies

Chapter 3
Responses of Most Vital and Less-Vital Organs to Oxygen and CO_2 Supply

3.1 Introduction

It is well known that under emergency situations the autonomic nervous system (ANS), mainly its sympathetic branch including the adrenal gland, dominates the compensatory mechanisms of the body related to O_2 deficiency (Fig. 3.1). The rapid compensatory reaction to a decrease in blood volume (hypovolemia), for example, includes redistribution of blood flow (and consequently O_2) to various organs, giving preference to the most vital organs in the body, namely the brain, heart, and adrenal glands (Ganong 1991; Guyton 1991a, b; Barbara 1994; Scher 1996). The less vital organs and tissues are hypoperfused and O_2 supply decreases (Harbrecht et al. 2004). Nevertheless, only very few studies had proved the concept of blood-flow redistribution developed under an emergency situation in animal models (Kuwahira et al. 1993; Licker et al. 1998; Levy et al. 2003). Many critical conditions are characterized by tissue hypoxia, which often leads to "cytopathic hypoxia" (Hotchkiss and Karl 1992; Marik and Varon 1998; Ince and Sinaasappel 1999; Fink 2001, 2002). Since most of the oxygen in the body is utilized by the mitochondria (Waltemath 1970) to produce ATP, it is obvious that mitochondrial function plays a crucial role in tissue and organ vitality. Recently it has been indicated that optimization of oxygen delivery is the best method for the prevention and the only method for the treatment of common intensive care syndromes such as sepsis, multiple organ dysfunction syndrome (MODS), and acute lung injury (Blinman and Maggard 2000). Consequently, a vast amount of resources in critical care research is directed at identifying and analyzing generic markers of incomplete resuscitation(Dantzker 1993; Mutlu et al. 2003; Tamion et al. 2003; Varela et al. 2003; Miami Trauma Clinical Trials G 2005). Today, the most familiar parameters being monitored in ICUs and operating rooms include pCO_2, pH, pO_2, Cyt aa3, and TBF, some of which are monitored in the blood and some in various tissues (Beale and Bihari 1993; Cairns et al. 1997; Marik and Varon 1998; Miller et al. 1998; Ackland et al. 2000; Boldt 2002). However, most of these techniques are often

© The Author(s), under exclusive license to Springer Nature Switzerland AG 2025

A. Mayevsky et al., *Body Oxygen Homeostasis and Mitochondrial Function*,
https://doi.org/10.1007/978-3-031-94113-9_3

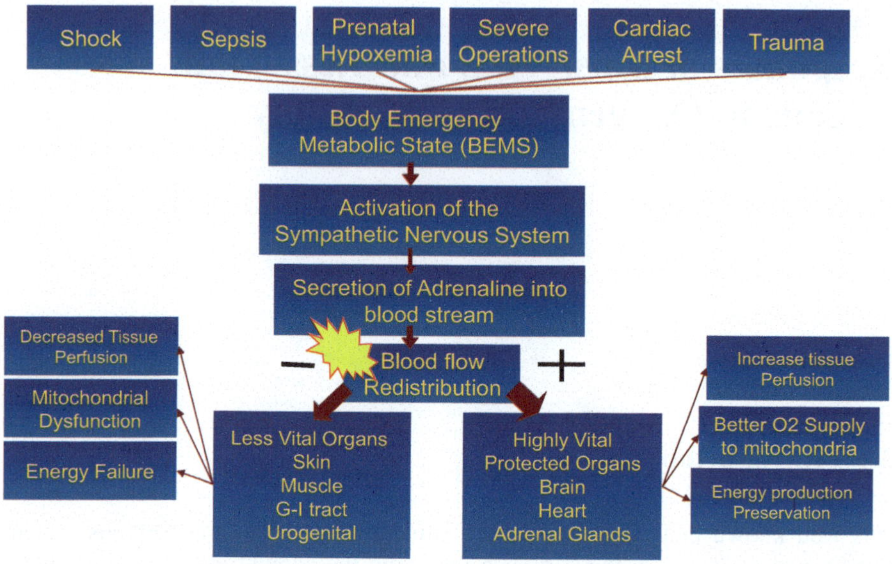

Fig. 3.1 Schematic representation of body emergency metabolic states(BMES) developing in various pathological conditions, leading to redistribution of the blood flow toward the most vital organs at the expense of less vital organs. (Mayevsky 2015)

insensitive, nonspecific, and show abnormality only at a very late stage of disease (Dantzker 1993). In view of this, techniques that directly appraise the tissue energetic state would be an optimal approach for the evaluation of tissue integrity (Dantzker 1997). The mitochondrial NADH redox state is the most sensitive parameter of oxygen deficiency and is well correlated with tissue pO_2 level under oxygen-deficiency conditions (Mayevsky et al. 1980, 1992, 1998). Therefore, in the present study, we aimed to simultaneously monitor TBF and mitochondrial NADH fluorescence in the brain, a most vital organ, and the small intestine, a less vital organ, under oxygen deficiency (anoxia and hypoxia) and hypercapnia. Our hypothesis was that by monitoring the small intestine together with the brain using this unique monitoring device in different acute stressful conditions, early detection of a systemic impaired metabolic state of the body can be accomplished, assuming that this model will help to characterize the "brain-sparing effect" developed under various emergency situations.

3.2 Methods

The Two-Site MultiParametric (TSMP) Monitoring System

Monitoring of the rat brain and small intestine was performed using a TSMP monitoring system developed in our laboratory. Each channel of this monitoring device contains a bundle of optical fibers for NADH redox state monitoring using the

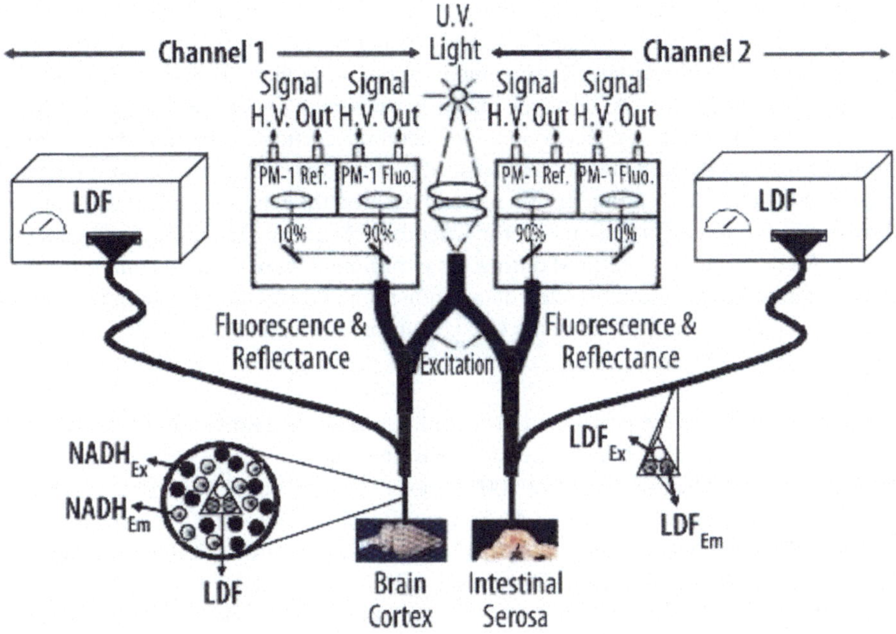

Fig. 3.2 Schematic presentation of the multisite-multiparametric (MSMP) monitoring system and its location on the brain surface and intestinal serosa. Tissue blood flow (TBF) was monitored using the LDF technique: each bundle contains an excitation fiber (LDF_{ex}) and two emission fibers (LDF_{em}). Mitochondrial NADH was monitored by the fluorometric technique using a two-channel DC fluorometer-reflectometer and a bundle of optical fibers for tissue excitation ($NADH_{ex}$) and emission ($NADH_{em}$). The LDF fibers are located together with the NADH fibers, enabling simultaneous monitoring of TBF and mitochondrial NADH redox state from the same location on the tissue. (Tolmasov et al. 2007)

fluorometric technique and another bundle of fibers for TBF monitoring using laser Doppler flowmetry (Figs. 3.2 and 2.14). The diameter of the probe (including all fibers) is 3 mm. The principle of NADH monitoring from the surface of the tissue (up to 1 mm in depth) is that excitation light (366 nm) passes from the fluorometer to the tissue via a bundle of optical fibers made of quartz. The emitted light (450 nm fluorescence), along with the reflected light (366 nm reflectance), is transmitted to the fluorometer via another bundle of fibers (Mayevsky 1993). The emitted light passes through appropriate filters in order to differentiate between the 366 nm reflectance and the NADH fluorescence (450 nm). In addition, a specific filter is used to prevent red light (laser Doppler flowmeter) from interfering with mitochondrial NADH monitoring. Changes in the 366 nm reflected light are correlated to changes in tissue blood volume and therefore serve as a correct tool for the hemodynamic artifacts in NADH monitoring (Mayevsky 1984). The fluorometer is calibrated in such a way that a range of 500 mV is equal to a change of 100% in the monitored signal. The corrected fluorescence (NADH) is obtained by subtracting the reflectance signal from the fluorescence signal at a 1:1 ratio (Mayevsky 1984). In general, an inverse linear correlation was clearly demonstrated between the changes in NADH level and tissue metabolic activity. Namely, under conditions in which the oxygen supply to the tissue is prevented (hypoxia, ischemia) and the metabolic activity decreases, the NADH level increases, whereas under conditions in which tissue activity is increased (such as under cerebral cortical spreading depression), NADH is oxidized and its level decreases (Mayevsky and Rogatsky 2007). TBF was monitored using a laser Doppler flowmeter, based on the Doppler shift reflecting the flow of red blood cells in the tissue for a depth of 1–2 mm (Mayevsky and Chance 1982). All the signals monitored during the experiment were digitized and transmitted to a multichannel computerized data acquisition and recording system (Labview A/D software, National Instruments Co, USA) for further analysis.

Animal Preparation

All experiments were performed in accordance with the Guidelines of the Animal Care Committee of Bar Ilan University. Wistar male rats (250–300 g) were anesthetized by IP injection (0.3 ml/100 g body weight) of Equithesin (each ml containing 9.72 mg pentobarbital, 42.51 mg chloral hydrate, 21.25 mg magnesium sulfate, 44.34% w/v propylene glycol, 11.5% alcohol, and distilled water). During the entire experimental period, anesthesia was maintained by 0.1 ml Equithesin injections every 30 min and the rats were breathing spontaneously. The rats were placed on a warming plate and body temperature was maintained at 37 °C. Polyethylene catheters were introduced into the femoral vein for drug administration and into the femoral artery for systemic blood pressure monitoring. To prepare for brain monitoring, the rat was placed in a special head holder seen in Fig. 3.3a. A 6 mm diameter hole was drilled in the parietal bone and the bone was removed (the dura mater remained intact). A special light guide was then placed above the brain surface. The optical fiber probes of the TSMP device were later placed inside the holder. Two stainless steel screws were inserted into the parietal bone for better fixation of the TSMP holder to the skull surface by dental acrylic cement. Then the rat was placed on its back and a 2 cm incision was performed in the abdomen. A small segment of the small intestine was exposed, and the monitoring probe was

Connection Model	Connection Model figure	Tissue/Organ
A. Cementation	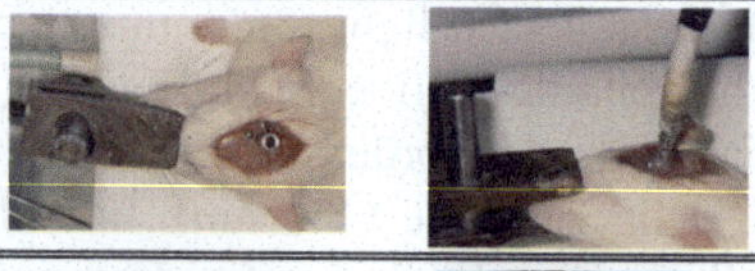	Brain
B. Adhesion	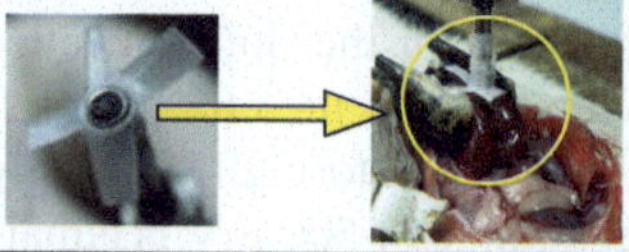	Soft Tissue & Visceral Organs

Fig. 3.3 The two techniques used to connect the monitored organ to the fluorometer

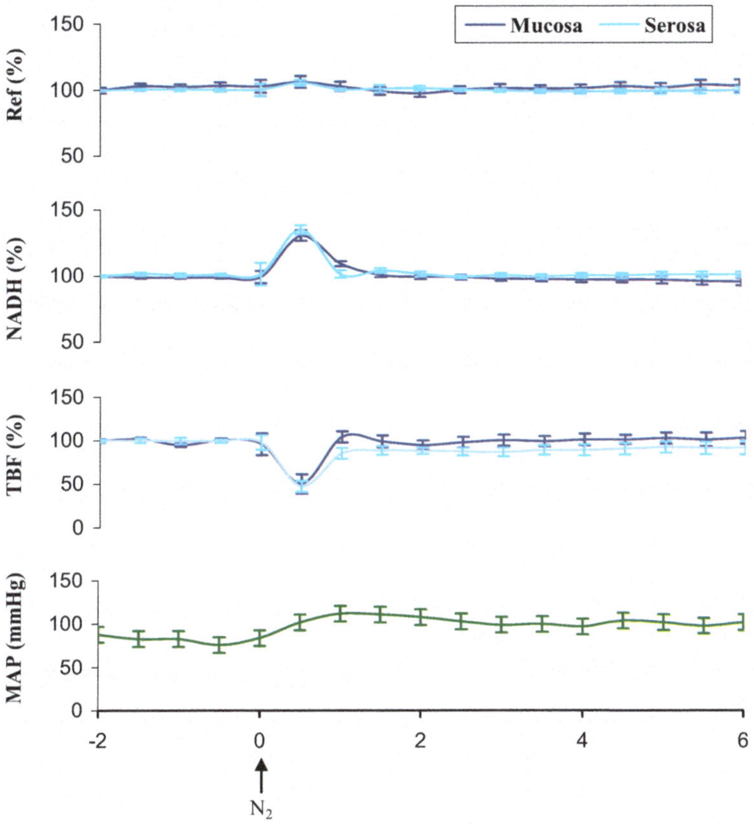

Fig. 3.4 The responses of the MAP (green), mucosa (purple), and serosa (light blue) during short anoxia ($n = 8$, mean ± S.E). *Ref* reflectance, *NADH* mitochondrial NADH redox state, *TBF* tissue blood flow, *MAP* mean arterial pressure

placed on the intestinal serosa using a micromanipulator (Fig. 3.3b) and then connected to the intestinal serosa by cyanoacrylate adhesive (Arvidsson et al. 1988).

The decision to measure the various parameters in the serosa side was based on a short preliminary study in which the measurements of the various parameters were done in the serosa as well as in the mucosal side of the small intestine.

The first experimental model involved short anoxia, i.e., inhalation of 100% N_2 until the rat stopped breathing. The second model involved IV injection of 6 μg/kg epinephrine. In both models, the responses of the serosa and mucosa were compared.

Figure 3.4 presents the responses of the serosa and mucosa to short anoxia. Mean arterial pressure (MAP) increased, and both layers of the intestine, the serosa and mucosa, responded by a decrease in TBF and an increase in NADH. No significant differences were observed between the two layers. Figure 3.5 represents the responses of the serosa and mucosa to IV injection of epinephrine. As is seen,

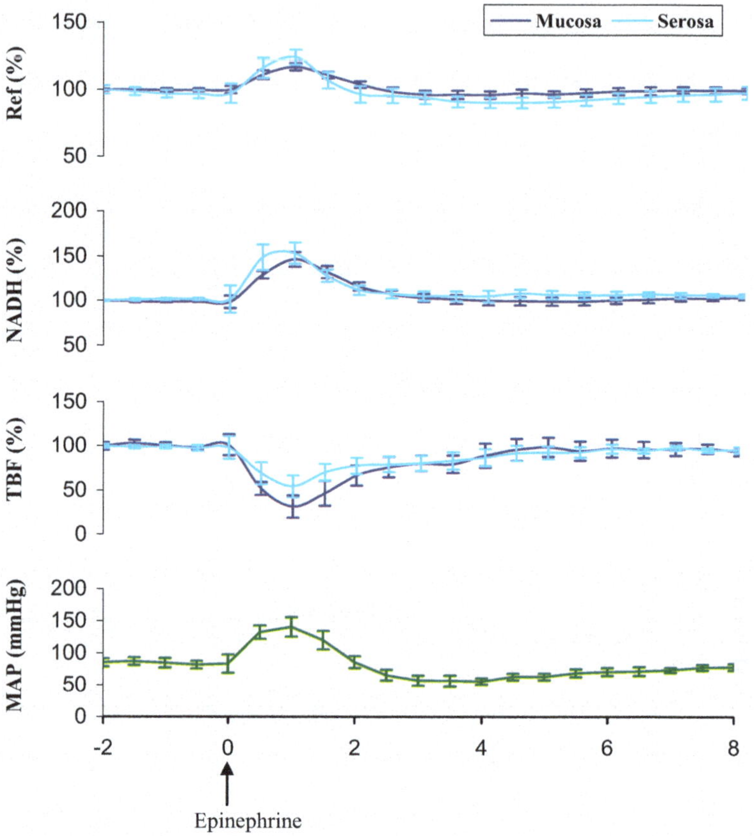

Fig. 3.5 The responses of the MAP (green), mucosa (purple), and serosa (light blue) during injection of epinephrine ($n = 8$, mean ± S.E). *Ref* reflectance, *NADH* mitochondrial NADH redox state, *TBF* tissue blood flow, *MAP* mean arterial pressure

both intestinal layers responded similarly to epinephrine. MAP increased, TBF decreased, and both the reflectance and NADH increased. Furthermore, no significant differences were observed between the two layers.

According to these results, showing no significant differences between the serosa and mucosa, we decided to place the intestinal probe on the serosa, since it is less invasive and easier to manipulate. A schematic presentation of the operated rat connected to the monitoring system is shown in Fig. 3.6.

Experimental Protocols

After surgery and fixation of the TSMP system to the brain and to the intestinal serosa, the following protocols were implemented:

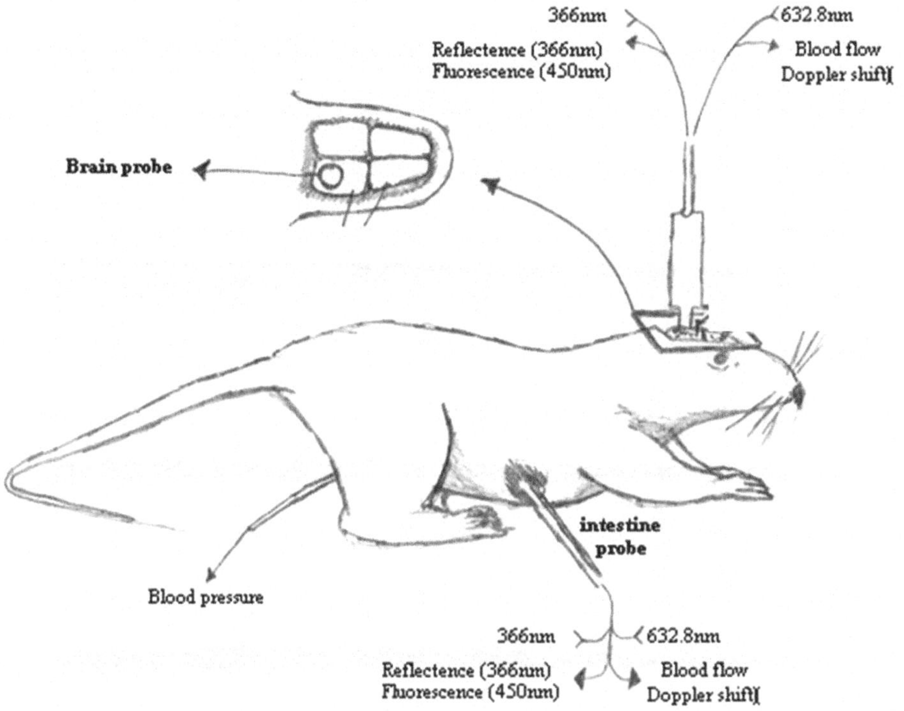

Fig. 3.6 Schematic presentation of the location of the TSMP on the brain and the serosa of the small intestine simultaneously

Control Experiments

In order to ensure a reliable monitoring of both organs, we run control experiments in which both organs and MAP were monitored continuously for almost 1.5 h. As presented in Fig. 3.7, all the parameters remained steady through the entire experimental period. Furthermore, there were no differences in the levels of all the parameters between the brain and small intestine, except for the final 5 min, in which the TBF and the reflectance in the brain were slightly higher (115 ± 12% and 110 ± 3%, respectively) than those of the intestine.

Anoxia

All animals were exposed to a short anoxia (30 s) induced by 100% nitrogen inhalation. This procedure was performed to ensure the correct placement of the monitoring probe on the monitoring site (the brain and intestinal serosa) as well as to test the organs' response to a complete lack of oxygen (Mayevsky and Chance 1982).

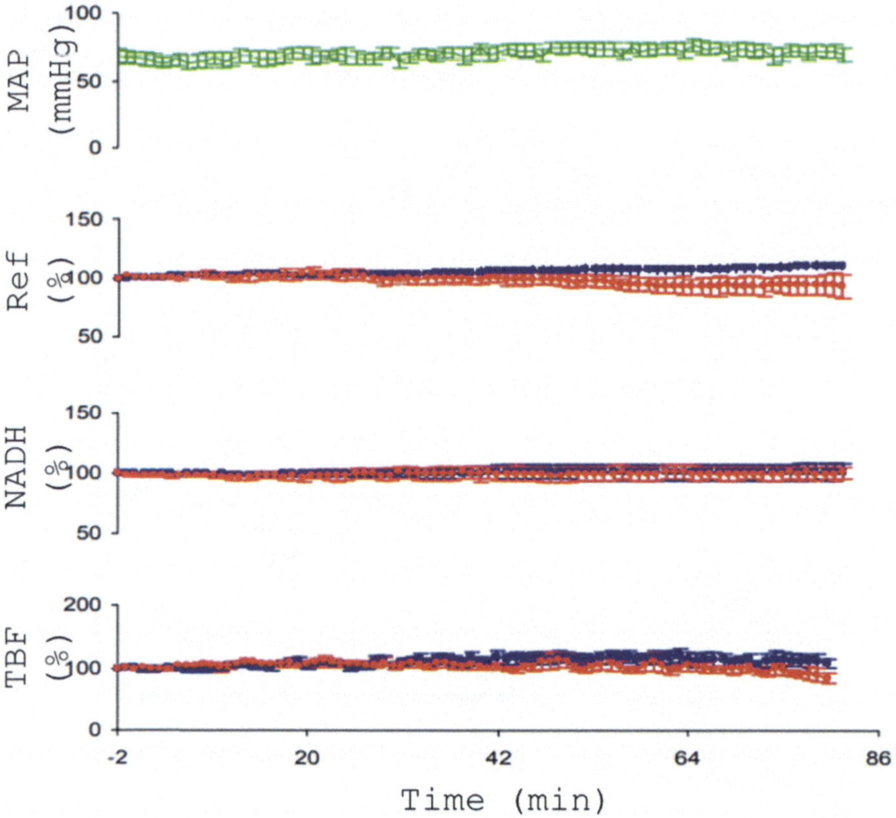

Fig. 3.7 The brain (blue) and small intestine (red) under normal conditions. *MAP* Mean Arterial Pressure, *REF* Reflectance, *NADH* mitochondrial NADH, *TBF* Tissue Blood Flow. ($N = 10$)

Hypoxia, Hypercapnia, and Hyperoxia

Three groups of 10 rats each were exposed to the following gas mixtures for a period of 20 min; thereafter, the rats were allowed to breathe normal air for 90 min. At the end of these periods, the rats were sacrificed by pure N_2 inhalation.

Hypoxia: 12% O_2 + 87% N_2 + 1% CO_2.
Hypercapnia: 21% O_2 + 69% N_2 + 10% CO_2.
Hyperoxia: 100% O_2.

The main limitation in this study is the inability to withdraw blood for gas analysis in the same rats that were monitored. This is because the femoral artery (which is the main blood vessel commonly used for blood withdrawal in a rat model) is near the intestine, and any manipulation near the intestinal monitoring site yielded artifacts in the intestinal monitored signals. Furthermore, using another blood vessel for blood withdrawal, such as the carotid artery, will lead to altered cerebral blood supply. Moreover, since the present study is based on the differences in the responses of the brain and intestine to systemic alterations, it is clear that under the various protocols used, both organs are exposed to the same pO_2, pCO_2, and pH levels. Nevertheless, blood pO_2, pCO_2, and pH were evaluated using a group of rats in another study performed previously in our laboratory. In the same species of Wistar rats, the same protocol, namely, spontaneous breathing of similar mixtures of gases for similar durations, was applied (Meilin et al. 1996). In addition, we monitored arterial blood pressure in order to obtain important information on the cardiovascular state of the rat.

Statistical Analysis

The two-tailed paired Student's *t*-test was used to examine the differences between the responses of the brain and intestinal serosa to various oxygen and CO_2 concentrations at specific time points (in anoxia for each second and in the rest of the groups for each minute). The two-tailed unpaired Student's *t*-test was used to examine the significance of changes in each parameter at specific points after the experimental period, compared with its first baseline value. A value of $P < 0.05$ was considered to be significant. All values are presented as the average ± *S.E.* In NADH calculations, the basal level before perturbation was considered 100%, NADH reduction is shown by NADH increase above 100%, and NADH oxidation is indicated by NADH decrease below 100%. TBF at death was considered 0%, whereas the range of signal from the basal level to the death level was considered as 100%. MAP values are presented as the changes in mmHg.

3.3 Results

In order to test the ability of the multisite-multiparametric (MSMP) device to detect changes in the hemodynamic and metabolic states of the brain and the intestine simultaneously in a rat model, we used several experimental protocols. Exposure of the rat to various concentrations of oxygen or CO_2 enabled us to test the differences in the responses of these two organs to oxygen deficiency or surplus.

3.3.1 Effects of Anoxia

As seen in Fig. 3.8 and Table 3.1, 100% N_2 inhalation for 30 s caused a decrease of 10 ± 8 mmHg in MAP, followed by an increase of approximately 140% in NADH in both organs. With respect to blood flow, a decrease of 40 ± 11% ($P < 0.01$) was observed in the intestine, while in the brain, CBF increased to the level of 315 ± 53% ($P < 0.01$). Moreover, the kinetics of response in the two organs was different. The time needed for CBF to reach maximum hyperemia was 120 s, whereas in the intestine the maximum decrease in TBF was recorded 40 s after the beginning of anoxia. During the recovery phase, when the rat breathed normal air, NADH in the brain decreased (oxidized) immediately, whereas in the intestine NADH decreased

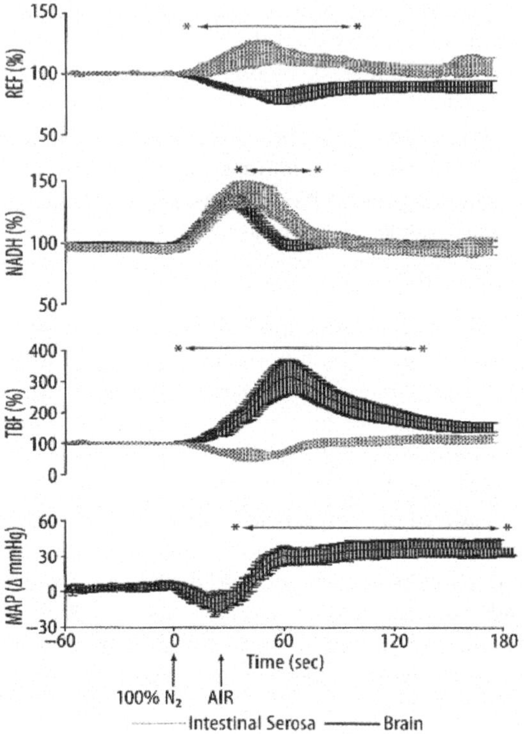

Fig. 3.8 The responses of the brain (blue) and small intestine (red) to short anoxia. Abbreviations are as above. Arrows show the period of time in which there were significant differences between the two organs for each minute. For MAP, the significance is related to the basal level. ($N = 9$). (*) $P < 0.05$. (Barbiro-Michaely et al. 2007)

Table 3.1 The major (min/max) changes in the systemic blood pressure and cerebral and intestinal parameters during the induction of anoxia, hypoxia, and hypercapnia (Barbiro-Michaely et al. 2007)

Perturbation	MAP change	Parameter	Brain	Intestine
Anoxia	−10 ± 8	TBF	315 ± 53	60 ± 11
		NADH	134 ± 3.8	139 ± 10
		REF	81.5 ± 5.7	117 ± 9.7
Hypoxia	−34 ± 4	TBF	140 ± 12	56 ± 13
		NADH	119 ± 2.8	151 ± 14
		REF	92 ± 2	119 ± 8.2
Hypercapnia	+13 ± 14	TBF	143 ± 11	113 ± 12
		NADH	90.3 ± 1.3	93 ± 1.6
		REF	121 ± 2.7	90 ± 7.6

The maximum and minimum changes (mean ± S.E.) in *TBF* tissue blood flow, *NAD* mitochondrial NADH redox state, *REF* tissue reflectance, and *MAP* mean arterial blood pressure in the brain and intestinal wall during the induction of three models of acute stress (anoxia, hypoxia, and hypercapnia). The basal level for each parameter was 100%, and the changes were monitored within the time the rats were exposed to the specific gas mixtures according to the protocol only 20 s later. Also, during the recovery phase, CBF continued to increase for 30 s, even though the rats were breathing air, and the recovery lasted for 90 s (until CBF reached its basal level). In contrast, TBF in the intestine returned to its baseline level 30 s after air inhalation. Changes in the reflectance, traced in both organs, were inversely correlated to changes in TBF.

Additionally, the changes in the TBF in both organs were significantly correlated with the changes in the mitochondrial NADH level. As presented in Fig. 3.9, the hyperemia in the cerebral tissue was associated with NADH level elevation ($r = 0.9432$, $P < 0.01$, $n = 30$) due to complete oxygen deficiency, while in the intestine NADH was elevated also due to a decreased TBF ($r = 0.9988$, $P < 0.01$, $n = 30$). A significant correlation was also detected between changes in the blood pressure and the level of TBF in both organs. As presented in Fig. 3.10, a decrease in MAP is accompanied by hyperemia in the brain ($r = 0.7813$, $P < 0.01$, $n = 30$) and deterioration of blood supply to the small intestine ($r = 0.9154$, $P < 0.01$, $n = 30$).

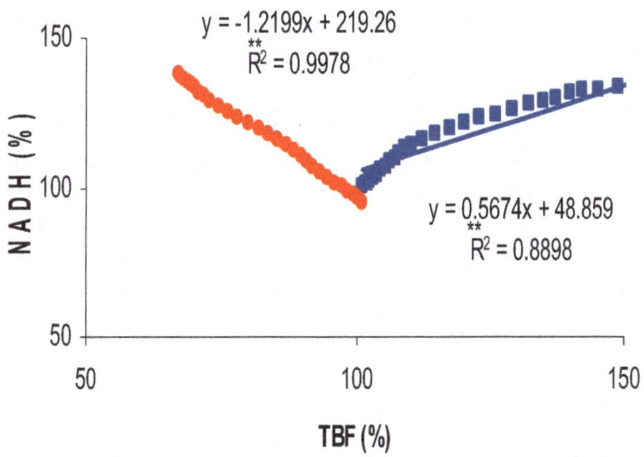

Fig. 3.9 The correlation between tissue blood flow (TBF) and mitochondrial NADH measured in the rat brain (blue) and in the small intestine (red) under anoxia. (N = 9). The data used for the present correlation were obtained from the time range (0–30 s). (**) $P < 0.01$

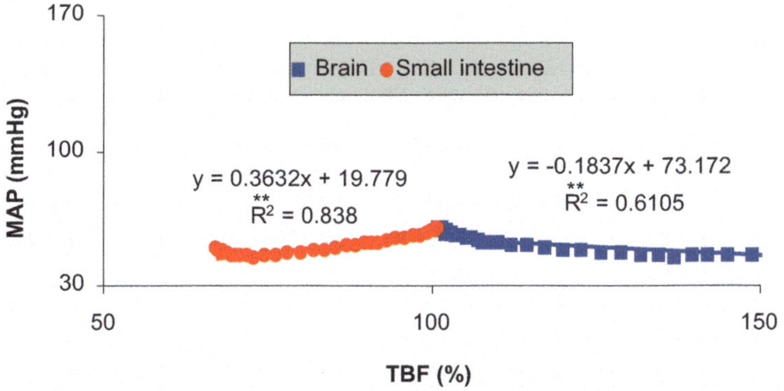

Fig. 3.10 The correlation between mean arterial pressure (MAP) and tissue blood flow (TBF) measured in the rat brain (blue) and in the small intestine (red) under anoxia. ($N = 9$). The data used for the present correlation were obtained from the time range (0–30 s). (**) $P < 0.01$

3.3.2 *Effects of Hypoxia*

Previous studies in our laboratory showed that by exposing rats ($n = 6$) (breathing spontaneously) to a gas mixture of 12% O_2: 1% CO_2: 87% N_2 (hypoxia), pO_2 decreased to 54 ± 4 mmHg, pCO_2 decreased to 30 ± 3 mmHg, and pH increased to 7.27 ± 0.03 (Table 3.2). The decrease in pCO_2 and the increase in pH level were probably caused by the spontaneous hyperventilation that results from exposure to the low oxygen level in the inspired air. When testing the effects of hypoxia on the brain and intestinal metabolism, the following results were recorded (Fig. 3.8 and Table 3.2): MAP decreased by 34 ± 4 mmHg ($P < 0.001$) immediately after the rats started breathing the low-oxygen mixture. This change was followed by a decrease in TBF in the intestine to a level of 56 ± 13% ($P < 0.05$) and a tendency of CBF increase to a level of 140 ± 12%. Simultaneously, the reflectance in the intestine increased (119 ± 8%, $P < 0.05$), while in the brain it decreased (92 ± 2%, $P < 0.01$). The changes in the blood supply to the organs produced corresponding changes in the levels of mitochondrial NADH. In the intestine, NADH increased to a level of 151 ± 14% ($P < 0.01$) and remained at this level throughout the entire hypoxic period, while in the brain NADH reached a level of 119 ± 2.8% ($P < 0.001$). When the rats started breathing air, all the parameters returned to the basal level within 2 min in the intestine, whereas in the brain hyperemia was observed for 11 min, followed by full recovery. With reference to NADH and reflectance, full recovery was observed immediately after the rats started breathing air (Fig. 3.11).

The correlation between TBF and NADH in both organs during the hypoxic period is presented in Fig. 3.12a. As seen, hyperemic levels of CBF were associated with relatively stable level of NADH ($r = 0.7547$, $P < 0.01$, $n = 20$); however, in the small intestine, low levels of TBF were associated with high NADH levels ($r = 0.9172$, $P < 0.01$, $n = 20$). Additionally, decrease of MAP significantly correlated with the blood-flow deterioration in the small intestine ($r = 0.7004$, $P < 0.01$, $n = 20$), whereas changes in this systemic parameter were not correlated with the blood supply to the brain ($r = 0.3613$, n.s., $n = 20$) (Fig. 3.12b).

Table 3.2 Effects of hypoxia (12% O_2, 1% CO_2, 87% N_2) and hypercapnia (100% CO_2, 21% O_2, 69% N_2) on sample blood parameters (Barbiro-Michaely et al. 2007)

	Control (before hypoxia)	Hypoxia	Control (before hypercapnia)	Hypercapnia
pO_2	108.27 ± 5.59	54.39 ± 3.88***	113.83 ± 7.65	160.49 ± 7.7***
pCO_2	36.42 ± 2.82	30.04 ± 6.83	37.47 ± 6.83	69.94 ± 5.95**
pH	7.22 ± 0.03	7.27 ± 0.03***	7.26 ± 0.05	7.07 ± 0.05***
% sat.	95.39 ± 0.84	77.42 ± 4.63**	96.43 ± 0.52	97.64 ± 0.13

The levels of pO_2, pCO_2, pH and the saturation of arterial blood before and following hypoxia or hypercapnia induction, Hypoxia: the rats were exposed to a gas mixture of 12% O_2 + 1% CO_2 + 87% N_2 for 20 min and under hypercapnia the rats were exposed to a gas mixture of 10% CO_2 + 21% O_2 + 69% N_2 for 20 min. Significance between control levels and the levels measured under treatment is presented as follows: (*) $P < 0.05$, (**) $P < 0.01$, (***) $P < 0.001$

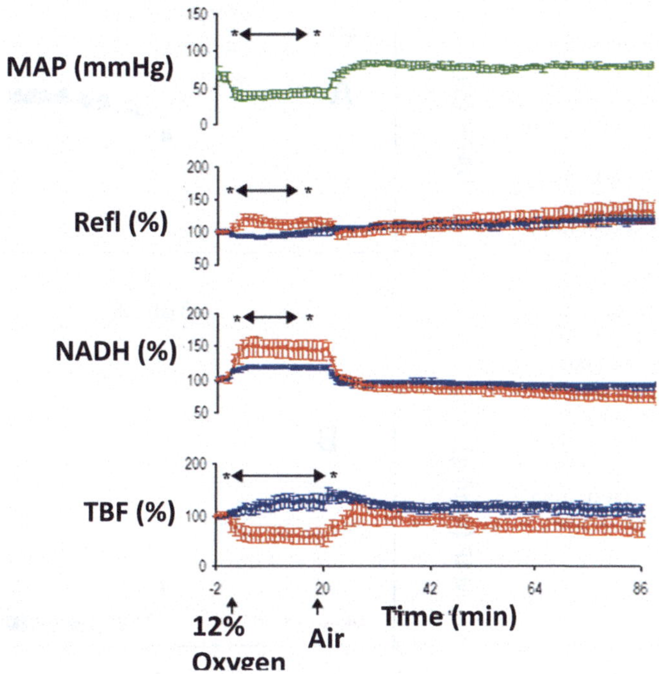

Fig. 3.11 The responses of the brain cortex and intestinal serosa (mean ± S.E.) to hypoxia induced by exposure to a low oxygen mixture: 12% O_2 + 87% N_2 + 1% CO_2. *Ref* reflectance, *NADH* mitochondrial NADH redox state, *TBF* tissue blood flow, *MAP* mean arterial pressure. $n = 10$, (*) $P < 0.05$. (Barbiro-Michaely et al. 2007)

3.3.3 Effects of Hypercapnia

Under hypercapnia induced by 10% CO_2 in air (in another group of 6 rats), pCO_2 increased to approximately 70 ± 6 mmHg and pO_2 increased to 160.5 ± 7.7 mmHg (Table 3.2). In the group of rats that were monitored by the TSMP system (Fig. 3.13), hypercapnia induced the following results: There were no significant differences between the brain and intestine, except for the difference in reflectance at the 17th min, when it increased in the brain to 106 ± 2% and remained constant in the intestine. When testing the effects of hypercapnia on each organ separately, the following results were obtained: In the brain, hypercapnia caused a significant increase in CBF to a level of 143 ± 11% ($P < 0.05$) 6 min after hypercapnia induction. The hyperemia continued for approximately 60 min. However, mitochondrial NADH was decreased from the very beginning of hypercapnia until the end of the experiment. Considering systemic changes, a significant MAP increase of 13 ± 4 mmHg was seen 3 min after hypercapnia induction, and it continued to increase for 17 min. Hypercapnia had no significant changes in intestinal TBF.

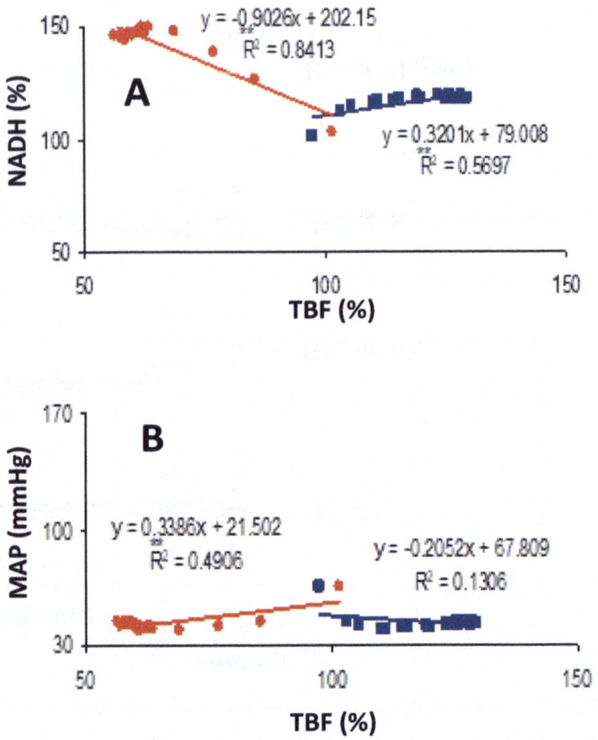

Fig. 3.12 (**a**) The correlation between tissue blood flow (TBF) and NADH measured in the rat brain (blue) and in the small intestine (red) under hypoxia. ($N = 9$). The data used for the present correlation were taken from the time range (0–20 min). (**) $P < 0.01$. (**b**) The correlation between tissue blood flow (TBF) and MAP measured in the rat brain (blue) and in the small intestine (red) under hypoxia. (N = 9). The data used for the present correlation were taken from the time range (0–20 min). (**) $P < 0.01$

However, a significant decrease in NADH was observed 4 min after the rats were exposed to high CO_2 concentrations. The oxidation of mitochondrial NADH continued for 10 min.

As seen in Fig. 3.14a, brain hyperemia was significantly correlated with NADH oxidation ($r = 0.9333$, $P < 0.01$, $n = 20$), while in the small intestine the level of the blood flow was not correlated with the NADH level ($r = 0.3074$, ns., $n = 20$). In addition, the analysis of the correlation between the blood pressure and TBF indicates a statistically significant linear relationship ($P < 0.01$) detected in both organs (Fig. 3.14b). Thus, the elevation of blood pressure accompanied hyperemia in the brain ($r = 0.8624$, $P < 0.01$, $n = 20$) and correlated with the TBF level in the small intestine ($r = 0.7082$, $P < 0.01$, $n = 20$).

3.3.4 *Effects of Hyperoxia*

Hyperoxia is a condition in which oxygen is in surplus. We wanted to test the responses of the brain versus the intestine to a condition in which oxygen is available in high concentrations. Figure 3.15 demonstrates the responses of both organs to hyperoxia (100% O_2). No significant differences between the two organs were

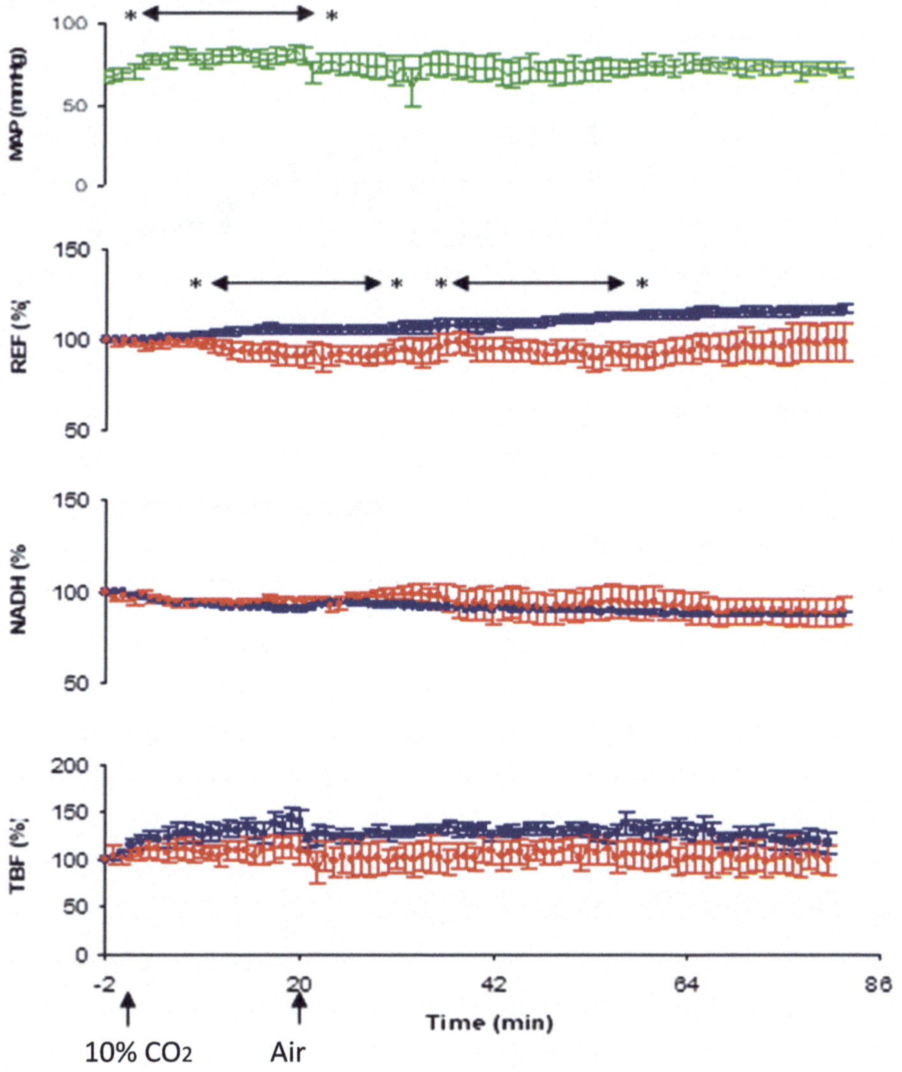

Fig. 3.13 The responses of the brain cortex and intestinal serosa (mean ± S.E.) to hypercapnia induced by exposure to a high CO_2 mixture: 21% O_2 + 69% N_2 + 10% CO_2. *Ref* reflectance, *NADH* mitochondrial NADH redox state, *TBF* tissue blood flow, *MAP* mean arterial pressure. $n = 10$, (*) $P < 0.05$. (Barbiro-Michaely et al. 2007)

observed. MAP increased by 16 ± 4.7 mmHg ($P < 0.001$) immediately after the rats started breathing pure oxygen. When testing the effects of hyperoxia on each organ separately, the following results were obtained. The elevation of the MAP level was not accompanied by any changes in the blood supply of both organs. However, the mitochondrial NADH was significantly oxidized in both organs during the entire period of exposure to 100% O_2, reaching the levels of 93 ± 1.5% in the brain

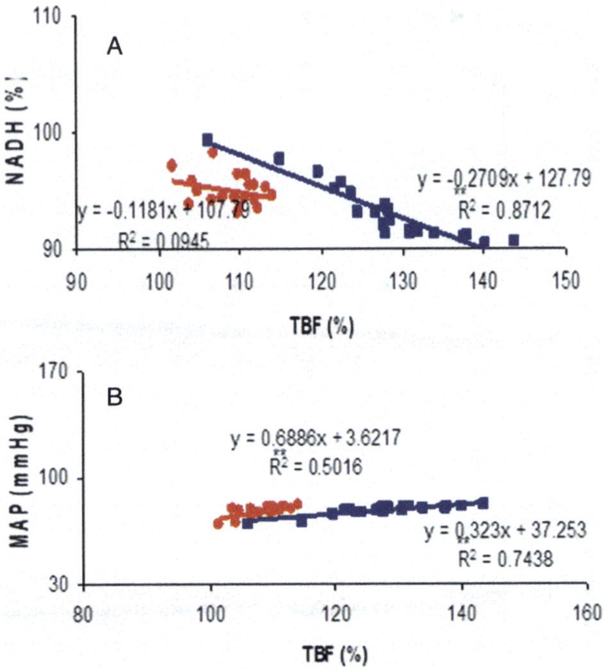

Fig. 3.14 (**a**) The correlation between tissue blood flow (TBF) and NADH redox state (NADH) measured in the rat brain (blue) and in the small intestine (red) under hypercapnia. ($N = 9$). The data used for the present correlation were taken from the time array (0–20 min). (**) $P < 0.01$. (**b**) The correlation between mean arterial pressure (MAP) and tissue blood flow (TBF) measured in the rat brain (blue) and in the small intestine (red) under hypercapnia. ($N = 9$). The data used for the present correlation were taken from the time array (0–20 min). (**) $P < 0.01$

($P < 0.01$) and 95 ± 0.6% in the intestine ($P < 0.01$). With reference to the reflectance, this parameter was stable through the entire experimental period in both organs.

There was a significant linear correlation between TBF and NADH (Fig. 3.16) in both organs (brain: $r = 0.5656$, $P < 0.01$, $n = 20$ and intestine: $r = 0.5158$, $P < 0.01$, $n = 20$), namely, as TBF increased, NADH was oxidized and the opposite.

3.4 Discussion

The aim of the present study was to establish a new small animal model that would yield conditions similar to some of the emergency situations developed in patients. We used a monitoring system that provides real-time data on microcirculatory blood flow and volume as well as the intramitochondrial redox state of NADH in rats exposed to systemic perturbations. Under critical conditions, blood flow to the gastrointestinal (GI) tract is selectively and markedly reduced, and the small intestine is one of the first organs to be affected (Neviere et al. 2002), making it ideal to serve as a surrogate organ to the brain for deterioration of body oxygenation. Additionally, the GI tract is relatively easy to access in experimental as well as clinical situations and provides reliable information.

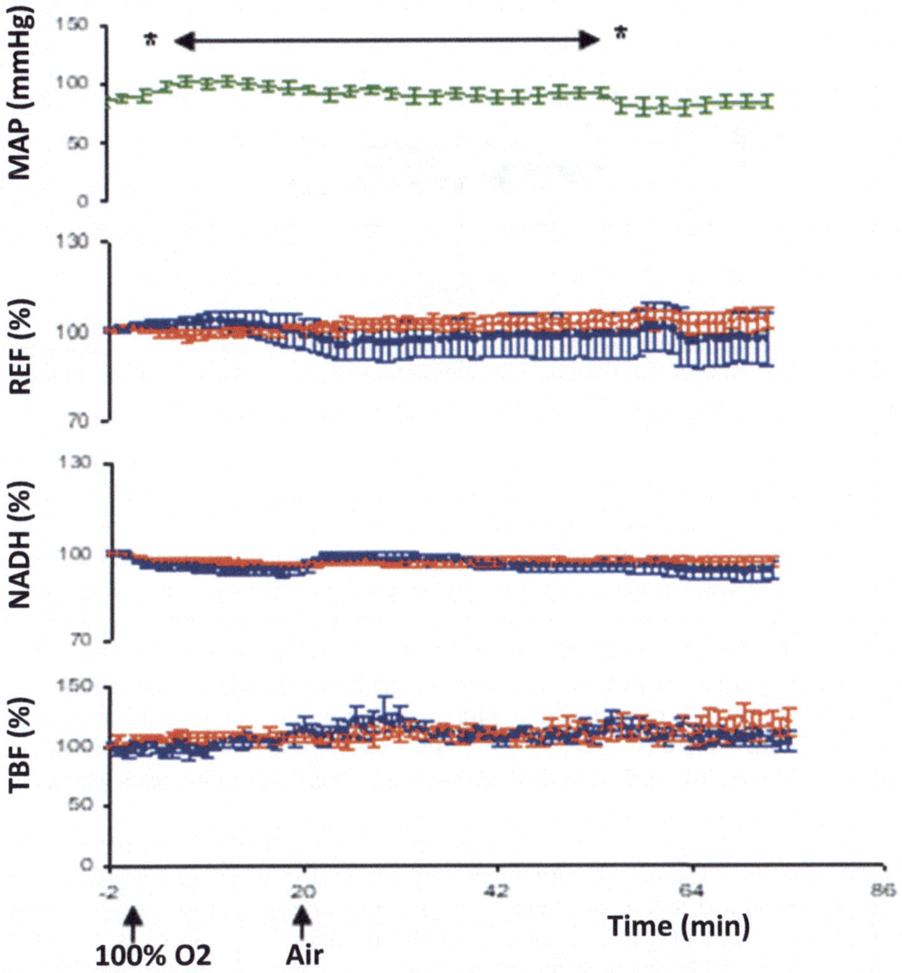

Fig. 3.15 The responses of the brain cortex and intestinal serosa (mean ± S.E.) to hyperoxia induced by exposure to 100% oxygen. *Ref* reflectance, *NADH* mitochondrial NADH redox state, *TBF* tissue blood flow, *MAP* mean arterial pressure. $n = 10$, (*) $P < 0.05$

The three monitored parameters represent the behavior of the microcirculatory hemodynamics and its coupling to intracellular mitochondrial function. In order to evaluate tissue oxygenation under anoxia, hypoxia, hypercapnia, and hyperoxia, we used the intracellular marker located in the mitochondria, which are responsible for generating most of the ATP in the tissue. As shown in Table 3.2, the perturbations used led to changes in systemic arterial blood parameters. The two monitored organs were exposed to the same internal environment, but their responses were very different. We believe that the use of electrodes for tissue pO_2 measurements is less accurate than the evaluation of mitochondrial NADH redox state, since it is very

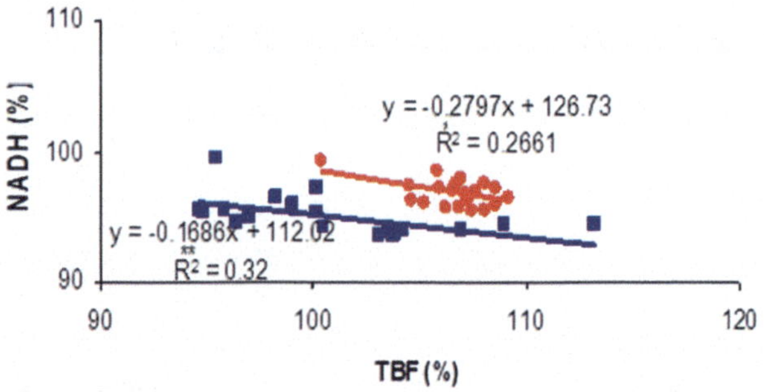

Fig. 3.16 The correlation between tissue blood flow (TBF) and NADH redox state measured in the rat brain (blue) and in the small intestine (red) under hyperoxia. ($N = 9$). The data used for the present correlation were taken from the time array (0–20 min). (*) $P < 0.05$. (**) $P < 0.01$. (Barbiro-Michaely et al. 2007)

sensitive to TBF and provides an average level of oxygenation of the microcirculatory blood and extracellular and intracellular compartments. Tissue microcirculatory blood flow, together with NADH redox state, provides, to the best of our knowledge, the most important information to describe the metabolic activity in the tissue. Since we compared the responses of two different organs in the same body, the absolute level of pO_2 is less important. By measuring the NADH redox state, we bypass the need for pO_2 values, which are also very difficult to calibrate in vivo due to O_2 consumption by the electrode itself. Moreover, we do not compare our results to previous studies in which similar perturbations were tested, but rather compare the responses of these two organs (brain and intestine) to various conditions in which the rat breathes different gas mixtures.

The systemic effect of hypoxia, for example, is the same in the blood that flows to the brain or the intestine. The difference in the responses of the two organs is the main question, and indeed our multiparametric monitoring approach may provide the answer. When pO_2 is measured, the question that always arises is what the physiological interpretation of the 30 mmHg or 10 mmHg is. In our monitoring approach, the intracellular activity of the mitochondria is assessed, and the critical levels for intact activity of the tissue can be evaluated using these data. We know, for example, that an increase in NADH by 50% or more is very significant to the function of the mitochondria. In the past, we compared tissue pO_2 with NADH and found that NADH is much more significant (Mayevsky et al. 1980). The same was found when we compared HbO_2 oxygenation with NADH. This parameter is also very sensitive to changes in blood flow at the microcirculation level (Mayevsky et al. 1992).

A negative balance between O_2 supply and O_2 demand has a crucial role in many pathophysiological states. However, under critical conditions, the autonomic nervous system redistributes the supply of blood, and more oxygen is provided to the most vital organs compared with less vital ones. This protective mechanism was demonstrated in the present study using real-time monitoring of tissue energy metabolism. Our results showed that when the oxygen supply to the tissue is impaired (under hypoxia), the responses of TBF and mitochondrial NADH in the brain were different from those in the intestine. Under oxygen deficiency, the mechanisms of brain autoregulation became activated, leading to an increase in CBF due to vasodilatation of small arterioles. This phenomenon is well known in mammals, e.g., rats, where the increase in CBF under hypoxia is mediated by the rostral ventrolateral reticular nucleus of the brainstem (Nunn 1987; Xu et al. 2004). On the other hand, in the small intestine, where no autoregulation mechanisms exist, hypoxia caused a significant decrease in TBF, which correlated with the decrease in MAP. Consequently, oxygen supply was interrupted and mitochondrial dysfunction developed, leading to a significant increase in NADH in the intestine. Although NADH also increased in the brain during hypoxia, this increase was significantly lower than the increase observed in the intestine, indicating the ability of autoregulatory mechanisms to improve tissue oxygenation rather than their ability to prevent energy failure. It is important to emphasize that although the brain and intestine differ in their optical characteristics, a comparison of the NADH levels in these two organs is possible since the responses to a complete lack of oxygen, as observed under anoxia or following death (results not shown), are identical in both organs. NADH monitoring using the fluorometric technique in these two organs is not affected by the differences in their optical nature due to the calibration technique, which brings the signal to the same baseline. Furthermore, the fact that NADH increased significantly in both organs, although TBF responded inversely may point out the risk involved in the evaluation of organ viability by monitoring only TBF.

Contrasting responses of TBF in these two organs were also observed under anoxia, where CBF in the brain increased while TBF in the intestine decreased. Nevertheless, NADH increased to the same level in both organs due to the complete depletion of oxygen from the blood. The advantage of the state of the brain following anoxia is observed during the recovery phase, in which mitochondrial NADH was fully oxidized earlier than the NADH in the intestine.

Under hypercapnia, CBF increased due to the elevation in pCO_2. NADH became oxidized before the increase in blood supply, probably due to the increase in respiratory rate (which was observed in real time), leading to an increase in oxygen supply to both tissues. This was also observed in the hypercapnia group presented in a previous study (Meilin et al. 1996), where hypercapnia of 20 min increased pO_2 by 47 mmHg, as well as in another previous study (Kallinen et al. 1991).

Consequently, NADH was oxidized in the intestine as well, although only for a short duration, since no increase in TBF occurred. In contrast, the blood supply increased in the brain in addition to the increase in the respiratory rate; therefore, NADH oxidation lasted longer.

Under hyperoxia, the only change observed was the oxidation of NADH in both organs, probably due to the increase in blood oxygenation. The response of CBF to high oxygen tension, as indicated in previous studies, seems to be heterogeneous for several reasons: the duration of hyperoxia, artificial ventilation versus spontaneous breathing, CBF monitoring techniques, and the anesthetic used. Transient hyperoxia (2 min) caused an increase in CBF due to hypoventilation, which increased $PaCO_2$ levels (Sicard and Duong 2005). In contrast, chronic hyperoxia is suggested to lead to vasoconstriction of small arterioles in the brain and thus to a decrease in CBF. Other studies, however, showed that hyperoxia does not alter CBF (Duong et al. 2001). With regard to monitoring techniques, using laser Doppler flowmetry showed no significant changes in CBF during hyperoxia (Wolf et al. 1997), although other blood-flow monitoring techniques, such as microspheres and H_2 clearance, showed a decrease in CBF under exposure to high pO_2 levels (Davidson et al. 1986).

Since many critical conditions treated in intensive care units and operating rooms involve an imbalance between oxygen supply and oxygen demand, which may lead to a general breakdown of homeostasis, it is crucially important to develop a method for the early and reliable detection of these conditions. Such early detection is necessary to mitigate adverse consequences, including ATP catabolism, the production of reactive oxygen metabolites, and the activation of the inflammatory processes, which can lead to progressive cellular dysfunction and cell death (Grum 1993). Moreover, during the last decades, it has become clear that traditional global methods for the estimation of tissue hypoxia, such as lactate levels and mixed venous pO_2 measurements, are nonspecific and insensitive in detecting regional abnormalities (Dantzker 1997). We believe that the use of the MSMP system in animal models that mimic clinical situations will enable the detection of partial or complete tissue hypoxia at an early phase, before irreversible damage occurs.

3.5 Conclusion

This study expands on and supports our previous work in which we simultaneously monitored TBF and mitochondrial NADH redox state in four organs, i.e., the brain, kidney, liver, and the testis, in rat models of anoxia and cardiac arrest (Kraut et al. 2003) as well as after norepinephrine injection (Chance and Legallias 1959). Moreover, since intestinal blood flow as well as mitochondrial NADH levels showed a significant response to oxygen deficit, the monitoring of the intestinal wall, or any other less vital organ, could be used for the early detection of whole-body deterioration.

References

Ackland G, Grocott MP, Mythen MG (2000) Understanding gastrointestinal perfusion in critical care: so near, and yet so far. Crit Care 4(5):269–281

Arvidsson D, Svensson H, Haglund U (1988) Laser-Doppler flowmetry for estimating liver blood flow. Am J Phys 254(4 Pt 1):G471–G476

Barbara A (1994) Shock. In: Schwartz SI (ed) Principles of surgery. McGraw-Hill, New York, pp 101–122

Barbiro-Michaely E, Tolmasov M, Rinkevich-Shop S, Sonn J, Mayevsky A (2007) Can the "brain-sparing effect" be detected in a small-animal model? Med Sci Monit 13(10):Br211–Br219

Beale R, Bihari D (1993) Multiple organ failure: the pilgrim's progress. Crit Care Med 21(2 Suppl): S1–S3

Blinman T, Maggard M (2000) Rational manipulation of oxygen delivery. J Surg Res 92(1): 120–141

Boldt J (2002) Clinical review: hemodynamic monitoring in the intensive care unit. Crit Care 6(1): 52–59

Cairns CB, Moore FA, Haenel JB, Gallea BL, Ortner JP, Rose SJ, Moore EE (1997) Evidence for early supply independent mitochondrial dysfunction in patients developing multiple organ failure after trauma. J Trauma 42(3):532–536

Chance B, Legallias V (1959) Differential microfluorimeter for the localization of reduced pyridine nucleotide in living cells. Rev Sci Instr 30(8):732–735

Dantzker DR (1993) Adequacy of tissue oxygenation. Crit Care Med 21(2 Suppl):S40–S43

Dantzker DR (1997) Monitoring tissue oxygenation. The search for the grail. Chest 111(1):12–14

Davidson D, Stalcup SA, Mellins RB (1986) Systemic hemodynamics affecting cardiac output during hypocapnic and hypercapnic hypoxia. J Appl Physiol (1985) 60(4):1230–1236

Duong TQ, Iadecola C, Kim SG (2001) Effect of hyperoxia, hypercapnia, and hypoxia on cerebral interstitial oxygen tension and cerebral blood flow. Magn Reson Med 45(1):61–70

Fink MP (2001) Cytopathic hypoxia. Mitochondrial dysfunction as mechanism contributing to organ dysfunction in sepsis. Crit Care Clin 17(1):219–237

Fink MP (2002) Cytopathic hypoxia. Is oxygen use impaired in sepsis as a result of an acquired intrinsic derangement in cellular respiration? Crit Care Clin 18(1):165–175

Ganong WF (1991) Circulation through special regions. Appleton and Lange medical book. Rev Med Physiol:562–582

Grum CM (1993) Tissue oxygenation in low flow states and during hypoxemia. Crit Care Med 21(2 Suppl):S44–S49

Guyton AC (1991a) Circulatory shock and physiology of its treatment. In: Guyton AC (ed) Textbook of medical physiology, 8th edn. W.B. Saunders Company, Philadelphia, pp 263–271

Guyton AC (1991b) General principles of gastrointestinal function-motility, nervous control, and blood circulation. In: Guyton AC (ed) Textbook of medical physiology, 8th edn. W.B. Saunders Company, Philadelphia, pp 688–697

Harbrecht BG, Alarcon LH, Peitzman AB (2004) Management of shock. In: Moore EE, Feliciano DV, Mattox KL (eds) Trauma, 5th edn. McGraw-Hill, New York, pp 201–226

Hotchkiss RS, Karl IE (1992) Reevaluation of the role of cellular hypoxia and bioenergetic failure in sepsis. JAMA 267:1503–1510

Ince C, Sinaasappel M (1999) Microcirculatory oxygenation and shunting in sepsis and shock. Crit Care Med 27:1369–1377

Kallinen J, Didier A, Miller JM, Nuttall A, Grenman R (1991) The effect of CO_2- and O_2-gas mixtures on laser Doppler measured cochlear and skin blood flow in Guinea pigs. Hear Res 55(2):255–262

Kraut A, Barbiro-Michaely E, Zurovsky Y, Mayevsky A (2003) Multiorgan monitoring of hemodynamic and mitochondrial responses to anoxia and cardiac arrest in the rat. Adv Exp Med Biol 510:299–304

Kuwahira I, Heisler N, Piiper J, Gonzalez NC (1993) Effect of chronic hypoxia on hemodynamics, organ blood flow and O_2 supply in rats. Respir Physiol 92(2):227–238

Levy B, Mansart A, Bollaert PE, Franck P, Mallie JP (2003) Effects of epinephrine and norepinephrine on hemodynamics, oxidative metabolism, and organ energetics in endotoxemic rats. Intensive Care Med 29(2):292–300

Licker M, Schweizer A, Hohn L, Morel DR (1998) Haemodynamic and metabolic changes induced by repeated episodes of hypoxia in pigs. Acta Anaesthesiol Scand 42(8):957–965

Marik PE, Varon J (1998) The hemodynamic derangements in sepsis: implications for treatment strategies. Chest 114:854–860

Mayevsky A (1984) Brain NADH redox state monitored *in vivo* by fiber optic surface fluorometry. Brain Res Rev 7:49–68

Mayevsky A (1993) Biochemical and physiological activities of the brain as *in vivo* markers of brain pathology. In: Bernstein EF, Callow AD, Nicolaides AN, Shifrin EG (eds) Cerebral revascularization. Med-Orion Publication, pp 51–69

Mayevsky A (2015) Mitochondrial function *in vivo* evaluated by NADH fluorescence. Springer International, Cham

Mayevsky A, Chance B (1982) Intracellular oxidation-reduction state measured *in situ* by a multicannel fiber-optic surface fluorometer. Science 217:537–540

Mayevsky A, Rogatsky G (2007) Mitochondrial function *in vivo* evaluated by NADH fluorescence: from animal models to human studies. Am J Physiol Cell Physiol 292:C615–C640

Mayevsky A, Lebourdais S, Chance B (1980) The interrelation between brain PO_2 and NADH oxidation- reduction state in the gerbil. J Neurosci Res 5:173–182

Mayevsky A, Frank K, Muck M, Nioka S, Kessler M, Chance B (1992) Multiparametric evaluation of brain functions in the Mongolian gerbil *in vivo*. J Basic Clin Physiol Pharmacol 3:323–342

Mayevsky A, Meilin S, Manor T, Ornstein E, Zarchin N, Sonn J (1998) Multiparametric monitoring of brain oxygen balance under experimental and clinical conditions. Neurol Res 20:S76–S80

Meilin S, Dekel N, Granot E, Kadusi R, Zarchin N, Mayevsky A (1996) A new brain function multiprobe for measuring cerebral function. Israel J Med Sci 32(suppl):S23

Miami Trauma Clinical Trials, G (2005) Splanchnic hypoperfusion-directed therapies in trauma: a prospective, randomized trial. Am Surg 71(3):252–260

Miller PR, Kincaid EH, Meredith JW, Chang MC (1998) Threshold values of intramucosal pH and mucosal-arterial CO_2 gap during shock resuscitation. J Trauma 45(5):868–872

Mutlu GM, Mutlu EA, Factor P (2003) Prevention and treatment of gastrointestinal complications in patients on mechanical ventilation. Am J Respir Med 2(5):395–411

Neviere R, Chagnon JL, Teboul JL, Vallet B, Wattel F (2002) Small intestine intramucosal PCO_2 and microvascular blood flow during hypoxic and ischemic hypoxia. Crit Care Med 30(2): 379–384

Nunn JE (1987) Hypoxia. In: Nunn JE (ed) Applied respiratory Physiology. Butterworths, London, pp 471–477

Scher A (1996) Control of arterial blood pressure: measurement of pressure and flow. In: Ruch TC, Patton HD (eds) Physiology & Biophysices. W.B. Saunders Company, Philadelphia, p 667

Sicard KM, Duong TQ (2005) Effects of hypoxia, hyperoxia, and hypercapnia on baseline and stimulus-evoked BOLD, CBF, and $CMRO_2$ in spontaneously breathing animals. NeuroImage 25(3):850–858

Tamion F, Richard V, Sauger F, Menard JF, Girault C, Richard JC, Thuillez C, Leroy J, Bonmarchand G (2003) Gastric mucosal acidosis and cytokine release in patients with septic shock. Crit Care Med 31(8):2137–2143

Tolmasov M, Barbiro-Michaely E, Mayevsky A (2007) Simultaneously multiparametric spectroscopic monitoring of tissue viability in the brain and small intestine. – art. no. 64341N. Opt Tomogr Spectrosc Tissue VII 6434:N4341–N4341

Varela JE, Cohn SM, Diaz I, Giannotti GD, Proctor KG (2003) Splanchnic perfusion during delayed, hypotensive, or aggressive fluid resuscitation from uncontrolled hemorrhage. Shock 20(5):476–480

Waltemath CL (1970) Oxygen, uptake, transport, and tissue utilization. Anesth Analg 49(1): 184–203

Wolf T, Lindauer U, Villringer A, Dirnagl U (1997) Excessive oxygen or glucose supply does not alter the blood flow response to somatosensory stimulation or spreading depression in rats. Brain Res 761(2):290–299

Xu K, Puchowicz MA, LaManna JC (2004) Renormalization of regional brain blood flow during prolonged mild hypoxic exposure in rats. Brain Res 1027(1–2):188–191

Chapter 4
Effects of Vasoactive Mediators and Anesthesia on Brain and Small Intestine

4.1 Introduction

4.1.1 The Effects of Epinephrine

Catecholamines, such as epinephrine, are the most important circulating endogenous vasoactive substances that influence the brain and the intestinal blood flow (IBF) (Takala 1996). The central adrenergic neurons have several functions in the brain: they enhance the cerebral vascular tone by acting on α-receptors sites (Strandgaard and Paulson 1984) and influence the metabolic rate of the brain tissue by acting on β-receptors sites on the cell membrane (Kogure et al. 1979). Additionally, catecholamines modulate the response of the cerebral cortex to an increase in the brain metabolic demand. Norepinephrine, for example, may mediate its effect by potentiating Na^+/K^+-ATP'ase or through its effects on vascular reactivity (Chance and Legallias 1959).

In contrast to the brain, the sympathetic nervous activity reduces the IBF by increasing the vascular resistance of its arteries and arterioles (Takala 1996). The receptor profile of epinephrine is complex, and its pharmacological effects are dose-dependent (Levy et al. 2003). At the low doses, epinephrine acts predominantly on the peripheral β1- and β2-adrenergic receptors (Levy et al. 2003). However, as the dose of epinephrine is increased, the α1-adrenergic receptor-mediated vasoconstrictor effects predominate (Levy et al. 2003). Moreover, epinephrine exhibits major metabolic effects and increases glucose and lactate concentrations through an increase in hepatic glucose output. Due to its significant role in the redistribution of blood flow during critical conditions, the injection of epinephrine in laboratory animals serves as a convenient model for the investigation of various critical situations in patients.

© The Author(s), under exclusive license to Springer Nature Switzerland AG 2025

A. Mayevsky et al., *Body Oxygen Homeostasis and Mitochondrial Function*,
https://doi.org/10.1007/978-3-031-94113-9_4

4.1.2 The Effects of Nitric Oxide

Nitric oxide (NO) may be an important modulator of sympathetic tone (Jordan et al. 2001). NO is a mediator of a diverse array of physiological processes, including vasodilator tone, neurotransmission, and prevention of platelet aggregation, and it is strongly implicated in the pathophysiology of sepsis, including circulatory abnormalities (Boughton-Smith et al. 1993). The early hypotensive response in rats injected with an endotoxin is associated with a marked increase in constitutive NO production (Boughton-Smith et al. 1993). NO increases the intracellular levels of cGMP, leading to vascular smooth muscle relaxation (Boughton-Smith et al. 1993). Under normal physiological circumstances, this allows for microvascular adaptation to local requirements (Boughton-Smith et al. 1993). However, an excessive production of NO leads to a loss of this control, resulting in a general vasodilated state, accompanied by hypotension that is often refractory to fluid and vasopressors (Boughton-Smith et al. 1993). By inhibiting NO synthesis, the circulation can be restored and the hyporesponsiveness to vasopressors reversed (Boughton-Smith et al. 1993). NO is also implicated in vessel integrity (Boughton-Smith et al. 1993). The generation of NO results in an increased vascular permeability to plasma proteins in both the colon and jejunum, 3 h after an endotoxin challenge increased permeability, which could be prevented by pretreatment with dexamethasone (which inhibits inducible NO synthase) and by post-treatment with L-NAME (Boughton-Smith et al. 1993). These circulatory abnormalities are accompanied by a blood flow redistribution and are considered to be the putative cause of altered oxygen metabolism during sepsis (Nelson et al. 1987; Gutierrez 1993; Maginniss et al. 1994; Crouser et al. 2002). The decreased blood flow to the intestines, for example (Weiss and Cohen 1978; Spain et al. 1994), has been consistently documented in experimental septic shock, contrasting with normal or increased flows generally reported for skeletal muscle (Lubbe et al. 1992; Hershey and Bond 1993).

In addition to the interorgan redistribution of the blood flow, intraorgan redistribution also occurs, as demonstrated in the intestine (Vallet et al. 1994; Revelly et al. 1996). These alterations may favor the development of localized zones of ischemia, resulting in an imbalance between regional O_2 demand and delivery. However, a growing body of evidence indicates that tissue ischemia in sepsis may not entirely account for the metabolic abnormalities induced by sepsis at the cellular level. In several experimental studies, tissue anaerobiosis and impaired energy metabolism were found to develop, in spite of normal tissue perfusion and O_2 tension (van Lambalgen et al. 1984; Astiz et al. 1988; VanderMeer et al. 1995; Revelly et al. 2000). For example, Astiz and collaborators (1988) showed that ATP levels in the skeletal muscle of septic rats significantly decreased despite normal values of muscle pO_2. This functional cellular energy starvation is termed "cytopathic hypoxia"and is thought to be one of the major causes of body deterioration under sepsis (Crouser et al. 2002).

4.1.3 The Effects of Sepsis

Sepsis is a systemic inflammatory response associated with the development of progressive damage in multiple organs, and is a leading cause of patient mortality in intensive care units (Feihl et al. 2001). Sepsis includes a heterogeneous class of syndromes such as tachycardia, tachypnea, fever (or hypothermia), and leukocytosis (or leukopenia) (Brun-Buisson et al. 1996; Sands et al. 1997) caused by a systemic inflammatory response to infection. This pathophysiological condition develops due to the invasion of bacteria, viruses, or parasites that release various toxic products into the blood (Parrillo 1993; Glauser et al. 1994). These microorganisms and their products are able to activate the systemic host defenses, comprising both humoral and cellular components that amplify the inflammatory response (Marsh and Wewers 1996). Complex interactions between these different mediators produce profound pathophysiological alterations, which ultimately lead to diffuse tissue injury and to the progressive deterioration of organ function. The fall in organ functionality is called the multiple organ dysfunction syndrome (MODS) (Fink and Heard 1990; Parrillo 1993; Beal and Cerra 1994; Astiz et al. 1995; Fink 2002). In the pathophysiology of sepsis, the barrier function of the intestine may play a central role (Fink 1991). The mucosa serves as an important mechanical barrier that helps in preventing systemic contamination by microbes and microbial products that are normally present within the lumen of the intestine (Fink 1991). There are two underlying mechanisms that may be important in the pathogenesis of septic shock—microbial translocation of bacteria and/or yeast into the bloodstream and an increase in mucosal permeability to lipopolysaccharide (LPS) and other microbial toxins (Fink 1991). An additional mechanism that may play a role in generating the septic response is the involvement of NO (Ackland et al. 2000).

In contrast to sepsis, in experimental animals and human subjects, the circulatory abnormalities in the gastrointestinal (GI) tract during hemorrhagic shock may play a major part in producing metabolic dysfunction (Fink 1991).

4.2 Methods

The main basic methods used in this section have been described in Sect. 3.2.

4.2.1 The Effects of Various Epinephrine Concentrations

A group of nine rats was injected with Epinephrine (2, 4, 6, 8, and 10 μg/100 g body weight [BW], IV) with 15-min intervals between the injections.

The animals in the control group ($N = 8$) were injected with saline at the same volume as well as the same intervals.

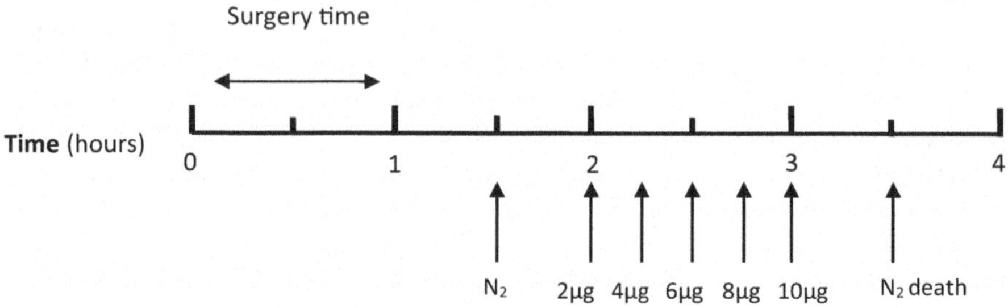

4.2.2 The Effects of Anesthesia on the Response to Epinephrine

A group of nine rats was injected with epinephrine (epinephrine hydrochloride 4-[1-Hydroxy-2-(methylamino) ethyl]-1,2-benzenediol) $C_9H_{13}NO_3 \cdot HCL$, N-51 K1096, Sigma) 6 μg/100 g IV twice. The first injection of Epinephrine was performed when the animals were under mild anesthesia. Thereafter, these animals were monitored for 15 min. At the end of this period, the animals were deeply anesthetized by pentobarbitone (Pentobarbitone Sodium 200 mg/ml, Lot 170035, CTS, Israel), which was given as follows: one bolus of 30 mg/kg pentobarbitone followed by an IV infusion of 4 mg/kg pentobarbitone for 10 min. In the midterm of the pentobarbitone infusion, when the animals were under deep anesthesia, the second dose of epinephrine (6 μg/kg) was applied.

The responses of the brain and small intestine to the injection of epinephrine under mild anesthesia were used as a control for the deep anesthesia treatment.

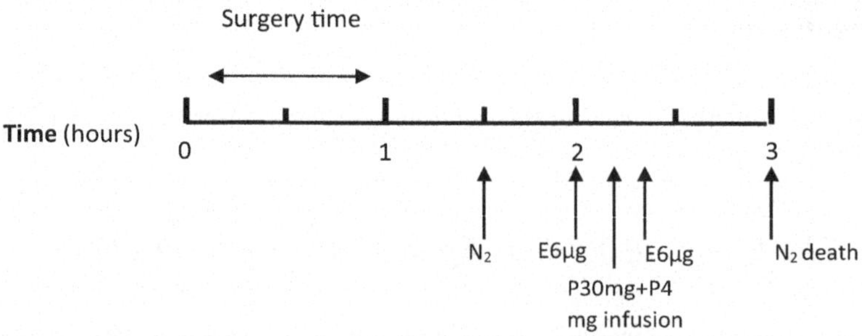

4.2.3 The Effects of L-NAME and Nitrite

The experimental group ($N = 14$) was IV injected with N-omega-nitro-L-arginine methyl ester (L-NAME N-5751-1G, Sigma) (50 mg/kg). Then, the rats were monitored for 30 min. At the end of this monitoring period, the rats were IV infused for 10 min by Sodium nitrite ($NaNO_2$ 99.99 + %, Sigma) 20 mg/kg. The monitoring proceeded for another 60 min after the infusion of nitrite.

Two control groups were tested in this protocol. The control for nitrite in which the rats ($n = 8$) were injected with L-NAME (50 mg/kg) and infused with saline for10 min. The control for L-NAME in which the rats ($n = 8$) were injected with saline and infused with Nitrite (20 mg/kg) for 10 min.

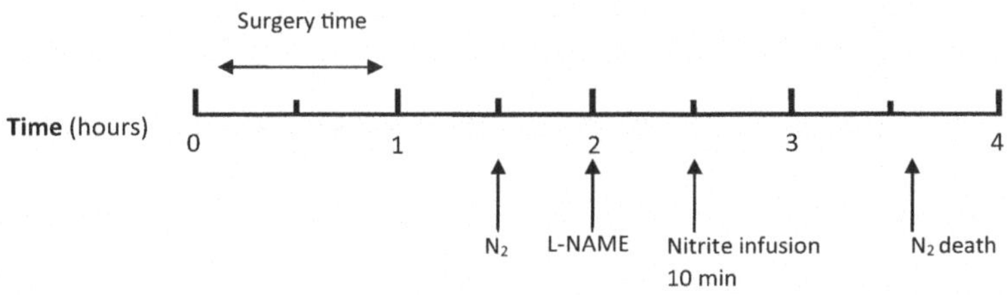

4.2.4 The Animal Model of Sepsis

A group of 10 rats was IP injected with lipopolysaccharide (LPS, *E. coli* stereotype 0111:B4, Sigma) (30 mg/kg). Thereafter, these animals were monitored for 200 min or until spontaneous death occurred.

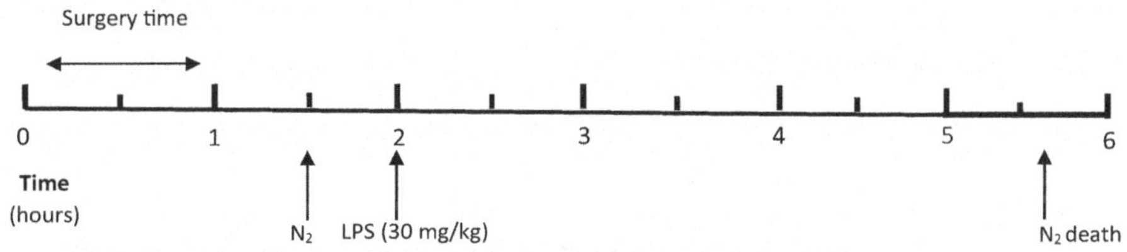

4.3 Results

The Effects of Various Vasoactive Mediators

This part is dedicated to testing the responses of the brain and the small intestine to the injection of several substances known for their vasoactive effects.

4.3.1 *The Responses to Increasing Concentrations of Epinephrine*

In order to test the responses of the brain and the small intestine to increasing doses of Epinephrine, we first examined a control group where the animals were injected (five times) with saline (0.1 ml/100 g) at intervals of 15 min. Figure 4.1 presents the responses of both organs to one session of saline injection, although the same responses were detected in all five sessions of saline injection. As observed, the injection of saline caused no changes in any of the monitored parameters in both organs, nor did it change the Mean Arterial Pressure (MAP) level.

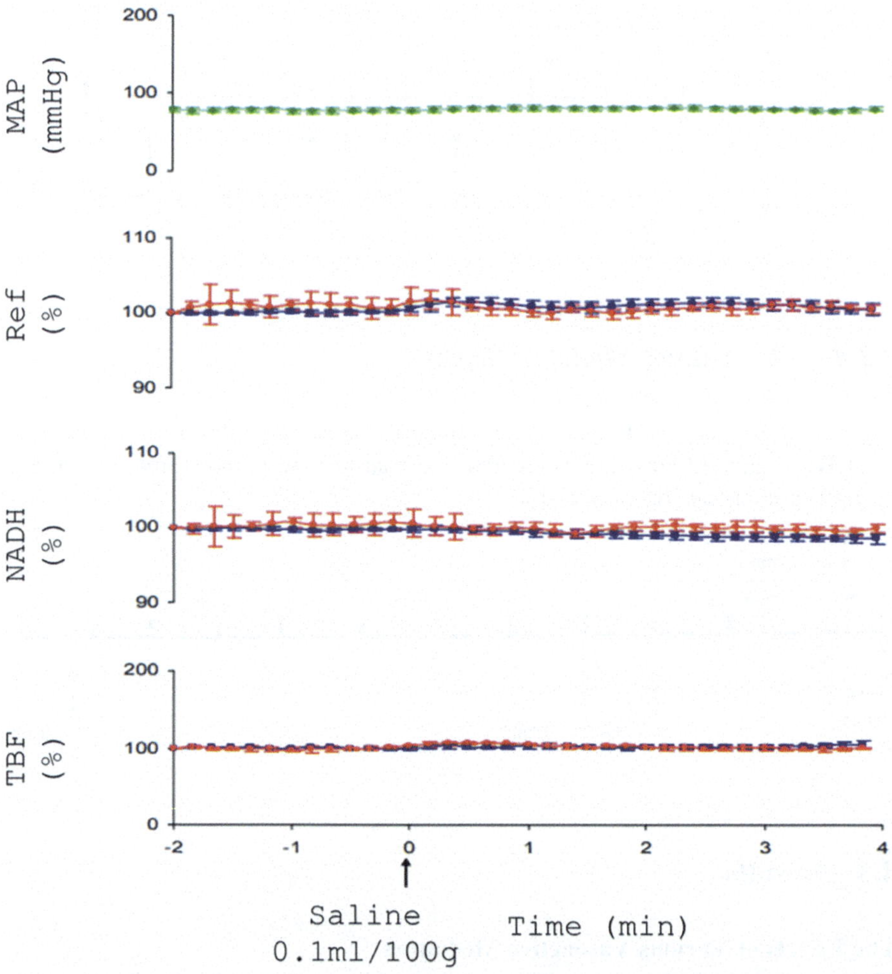

Fig. 4.1 The responses of the brain (blue) and small intestine (red) to the first injection of Saline (0.1 ml/100 g IV) *MAP* Mean Arterial Pressure, *REF* Reflectance, *NADH* Mitochondrial NADH, *TBF* Tissue Blood Flow, ($N = 8$)

4.3.1.1 Epinephrine (2 μg/100 g BW, IV)

Figure 4.2 presents the effect of the epinephrine injection at 2 μg/100 g BW on the brain and small intestine in a group of nine rats. The pattern of response of all the parameters to the epinephrine injection was similar for all doses; however, the degrees of change were different. As seen, there was a clear difference in the responses of the two organs. Immediately following the epinephrine injection, MAP increased to the level of 103 ± 8 mmHg ($P < 0.05$) and remained high for

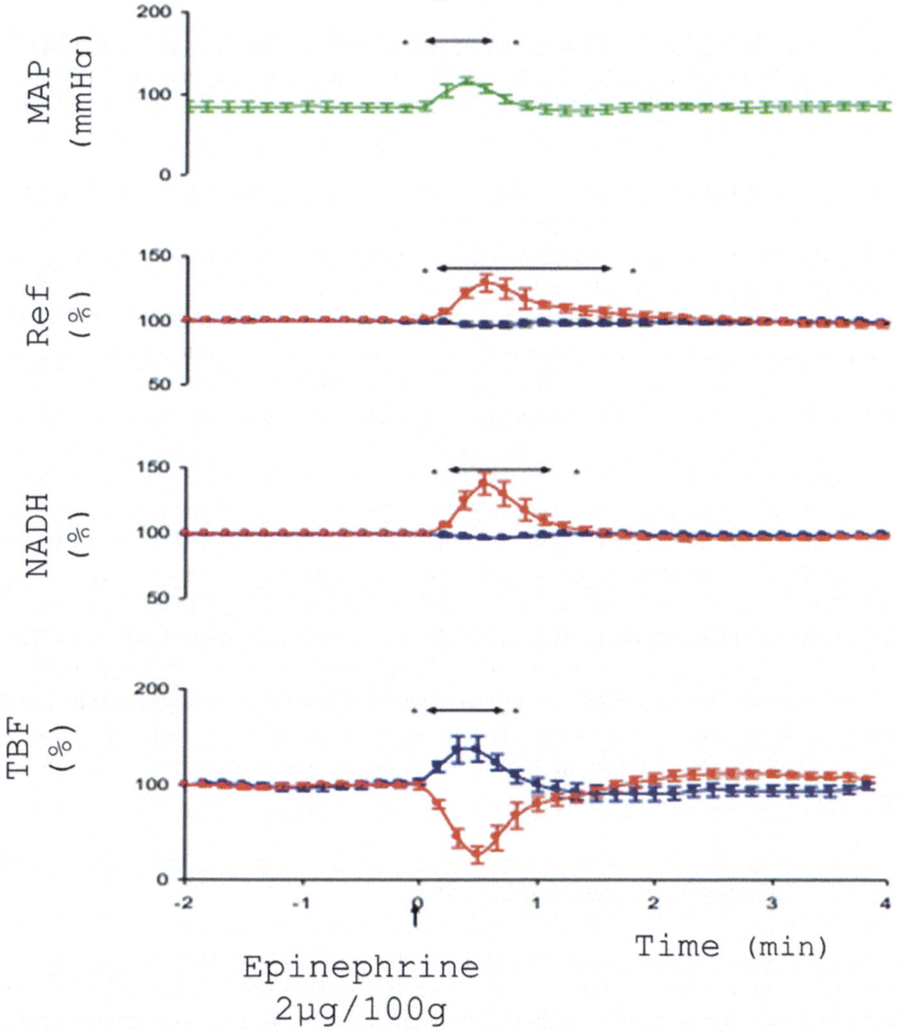

Fig. 4.2 The responses of the brain (blue) and small intestine (red) to Epinephrine injection (2 μg/100 g BW, IV). Abbreviations are as in Fig. 4.1. Arrows represent the period of time in which the differences between the two organs were significant for each minute. ($N = 9$) (*) $P < 0.05$

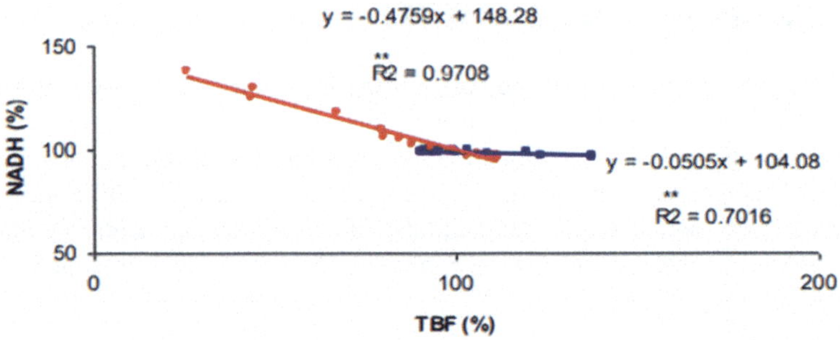

Fig. 4.3 The correlation between tissue blood flow (TBF) and NADH redox state (NADH) measured in the rat brain (blue) and in the small intestine (red) under epinephrine (2 μg/100 g BW). The data used for the present correlation were taken from the time range (0–4 min). (**) $P < 0.01$; (*) $P < 0.05$; ($N = 9$)

30 s. This hypertension was associated with an increase in cerebral blood flow (CBF) up to 138 ± 13% ($P < 0.05$), while the IBF decreased to 26 ± 9% ($P < 0.001$). These changes in the blood supply yielded significant differences between the two organs for 30 s. Consequently, NADH in the brain was oxidized (97 ± 1%, $P < 0.05$) while in the small intestine NADH was reduced (138 ± 8%, $P < 0.01$). The level of NADH in the brain was significantly different from its level in the small intestine for 40 s of monitoring. Regarding the reflectance, an increase was recorded in the small intestine (129 ± 7%, $P < 0.01$), whereas in the brain the reflectance decreased to 96 ± 2%, leading to a significant difference between the two organs for 1 min of monitoring.

Further analysis of the present results revealed that the changes in tissue blood flow (TBF) were strongly correlated with the changes in NADH (Fig. 4.3). As Fig. 4.3 shows, NADH remained steady (100%) through a wide range of CBF levels (80–150%), whereas in the intestine the low levels of TBF (26 ± 9%) were correlated with high NADH values (138 ± 8%) ($P < 0.001$) and high levels of TBF (138 ± 13%) were associated with oxidized NADH levels (97 ± 1%) ($P < 0.001$).

The systemic changes in MAP were strongly correlated with changes in the blood supply to both organs (Fig. 4.4). The elevation of the blood pressure induced a decline in TBF ($P < 0.01$) in the small intestine and hyperemia in the brain ($P < 0.01$).

4.3.1.2 Epinephrine (4 μg/100 g B.U, IV)

Figure 4.5 shows the responses of both organs to Epinephrine (4 μg/100 g IV) injection. The injection of Epinephrine (4 μg/100 g BW) immediately elevates the MAP to 118 ± 8 mmHg ($P < 0.001$). This high level of MAP was observed during 1 min of monitoring. As a result, the blood supply to the brain increased to 174 ± 22% ($P < 0.05$) at the 30th second following the injection, while the IBF decreased at this time point to 11 ± 3% ($P < 0.001$). These changes in CBF were

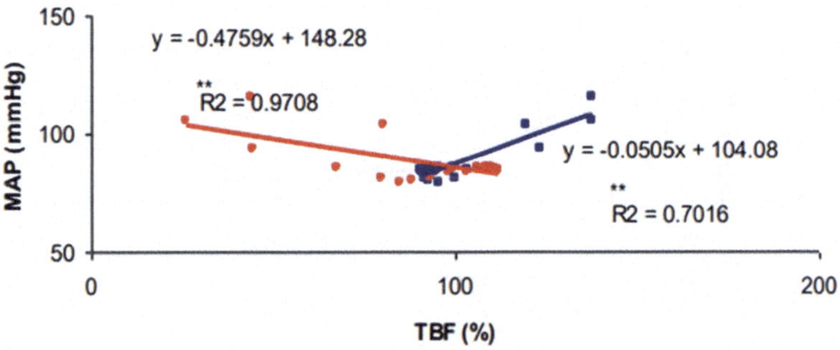

Fig. 4.4 The correlation between tissue blood flow (TBF) and mean arterial pressure (MAP) measured in the rat brain (blue) and in the small intestine (red) under epinephrine (2 μg/100 g BW). The data used for the present correlation were taken from the time range (0–4 min). (**) $P < 0.01$; (*) $P < 0.05$; ($N = 9$)

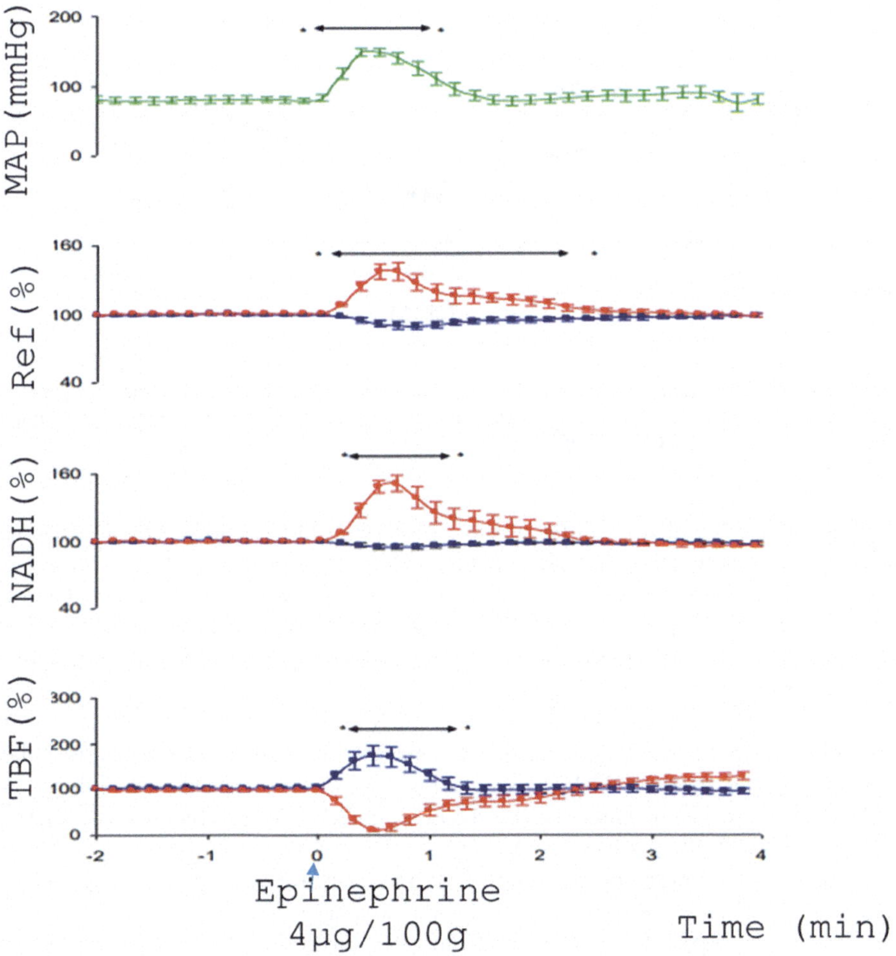

Fig. 4.5 The responses of the brain (blue) and small intestine (red) to Epinephrine injection (4 μg/100 g BW, IV). Abbreviations are as above. Arrows represent the period of time in which the differences between the two organs were significant for each minute. (N = 9) (*) $P < 0.05$

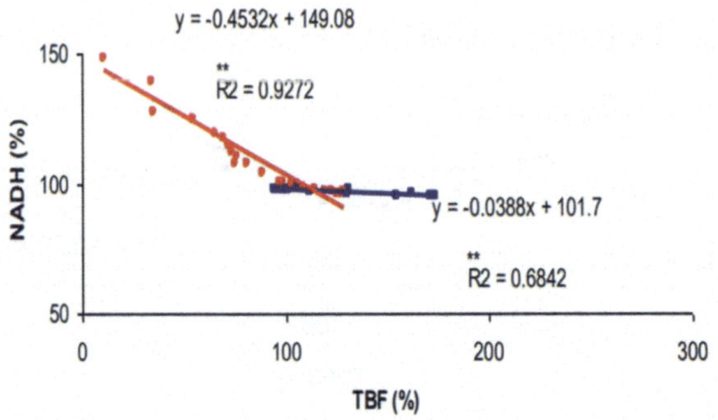

Fig. 4.6 The correlation between tissue blood flow (TBF) and NADH redox state (NADH) measured in the rat brain (blue) and in the small intestine (red) under epinephrine (4 μg/100 g BW). The data used for the present correlation were taken from the time range (0–4 min). (**) $P < 0.01$; (*) $P < 0.05$; ($N = 9$)

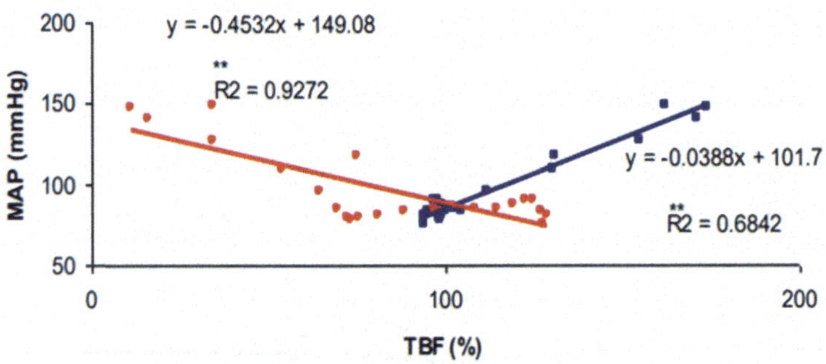

Fig. 4.7 The correlation between mean arterial pressure (MAP) and tissue blood flow (TBF) measured in the rat brain (blue) and in the small intestine (red) under epinephrine (4 μg/100 g BW). The data used for the present correlation were taken from the time range (0–4 min). (**) $P < 0.01$; (*) $P < 0.05$; ($N = 9$)

significantly different from those of the IBF for more than 1 min. The increase in CBF was accompanied by NADH oxidation (95 ± 2%, $P < 0.05$), while in the small intestine NADH was reduced (148 ± 5%, $P < 0.01$). The reflectance increased in the small intestine (137 ± 7%, $P < 0.001$), whereas in the brain it decreased to 92 ± 3% ($P < 0.01$), leading to a significant difference between the two organs during 2 min of monitoring.

Here also, the correlation between TBF and NADH (Fig. 4.6) exhibited constant levels of NADH (around 100%) for a vast range of CBF (90–180%), whereas in the intestine the low levels of TBF (down to 11 ± 3%) were correlated with high levels of NADH (up to 148 ± 5%) ($P < 0.01$) and high TBF (174 ± 22%) was associated with low NADH (96 ± 2%) ($P < 0.01$).

In addition, a correlation between MAP and TBF in both organs was detected. The high level of MAP correlated with brain hyperemia ($P < 0.01$) and the low levels of blood supply to the small intestine ($P < 0.01$) (Fig. 4.7).

4.3.1.3 Epinephrine (6 µg/100 g BW, IV)

Figure 4.8 shows the responses of both organs to epinephrine (6 µg/100 g IV) injection. As seen, the two organs responded significantly different to the Epinephrine injection. Immediately after the epinephrine injection (6 µg/100 g), the MAP increased to the level of 126 ± 9 mmHg ($P < 0.01$), which remained high for 2 min of monitoring. The hypertension resulted in the elevation of CBF to 130 ± 6% ($P < 0.01$), while the IBF decreased to 46 ± 17% ($P < 0.01$).

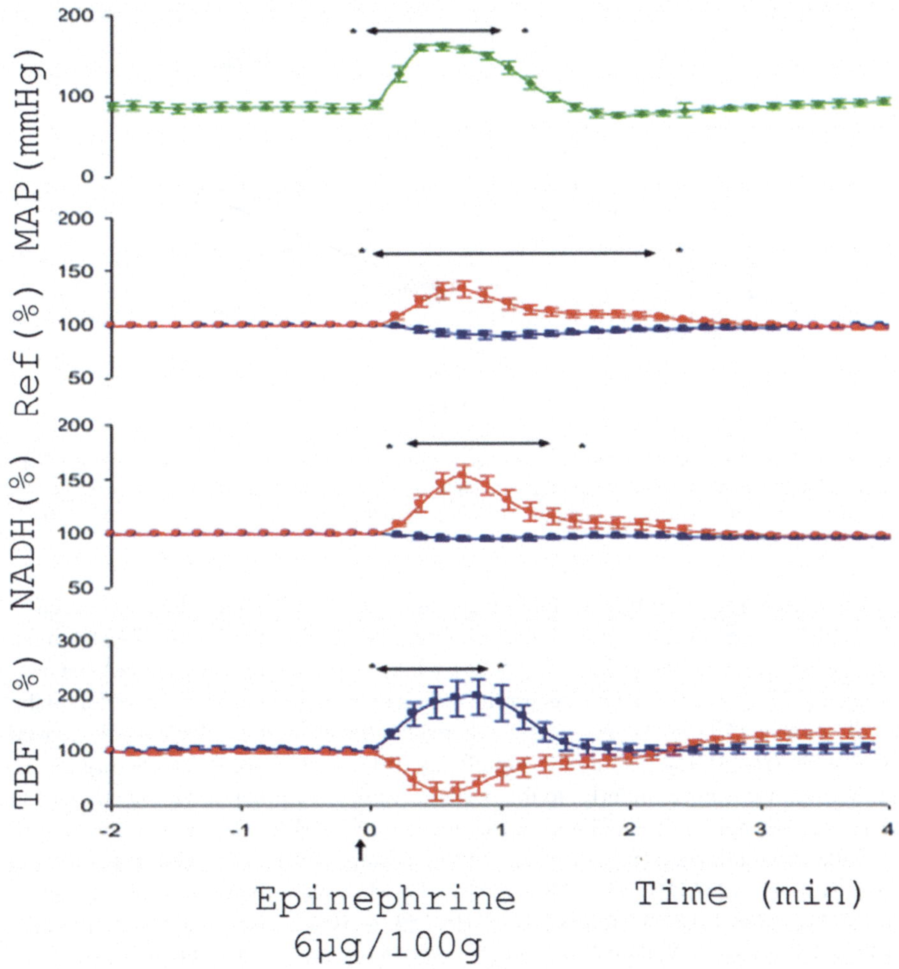

Fig. 4.8 The responses of the brain (blue) and small intestine (red) to Epinephrine injection (6 µg/100 g BW, IV). Abbreviations are as above. Arrows represent the period of time in which the differences between the two organs were significant for each minute. ($N = 9$)

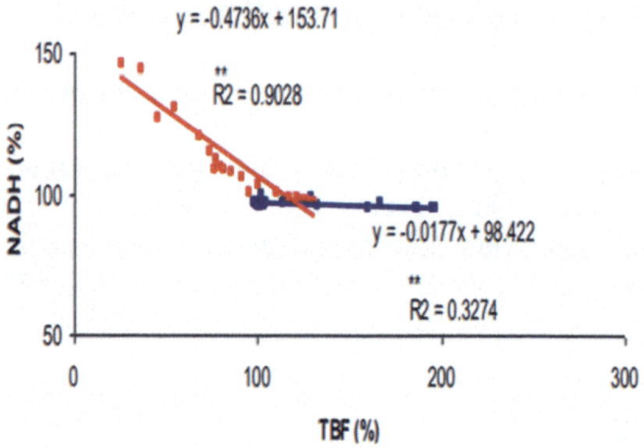

Fig. 4.9 The correlation between tissue blood flow (TBF) and NADH redox state (NADH) measured in the rat brain (blue) and in the small intestine (red) under epinephrine (6 μg/100 g BW). The data used for the present correlation were taken from the time range (0–4 min). (**) $P < 0.01$; (*) $P < 0.05$; ($N = 9$)

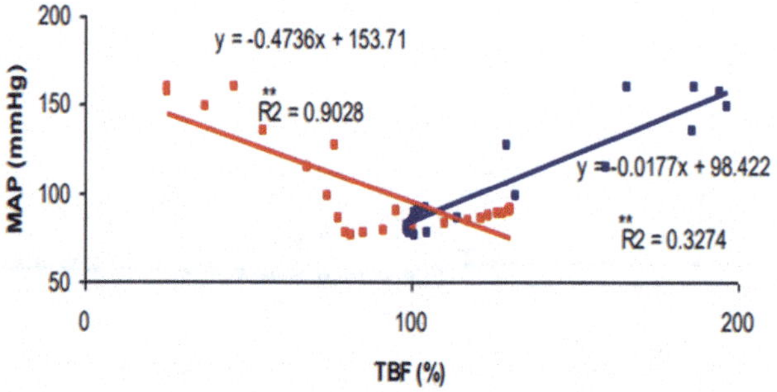

Fig. 4.10 The correlation between mean arterial pressure (MAP) and tissue blood flow (TBF) measured in the rat brain (blue) and in the small intestine (red) under epinephrine (6 μg/100 g BW). The data used for the present correlation were taken from the time range (0–4 min). (**) $P < 0.01$; (*) $P < 0.05$; ($N = 9$)

These changes in the blood supply were significantly different when compared to the small intestine during 1 min of the experimental period. As a result, NADH in the brain was oxidized (94 ± 2%, $P < 0.05$) while in the small intestine NADH was reduced (152 ± 9%, $P < 0.01$). As for the reflectance, an increase was recorded in the small intestine (133 ± 7%, $P < 0.01$), whereas in the brain the reflectance decreased to 89 ± 2% ($P < 0.01$), leading to a significant difference between the two organs for 2 min of monitoring. Similarly, to the previous doses of Epinephrine, the changes in TBF were strongly correlated with the changes in NADH levels, in the brain as well as in the small intestine (Fig. 4.9). The increased brain TBF (197 ± 34%) resulted in a steady level of NADH (94 ± 2%) ($P < 0.01$), whereas in the small intestine, when the TBF (25 ± 17%) decreased, NADH (153 ± 10%) rose, and when the TBF (132 ± 16%) rose NADH (97 ± 1%) ($P < 0.01$) became oxidized (Fig. 4.9).

Figure 4.10 demonstrates the correlation between MAP and TBF in both organs. The blood pressure elevation induced a decline in the blood supply to the small intestine. Conversely, a high level of TBF in the brain was observed.

4.3.1.4 Epinephrine (8 μg/100 g BW, IV)

Figure 4.11 shows the responses of both organs to epinephrine injection (8 μg/100 g IV). There was detected a significantly different response of the brain as compared to that of the small intestine. The epinephrine injection (8 μg/100 g) immediately elevates the level of MAP to 117 ± 7 mmHg ($P < 0.01$). MAP remained high during 1.2 min of monitoring. The hypertension resulted in CBF increasing to 131 ± 7% ($P < 0.05$), while the IBF was characterized by a decrease to the level

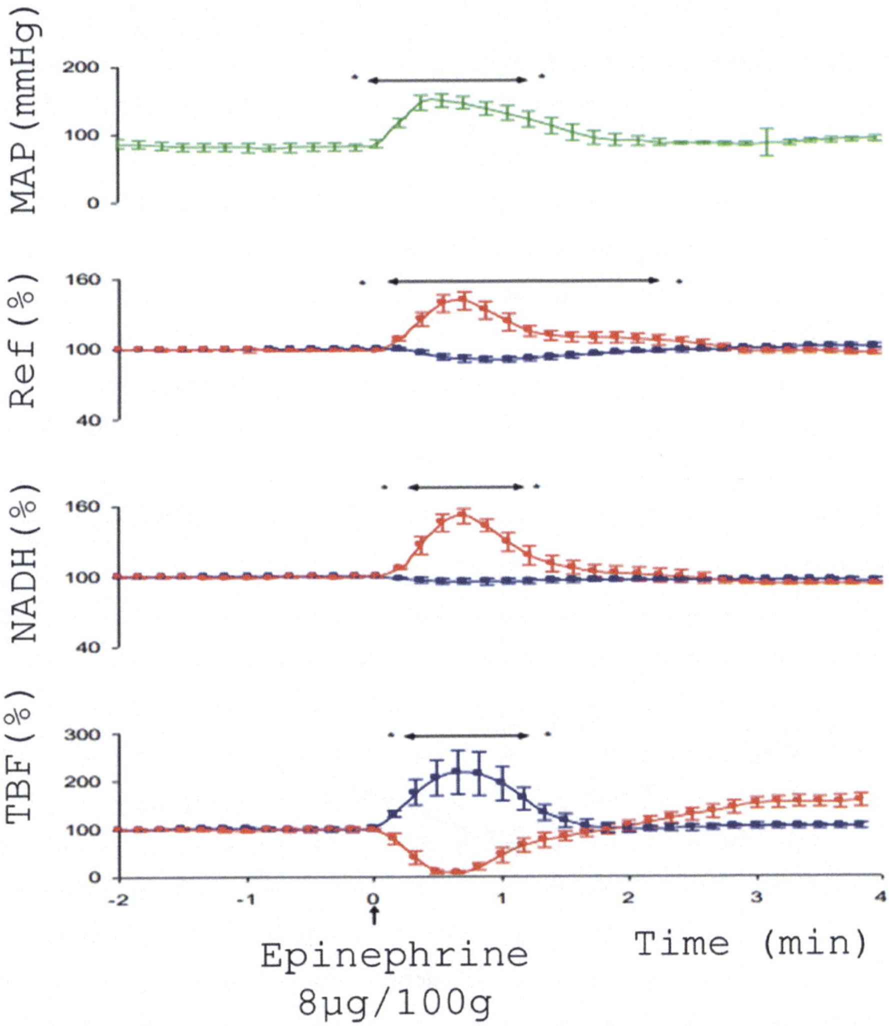

Fig. 4.11 The responses of the brain (blue) and small intestine (red) to Epinephrine injection (8 μg/100 g BW, IV). Abbreviations are as above. Arrows represent the period of time in which the differences between the two organs were significant. ($N = 9$) (*) $P < 0.05$

of 39 ± 14% ($P < 0.01$). These changes in the blood supply yielded significant differences between the two organs for 30 s. As a result, NADH in the brain was oxidized (96 ± 1%, $P < 0.05$) almost at the second minute of monitoring, while in the small intestine NADH was reduced (126 ± 8%, $P < 0.01$) immediately after the epinephrine injection. Regarding the reflectance, its increase was recorded in the small intestine (126 ± 6%, $P < 0.01$) immediately following the treatment, whereas in the brain the reflectance decreased to 96 ± 1% ($P < 0.05$), leading to a significant difference between the two organs for 2 min of monitoring.

A strong correlation between TBF and NADH in both organs was observed. As seen in Fig. 4.12, although TBF increased in the brain, NADH remained steady. In the intestine, when the TBF decreased, NADH increased, and when TBF increased, NADH became oxidized (Fig. 4.12).

As Fig. 4.13 indicates, the increase in MAP correlated with the brain hyperemia and the low level of blood flow in the small intestine.

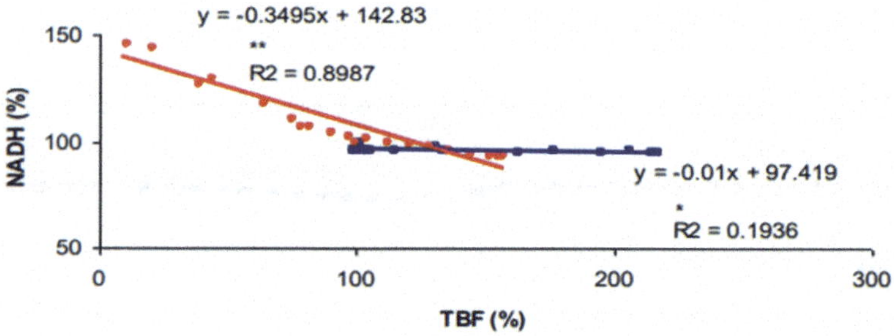

Fig. 4.12 The correlation between tissue blood flow (TBF) and NADH redox state (NADH) measured in the rat brain (blue) and in the small intestine (red) under Epinephrine (8 μg/100 g BW). The data used for the present correlation were taken from the time range (0–4 min). (**) $P < 0.01$; (*) $P < 0.05$; ($N = 9$)

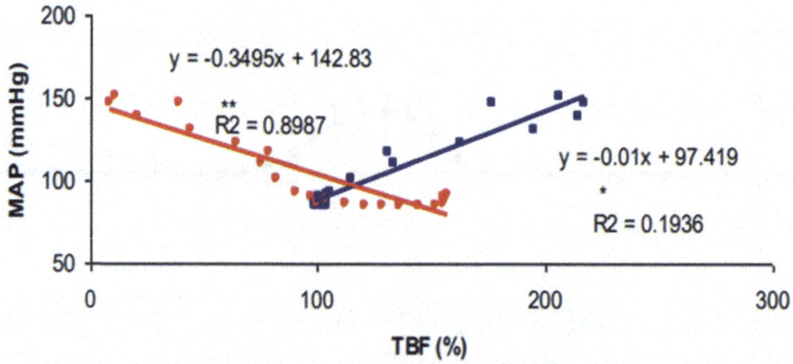

Fig. 4.13 The correlation between the mean arterial pressure (MAP) and tissue blood flow (TBF) measured in the rat brain (blue) and in the small intestine (red) under epinephrine (8 μg/100 g BW). The data used for the present correlation was taken from the time range (0–4 min). (**) $P < 0.01$; (*) $P < 0.05$; ($N = 9$)

4.3.1.5 Epinephrine (10 μg/100 g BW, IV)

Figure 4.14 illustrates the responses of both organs to epinephrine injection (10 μg/100 g IV). Following the epinephrine injection (10 μg/100 g), MAP significantly increased to the level of 127 ± 8 mmHg ($P < 0.01$) and remained high for 1.2 min. This hypertension was associated with an increase of CBF up to 139 ± 7%

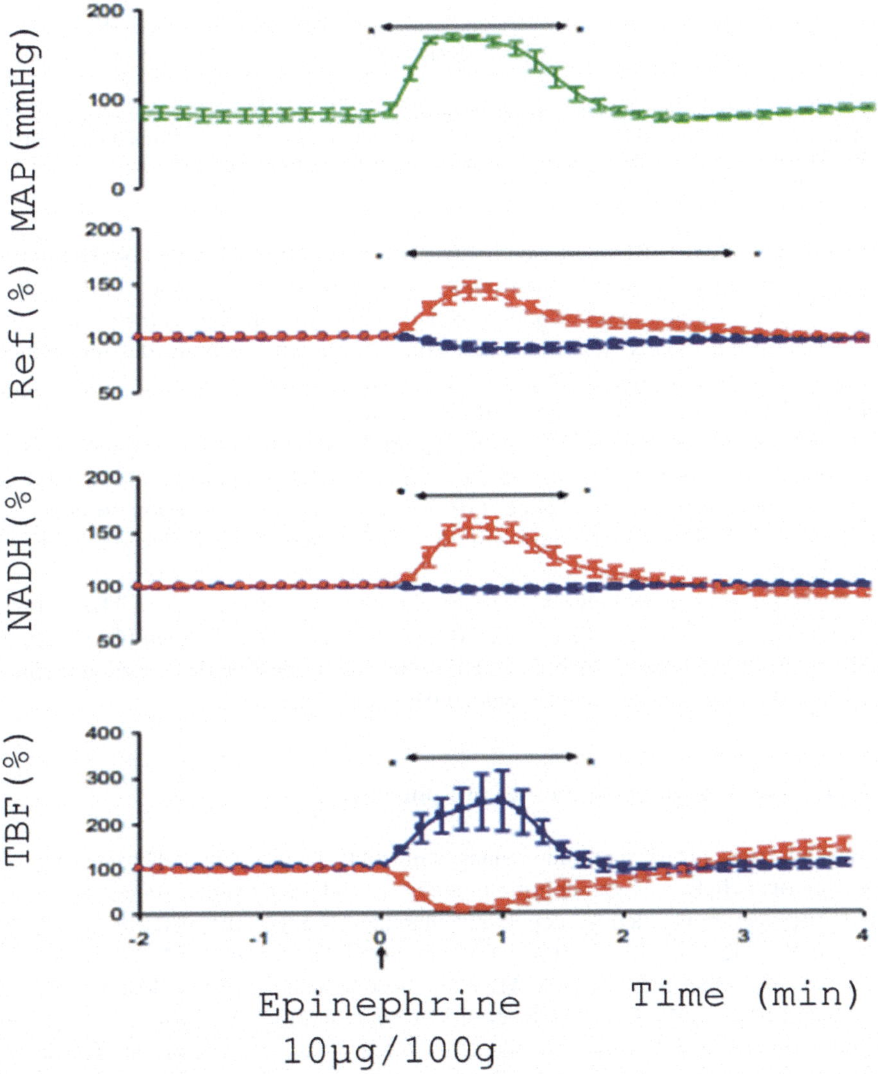

Fig. 4.14 The responses of the brain (blue) and small intestine (red) to Epinephrine injection (10 μg/100 g BW, IV). Abbreviations are as above. Arrows represent the period of time in which the differences between the two organs were significant for each minute. ($N = 9$) (*) $P < 0.05$

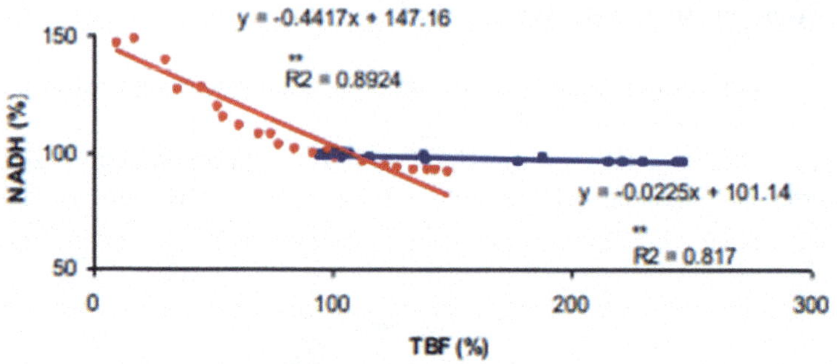

Fig. 4.15 The correlation between tissue blood flow (TBF) and NADH redox state (NADH) measured in the rat brain (blue) and in the small intestine (red) under epinephrine (10 μg/100 g BW). The data used for the present correlation were taken from the time range (0–4 min). (**) $P < 0.01$; ($N = 9$)

($P < 0.01$), while the IBF decreased to 36 ± 6% ($P < 0.01$). Although NADH in the brain remained stable, in the small intestine NADH became elevated (126 ± 9%, $P < 0.01$). An increase in reflectance was recorded in the small intestine (126 ± 6%, $P < 0.01$), immediately after the treatment, whereas in the brain the reflectance decreased at this time point to 96 ± 2% ($P < 0.01$), leading to a significant difference between the two organs for 2.5 min of monitoring.

In addition, the changes in TBF strongly correlated with the changes in NADH in both organs (Fig. 4.15). As seen in Fig. 4.17, NADH remained steady (~100%) through a wide range of TBF changes (90–250%). In the intestine, upon the decrease in the TBF, NADH increased, and with the TBF increase NADH became oxidized (Fig. 4.15).

The systemic changes in blood pressure significantly correlated with TBF in both organs (Fig. 4.16). Thus, an increase in MAP resulted in a high level of blood supply to the brain. In contrast to the brain, TBF in the small intestine decreased and when the level of blood pressure deteriorated, TBF became elevated.

4.3.1.6 The Maximum Responses to Epinephrine

Figure 4.17 presents the maximum changes in MAP, TBF, and NADH following the injection of each dose of epinephrine in the brain and the intestine. As it is seen, the MAP response showed a dose-dependent character. Namely, the increase from 2 μg/100 g BW epinephrine to 4 μg/100 g BW enhanced the response of MAP. However, there was no further increase in MAP, even though the epinephrine dose was raised to 6, 8 and 10 μg/100 g BW. The level of MAP was significantly lower ($P < 0.05$) at 2 and 4 μg/100 g BW versus all the other three doses of epinephrine. The dose-dependent response pattern at the systemic level was also observed in the response of the blood flow in the brain. Namely, the peak of CBF responses to epinephrine at 2 μg/100 g (154 ± 13%, $P < 0.05$) was lower than the peak values of CBF after 4 μg/100 g (187 ± 21%, $P < 0.05$) and 6 μg/100 g (215 ± 32%, $P < 0.05$). In addition, the

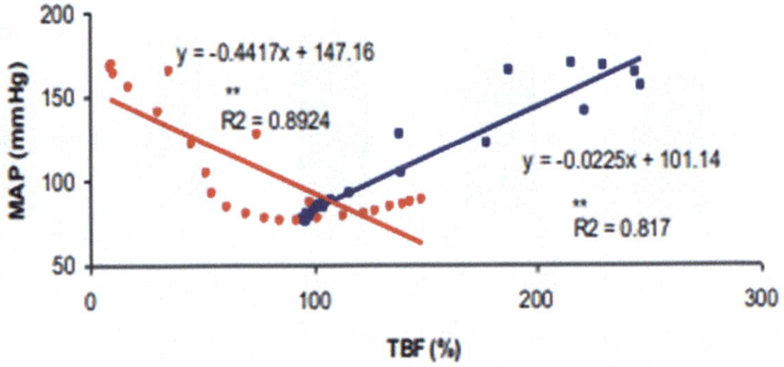

Fig. 4.16 The correlation between mean arterial pressure (MAP) and tissue blood flow (TBF) measured in the rat brain (blue) and in the small intestine (red) under epinephrine (10 μg/100 g BW). The data used for the present correlation were taken from the time range (0–4 min). (**) $P < 0.01$; (*) $P < 0.05$; ($N = 9$)

maximal response under 4 μg/100 g was significantly different when compared to the maximum change induced by 6 μg/100 g epinephrine. In contrast to the brain, the IBF decreased at the same level under all the doses of epinephrine injection.

Consequently, the intestinal NADH responded in the same manner under all the doses of epinephrine, whereas in the brain the peak level of NADH after 2 μg/100 g epinephrine (95 ± 0.5%, $P < 0.05$) was significantly different when compared to the maximum change that was monitored after 6 μg/100 g epinephrine (90 ± 1.6%, $P < 0.05$). Analysis of the brain reflectance revealed significant differences between 2 μg/100 g (93 ± 2%, $P < 0.01$) and all the other doses (4 μg/100 g, 88 ± 2%, $P < 0.01$; 6 μg/100 g, 87 ± 2%, $P < 0.01$; 8 μg/100 g, 88 ± 2%, $P < 0.01$; 10 μg/100 g, 86 ± 3%, $P < 0.01$). However, in the intestine, a significant difference was observed only between the doses of 2 μg/100 g (137 ± 7%, $P < 0.05$) and 10 μg/100 g (148 ± 6%, $P < 0.05$).

4.3.2 The Effects of Anesthesia on the Responses to Epinephrine

In order to test the effect of anesthesia on the responses of the brain and the intestine to epinephrine, we chose the dose of 6 μg/100 g BW, since this was the dose that yielded the maximum response of MAP and TBF. A group of nine rats was injected with pentobarbitone (as detailed in the Methods section) to induce deep anesthesia. The systemic changes during the first 2 min post epinephrine injection were similar under both levels of anesthesia. However, at the last 2 min, MAP in the mild anesthesia group was significantly higher (75 ± 4 mmHg, $P < 0.05$) than the MAP in the deeply anesthetized group (58 ± 4 mmHg, $P < 0.05$) due to an increase of this parameter at the beginning of the second minute post epinephrine injection.

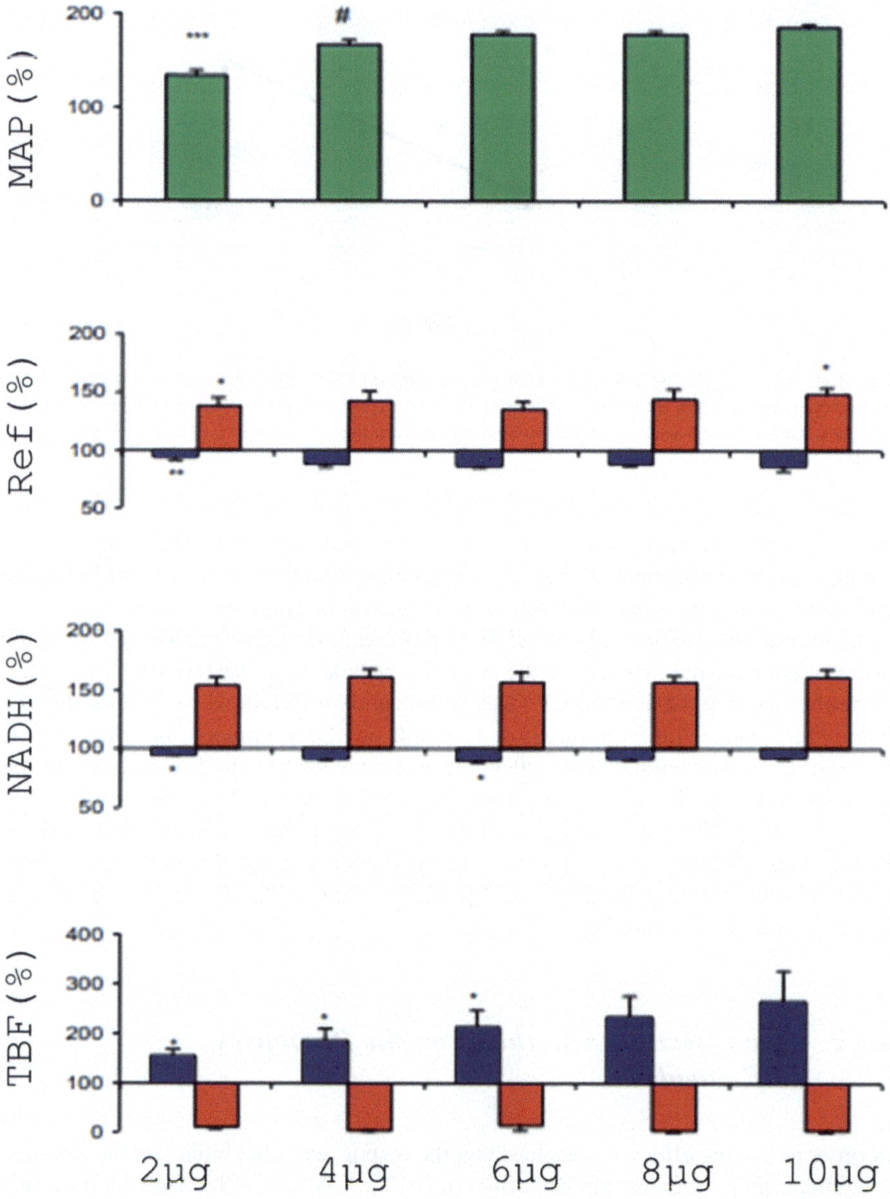

Fig. 4.17 Peak values of the response of the brain (blue) and small intestine (red) to increasing doses of Epinephrine (2–10 µg/100 g BW, IV). Abbreviations are as above, (N = 9), (*) $P < 0.05$—significant differences between peaks, (**) $P < 0.05$—significant differences between 2 µg/100 g BW and all the other doses, (***) $P < 0.001$—significant differences between 2 µg/100 g BW and all the other peaks, (#) $P < 0.05$—significant differences between 4 µg/100 g and all the other doses

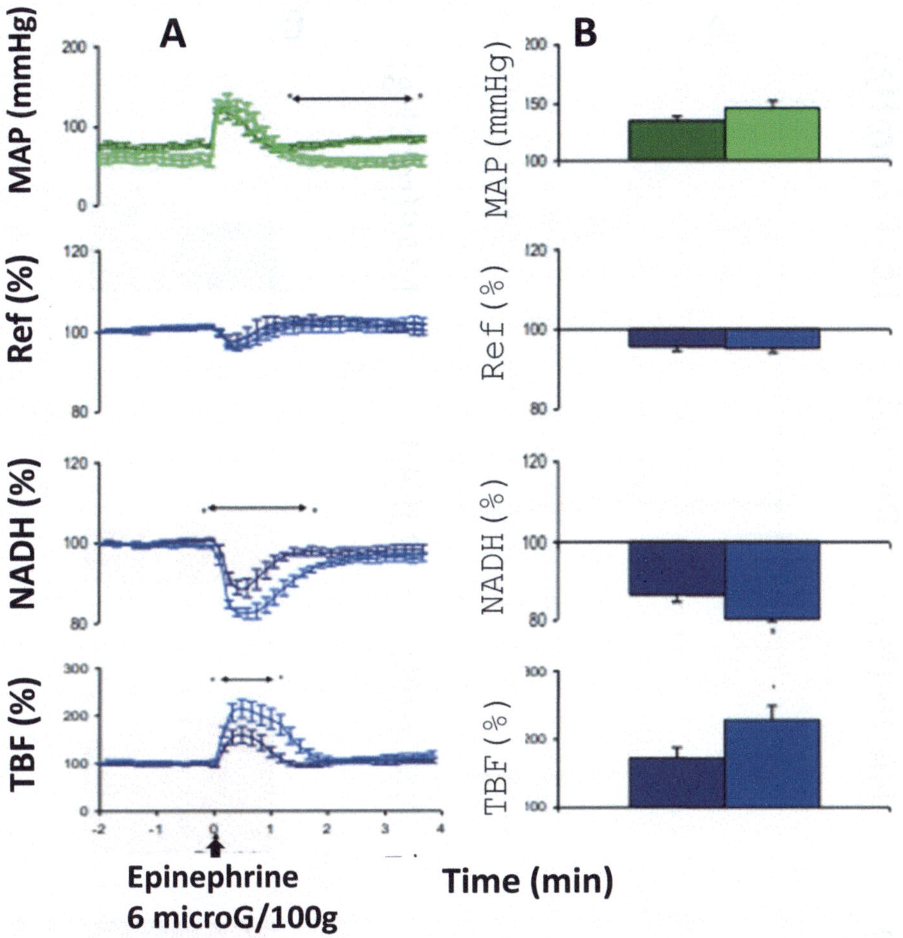

Fig. 4.18 The responses of the brain to Epinephrine (6 μg/100 g BW) under mild anesthesia (blue) and deep anesthesia (light blue). (**a**) The whole response to epinephrine. (**b**) The peak values of the response to epinephrine injection. For MAP: mild anesthesia (green) and deep anesthesia (light green). Arrows show the period of time in which significant differences between the two responses were detected. Abbreviations are as above. ($N = 9$). (*) $P < 0.05$

Considering the hemodynamic and metabolic changes in the deeply anesthetized rats, the cerebral hyperemia (228 ± 21% vs. 172 ± 15%; $P < 0.05$) and NADH oxidation (80 ± 1%; $P < 0.05$) were larger than in the mildly anesthetized rats (86 ± 1%, $P < 0.05$) (Fig. 4.18). Nevertheless, the response pattern to the epinephrine injection was similar under these two levels of anesthesia. On the other hand, in the intestine, there were no differences between these two levels of anesthesia (Fig. 4.19). The IBF in the mildly and deeply anesthetized rats decreased to the

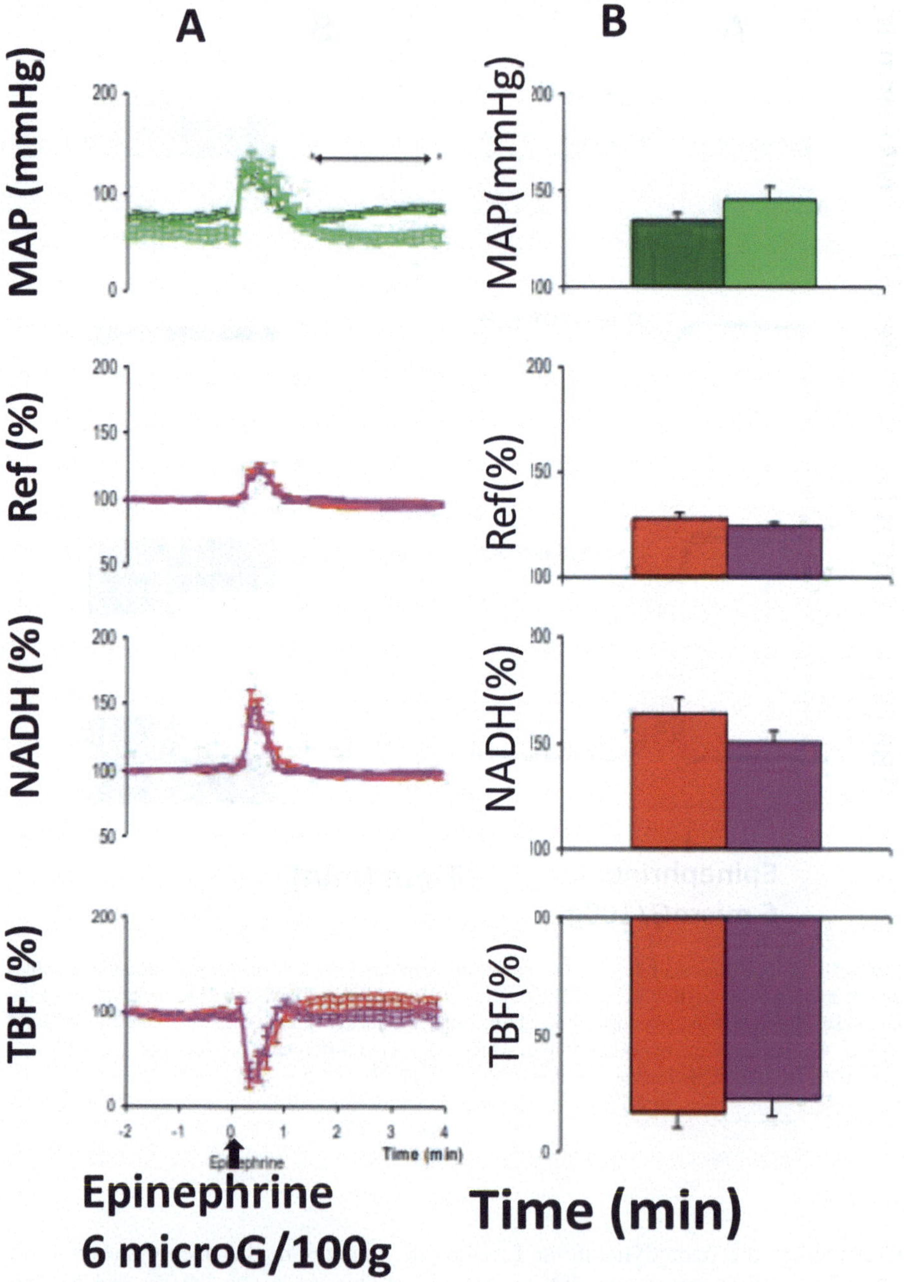

Fig. 4.19 The responses of the small intestine to Epinephrine (6 μg/100 g BW) under mild anesthesia (red) and deep anesthesia (pink). (**a**) The whole response to epinephrine. (**b**) The peak values of response to epinephrine injection. For MAP: mild anesthesia (green) and deep anesthesia (light green). Arrows show the period of time in which significant differences between the two responses were detected. Abbreviations are as above. ($N = 9$). (*) $P < 0.05$

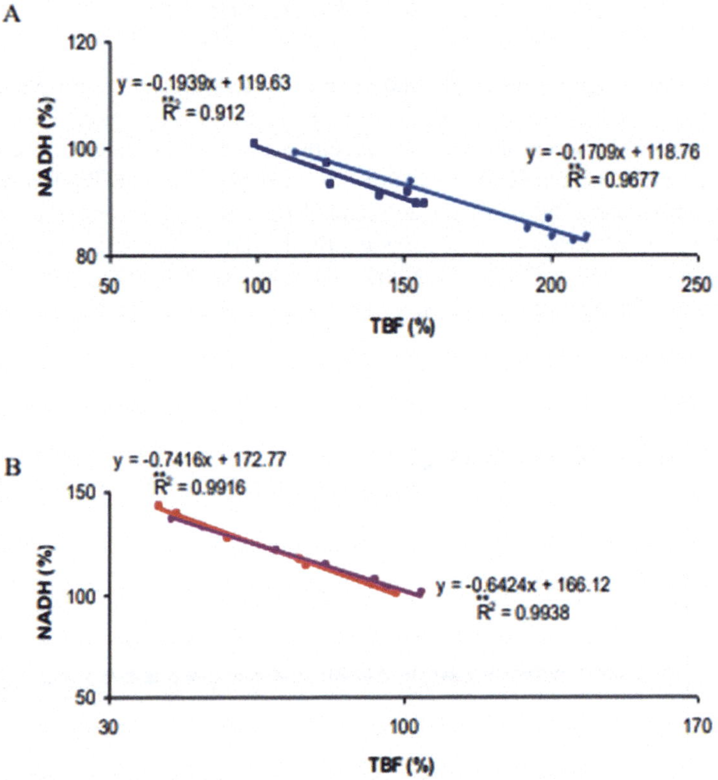

Fig. 4.20 The correlation between tissue blood flow (TBF) and NADH redox state (NADH) measured in the rat brain (**a**) and in the small intestine (**b**) in mild anesthesia and under deep anesthesia following epinephrine injection (6 μg/100 g BW). The data used for the present correlation were taken from the time array (0–1 min). (**) $P < 0.01$; (*) $P < 0.05$; ($N = 9$)

level of 42 ± 14% and 45 ± 15%, respectively ($P < 0.01$). Consequently, NADH increased to the level of 143 ± 8% and 137 ± 8%, respectively ($P < 0.01$). The intestinal reflectance increased in both groups to the level of approximately 120% ($P < 0.01$).

The correlation between TBF and NADH in the brain showed similar slopes, although in the deeply anesthetized rats the CBF reached the very high level of 213 ± 20%, whereas in the mildly anesthetized rats the maximal CBF was 157 ± 17% (Fig. 4.20a). On the other hand, in the intestine (Fig. 4.20b), the correlation between the blood supply and the mitochondrial NADH redox state was similar under both anesthesia levels.

4.3.3 The Responses to Pentobarbitone Injection

In order to test the responses of the brain and the small intestine to deep anesthesia, pentobarbitone was injected (as described in the Methods section). The level of changes induced by deep anesthesia was similar in both organs. As presented in Fig. 4.21, MAP decreased to 57 ± 3% ($P < 0.01$) immediately following the injection. This deterioration was accompanied by a decrease in CBF to 92 ± 3% ($P < 0.01$) and IBF to 92 ± 7%. Consequently, NADH in the brain was elevated to 103 ± 1% ($P < 0.05$) and in the small intestine it increased to 102 ± 1% ($P < 0.05$). The reflectance in the brain remained steady through the entire period of monitoring,

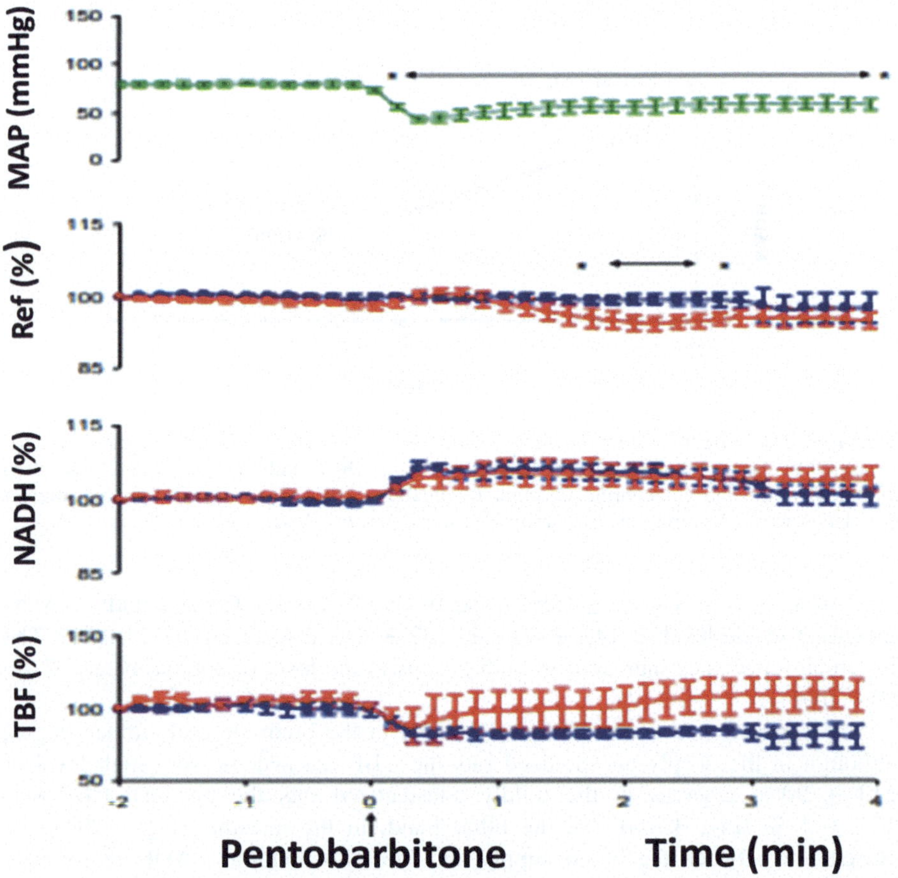

Fig. 4.21 The responses of the brain (blue) and small intestine (red) to pentobarbitone (one bolus of 30 mg/kg followed by an IV infusion of 4 mg/kg pentobarbitone for 10 min). Abbreviations are as above. Arrows show the period of time in which significant differences between the two organs were detected for each minute. ($N = 9$). (*) $P < 0.05$

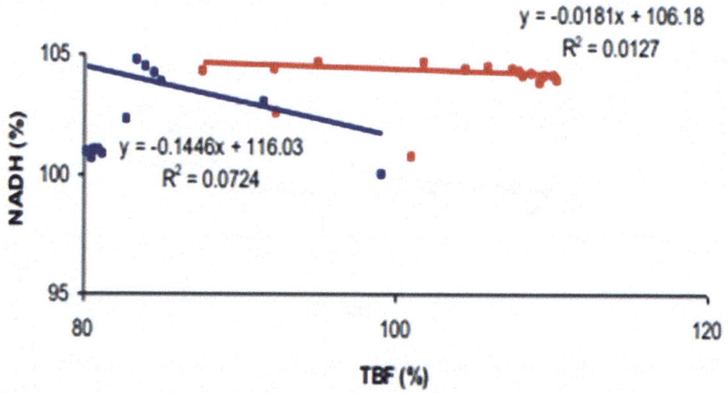

Fig. 4.22 The correlation between tissue blood flow (TBF) and NADH redox state (NADH) measured in the rat brain (blue) and small intestine (red) responses to pentobarbitone (one bolus of 30 mg/kg pentobarbitone followed by an IV infusion of 4 mg/kg pentobarbitone for 10 min). The data used for the present correlation were taken from the time range (0–4 min); ($N = 9$)

while in the intestine it started to decrease about 50 s after pentobarbitone was injected and reached a minimum level of 95 ± 2% ($P < 0.05$) 2 min post injection. Moreover, the blood flow in the brain responded differently from the intestine. The CBF decreased to 82 ± 6% ($P < 0.05$) and remained low until the end of the experiment. In contrast, IBF showed no significant changes through the entire experiment, although there was a trend of decrease immediately after the injection of pentobarbitone and an immediate recovery with a hyperemic tendency (although not statistically significant), presumably due to the large variability. Consequently, the NADH level in the small intestine increased only during the first minute post pentobarbitone injection, whereas in the brain NADH was elevated through the first 3 min post pentobarbitone injection.

Figure 4.22 presents the correlation between TBF and NADH levels in both organs under deep anesthesia. As it is seen, there was no correlation between these two parameters. Oppositely, the MAP correlates with TBF in both organs (Fig. 4.23). The low level of blood pressure induced a decrease in the blood flow in the brain and in the small intestine, while the increasing MAP level was accompanied by a recovery of the blood supply to both organs.

4.3.4 *The Effects of Nitric Oxide (NO)*

In order to test the involvement of NO in the metabolism of the brain and intestine, three groups of rats were monitored, where two served as controls and the third was the experimental group. The effect of low levels of NO in the body was tested in a control group of animals that were injected with L-NAME (50 mg/kg IV) and infused with saline for 10 min. As Fig. 4.24 demonstrates, in the absence of an

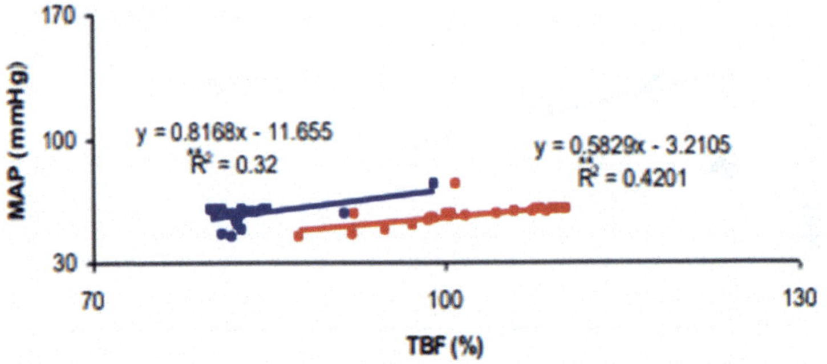

Fig. 4.23 The correlation between mean arterial pressure (MAP) and tissue blood flow (TBF) measured in the rat brain (blue) and in the small intestine (red) under pentobarbitone (one bolus of 30 mg/kg followed by an IV infusion of 4 mg/kg pentobarbitone for 10 min). The data used for the present correlation were taken from the time range (0–4 min). (**) $P < 0.01$; (*) $P < 0.05$; ($N = 9$)

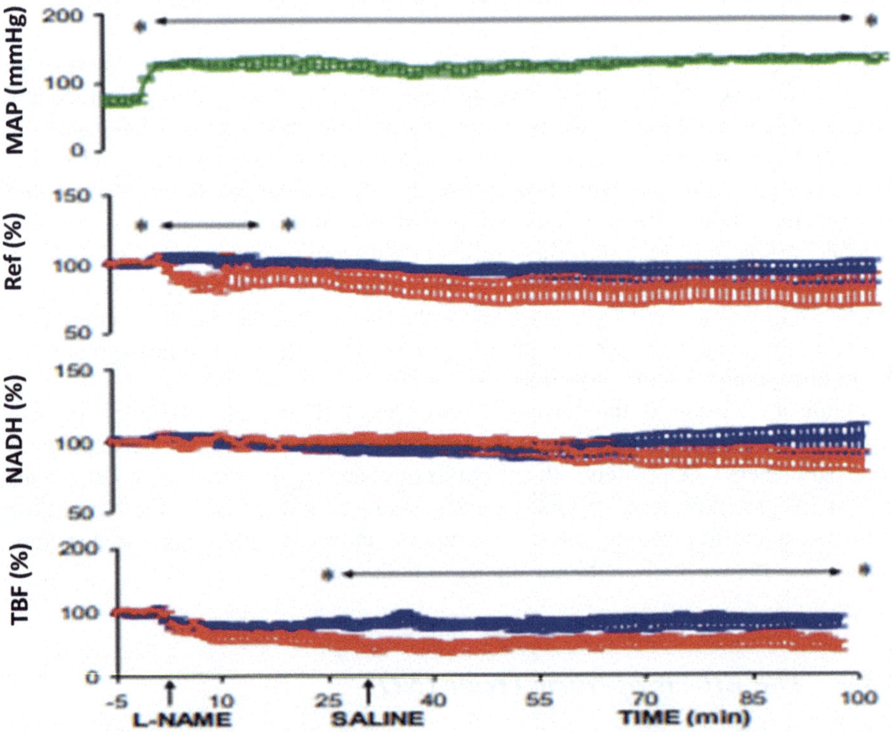

Fig. 4.24 The responses of the brain (blue) and small intestine (red) to L-NAME injection (50 mg/kg.) and Saline infusion (10 min). Abbreviations are as above. Arrows show the period of time in which differences between the two organs were detected for each minute. ($N = 8$). (*) $P < 0.05$ (Tolmasov et al. 2009)

alternative source of NO, the level of MAP remained high during the entire period of monitoring (133 ± 2 mmHg, $P < 0.05$). The TBF in both organs decreased to the same level (78 ± 6%, $P < 0.05$); however, from the 20th min post L-NAME, the CBF began to recover, while the IBF remained low, inducing a significant difference between the levels of TBF in the two organs from the 25th min until the end of the experiment.

There were no changes in the NADH level and the reflectance was not affected by the saline infusion, although the onset of the L-NAME injection and the following 10 min of monitoring yielded a small decrease in the intestine (85 ± 4%, $P < 0.05$), whereas the reflectance in the brain remained stable (101 ± 3%, $P < 0.05$).

4.3.4.1 The Effects of Nitrite (Control 2)

In this control group, the rats were injected with saline (IV) and thereafter infused with Nitrite (20 mg/kg IV 10 min).

As presented in Fig. 4.25, the MAP was not significantly affected by the Nitrite infusion, although a trend of decrease was observed. Nitrite induced no significant changes in any of the other parameters, though the reflectance showed a tendency of decrease in the intestine and a tendency of increase in the brain. The correlation between TBF and NADH (Fig. 4.26) showed that the increased blood supply to the brain was significantly correlated with the oxidation of NADH ($P < 0.01$), whereas in the intestine, no correlation between TBF and NADH was observed.

4.3.4.2 The Effects of L-NAME and Nitrite

In order to test the involvement of NO in the regulation of blood flow in both organs, L-NAME was used—a well-known nonspecific inhibitor of nitric oxide synthase (NOS) following the injection of nitrite, which is an exogenous source of NO. The injection of L-NAME yielded a significant increase in MAP up to 121 ± 2 mmHg ($P < 0.05$), which remained high during 38 min of monitoring (Fig. 4.27). In continuation, after 8 min of nitrite infusion, the level of MAP started to decrease and remained low for another 28 min of monitoring (80 ± 8 mmHg, $P < 0.05$). Further monitoring showed a recovery of MAP, indicating that the effect of nitrite was terminated and only the effect of L-NAME remained (108 ± 4 mmHg, $P < 0.05$). Although the injection of L-NAME decreased the TBF in the brain and small intestine at the same level, their responses to Nitrite were significantly different. The infusion of Nitrite immediately induced an elevation of CBF (108 ± 8%, $P < 0.05$), which remained significantly higher as compared to the IBF level (65 ± 9%, $P < 0.05$), from this point onward. In spite of the differences between these two organs with respect to TBF, NADH remained stable through the entire experiment in both organs. In addition, the reflectance in the small intestine tissue was significantly lower (81 ± 5%, $P < 0.05$) during the last 46 min of monitoring as compared to its level in the brain (94 ± 4%, $P < 0.05$).

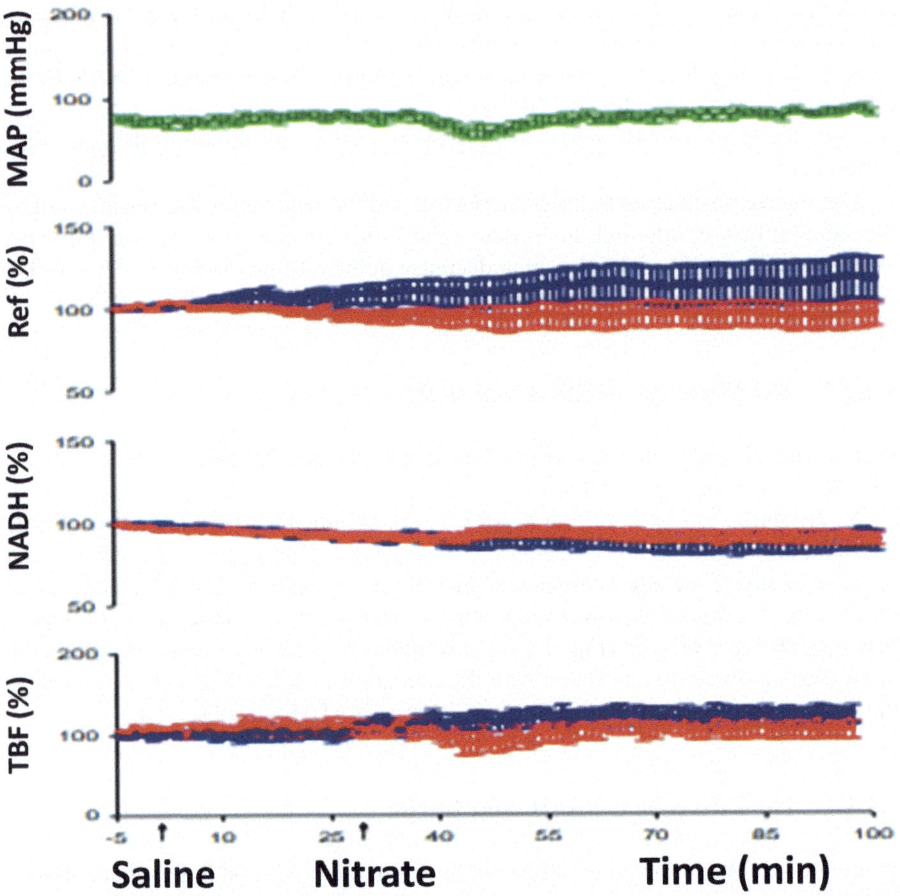

Fig. 4.25 The responses of the brain (blue) and small intestine (red) to Saline injection and Nitrite infusion (20 mg/kg 10 min). Abbreviations are as above. Arrows show the period of time in which differences between the two organs were detected for each minute. ($N = 8$). (Tolmasov et al. 2009)

Analysis of correlation between the blood flow and NADH revealed that after L-NAME injection, the TBF decrease in the brain to 80 ± 4% ($P < 0.01$) significantly correlated to NADH oxidation (93 ± 3%) (Fig. 4.28a), as well as the deterioration in TBF in the small intestine (69 ± 10%, $P < 0.01$) (Fig. 4.28a) correlated with a decrease in NADH (94 ± 3%). Similarly, a correlation between the same parameters was observed in both organs under Nitrite infusion. Thus, the TBF recovery in the brain to 108 ± 8% ($P < 0.01$) (Fig. 4.28b) was accompanied by a further oxidation of NADH (87 ± 4%). In the small intestine, the TBF level of 67 ± 10% (Fig. 4.28b) was significantly correlated with the NADH level of 94 ± 3%.

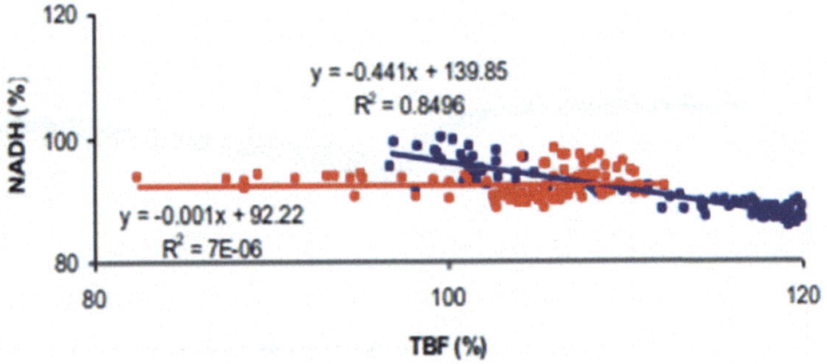

Fig. 4.26 The correlation between tissue blood flow (TBF) and NADH redox state (NADH) measured in the rat brain (blue) and small intestine (red) in their responses to Nitrite infusion (20 mg/kg 10 min). The data used for the present correlation were taken from the time range (0–100 min). (**) $P < 0.01$; (*) $P < 0.05$; ($N = 8$)

Figure 4.29a presents the correlation between the blood pressure and TBF in both organs. Following the injection of L-NAME, the blood pressure increase correlated with the deterioration of the blood supply to the brain and the small intestine. However, following the injection of Nitrite, no correlation between these parameters was detected (Fig. 4.29b).

4.3.5 *Responses to Sepsis*

A group of 10 rats was injected intravenously with LPS (30 mg/kg) and monitored during 5 h or until death. In the control group, 12 rats were injected with 0.1 ml/100 g saline and monitored for a similar time period. In the septic group, 50% of the rats spontaneously died within the monitored period, while the others showed extreme deterioration in all the parameters, except for the reflectance in the brain.

Figure 4.30 displays the parameter changes in the brain and the small intestine under sepsis compared to the control group. As seen, MAP in the septic animals was significantly lower than the MAP in the control animals through almost the entire monitored period ($P > 0.001$). While the MAP in the control group was stable, in the septic rats the MAP immediately decreased to the level of 74.9 ± 8.2% ($P < 0.05$), followed by a minor increase to the level of 85.8 ± 6.7%. From this point onward, the blood pressure in the septic rats further decreased to the minimum levels of 26.8 ± 6.2%, ($P < 0.001$) at the end of the experimental period. The CBF in the septic rats significantly decreased ($P < 0.001$) from time 5% and eventually reached the level of 13.6 ± 11.5% ($P < 0.001$) at the end of the experiment. In contrast, in the intestine, the TBF remained steady. These changes in CBF were accompanied by an elevation of the reflectance and NADH in the brain in both groups. The cerebral reflectance monitored in the septic and control animals was similar. However, the

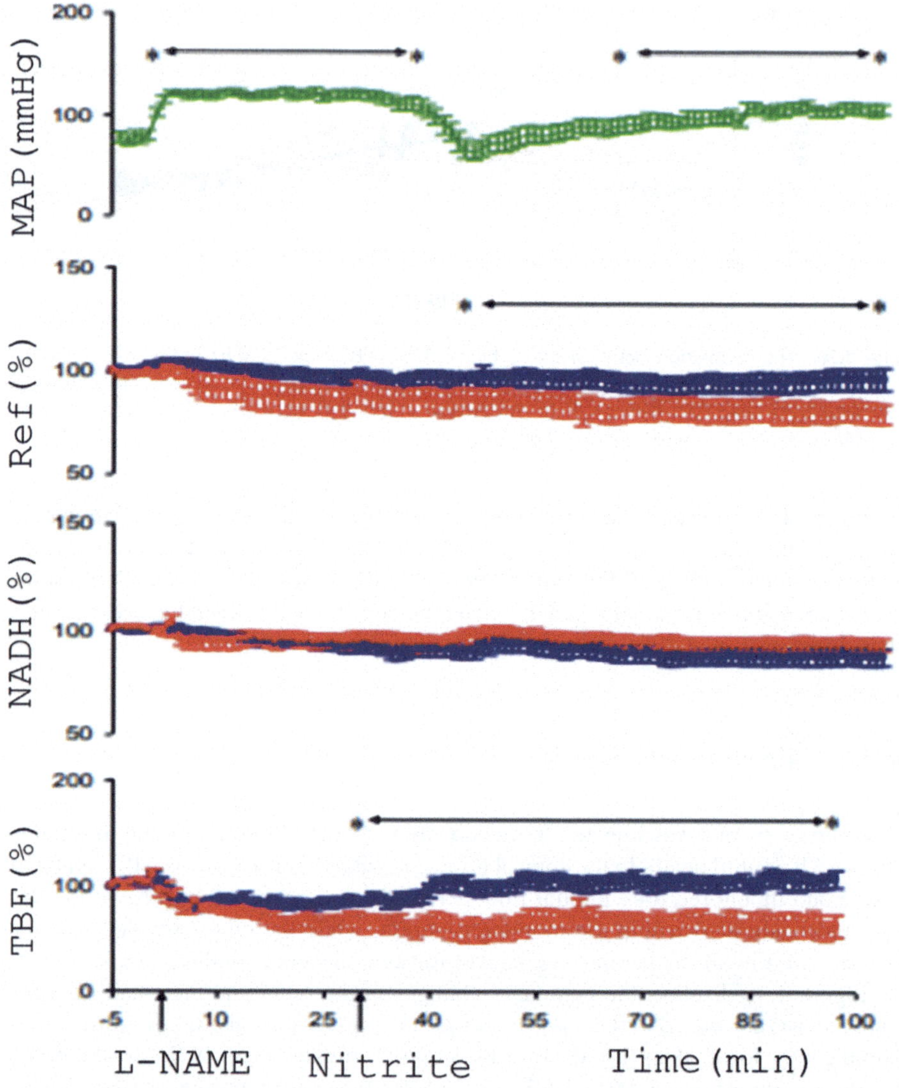

Fig. 4.27 The responses of the brain tissue (blue) and the small intestine (red) to L-NAME injection (50 mg/kg) and Nitrite infusion (20 mg/kg, 10 min) *REF* Reflectance, *NADH* NADH-corrected, *TBF* Tissue Blood Flow, and *MAP* Mean Arterial Pressure. Arrows show the period of time in which significant differences between the two organs were detected. ($N = 14$). (*) $P < 0.05$ (Tolmasov et al. 2009)

reflectance changes in the septic group were statistically significant when compared to their baseline values ($P < 0.001$). The monitoring of cerebral NADH in the control group showed no significant differences compared to the baseline values, while in the septic group NADH significantly increased after 60% of the

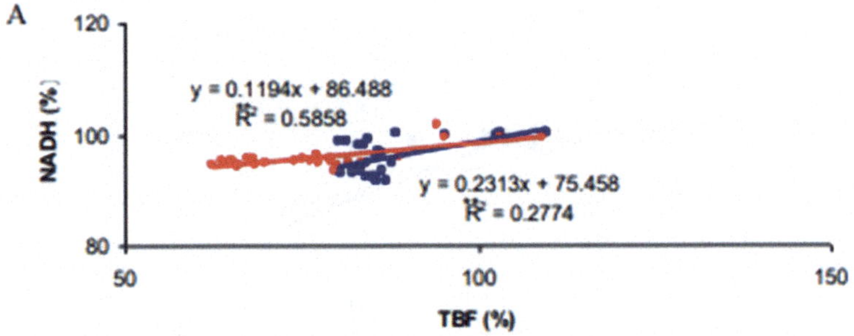

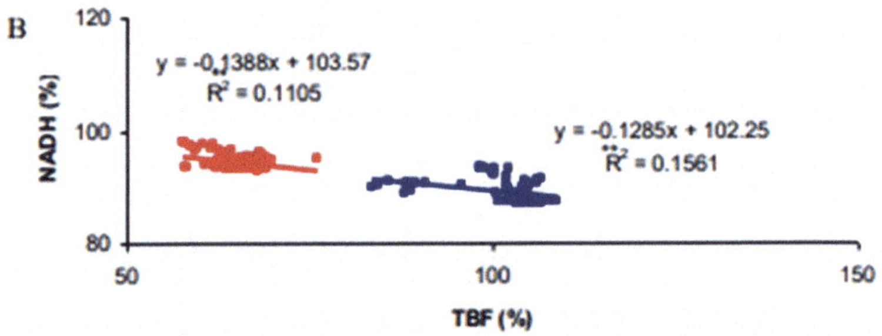

Fig. 4.28 The correlation between tissue blood flow (TBF) and NADH redox state (NADH) (50 mg/kg) (**a**). The data used for the present correlation were taken from the time range (0–30 min) and Nitrite infusion (20 mg/kg, 10 min) (**b**). The data used for the present correlation were taken from the time range (31–100 min). (**) $P < 0.01$; (*) $P < 0.05$; ($N = 14$)

experimental period. Furthermore, the NADH level monitored in the septic group at the end of the experiment (52 ± 9%, $P < 0.001$) was significantly higher than the NADH level in the control group (10.2 ± 5.4%). Significant differences between the two groups were seen from 85% of time post LPS/Saline injections until the end of the experiment ($P < 0.01$). The IBF was significantly different between the two groups in the second half of the experiment ($P < 0.001$). The IBF in the control group remained stable, while in the septic group the IBF continuously decreased, down to the level of 26.7 ± 16.4% (45% in time) post LPS injection. The IBF reached a minimum level of 8.9 ± 2.2% at the end of the experiment ($P < 0.001$).

The changes in the IBF were accompanied by an elevation of NADH in both groups in a similar manner; however, the degree of change in the last 10% of the experiment was significantly different ($P < 0.05$). NADH increased to the level of 16.4 ± 6.4% in the control group, while the maximal increase to 73.3 ± 20% was seen under sepsis ($P < 0.001$). This elevation of NADH in the septic group was detected in the last 35% of the experimental period. The reflectance in the small intestine showed opposite responses in the control group versus the septic group,

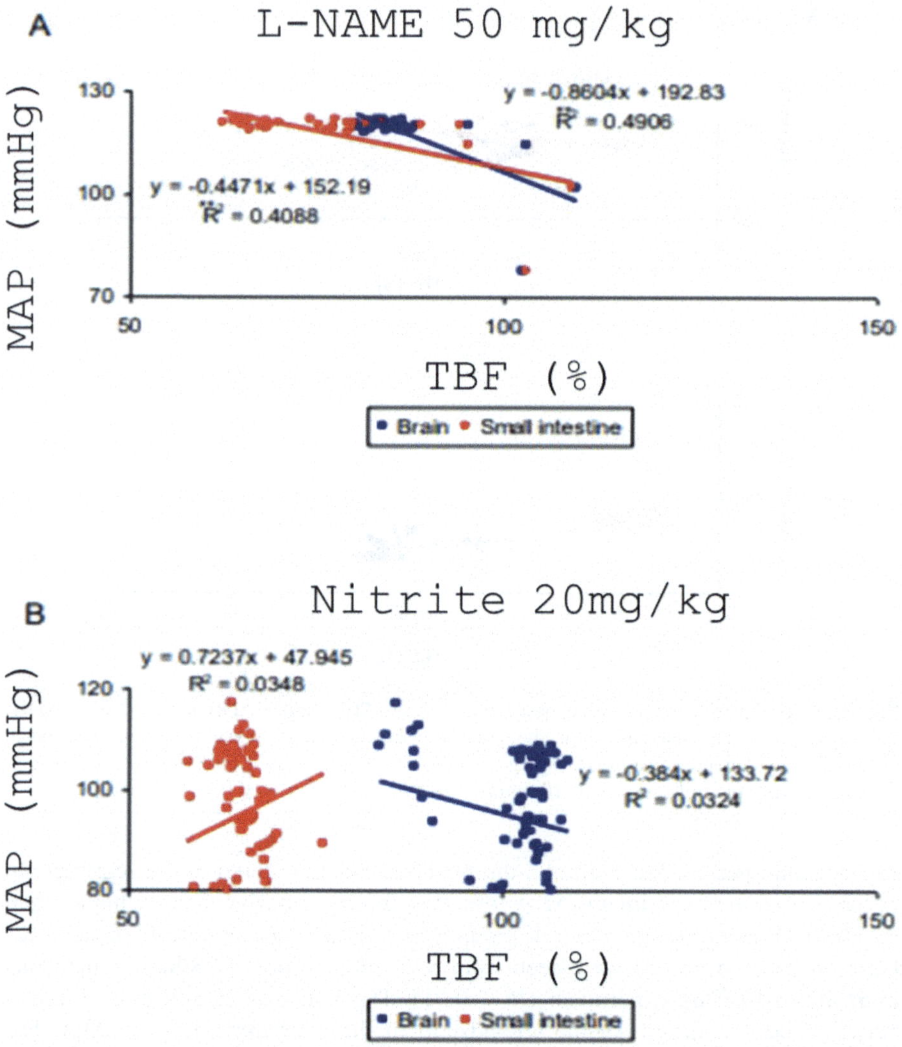

Fig. 4.29 The correlation between mean arterial pressure (MAP) and tissue blood flow (TBF) measured in the rat brain (blue) and small intestine (red) in their responses to L-NAME injection (50 mg/kg) (**a**). The data used for the present correlation were taken from the time range (0–30 min) and Nitrite infusion (20 mg/kg, 10 min) (**b**). The data used for the present correlation were taken from the time range (31–100 min). (**) $P < 0.01$; (*) $P < 0.05$; ($N = 14$)

with a significant difference at the last 20% of the experimental period ($P < 0.05$). At the end of the experiment, the reflectance in the control animals significantly increased by 6.9 ± 9.8%, whereas in the septic group the reflectance decreased by 30.4 ± 11.5% ($P < 0.05$).

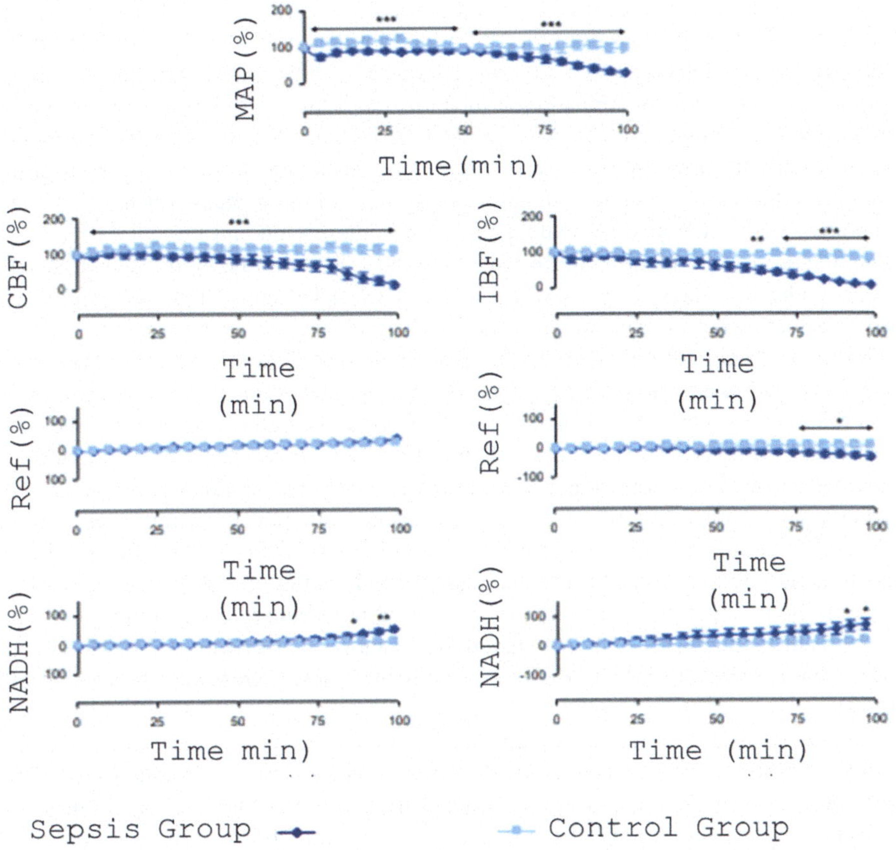

Fig. 4.30 Changes in the brain (left) and the small intestine (right), together with systemic blood pressure (MAP), following IV injections of LPS (Sepsis Group, $N = 10$), and Saline (Control Group, $N = 12$). Abbreviations are as above. MAP was calculated using a baseline value of 100% as the control value. The injections were given at time = 0%, while time = 100% represents the maximum 5 h of monitoring

4.4 Discussion

4.4.1 The Effects of Various Vasoactive Mediators on the Metabolism of the Brain and Small Intestine

The aim of this study was to employ simultaneous multiple organ monitoring in real time in “vital” and “less vital” organs, following the injection of various vasoactive compounds whose physiological effects are related to the sympathetic system activity.

The sympathetic nervous system and adrenal medulla act together to preserve homeostasis by modifying the blood flow distribution, and altering metabolism to improve oxygen delivery to vital organs (Bao et al. 1997). Catecholamines, such as epinephrine released from the adrenal medulla, contribute to this sympatho-mimetic effect (De Champlain and Van Ameringen 1972). Hence, brain monitoring simultaneously with the small intestine, following epinephrine injection, has a great significance for the early detection of whole body deterioration based on the different responses to the activated sympathetic system in these vital and less vital organs.

In the present study, epinephrine injection induced a significant increase in MAP levels under all doses (Figs. 4.4, 4.7, 4.10, 4.13, 4.16, and 4.19). As previously indicated, epinephrine can increase the MAP, primarily by increasing the cardiac index and stroke volume with more modest increases in the systemic vascular resistance and heart rate (Holmes et al. 1980). The elevation of the blood pressure was followed by a significant increase in the CBF and a deterioration in the TBF of "less vital" organs such as the small intestine. The same result was previously obtained in our laboratory when the brain was monitored simultaneously with the kidney, liver, and testis, following injection of norepinephrine (Kraut et al. 2004). The ability of epinephrine to improve the blood flow to the brain likely derives from its alpha-adrenergic activity (Holmes et al. 1980; Kraut et al. 2004). Several investigators showed that the destruction of the central noradrenergic system produces an increase in resting CBF and CO_2 reactivity (Sharma and Gulati 1994). However, others have found no effect of this procedure on these parameters (Onesti et al. 1989).

In the small intestine, the sympathetic nervous activity reduces the IBF by increasing the vascular resistance of the arteries and arterioles (Takala 1996). The low intestinal perfusion caused a significant increase in NADH, whose high level reflects the deterioration in the energetic state of the tissue. The high level of reflectance in the small intestine is due to the decrease in the blood volume caused by a decrease in the blood flow and results in changes in the optical features of the tissue. In addition, the blood pressure responded to epinephrine in a dose-dependent manner (Fig. 4.17). Significant differences in MAP were noted between 2 μg/100 g epinephrine and all the other doses. The dose-dependent response of MAP probably reflects a selective increase in the cardiac index and stroke volume. The mechanism responsible for the rise in systolic blood pressure that occurs after epinephrine administration relates to its positive inotropic and chronotropic actions on the heart (predominantly β_1-receptors) and the vasoconstriction induced in many vascular beds (alpha-receptors) (Katzung 1992). The increase in the cerebral blood supply under various concentrations of epinephrine reflects the ability of the central protection mechanism to activate itself and thus to preserve the blood supply to the brain, while the "less vital" organs, like the small intestine, undergo vasoconstriction when alpha-stimulation predominates (Edvinsson et al. 1978; Sharma and Gulati 1994; Takala 1996). Consequently, the NADH level in the small intestine increased similarly under all doses of epinephrine. The high level of NADH is due to a low blood supply. Analysis of the brain reflectance revealed a significant difference between 2 μg/kg and all the other doses. This selective decrease in the brain reflectance is due to a dose-dependent elevation of the blood volume, and it results in changes in the optical features of the tissue. However, in the intestine, a significant difference was observed only between the doses of 2 μg/100 g and 10 μg/100 g. This mainly reflects a comparable level of changes in the blood volume under intermediate doses of epinephrine. It is important to clarify that all the changes in TBF and NADH parameters presented in this study were a direct result of epinephrine injection, since the results for the control group (five saline injections) showed no effect on TBF and NADH in the brain as well as in the small intestine (Fig. 4.1).

The results of the present study demonstrate a high correlation between the TBF (O_2 supply) and tissue energetic state under all doses of epinephrine in both organs (Figs. 4.3, 4.6, 4.7, 4.12, and 4.15). However, although TBF increased in the brain, NADH remained stable. A possible reason for this is that the normoxic brain is well oxidized in the normal perfusion levels; thus, an increase in TBF has no further beneficial effect on mitochondrial NADH levels (Kraut et al. 2004). However, the decrease of the IBF was accompanied by NADH elevation, which manifested the deterioration of the energetic metabolism. The present study demonstrated a high correlation between MAP and TBF parameters (Figs. 4.4, 4.7, 4.10, 4.13, and 4.16) under all doses of epinephrine, in both organs. This is explained by the fact that the decrease in IBF is mostly influenced by systemic changes. However, the brain has its autoregulatory mechanisms, which preserve the CBF under various MAP levels (Strandgaard et al. 1973; Strandgaard and Paulson 1984). The underlying mechanism of CBF autoregulation is based on the difference in the resistance of large blood vessels versus small vessels during blood pressure changes (Strandgaard and Paulson 1984). The sympathetic nerves, when stimulated, constrict the larger vessels, whereas cerebral autoregulation predominantly involves changes in the resistance of smaller vessels (Strandgaard and Paulson 1984).

The brain, a vital organ, showed an increase in TBF with no damage to its mitochondrial function and energetic state, while in the small intestine, which is defined as "less vital" organ (relatively to the brain), the TBF decreased and NADH increased. These results may imply that the small intestine is less important in stress conditions. Here, the question arises whether the inhibition of the sympathetic system induced by deep anesthesia affects the response to epinephrine of both TBF and NADH in the two organs.

Anesthesia may produce a considerable bias in experimental medicine, particularly in conditions of stress (Kerger et al. 1997). Barbiturates are commonly used as anesthetic agents both in clinical and experimental conditions (Kerger et al. 1997). As previously indicated, barbiturates decreased the central sympathetic tone (Feldberg and Guertzenstein 1972). In the present study, immediately after pentobarbitone intravenous injection, a decrease of 30 mmHg in MAP was developed (Fig. 4.21). Several investigators achieved the same result in rabbits (Korner et al. 1968). However, others have found no effect of this anesthetic agent on the blood pressure following intravenous treatment, whereas an injection of a few milligrams of pentobarbitone sodium into the cerebral ventricles of anesthetized cats, under artificial ventilation, regularly produced a fall in MAP (Feldberg and Guertzenstein 1972). The difference may be related to the species or, more likely, to alterations in the central neural control of blood pressure induced by mild anesthesia. The systemic changes observed in our study are presumably due to cardiac output depression, a main cause of MAP reduction and an important factor in microvascular blood flow decrease leading to oxygen delivery impairment (Kerger et al. 1997). Indeed, the present results showed that the decrease in blood pressure was followed by a deterioration of CBF. The sensitivity of CBF to pentobarbitone injection primarily indicates that this vasopressor agent has a central mechanism of activity (Lowry and Fillenz 2001). However, the IBF remained stable under deep anesthesia. This was possibly due to the fact that the small intestine is a "less vital" organ whose blood supply decreased already following mild anesthesia; thus, a further decrease in MAP induced by the deep anesthesia has no additional effects on IBF levels. Consequently, the elevation of NADH in the small intestine was equal under both levels of anesthesia, whereas in the brain, deep anesthesia induced a decrease in tissue metabolism (Lowry and Fillenz 2001), which was expressed by the larger response of CBF and NADH oxidation after adrenaline injection in the deep anesthesia group versus the low anesthesia group. In addition, the kinetics of the response were similar in the intestine under the two anesthesia levels, whereas in the brain the deeper anesthesia induced a prolonged response of CBF, NADH, and reflectance, due to the suppressor effect of pentobarbitone on the catecholamines in the brain (Dalley et al. 1998). As a result, the cerebral vaso activity was depressed. The present results showed that TBF was not correlated with NADH in both organs (Fig. 4.22). This is probably due to a low level of NADH increase observed under deep anesthesia. In contrast, MAP correlates with TBF in both organs (Fig. 4.23). This is due to the strong reduction effect of pentobarbitone on the blood pressure and vasomotor tone (Feldberg and Guertzenstein 1972).

In order to test the response of the brain and small intestine under deep anesthesia to the activated sympathetic system, epinephrine injection was performed. The responses of the brain TBF and NADH to Epinephrine under deep anesthesia were more significant than those measured in the awake state (Fig. 4.20). This is explained by the fact that deep anesthesia lowers the level of MAP, and causes the elevation of adenosine, which is one of the mediators of CBF autoregulation that activates through the vasodilatation of small arteries (Strandgaard and Paulson 1984), leading to a significant response of CBF and NADH to epinephrine in the brain under deep anesthesia versus mild anesthesia. In contrast to the brain, in the small intestine, there were no differences between the responses of these parameters to epinephrine under different levels of anesthesia (Fig. 4.21). However, the IBF as well as intestinal metabolism are similar under mild and deep anesthesia; therefore, the response to epinephrine was similar under both conditions. In addition, the brain, as a vital organ, has an extremely high level of metabolism, while the small intestine, a "less vital" organ, has a low metabolic activity. As a result, the brain is much more sensitive to a decrease in the level of metabolism induced by deep anesthesia when compared to the small intestine. Our results showed that the level to which the blood pressure recovered in the deep anesthesia group was lower than that in the mild anesthesia group.

The greater reduction in blood pressure under deep anesthesia during recovery is probably related to the inhibitory effect of barbiturates on NOS activity (Castillo et al. 1999). This possibly occurs because epinephrine produces systemic vasodilatation through the release of NO (Hemida et al. 1995), whose synthesis is impaired by barbiturates (Castillo et al. 1999). Thus, a deficit of this vasodilator compound is the cause of strong vasoconstrictions and a blood pressure reduction. An additional explanation for the different levels of blood pressure recovery is that this systemic parameter under deep anesthesia was lower than it is under mild anesthesia. It appears important to evaluate the hemodynamic and metabolic changes induced by NO inhibitors and donors in the brain, monitored simultaneously with "less vital" organs such as the small intestine.

In order to test the involvement of NO in the regulation of the blood flow to the brain and the small intestine, L-NAME was utilized. This is a well-known nonspecific inhibitor of NOS and nitrite, which is an alternative exogenous source of NO.

In animals injected with L-NAME and infused with nitrite, the MAP was significantly elevated immediately after the L-NAME injection (Fig. 4.27). This is due to the e-NOS inhibitory effect of L-NAME, which is responsible for vasoconstriction of regional blood vessels (Licker et al. 1998; Rifkind et al. 2007). Additionally, experimental studies in animals have provided convincing evidence that the neuronal isoform of NO synthase is an important component of the transduction pathways that tonically inhibit the sympathetic outflow from the brain stem (Ye et al. 1997). Thus, the hypertensive effect of L-NAME in rats could be partly due to the activation of central noradrenergic neurons (Ye et al. 1997). Analysis of MAP responses to nitrite revealed that this systemic parameter immediately decreased under conditions of an enhanced increase in the NO level. This is due to the reduction of nitrite by red cells and the production of NO, which causes systemic vasodilatation (Tsuchiya et al. 2005; Rifkind et al. 2007). Further monitoring showed a recovery of MAP, indicating that the effect of nitrite was terminated and only the L-NAME effect remained. The relatively short effect of nitrite may be explained by a rapid decline of the nitrite level in the plasma after the intravenous injection, due to the transport of nitrite into erythrocytes where it can bind to hemoglobin, its transport into tissues where it is oxidized to nitrate and the excretion of nitrite into the urine (Kohn et al. 2002). The hypertensive effect of L-NAME was accompanied by a reduction of the blood supply to both the brain and the small intestine. This is due to the vasopressor effects of L-NAME on regional hemodynamics (Tsuchiya et al. 2005; Rifkind et al. 2007). Interestingly, the infusion of nitrite immediately induced an elevation in CBF, whose level remained significantly higher when compared to the level of IBF, from this point further. The difference between CBF and IBF may imply that the blood supply to the brain, a vital organ, is more affected by the level of NO in the body than IBF. A possible explanation for this variation relies on the major difference in oxygen consumption and flow characteristics of vital and "less vital" organs. The brain is a highly perfused vital organ, whose metabolic activity is characterized by high oxygen consumption;

therefore, it is more sensitive to the release of NO in the bloodstream as compared to the small intestine. Moreover, the role of NO in the cerebral tissue is well documented. NO participates in both the regulation of the basal tone of cerebral microvessels and the autoregulation of CBF (Kobari et al. 1994), although other studies have found no effect of NO on the CBF autoregulation in anesthetized rats (Buchanan and Phillis 1993).

In spite of the differences in TBF between these two organs, the level of NADH in the brain was similar to the NADH level in the intestine. This probably reflects a situation in which the decrease of blood supply to the intestine does not affect mitochondrial NADH levels, since intestinal metabolism is relatively low, as the intestine is a "less vital" organ. The results of this study show a significant correlation between the TBF and the energetic state in both organs under NO deficit, as well as under increased NO levels. This demonstrates the important role of NO in the regulation of blood supply to the brain and the small intestine. However, the blood supply to these two organs was not correlated with blood pressure levels under NO elevation. In contrast, a significant correlation between systemic changes and the blood flow was observed in both organs under NO deficit. This may be explained by the relatively short effect of nitrite infusion on the level of blood pressure. In addition, the reflectance in the intestine was significantly different during the last 46 min of monitoring, when compared to the level of reflectance in the brain. This derives from differences in the blood volume.

The effect of the low NO levels in the body was tested in a control group of animals which were injected with L-NAME and infused with saline (Fig. 4.24). The findings showed that the level of MAP remained high during the entire period of monitoring in the absence of an alternative source of NO. With regard to CBF and IBF following L-NAME injection, the results for this control group emphasized the great sensitivity of CBF to a low level of NO. Namely, following L-NAME application, the CBF was restored to its baseline level only 61 min post injection, while after the nitrite infusion, the CBF stabilized only within 5 min. In contrast, the small intestine did not respond to the elevation of NO concentration following nitrite infusion.

In addition, the elevation of NO concentration may be responsible for changes in the reflectance. Thus, the reflectance of both organs—the brain and the small intestine—was not affected by saline, while after nitrite infusion different responses of these two organs were observed. The brain reflectance remained stable, while the reflectance in the small intestine tissue decreased.

Another control group was tested, where the rats were injected with saline and infused with nitrite. The MAP (Fig. 4.25) was not affected by nitrite infusion and remained stable during the entire monitoring period. Furthermore, both organs—the brain and the small intestine—responded in the same way in all the parameters. We believe that the amount of exogenous NO is minor; thus, when given with no other inhibition of NOS, its effect could not be seen.

Conclusion

- The present stress condition study shows that the blood pressure and CBF rise with increased doses of epinephrine in a dose-dependent manner.
- In contrast, IBF responded to epinephrine by a deterioration of its level. The blood supply to the small intestine decreased to the same level under all epinephrine doses.
- The increase in the cerebral blood supply under various concentrations of epinephrine reflects the ability of the central protection mechanism to activate and thus preserve the blood supply to the brain, while the "less vital" organs, such as the small intestine, undergo vasoconstriction when alpha-stimulation predominates.
- The brain, a vital organ, showed an increase in TBF with no damage to its mitochondrial function and energetic state, while in the small intestine, defined as a "less vital" organ (relatively to the brain), TBF decreased and NADH increased.
- Anesthesia induced by barbiturates, such as pentobarbitone, may produce a considerable distortion in experimental medicine, particularly in conditions of stress. As previously indicated, barbiturates decreased the central sympathetic tone.
- In the present study, the MAP decrease of 30 mmHg developed immediately after pentobarbitone intravenous injection.
- Pentobarbitone induced a deterioration of CBF, while IBF remained stable. The high sensitivity of CBF to pentobarbitone injection primarily indicates that this vasopressive agent has a central mechanism of activity.
- Consequently, the elevation of NADH in the small intestine was equal under both levels of anesthesia, whereas in the brain, deep anesthesia induced a decrease in tissue metabolism.
- The activation of the sympathetic system by an injection of epinephrine under deep anesthesia demonstrated that the CBF and metabolic state were more significant than those measured in the awake state, whereas there were no differences between the responses of these parameters to epinephrine under different levels of anesthesia in the small intestine.
- The blood supply to the brain, as a vital organ, is more affected by the NO level in the body than IBF.

4.4.2 Responses to Sepsis

Since sepsis is one of the frequently observed conditions in critical care medicine, we tested the responses of the brain and the small intestine to sepsis, which was induced by the injection of LPS. The responses of each organ under sepsis were compared to the responses of the same organ in normal healthy rats, showing that the intestine was remarkably affected by the development of sepsis as compared to the brain.

LPS are the bacterial components that mainly induce sepsis and are also termed endotoxins (Ivanov and Zilberter 2011). The presence of these components in the bloodstream is the main cause for the induction of a systemic inflammatory response (Van Amersfoort et al. 2003; Ivanov and Zilberter 2011) which produces a severe disruption of microcirculation and consecutive tissue hypoxia. This hypoxia is considered to be a key factor in the development of organ dysfunction and multiple organ failure (Koch et al. 2001).

The progression of sepsis is marked by changes in the relationship between oxygen supply and oxygen consumption in various organs (Ivanov and Zilberter 2011). As the oxygen supply is relatively short, compensatory mechanisms redirect the blood supply to the vital organs, such as the heart and the brain, leaving the "less vital" organs in oxygen debt. The most vulnerable organ in this context is the intestine (Ivanov and Zilberter 2011). The ability to monitor the changes in the blood flow and the tissue energy state of the small intestine, simultaneously with the brain, could provide valuable insights for the early detection of sepsis.

When defining the pathogenesis of sepsis, the element of time is crucial and must be taken into account. Given that, the nature of early sepsis is different from that of late sepsis.

In the present study, we divided the hemodynamic and metabolic events that occurred in response to the injection of LPS into four different periods: the "Initial Response" (phase I), "Static Phase" (phase II), and "Primary and Secondary Deterioration" (phases III and IV). The first two phases represent the early stage of sepsis, while phases III and IV represent late sepsis.

The results showed that LPS injection induced hypotension. The deterioration of the blood pressure was gradual and moderate.

Such hypotension may result from various mechanisms, such as systemic vasodilatation, cardiac dysfunction, and central and/or peripheral deregulation (Anning et al. 1999). These massive cardiovascular alterations are likely due to the overproduction of NO, whose role in modulating the vascular response to sepsis is well known (Anning et al. 1999). As previously indicated, the selective inhibition of inducible nitric oxide synthase (iNOS) corrected the sepsis-induced hypotension, in most of the studies, by increasing the systemic vascular resistance (Hauser et al. 2005).

The circulatory abnormalities alter the oxygen delivery from organs with low metabolic requirements to organs with higher demands and are considered to be the putative cause of altered O_2 metabolism during sepsis (Gutierrez 1993; Maginniss et al. 1994; Crouser et al. 1999; Ince 2005). Our results showed a greater decrease in the blood flow in the small intestine as compared to the brain, during most of the septic periods (phases I-III). This may imply that the brain, as a vital organ, is better protected than the small intestine, especially in the early stages of sepsis.

It has been consistently documented that the blood supply to the intestine decreases in experimental sepsis (Whitworth et al. 1989), whereas normal or increased flows are monitored in skeletal muscle (Lubbe et al. 1992; Hershey and Bond 1993).

Additionally, in the sepsis group, the small intestine exhibited a significant blood flow decrease earlier than the brain. Indeed, splanchnic hypoperfusion occurs early in the shock, and may even take place before the usual shock indicators emerge (Whitworth et al. 1989).

It appears that in the early stages of sepsis, the microcirculatory function is responsible for the maintenance of systemic circulation by redistributing the blood flow to the vital organs, such as the brain, while the blood flow in the "less vital" organs, such as the small intestine, decreases.

In addition to this interorgan redistribution of the blood flow, intraorgan redistribution also occurs, as others demonstrated for the intestine (Vallet et al. 1994; Revelly et al. 1996). It is likely that the NO system is a central component in this autoregulatory control of microcirculatory patency (Ince 2005). This system is severely disturbed in sepsis by a heterogeneous expression of iNOS in different areas of organ beds, resulting in a pathological shunting of flow (Ince 2005). Since iNOS is not expressed homogenously in organ systems, areas lacking iNOS have less NO-induced vasodilatation and become under-perfused (Ince 2005).

However, the regulation of microcirculation in the brain under sepsis is a matter of controversy. Although NO is a potent cerebral vasodilator, the brain blood flow is obviously regulated by mechanisms other than NO during sepsis (Takala et al. 2003). The additional factor influencing CBF is the tissue pCO_2, whose level increases in the cerebral cortex under sepsis (Sato et al. 1998). This increase of pCO_2 is responsible for tissue acidosis, whose ability to dilate the cerebral vessels in an attempt to increase the blood flow is recognized (Strandgaard and Paulson 1984).

The microcirculatory alterations may favor the development of localized zones of ischemia, resulting in an imbalance between oxygen demand and delivery. It is likely that the progression from an early to severe sepsis is accompanied or even possibly caused by a deficit of the oxygen supply, which leads to mitochondrial dysfunction over time (Ince 2005).

The present findings showed that, in the sepsis group, the decrease in CBF did not influence the energy status of the brain, as the brain tissue demonstrated stress later than expected, namely in the secondary deterioration phase.

It might be expected that the drop in the blood flow would result in an oxygen deficit at the cellular level and a consequent impaired cellular respiration, that is, an inhibition of the oxidative phosphorylation process in the mitochondria. It appears that in sepsis, at least until the late stages, the vitality of the brain is not influenced by the reduced perfusion (Hotchkiss and Karl 1992). When the decrease in the blood flow is gradual and moderate, as observed in the brain, a new baseline for the flow is achieved. The body has the ability to adapt to this new flow level. Hence, a delay in the cellular energy reduction may arise. This may imply that the auto-vaso-regulation mechanism is capable of preventing the development of oxygen deficit in the brain under sepsis (Lush and Kvietys 2000).

In contrast to the brain, the small intestine appears to be very sensitive to the blood flow deterioration under sepsis, due to the unique microvascular structure of the intestine, with a counter-current blood flow in the intestinal microvilli (Takala 1996; De Backer et al. 2002). The low pO_2 may make this intestinal villus susceptible to tissue hypoxia (Takala 1996). It may lead to microvascular shunting, when the arteriolar pO_2 is lower than the venular pO_2 (Ince and Sinaasappel 1999). The result of the shunting is a heterogeneity of the microvascular blood flow, which may account for the alterations in oxygen extraction capabilities, as seen in sepsis (De Backer et al. 2002). Therefore, the splanchnic microvasculature is prone to injury because of an oxygen extraction deficit (Ince 2005) and because of the fact that splanchnic vasoconstriction is a common event in conditions related to hemodynamic disturbances (Fink 1991).

A decrease in the IBF is of particular importance, since it may lead to an increase in the intestinal permeability and to a breakdown of the gastrointestinal barrier function, which may allow for the translocation of bacteria and bacterial toxins from the luminal site into the systemic circulation, thereby initiating or perpetuating septic events (Fink 1993). These pathogenic effects of sepsis may lead to organ failure (Ince 2005). Hence, the failure of organs, such as the gut in general and the small intestine in particular, is considered to be the "driving force" of a septic multiorgan failure condition (Ince 2005).

The decrease in the intestinal flow, observed in the sepsis group, impaired the oxygen supply to this "less vital" organ. However, the increase in NADH is even more indicative of the importance of a metabolic failure in the intestine under sepsis conditions.

Brealey et al. (2002) showed that mitochondrial dysfunction indeed plays an important role in sepsis (Ince 2005). During sepsis, ultrastructural injury to the mitochondria commonly develops, impairing ATP biosynthesis. This phenomenon is known as "cytopathic hypoxia." Several mechanisms may be responsible for the "cytopathic hypoxia"in sepsis: dysfunction of the pyruvate dehydrogenase enzyme complex, inhibition of the mitochondrial tricarboxylic acid cycle enzymes or inhibition of the electron transport chain, and activation of the nuclear enzyme poly-(ADP-ribosyl-polymerase) (PARP) that leads to a massive depletion of NAD^+/NADH (Fink 1997, 2001). Since NADH is the main reducing equivalent used to support oxidative phosphorylation, the activation of PARP may result in a marked impairment of aerobic respiration (Fink 1997).

Impaired intestinal oxidative phosphorylation occurred as early as 5 min post endotoxin injection, with a significant fall in the intestinal cytochrome a,a_3 redox state, which was dose-dependent and correlated with falls in the blood pressure and IBF (Schaefer et al. 1991).

The present findings suggested that, in the small intestine (i.e., in a "less" vital organ), the energetic metabolism under sepsis deteriorated before this occurred in the brain.

In view of the fact that the small intestine demonstrated early signs of circulatory abnormalities, followed by cellular injury and stress, the monitoring of these parameters in the gut under sepsis might permit an early detection of the metabolic changes in the whole body.

In the present study, the microcirculatory dysfunction induced by sepsis is accompanied by the cellular energy metabolism deterioration, whose progression is associated with the inhibition of mitochondrial respiration and reduced availability of substrates (Fink 2002). It is becoming clear that sepsis is a disease of the microcirculation, which causes mitochondrial dysfunction. Hence, this pathophysiological condition has been termed Microcirculatory and Mitochondrial Dysfunction Syndrome (MMDS) (Ince 2005). It is likely that the development of MMDS is an indication of a progression from an early to severe sepsis (Ince 2005).

Conclusion

- The largest drop in blood pressure occurred in the last phase in the sepsis group.
- In the sepsis group, there was an interorgan redistribution of the blood flow away from the small intestine.
- In the sepsis group, the brain experienced a late appearance of metabolic failure, regardless of tissue perfusion abnormalities.

References

Ackland G, Grocott MP, Mythen MG (2000) Understanding gastrointestinal perfusion in critical care: so near, and yet so far. Crit Care 4(5):269–281

Anning PB, Sair M, Winlove CP, Evans TW (1999) Abnormal tissue oxygenation and cardiovascular changes in endotoxemia. Am J Respir Crit Care Med 159(6):1710–1715

Astiz M, Rackow EC, Weil MH, Schumer W (1988) Early impairment of oxidative metabolism and energy production in severe sepsis. Circ Shock 26(3):311–320

Astiz ME, DeGent GE, Lin RY, Rackow EC (1995) Microvascular function and rheologic changes in hyperdynamic sepsis. Crit Care Med 23(2):265–271

Bao G, Metreveli N, Li R, Taylor A, Fletcher EC (1997) Blood pressure response to chronic episodic hypoxia: role of the sympathetic nervous system. J Appl Physiol 83(1):95–101

Beal AL, Cerra FB (1994) Multiple organ failure syndrome in the 1990s. Systemic inflammatory response and organ dysfunction. JAMA 271(3):226–233

Boughton-Smith NK, Evans SM, Whittle BJ, Moncada S (1993) Induction of nitric oxide synthase in rat intestine and its association with tissue injury. Agents Actions 38. Spec No: C125-126

Brealey D, Brand M, Hargreaves I, Heales S, Land J, Smolenski R, Davies NA, Cooper CE, Singer M (2002) Association between mitochondrial dysfunction and severity and outcome of septic shock. Lancet 360(9328):219–223

Brun-Buisson C, Doyon F, Carlet J (1996) Bacteremia and severe sepsis in adults: a multicenter prospective survey in ICUs and wards of 24 hospitals. French Bacteremia-Sepsis Study Group. Am J Respir Crit Care Med 154(3 Pt 1):617–624

Buchanan JE, Phillis JW (1993) The role of nitric oxide in the regulation of cerebral blood flow. Brain Res 610(2):248–255

Castillo C, Asbun J, Escalante B, Villalon CM, Lopez P, Castillo EF (1999) Thiopental inhibits nitric oxide production in rat aorta. Can J Physiol Pharmacol 77(12):958–966

Chance B, Legallias V (1959) Differential microfluorimeter for the localization of reduced pyridine nucleotide in living cells. Rev Sci Instr 30(8):732–735

Crouser ED, Julian MW, Dorinsky PM (1999) Ileal VO(2)-O(2) alterations induced by endotoxin correlate with severity of mitochondrial injury. Am J Respir Crit Care Med 160(4):1347–1353

Crouser ED, Julian MW, Blaho DV, Pfeiffer DR (2002) Endotoxin-induced mitochondrial damage correlates with impaired respiratory activity. Crit Care Med 30(2):276–284

Dalley JW, Parker CA, Wulfert E, Hudson AL, Nutt DJ (1998) Potentiation of barbiturate-induced alterations in presynaptic noradrenergic function in rat frontal cortex by imidazol(in)e alpha2-adrenoceptor agonists. Br J Pharmacol 125(3):441–446

De Backer D, Creteur J, Preiser JC, Dubois MJ, Vincent JL (2002) Microvascular blood flow is altered in patients with sepsis. Am J Respir Crit Care Med 166(1):98–104
De Champlain J, Van Ameringen MR (1972) Regulation of blood pressure by sympathetic nerve fibers and adrenal medulla in normotensive and hypertensive rats. Circ Res 31(4):617–628
Edvinsson L, Hardebo JE, MacKenzie ET, Owman C (1978) Effect of exogenous noradrenaline on local cerebral blood flow after osmotic opening of the blood-brain barrier in the rat. J Physiol 274:149–156
Feihl F, Waeber B, Liaudet L (2001) Is nitric oxide overproduction the target of choice for the management of septic shock? Pharmacol Ther 91(3):179–213
Feldberg W, Guertzenstein PG (1972) A vasodepressor effect of pentobarbitone sodium. J Physiol 224(1):83–103
Fink MP (1991) Gastrointestinal mucosal injury in experimental models of shock, trauma, and sepsis. Crit Care Med 19(5):627–641
Fink MP (1993) Adequacy of gut oxygenation in endotoxemia and sepsis. Crit Care Med 21(2 Suppl):S4–S8
Fink M (1997) Cytopathic hypoxia in sepsis. Acta Anaesthesiol Scand Suppl 110:87–95
Fink MP (2001) Cytopathic hypoxia. Mitochondrial dysfunction as mechanism contributing to organ dysfunction in sepsis. Crit Care Clin 17(1):219–237
Fink MP (2002) Cytopathic hypoxia. Is oxygen use impaired in sepsis as a result of an acquired intrinsic derangement in cellular respiration? Crit Care Clin 18(1):165–175
Fink MP, Heard SO (1990) Laboratory models of sepsis and septic shock. J Surg Res 49(2): 186–196
Glauser MP, Heumann D, Baumgartner JD, Cohen J (1994) Pathogenesis and potential strategies for prevention and treatment of septic shock: an update. Clin Infect Dis 18(Suppl 2):S205–S216
Gutierrez G (1993) Regional blood flow and oxygen transport: implications for the therapy of the septic patient. Crit Care Med 21(9):1263–1264
Hauser B, Bracht H, Matejovic M, Radermacher P, Venkatesh B (2005) Nitric oxide synthase inhibition in sepsis? Lessons learned from large-animal studies. Anesth Analg 101(2):488–498
Hemida MR, Brum JM, Estafanous FG, Khairallah PA, Shamloula M, El-Kasstawy B (1995) Role of nitric oxide in systemic hemodynamic responses to dobutamine, epinephrine, and amrinone. J Cardiothorac Vasc Anesth 9(6):627–633
Hershey JC, Bond RF (1993) Oxygen extraction and vascular dilation are dependently increased in skeletal muscle during canine endotoxemia. Circ Shock 40(2):132–138
Holmes HR, Babbs CF, Voorhees WD, Tacker WA Jr, de Garavilla B (1980) Influence of adrenergic drugs upon vital organ perfusion during CPR. Crit Care Med 8(3):137–140
Hotchkiss RS, Karl IE (1992) Reevaluation of the role of cellular hypoxia and bioenergetic failure in sepsis. JAMA 267:1503–1510
Ince C (2005) The microcirculation is the motor of sepsis. Crit Care 9(Suppl 4):S13–S19
Ince C, Sinaasappel M (1999) Microcirculatory oxygenation and shunting in sepsis and shock. Crit Care Med 27:1369–1377
Ivanov A, Zilberter Y (2011) Critical state of energy metabolism in brain slices: the principal role of oxygen delivery and energy substrates in shaping neuronal activity. Front Neuroenerg 3:9
Jordan J, Tank J, Stoffels M, Franke G, Christensen NJ, Luft FC, Boschmann M (2001) Interaction between beta-adrenergic receptor stimulation and nitric oxide release on tissue perfusion and metabolism. J Clin Endocrinol Metab 86(6):2803–2810
Katzung BG (1992) Basic & clinical pharmacology, 5th edn. Prentice-Hall International Inc, Upper Saddle River
Kerger H, Saltzman DJ, Gonzales A, Tsai AG, van Ackern K, Winslow RM, Intaglietta M (1997) Microvascular oxygen delivery and interstitial oxygenation during sodium pentobarbital anesthesia. Anesthesiology 86(2):372–386
Kobari M, Fukuuchi Y, Tomita M, Tanahashi N, Takeda H (1994) Role of nitric oxide in regulation of cerebral microvascular tone and autoregulation of cerebral blood flow in cats. Brain Res 667(2):255–262

Koch T, Geiger S, Ragaller MJ (2001) Monitoring of organ dysfunction in sepsis/systemic inflammatory response syndrome: novel strategies. J Am Soc Nephrol 12(Suppl 17):S53–S59

Kogure K, Scheinberg P, Kishikawa H, Utsunomiya Y, Busto R (1979) Adrenergic control of cerebral blood flow and energy metabolism in the rat. Stroke 10(2):179–184

Kohn MC, Melnick RL, Ye F, Portier CJ (2002) Pharmacokinetics of sodium nitrite-induced methemoglobinemia in the rat. Drug Metab Dispos 30(6):676–683

Korner PI, Uther JB, White SW (1968) Circulatory effects of chloralose-urethane and sodium pentobarbitone anaesthesis in the rabbit. J Physiol 199(2):253–265

Kraut A, Barbiro-Michaely E, Mayevsky A (2004) Differential effects of norepinephrine on brain and other less vital organs detected by a multisite multiparametric monitoring system. Med Sci Monit 10(7):BR215–BR220

Levy B, Mansart A, Bollaert PE, Franck P, Mallie JP (2003) Effects of epinephrine and norepinephrine on hemodynamics, oxidative metabolism, and organ energetics in endotoxemic rats. Intensive Care Med 29(2):292–300

Licker M, Boussairi H, Hohn L, Morel DR (1998) Role of nitric oxide in the regulation of regional blood flow and metabolism in anaesthetized pigs. Acta Physiol Scand 163(4):339–348

Lowry JP, Fillenz M (2001) Real-time monitoring of brain energy metabolism in vivo using microelectrochemical sensors: the effects of anesthesia. Bioelectrochemistry 54(1):39–47

Lubbe AS, Garrison RN, Cryer HM, Alsip NL, Harris PD (1992) EDRF as a possible mediator of sepsis-induced arteriolar dilation in skeletal muscle. Am J Phys 262(3 Pt 2):H880–H887

Lush CW, Kvietys PR (2000) Microvascular dysfunction in sepsis. Microcirculation 7(2):83–101

Maginniss LA, Connolly H, Samsel RW, Schumacker PT (1994) Adrenergic vasoconstriction augments tissue O_2 extraction during reductions in O_2 delivery. J Appl Physiol (1985) 76(4): 1454–1461

Marsh CB, Wewers MD (1996) The pathogenesis of sepsis. Factors that modulate the response to gram-negative bacterial infection. Clin Chest Med 17(2):183–197

Nelson DP, Beyer C, Samsel RW, Wood LD, Schumacker PT (1987) Pathological supply dependence of O_2 uptake during bacteremia in dogs. J Appl Physiol (1985) 63(4):1487–1492

Onesti ST, Strauss RC, Mayol B, Solomon RA (1989) The effects of norepinephrine depletion on cerebral blood flow in the rat. Brain Res 477(1–2):378–381

Parrillo JE (1993) Pathogenetic mechanisms of septic shock. N Engl J Med 328(20):1471–1477

Revelly JP, Ayuse T, Brienza N, Fessler HE, Robotham JL (1996) Endotoxic shock alters distribution of blood flow within the intestinal wall. Crit Care Med 24(8):1345–1351

Revelly JP, Liaudet L, Frascarolo P, Joseph JM, Martinet O, Markert M (2000) Effects of norepinephrine on the distribution of intestinal blood flow and tissue adenosine triphosphate content in endotoxic shock. Crit Care Med 28(7):2500–2506

Rifkind JM, Nagababu E, Barbiro-Michaely E, Ramasamy S, Pluta RM, Mayevsky A (2007) Nitrite infusion increases cerebral blood flow and decreases mean arterial blood pressure in rats: a role for red cell NO. Nitric Oxide 16:448–456

Sands KE, Bates DW, Lanken PN, Graman PS, Hibberd PL, Kahn KL, Parsonnet J, Panzer R, Orav EJ, Snydman DR, Black E, Schwartz JS, Moore R, Johnson BL Jr, Platt R, G. Academic Medical Center Consortium Sepsis Project Working (1997) Epidemiology of sepsis syndrome in 8 academic medical centers. JAMA 278(3):234–240

Sato Y, Weil MH, Tang W (1998) Tissue hypercarbic acidosis as a marker of acute circulatory failure (shock). Chest 114(1):263–274

Schaefer CF, Biber B, Lerner MR, Jobsis-VanderVliet FF, Fagraeus L (1991) Rapid reduction of intestinal cytochrome a,a3 during lethal endotoxemia. J Surg Res 51(5):382–391

Sharma AC, Gulati A (1994) Effect of diaspirin cross-linked hemoglobin and norepinephrine on systemic hemodynamics and regional circulation in rats. J Lab Clin Med 123(2):299–308

Spain DA, Wilson MA, Bar-Natan MF, Garrison RN (1994) Role of nitric oxide in the small intestinal microcirculation during bacteremia. Shock 2(1):41–46

Strandgaard S, Paulson OB (1984) Cerebral autoregulation. Stroke 15(3):413–416

Strandgaard S, Olesen J, Skinhoj E, Lassen NA (1973) Autoregulation of brain circulation in severe arterial hypertension. Br Med J 1(5852):507–510

Takala J (1996) Determinants of splanchnic blood flow. Br J Anaesth 77(1):50–58

Takala J, Booke M, Westphal M, Hinder F, Traber LD, Traber DL (2003) Cerebral blood flow is not altered in sheep with Pseudomonas aeruginosa sepsis treated with norepinephrine or nitric oxide synthase inhibition. Anesth Analg 96(4):1122–1128

Tolmasov M, Barbiro-Michaely E, Mayevsky A (2009) The involvement of nitric oxide in the hemodynamic and metabolic activities of the brain and small intestine. SPIE BiOS: Biomedical Optics A. Mahadevan-Jansen, T. Vo-Dinh and W. S. Grundfest, International Society for Optics and Photonics: 716906-716906-716907

Tsuchiya K, Kanematsu Y, Yoshizumi M, Ohnishi H, Kirima K, Izawa Y, Shikishima M, Ishida T, Kondo S, Kagami S, Takiguchi Y, Tamaki T (2005) Nitrite is an alternative source of NO *in vivo*. Am J Physiol Heart Circ Physiol 288(5):H2163–H2170

Vallet B, Lund N, Curtis SE, Kelly D, Cain SM (1994) Gut and muscle tissue PO_2 in endotoxemic dogs during shock and resuscitation. J Appl Physiol (1985) 76(2):793–800

Van Amersfoort ES, Van Berkel TJ, Kuiper J (2003) Receptors, mediators, and mechanisms involved in bacterial sepsis and septic shock. Clin Microbiol Rev 16(3):379–414

van Lambalgen AA, Bronsveld W, van den Bos GC, Thijs LG (1984) Distribution of cardiac output, oxygen consumption and lactate production in canine endotoxin shock. Cardiovasc Res 18(4): 195–205

VanderMeer TJ, Wang H, Fink MP (1995) Endotoxemia causes ileal mucosal acidosis in the absence of mucosal hypoxia in a normodynamic porcine model of septic shock. Crit Care Med 23(7):1217–1226

Weiss HR, Cohen JA (1978) Changes in small vessel blood content of the rat heart induced by hypercapnic, hyperoxic or asphyxic conditions. Cardiology 63(4):199–207

Whitworth PW, Cryer HM, Garrison RN, Baumgarten TE, Harris PD (1989) Hypoperfusion of the intestinal microcirculation without decreased cardiac output during live Escherichia coli sepsis in rats. Circ Shock 27(2):111–122

Ye S, Nosrati S, Campese VM (1997) Nitric oxide (NO) modulates the neurogenic control of blood pressure in rats with chronic renal failure (CRF). J Clin Invest 99(3):540–548

Chapter 5
Effects of Hemorrhage and Resuscitation

5.1 Introduction

5.1.1 Homeostasis

The maintenance of a relatively stable internal environment is defined as Homeostasis (Sherwood 1997b). Usually, it is maintained by a negative-feedback mechanism; this means that the response is negative to the initial stimulus. Very often, the single feedback mechanism acts as a part of a greater network of controls rather than operating in isolation (Guyton 1999a). The combination of several mechanisms can act either antagonistically or synergistically. The more vital a parameter is, the more systems the body will activate in order to regulate it. The homeostasis is kept within the body levels of organization, including the cell, the tissue, the organ, the body systems, and the entire body. The price for this complicated regulation is energy (Boulpaep and Boron 2003a).

The regulation of body function is divided into two major systems; the nervous system and the hormonal system (Guyton 1999a; Levin et al. 1999). The *nervous system* includes sensory receptors that are dispersed all over the body and detect the environmental state. Those receptors pass the information to the central nervous system (CNS) that makes the decision on how to react to the signals and then passes it toward the target using the motor nervous system. A large part of those processes is done by the autonomic nervous system (ANS) (Guyton 1999a; Levin et al. 1999; Levin 2006). The *hormonal system* includes eight endocrine glands that secrete hormones. The hormones, with the assistance of the extracellular fluid, are transferred through the entire body. Both systems complement each other (Guyton 1999a). As long as normal conditions are maintained in the internal environment, the entire body and its organs continue to function properly. When one part of this "puzzle" loses its ability to keep the optimal status—all of the body suffers (Sher-

© The Author(s), under exclusive license to Springer Nature Switzerland AG 2025
A. Mayevsky et al., *Body Oxygen Homeostasis and Mitochondrial Function*,
https://doi.org/10.1007/978-3-031-94113-9_5

wood 1997b; Guyton 1999a). There are a large number of control systems in the body; among them, there is the circulatory system, which plays a role in maintaining the homeostasis of the body (Guyton 1999a).

5.1.2 The Circulatory System

The major role of the circulatory system is the distribution of dissolved gases and other molecules for nutrition, growth, and repair (Pries et al. 1996; Boulpaep 2003; den Uil et al. 2008). Delivering oxygen is considered to be the most important function of this system (Duling 1972; Torres et al. 2004a, b). The secondary role includes taking part in the signaling pathway by transferring hormones and neurotransmitters, dissipating heat all over the body, and assisting the immune system by mediating inflammatory and host defense response (Boulpaep 2003; den Uil et al. 2008). By transporting and distributing the blood and its contents via the blood vessels according to the body's needs, the circulatory system contributes to homeostasis (Sherwood 1997a).

Blood flow is a rather ubiquitous phenomenon. Its tendency to remain constant despite changes in arterial perfusion pressure is defined as blood flow autoregulation. There are three essential mechanisms, including the neurogenic, myogenic, and metabolic regulations (Johnson 1986, 1988; Klabunde 2004). The *neuronal mechanism* includes the baroreceptor reflex and the chemoreceptor reflex. A rise in pressure stretches the baroreceptors and causes them to transmit signals into the CNS, then a signal is transmitted back to reduce the pressure toward normal. Whenever the arterial pressure falls, it causes the flow to diminish and the chemoreceptors become activated due to a decrease in oxygen and an excess of other metabolites like CO_2 (Guyton 1999b; Boulpaep and Boron 2003b). The *myogenic hypothesis* states that the small arteries and arterioles respond to intravascular pressure as a stimulus. When a lumen of blood vessels is suddenly expanded following an elevation in the intravascular pressure, the smooth muscle responds by contracting. In addition, vascular resistance is directly related to the ratio of oxygen supply to oxygen demand. In case there is an increase in oxygen demand, the vascular resistance will decrease and blood flow will increase (Duling 1972; Johnson 1986). According to the *metabolic hypothesis*, blood flow and tissue metabolism are tightly coupled; thus, the decrease in the O_2 content of the blood leads to increase in blood flow and vice versa. A reduction in arterial inflow or an increase in the activity of cells causes a buildup of vasodilator metabolites in the tissue (Duling 1972; Johnson 1986, 1988; Klabunde 2004; Singel and Stamler 2005). It is clear that more than one mechanism is involved in autoregulation, depending upon the vascular bed of each organ (Johnson 1986, 1988; Gjedde 2005).

5.1.3 The Importance of Oxygen

Mammalian life and metabolic processes that maintain cellular integrity depend on a continuous supply of oxygen to sustain aerobic metabolism (Grum 1993; Treacher and Leach 1998). The microcirculation must ensure this adequate oxygen delivery (DO_2) to meet the oxygen demands of every cell within an organ (Grum 1993; Ellis et al. 2005). DO_2 is defined as the product of cardiac output (CO) and arterial O_2 content. At normal or high levels of DO_2, O_2 uptake is independent of DO_2 because changes in DO_2 result in reciprocal changes in tissue O_2 extraction ratio (Huang 2005). The oxygen takes a crucial part in the formation of ATP (Adenosine Tri Phosphate) by serving as the terminal electron acceptor (Acker and Acker 2004). Most of the energy is generated at the mitochondrion primarily (above 90%) through oxidative phosphorylation at the electron transfer chain (Boulpaep and Boron 2003b; Rixen and Neugebauer 2004). During this process, NADH (Reduced Nicotinamide Adenine Dinucleotide), which is produced by the oxidation of nutrients, such as glucose, transfers electrons to molecular oxygen via other complexes. The passage of electrons causes energy release that is stored in the form of proton gradient, which is used by the ATP synthase to create ATP from ADP (Adenosine 5′-diphosphate) and phosphate (Saraste 1999). Inadequate DO_2 to the mitochondrion impairs the production of ATP since there is not enough oxygen to receive the electrons at the end of the chain. Therefore, the other components stay reduced. Since NADH is the first component of the chain to transfer electrons, its concentration is elevated in proportion to the decrease in DO_2 (MacKenzie et al. 1979; Grum 1993; Carden and Granger 2000). As a result, the cell's ability to perform chemical and mechanical work is compromised. When DO_2 is severely compromised and gets below the tissue oxygen consumption (VO_2), the cellular metabolism is impaired, leading to acidosis and eventual cell death (Grum 1993; Cryer et al. 2005).

Among the physiological and biochemical activities that are common to all tissues, the most prominent are the different aspects of the cellular and microcirculatory activities regarding energy metabolism and oxygen balance (Leach and Treacher 1998; Mayevsky et al. 2001).

5.1.4 Vital Organ Versus Less Vital Organ

In cases of severe stress to the body, like hypoxemia and hypotension, the metabolic state of the total body will deteriorate and energy failure will develop. As a response, an autoregulation mechanism will be activated toward the redistributions of blood flow to the "most vital organ" such as the brain, heart, and adrenal gland, whereas the "less vital" and peripheral organs and tissues like the gastrointestinal tract, muscle, and skin will undergo vasoconstriction and a decrease in blood flow and oxygen supply will appear. The perfusion of "vital organs" is maintained at the expense of perfusion to the "less vital" organs (Kozuka 1995; Gruhn et al. 2001; Jakob 2002).

5.1.4.1 Cerebral Metabolism and Autoregulation

Although the normal weight of the human adult brain is only 2% of body weight, the cerebral blood flow (CBF) accounts for 15% of the resting CO, reflecting the high cerebral metabolic rate of oxygen. The brain consumes on average 3.3 ml O_2/min/100 g, which is about 20% of the total body consumption (Ganong 1991). The CBF is dynamic and the blood supply changes depending on which brain area is most active in a given moment (Josko and Gwozdz 2003). Cerebral autoregulation is the maintenance of CBF in consideration of the varying Cerebral Perfusion Pressure (CPP) in the range of 50–170 mmHg (Aaslid et al. 1989; O'Sullivan et al. 2007). The brain is especially sensitive to circulatory changes that reduce oxygen and glucose delivery (Johnson 1986; Gruhn et al. 2001). The autoregulation has a rapid response which is initiated within a few seconds after a change in the resistance of vessel transmural pressure occurs and largely completed within 15–30 s (Aaslid et al. 1989; Solenski et al. 2002). During blood pressure changes, the sympathetic nerves, when stimulated, constrict the larger vessels, whereas autoregulation is mainly a function of the smaller resistance vessels (Strandgaard and Paulson 1984). Normal resting CBF is approximately 50 ml/min/100 g brain tissue and is kept constant within a range of 50–160 mmHg of mean arterial pressure (MAP); there is no full agreement regarding the values of the autoregulation limits among the researchers (Novak et al. 1998; Gruhn et al. 2001; Lavi et al. 2003). When arterial blood pressure rises, the cerebral arterioles constrict and when blood pressure falls or intracranial pressure rises, they dilate, keeping CBF constant (Strandgaard 1976). Neuron cell function and viability, and thus structural and functional integrity of the brain, rely heavily on a constant oxygen and glucose supply. A large proportion of the neuronal ATP generated during oxidative phosphorylation is required to maintain ion homeostasis and membrane potential. Any disturbance in this supply becomes life-threatening and may result in severe loss of brain function. In particular, reductions in oxygen availability caused by systemic or local blood circulation irregularities cannot be tolerated for longer periods because the energy supply to the brain provided by anaerobic glycolysis is insufficient (Acker and Acker 2004). A failure of autoregulation is implicated in a variety of diseases, such as stroke and trauma (Aaslid et al. 1989; Payne and Tarassenko 2006). There are two separate thresholds for ischemic neuronal damage. Electrocerebral activity diminishes at flows below 20 ml/min/100 g, whereas membrane failure and cell death occur only when CBF falls below ~10 ml/min/100 g. Values between this range have been correlated with reversible neurological deficits (Powers et al. 1985).

5.1.4.2 Splanchnic Metabolism and Autoregulation

Normally, the total hepatosplanchnic blood flow is approximately 20–30% of CO. The splanchnic oxygen consumption at rest is approximately 20–35% of whole body VO_2. When the metabolic demands of the splanchnic region increase, blood flow to the relevant organs increases in proportion to the increase in regional VO_2 (Takala 1996). Under conditions of low systemic blood flow, hypotension or low CO, the blood flow to the splanchnic region is reduced in proportion to systemic blood flow and perfusion (Jakob et al. 2001, Jakob 2002). If the oxygen extraction is very low, an increase in blood flow will precede in order to increase oxygen extraction. In extreme conditions, like under hypoxia, splanchnic oxygen extraction may increase to close to 90%. Probably because of its important role in the regulation of blood volume (BV) and redistribution of blood flow, the splanchnic region has a high capacity of adaptation to reduced blood flow, which is achieved by its ability to increase oxygen extraction (Boquist 1972). Selective vasoconstriction of the afferent mesenteric arterioles serves to sustain vascular resistance and therefore to maintain arterial blood pressure (Jakob 2002). If blood flow to the splanchnic tissues is sufficiently low, ischemia, tissue damage, and necrosis may develop. In case this low flow condition is kept for long term, it may also cause multiple organ failure and death (Takala 1996; Carden and Granger 2000; Jakob 2002).

5.1.5 Hemorrhage

The average adult BV represents 7% of total body weight. BV varies with age and physiological state. While infants have a BV of 8–9% of body weight, older individuals have less than 7% (Gutierrez et al. 2004). Hemorrhage is defined as loss of circulating volume and is a natural consequence of injury and surgical procedure (Peitzman et al. 1995). Life-threatening decrease in blood pressure is often associated with a state of shock—a condition of inadequate tissue perfusion, in which the delivery of oxygen to tissue and cells is insufficient to maintain normal aerobic metabolism (Gutierrez et al. 2004; Harbrecht et al. 2004). Once a critical threshold in the shock state is reached, dictated by the severity or duration of shock, simple resuscitation is not likely to reverse the pathologic cascade of events and decompensated shock is developed (Peitzman et al. 1995).

During hemorrhage VO_2 is sufficient to maintain function of vital organs, while during hemorrhagic shock VO_2 declines (Fig. 5.1). Throughout severe and sustained hemorrhagic shock, a decrease in DO_2 eventually overwhelm the microvascular response, till the mitochondria cannot sustain aerobic metabolism, this phase is referred to as decompensated shock (Peitzman et al. 1995; Gutierrez et al. 2004). The state in which VO_2 starts to decline is referred to as critical DO_2 (DO_{2crit}) (Gutierrez et al. 2004). Below DO_{2crit}, oxygen extraction cannot increase enough to sustain O_2 uptake and VO_2 becomes supply dependent. Then, the tissue begins to consume energy stores preferentially in the form of phosphocreatine, eventually turning to anaerobic glycolysis to generate ATP, resulting in lactic acidosis (Huang 2005; Torres Filho et al. 2005).

Under conditions of hemorrhage and compensated shock, cellular functions are maintained as long as the combined yield of aerobic and anaerobic sources of energy provides sufficient ATP for protein synthesis and contractile processes.

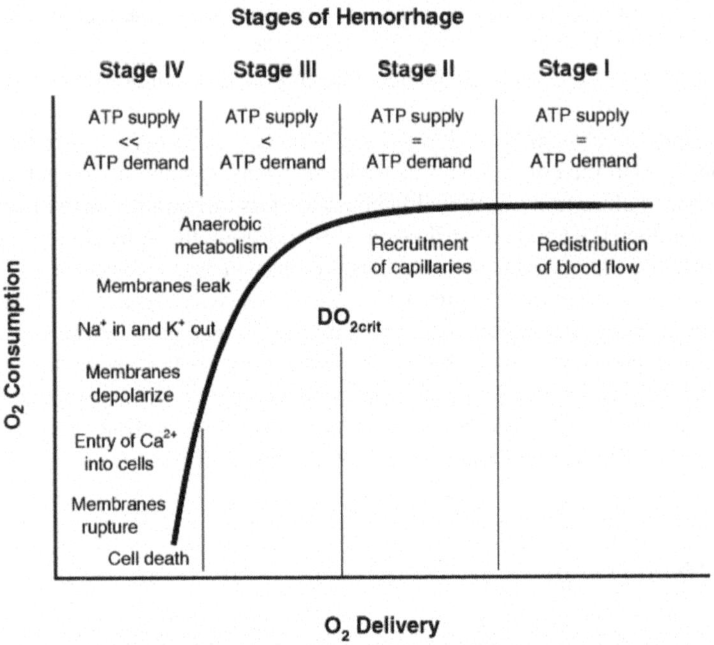

Fig. 5.1 Stages of hemorrhage. Changes in oxygen consumption shown as a function of oxygen delivery and the relationship of these parameters to the different stages of hemorrhage, together with the changes in cellular membrane integrity. DO_{2crit}-critical oxygen delivery. (Reprinted with kind permission of Springer Science + Business Media) (Gutierrez et al. 2004)

Decompensated shock results in irreversible tissue damage occurring when the mitochondrion become depressed and dysfunctional and the combined aerobic and anaerobic supplies of ATP are not sufficient to maintain cellular function (Gutierrez et al. 2004; Payne and Tarassenko 2006).

5.1.5.1 The Regulation Mechanisms Activated by Hemorrhage

The homeostatic response to hemorrhage is initiated by the loss of circulatory blood volume, followed by a decrease in CO and blood pressure. It includes three mechanisms: autonomic nervous system, hormonal system, and local microcirculatory mechanisms (Gutierrez et al. 2004; Harbrecht et al. 2004; Payne and Tarassenko 2006).

Autonomic Nervous System Regulation

The nervous system responds immediately to loss of circulatory blood volume with sympathetically mediated vasoconstriction, which rapidly reduces the capacitance of the circulatory system (Peitzman et al. 1995; Harbrecht et al. 2004). The high-

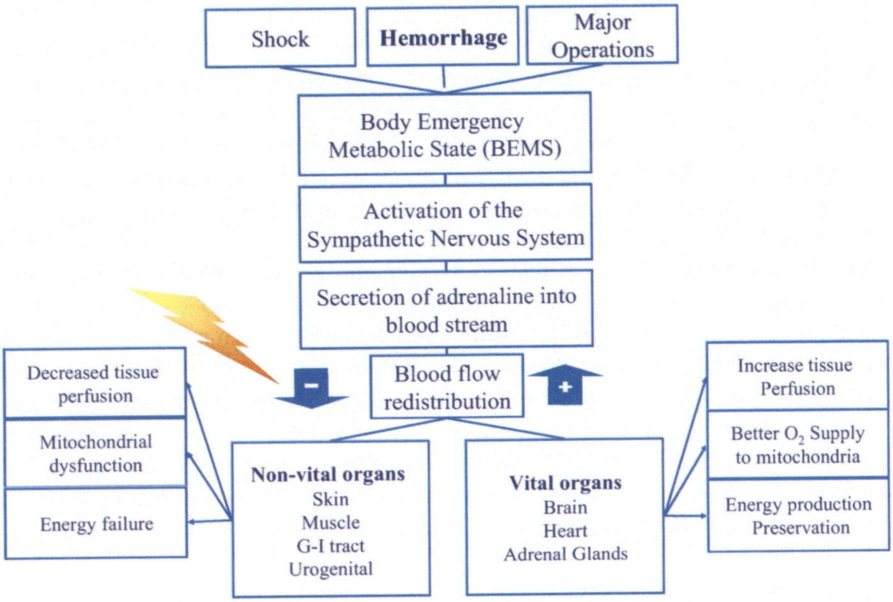

Fig. 5.2 Schematic presentation of body emergency metabolic states (BEMS) developed in various pathological conditions, leading to redistribution of the blood flow toward the most vital organs at the expense of less vital organs. (Modified from Mayevsky and Barbiro-Michaely 2013)

pressure baroreceptors in the aortic arch and carotid sinus respond immediately to a decrease in blood pressure by adjusting the sympathetic tone. The decrease in effective circulating blood volume or blood pressure causes a release of the chronic inhibition of the vasoconstrictor centers in the brainstem imposed by the baroreceptors and increased chemoreceptor stimulation of vasomotor centers (Peitzman et al. 1995). An increased heart rate is caused by the inhibition discharge of the afferent vagal fibers and a release of norepinephrine (NE) from postganglionic sympathetic fibers (Sander-Jensen et al. 1986; Schertel et al. 1991). Further blood loss mediates signals to the hypothalamus, causing a release of corticotropin and the release of vasopressin from the neurohypophsis (Peitzman et al. 1995). The sympathetic response to acute loss of circulatory blood volume has several effects. Arterioles in most vascular beds constrict increasing total peripheral vascular resistance (Peitzman et al. 1995; Torres et al. 2004b) and venous capacitance is decreased, resulting in accelerated venous return to the heart (Peitzman et al. 1995). The aim of these compensatory mechanisms is to increase CO and maintain perfusion pressure of the life-preserving circulation of the "vital organs" at the expense of the "less vital organs" (Fig. 5.2) (Pries et al. 1996; Vagnozzi et al. 1997; Zakaria et al. 2005).

Hormonal Regulation

Acute hemorrhage also initiates hormonal responses. Plasma levels of glucagon, growth hormone, cortisol, and adrenocorticotropic hormone (ACTH) increase (Wattanasirichaigoon et al. 1999; Harbrecht et al. 2004; Payne and Tarassenko

2006). The rennin-angiotensin-aldosterone axis is stimulated with release of the vasoconstrictive hormone angiotensin II (Mayevsky and Ziv 1991; Harbrecht et al. 2004; Payne and Tarassenko 2006; Peng et al. 2006). Similarly, release of vasopressin (antidiuretic hormone) also occurs after hemorrhage, resulting in increased water absorption in the distal tubule of the kidney. Vasopressin also induces splanchnic vasoconstriction (Woolf 1986; Peitzman et al. 1995; Friesenecker et al. 2006). Growth hormone and glucagon oppose the effects of insulin by promoting gluconeogenesis, lipolysis, and glycogenolysis. The resultant hyperglycemia is stimulated further by inhibition of insulin release by epinephrine and NE and by increasing tissue resistance to insulin. The hyperglycemia and increased blood osmolality tend to shift fluid from the cells and interstitium into the intravascular space and thus help maintain circulating volume (Woolf 1986; Peitzman et al. 1995; Harbrecht et al. 2004). The sympathetic vasoconstrictor mechanisms are augmented by these hormonal effects in response to hemorrhage. The summation of the neuroendocrine response is to maximize cardiac function, conserve salt and water for the maintenance of circulating blood volume, and to provide nutrients and oxygen to the most vital organs like the brain and the heart (Peitzman et al. 1995; Harbrecht et al. 2004).

Local Microcirculatory Regulation

Another regulation mechanism in a response to hemorrhage is capillary recruitment. During normal conditions, only a fraction of capillaries are usually open to accommodate the passage of erythrocytes, whereas the remaining capillaries allow only the passage of plasma. During hemorrhage, the number of open capillaries increases in proportion to the degree of tissue hypoxia. Capillary recruitment shortens the diffusion distance from red blood cells to the surrounding tissue and increases the capillary surface area available for oxygen diffusion. The overall effect of capillary recruitment is the maintenance of tissue oxygen flux at a lower capillary oxygen tension, which is a vital response in organs on the edge of hypoxia (Gutierrez et al. 2004; Zhu et al. 2007).

5.1.5.2 The Responses of the Brain and Intestine to Hemorrhage

Cerebral Regulation

The relationship between CPP and cerebral circulation is of paramount importance in situations as diverse as hemorrhagic shock (MacKenzie et al. 1979). During hemorrhage (Fig. 5.2), CBF is well maintained as is cerebral oxygen transport, in part because of the changes in sympathetic outflow. The intense triggering of sympathetic signals is activated when the arterial blood pressure falls below 50 mm Hg and is maximally stimulated when the systolic blood pressure is less than 15 mm Hg (Peitzman et al. 1995). Autoregulation of CBF is thought to maintain essential cerebral perfusion and sustain vital CNS functions (Strandgaard and

Paulson 1984; Peitzman et al. 1995). VO_2 is significantly increased, although DO_2 remains constant due to an increase in oxygen extraction. The increase in VO_2 is thought to be a result of increased excitation of neurons responsible for cardiovascular regulation (Chen et al. 1984). Hemorrhage causes the release of various circulating vasoactive substances, including catecholamines and renin-angiotensin. The catecholamines can cause an increase in cerebral oxygen consumption and blood flow rates, and angiotensin II can stimulate the sympathetic neurons in the brain (Chen et al. 1984; Strandgaard and Paulson 1984). When severe shock is formed, it leads to a decrease in CBF and VO_2 despite the increase in oxygen extraction, leading to cerebral hypoxia and dysfunction and finally irreversible shock syndrome (Chen et al. 1984). Previous studies have found that arteriolar dilation continued as the pressure fell and reached their maximum constriction MAP of 40 mmHg. Under lower levels of MAP, the vessels collapsed (MacKenzie et al. 1979). During hemorrhage, regional differences in CBF occur due to alpha-adrenergic activity (Strandgaard and Paulson 1984) and may contribute to the patchy areas of neurologic injury seen on histologic evaluation after ischemia (Chen et al. 1984; Peitzman et al. 1995). Differences in regional glucose use also occur during hemorrhage, with the brainstem and midbrain manifesting increased glucose use (Chen et al. 1984; Peitzman et al. 1995). Hemorrhage also produces changes in the electrical and functional activity of the nervous system. Electrical activity of the cortex and other areas of the brain is depressed by hypotension. However, these changes are reversible in patients with mild shock, whereas severe shock produces changes that are not reversible when resuscitation to an adequate blood pressure is induced. Similarly, reflex activity and some selected neurological functions are also depressed by hypotension (Peitzman et al. 1995).

Intestinal Regulation

One of the primary reflex adjustments to hemorrhage involves intense total peripheral resistance; however, the extent of increase varies between the different peripheral beds (Fig. 5.2) (Tomlinson et al. 1993; Vajda et al. 2004b). Thus, alterations in splanchnic blood flow occur mainly due to vasoconstriction and redistribution of blood flow from the intestine to the brain and heart (Vatner 1974). These responses are mediated by epinephrine and NE, which are released from the adrenal medulla and through direct effects of local sympathetic activity on blood vessel walls (Peitzman et al. 1995; Payne and Tarassenko 2006). When α-adrenergic receptors on postcapillary mesenteric venules and veins are stimulated, the resulting autotransfusion will improve the performance of the heart by increasing cardiac filling. Selective vasoconstriction of the afferent mesenteric arterioles serves to sustain systemic vascular resistance and therefore to maintain arterial blood pressure (Jakob 2002). The markedly disproportionate response of the mesenteric resistance vessels is largely independent of the sympathetic nervous system and variably related to vasopressin, but mediated primarily by the renin-angiotensin axis (Pries et al. 1996; Jakob 2002; Payne and Tarassenko 2006). The response of the intestinal microvasculature to hemorrhage is complex. Large arterioles constrict, whereas smaller, premucosal arterioles dilate, a response mediated in part by prostaglandins. Thus, vasodilatory prostaglandins are hypothesized to be an important compensatory response of the intestinal vasculature to hemorrhage (Peitzman et al. 1995). The loss of blood volume causes a decrease in DO_2 to peripheral tissues due to a decrease in CO (Torres et al. 2004a). Within the intestine, there is a redistribution of blood flow from tissues with low oxygen demand toward tissues with high oxygen demand. This is an important mechanism for maintaining VO_2 independent from DO_2 levels (Krejci et al. 2000).

5.1.6 Decompensated Hemorrhagic Shock

5.1.6.1 The Systemic Level

During decompensated shock, a progressive vascular vasodilatation and a continuous decrease in peripheral vascular resistance despite augmentation of sympathetic nervous activity occur and are referred to as vasodilatory shock (Pieber et al. 1999; Md et al. 2003; Friesenecker et al. 2006). Decompensated shock can evolve from either hemorrhage with no further resuscitation or followed by resuscitation. The transition from the compensatory to the decompensatory phase of hemorrhagic shock is associated with vascular hyporeactivity to vasoconstrictors, including NE, epinephrine, catecholamines, vasopressin, and angiotensin II (Pieber et al. 1999; Friesenecker et al. 2006). Especially after hemorrhage, the ability of the organism to maintain blood pressure and organ perfusion is limited by a progressive loss of responsiveness of the vasculature to the compensatory release of vasoconstrictors. As a result, blood pressure further declines and becomes refractory to volume substitution, which irreversibly leads to death (Mayevsky and Ziv 1991; Ellis et al. 2005; Peng et al. 2006). The mechanisms involved in peripheral vascular failure are still unclear, though NO has been identified as a major mediator of vascular hyporesponsiveness to adrenoreceptor activation in hemorrhage. However, the impact of NO on other pressor systems such as the renin-angiotensin axis has not yet been revealed. Moreover, it is unknown whether hyporesponsiveness simultaneously and equally affects several vasopressor systems (Pieber et al. 1999; Cuzzocrea et al. 2002; Md et al. 2003; Friesenecker et al. 2006).

5.1.6.2 The Cellular Level—Mitochondrial Response

It has been emphasized that the initial site to be affected following ischemia is the mitochondria (Chance et al. 1973; Jassem et al. 2002; Lu et al. 2005). Deranged metabolism occurring during decompensated shock can lead to a rapid onset in overproduction of ROS, which may destroy the cellular protein machinery and the genome of the host cell. This compromises the ability to restore normal function and repair from damage (Peitzman et al. 1995; Harbrecht et al. 2004). Failure of membrane-associated ion transport pumps (Fig. 5.1), in particular those associated with the regulation of calcium and sodium, results in the loss of membrane integrity and in cellular swelling (Levin et al. 1999; Jassem et al. 2002; Harbrecht et al. 2004; Hubbard et al. 2004; Payne and Tarassenko 2006). If the mitochondrion becomes dysfunctional, there are feed-forward responses such as fission and increased ROS production that have deadly consequences for the host cell (Jassem et al. 2002; Hubbard et al. 2004). If homeostasis is not restored and ATP supplies become insufficient, there can be ripple effects from the compromised mitochondria that may lead to deranged gene expression, inactivation of mediator substances, and loss of function for receptors and signaling pathways (Peitzman et al. 1995; Jassem et al. 2002; Harbrecht et al. 2004; Hubbard et al. 2004).

5.1.7 The Different Experimental Models for Hemorrhage

Extensive efforts have been made to try to clarify the pathophysiologic mechanisms associated with severe hemorrhage. A broad variety of experimental conditions have been established; however, translating these experiments to the clinic is a challenge, since they are barely identical to real-life situations (Rixen and Neugebauer 2004; Lomas-Niera et al. 2005). The uncontrolled hemorrhage, which reflects the clinical setting of a trauma victim, is difficult to be analyzed in research and therefore is less common than controlled hemorrhage. The latter model offers a better management of the degree of shock induced and therefore easier to be analyzed, but is clinically irrelevant since it does not resemble the situation observed in trauma patients (Rixen and Neugebauer 2004; Lomas-Niera et al. 2005). Controlled hypotension is widely used in order to study the metabolic and histographic changes, the survival rates, and the effect of different approaches of therapeutic interventions. The controlled hemorrhage models are based on two methods; hypovolemic hemorrhage and hemorrhagic hypotension (Lomas-Niera et al. 2005; Peng et al. 2006). Hypovolemic hemorrhage is carried out using a fixed volume of blood removed based on the weight of the animals. It allows the assessment of compensatory mechanisms and enables to elucidate the specific hemodynamic response to a fixed volume of blood loss. Nevertheless, the response to a given amount of blood loss differs among individuals, and therefore a large variability may result. Hemorrhagic hypotension is based on the amount of blood needed to obtain in order to hold a fixed pressure for a specified period. It answers the variability problem and enables the regulation of the intensity of hypotensive shock carried out based on physiological and end-organ injury outcome, which is an important consideration in designing therapeutic interventions, though it is less clinically comparable (Rixen and Neugebauer 2004; Lomas-Niera et al. 2005).

The clinical data gathered from the hemorrhage models is heterogeneous and confounded by several effectors. All models require anesthesia, which affects the cardiovascular function and metabolic demand and has diverse effects on the circulating and resident immune cell population (Rixen and Neugebauer 2004; Lomas-Niera et al. 2005; Peng et al. 2006). Heparin is frequently used in order to anticoagulate the catheters and reservoir of shed blood. However, high doses interfere with the normal regulation mechanism. The gender affects the survival rate due to estrogen, which highly activates the immune system and decreases gut injury. Advanced age, on the other hand, increases immune suppression (Rixen and Neugebauer 2004; Lomas-Niera et al. 2005).

5.1.8 Cerebral Ischemia

5.1.8.1 Incomplete Global Cerebral Ischemia

Cerebral blood supply in mammals includes four main vessels: two carotid arteries and two vertebral arteries which, together with communicating arteries, create the circle of Willis (Hendrikse et al. 2001). This special structure is responsible for the continuity of CBF even under incomplete global ischemic conditions (Mayevsky 1976).

Unilateral Carotid Occlusion

Unilateral carotid occlusion (UCO) is one of the models of incomplete global cerebral ischemia which mimics clinical situations, such as stenosis or atherosclerosis, that increase the probability of cerebral insults, such as stroke (Hendrikse et al. 2001; Dieter and Laird 2005). UCO by itself does not appear to immediately affect the brain circulation and does not induce irreversible damage to the cerebral tissue (Hendrikse et al. 2001; Mallick et al. 2004). Blood supply to the entire brain remains intact due to the compensation by the other arteries and the Circle of Willis (Hendrikse et al. 2001; Mallick et al. 2004). UCO appears to produce late effects on the brain, such as over-developed anastomoses (Coyle and Panzenbeck 1990) and metabolic oscillations in the blood volume and mitochondrial NADH (Mayevsky 1984; Mayevsky and Ziv 1991). These phenomena are restricted to the hemisphere ipsilateral to the occluded artery and appear only 24 h after UCO, due to changes in blood circulation and oxygen supply, which occur under various ischemic insults (Mayevsky 1976; Mayevsky and Rogatsky 2007).

Bilateral Carotid Occlusion

As a result of bilateral carotid occlusion (BCO), there is a decrease of around 80% of CBF (Mayevsky 1976). This decrease affects especially the forebrain. However, the blood continues to be supplied to the brain via the vertebral arteries due to the circle of Willis (Tomlinson et al. 1993).

Occlusion of both carotid arteries interrupts the energetic balance in the brain, leading to a decrease in ATP and GTP and an increase in AMP. In addition, a decrease in other metabolites such as adenosine nucleotides and PCr occurs, pointing toward an injury in the energetic metabolism (Vagnozzi et al. 1997). The ATP and

PCr levels continue to decrease in accordance with the length of the ischemic period (Kozuka 1995). The mitochondrial NADH level immediately increases following the occlusion. However, it returns to its basal level a few minutes later, due to the remnant blood supply (Mayevsky et al. 1990). Simultaneously with the increase in ANDH level, an increase in extracellular K^+ occurs (Mayevsky 1984). These phenomena emerge due to a disruption in the integrity of the electron transport chain, leading to a lack of ATP in the tissue (Mayevsky et al. 1990).

5.1.8.2 Focal Cerebral Ischemia

Ischemic stroke is a major cause of death and adult disability in many countries (Zhan and Yang 2006). Focal ischemia is characterized by reduced CBF in the cortex, which leads to disruption in oxygen balance (Sherwood 1997b; Levin et al. 1999), therefore leading to interruption of extracellular ion balance, increased lactate levels, decrease in extracellular pH, accumulation of Glutamate and free radicals, and suppression of spontaneous electrical activity (Brun-Buisson et al. 1996; Takala 1996; Levin et al. 1999).

Focal ischemia in the brain creates a gradual injury, which spreads from severe injury in the center of the lesion, known as the core, toward moderate damage in periphery regions, known as the penumbra (Lipton 1999). The severity of the injury is also known to be influenced by the duration of blood shortage, affecting the damage occurring during the ischemic period, as well as reperfusion injury after restoration of blood supply (Hossmann and Traystman 2009).

A decrease in blood supply leads to the shortage of oxygen in the brain, affecting the intracellular respiratory chain activity in the mitochondria. As a result, ATP synthesis is reduced and mitochondrial NADH levels increase (Mayevsky 1984; Sims and Muyderman 2010). Due to its high sensitivity to intracellular oxygenation, NADH levels can serve as an indicator of mitochondrial function and tissue metabolic state (Mayevsky 1984, 1993).

5.1.8.3 Cell Responses to Ischemia and Reperfusion

Ischemic stress causes cells to lose viability (ischemic injury). Typically, cell death occurs within a few hours, sometimes within minutes. Onset of cell death is the consequence of acute and profound disruption of cellular metabolism, leading most often to ATP depletion, ion disregulation, cellular swelling, and activation of degradative enzymes (Lemasters et al. 1997). The restoration of blood flow following a period of ischemia can place ischemic organs at risk of further cellular necrosis and thereby limit the recovery of function. This situation is defined as ischemic-reperfusion injury (IRI). The reperfusion injury frequently exceeds the original ischemic injury (Carden and Granger 2000; Cerqueira et al. 2005). The endothelial cells appear to be particularly vulnerable to the deleterious effects of both ischemia and reperfusion. Prolonged ischemia is known to alter membrane potential, interrupt the distribution of ions, increase intracellular volume, decrease membrane fluidity, and impair the cytoskeletal organization of endothelial cells (Cerqueira et al. 2005; Lu et al. 2005). These changes can be accompanied by depletion of energy stores, a diminished production of certain bioactive agents, such as prostacyclin and nitric oxide (NO), and an accelerated production of others like endothelin and thromboxane A2 (Jassem et al. 2002; Cerqueira et al. 2005). Likewise, some genes are induced (adhesion molecules and cytokines), while others (cNOS, thrombomodulin) are suppressed in hypoxic endothelial cells (Carden and Granger 2000). Consequently, in the initial period after reperfusion, severe endothelial cell dysfunction occurs rapidly and can develop without obvious morphological cell injury. Those morphological alterations that accompany reperfusion after prolonged ischemia generally include cell swelling, lifting of endothelial cells from the underlying basement membrane, and the adherence of activated leukocytes (mainly neutrophils) to the endothelial cell surface (Chance et al. 1973; Cerqueira et al. 2005). Profound changes in the production of reactive oxygen species (ROS) and NO are observed early after reperfusion (Jassem et al. 2002; Hubbard et al. 2004). A further devastating consequence of tissue reperfusion is the development of damage in organs uninvolved in the initial ischemic insult, known as multiple organ failure (MOF). MOF is associated with intense splanchnic vasoconstriction and relative mesenteric ischemia. Even short periods of enteric ischemia are associated with loss of intestinal mucosal barrier integrity (increased intestinal permeability) and the movement or "translocation" of intestinal microflora and lipopolysaccharide (LPS) from the intestinal lumen to mesenteric lymph nodes and portal blood. During reperfusion, bacterial products and other mediators are released into the circulation and may cause macrophage activation and the expansion of inflammatory cytokines like TNFα, capable of activating vascular endothelium in virtually any organ (Morini et al. 2000; Mallick et al. 2004; Cerqueira et al. 2005). Furthermore, inflammatory mediators released as a consequence of reperfusion may activate circulating leukocytes and vascular endothelial cells, enhance the expression of adhesion molecules, and promote neutrophil–endothelial cell interactions in multiple vascular beds (Carden and Granger 2000; Cerqueira et al. 2005). Usually, MOF is observed within 24–72 h after the initiation of the ischemic event and is expressed by the development of respiratory insufficiency (Carden and Granger 2000).

5.2 Methods

5.2.1 The Multisite-Multiparametric Monitoring System (MSMP)

In this study, the MSMP was used for monitoring two organs simultaneously—the brain and the small intestine (Fig. 5.3).

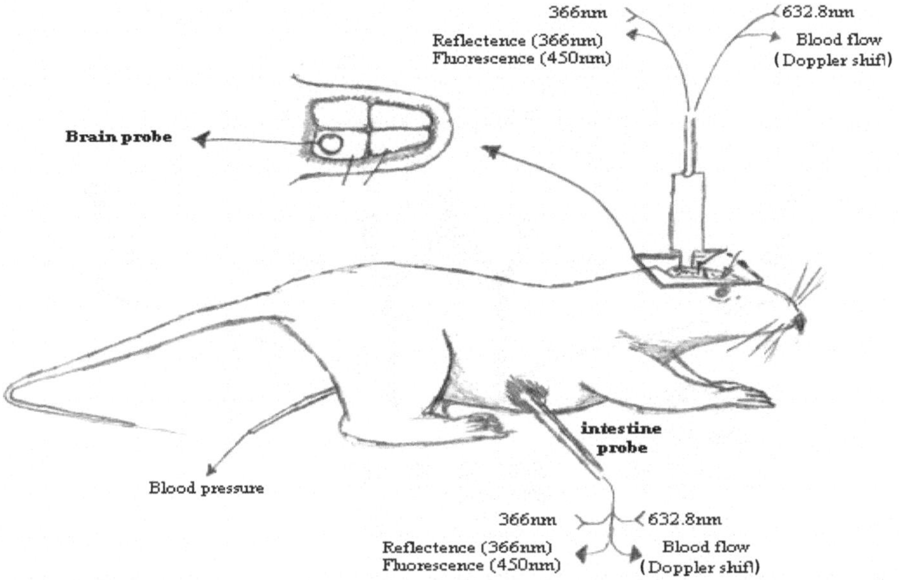

Fig. 5.3 Schematic presentation of the location of the MSMP probes on the brain and the small intestine simultaneously

The MSMP monitors both NADH redox state using the fluorometry technique and tissue blood flow (TBF) using the Laser Doppler Flowmeter (LDF) simultaneously and from the same location on each organ (Fig. 5.4) (Kraut et al. 2004; Barbiro-Michaely et al. 2007).

5.2.1.1 NADH Redox State Fluorometry

The principle of NADH measurements is that excitation light at 366 nm passes through a bundle of optical fibers made of quartz to the tissue. Both the emitted light at 450 nm and the reflected light at 366 nm are transferred back to the fluorometer via another bundle of optical fibers containing appropriate filters (Mayevsky et al. 1999; Kraut et al. 2004; Barbiro-Michaely et al. 2007). The changes in the reflected light are correlated to changes in tissue blood volume and also serve for correction of hemodynamic artifacts appearing in the NADH measurements (Barbiro-Michaely et al. 2007; Mayevsky and Chance 2007; Mayevsky and Rogatsky 2007). The changes in the fluorescence and reflectance signals are calculated relative to the calibrated signals under normoxic conditions (Mayevsky 1993; Mayevsky and Chance 2007). Those changes are presented in a percentage scale of 0–200% where 100% represents the normoxic condition.

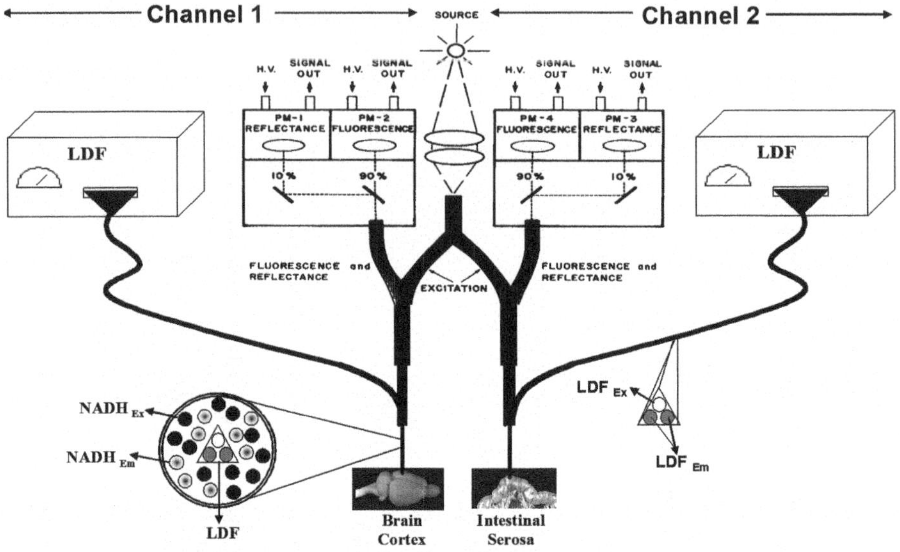

Fig. 5.4 The Multisite-Multiparametric Monitoring System (MSMP). Schematic presentation of the probe containing optical fibers for NADH monitoring (NADH mitochondrial NADH redox state, Ex Excitation, Em Emission, H.V. High Voltage) and for Laser Doppler monitoring (LDF) (Barbiro-Michaely et al. 2007)

5.2.1.2 Microcirculatory Blood Flow

TBF was monitored using a LDF (Perimed® Periflux PF 2B), which passes monochromatic red light at 632.8 nm (from a He-Ne tube) through a bundle of optical fibers (Corbett et al. 2000). The principle underlying LDF is that light undergoes a Doppler shift when it collides with moving red blood cells. The light that returns from tissue is separated into the shifted light that is influenced by flow changes and to the reflected light which serves as an indirect indicator for blood volume (Corbett et al. 2000; Barbiro-Michaely et al. 2007; Mayevsky and Chance 2007). The changes in TBF signals are presented as relative flow values by a percentage scale of 0–200%, while 100% represents the normoxic condition and 0% represents "no flow"—meaning death.

5.2.2 Animal Preparation

All experiments are performed in accordance with the Animal Care Committee of Bar-Ilan University Guidelines. Wister male rats (220–300 g) were anesthetized by an IP injection of 0.3 ml/100 g of Equithesin (each ml contains: pentobarbital 9.72 mg; chloral hydrate 42.51 mg; magnesium sulfate 21.25 mg; propylene glycol

44.34% w/v; alcohol 11.5%; and water). During the entire experiment, the rats were maintained anesthetized by an addition of 0.1 ml Equithesin every half an hour. In addition, a heating pad was placed under the rat to maintain body temperature at 37 ° C, and a rectal thermistor probe (Yellow Springs Instruments Co.Inc.) was inserted for body temperature measurements.

5.2.2.1 Cannulations Preparation

The femoral arteries, one femoral vein, and the tail artery were exposed and cannulated with polyethylene catheters (PE50). Both femoral arteries were cannulated for blood withdrawal. The femoral vein was cannulated for resuscitation. The tail artery was cannulated for blood pressure monitoring, via an UTR-disposable pressure transducer (Biometrix, Israel) which converts the pressure signals into electrical signals.

5.2.2.2 Brain Craniotomy

A midline incision was made in the skin, exposing the skull. Then, three holes were drilled in the skull. The first hole (3.5 mm) was drilled on the parietal bone for fixation of a cannula in which a monitoring probe was inserted. The other two holes (1 mm) were drilled in the frontal bone—distal to the Bregma and from both sides of the central sinus and screws were screwed in for better fixation of the cannula to the skull. Then the whole complex was fixed using dental acrylic cement. The probe was inserted into the cannula, and a black cloth was placed around to avoid room light from entering the monitored site and causing artifacts.

5.2.2.3 Small Intestine Preparation

For the exposure of the small intestine, an abdominal midline incision below the rib cage was created. Then, a small part of the ileum was exposed and placed above a special plastic holder. Afterward, the probe was placed on the ileum's serosa using a micromanipulator, with the assistance of a small piece of parafilm and cyanoacrylate adhesive. Finally, a slice of parafilm was placed around the probe and over the incision to prevent dehydration, and a black cloth was placed above to avoid room light from entering the monitored site and causing artifacts.

5.2.2.4 Middle Cerebral Artery Occlusion (MCAO)

The skull was exposed by a midline incision, and a round hole was drilled in the right parietal bone for a cannula, in which one probe of the MSMP system will be placed, 1.5 mm posterior and 1.5 mm lateral to the Bregma, which according to previous studies is equivalent to the penumbra of the ischemic area (Lipton 1999; Murphy et al. 2008). The cannula was placed epidurally on the right hemisphere and fixated to the skull using dental acrylic cement, together with additional screws for better fixation.

Afterward, the rat was turned over on its back, and right Middle Cerebral Artery Occlusion (MCAO) surgery was performed according to Longa et al. (1989), with several adaptations. A 4/0 nylon filament, with a heated 37 nm diameter bulb head covered with Poly-L-Lysine was inserted through a puncture to the external carotid artery onto the common carotid artery. After 60 min of occlusion, the nylon filament was removed and the tissue was reperfused. The rat was left in a cage with free access to food and water for 24 h before the monitoring began.

5.2.3 *Experimental Protocols*

5.2.3.1 General Protocol

At the beginning of each trial, the rat is being ventilated by 100% N_2 (short anoxia) until breathing is arrested (~30 s) in order to make sure the probes are placed correctly on the various organs, that all parameters act as expected, and the viability of the tissues. At the end of the experiments, the rats were sacrificed by inhalation of 100% N_2.

All permanent perturbations (namely the start point, short anoxia, and death) will be written with no extra styling, while the specific perturbations for each protocol will be written bolded and italic.

5.2.3.2 Development of the Monitoring Model—Serosa and Mucosa Monitoring

Protocol 1—Short Anoxia

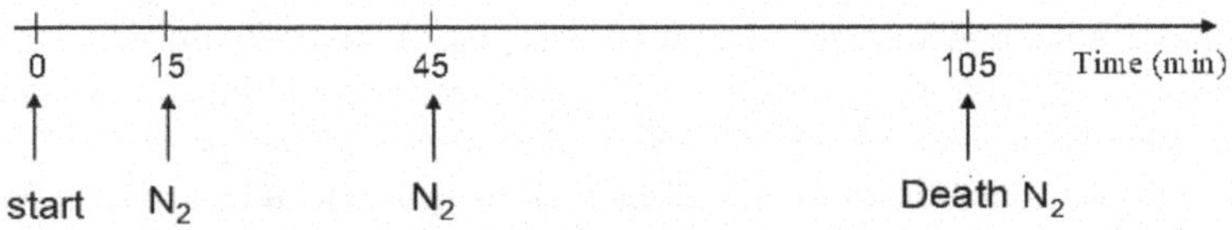

The rat was ventilated with 100% N_2 (anoxia) until breathing was arrested. Then, if the rat did not spontaneously breathe, it was resuscitated by inhalation of air directly to the respiratory airways. Monitoring proceeded for another hour ($n = 7$).

Protocol 2—IV Injection of Epinephrine 6 μg/Kg

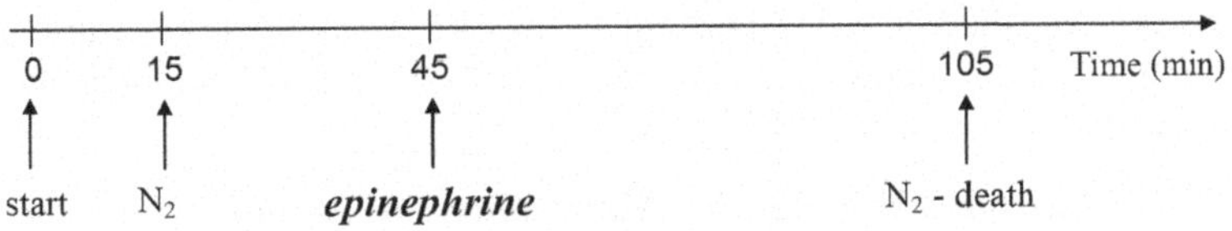

The rat was injected IV with Epinephrine 6 μg/Kg (Epinephrine hydrochloride 4-[1-Hydroxy-2-(methylamino) ethyl]-1,2-benzenediol) $C_9H_{13}NO_3 \cdot HCL$, N-51K1096, Sigma) and was further monitored for another hour ($n = 7$).

5.2.3.3 Hemorrhage Models—Brain and Small Intestine Monitoring

Protocol 3—Normotensive Group

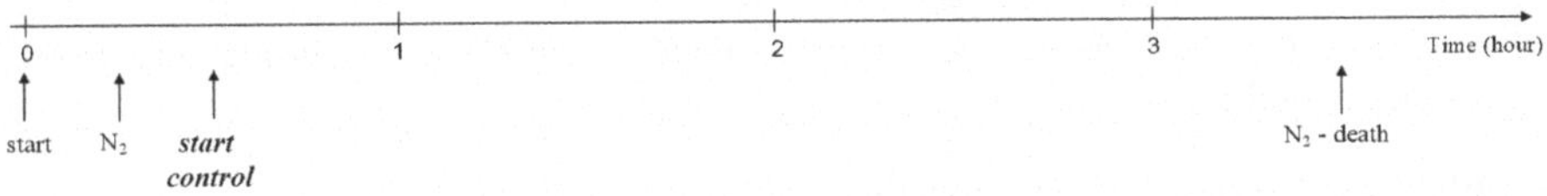

The rats were operated and then were monitored for 3 h with no perturbation being carried out ($n = 4$).

Protocol 4—Uncontrolled Hypotension

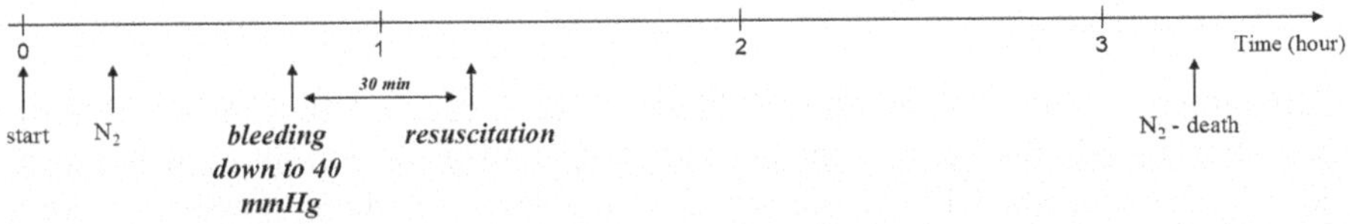

Blood was rapidly withdrawn until blood pressure reached 40 mmHg (~2 min). Then after, there was a period of 30 min, in which no perturbation was done. During resuscitation, the shed blood was infused back to the rat (~3 min') and monitoring proceeded for another 2 h ($n = 9$).

Protocol 5—Controlled Hypotension Maintained for 15 min

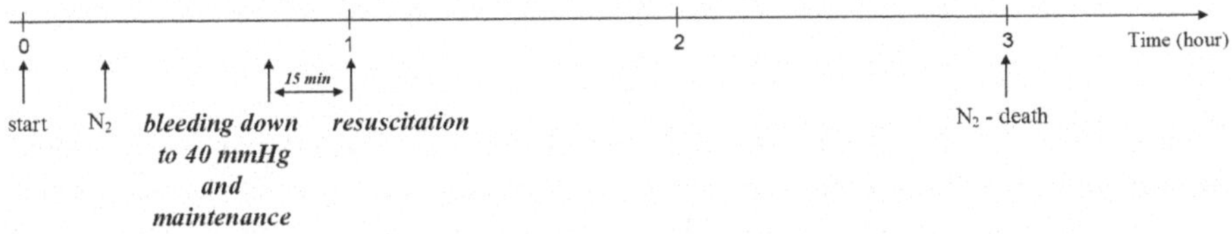

Blood was rapidly withdrawn until blood pressure reached 40 mmHg (~2 min). Then after, blood pressure was maintained at 40 mmHg for 15 min by withdrawal and reinfusion of shed blood according to the level of blood pressure. During resuscitation, the shed blood was reinfused back to the rat (0.1 ml/min) and monitoring proceeded for another 2 h ($n = 9$).

Protocol 6—Controlled Hypotension Maintained for 30 min

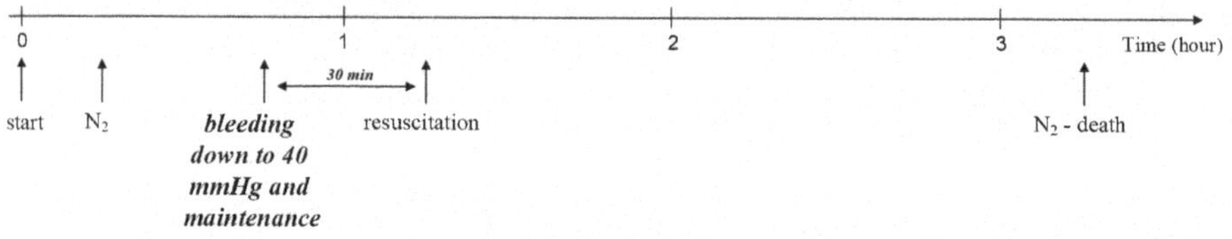

Blood was rapidly withdrawn until blood pressure reached 40 mmHg (~2 min). Then after, blood pressure was maintained at 40 mmHg for 30 min by withdrawal and reinfusion of shed blood according to the level of blood pressure. During resuscitation, the shed blood was reinfused back to the rat (0.1 ml/min) and monitoring proceeded for another 2 h ($n = 12$).

Protocol 7—Controlled Hypotension Maintained for 60 min

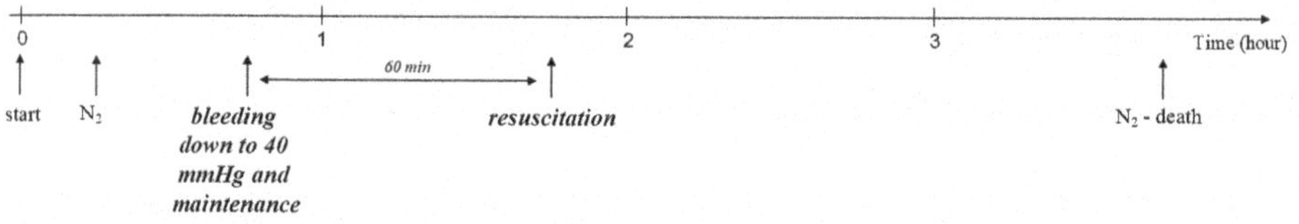

Blood was rapidly withdrawn until blood pressure reached 40 mmHg (~2 min). Then after, blood pressure was maintained at 40 mmHg for 60 min by withdrawal and reinfusion of shed blood according to the level of blood pressure. During resuscitation, the shed blood was reinfused back to the rat (0.1 ml/min) and monitoring proceeded for another 2 h ($n = 13$).

Protocol 8—Graded Hemorrhage in 10-minute Intervals Maintained for 30 min

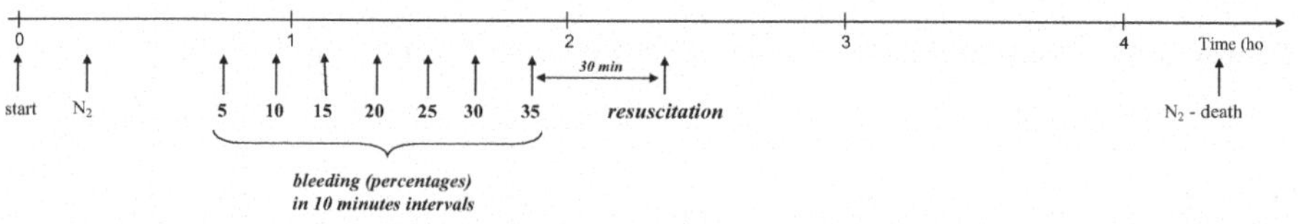

Five percent of the rat's initial total blood volume was withdrawn in 10-minute intervals until reaching a blood pressure of 40 mmHg. Afterward, the rat was left without any perturbation for 30 min. Then after, during resuscitation, the shed blood was reinfused back to the rat (0.1 ml/min) and monitoring proceeded for another 2 h ($n = 9$).

Protocol 9—Graded Hemorrhage in 5-minute Intervals Maintained for 30 min

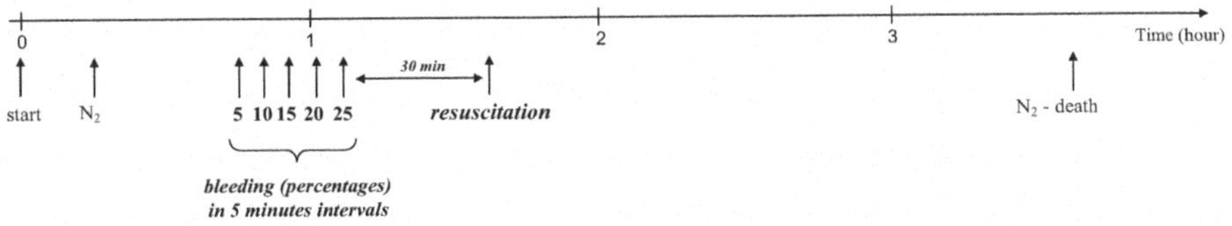

Five percent of the rat's initial total blood volume was withdrawn in 5-minute intervals until reaching a blood pressure of 40 mmHg. Afterward, the rat was left without any perturbation for 30 min. Then after, resuscitation began, the shed blood was reinfused back to the rat (0.1 ml/min) and monitoring proceeded for another 2 h ($n = 10$).

5.2.3.4 Resuscitation Models—Brain and Small Intestine Monitoring

Protocol 10—Controlled Hypotension Maintained for 15 min Followed by a Saline Resuscitation

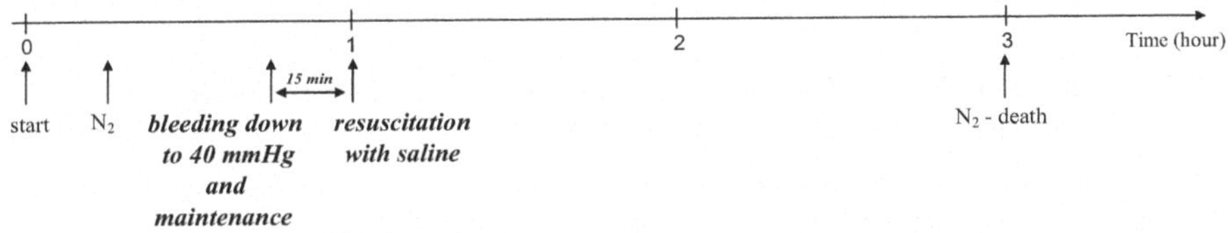

Blood was rapidly withdrawn until blood pressure reached 40 mmHg (~2 min). Then after, blood pressure was maintained at 40 mmHg for 15 min by withdrawal and reinfusion of shed blood according to the level of blood pressure. During resuscitation, the normal saline was reinfused back to the rat (0.1 ml/min) and monitoring proceeded for another 2 h ($n = 12$).

Protocol 11—Controlled Hypotension Maintained for 30 min Followed by a Shed Blood and Saline Resuscitation

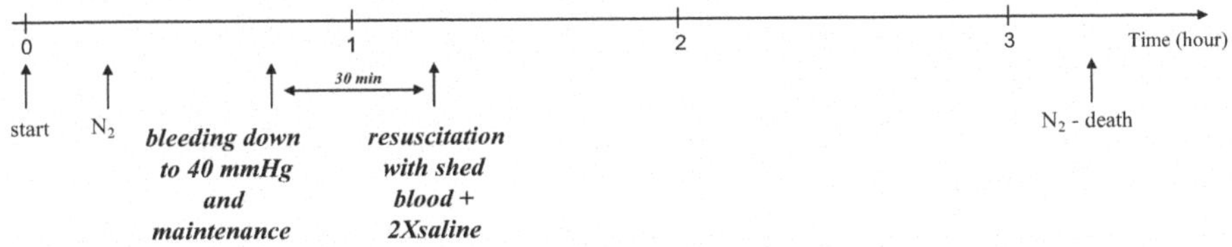

Blood was rapidly withdrawn until blood pressure reached 40 mmHg (~2 min). Then after, blood pressure was maintained at 40 mmHg for 15 min by withdrawal and reinfusion of shed blood according to the level of blood pressure. During resuscitation, the shed blood + 2X normal saline were reinfused back to the rat (0.1 ml/min) and monitoring proceeded for another 2 h ($n = 12$).

Protocol 12—Controlled Hypotension Maintained for 15 min Followed by Blood Substitute Resuscitation

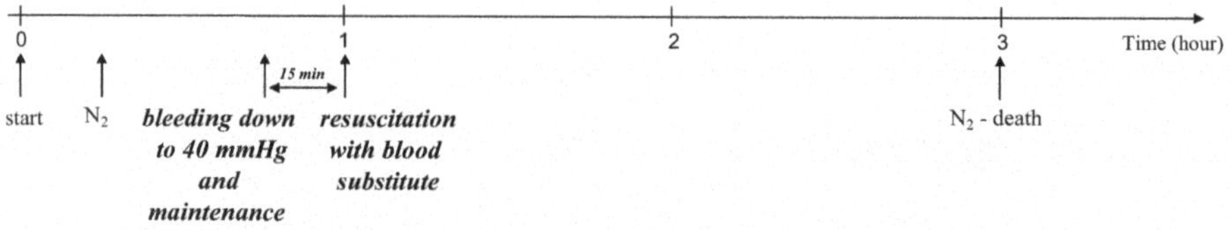

Blood was rapidly withdrawn until blood pressure reached 40 mmHg (~2 min). Then after, blood pressure was maintained at 40 mmHg for 15 min by withdrawal and reinfusion of shed blood according to the level of blood pressure. During resuscitation, a blood substitute (Glutaraldehyde-polymerized porcine hemoglobin) was reinfused back to the rat (0.1 ml/min) and monitoring proceeded for another 2 h ($n = 12$).

The blood substitutes components: hemoglobin concentration ~105–115 mg/mL; methemoglobin <5%; colloid osmotic pressure 20 ± 3 mmHg; P50 – 40 ± 5 mmHg; mean molecular weight 550–850 kDa; pH 7.35–7.45; electrolyte osmotic pressure: 285–300 mOsm.

5.2.3.5 Hemorrhage Under Cerebral Ischemia—Brain and Small Intestine Monitoring

Protocol 13—Unilateral Carotid Occlusion Control Group

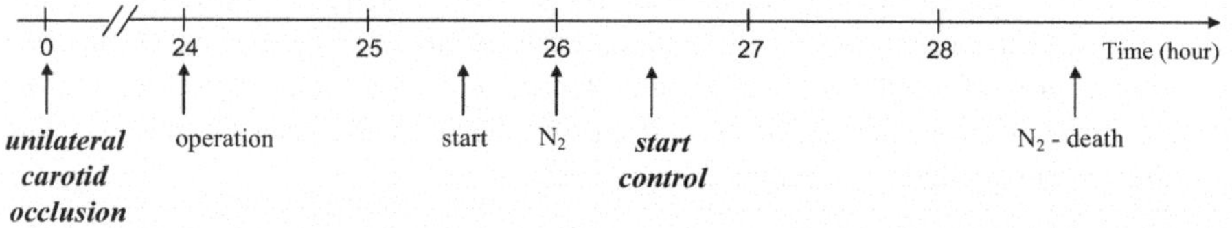

Twenty-four hours before the experiment was initiated, unilateral carotid occlusion was carried out, and the rat was left to recover with free access to water and food. On the day after, the rat was operated and monitored continuously for 3 h with no further perturbation ($n = 4$).

Protocol 14—Controlled Hypotension Maintained for 15 min Under Unilateral Carotid Occlusion

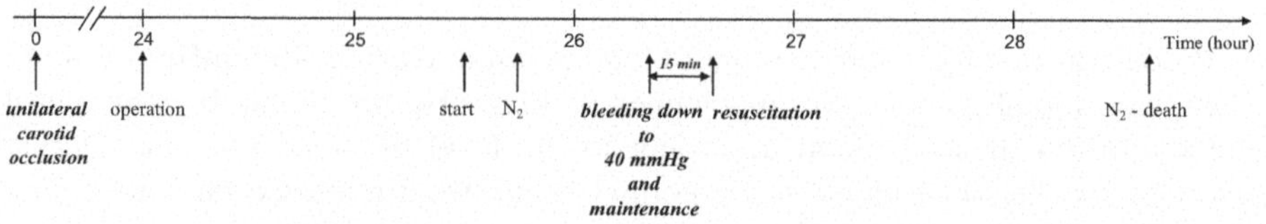

Twenty-four hours before the experiment was initiated Unilateral carotid artery occlusion (UCO) was carried out and the rat was left to recover with free axcess to water and food. On the next day, after the operation and short anoxi blood was rapidly withdrawn until blood pressure reached 40 mmHg (~2 min). Then after, blood pressure was maintained at 40 mmHg for 15 min by withdrawal and reinfusion of shed blood according to the level of blood pressure. During resuscitation, the shed blood was reinfused back to the rat (0.1 ml/min) and monitoring proceeded for another 2 h ($n = 9$).

Protocol 15—Bilateral Carotid Occlusion Control Group

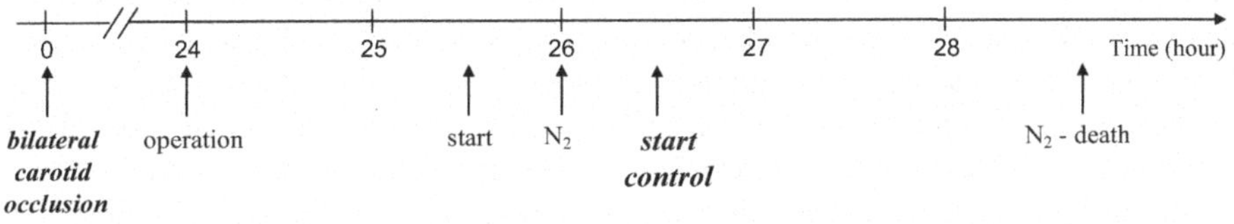

Twenty-four hours before the experiment was initiated, both carotid arteries were occluded and the rat was left to recover with free axcess to water and food. On the day after, the rat was operated and monitored continuously for 3 h with no further perturbation ($n = 4$).

Protocol 16—Controlled Hypotension Maintained for 15 min Under Bilateral Carotid Occlusion

Twenty-four hours before the experiment was initiated, both carotid arteries were occluded and the rat was left to recover with free axcess to water and food. On the next day, after the operation and short anoxia, blood was rapidly withdrawn until blood pressure reached 40 mmHg (~2 min). Then after, blood pressure was maintained at 40 mmHg for 15 min by withdrawal and reinfusion of shed blood according to the level of blood pressure. During resuscitation, the shed blood was reinfused back to the rat (0.1 ml/min) and monitoring proceeded for another 2 h ($n = 9$).

Protocol 17—Middle Cerebral Artery Occlusion Control Group

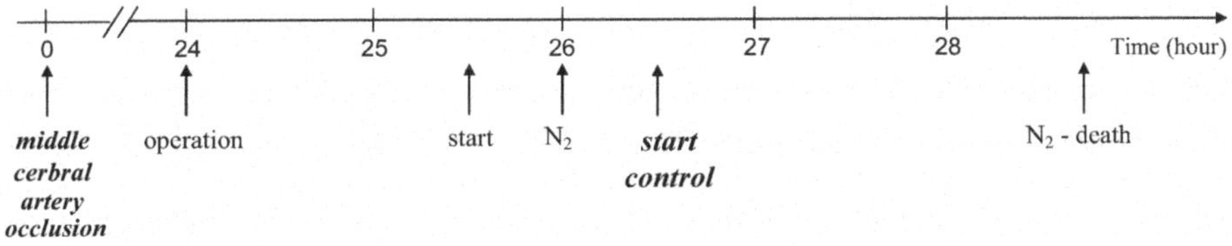

Twenty-four hours before the experiment was initiated, the middle cerebral artery was occluded and the rat was left to recover with free axcess to water and food. On the day after, the rat was operated and monitored continuously for 3 h with no further perturbation ($n = 4$).

Protocol 18—Controlled Hypotension Maintained for 15 min Under Middle Cerebral Artery Occlusion

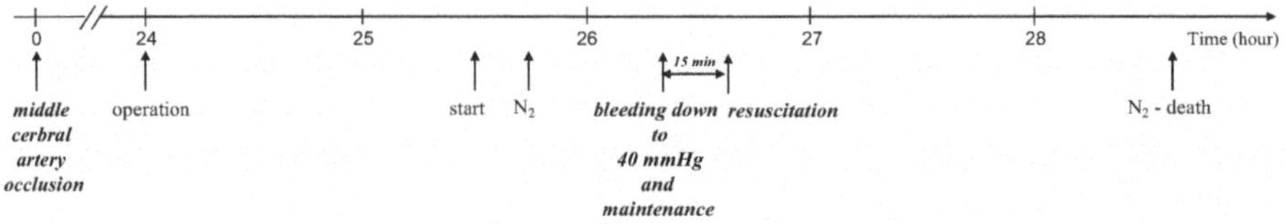

Twenty-four hours before the experiment was initiated middle cerebral artery was occluded, and the rat was left to recover with free axcess to water and food. On the next day, after the operation and short anoxia, blood was rapidly withdrawn until blood pressure reached 40 mmHg (~2 min). Then after, blood pressure was maintained at 40 mmHg for 15 min by withdrawal and reinfusion of shed blood according to the level of blood pressure. During resuscitation, the shed blood was reinfused back to the rat (0.1 ml/min) and monitoring proceeded for another 2 h ($n = 7$).

5.2.4 *Data Collection and Analysis*

5.2.4.1 Data Collection

The data recorded simultaneously from nine channels during the experiment was transferred to a multichannel computerized acquisition and recording system (Labview A/D software, National Instruments Co., USA) and to a Grass polygraph paper recorder (7D model, Grass Instruments) for further analysis. The data for TBF, reflectance, and NADH were received in a frequency of 100 Hz, from which 120 values per second were sampled. The data was converted from analog to digital form. The data from the Labview was calculated and transferred to Excel (Microsoft, USA) for further analysis. The data for MAP was sampled by 120 values per minute.

5.2.4.2 Data Processing

The raw data recorded in both the computer and the polygraph were collected from all the animals and sorted by parameters. The values of NADH, TBF, and reflectance were converted from mV to percentages. For both TBF and NADH, the baseline was calibrated to 100%, though for TBF, 0% was referred to as no flow (death) while for NADH, a range of 500 mV was referred to as a 100% change. The MAP was converted from mV to mmHg.

5.2.4.3 Statistical Analysis

All statistical analysis was performed in coordination with a professional statistician. The statistical analysis included Student's T-test, ANOVA analysis, and Pearson correlation. The paired two-tailed Student's T-test was used in order to evaluate the significance of change for each parameter regarding its basal level and the differences between the two organs in the same group for each minute. The unpaired two-tailed Student's T-test was used to evaluate the significance of differences between the responses of the brain and small intestine in the different groups. One-way ANOVA analysis was used in order to evaluate the differences between the maximum responses of several parameters in the various models. In case of significant difference, the Fisher post hoc test was carried out. ANOVA repeated measures analysis was used in order to evaluate the differences between the responses of the parameters in the various models during a specific period. An unpaired two-tailed Student's T-test was carried out for each minute during that period in order to identify those significant episodes. ANCOVA analysis was carried out in order to evaluate the differences between the bleeding periods of the various models. Pearson correlation coefficient analysis was used in order to find the relationship between the different parameters during the various periods of the experiment. Three levels of significance were determined $p < 0.05$, $p < 0.01$, and $p < 0.001$, and they were marked by 1–3 stars, respectively.

5.3 Results

5.3.1 Development of the Monitoring Model

During the development of the hemorrhage model, the question arose where to place the probe for intestinal monitoring. To resolve it, we tested two experimental models, in which one probe was placed on the serosa and the other on the mucosa. The first experimental model involved short anoxia, i.e., inhalation of 100% N_2 until the rat stopped breathing. The second model involved IV injection of 6 μg/kg epinephrine. In both models, the responses of the serosa and mucosa were compared.

5.3.1.1 Responses of Serosa and Mucosa to Short Anoxia

Figure 5.5 presents the responses of the serosa and mucosa to short anoxia. MAP increased, and both layers of the intestine, the serosa and mucosa, responded by a decrease in TBF and increase in NADH. No significant differences were observed between the two layers.

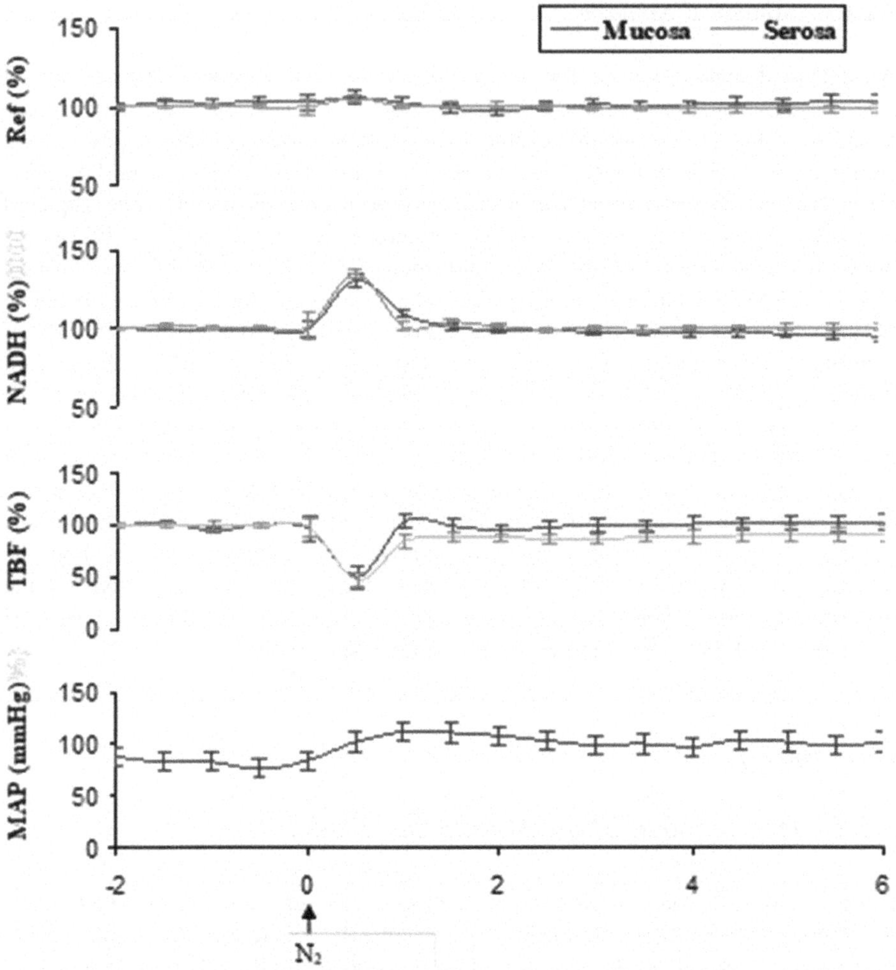

Fig. 5.5 The responses of the MAP, mucosa (dark gray) and serosa (light gray) during short anoxia ($n = 8$, mean ± S.E). Ref reflectance, NADH mitochondrial NADH redox state, TBF tissue blood flow, MAP mean arterial pressure

5.3.1.2 Responses of Serosa and Mucosa to Epinephrine Injection

Figure 5.6 represents the responses of the serosa and mucosa to IV injection of epinephrine. As it is seen, both intestinal layers responded similarly to epinephrine. MAP increased, TBF decreased, and both the reflectance and NADH increased. Furthermore, no significant differences were observed between the two layers.

According to these results, showing no significant differences between the serosa and mucosa, we decided to place the intestinal probe on the serosa, since it is less invasive and easier to manipulate.

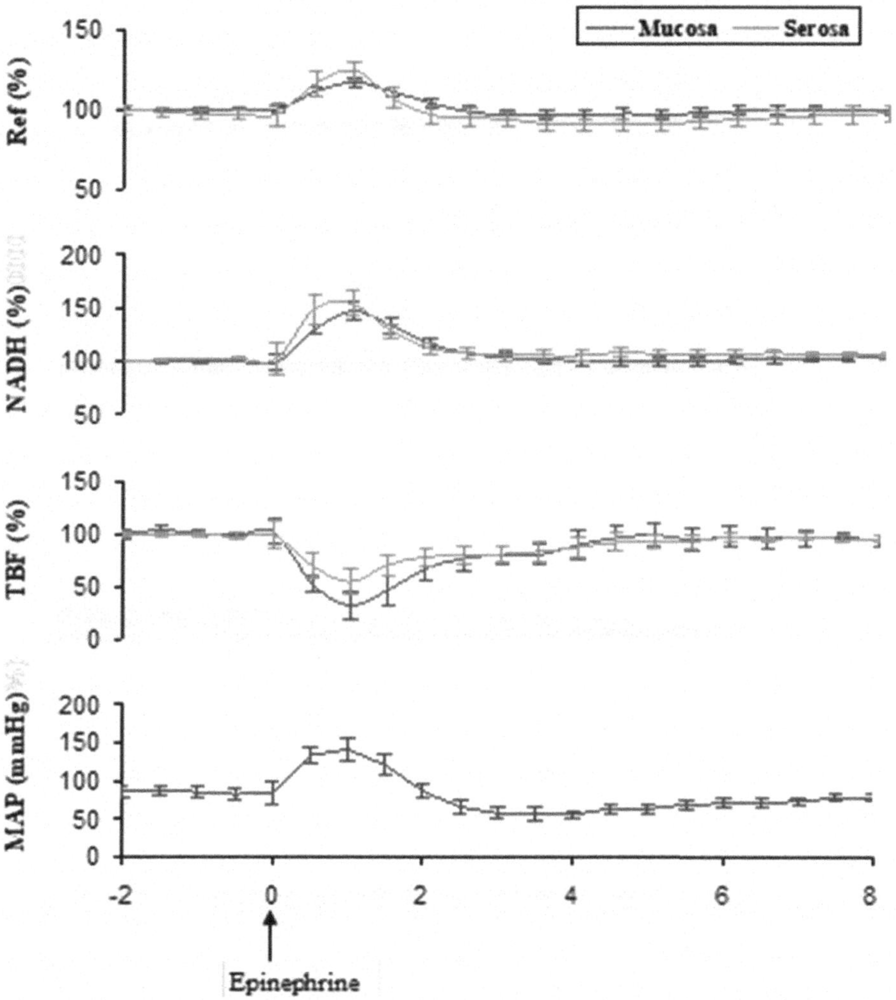

Fig. 5.6 The responses of the MAP, mucosa (dark gray) and serosa (light gray) during injection of epinephrine ($n = 8$, mean ± S.E). Ref reflectance, NADH mitochondrial NADH redox state, TBF tissue blood flow, MAP mean arterial pressure

5.3.1.3 Monitoring of a Vital and Less Vital Organ in Normotensive Rats

Figure 5.7 represents the responses of a normotensive group (control monitoring of the brain and the intestine), including four rats. This group was monitored for 180 min, with no perturbations induced during the entire monitoring period. For the entire trial, there were no significant changes in any of the parameters, or only every minute ones in comparison to the baseline and between the two organs.

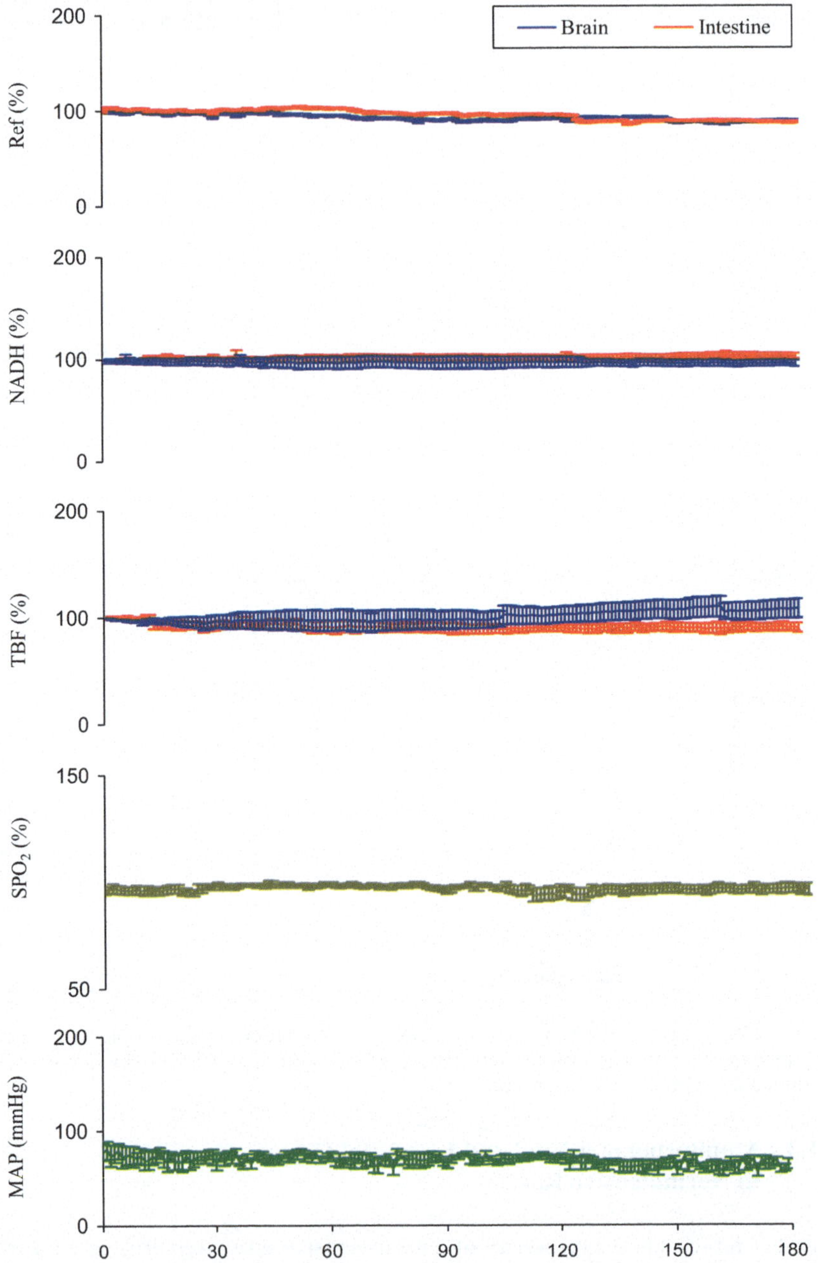

Fig. 5.7 The changes in all the parameters in both organs: the brain (blue) and the small intestine (red) in the normotensive group ($n = 4$). Ref reflectance, NADH mitochondrial NADH redox state, TBF tissue blood flow, MAP mean arterial pressure (green)

5.3.2 *Uncontrolled Hemorrhage*

5.3.2.1 Effects of Uncontrolled Hemorrhage on Brain and Intestine

Analogical results (figure not shown) of an experiment where 40 mmHg hypotension was obtained, followed by no further perturbations for 30 min. This protocol will be henceforth termed *uncontrolled hypotension.* This level of hypotension was induced by bleeding out 6.51% of the rat's total blood volume.

As observed during the uncontrolled bleeding phase, MAP decreased sharply to 40 mmHg followed by several recovery steps until reaching basal levels within 10 min. The brain and the intestine showed different responses to the hypotension state. The intestine immediately responded to the bleeding in all of the parameters, whereas the brain showed only slight changes. TBF in the intestine decreased immediately after bleeding started. Then, after bleeding had stopped, intestinal blood flow (IBF) recovered to hyperemic levels and later returned to the initial level. After these changes, the reflectance and NADH increased, followed by a decrease to basal levels. The brain responded quite differently: TBF was elevated slowly in the course of time, while the reflectance slightly decreased over time and NADH remained relatively stable.

During resuscitation, the MAP increased above the basal level and remained almost completely stable around this level until the end of the experiment. The intestinal responses to resuscitation exhibited a further recovery of all the parameters toward the basal levels. TBF and reflectance returned to the basal levels, and NADH slightly decreased, followed by full recovery. In the brain, TBF continued to increase until stabilizing at high values. In contrast, the reflectance further decreased, and only after about 15 min, started to increase above the basal levels. NADH remained stable during the resuscitation, and decreased slightly 15 min afterward.

Figure 5.8 represents the responses of nine rats to uncontrolled hypotension for 30 min. In this group, hypotension of 40 mmHg was achieved by bleeding out $12 \pm 3\%$ of the rat's total blood volume.

During the bleeding, MAP immediately decreased (44 ± 5 mmHg, $p < 0.01$) and remained significantly low for 10 min, followed by full recovery within the hemorrhage phase. The intestinal responses were greater and faster than those of the brain. After bleeding, TBF in the intestine immediately decreased by $21 \pm 5\%$ ($p < 0.01$) and later almost fully recovered. In contrast, TBF in the brain increased by $28 \pm 13\%$ ($p < 0.05$) only several minutes after the hypotension induction and remained high for the entire period. Intestinal reflectance increased by $8 \pm 3\%$ ($p < 0.05$) and then decreased back to its basal level while the cerebral reflectance remained stable. Intestinal NADH immediately increased ($13 \pm 5\%$, $p < 0.05$), followed by full recovery, and cerebral NADH did not change significantly.

During resuscitation and afterward, MAP remained relatively stable. TBF in the intestine transiently recovered, reaching the high levels of $115\% \pm 9$ followed by a gradual decrease by $35.5 \pm 10\%$ to very low levels. In comparison, TBF in the brain

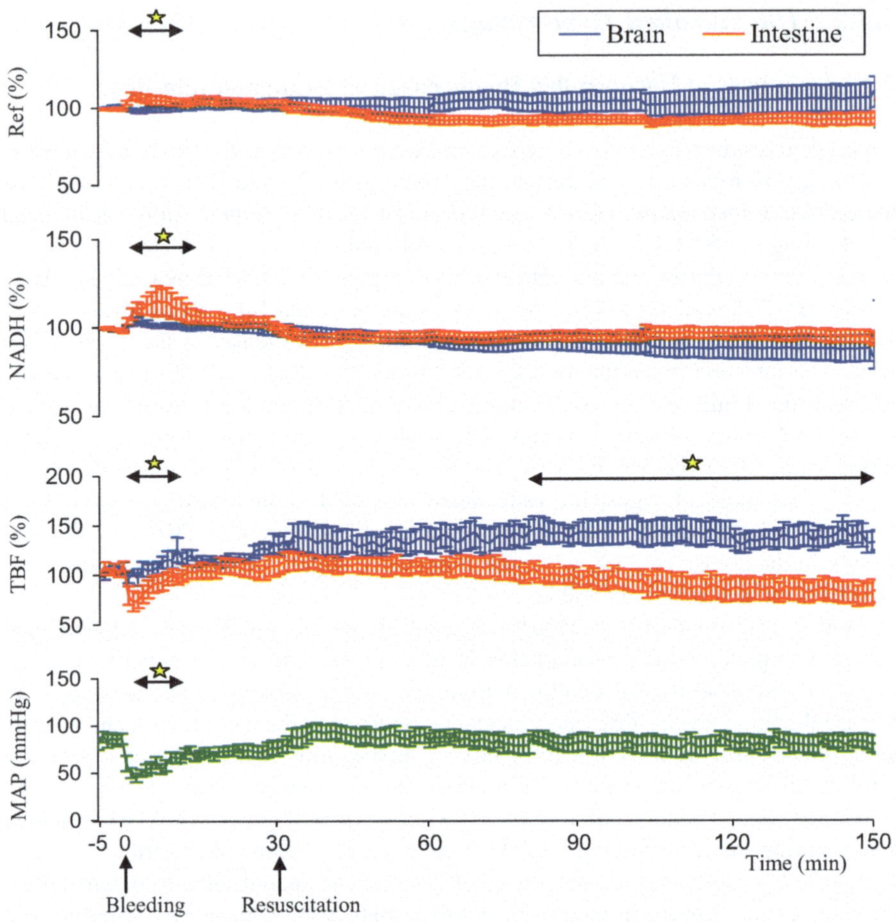

Fig. 5.8 The responses of the brain (blue), small intestine (red), and MAP (green) to uncontrolled hypotension for 30 min ($n = 9$, mean ± S.E). The arrows indicate the period in which significant differences were found between the two organs for each minute. The asterisks show the level of significance: $*p < 0.05$. Ref reflectance, NADH mitochondrial NADH redox state, TBF tissue blood flow, MAP mean arterial pressure

slowly increased, reaching a level of 147 ± 0.12%. About 20 min after resuscitation, the intestinal reflectance decreased by 11 ± 2% ($p < 0.05$) and remained significantly low for an hour, while the cerebral reflectance remained stable. However, NADH in both organs remained stable, with no significant changes.

The comparison between the organs reveals significant differences in all of the parameters during the bleeding session ($p < 0.05$). TBF exhibited major differences between the organs during the bleeding session and 1 h after resuscitation ($p < 0.05$).

5.3.2.2 Uncontrolled Hemorrhage Compared to Normotensive Conditions

For the comparison of the hypotensive and normotensive groups (no perturbation), 5 points were selected within the experimental period for each group, as follows: The first point is the basal level, the second is the end of the hypotension period (maximum response to hypotension), the third is when MAP reaches its maximal level after resuscitation (end of reperfusion), the fourth occurs 1 h after resuscitation (rec_0), and the fifth is 2 h after resuscitation (rec_1).

Figure 5.9 presents a comparison of the MAP, brain, and intestinal responses, to uncontrolled hypotension and normotension. For this group, the second point was replaced with the last point of 40 mmHg, since in this group, after reaching 40 mmHg, MAP gradually increased and all the other parameters responded within the hypotension period.

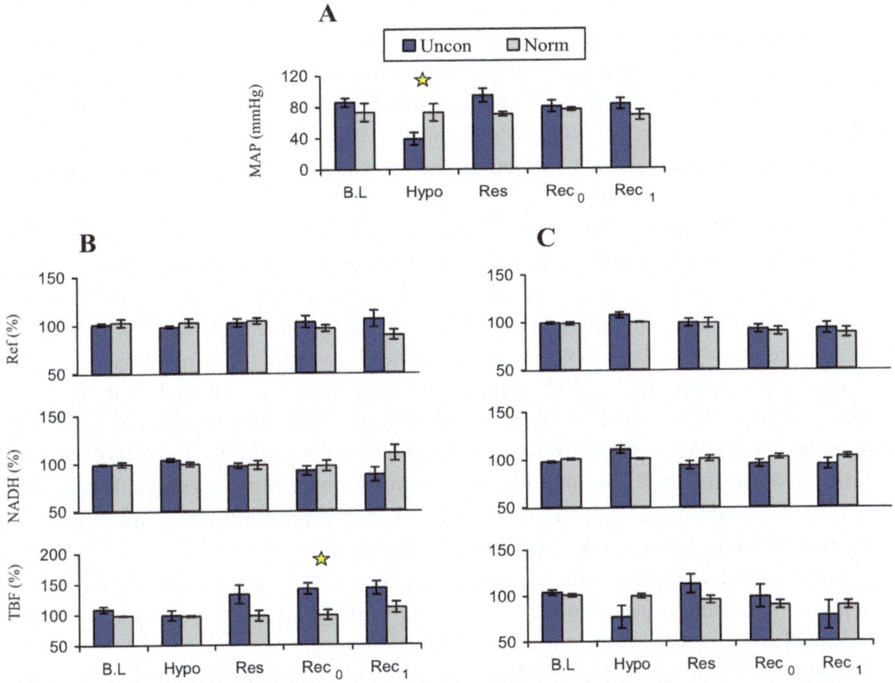

Fig. 5.9 Comparison of the MAP (**a**), cerebral (**b**), and intestinal (**c**) responses in the uncontrolled hypotension group (Uncon-purple, $n = 9$) and normotensive group (Norm-gray, $n = 4$, mean ± S. E). The asterisk represents the level of significance—$p < 0.05$. B.L. basal level, Hypo end of the hypotension period, Res end of resuscitation, Rec_0 1 h following resuscitation, Rec_1 2 h following resuscitation, Ref reflectance, NADH mitochondrial NADH redox state, TBF tissue blood flow, MAP mean arterial pressure

Significant differences between the uncontrolled hypotension group and the normotensive group were observed in MAP (Fig. 5.5a) during the hypotension period, with lower levels shown in the uncontrolled hypotension group ($p < 0.05$).

CBF (B) in the uncontrolled hypotension group increased, yet, only an hour after resuscitation, it was significantly different from the TBF level in the normotensive group ($p < 0.05$). No significant differences between the groups were found in either cerebral reflectance or NADH.

None of the intestinal parameters demonstrated significant differences between the two groups (C).

5.3.3 Controlled Hemorrhage

5.3.3.1 Controlled Hemorrhage Maintained for 15 Min

Comparison Between Brain and Intestine

Typical results of controlled hypotension maintained for 15 min (figure not shown). The level of 40 mmHg hypotension was achieved due to bleeding of 29% of the rat's total blood volume.

Following bleeding, MAP immediately decreased to the level of 40 mmHg, showing several attempts of recovery, which were blocked by another session of bleeding. The SpO_2, on the other hand, was not altered. In both organs, TBF decreased following bleeding, yet in the intestine for a greater amount. The cerebral reflectance decreased, while the intestinal reflectance increased. Both cerebral and intestinal fluorescence increased. However, the change in the intestine was greater. Nevertheless, the NADH in both organs presented a very similar response.

Following resuscitation, MAP fully recovered, reaching higher levels than the baseline, while SpO_2 remained stable. Both organs' TBF fully recovered, reaching hyperemic levels. While the CBF remained stable for the entire trial, the IBF showed a short decrease followed by an additional increase reaching hyperemic levels, and then it slowly decreased back to the basal level. All other parameters, as well, fully recovered and remained stable till the end of the trial.

Figure 5.10 demonstrates the responses of nine rats to controlled hypotension, which was maintained for 15 min. MAP of 40 mmHg was achieved by an average withdrawal of 30 ± 2% of the rats' total blood volume.

During the hypotension phase, MAP decreased reaching the low level of 39 ± 3% mmHg ($p < 0.005$), whereas SpO_2 was not altered. Both the cerebral and intestinal TBF decreased, reaching the low levels of 82 ± 2% ($p < 0.01$) and 64 ± 5% ($p < 0.005$), respectively. The cerebral reflectance was not altered while the intestinal reflectance increased, reaching the high level of 111 ± 2% ($p < 0.01$). Both organs' fluorescence increased, the brain by 9 ± 3% ($p < 0.05$) and the intestine by 28 ± 8 ($p < 0.05$). In accordance with that, the NADH in both organs increased as well, reaching high levels of 111 ± 4% ($p < 0.05$) in the brain and 118 ± 8% ($p < 0.05$) in the intestine.

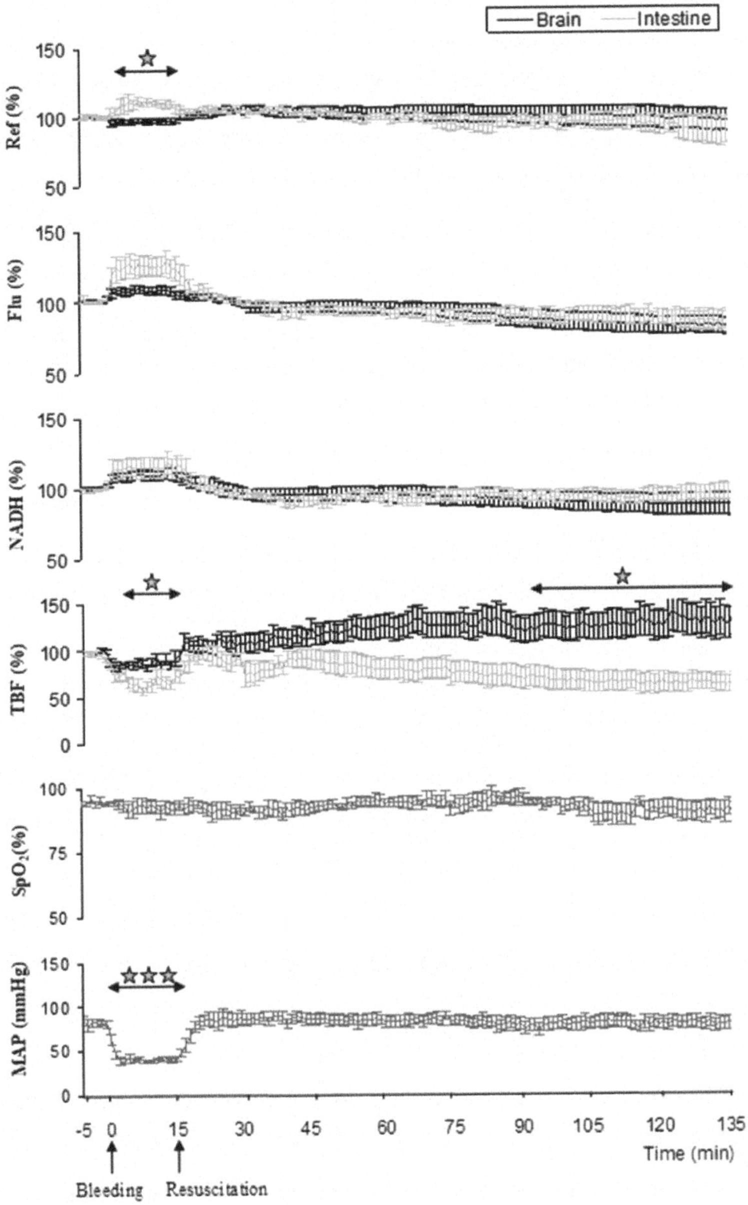

Fig. 5.10 The responses of the brain (black), small intestine (gray), and MAP to controlled hypotension for 15 min in a group—repeated ($n = 9$, mean ± S.E). The arrows represent the period in which significant differences were found between the two organs (in each minute) and the asterisks indicate levels of significance: $*p < 0.05$, $**p < 0.01$, and $***p < 0.001$. Asterisks in Ref, Flu, NADH, and TBF mark significance between brain and intestine; asterisks in SpO_2 and MAP mark significance compared to baseline. Ref reflectance, NADH mitochondrial NADH redox state, TBF tissue blood flow, MAP mean arterial pressure

Following resuscitation, MAP fully recovered and remained stable for the entire trial, while the SpO_2 remained stable. The TBF in both organs was fully recovered, yet while the CBF continued to be elevated, reaching the hyperemic level of 130 ± 18 (not significant due to a high standard error), IBF immediately deteriorated reaching the low level of $60 \pm 9\%$ ($p < 0.05$). The reflectance in both organs fully recovered and remained stable until the end of the trial. The fluorescence in both organs fully recovered, as well, yet in the brain it continued to decrease, reaching the level of $81 \pm 6\%$ ($p < 0.05$). Both the cerebral and intestinal NADH were fully recovered.

A comparison of the organs' response revealed significant differences in the TBF ($p < 0.01$) and the reflectance ($p < 0.01$) during the bleeding session. Following resuscitation, significant differences between the organs were found only for TBF ($p < 0.05$) within the last 40 min of the trial.

Controlled Hemorrhage for 15 Min Compared to Normotensive Conditions

As Fig. 5.11a shows, significant differences were found for MAP during hypotension ($p < 0.01$) and resuscitation ($p < 0.05$), as well as during the entire recovery phase, between the hypotension group and the normotensive group.

In the brain (Fig. 5.11b), none of the parameters presented a significant difference between the normotensive group and the controlled hypotension group.

In the intestine (Fig. 5.11c), the TBF presented a significant difference during the hypotension ($p < 0.01$) and following resuscitation ($p < 0.01$). In the reflectance, a significant difference was found following resuscitation ($p < 0.05$). At the fluorescence, it was found during the hypotension phase ($p < 0.05$) and following resuscitation ($p < 0.05$), while for NADH it was found only following resuscitation ($p < 0.05$).

5.3.3.2 Controlled Hemorrhage Maintained for 30 Min

Comparison Between Brain and Intestine

Typical results of controlled hypotension at 40 mmHg maintained for 30 min (figure not shown). Hypotension was induced by bleeding out 34.8% of the rat's total blood volume.

As hemorrhage started, MAP sharply decreased, whereas SpO_2 remained relatively stable. Both organs' TBF decreased; however, the IBF decreased to a greater extent. The reflectance in the brain remained relatively stable, while in the intestine it sharply and greatly increased. Both organs' fluorescence increased, yet in the intestine to a greater extent. The NADH in the brain sharply increased, while in the intestine it remained stable.

Following resuscitation, MAP presented a steps recovery. CBF recovered to hyperemic levels followed by stabilization at basal levels, whereas IBF fully

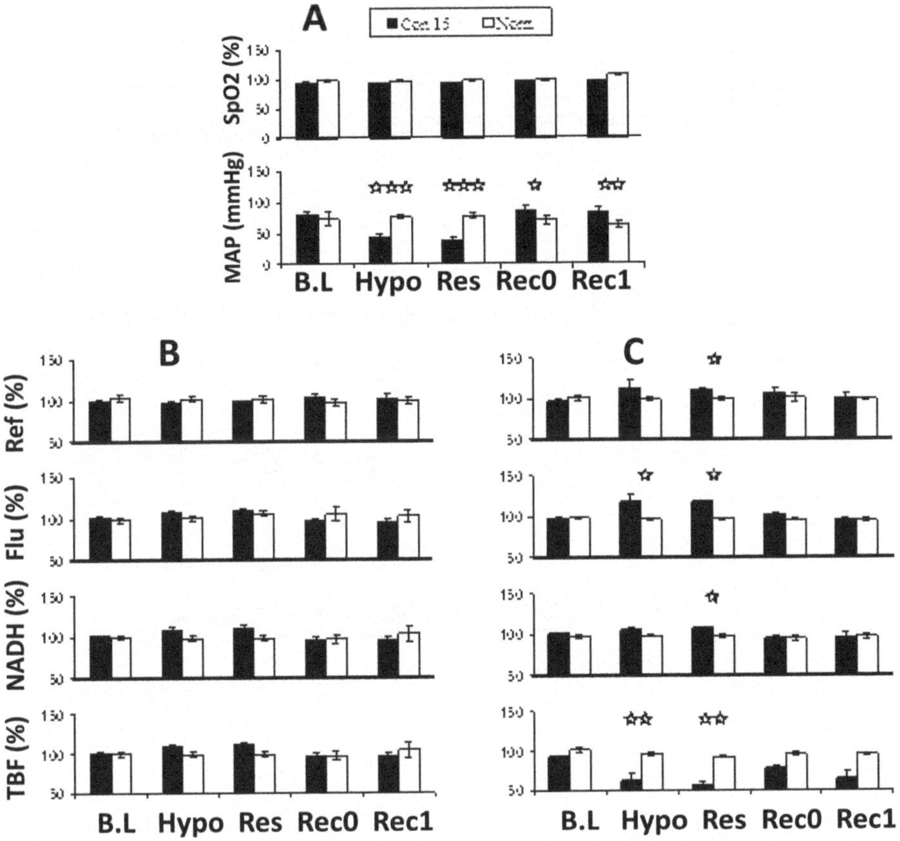

Fig. 5.11 Comparison of the MAP (**a**), cerebral (**b**), and intestinal (**c**) responses in the controlled hypotension group for 15 min (Con 15-black, $n = 7$) and normotensive group (Norm-white, $n = 4$, mean ± S.E)—repeated. The asterisks represent the levels of significance in relation to the control rats: $*p < 0.05$ and $***p < 0.005$. B.L. basal level, Hypo the end of the hypotension period, Res the end of resuscitation, Rec_0 1 h following resuscitation, Rec_1 2 h following resuscitation, Ref reflectance, NADH mitochondrial NADH redox state, TBF tissue blood flow, SpO_2 Saturation of peripheral oxygen, MAP mean arterial pressure

recovered, followed by a later deterioration. Cerebral reflectance remained stable, and intestinal reflectance fully recovered. Cerebral fluorescence and NADH fully recovered, whereas intestinal fluorescence and NADH only partially recovered.

Figure 5.12 exhibits the average response of nine rats to controlled hypotension to 40 mmHg maintained for 30 min. The hypotension was induced by bleeding 35 ± 6% of the rats' total blood volume.

During the bleeding session, MAP decreased reaching the level of 38.6 ± 2 mmHg ($p < 0.005$), whereas SpO_2 remained stable. Both organs' TBF decreased, in the brain reaching the level of 63.5 ± 4% ($p < 0.005$) and in the intestine reaching the level of 56.2 ± 10% ($p < 0.005$). Reflectance and fluorescence

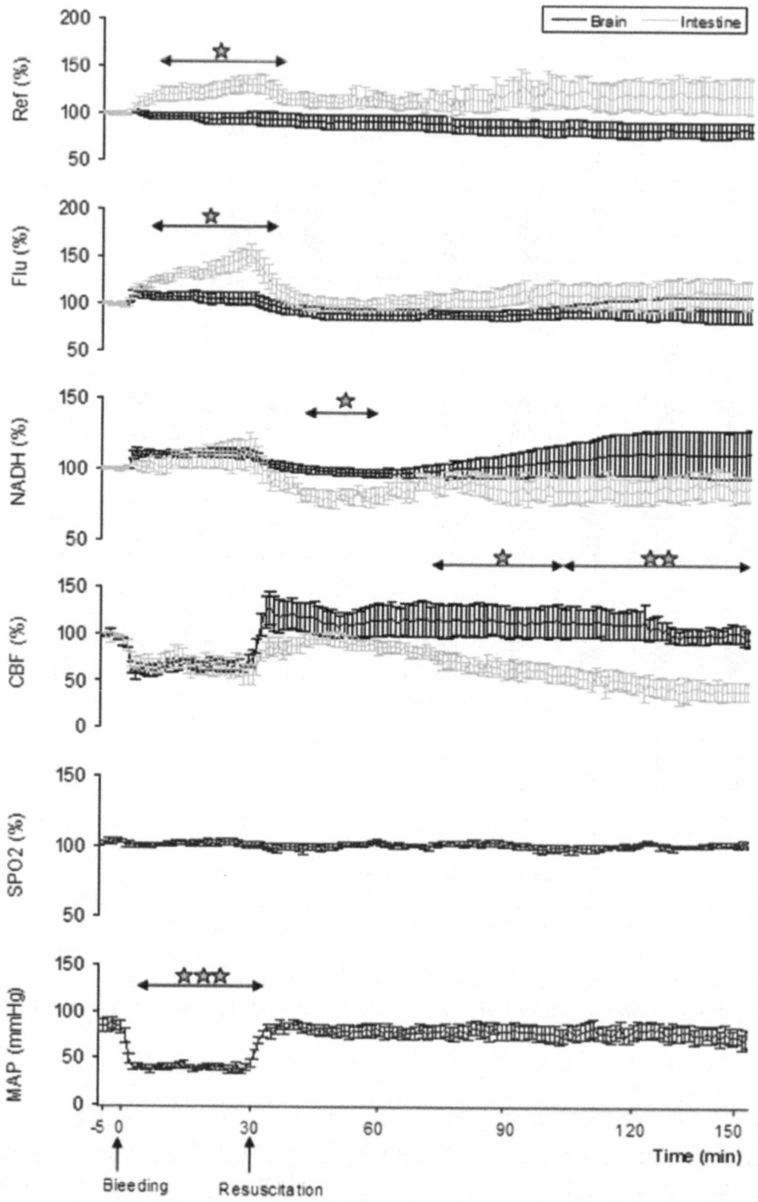

Fig. 5.12 The responses of the brain (black), small intestine (gray) SpO_2, and MAP to controlled hypotension for 30 min in a group—repeated ($n = 7$, mean ± S.E). The arrows represent the period in which significant differences were found between the two organs for each minute, and the asterisks represent levels of significance: $*p < 0.05$, $**p < 0.01$, and $*** p < 0.001$. Asterisks in Ref, Flu, NADH, and TBF mark significance between brain and intestine; asterisks in MAP mark significance compared to baseline. Ref reflectance, NADH mitochondrial NADH redox state, TBF tissue blood flow, SpO_2 Saturation of peripheral oxygen, MAP mean arterial pressure

in the brain remained stable, while in the intestine they were elevated to 131.2 ± 6% ($p < 0.01$) and 152.3 ± 10% ($p < 0.005$), respectively. Both the cerebral and intestinal NADH were elevated, yet only in the brain significantly (111.6 ± 3%, $p < 0.05$).

Following resuscitation, MAP fully recovered (85.9 ± mmHg) and remained steady for the entire trial. SpO_2 was not affected by the resuscitation and remained stable for the entire time left. Blood flow in both organs fully recovered, but while CBF at first showed a trend, yet not significant, hyperemia (127 ± 18%), following stabilization at basal levels, IBF reached the level of 95.1 ± 1% and then deteriorated again to the level of 38.4 ± 9% ($p < 0.005$). The cerebral reflectance and fluorescence were not altered, while these parameters fully recovered in the intestine.

Controlled Hemorrhage for 30 Min Compared to Normotensive Conditions

Figure 5.13 exhibits a comparison between the controlled hypotension group maintained for 30 min and the normotensive group. MAP was significantly different ($p < 0.05$) following hypotension, while SpO_2 showed no differences at any point. Brain monitoring was stable, whereas the intestinal parameters presented differences as follows: in IBF and reflectance during hypotension ($p < 0.01$), in the fluorescence during hypotension and following resuscitation, in the NADH following resuscitation.

5.3.3.3 Controlled Hemorrhage Maintained for 60 Min

Comparison Between Brain and Intestine

Figure 5.14 represents analogical results for controlled hypotension at 40 mmHg, maintained for 60 min. The hypotension was induced by bleeding out 47% of the rats' total blood volume.

As hemorrhage started, MAP rapidly decreased, and so did IBF, which remained very low for the entire period. The reflectance gradually increased while NADH increased in two steps. The brain responded to the hemorrhagic insult in a moderate manner. TBF decreased, but very rapidly recovered to its basal level. Accordingly, reflectance increased and NADH also gradually increased.

Following resuscitation, MAP increased above the basal levels, followed by full recovery. IBF responded similarly, although it showed only partial recovery and about 30 min after resuscitation reached significantly low levels. Intestinal reflectance also showed a partial recovery and intestinal NADH decreased rapidly; nevertheless, its levels remained high above the basal levels. CBF, reflectance, and NADH were fully recovered.

Under controlled 40 mmHg hypotension maintained for 60 min, only 7 rats out of 13 survived. Figure 5.10 shows the averaged responses of those rats. The hypotension was induced by bleeding out 45 ± 4% of the rats' total blood volume.

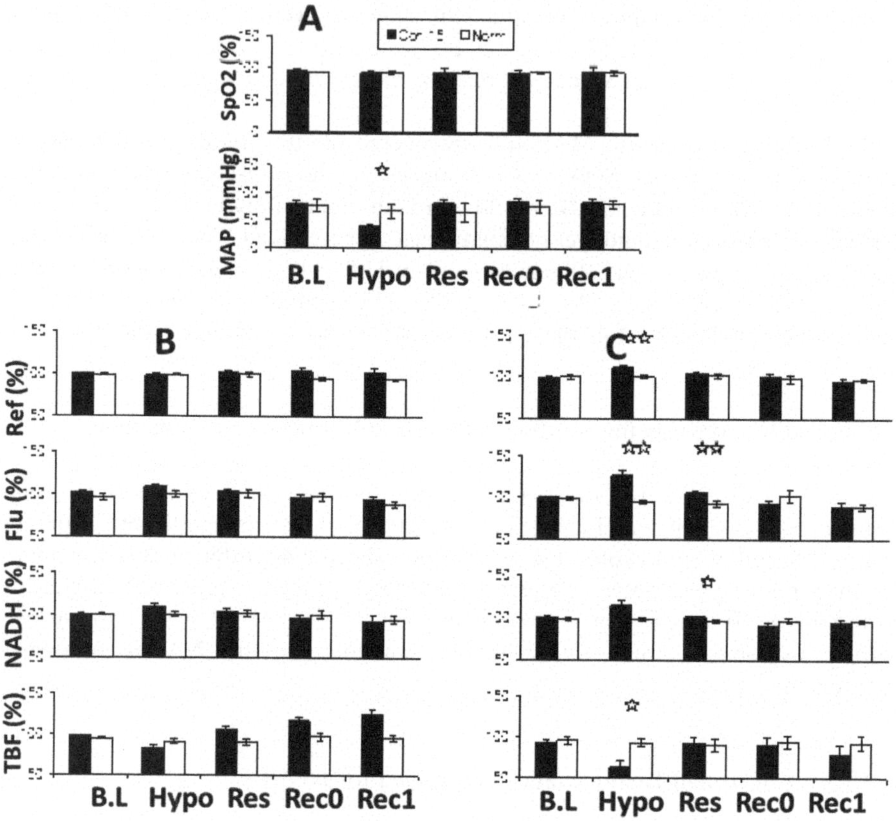

Fig. 5.13 Comparison of the MAP (**a**), cerebral (**b**), and intestinal (**c**) responses in the group with controlled hypotension maintained for 30 min—repeated (Con 30-black, Norm-gray, $n = 7$) and normotensive group (Norm-white, $n = 4$, mean ± S.E). Asterisks represent the levels of significance: $*p < 0.05$, $**p < 0.01$, and $***p < 0.001$. B.L. basal level, Hypo end of the hypotension period, Res end of resuscitation, Rec0 1 h after resuscitation, Rec1 2 h after resuscitation, Ref reflectance, NADH mitochondrial NADH redox state, TBF tissue blood flow, SpO_2 Saturation of Peripheral Oxygen, MAP mean arterial pressure

During bleeding, MAP sharply decreased by 51 ± 0.5 mmHg ($p < 0.001$). Both TBFs decreased rapidly and stabilized at low levels. However, while IBF significantly decreased by 61 ± 6% ($p < 0.001$), CBF significantly decreased only by 45 ± 5.5% ($p < 0.01$). Intestinal reflectance showed a trend of increase, while the cerebral reflectance increased by 5 ± 2% ($p < 0.05$) and decreased back to its basal level. In comparison to the intestinal NADH, which slowly increased up to 228 ± 24% ($p < 0.01$), cerebral NADH increased only to a level of 142 ± 5% ($p < 0.01$). The maximum levels of NADH in both organs were monitored at the end of the hemorrhagic period and were also associated with the maximum changes in TBF.

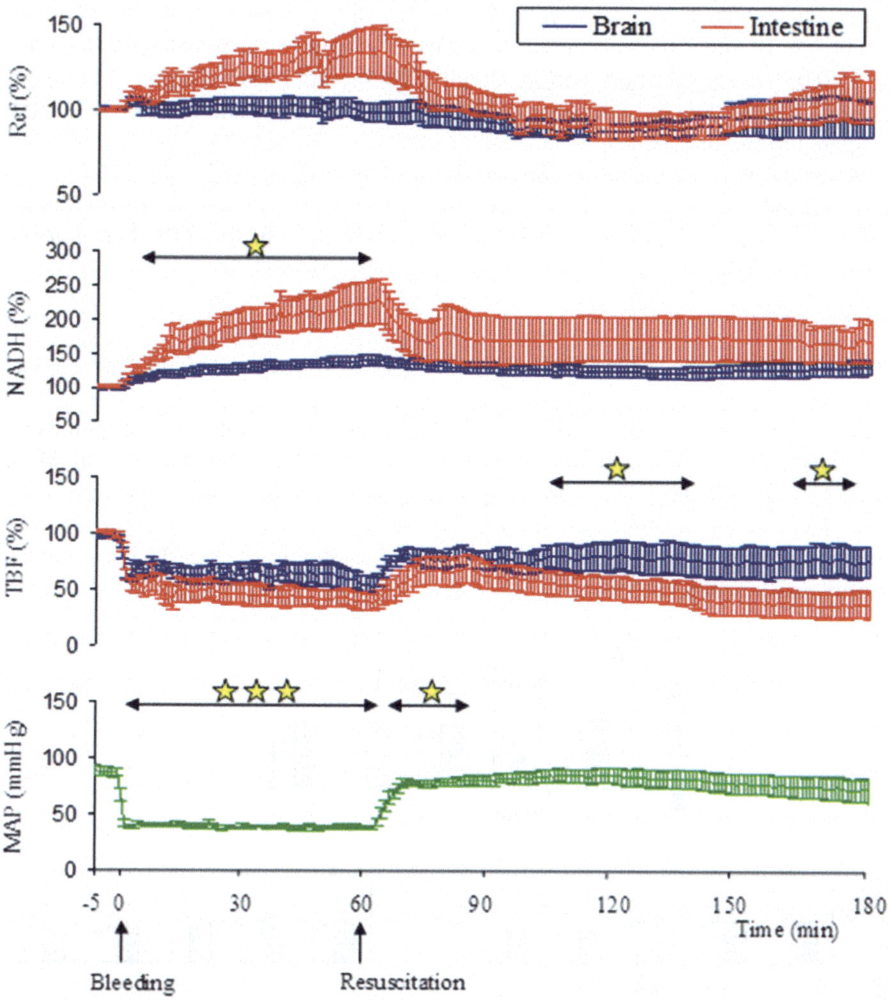

Fig. 5.14 The responses of the brain (blue), small intestine (red), and MAP to controlled hypotension for 60 min in a group ($n = 7$, mean ± S.E). The arrows represent the period in which significant differences were found between the two organs and the asterisks represent significance levels: $*p < 0.05$, $**p < 0.01$, and $***p < 0.001$. Asterisks in Ref, NADH, and TBF mark significance between brain and intestine; asterisks in MAP mark significance compared to baseline. Ref reflectance, NADH mitochondrial NADH redox state, TBF tissue blood flow, MAP mean arterial pressure

Following resuscitation, MAP increased up to the basal level. The intestine and the brain responded differently to resuscitation. IBF significantly increased by 30 ± 9% ($p < 0.05$), reaching a level of 69%, followed by a secondary decrease down to 37 ± 11% ($p < 0.01$). Blood flow in the brain increased by 26.5 ± 8% ($p < 0.01$) and stabilized at a level of 82%. The intestinal reflectance decreased sharply below its basal level, followed by an increase back to the basal level. Cerebral reflectance showed nearly no changes except for a decrease of 12 ± 5% ($p < 0.05$) during a short period, about an hour after resuscitation. Following resuscitation, NADH in both organs only partially recovered and remained elevated compared to the basal level.

When the two organs are compared during bleeding, a significant difference is observed only with respect to NADH levels ($p < 0.05$). Following resuscitation, there were only two episodes of significant differences in the TBF between the organs ($p < 0.05$).

Controlled Hemorrhage for 60 Min Compared to Normotensive Conditions

Figure 5.15 presents the responses of the group with controlled hypotension maintained for 60 min and the normotensive group. During Hypotension, MAP (A) in the controlled hypotension group decreased significantly in comparison to the MAP level in the normotensive group. Following resuscitation, MAP in the controlled hypotension group increased significantly as opposed to the MAP level in the normotensive group.

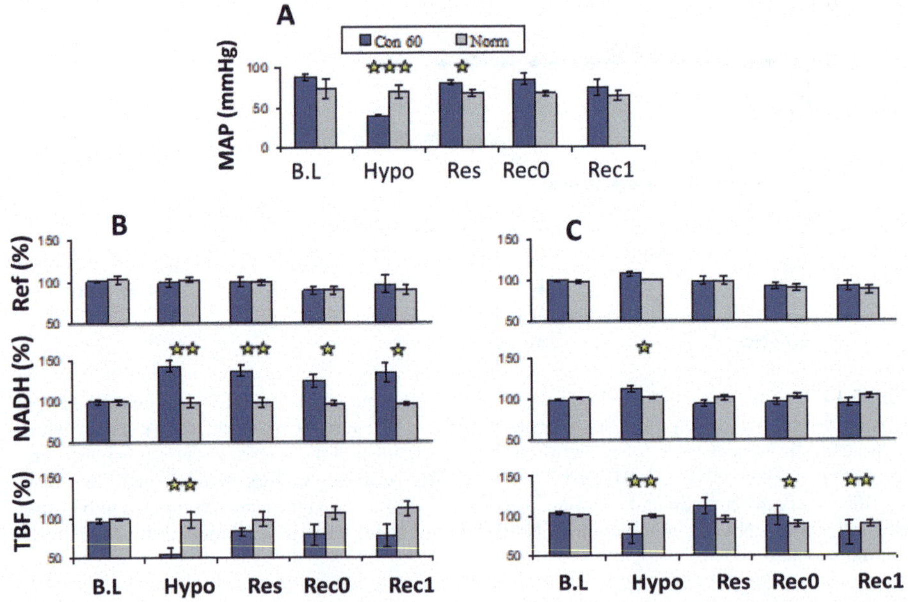

Fig. 5.15 Comparison of the MAP (**a**), cerebral (**b**), and intestinal (**c**) responses in the group with controlled hypotension maintained for 60 min (Con 60-purple, $n = 7$) and normotensive group (Norm-black, $n = 4$, mean ± S.E.). The asterisks represent the levels of significance: *$p < 0.05$, **$p < 0.01$, and ***$p < 0.001$. B.L. basal level, Hypo end of the hypotension period, Res end of resuscitation, Rec_0 1 h after resuscitation, Rec_1 2 h after resuscitation, Ref reflectance, NADH mitochondrial NADH redox state, TBF tissue blood flow, MAP mean arterial pressure

Cerebral TBF demonstrated significant differences during hypotension (Fig. 5.15b). Both groups presented the same levels of cerebral reflectance. NADH level in the controlled hypotension group was significantly higher than in the normotensive group, from the hypotension onset until the end of the experiment ($p < 0.05$–$p < 0.01$).

Intestinal TBF was significantly lower in the controlled hypotension group as compared to the normotensive group (Fig. 5.15c). Following resuscitation, TBF increased and became significantly high through the first hour of recovery ($p > 0.05$), but it deteriorated again within the second hour of recovery ($p < 0.01$). The intestinal reflectance was similar in both groups. Intestinal NADH level was significantly higher in the controlled hypotension group than in the normotensive group in the hypotension phase ($p < 0.05$).

Comparison Between the Surviving and Non-Surviving Rats

Under controlled hypotension maintained for 60 min, 6 rats out of 13 did not survive: two rats died within the hypotension phase, one during the resuscitation phase, and the other three during the recovery phase. ANCOVA analysis revealed no effect of the group (either survivors or non-survivors) on the time interval until reaching 40 mmHg, but only an effect of time ($F(1,44) = 93.308$, $p < 0.001$).

Hypotension Phase

MAP Responses
As Fig. 5.16a demonstrates, the initial MAP of both groups was the same. During hypotension, MAP reached almost the same level in both the surviving and non-surviving rats. Although in the surviving rats, MAP remained steady around 40 ± 2 mmHg, in the non-surviving rats, a biphasic decrease appeared. After an initial decrease, MAP stabilized around 46 ± 7 mmHg, and at 60% of time, MAP decreased again and remained low at the level of 35 ± 3 mmHg until the end of the experiment.

Cerebral Responses
In the brain (Fig. 5.16b), TBF demonstrated the same trend for both groups, although the final level of TBF in the surviving rats was lower than among the non-survivors (55 ± 7% and 72 ± 14%, respectively). In both groups, the reflectance remained relatively stable around the basal level. NADH was elevated in both groups, with a larger increase in the surviving rats (141 ± 7% in comparison to 122 ± 12%).

Intestinal Responses
In the intestine (Fig. 5.16c), TBF decreased in both groups, though the decrease was larger in the non-surviving group as compared to the survivors (27 ± 7% and 39 ± 8%, respectively). Both groups demonstrated an elevation in the reflectance, yet, in the non-surviving rats, it increased to a higher level (158 ± 39%) than in the surviving animals (134 ± 14%). NADH increased as well; however, in the surviving rats this elevation was greater (226 ± 32% in comparison to 186 ± 35%).

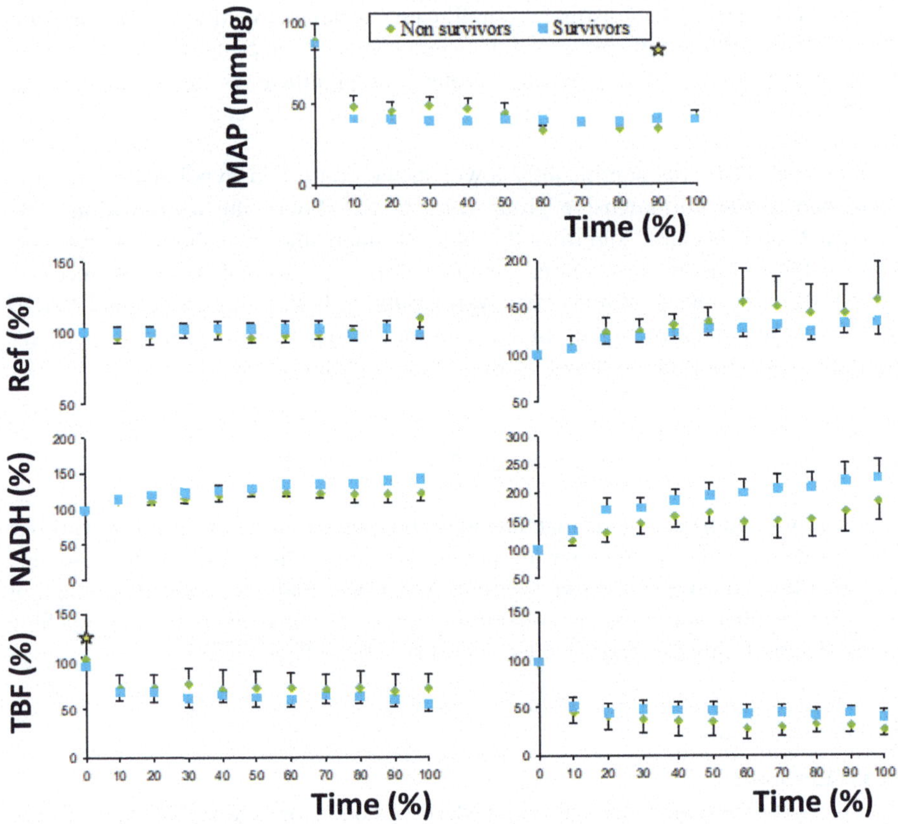

Fig. 5.16 Comparison between the responses of the MAP (**a**), the brain (**b**), and intestine (**c**) in the surviving rats ($n = 7$, squares) and the non-surviving rats ($n = 6$, circles) during the 60 min hypotension period (time was normalized with 10% intervals, mean ± S.E.). The asterisks represent the level of significance: $*p < 0.05$, Ref reflectance, NADH mitochondrial NADH redox state, TBF tissue blood flow, MAP mean arterial pressure

Resuscitation Phase

MAP Responses

Figure 5.17a, representing the normalized resuscitation time, indicates that during the first 30% of time, both groups showed an elevation of MAP with the same pattern. However, since 40% of the resuscitation period, significant differences were observed between the responses ($p < 0.05$). In the surviving group, MAP remained stabilized at a high level of 81 ± 4 mmHg, whereas in the non-surviving group, MAP failed to remain high and decreased to 50 ± 11 mmHg.

Cerebral Responses

CBF (Fig. 5.17b) increased in the non-surviving group during the first 10% of time, followed by a gradual decrease to 67 ± 17%. In the surviving rats, TBF gradually increased during the first 40% of the time and remained stable around 76 ± 8% for

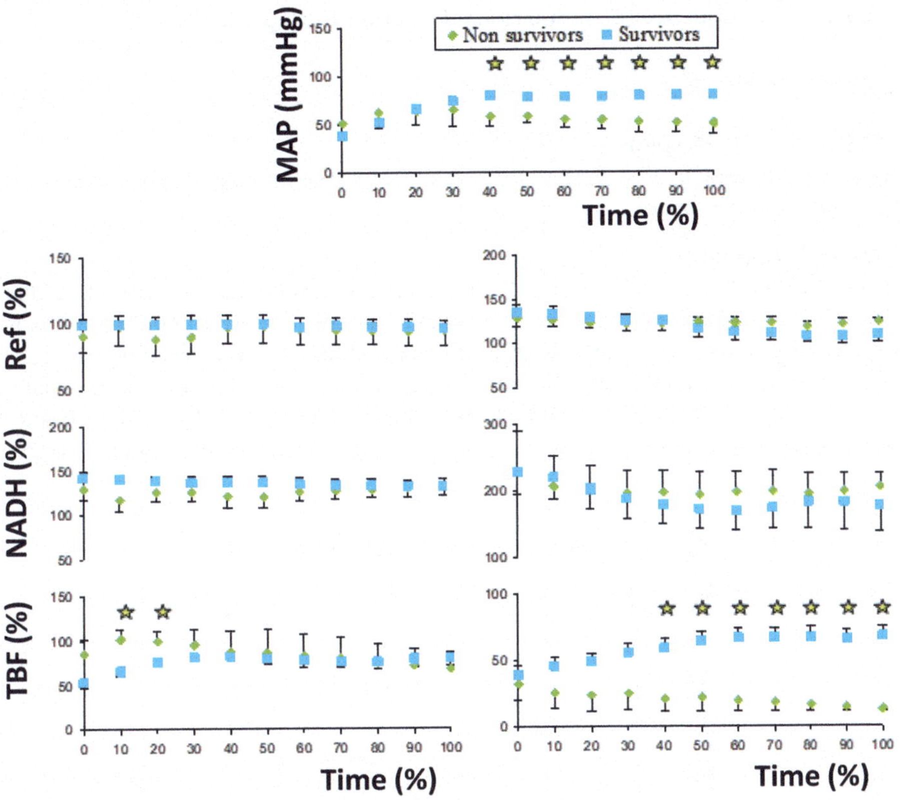

Fig. 5.17 Comparison between the responses of the MAP (**a**), the brain (**b**), and intestine (**c**) in the surviving rats ($n = 7$, squares) and the non-surviving rats ($n = 3$, circles) during the resuscitation period (time was normalized with 10% intervals, mean ± S.E.). The asterisks represent the level of significance: *$p < 0.05$. Ref reflectance, NADH mitochondrial NADH redox state, TBF tissue blood flow, MAP mean arterial pressure

the rest of the experiment. During 10–20% of time, differences were significant ($p < 0.05$). Both the reflectance and NADH reached the same level, although in the surviving rats, the reflectance and NADH were more stable.

Intestinal Responses

In the intestine (Fig. 5.17c), TBF of the surviving rats increased up to 67 ± 12%, while the TBF of the non-surviving group decreased down to 13 ± 2%, leading to a significant difference between the two groups in the last 60% of the experimental period ($p < 0.05$). The reflectance decreased with time in the surviving rats down to 108 ± 8%, whereas in the non-surviving rats it decreased to 122 ± 10% and remained so until the end of the experiment. In both groups, NADH decreased during the first 40% of time, and then it remained steady, though the non-surviving rats maintained a higher level (198 ± 25%) than the surviving rats (176 ± 39%).

5.3.3.4 The Influence of Hemorrhage Duration in Repeated Groups

Hypotension Phase

MAP and SpO_2 Responses

As it can be seen in Fig. 5.18a and Table 5.1, no significant levels were found between the three models.

Cerebral Responses

As can be seen from Fig. 5.18b, the 15-minute model responded differently from the two other models, which reacted at the same pattern. The 15-minute model was significantly lower than the 30-minute model 90% of the bleeding sessions ($p < 0.01$–$p < 0.001$) and for the entire period from the 60-minute model ($p < 0.05$, $p < 0.001$). The NADH level in the 60-minute model was significantly higher than the other models during 30–100% of the bleeding session ($p < 0.05$–$p < 0.01$).

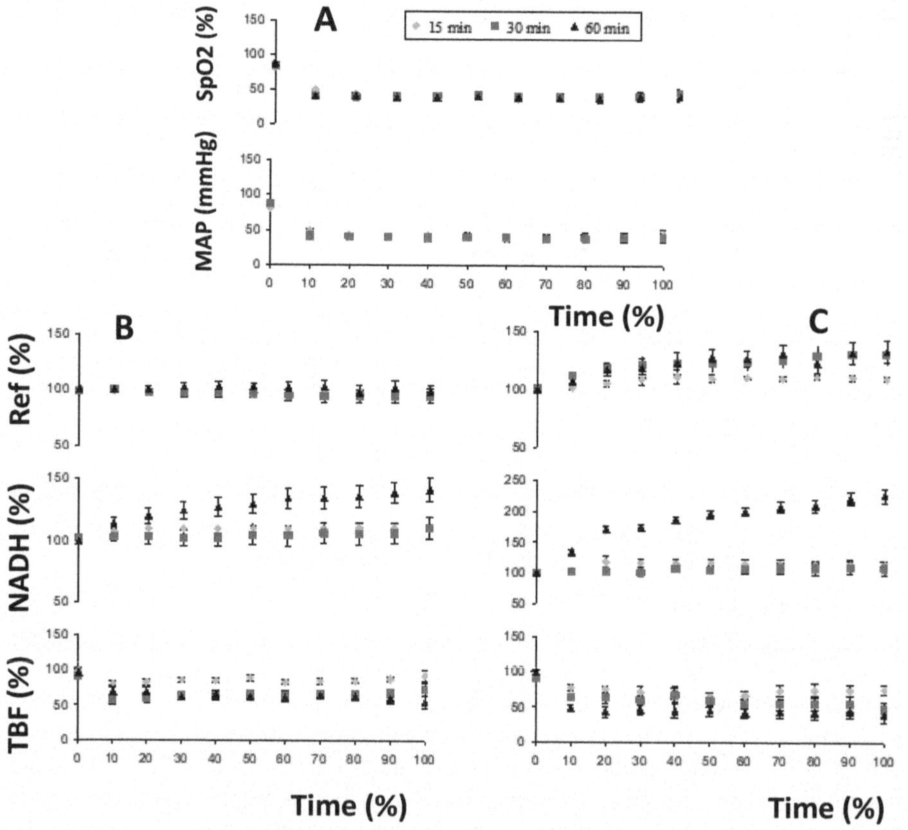

Fig. 5.18 The effect of the duration of hypotension during the hypotension periods on the MAP (**a**), the brain (**b**), and intestine (**c**) in the 15-minute model ($n = 9$, means ± SE), 30- and 60-minute models ($n = 7$, means ± SE)—repeated. Ref reflectance, NADH mitochondrial NADH redox state, TBF tissue blood flow, MAP mean arterial pressure

Table 5.1 Significance levels of variance between the different models of hypotension, during the hypotension period—repeated

Parameter	Sig.	15 min vs. 30 min		15 min vs. 60 min		30 min vs. 60 min	
		Brain	Intestine	Brain	Intestine	Brain	Intestine
REF	p<0.05		70, 100		70		
NADH	p<0.05			30-50		30-100	
	p<0.001			60-100	10-100		10-100
TBF	p<0.05		100	10-50	10-20, 60-100		
	p<0.01	60-90					
	p<0.001	10-50		90-100			
MAP							
SpO_2							

The data represent periods of significant differences between the various models of controlled hypotension. The hypotension duration is normalized with 10% intervals
Ref reflectance, *NADH* mitochondrial NADH redox state, *TBF* tissue blood flow, *MAP* mean arterial pressure

Intestinal Responses
No significant differences were observed for IBF between the different models (Fig. 5.18c and Table 5.1). As for the reflectance, all models responded similarly, while the 15-minute model was significantly different only during two short episodes ($p < 0.05$). The most clear difference between the models was for NADH level in the 60-minute model. The NADH level in this model was greater beginning on the 10% of the session until the end ($p < 0.05$–$p < 0.01$).

Resuscitation Phase

MAP and SpO_2 Responses
As can be seen from Fig. 5.19a and Table 5.2, all models presented the same pattern for MAP. Yet, MAP level in the 60-minute model was significantly lower than the 30-minute model during the baseline ($p < 0.05$). As for SpO_2, no significant difference was observed between any of the models.

Cerebral Responses
The CBF in all models responded very similarly (Fig. 5.19b and Table 5.2) yet the CBF level in the 60-minute model was lower and therefore significantly different from the 30-minute model during the entire period ($p < 0.05$–$p < 0.01$) and significantly different from the 30-minutes model during the first 15 min ($p < 0.05$). However, no differences in the reflectance were found. In accordance with CBF, the NADH level in the 60-minute model was significantly higher than the two other models for the entire period ($p < 0.05$–$p < 0.001$).

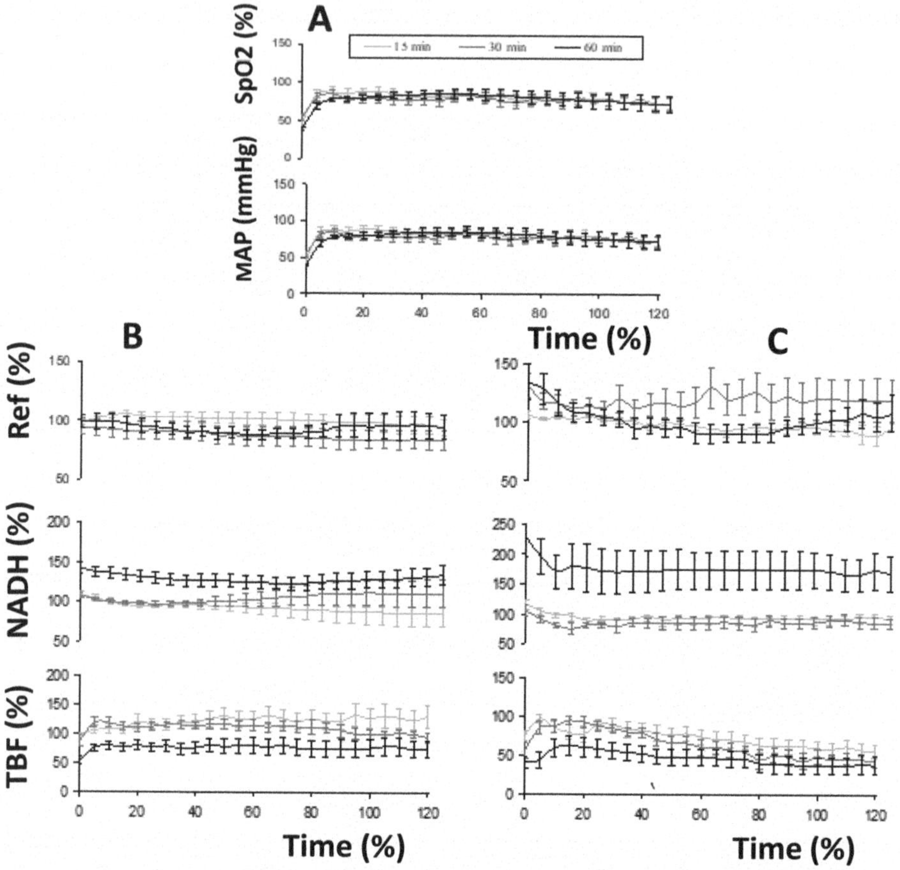

Fig. 5.19 The effect of the duration of hypotension during the resuscitation and recovery periods on MAP (**a**), the brain (**b**), and intestine (**c**) in the 15-minute model ($n = 9$, means ± SE), 30- and 60-minute models ($n = 7$, means ± SE). Ref reflectance, NADH mitochondrial NADH redox state, TBF tissue blood flow, MAP mean arterial pressure

Intestinal Responses

As can be seen from Fig. 5.19c and Table 5.2, the IBF in the 15-minute model and the 30-minute model responded similarly to the resuscitation. During the first 30 min, the IBF in the 60-minute model was significantly lower than the two other models ($p < 0.05$, $p < 0.001$ in the 15-minute model and $p < 005$–$p < 0.01$ in the 30-minute model). The only significant difference in the reflectance was between the 15-minute model and the 30-minute model during the beginning of resuscitation ($p < 0.05$). However, the 60-minute model was significantly different from the other models during the entire period ($p < 0.05$–$p < 0.001$).

Table 5.2 Significance levels of variance between the different models of hypotension, during the recovery period

Parameter	Sig.	15 min vs. 30 min		15 min vs. 60 min		30 min vs. 60 min	
		Brain	**Intestine**	**Brain**	**Intestine**	**Brain**	**Intestine**
Reflectance	p<0.05		0				
NADH	p<0.05		10		20-120	40-50	30-120
	p<0.01			40-120	0-10		
	p<0.001			0-30		0-40	0-20
TBF	p<0.05			10-120	25-30, 45	0-15	15-30
	p<0.01			0-5			5
	p<0.001				0-5		
MAP	p<0.05					0	
SpO2							

The data represent periods of significant differences between the various models of controlled hypotension
Ref reflectance, *NADH* mitochondrial NADH redox state, *TBF* tissue blood flow, *MAP* mean arterial pressure

5.3.3.5 TBF and NADH Levels During Different Hemorrhage Durations

As can be observed from Fig. 5.20a, the CBF in the 15-minute model was significantly different from the 30- and 60-minute models ($p < 0.001$).

In IBF (Fig. 5.20b), a significant difference was observed only between the 15- and 60-minute model ($p < 0.001$).

As can be seen from Fig. 5.20c, d, the 60-minute model was significantly different from the two other models ($p < 0.001$).

5.3.3.6 Survival Rates in Different Hemorrhage Durations

Figure 5.21 exhibits the effect of the duration of hypotension on the survival rates as observed in the repeated groups. In the 15- and 30-minute model, 100% of the rats survived, whereas in the 60-minute model, only 54% survived.

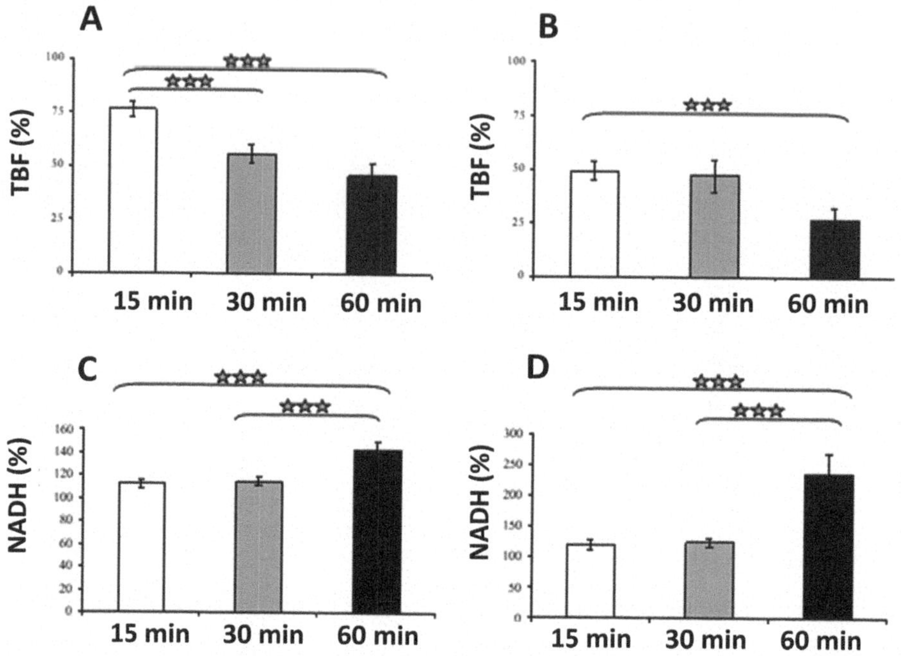

Fig. 5.20 The effect of duration of hypotension on the TBF and NADH levels during hypotension. (**a**) the effect on cerebral TBF, (**b**) the effect on intestinal blood flow, (**c**) the effect on cerebral NADH, and D—the effect on intestinal NADH (mean ± S.E.). White—15-minute model, gray—30-minute model, and black—60-minute model. The asterisks represent levels of significance: $*p < 0.05$, $**p < 0.01$, and $***p < 0.001$

Fig. 5.21 The effect of the duration of hypotension on the survival rates in various models—repeated groups. 15 min—white, 30 min—gray, and 60 min—black

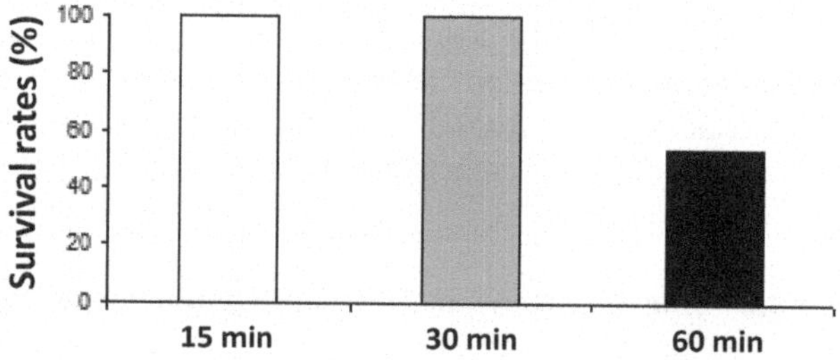

5.3.4 *Comparison Between Uncontrolled and Controlled Hemorrhage*

5.3.4.1 MAP Responses

During hemorrhage, a very high significant difference was observed between the MAP in the two models ($p < 0.01$, Fig. 5.22). In the controlled hypotension group,

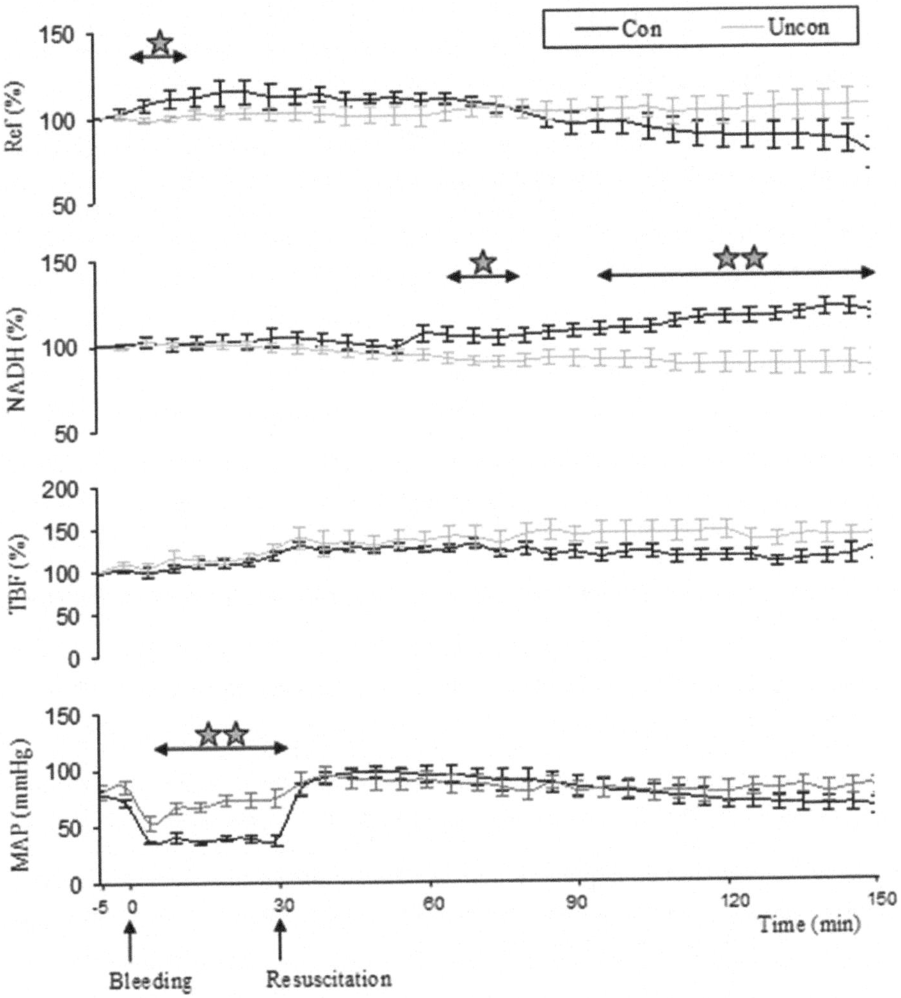

Fig. 5.22 Comparison of the responses of MAP and the brain to controlled (CON—black, $n = 7$) and uncontrolled hypotension for 30 min (UNCON-gray, $n = 9$, mean ± S.E). The asterisks present the levels of significance—$*p < 0.05$, and $**p < 0.01$. Ref reflectance, NADH mitochondrial NADH redox state, TBF tissue blood flow, MAP mean arterial pressure

MAP remained low, at a level of 40 mmHg through the entire 30 min, whereas in the uncontrolled hypotension group, immediately after MAP reached its minimal level, it increased to the basal level.

5.3.4.2 Cerebral Responses

Both groups showed an elevation of TBF levels; in the controlled hypotension group, it increased up to 129 ± 4% and stabilized, whereas under uncontrolled hypotension, it increased to a level of 140 ± 13% and then stabilized. The reflectance increased in the controlled hypotension group ($p < 0.05$) through the bleeding phase and then decreased through the rest of the experiment, reaching the level of 76% ± 9. In the uncontrolled hypotension group, the reflectance remained steady through the entire experimental period. NADH was also steady in the uncontrolled hypotension group for the first 60 min, followed by a decrease to 87 ± 7%, while in the second group, after the first hour in which NADH was steady, it increased up to 118 ± 4%, yielding significant differences between NADH levels in these groups ($p < 0.05$ and $p < 0.01$).

5.3.4.3 Intestinal Responses

Figure 5.23 demonstrates significant differences, during hypotension, between the controlled and uncontrolled groups ($p < 0.01$). TBF showed the same trend of change, namely, a biphasic increase (followed by a decrease) through the bleeding phase, followed by another increase during the resuscitation, and a decrease further on. However, TBF levels were significantly different between the two groups during the last 20 min of the hemorrhage period ($p < 0.05$). The reflectance decreased gradually in the uncontrolled hypotension group, whereas in the controlled hypotension group it decreased in the first minutes of bleeding and then increased in a biphasic manner leading to significant differences between the groups within the 50–150 min of the experiment ($p < 0.05$ and $p < 0.01$), with minor intervals of nonsignificant difference. In the bleeding phase, NADH increased in both groups; however, in the uncontrolled hypotension rats, it recovered within the bleeding phase, leading to significant differences between the groups during 15–25 min ($p < 0.01$). Following resuscitation, NADH in both groups decreased. However, while in the uncontrolled hypotension group NADH remained stable for the entire experiment, in the controlled hypotension group it increased during the 30 min after resuscitation and then remained stable.

5.3.4.4 Maximum Responses of TBF and NADH

As can be seen in Fig. 5.24, no significant differences between the models were found for any of the parameters.

5.3.4.5 Survival Rates

During uncontrolled hypotension, 100% of the rats survived, while during the controlled hypotension, only 58% survived (Fig. 5.25).

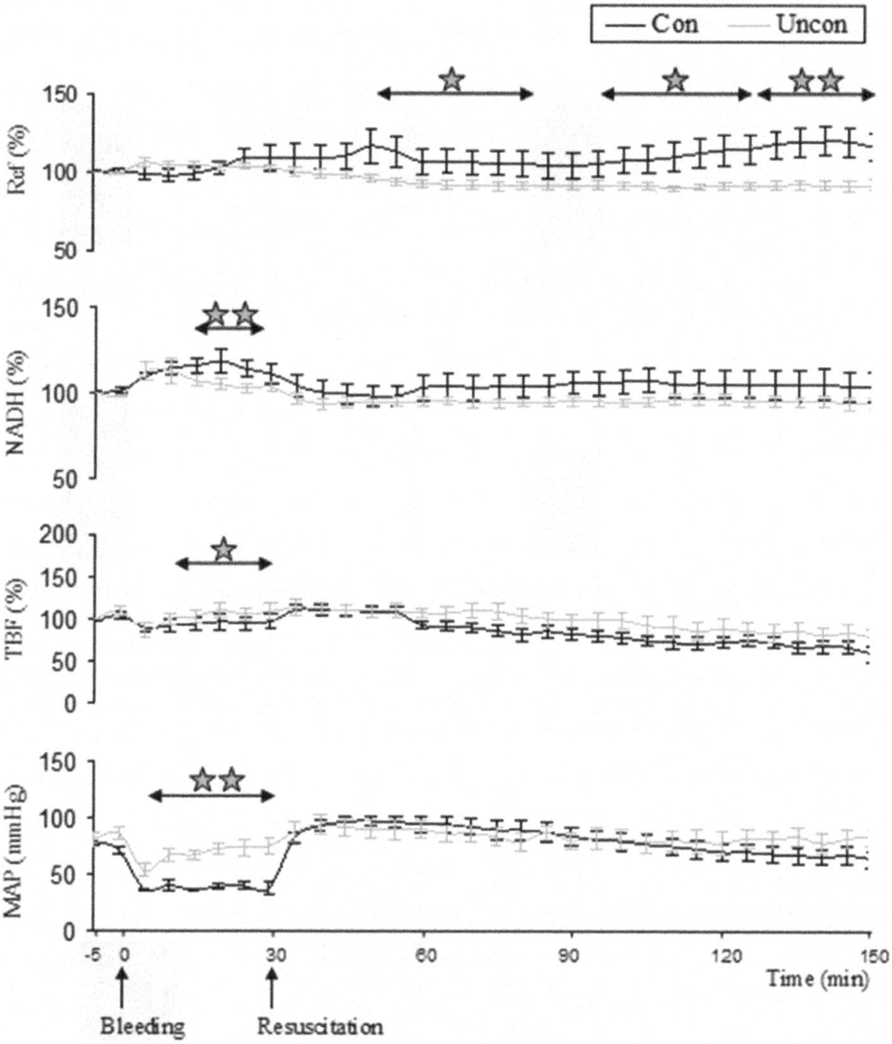

Fig. 5.23 Comparison of the responses of MAP and the intestine to controlled (CON—black, $n = 7$) and uncontrolled hypotension for 30 min (UNCON-gray, $n = 9$). The asterisks represent levels of significance—$*p < 0.05$, and $**p < 0.01$. Ref reflectance, NADH mitochondrial NADH redox state, TBF tissue blood flow, MAP mean arterial pressure

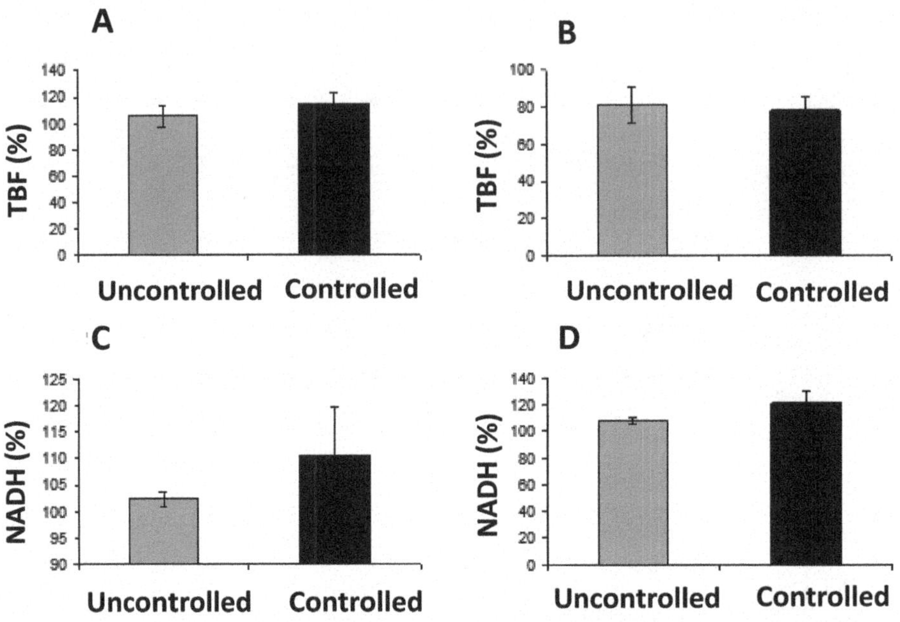

Fig. 5.24 The effect of uncontrolled (gray) and controlled (black) hypotension on the TBF and NADH levels during hypotension. (**a**) the effect on cerebral TBF, (**b**) the effect on intestinal blood flow, (**c**) the effect on cerebral NADH, and (**d**) the effect on intestinal NADH (mean ± S.E.). TBF tissue blood flow, NADH mitochondrial NADH redox state

Fig. 5.25 Survival rates in the uncontrolled hypotension group (gray) and controlled hypotension group (black)

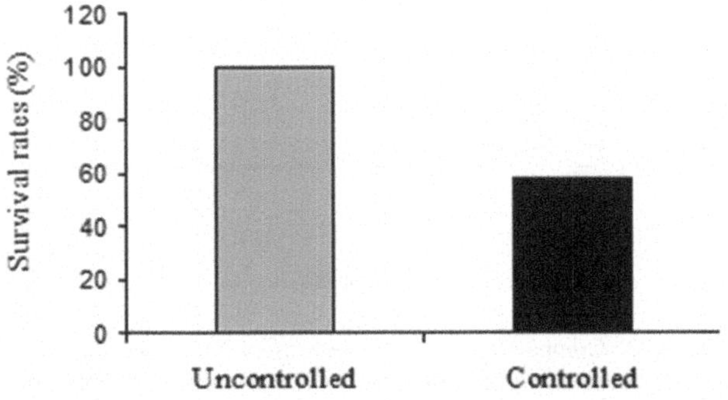

5.3.5 *Graded Hemorrhage*

5.3.5.1 Graded Hemorrhage in 10-Minute Intervals

Comparison Between Brain and Intestine

In this group, in each withdrawal of blood, 5% of the rat's initial total blood volume was removed, with 10-minute intervals, until MAP reached a level of 40 mmHg. The percentage of hemorrhage (namely 5%, 10%, etc.) is determined based on the overall blood volume being removed up to this perturbation.

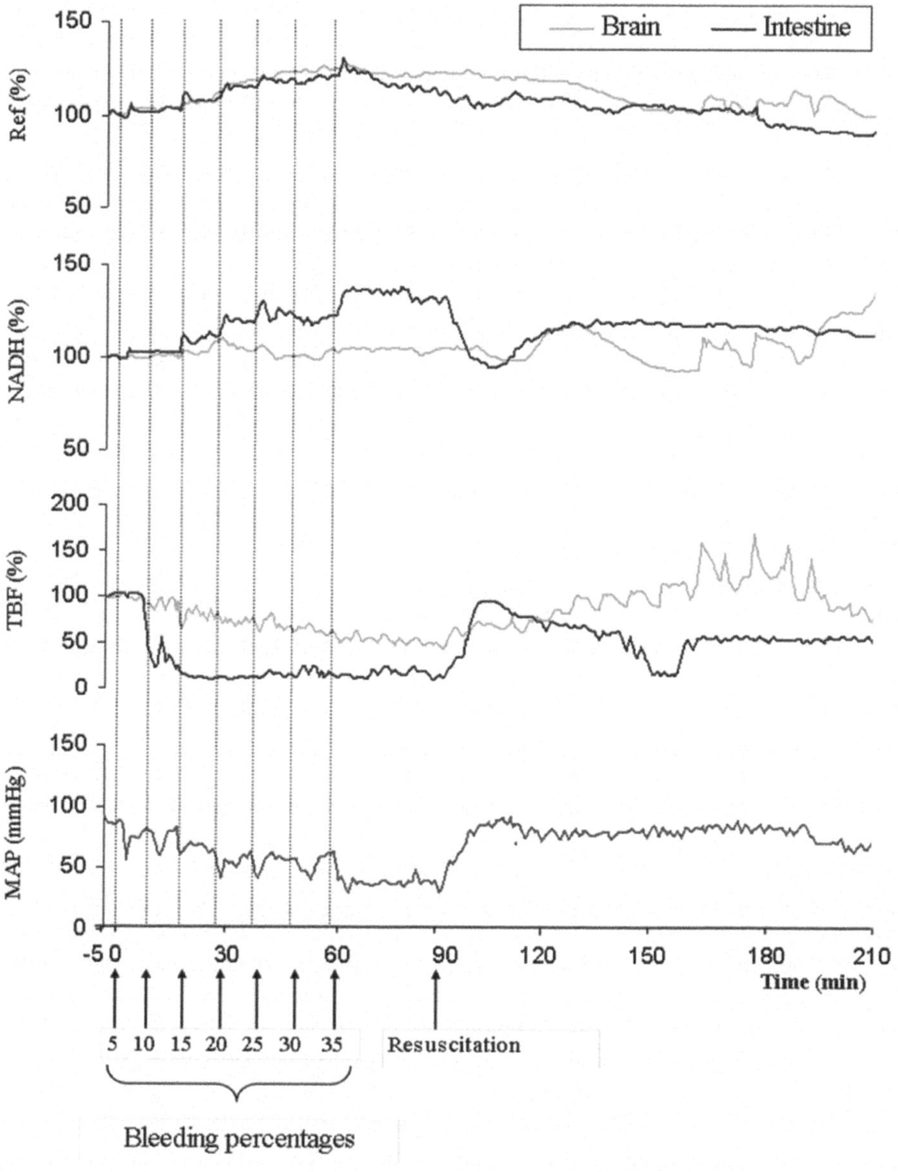

Fig. 5.26 Typical responses of the brain (gray), small intestine (black), and MAP to graded hemorrhage in 10-minute intervals in a single rat. Ref reflectance, NADH mitochondrial NADH redox state, TBF tissue blood flow, MAP mean arterial pressure

Figure 5.26 presents typical responses of the brain and intestine to this graded hemorrhage. As seen, during the 5% hemorrhage, MAP sharply decreased whereas all other parameters remained relatively steady.

Following this withdrawal, MAP partially recovered, but decreased again during the second bleeding phase (10% hemorrhage). In accordance TBF in both organs decreased, yet differently. The CBF decreased moderately, whereas IBF decreased very sharply. However, neither the reflectance nor the NADH in both organs was altered.

Later on, both MAP and blood flows attempted to recover, but the third session of bleeding (the 15% hemorrhage) decreased MAP and CBF and totally deteriorated IBF. Correspondingly, the reflectance in both organs slightly decreased, whereas NADH increased only in the intestinal tissue.

Once again, the MAP partially recovered until the next bleeding, 20% hemorrhage, through which it decreased again. CBF continued to decrease, while IBF remained very low. The trend of elevation in the reflectance in both organs and in the intestinal NADH continued, and the level of cerebral NADH was elevated as well.

Following the 5th session of bleeding (25% hemorrhage) MAP decreased again. The bloodflows patterns remained, where the CBF continued to decrease, and the level of IBF stayed extremely low. Again, the reflectance in both the brain and intestine and the intestinal NADH continued to increase, whereas the cerebral NADH remained relatively stable.

Before and during the 30% bleeding, all the trends remained, except for the cerebral NADH level, which slightly decreased.

During the 35% hemorrhage and during the following 30 min the MAP remained steady around 40 mmHg, both blood flows remained decreased, the reflectance in both organs increased for a short period and then decreased, and cerebral NADH showed a slight increase with stabilization on that level, whereas, the intestinal NADH showed a significant large increase with stabilization at this elevated level.

Following resuscitation, MAP fully recovered for a short period, followed by a slight decrease and stabilization on that level for about 70 min, with another decrease at the end of the experiment. The IBF also temporarily recovered, but for a shorter period with a biphasic deterioration and a partial recovery an hour before the end of the trial. The CBF, on the contrary, showed full recovery with a long hyperemic phase. Both the cerebral and intestinal reflectance fully recovered, while the NADH in both organs showed a short phase of recovery followed by an increase approximately 15 min after resuscitation began.

As indicated in the methods section, our endpoint of hemorrhage was reaching a MAP level of 40 mmHg by intermittent bleeding of 5% of the rat's blood volume in each bleeding session. This special model was involved with differences in the kinetics of the decrease of MAP, due to different vasoreactivity of the blood vessels, yielding large variance within the group. In order to overcome this issue, we omitted the bleeding sessions, leaving only the auto-recovery phases within the hemorrhage protocol, as well as the resuscitation phase followed by 2 h of recovery (Fig. 5.27).

In this group of graded hemorrhage in 10-minute intervals 8 out of 9 rats survived. As can be seen from Fig. 5.27, during the 5% of bleeding, MAP decreased from 98 ± 3 mmHg to 66 ± 3 mmHg ($p < 0.001$). However, neither the cerebral nor the IBFs, as well as the reflectance in these organs, were significantly changed. Nevertheless, intestinal NADH increased to a level of $104 \pm 1\%$ ($p < 0.05$) while cerebral NADH remained stable.

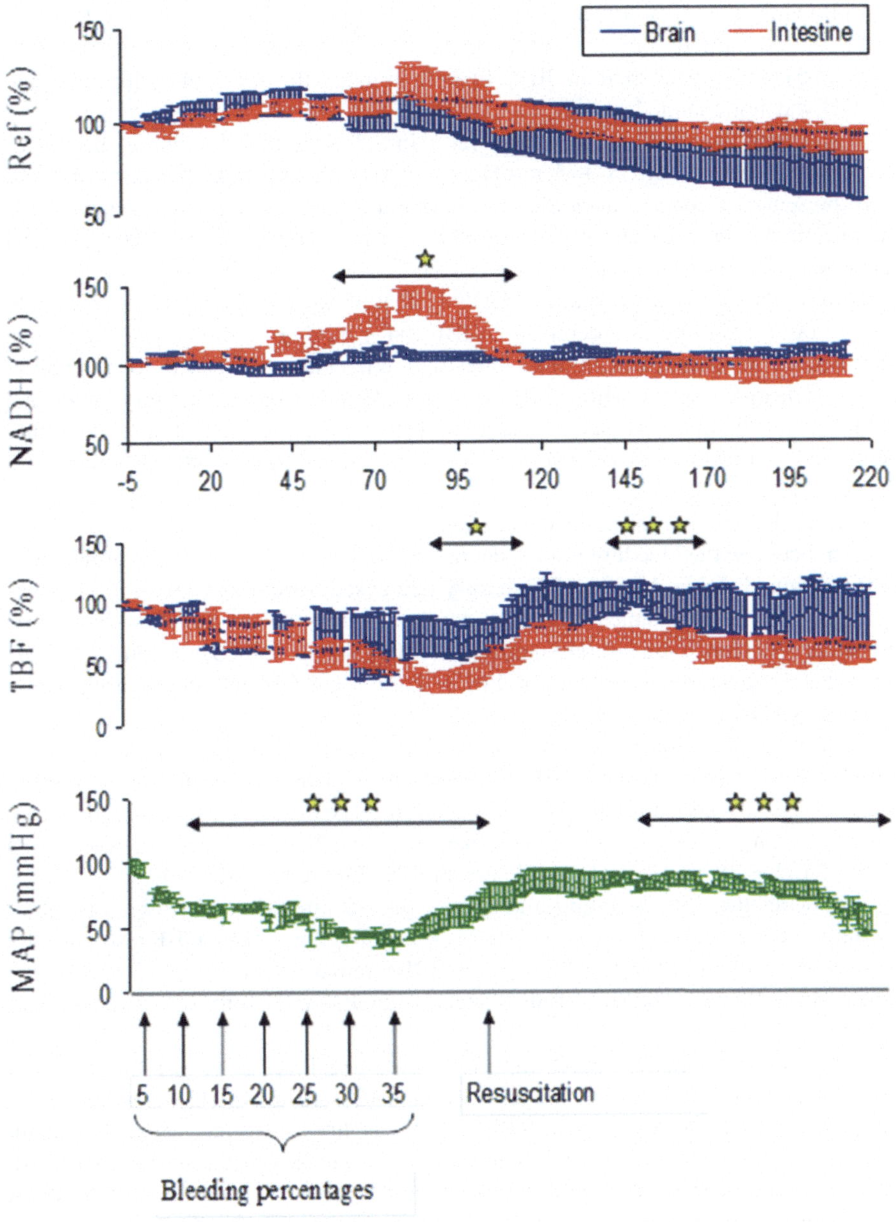

Fig. 5.27 The changes in all parameters in both organs: the brain (blue) and the small intestine (red) under graded hemorrhage in 10-minute intervals in a group ($n = 8$, mean ± S.E). The arrows represent the period in which significant differences were found between the two organs for each minute, and the asterisks represent levels of significance: *$p < 0.05$ and *** $p < 0.001$. Asterisks in Ref, NADH, and TBF mark significance between brain and intestine; asterisks in MAP mark significance compared to baseline. Ref reflectance, NADH mitochondrial NADH redox state, TBF tissue blood flow, MAP mean arterial pressure

Following the 10% of hemorrhage, MAP further decreased to a level of 59.58 ± 5.4 mmHg ($p < 0.001$), leading to a decrease in CBF (70 ± 9%, $p < 0.05$) with no change in IBF. In both organs, the level of reflectance and NADH was not changed.

While bleeding until 15% of the rat's initial total blood volume, the MAP decreased to a level of 53 ± 5 mmHg ($p < 0.001$), and both the intestinal and CBF presented a trend of decrease. The cerebral reflectance remained steady while the intestinal reflectance slightly increased to a level of 107 ± 3% ($p < 0.05$). NADH remained stable in both organs.

During the 20% hemorrhage, MAP decreased to a level of 42 ± 7 mmHg ($p < 0.001$), resulting in a decrease in IBF (64 ± 11%, $p < 0.05$) with no change in CBF. In accordance with that, intestinal reflectance increased to a level of 110 ± 3% ($p < 0.05$) while CBF was not altered. Correspondingly, intestinal NADH immediately increased to a level of 119 ± 3% ($p < 0.01$), but then slightly decreased to a level of 114 ± 6%, which was no longer significantly different from the basal level.

During the 25% hemorrhage MAP, which slightly increased during the 10-minute interval between the bleeding points, decreased back to a level of 43 ± 4 mmHg. The CBF remained stable, but the IBF sharply decreased, reaching a level of 55 ± 9% ($p < 0.001$). In accordance with that, cerebral reflectance remained stable, while intestinal reflectance moderately increased (110 ± 4%, $p < 0.05$). Intestinal NADH immediately increased to a level of 118 ± 4% ($p < 0.01$) and stayed high, while cerebral NADH remained steady.

Following the 30% bleeding, the MAP again decreased to a level of 42 ± 5 mmHg ($p < 0.001$). IBF continued to decrease (47 ± 6%, $p < 0.001$), whereas CBF remained relatively stable. Both the cerebral and intestinal reflectance were not changed. In accordance with that, intestinal NADH sharply increased to a level of 132 ± 5% ($p < 0.01$), while cerebral NADH remained stable.

Following the 35% bleeding and during the 30 min of no perturbation, MAP partially recovered, reaching a level of 75 ± 9 mmHg ($p < 0.5$). CBF also partially recovered, reaching a level of 86 ± 12, and IBF remained steady on relatively low levels (45 ± 8%, $p < 0.001$). Both reflectance and NADH fully recovered in both organs.

Following resuscitation, MAP fully recovered to a level of 87 ± 89 mmHg, but then it decreased back down to the level of 35 ± 12 mmHg, which was significantly lower from the basal level ($p < 0.001$). CBF, at first, fully recovered, but about 10 min after resuscitation ended, it presented a re-decreased pattern (84 ± 22%). IBF also tried to recover, yet only partially reaching a level of 76 ± 9% ($p < 0.05$), but it also deteriorated after resuscitation ended reaching the level of 57 ± 10% ($p < 0.001$). Intestinal reflectance fully recovered in opposed to cerebral reflectance, which had a pattern of gradual deterioration reaching a level of 74 ± 19%. The NADH level in both organs remained stable at a basal level.

A comparison between the two organs reveals significant differences in TBF from the point of 35% bleeding until resuscitation began, namely, during the 30 min of no perturbation ($p < 0.05$) and for 15 min within the recovery phase (about 10 min after resuscitation ended) ($p < 0.001$). Another significant difference can be found in the NADH level, from the end of the period after the 20% hemorrhage until almost the end of the no perturbation period.

Bleeding Sessions

In order to evaluate the effect of each session of bleeding on the brain and intestine, we collected the maximum responses of each parameter and three other important points: the 30 min of no perturbation, the end of resuscitation, and the end of the entire experiment.

As presented in Fig. 5.28, during the 5% hemorrhage, MAP decreased to a level of 71 ± 5 mmHg ($p < 0.001$), yet neither the intestinal nor CBF was significantly changed. In accordance with that, the reflectance and NADH levels were not altered either.

During the second session of bleeding (10% hemorrhage), MAP decreased to 54 ± 4 mmHg ($p < 0.001$), and in response, the IBF decreased to a level of 71 ± 10% ($p < 0.05$); however, CBF was not changed. The levels of the reflectance in both organs were not altered either. In accordance with TBF, intestinal NADH increased (110 ± 3%, $p < 0.05$), while cerebral NADH remained steady.

Throughout the 15% hemorrhage, the MAP slightly decreased (53 ± 4 mmHg, $p < 0.001$). CBF also decreased, reaching a level of 77 ± 7% ($p < 0.05$), and IBF slightly increased, obtaining a level of 73 ± 11% ($p < 0.05$), which was still significantly lower than the basal level. The reflectance and NADH levels in both organs remained unchanged.

Following the fourth session of bleeding (20% hemorrhage), MAP further decreased, reaching a level of 44 ± 3 mmHg. Consequently, IBF decreased reaching a level of 62 ± 10% ($p < 0.05$), whereas CBF increased and was no longer significantly lower than its initial level. Once again, the reflectance remained unchanged in both organs. Yet, intestinal NADH increased up to a level of 115 ± 6% ($p < 0.05$), whereas cerebral NADH stayed stable.

During the 25% bleeding session, MAP decreased to a level of 40 ± 3 mmHg, leading to a sharp decrease in IBF down to the level of 49 ± 8% ($p < 0.001$) with a significant increase of NADH up to the level of 131 ± 10% ($p < 0.05$).

The 30% bleeding session kept MAP on a level of 40 ± 3 mmHg, leading to a deterioration of all intestinal parameters with no change in the cerebral parameters. The blood flow decreased (47 ± 9%, $p < 0.001$), while both the reflectance and NADH levels were increased (122 ± 8%, $p < 0.05$ and 138 ± 4%, $p < 0.001$, respectively).

During the 35% bleeding session, MAP reached 40 ± 4 mmHg. Both organs maintained stable levels of blood flow and reflectance with an elevation in NADH levels. Cerebral NADH reached the level of 111 ± 4% ($p < 0.05$), while intestinal NADH obtained the high level of 148 ± 13% ($p < 0.05$).

Following the period of no perturbation, MAP level partially recovered, reaching a level of 48 ± 7 mmHg ($p < 0.001$). CBF showed a trend of decrease (65 ± 16%)

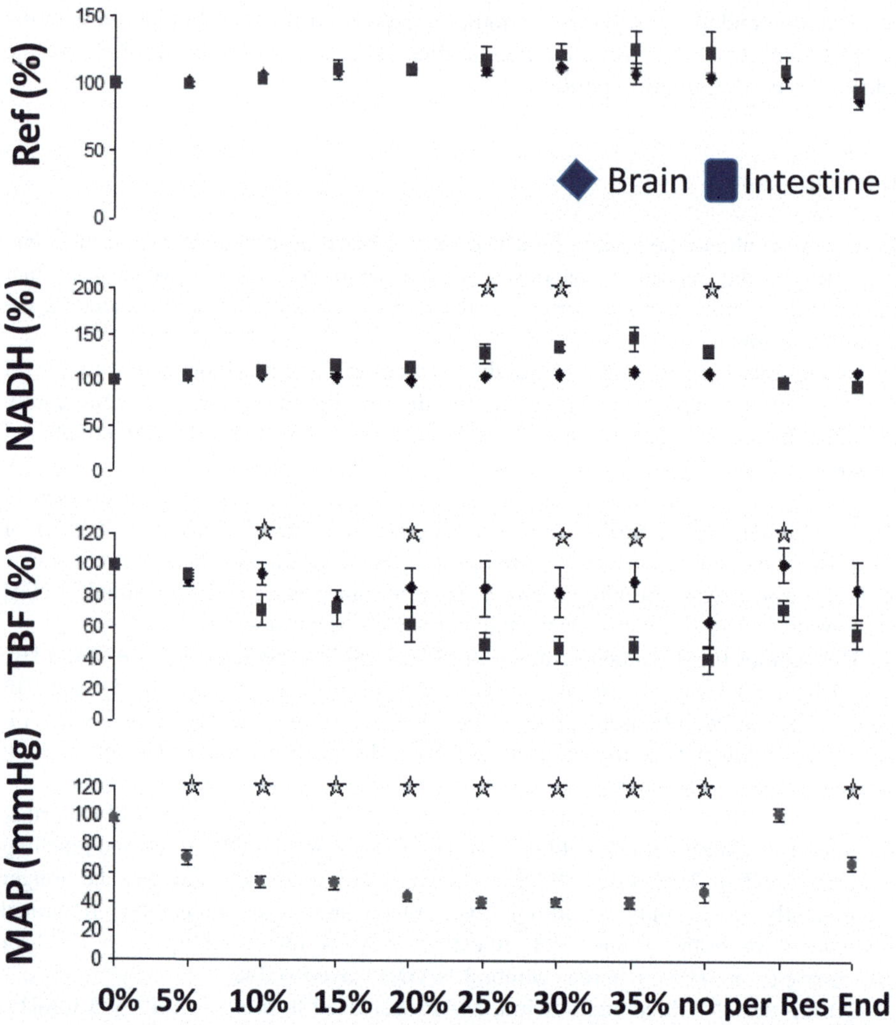

Fig. 5.28 The changes in all parameters in both organs: the brain (circles), the small intestine (squares), and MAP—mean arterial pressure during the bleeding under graded hemorrhage in 10-minute intervals ($n = 8$, mean ± S.E). The asterisks represent significant differences between the two. Ref reflectance, NADH mitochondrial NADH redox state, TBF tissue blood flow, no per the 30-minute period with no perturbation, Res resuscitation and End end of trial

and IBF decreased too (41 ± 9%, $p < 0.001$). The reflectance in both organs remained stable and so did the cerebral NADH, whereas intestinal NADH slightly decreased but remained significantly high (134 ± 6%, $p < 0.001$).

At the end of resuscitation, MAP fully recovered, obtaining a level of 102.25 ± 4.7 mmHg and so did the CBF (102 ± 11%), while IBF only partially recovered (73 ± 7%, $p < 0.01$). The other cerebral parameters remained steady (around basal levels), while the intestinal reflectance and NADH fully recovered ($111 \pm 11\%$ and $99 \pm 3\%$, respectively).

At the end of the trial, MAP was deteriorated again reaching a level of 68 ± 5 mmHg ($p < 0.001$), while all the other parameters remained relatively steady, except for IBF which decreased to a level of $56 \pm 8\%$ ($p < 0.001$) and CBF which showed a trend of decrease ($85 \pm 19\%$).

Comparison between the two organs revealed significant differences in TBF during the 10%, 20%, 30%, and 35% sessions of bleeding and at the end of resuscitation. Significant differences were also found in NADH level during the sessions of 25% and 30% hemorrhage and at the end of the 30-minute period of no perturbation.

Graded Hemorrhage in 10-Minute Intervals Compared to Normotensive Conditions

As can be seen from Fig. 5.29a, MAP in the hypotension group was significantly lower from the normotensive group during the entire hypotension phase ($p < 0.05$ and $p < 0.005$). Following resuscitation, MAP recovered, and there was no significant difference between both groups, which lasted for another hour post resuscitation. However, toward the end of the trial, MAP became significantly lower ($p < 0.01$) in the hypotensive group as compared to the normotensive group. On the other hand, SpO_2 presented no significant differences between the groups.

The cerebral parameters (Fig. 5.29b) also did not exhibit any significant changes between the two groups.

Intestinal TBF (Fig. 5.29c) in the hypotensive group exhibited significant differences from the normotensive group toward the end of the hypotension phase, namely 25% ($p < 0.01$), 30% ($p < 0.05$), 35% ($p < 0.005$), and the session of 30 min of auto recovery ($p < 0.005$). Following resuscitation, there was no longer a significant difference between both groups, but later on the TBF deteriorated again and a significant difference was observed during an hour following resuscitation and an hour later ($p < 0.05$). Intestinal reflectance exhibited no significant differences, probably due to high standard error. Intestinal NADH level was significantly higher in the hypotensive group during the bleeding sessions of 30% ($p < 0.005$), 35% ($p < 0.05$), and the 30 min of auto recovery ($p < 0.01$).

5.3.5.2 Graded Hemorrhage in 5-Minute Intervals

Comparison Between Brain and Intestine

Figure 5.30 presents typical responses to graded hemorrhage, induced every 5 min, in which 5% of the rat's initial total blood volume was removed. As presented, during the 5% hemorrhage, MAP sharply decreased and a distinct difference between the brain and the intestine was formed. While all cerebral parameters

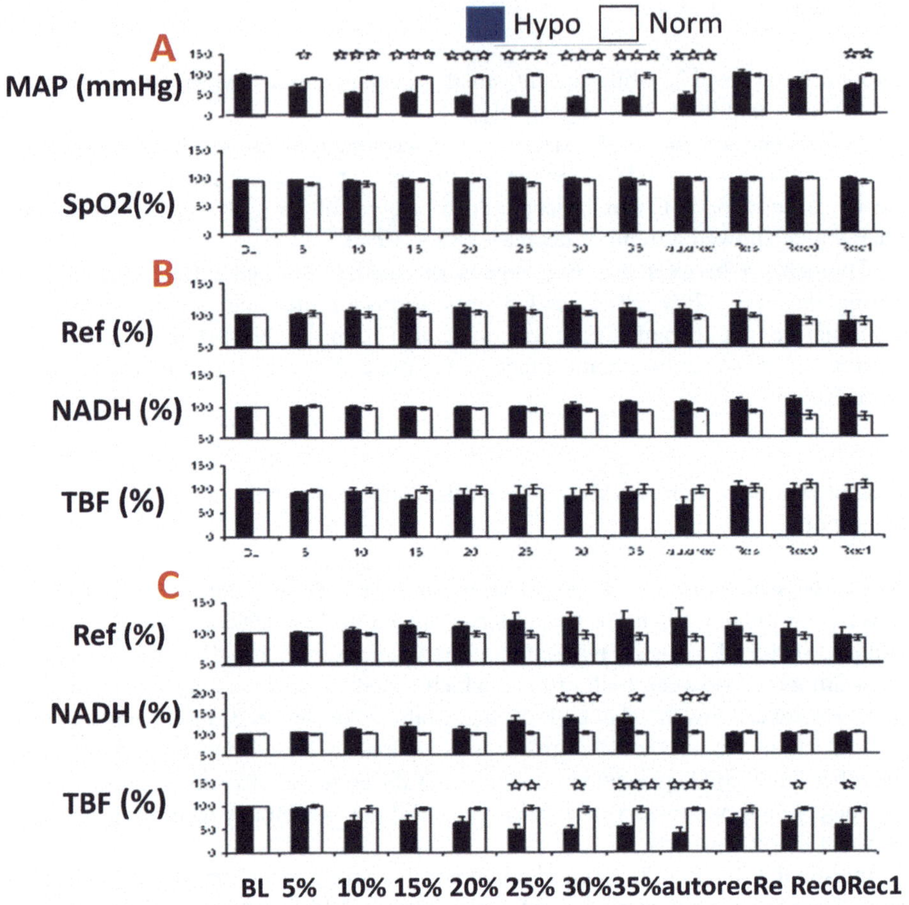

Fig. 5.29 Comparison of the MAP and SpO_2 (**a**), cerebral (**b**), and intestinal (**c**) responses in the group with graded hemorrhage in 10-minute intervals (Hypo-black, $n = 8$) and normotensive group (Norm-white, $n = 4$, mean ± S.E.). The asterisks represent the levels of significance: $^{*}p < 0.05$, $^{**}p < 0.01$, and $^{***}p < 0.001$. B.L. basal level, Hypo end of the hypotension period, Res end of resuscitation, Rec_0 1 h after resuscitation, Rec_1 2 h after resuscitation, Ref reflectance, NADH mitochondrial NADH redox state, TBF tissue blood flow, MAP mean arterial pressure

remained steady, the IBF immediately decreased, and in accordance the reflectance slightly increased and NADH was elevated too.

During the 5-minute interval, MAP partially recovered and so did the IBF. Yet, during the second session of bleeding (10% hemorrhage), both parameters decreased back, leading to an additional elevation in intestinal reflectance and NADH, whereas the cerebral parameters sustained a basal level.

Once again, during the 5-minute interval MAP was partially recovered. During the 15% hemorrhage, the MAP was slightly decreased, followed by full recovery, while the intestinal parameters also partially recovered, though immediately deteriorated and stabilized on lower levels, and the reflectance and NADH demonstrated a slight short-lasting decrease.

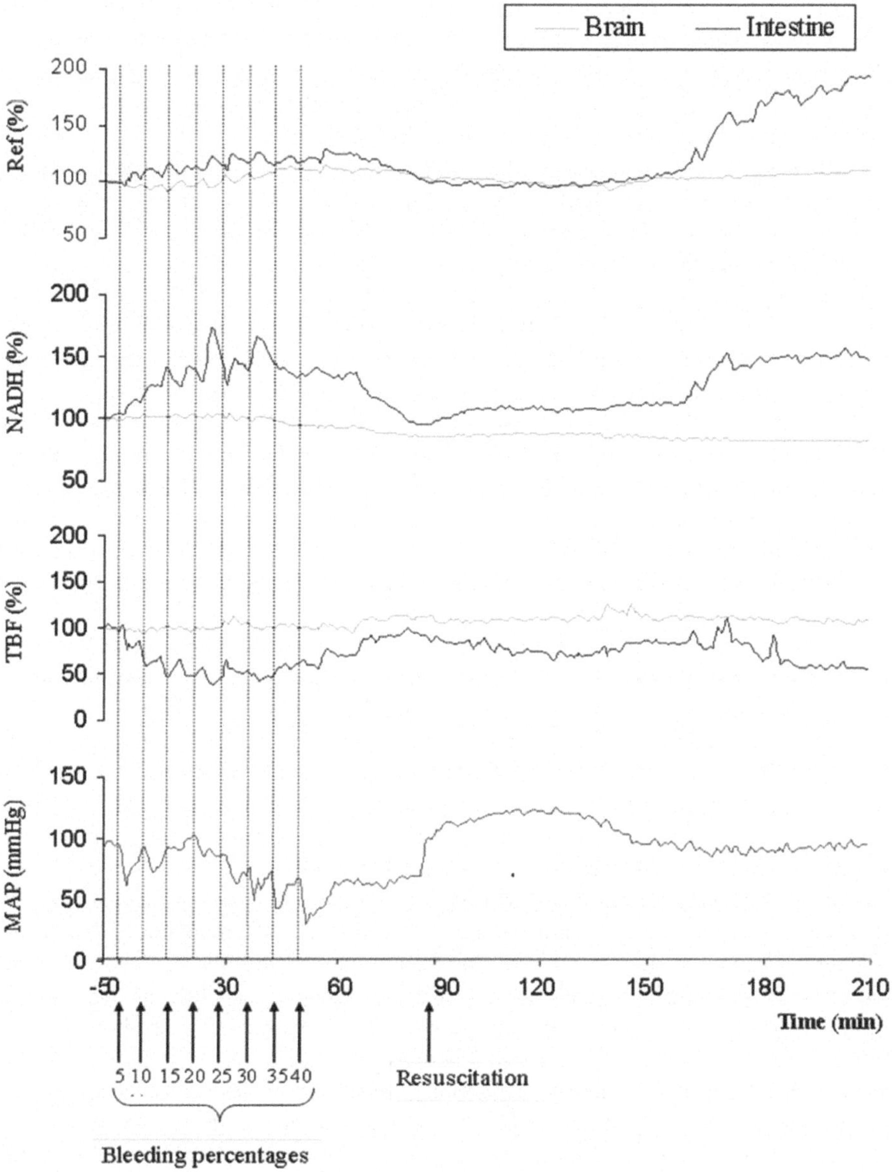

Fig. 5.30 Typical responses of the brain (gray), small intestine (black), and MAP to graded hemorrhage with 5-minute intervals in a single rat. Ref reflectance, NADH mitochondrial NADH redox state, TBF tissue blood flow, MAP mean arterial pressure

However, the 20% hemorrhage was a breaking point in this trial, MAP immediately decreased, and all intestinal parameters showed a maximal response, namely, IBF reached its lowest level, whereas the reflectance and NADH exhibited their highest level. Yet, the cerebral parameters remained relatively steady. Later on, the intestinal parameters fully recovered to their former level.

During the next hemorrhage (25%), MAP sharply decreased, IBF decreased as well, whereas CBF remained stable. Both cerebral and intestinal reflectance increased, and so did the intestinal NADH. After the hemorrhage sessions, MAP and the other parameters only slightly recovered.

Following the 30% hemorrhage, MAP dramatically decreased. IBF continued to decrease but moderately, while CBF remained stable. The reflectance in both organs increased, yet in the intestine, this increase was more intense. While the intestinal NADH sustained its elevation pattern, the cerebral NADH remained steady.

As in the former bleeding sessions, the MAP and the intestinal parameters showed a partial recovery during the intervals between bleeding. Then after during the subsequent hemorrhage (35%), MAP decreased back and IBF recovered very slightly. Both the cerebral and intestinal reflectance were elevated. However, the cerebral and intestinal NADH showed a decrease in their level. Following this hemorrhage MAP, IBF, and intestinal NADH were elevated, while all the other parameters remained stable.

In response to the 40% hemorrhage, MAP decreased and as a result IBF and NADH increased too. All the other parameters sustained a relatively steady state. During the following 30 min, at first MAP sharply increased for about 10 min, but then after it remained quite stable. IBF fully recovered, while CBF remained steady for almost the entire period, but about 10 min before resuscitation CBF was slightly elevated. The reflectance in both organs presented a similar pattern. At first, they were moderately elevated and then started to recover. The intestinal NADH fully recovered, while the cerebral NADH continued to oxidize.

During the resuscitation, MAP sharply increased and remained elevated for an hour, and then moderately decreased back to basal level. IBF moderately decreased with an occasional attempt to recover, and CBF remained slightly elevated. Both the cerebral and intestinal reflectance were fully recovered, yet while in the brain it remained around basal level, in the intestine it showed a large increase about 45 min before the end of the trial. The intestinal NADH also increased, yet in two phases, at first it showed a moderate increase for about 75 min, but then it sharply increased and stayed elevated. In contrast, the cerebral NADH continued to slowly decrease.

Figure 5.31 exhibits the average response to graded hemorrhage of 5% from the rat's initial blood volume with 5-minute intervals. In this group, only 7 out of 10 rats survived.

Following the 5% hemorrhage, MAP drastically decreased from 91.4 ± 2.8 mmHg to the level of 75 ± 5.6 mmHg ($p < 0.05$). As a result, IBF decreased as well, reaching a level of $92 \pm 1.4\%$ ($p < 0.001$), whereas CBF remained stable. In accordance, both the intestinal reflectance and NADH were elevated ($104.38 \pm 1.7\%$, $p < 0.05$ and 107.42 ± 1.5, $p < 0.001$, respectively), while the cerebral parameters maintained a basal level.

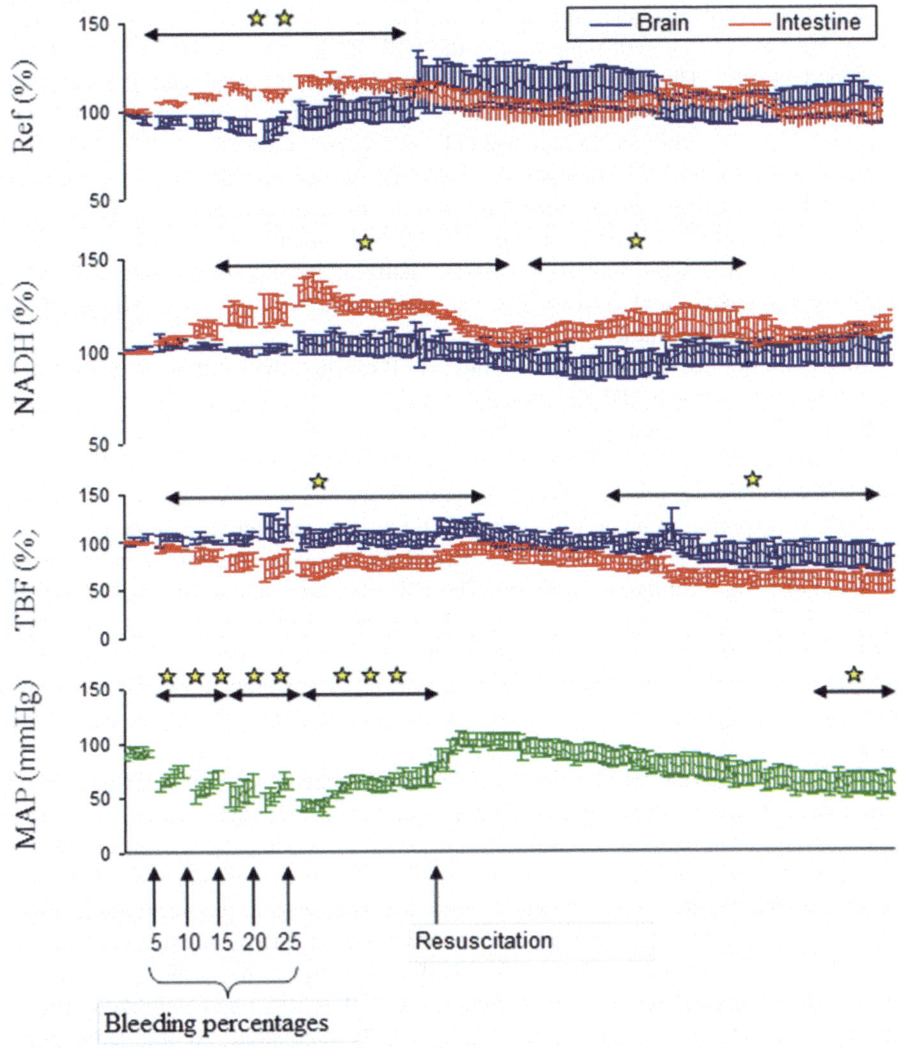

Fig. 5.31 The changes in all parameters in both organs: the brain (blue) and the small intestine (red) under graded hemorrhage in 5-minute intervals in a group ($n = 7$, mean ± S.E). The arrows represent the period in which significant differences were found between the two organs for each minute, and the asterisks represent levels of significance: *$p < 0.05$, ** $p < 0.01$, and ***$p < 0.001$. Asterisks in Ref, NADH, and TBF mark significance between brain and intestine; asterisks in MAP mark significance compared to baseline. Ref reflectance, NADH mitochondrial NADH redox state, TBF tissue blood flow, MAP mean arterial pressure

After the next bleeding (10%), MAP reached a level of 70 ± 6 mmHg. Consequently, IBF continued to decrease (90 ± 4%, $p < 0.05$), while CBF sustained its basal level. Again, the intestinal parameters were elevated while cerebral parameters remained stable. The intestinal reflectance reached the level of 108 ± 2% ($p < 0.05$) and the intestinal NADH increased to 113 ± 3% ($p < 0.05$).

As a result of the subsequent hemorrhage (15%), MAP further decreased (62 ± 10 mmHg, $p < 0.05$) and so did the IBF (82 ± 7%, $p < 0.05$), while CBF remained steady. The intestinal reflectance remained elevated and the cerebral reflectance remained stable. The cerebral NADH remained stable too, although the intestinal NADH continued to increase (119 ± 5%, $p < 0.05$).

In response to the 20% hemorrhage, the MAP further decreased (47 ± 8 mmHg, $p < 0.01$), but during the 5 min of no perturbation, it spontaneously recovered, reaching the level of 63 ± 5 mmHg ($p < 0.05$). As a result, IBF sustained its pattern of decrease, while CBF had an increasing pattern. In accordance with IBF, the intestinal reflectance and NADH also showed a pattern of increase. The cerebral parameters remained stable.

As a consequence of the 25% bleeding, MAP reached the level of 41 ± 7 mmHg ($p < 0.001$). As a result, IBF decreased (71 ± 7%, $p < 0.01$), while CBF maintained its basal level. Consequently, to the changes in IBF, both the reflectance and NADH in the intestine were elevated, though NADH showed a larger increase (115 ± 4%, $p < 0.001$ and 133 ± 6%, $p < 0.001$, respectively).

During the period of no perturbation, MAP partially recovered. At first, it showed a sharp increase (64 ± 4, $p < 0.001$) followed by a more moderate increase reaching the level of 69 ± 9 mmHg ($p < 0.05$). The IBF showed partial recovery (77 ± 6%, $p < 0.05$), while CBF remained steady. The intestinal reflectance fully recovered opposed to the cerebral reflectance, which showed a trend of increase. The intestinal NADH, like the other intestinal parameters, partially recovered and stabilized at the level of 124 ± 2% ($p < 0.001$), while the cerebral NADH remained steady at the basal level.

During resuscitation, MAP fully recovered and reached even higher levels than its initial level (103 ± 6 mmHg, $p < 0.05$), but later on it slowly deteriorated, reaching a level of 61 ± 10 mmHg ($p < 0.05$). IBF presented the same pattern as the MAP. At first, it partially recovered, reaching the level of 92 ± 6%, which was no longer significant. Then after, it was slowly deteriorated until reaching a very low level of 57 ± 11% ($p < 0.05$). The CBF showed an increasing pattern during resuscitation up to 115 ± 9%, followed by a trend of moderate decrease to 82 ± 13%. During resuscitation, cerebral reflectance was elevated (114 ± 12%) and stabilized around those levels for about an hour, afterward, it fully recovered. At the same time, intestinal reflectance fully recovered and stayed steady for the rest of the trial. While cerebral NADH maintained its basal level for the rest of the trial, intestinal NADH at first fully recovered, but then after presented a pattern of increase, it fully recovered until the rest of the trial.

Comparison of both organs revealed significant differences in TBF through all bleeding sessions, during the 30 min of no perturbation phase and an hour before the end of the trial ($p < 0.05$). In the reflectance, differences were observed during all bleeding sessions ($p < 0.01$). As for NADH, significant differences between the two organs were found since the 15% hemorrhage until about 30 min before the end of the trial ($p < 0.05$).

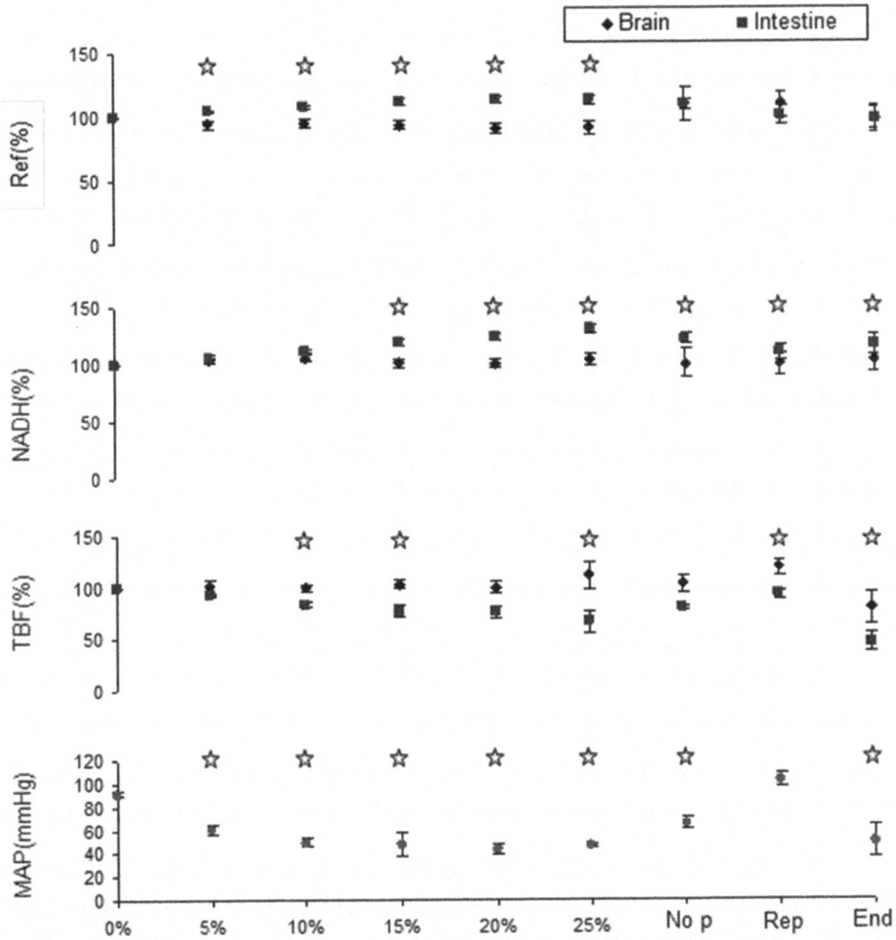

Fig. 5.32 The changes in all parameters in both organs: the brain (circles), the small intestine (squares), and MAP—mean arterial pressure during the bleeding at the graded hemorrhage with 5-minute interval models ($n = 8$, mean ± S.E). The asterisks represent significant differences between the two organs. Ref reflectance, NADH mitochondrial NADH redox state, TBF tissue blood flow, no per the 30-minute period with no perturbation, Res resuscitation, and End end of trial

Bleeding Sessions

Figure 5.32 presents the maximal response of the brain and the intestine to the act of hemorrhage. Throughout the 5% hemorrhage, MAP rapidly decreased from 92 ± 2 mmHg to 61 ± 4 mmHg ($p < 0.001$). Blood flow in both organs remained around the basal level. The reflectance and NADH in both organs were elevated (105 ± 0%, $p < 0.001$ and 106 ± 0%, $p < 0.001$, respectively) as opposed to the cerebral parameters, which remained stable.

During the second bleeding (10%), MAP further decreased (50 ± 3%, $p < 0.001$). Yet, only the intestinal parameters were altered, the blood flow continued to decrease

($84 \pm 3\%$, $p < 0.05$) and the reflectance and NADH increased ($108 \pm 0\%$, $p < 0.001$ and $112 \pm 3\%$, $p < 0.01$, respectively).

The 3rd hemorrhage (15%) was associated with a decrease in MAP to the level of 48 ± 10 mmHg ($p < 0.01$), sustained levels in the cerebral parameters and a further decrease in IBF ($77 \pm 5\%$, $p < 0.05$), and elevation in the reflectance ($112 \pm 0\%$, $p < 0.01$) and NADH ($120 \pm 5\%$, $p < 0.05$).

During the next withdrawal of blood (20%), MAP continued to decrease (43 ± 5 mmHg, $p < 0.001$). The cerebral parameters sustained their levels, while all the intestinal parameters were changed. The IBF decreased to a level of $76 \pm 67\%$ ($p < 0.05$), in accordance with both the reflectance and NADH increased ($113 \pm 3\%$, $p < 0.01$ and $123 \pm 6\%$, $p < 0.01$, respectively).

The last hemorrhage (25%) caused a decrease in MAP to a level of 40 ± 1 mmHg. CBF presented an increasing pattern, while the cerebral reflectance demonstrated a trend of decrease in accordance, and cerebral NADH remained stable. IBF showed a decrease pattern, the intestinal NADH showed an increase trend, and the reflectance maintained its high level ($113 \pm 4\%$, $p < 0.05$).

Following the 30 min of no perturbation, MAP partially recovered, reaching the level of 66 ± 5 mmHg ($p < 0.001$). The CBF and NADH maintained their levels, while cerebral reflectance had an elevation pattern. All intestinal parameters showed partial and even full recovery. IBF obtained a level of $79 \pm 6\%$ ($p < 0.05$), the reflectance decreased and fully recovered, and NADH decreased to the level of $122 \pm 2\%$ ($p < 0.001$).

At the end of the reperfusion session, MAP fully recovered and reached higher levels than the basal levels (103 ± 6 mmHg). CBF showed a trend of increase, whereas the cerebral reflectance and NADH maintained their levels. Both IBF and intestinal NADH partially recovered to $92 \pm 6\%$ and 110 ± 4, respectively. The intestinal reflectance fully recovered ($101 \pm 7\%$).

At the end of trial, MAP was totally deteriorated, reaching a level of 50 ± 13 mmHg ($p < 0.05$). Consequently, the blood flow in both organs decreased, the CBF trend of increase resulted in a level of $79 \pm 16\%$ and IBF reached a level of $46 \pm 9\%$ ($p < 0.01$). The reflectance in both organs maintained a basal level and so did the cerebral NADH, whereas the intestinal NADH showed an increasing pattern.

Comparison between the two organs revealed differences in TBF during the 10%, 15%, and 25% hemorrhage insults, following resuscitation and at the end of the trial ($p < 0.05$). In the reflectance, differences were observed as well, during all bleeding sessions, namely, 5%, 10%, 15%, 20%, and 25% hemorrhage ($p < 0.05$, $p < 0.01$, and $p < 0.00$, alternately). As for NADH differences were found from the 15% hemorrhage until the 25% hemorrhage, at the end of the 30 min of no perturbation and following resuscitation.

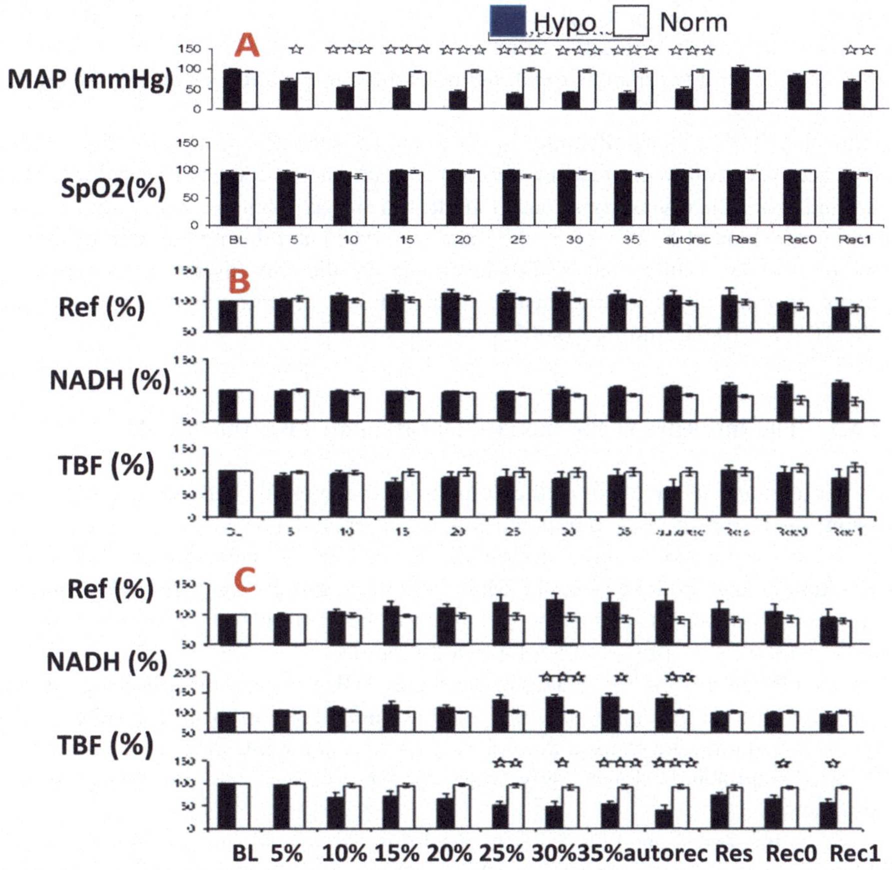

Fig. 5.33 Comparison of the MAP and SpO_2 (**a**), cerebral (**b**), and intestinal (**c**) responses in the group with graded hemorrhage in 5-minute intervals (Hypo-black, $n = 7$) and normotensive group (Norm-white, $n = 4$, mean ± S.E.). The asterisks represent the levels of significance: $*p < 0.05$, $**p < 0.01$, and $***p < 0.001$. B.L. basal level, Hypo end of the hypotension period, Res end of resuscitation, Rec_0 1 h after resuscitation, Rec_1 2 h after resuscitation, Ref reflectance, NADH mitochondrial NADH redox state, TBF tissue blood flow, MAP mean arterial pressure

Graded Hemorrhage in 5-Minute Intervals Compared to Normotensive Conditions

Figure 5.33 presents the responses of the group with graded hemorrhage in 5-minute intervals and the normotensive group. During Hypotension, MAP (Fig. 5.29a) in the hypotension group was significantly lower in all the bleeding sessions and the 30-minute interval before resuscitation compared to the normotensive group ($p < 0.01$ and $p < 0.005$). Following resuscitation, MAP in the hypotension group increased and there was no longer a difference compared to the normotensive group. However, later on, it decreased again until reaching a significant difference toward the end of the trial ($p < 0.05$). SpO_2, on the other hand, exhibited no significant difference during the entire trial.

All the Cerebral parameters exhibited no significant differences during the entire trial (Fig. 5.33b).

Intestinal TBF was significantly lower in the hypotension group only during the 2 h of recovery following resuscitation ($p < 0.05$ and $p < 0.01$, Fig. 5.33c). The intestinal reflectance was significantly higher in the hypotensive group during the bleeding sessions of 5–20% ($p < 0.05$ and $p < 0.01$) and during the auto recovery session ($p < 0.01$). Intestinal NADH level was significantly higher in the hypotension group than in the normotensive group during the entire hypotension phase ($p < 0.05$, $p < 0.01$, and $p < 0.005$)).

5.3.5.3 The Influence of the Interval Duration on TBF and NADH

In order to examine the effect of the duration of the intervals between two bleeding sessions on TBF responses to hypotension, we compared TBF level at 4 time points: the end of hypotension stage, following 30 min of no perturbation, following resuscitation, and on the end of the trial. Since both groups went through different total blood loss, we used only the points until the same blood volume was withdrawn, namely, 25% of the rat's total blood volume.

As can be seen from Fig. 5.34a, in the brain, TBF was significantly lower at the 10-minute interval group versus the 5-minute interval group, from a level of 15% blood loss and until the resuscitation phase, whereas in the intestine (Fig. 5.34b), the IBF was significantly lower only after 25% blood loss and until the end of resuscitation.

As can be seen from Fig. 5.35, in the brain, no significant differences at the NADH level were observed between both groups. Yet, in the intestine, following 15% blood loss and at the end of 30 min with no perturbation, NADH was significantly lower ($p < 0.05$), in the 10-minute intervals group.

5.3.6 Models of Resuscitation

5.3.6.1 Resuscitation with Saline

Comparison Between Brain and Intestine

During the hypotension session (figure not shown), MAP decreased to 40 mmHg and remained low for the entire period, while SpO_2 remained stable.

The blood flow in both organs decreased, yet in the intestine by a greater amount. Both organ reflectance and NADH increased in the same pattern, however, the NADH in the intestine increased to a greater extent.

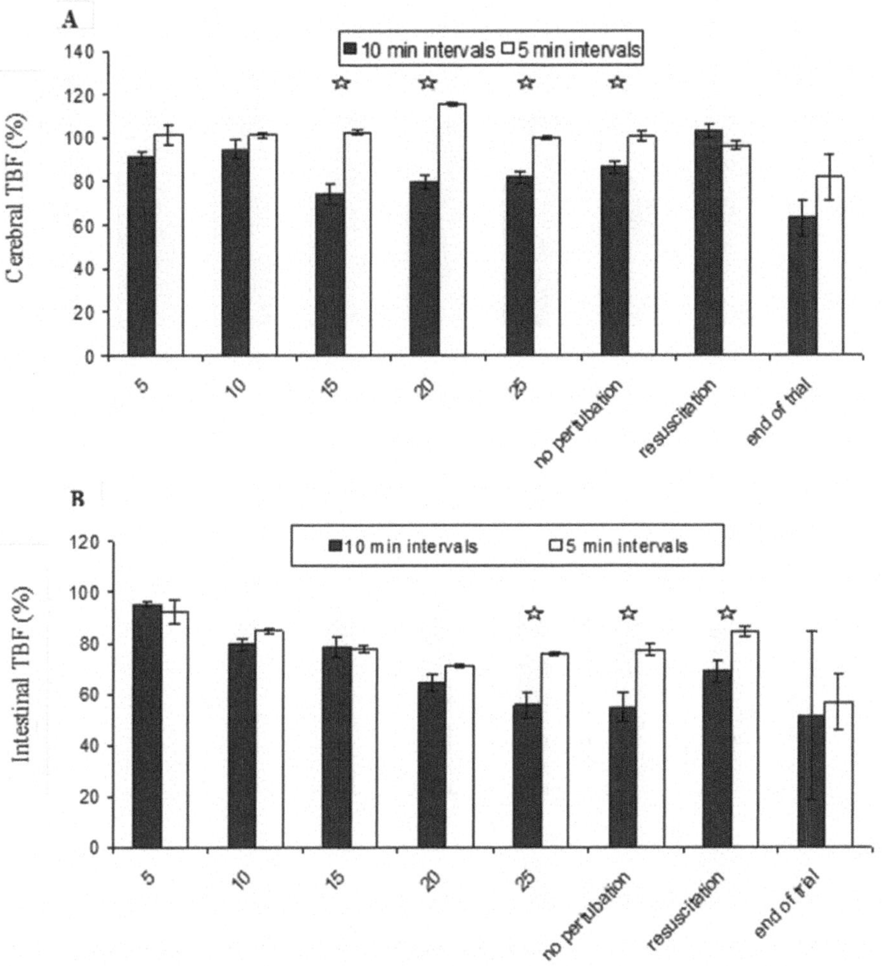

Fig. 5.34 The effect of the duration of the intervals between two bleeding sessions on the TBF levels. (**a**) the effect on cerebral TBF, (**b**) the effect on intestinal TBF (mean ± S.E.). Black—10-minute interval group, white —5-minute interval group. The asterisks represent significant level of $p < 0.05$

Following resuscitation with saline, MAP fully recovered and remained around basal level for the entire trial. Surprisingly, SpO_2 became immediately deteriorated and remained relatively low for the entire time. The CBF exhibited an immediate increase reaching hyperemic levels, followed by a sharp decrease and stabilization at low levels. On the other hand, the IBF made a short and only partial recovery before deteriorating again. Both organs' blood flow remained low for the entire trial.

The reflectance in both organs fully recovered in the same pattern, although the reflectance in the brain reached lower levels. The NADH in the brain fully recovered,

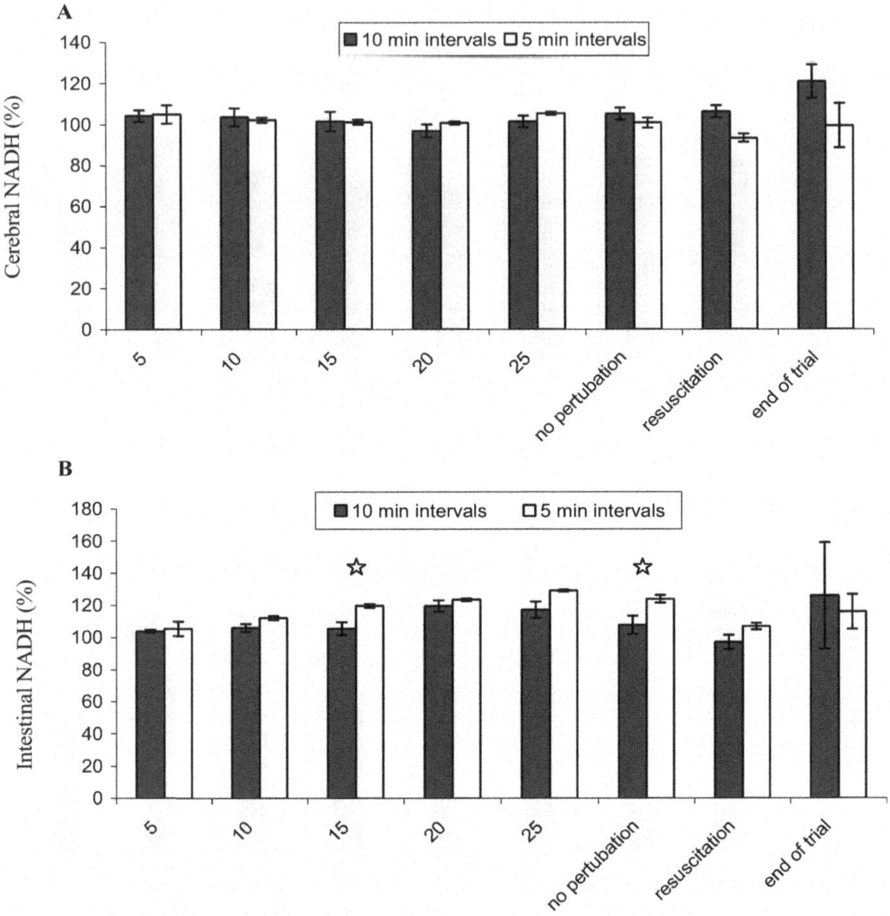

Fig. 5.35 The effect of the duration of the intervals between two bleeding sessions on the NADH levels. (**a**) The effect on cerebral NADH, (**b**) The effect on intestinal NADH (mean ± S.E.). Black—10-minute interval group, white—5-minute interval group. The asterisks represent a significant level of $p < 0.05$

reaching oxidized levels, while the intestinal NADH also recovered but slowly increased again.

Figure 5.36 shows the averaged responses of seven rats to controlled hypotension followed by saline resuscitation. The hypotension was induced by bleeding out $28 \pm 3\%$ of the rats' total blood volume.

During the bleeding session, MAP sharply decreased, reaching the low level of 44 ± 3 mmHg ($p < 0.005$), while the SpO_2 maintained its basal level.

TBF in both organs sharply decreased, reaching a very similar level; $74 \pm 9\%$ in the brain ($p < 0.05$) and 70 ± 10 ($p < 0.05$) in the intestine. Yet, only the intestinal

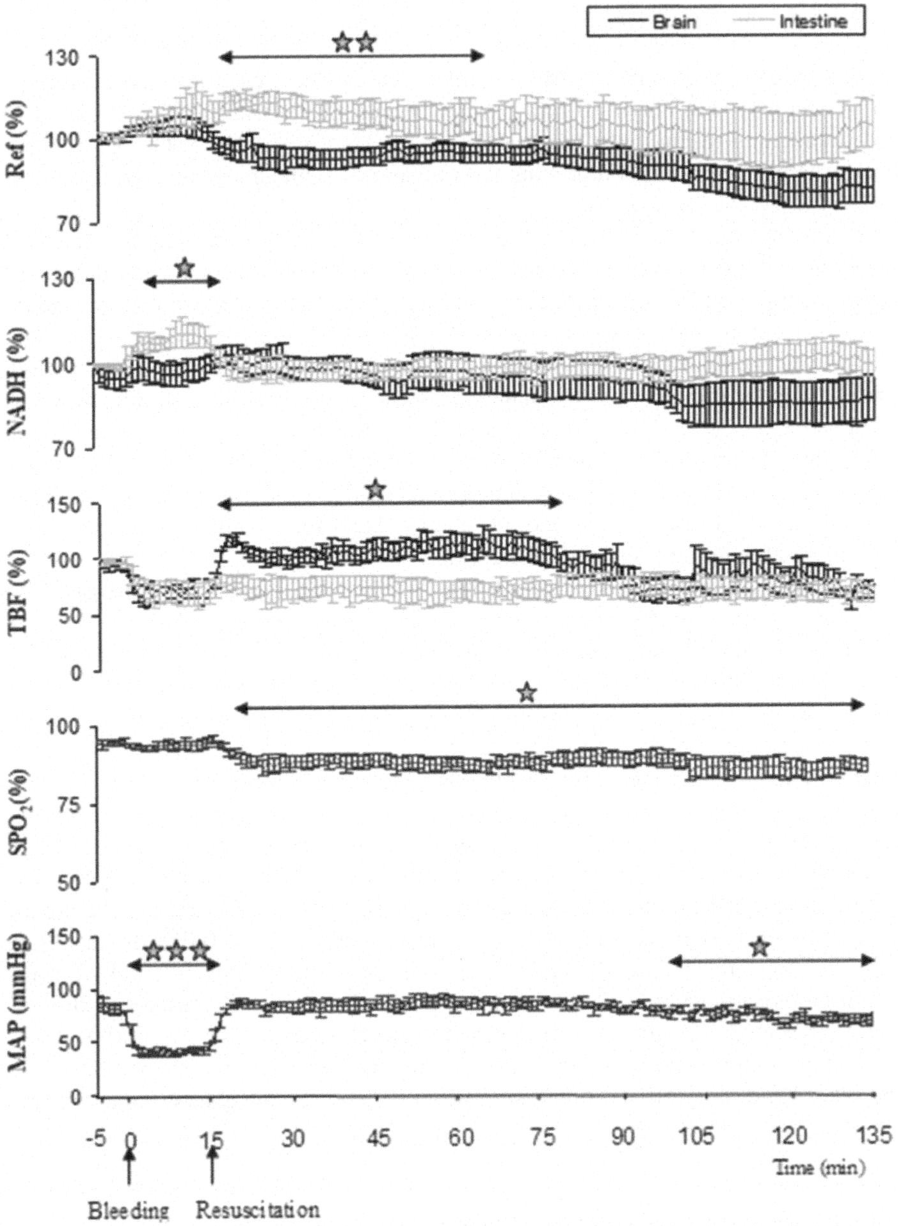

Fig. 5.36 The responses of the brain (black), small intestine (gray), SpO_2 (dark gray), and MAP (dark gray) to controlled hypotension for 15 min followed by resuscitation with saline in a group ($n = 7$, mean ± S.E.). The arrows represent the period in which significant differences were found between the two organs and the asterisks represent significance levels—*$p < 0.05$, **$p < 0.01$, and ***$p < 0.001$. Asterisks in Ref, Flu, NADH, and TBF mark significance between brain and intestine; asterisks in SpO_2 and MAP mark significance compared to baseline. Ref reflectance, NADH mitochondrial NADH redox state, TBF tissue blood flow, MAP mean arterial pressure

reflectance and NADH responded by increasing to a level of 110 ± 5% ($p < 0.05$) and 109 ± 4 ($p < 0.05$), respectively.

Following resuscitation, MAP fully recovered, reaching the level of 85 ± 5 mmHg, but about 30 min before the end of the trial, it slowly started to deteriorate until reaching the level of 69 ± 5 mmHg ($p < 0.05$). SpO_2, on the other hand, sharply deteriorated following the infusion of saline and stabilized at the low level of 86 ± 2% ($p < 0.05$).

Following the infusion of saline, the cerebral TBF immediately increased reaching the high level of 116 ± 5% ($p < 0.05$), but then after it sharply decreased until reaching a full recovery for about an hour, which was followed by a secondary decrease reaching the low level of 69 ± 9% ($p < 0.05$). The IBF, on the other hand, did not react at all to the saline infusion and remained very low for the entire period. In the brain, the reflectance immediately decreased until reaching a significantly low level (80 ± 5%, $p < 0.05$) about 30 min before the end of the trial. The intestinal reflectance, however, continued to increase until reaching the level of 113 ± 4% ($p < 0.05$), and then after it slowly deteriorated until the end of the trial. The cerebral NADH remained stable, while the intestinal NADH fully recovered and remained around basal level until the end of the trial.

In comparing the responses of the intestine and the brain during the bleeding episode, significant differences can be observed only in the NADH ($p < 0.05$). During resuscitation, significant changes were found in the TBF ($p < 0.05$) and reflectance ($p < 0.01$).

Resuscitation with Saline Compared to Normotensive Conditions

As can be seen from Fig. 5.37a, during both sessions of hypotension and 2 h following resuscitation, MAP was significantly lower in the controlled hypotension group ($p < 0.001$ and $p < 0.05$, respectively). However, SpO_2 presented a significant difference only an hour following resuscitation ($p < 0.05$). In the brain (Fig. 5.37b), only TBF at the end of the trial was significantly different from the normotensive group ($p < 0.05$). In the intestine, both the reflectance and fluorescence presented a significant difference following resuscitation ($p < 0.01$). The NADH presented a significant difference during the bleeding session ($p < 0.01$). On the other hand, the TBF exhibited significant differences for almost the entire trial ($p < 0.05$).

5.3.6.2 Resuscitation with Shed Blood and Saline

Comparison Between Brain and Intestine

Typical response to controlled hypotension at 40 mmHg was maintained for 30 min (figure not shown), followed by resuscitation with shed blood or saline. The hypotension was induced by bleeding out 34% of the rats' total blood volume.

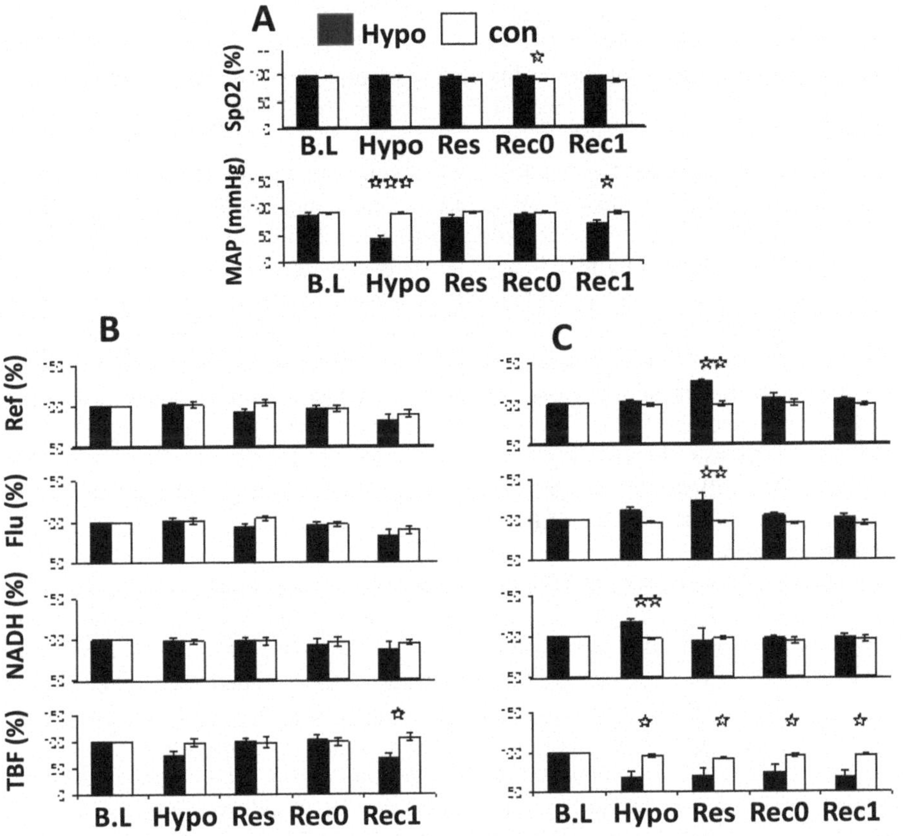

Fig. 5.37 Comparison of the MAP (**a**), cerebral (**b**), and intestinal (**c**) responses in the group with controlled hypotension maintained for 15 min followed by resuscitation with saline (Con 30-black, $n = 8$) compared to the normotensive group (Norm-white, $n = 4$, mean ± S.E). The asterisks represent the levels of significance: $^*p < 0.05$ and $^{***}p < 0.001$. B.L. basal level, Hypo end of the hypotension period, Res end of resuscitation, Rec_0 1 h after resuscitation, Rec_1 – 2 h after resuscitation, Ref reflectance, NADH mitochondrial NADH redox state, TBF tissue blood flow, MAP mean arterial pressure

As hemorrhage started, MAP rapidly decreased and remained low for the entire bleeding session, while SpO_2 remained stable. The cerebral parameters remained stable for the entire hypotension phase, while at the same time, the intestinal parameters were greatly altered. IBF rapidly decreased and remained low, while both intestinal reflectance and NADH were sharply increased and remained elevated.

Following resuscitation, MAP slowly recovered, yet about 30 min later, it started slowly deteriorating again. SpO_2 also increased, yet, on a much smaller scale and also slowly decreased. Cerebral TBF and NADH remained stable for the entire trial, while cerebral reflectance sharply increased and remained high for about 20 min, but then slowly decreased until reaching very low levels toward the end of the trial. IBF fully recovered, yet a half an hour later it started deteriorating again until stabilizing on a very low level. Both intestinal reflectance and NADH at first fully recovered and then increased again, yet in a different pattern. The reflectance immediately increased, while the NADH level at first was oxidized and then after it slowly increased.

Figure 5.38 presents the response of seven rats to controlled hypotension maintained for 30 min followed by a conventional resuscitation, meaning reperfusion of shed blood and twice the volume of saline. In this group, hypotension of 40 mmHg was achieved by bleeding 38 ± 1% of the rat's total blood volume.

During the hypotension period, MAP decreased to a level of 37 ± 1 mmHg ($p < 0.001$) while SpO_2 remained stable. In both tissues, blood flow decreased, yet the intestinal decrease was greater (89 ± 4%, $p < 0.05$ in the brain and 62 ± 3%, $p < 0.001$ in the intestine). However, only the intestinal reflectance and NADH increased in response (122 ± 7%, $p < 0.05$ and 127 ± 8%, $p < 0.05$, respectively). Although TBF was deteriorated in both organs, the mitochondrial response was different. This differentiation may point toward mitochondrial dysfunction in less vital organs versus the ability of the mitochondria of vital organs to cope with the hemorrhagic insult.

Following resuscitation, MAP increased above the basal level, reaching the level of 107 ± 5 mmHg, and SpO_2 remained steady except for a short episode of low levels about 40 min after reperfusion (88 ± 3, $p < 0.05$). Both the cerebral and intestinal TBF fully recovered, but while the CBF reached hyperemic levels (121 ± 6%, $p < 0.05$) and remained significantly high for about an hour after reperfusion, the IBF was deteriorated again about 40 min after reperfusion (60 ± 8%, $p < 0.01$). The intestinal reflectance fully recovered and remained stable for the entire trial, while the cerebral reflectance decreased, reaching the level of 80 ± 8% ($p < 0.01$). In accordance, the intestinal NADH also fully recovered and remained stable, whereas the cerebral NADH became oxidized (83 ± 4%, $p < 0.01$). This NADH oxidation further demonstrates the cerebral tissue's ability to cope with the hemorrhagic insult.

The comparison between the brain and the intestine revealed significant differences in all parameters during the bleeding session; TBF $p < 0.05$, reflectance $p < 0.01$, and NADH $p < 0.05$. The differences in TBF were significant for the entire trial, and the differences in reflectance became significant again toward the last 30 min of the trial.

Resuscitation with Shed Blood and Saline Compared to Normotensive Conditions

As can be seen from Fig. 5.39a, during the hypotension phase and following resuscitation, significant differences between the hypotensive and normotensive groups occurred ($p < 0.005$ and $p < 0.05$, respectively). Two hours after resuscitation, a significant difference between the two groups was exhibited ($p < 0.05$). SpO_2, on the other hand, presented no significant difference between the groups for

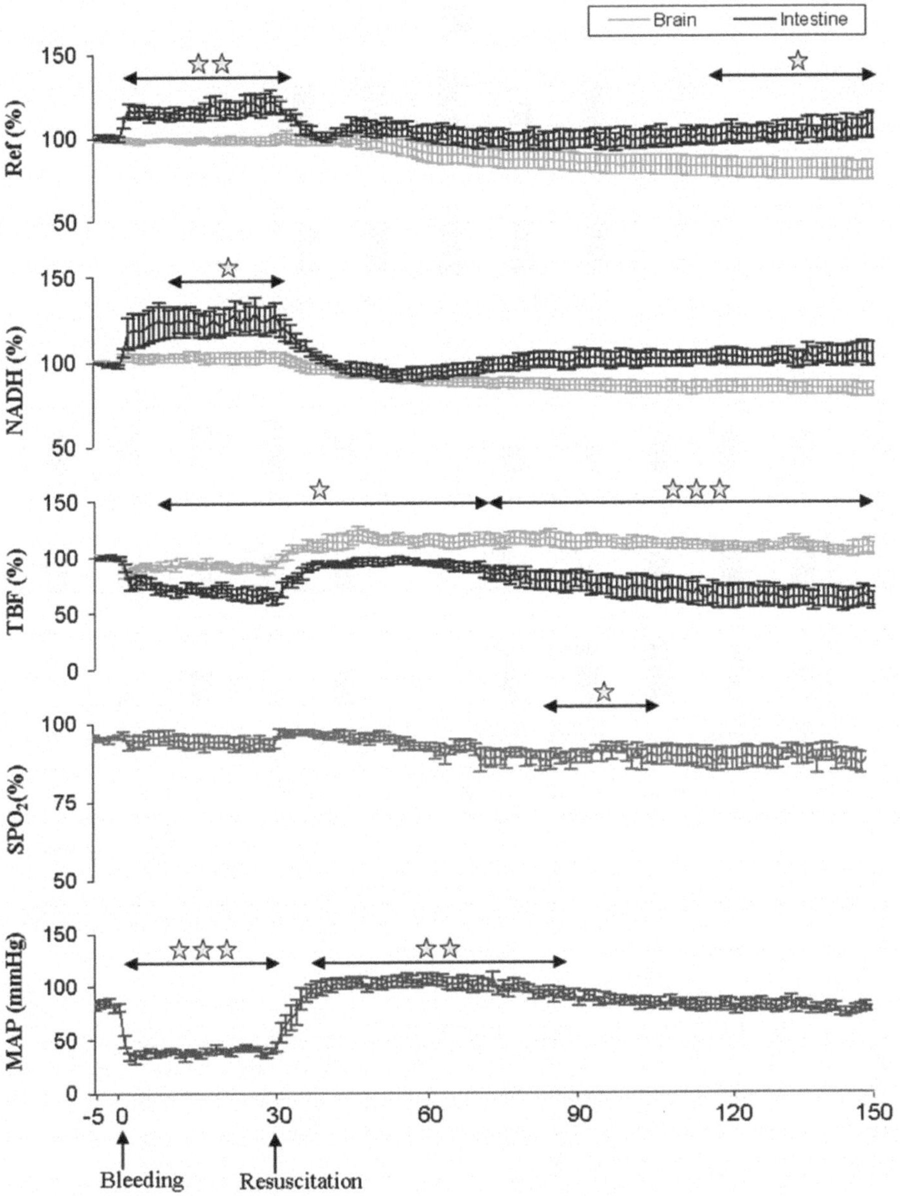

Fig. 5.38 The responses of the brain (gray), small intestine (black), SpO_2, and MAP to controlled hypotension for 30 minutes followed by a conventional resuscitation in a group ($n = 7$, mean ± S. E.). The arrows represent the period in which significant differences were found between the two organs, and the asterisks represent significance levels—$*p < 0.05$, $**p < 0.01$, and $***p < 0.001$. Asterisks in Ref, Flu, NADH, and TBF mark significance between brain and intestine; asterisks in SpO_2 and MAP mark significance compared to baseline. Ref reflectance, NADH mitochondrial NADH redox state, TBF tissue blood flow, MAP mean arterial pressure

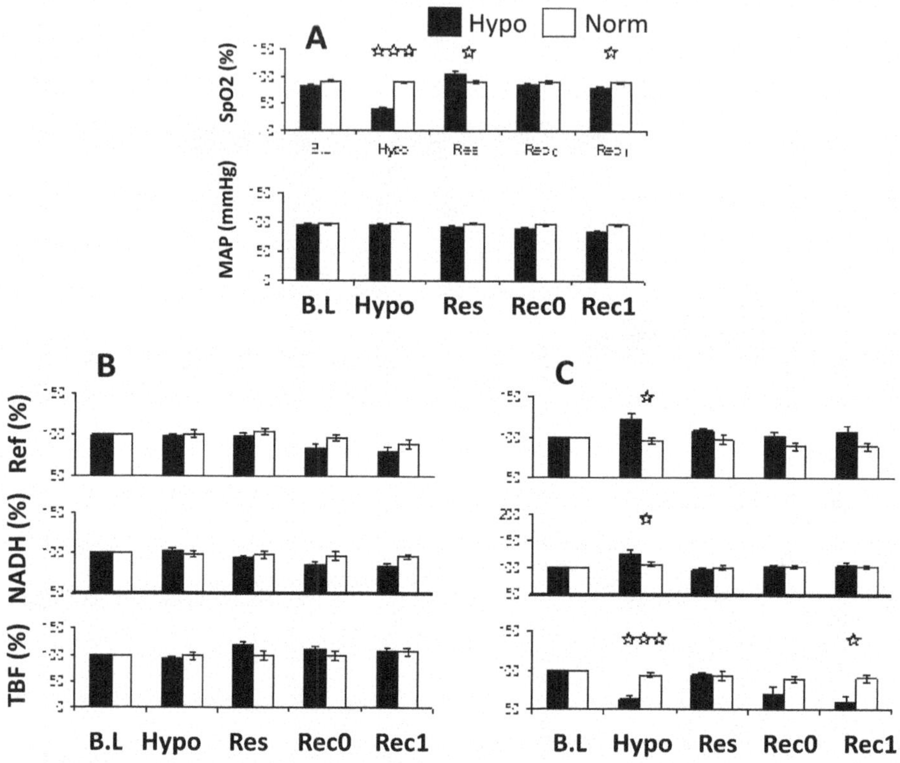

Fig. 5.39 Comparison of the MAP (**a**), cerebral (**b**), and intestinal (**c**) responses in the group with controlled hypotension maintained for 30 min followed by resuscitation with shed blood and saline (Con 30-black, $n = 7$) and normotensive group (Norm-white, $n = 4$, mean ± S.E). The asterisks represent the levels of significance**:** *$p < 0.05$ and ***$p < 0.001$. B.L. basal level, Hypo end of the hypotension period, Res end of resuscitation, Rec_0 1 h after resuscitation, Rec_1 2 h after resuscitation, Ref reflectance, NADH mitochondrial NADH redox state, TBF tissue blood flow, MAP mean arterial pressure

the entire trial. As for the brain (Fig. 5.39b), no significant differences were observed. While in the intestine (Fig. 5.39c), both the reflectance and NADH presented significant differences during the bleeding sessions ($p < 0.05$), and TBF presented significant differences during both the bleeding session ($p < 0.001$) and 2 h after resuscitation ($p < 0.05$).

5.3.6.3 Resuscitation with Blood Substitute

Blood Substitute Injections Compared to Saline Injections

Figure 5.40 exhibits the response of the systemic and local parameters to a short N_2 inhalation following a perfusion of either saline or blood substitute. The purpose of

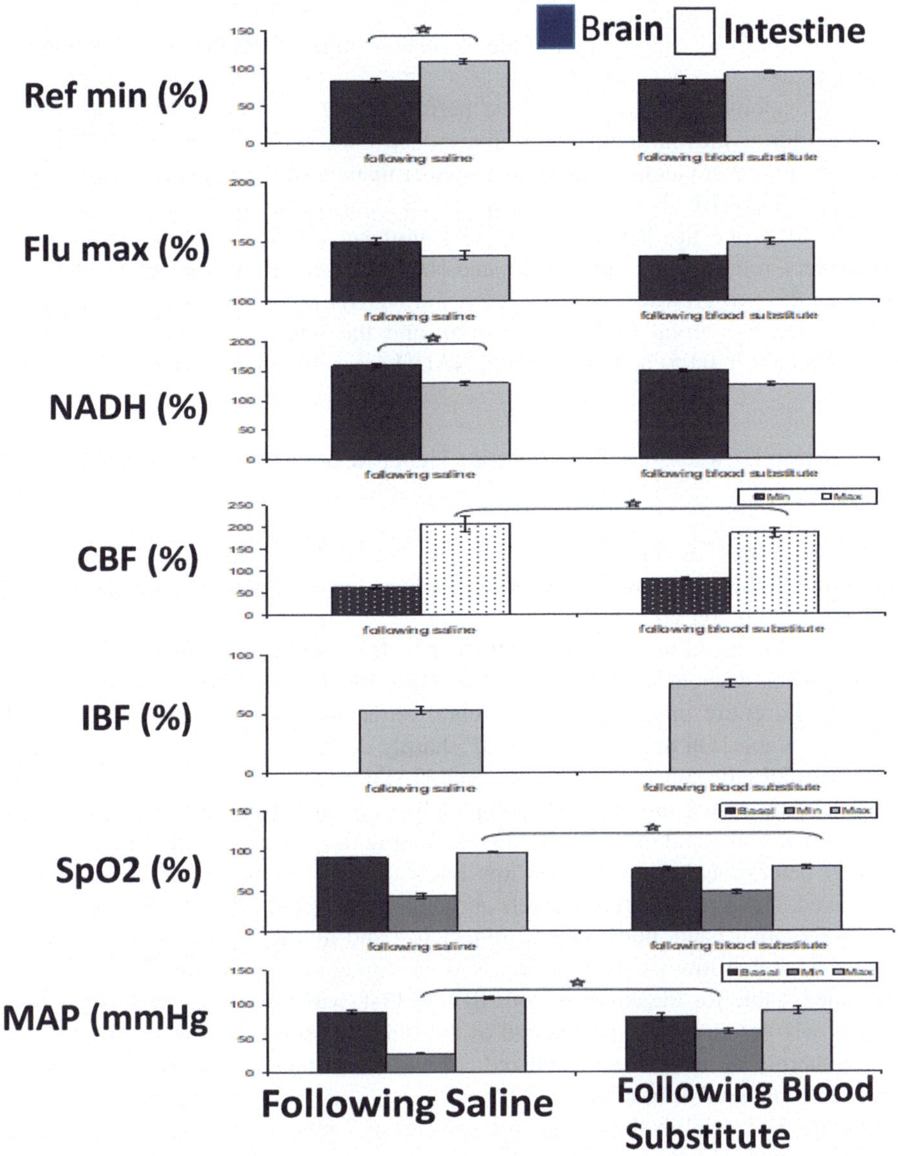

Fig. 5.40 The responses of the brain, small intestine, SpO_2, and MAP to N_2 inhalation following saline injection and blood substitute ($n = 6$, mean ± S.E.). The asterisks represent a significant difference, $*p < 0.05$. Ref reflectance, NADH mitochondrial NADH redox state, CBF cerebral blood flow, IBF Intestinal blood flow, SpO_2 Saturation of peripheral oxygen, MAP mean arterial pressure

this examination was to study the effect of this blood substitute on the rat's responses and vitality, as well as to find out if the blood substitute affects the optical properties of the tissues.

As a response to a blood substitute perfusion, the MAP decreased in a smaller manner (during the minimum phase) in response to the N_2 inhalation ($p < 0.05$). The SpO_2, on the other hand, increased in a smaller manner (during the maximum phase, $p < 0.05$). The CBF also increased in a smaller manner (during the maximum phase, $p < 0.05$), while the IBF presented no significant difference. None of the local parameters, reflectance, fluorescence, and NADH, presented a significant difference between the two groups. The reflectance and NADH presented a significant difference within the group between the brain and the intestine, while the intestinal reflectance ($p < 0.05$) and the cerebral NADH reached a higher level ($p < 0.05$).

Comparison Between Brain and Intestine Under Resuscitation with Blood Substitute

Typical response to controlled hypotension followed by resuscitation with blood substitute (figure not shown). Hemorrhage was induced by bleeding out 30 ± 2% of the rat's total blood volume.

During the bleeding session, MAP sharply decreased and remained low for the entire period, whereas SpO_2 stayed stable. The CBF slowly decreased and remained low for the entire time, while the cerebral reflectance, fluorescence, and NADH remained stable. On the other hand, IBF sharply decreased to a very low level, and the intestinal reflectance, fluorescence, and NADH increased.

Following resuscitation, MAP completely recovered and remained stable until 45 min before the end of the trial, when it sharply deteriorated, while SpO_2 became sharply deteriorated and remained low but stable for the entire trial. CBF slowly recovered, reaching hyperemic levels and remained high for the entire trial.

On the other hand, both cerebral reflectance and fluorescence sharply decreased and stabilized at low levels. Cerebral NADH slowly and lightly decreased too and remained stable for the entire period. IBF, at first, was fully recovered, but then it was slowly deteriorated until the end of the trial. Intestinal reflectance and fluorescence sharply decreased and stabilized at low levels. Intestinal NADH fully recovered and stabilized at basal level.

Figure 5.41 exhibits the averaged responses of 7 rats to bleeding of 28 ± 4% of the rat's total blood volume. During the hypotension phase, MAP decreased, reaching a low level of 38 ± 5 mmHg ($p < 0.001$). SpO_2, on the other hand, remained stable for the entire period. The blood flow in the brain decreased, reaching the low level of 82 ± 4% ($p < 0.05$), whereas the blood flow in the intestine decreased to a level of 73 ± 7% ($p < 0.05$). The cerebral reflectance remained stable, while the intestinal reflectance increased, reaching a high level of 111 ± 5% ($p < 0.05$). The cerebral fluorescence did not demonstrate a significant elevation, while the intestinal fluorescence increased, reaching the high level of 131 ± 5%

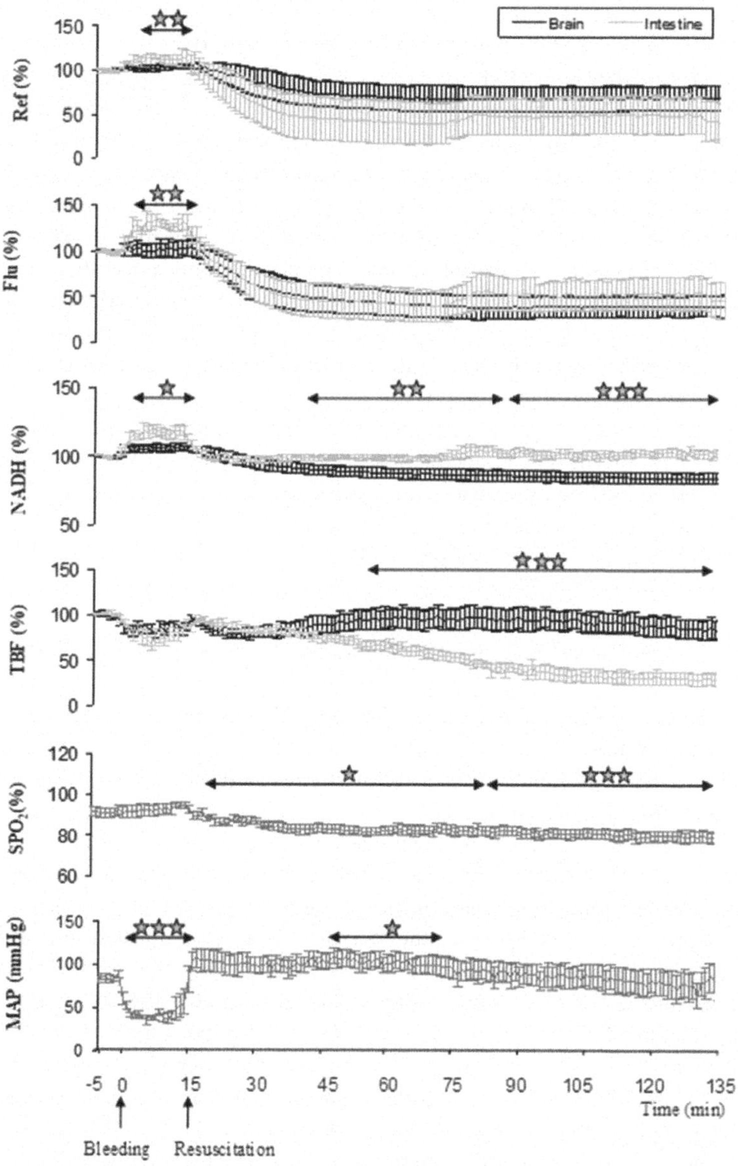

Fig. 5.41 The responses of the brain (black), small intestine (gray), SpO_2, and MAP to controlled hypotension for 15 min followed by a resuscitation with a blood substitute ($n = 7$, mean ± S.E.). The arrows represent the period in which significant differences were found between the two organs and the asterisks represent significance levels—$*p < 0.05$, $**p < 0.01$, and $***p < 0.001$. Asterisks in Ref, Flu, NADH, and TBF mark significance between brain and intestine; asterisks in SpO_2 and MAP mark significance compared to baseline. Ref reflectance, NADH mitochondrial NADH redox state, TBF tissue blood flow, SpO_2 Saturation of peripheral oxygen, MAP mean arterial pressure

($p < 0.001$). The NADH in the brain remained relatively stable, while the NADH in the intestine was elevated to a level of 119 ± 4% ($p < 0.01$).

During resuscitation, MAP fully recovered, reaching a high level of 105 ± 10 mmHg. Then after, it gradually decreased until reaching the basal level. On the other hand, the SpO_2 decreased during resuscitation, reaching the very low level of 79 ± 3% ($p < 0.001$) and remained stable for the entire trial. The CBF, at first, fully recovered, reaching the level of 95 ± 3%, but then after it was deteriorated again (79 ± 5%, $p < 0.05$) for about 35 min and then fully recovered again. The intestinal TBF also fully recovered at first, but immediately after sharply deteriorated, reaching the very low level of 29 ± 7% ($p < 0.001$). The NADH level in both organs fully recovered following the resuscitation, but while the intestinal NADH remained stable for the entire trial, the cerebral NADH was deteriorated again, reaching the level of 85 ± 4% ($p < 0.001$).

A comparison of the organs' responses revealed a significant difference, during the bleeding session, in the reflectance and fluorescence ($p < 0.01$) and in NADH level ($p < 0.05$). Following resuscitation, a difference was observed in TBF, about 35 min after resuscitation began ($p < 0.001$) and in NADH level, about 25 min after resuscitation began ($p < 0.01$ and $p < 0.001$).

Resuscitation with Blood Substitute Compared to Normotensive Conditions

As can be seen from Fig. 5.42a, MAP in the controlled hypotension was significantly lower than in the normotensive group ($p < 0.01$). As for SpO_2, in the controlled hypotension it was significantly lower following resuscitation ($p < 0.05$) and during an hour later and 2 h later ($p < 0.01$).

Figure 5.42b exhibits the brain's responses, while the CBF exhibits no significant differences. In both the reflectance and fluorescence following resuscitation, the controlled hypotension group was significantly lower ($p < 0.05$) and during an hour and 2 h later ($p < 0.001$). Yet, the cerebral NADH exhibited no significant difference. The intestinal responses are presented in Fig. 5.42c; during an hour ($p < 0.05$) and 2 h ($p < 0.001$) after resuscitation, the IBF was significantly lower in the controlled hypotension group. The controlled hypotension group was significantly lower in the reflectance following resuscitation and during an hour and 2 h later ($p < 0.01$). In the controlled group, the fluorescence response was significantly higher during the hypotension phase ($p < 0.05$) and lower following resuscitation ($p < 0.05$) and during an hour later ($p < 0.01$) and 2 h later ($p < 0.05$). And the NADH response was significantly higher during the hypotension phase ($p < 0.05$).

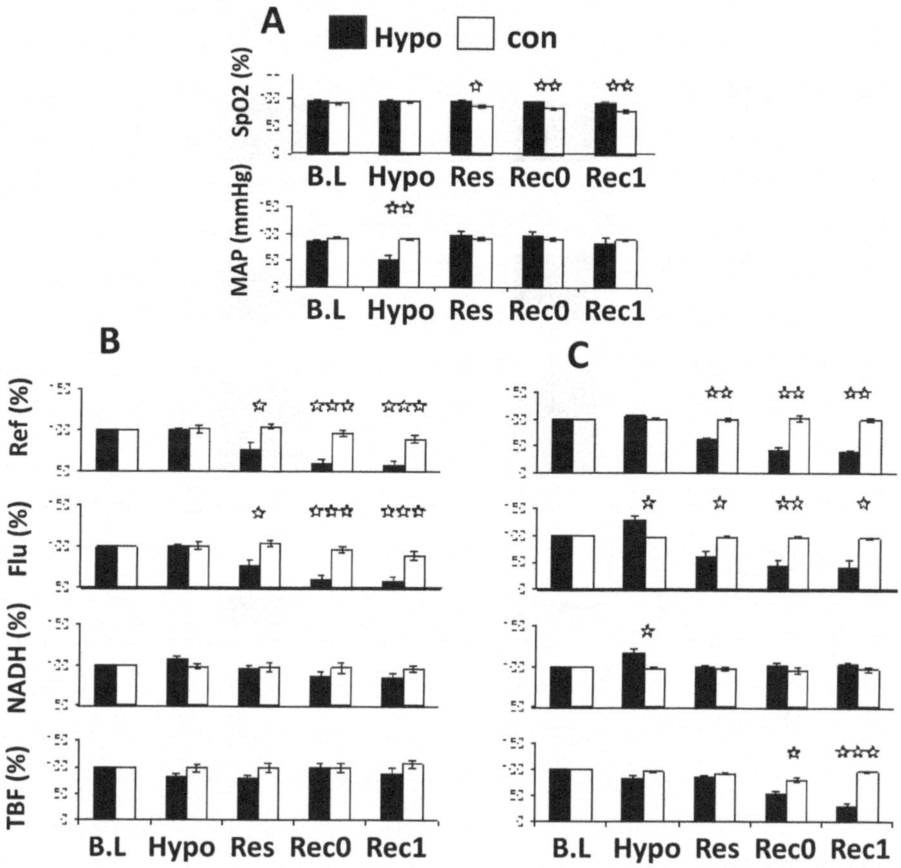

Fig. 5.42 Comparison of the MAP (**a**), cerebral (**b**), and intestinal (**c**) responses in the group with controlled hypotension maintained for 15 min followed by resuscitation with blood substitute (hypo, black, $n = 7$) and normotensive group (con-white, $n = 4$, mean ± S.E). The asterisks represent the levels of significance: $*p < 0.05$ and $***p < 0.001$. B.L. basal level, Hypo end of the hypotension period, Res end of resuscitation, Rec_0 1 h after resuscitation, Rec_1 2 h after resuscitation, Ref reflectance, NADH mitochondrial NADH redox state, TBF tissue blood flow, MAP mean arterial pressure

N_2 Inhalations Before Hemorrhage and 2 H Following Resuscitation with Blood Substitute

Figure 5.43 exhibits the systemic response and the response of both organs to a short N_2 inhalation before the hemorrhage insult and 2 h after resuscitation with the blood substitute (the end of the trial). The purpose of this check was to compare the vitality of the organs before and after the usage of blood substitute.

The only difference in MAP response was at the lowest level reached, which was significantly higher ($p < 0.05$) following the blood substitute usage. As for SpO_2, at

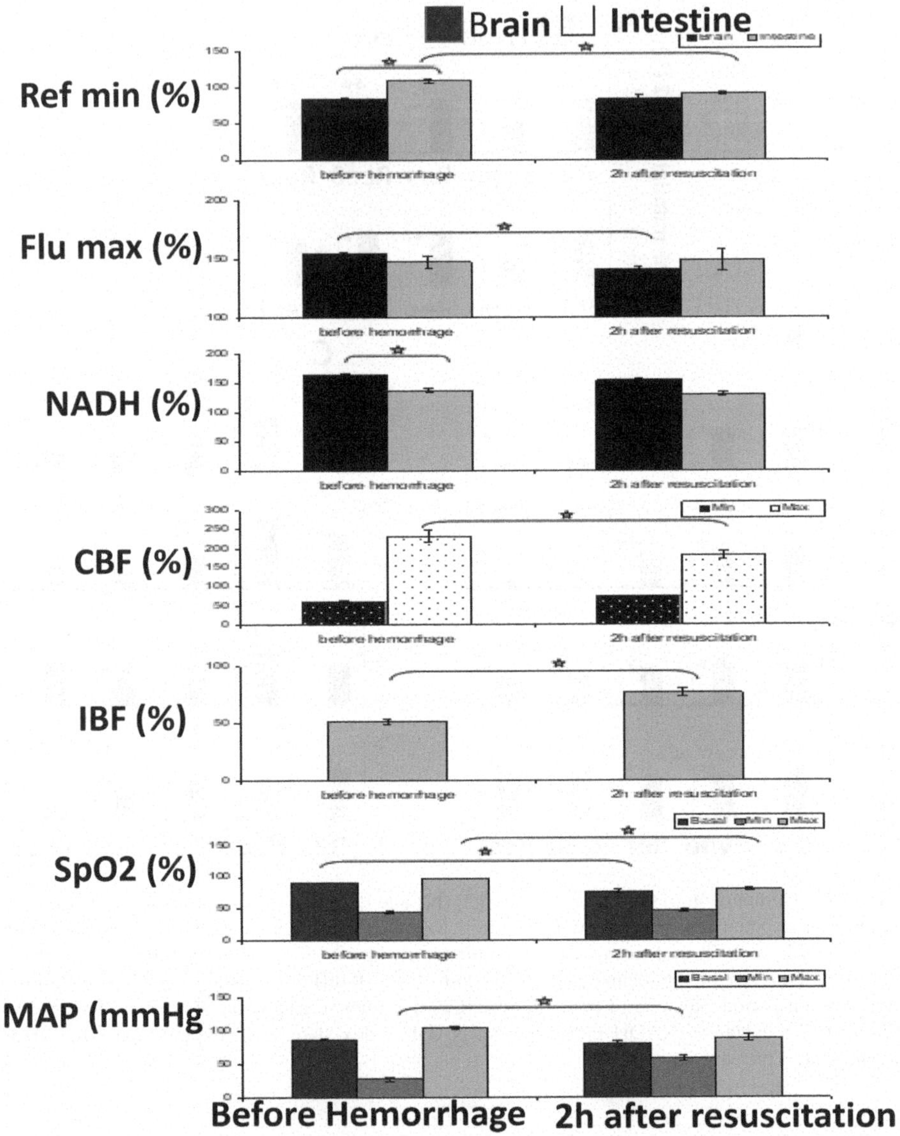

Fig. 5.43 The responses of the brain, small intestine, SpO_2, and MAP to an anoxic episode before the hemorrhage insult and at the end of the trial ($n = 7$, mean ± S.E.). The asterisks represent a significant difference. Ref reflectance, NADH mitochondrial NADH redox state, CBF cerebral blood flow, IBF Intestinal blood flow, SpO_2 Saturation of peripheral oxygen, MAP mean arterial pressure

the end of the trial, both the basal level and the maximal level (the level reached during the recovery from N_2 inhalation) were significantly lower ($p < 0.05$). The CBF presented a significant difference only for the hyperemic phase ($p < 0.05$), in which, following the usage of blood substitute, the hyperemic level was lower. The IBF response following blood substitute usage was significantly reduced ($p < 0.05$). The organs' reflectance response, before the hemorrhage, was smaller in the intestine ($p < 0.05$). Comparing the responses before the trial and afterward revealed a significant difference only in the intestine ($p < 0.05$). No significant difference between both organs was observed for fluorescence in both cases, yet the cerebral response following the blood substitute usage was significantly smaller ($p < 0.05$). As for NADH, the only significant response was between the organs before the hemorrhage, in which the brain reached a higher level ($p < 0.05$).

5.3.6.4 Comparison Between Different Methods of Resuscitation

In order to evaluate the influence of the method of resuscitation on the rats' ability to recover, we compared the results of the different groups, namely, resuscitation with saline, shed blood (blood), shed blood +2 times saline (blood + saline), and blood substitute (substitute). For the comparison, we chose four key points: the end of the hemorrhage, the end of the resuscitation, an hour after the beginning of resuscitation, and an hour later.

As can be seen from Fig. 5.44 and Table 5.3 at the end of the hemorrhage, no significant differences were observed between any of the groups.

Following resuscitation, the most significant differences were observed at the MAP level, while differences were observed between the shed blood + saline group to the groups including saline and blood substitute ($p < 0.001$) and between the shed blood group to the blood substitute group ($p < 0.001$). As for SpO_2, only between the shed blood + saline group to the blood substitute group a significant difference was observed ($p < 0.05$). In CBF, a significant difference was observed between the blood substitute group and the groups of saline and blood + saline ($p < 0.05$), while it reached a lower level. As for the IBF, the group of blood + saline was significantly different from the groups of blood and blood substitute ($p < 0.05$) by reaching a lower level. Only at the intestinal NADH, a significant difference was observed while the NADH level group of blood + saline was significantly higher than all the other groups; $p < 0.05$ at the saline and blood substitute groups and $p < 0.01$ at the blood group.

An hour following the beginning of resuscitation, the only differences were observed in SpO_2 and IBF. In SpO_2, the groups of blood and blood substitute were significantly different ($p < 0.01$), and in IBF the group of blood + saline and the group of blood substitute were significantly different ($p < 0.001$).

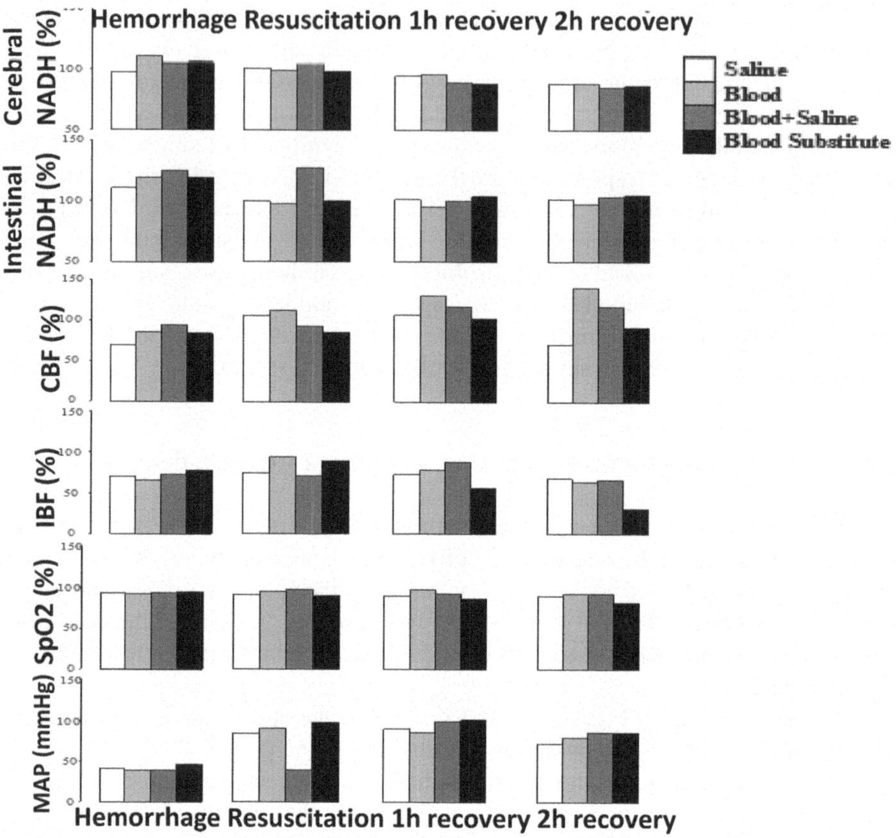

Fig. 5.44 The maximal responses of MAP, SpO2, TBF, and NADH in four time points; Hemorrhage (the end of the hemorrhage session); Resuscitation (the end of the resuscitation session); 1 h recovery (an hour after resuscitation began); and 2 h recovery (2 h after resuscitation began). These points were monitored in different models of resuscitation: infusion of saline (white), blood (pale gray), blood+saline (dark gray), or blood substitute (black)

Two hours following the beginning of resuscitation, significant differences were observed in SpO_2 between the group of blood and the group of blood substitute were significantly different ($p < 0.01$), while in the blood substitute the SpO_2 level was significantly lower. In CBF, the significant differences were observed between the saline group, which reached a significantly lower level, and the groups of blood ($p < 0.05$) and blood + saline ($p < 0.001$). Whereas in IBF, it was the group with blood substitute that was different from all the other groups by reaching a lower level ($p < 0.05$–$p < 0.001$).

Table 5.3 Significant differences in maximum responses following different models of resuscitation

Parameter	Comparison	Hemorrhage	Resuscitation	1h recovery	2h recovery
Cerebral NADH	sal vs. blood	*			
	sal vs. blood+sal				
	sal vs. substitute				
	blood vs. blood+sal				
	blood vs. substitute				
	blood+sal vs. substitute				
Intestinal NADH	sal vs. blood				
	sal vs. blood+sal		*		
	sal vs. substitute				
	blood vs. blood+sal		**		
	blood vs. substitute				
	blood+sal vs. substitute		*		
CBF	sal vs. blood				*
	sal vs. blood+sal	*			***
	sal vs. substitute		*		
	blood vs. blood+sal				
	blood vs. substitute		*		
	blood+sal vs. substitute				
IBF	sal vs. blood				
	sal vs. blood+sal				
	sal vs. substitute				***
	blood vs. blood+sal		*		
	blood vs. substitute				*
	blood+sal vs. substitute		*	***	**
SpO_2	sal vs. blood				
	sal vs. blood+sal				
	sal vs. substitute				
	blood vs. blood+sal				
	blood vs. substitute			**	
	blood+sal vs. substitute		*		*
MAP	sal vs. blood				
	sal vs. blood+sal		***		
	sal vs. substitute				
	blood vs. blood+sal				
	blood vs. substitute		***		
	blood+sal vs. substitute		***		

The level of significant difference between different models of resuscitation; infusion of saline (sal), blood, blood + saline (blood + sal), or blood substitute (substitute). The measured points were taken during Hemorrhage—the end of the hemorrhage session, Resuscitation—the end of the session, 1 h recovery—an hour after resuscitation began—,and 2 h recovery—2 h after resuscitation began. $*p < 0.05$, $**p < 0.01$, and $***p < 0.001$

5.3.7 *Hemorrhage Following Cerebral Ischemia*

5.3.7.1 Hemorrhage Following Unilateral Carotid Occlusion (UCO)

Unilateral Carotid Occlusion Control Group

Figure 5.45 presents the monitoring of rats that underwent permanent unilateral carotid occlusion (UCO) 24 h earlier ($n = 4$). This group was monitored for 140 min, with no perturbations induced during the entire monitoring period. For the entire trial, there were no significant changes in any of the parameters, not in comparison to the baseline and not when the two organs were compared.

Comparison Between Brain and Intestine During Hemorrhage Following UCO

Typical results of controlled hypotension maintained for 15 min in a rat that underwent permanent UCO 24 h earlier (figure not shown). The hypotension was induced by bleeding out 27% of the rat's total blood volume.

Following bleeding, MAP decreased sharply and remained low for the entire period, while SpO_2 remained stable. CBF decreased and remained low, and so did the IBF, although to a greater extent. While the cerebral reflectance remained stable, the intestinal reflectance increased and remained high for the entire bleeding session. Both organs' fluorescence and NADH increased, yet those parameters in the intestine altered to a much greater extent.

Following resuscitation, MAP fully recovered, but later on, it slowly decreased but remained stable. $SpPO_2$ remained stable for the entire period. All parameters in both organs recovered. The cerebral reflectance, at first, increased but then after it recovered. The fluorescence and NADH in the brain decreased after recovery while in the intestine, it increased.

During resuscitation, MAP fully recovered; however, about an hour later, it decreased again and stabilized on a level of 86 ± 4% ($p < 0.05$). As opposed to MAP, SpO_2 was not altered by the reperfusion. The Intestinal TBF partially recovered during the resuscitation session, reaching a significantly low level of 85 ± 6% ($p < 0.05$), yet immediately after it was gradually deteriorated again, reaching a low level of 56 ± 8% ($p < 0.001$), indicating that the intestine is somewhat affected by an ischemic injury. The CBF, on the other hand, remained unchanged. The intestinal reflectance fully recovered, while the cerebral reflectance at first increased (109 ± 3%, $p < 0.05$) but then fully recovered. Both the intestinal and cerebral NADH fully recovered; however, the cerebral NADH decreased further and stabilized at the low level of 87 ± 3% ($p < 0.05$).

In comparing the responses of the intestine and the brain during the bleeding episode, significant differences can be observed with reference to TBF ($p < 0.05$),

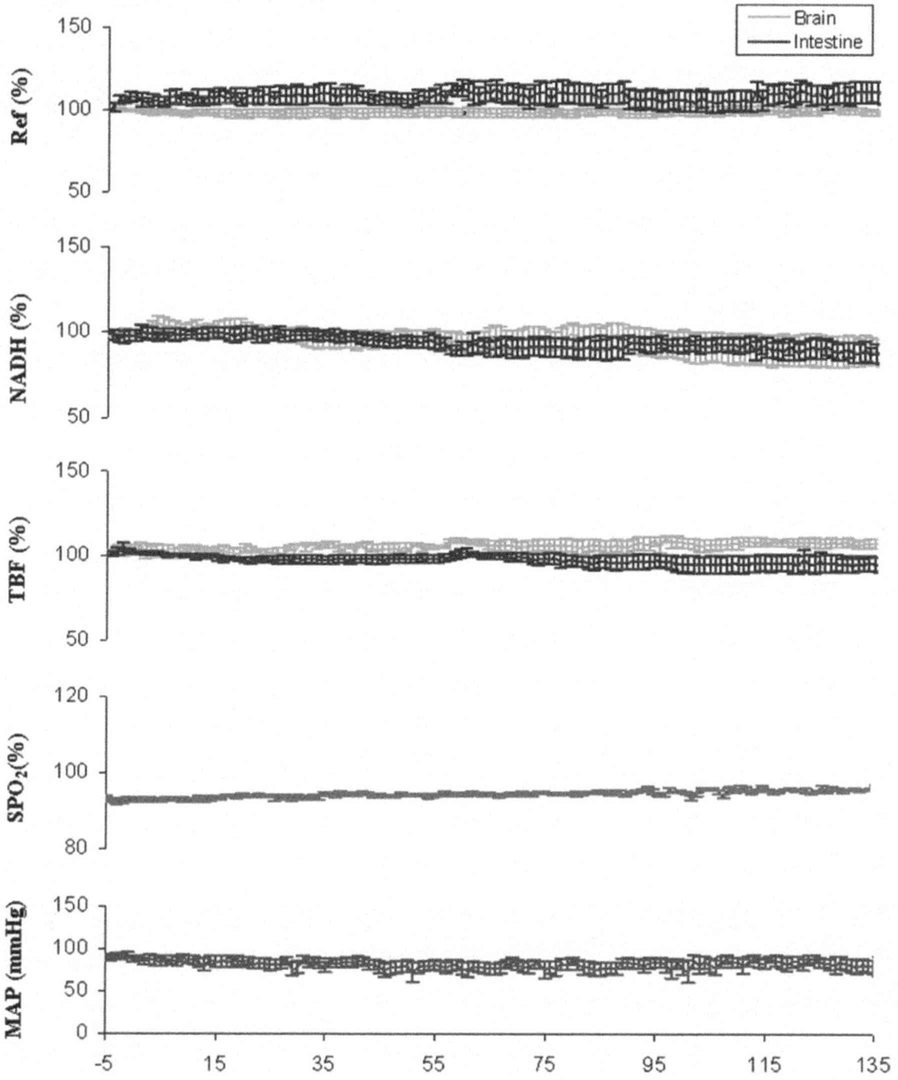

Fig. 5.45 Nontreatment monitoring of brain (gray) and the small intestine (black), SpO_2 and MAP 24 hours following permanent UCO ($n = 4$, mean ± S.E.). Ref reflectance, NADH mitochondrial NADH redox state, TBF tissue blood flow, MAP mean arterial pressure

reflectance ($p < 0.05$), and NADH ($p < 0.01$). During resuscitation, significant changes were found only in NADH ($p < 0.01$ followed by $p < 0.05$).

Figure 5.46 presents the responses of nine ischemic rats to controlled hypotension maintained for 30 min, 24 h following permanent UCO. In this group, hypotension of 40 mmHg was achieved by bleeding 28 ± 3% of the rat's total blood volume.

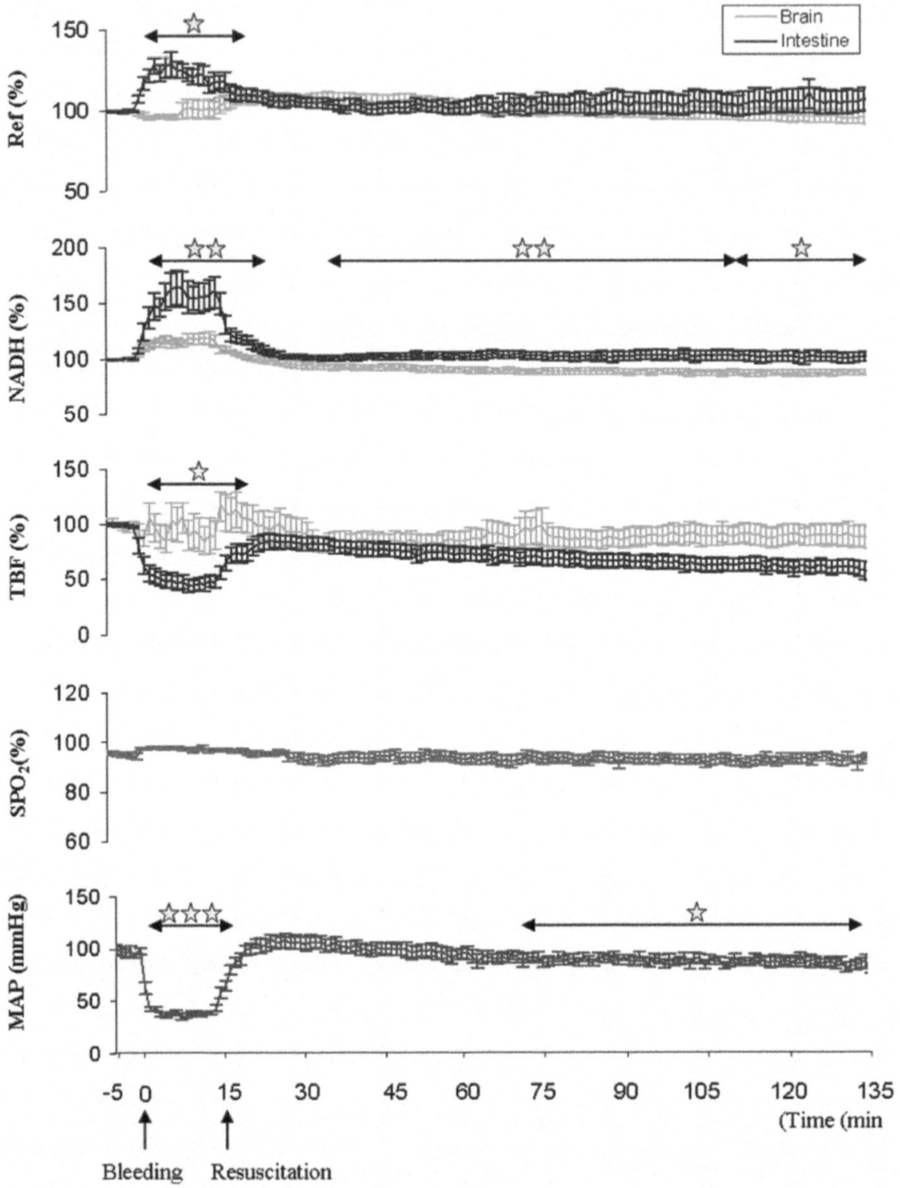

Fig. 5.46 The responses of the brain (gray), small intestine (black), SpO_2, and MAP to controlled hypotension for 15 min in a group of unilateral carotid occluded rats ($n = 9$, mean ± S.E.). The arrows represent the period in which significant differences were found between the two organs and the asterisks represent significance levels—$^*p < 0.05$, $^{**}p < 0.01$, and $^{***}p < 0.001$. Asterisks in Ref, NADH, and TBF mark significance between brain and intestine; asterisks in MAP mark significance compared to baseline. Ref reflectance, NADH mitochondrial NADH redox state, TBF tissue blood flow, MAP mean arterial pressure

During the bleeding session, MAP significantly decreased to a level of 39 ± 2 mmHg ($p < 0.001$) and SpO_2 remained stable. The tissue's blood flow responded differently. While Intestinal TBF immediately decreased (38 ± 4%, $p < 0.001$) and remained low for the entire period, the cerebral TBF showed no significant alterations, probably due to high variability (S.E = 12). This high variability might result from the presence of cerebral autoregulation mechanisms that are activated in order to prevent ischemic injury. The reflectance also exhibited a different pattern. The intestinal reflectance increased, achieving a level of 129 ± 7% ($p < 0.001$), whereas the cerebral reflectance decreased to a level of 96 ± 1% ($p < 0.05$). However, the NADH in both organs increased up to 64 ± 16% ($p < 0.01$) in the intestine and 18 ± 5% ($p < 0.01$) in the brain. These results indicate that the hemorrhagic insult caused mitochondrial dysfunction in both organs; the vital and the less vital organs. The fact that cerebral NADH was significantly altered in opposition to the CBF, highlights the importance of monitoring mitochondrial NADH in addition to TBF. Furthermore, the SpO_2, a systemic parameter, was not altered by the hemorrhagic insults, while the tissue parameters were altered. This emphasizes the high value of monitoring at the tissue level, giving reliable information on the body's metabolic state.

UCO With or Without Hemorrhage

As Fig. 5.47a demonstrates, during Hypotension, MAP levels were significantly lower in the ischemic controlled hypotension group as compared to the MAP level in the ischemic control group ($p < 0.001$). The SpO_2 and cerebral parameters exhibited no differences for the entire trial. (Fig. 5.47b).

Intestinal TBF in the hypotension group (Fig. 5.47c) was significantly lower during the hypotension ($p < 0.001$), yet, during resuscitation, it partially recovered followed by a significant deterioration an hour later ($p < 0.01$), which lasted till the end of the trial. Intestinal NADH was significantly higher during hypotension in the ischemic controlled hypotension group ($p < 0.05$).

Hemorrhage With or Without UCO

MAP Responses

As can be seen from Fig. 5.48, no significant differences were observed for MAP between the normally perfused rats and the rats that underwent unilateral carotid occlusion.

Cerebral Responses

During the hypotension, both groups reacted similarly, yet the cerebral parameters responded in a greater amount. The CBF in both groups decreased, yet in the UCO

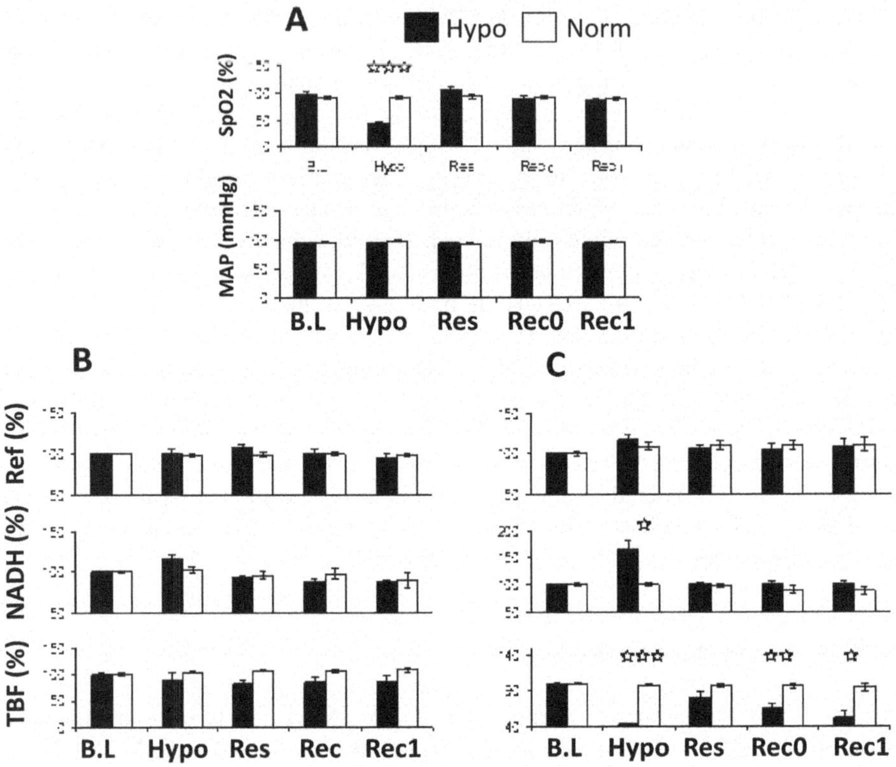

Fig. 5.47 Comparison of the MAP and SpO_2 (**a**), cerebral (**b**), and intestinal (**c**) responses in unilateral carotid occluded rats that underwent controlled hypotension (Hypo-black, $n = 9$) and unilateral carotid occluded rats alone (Norm-white, $n = 4$, mean ± S.E.). The asterisks represent the levels of significance: $*p < 0.05$, $**p < 0.01$, and $***p < 0.001$. B.L. basal level, Hypo end of the hypotension period, Res end of resuscitation, Rec_0 1 h following resuscitation, Rec_1 2 h following resuscitation, Ref reflectance, NADH mitochondrial NADH redox state, TBF tissue blood flow, MAP mean arterial pressure

group it was former. The reflectance reacted the same, while the NADH level in the UCO group was significantly higher ($p < 0.05$).

Following resuscitation, both groups' CBF recovered at first. But then after, the CBF in the normally perfused rats continued to decrease, while the CBF in the UCO group deteriorated again. Leading to a significant level of difference of $p < 0.05$. Both groups' reflectance decreased, yet in the UCO group it was more moderate ($p < 0.05$). The NADH level in the UCO rats fully recovered, while in the normally perfused rats it remained elevated ($p < 0.05$–$p < 0.01$).

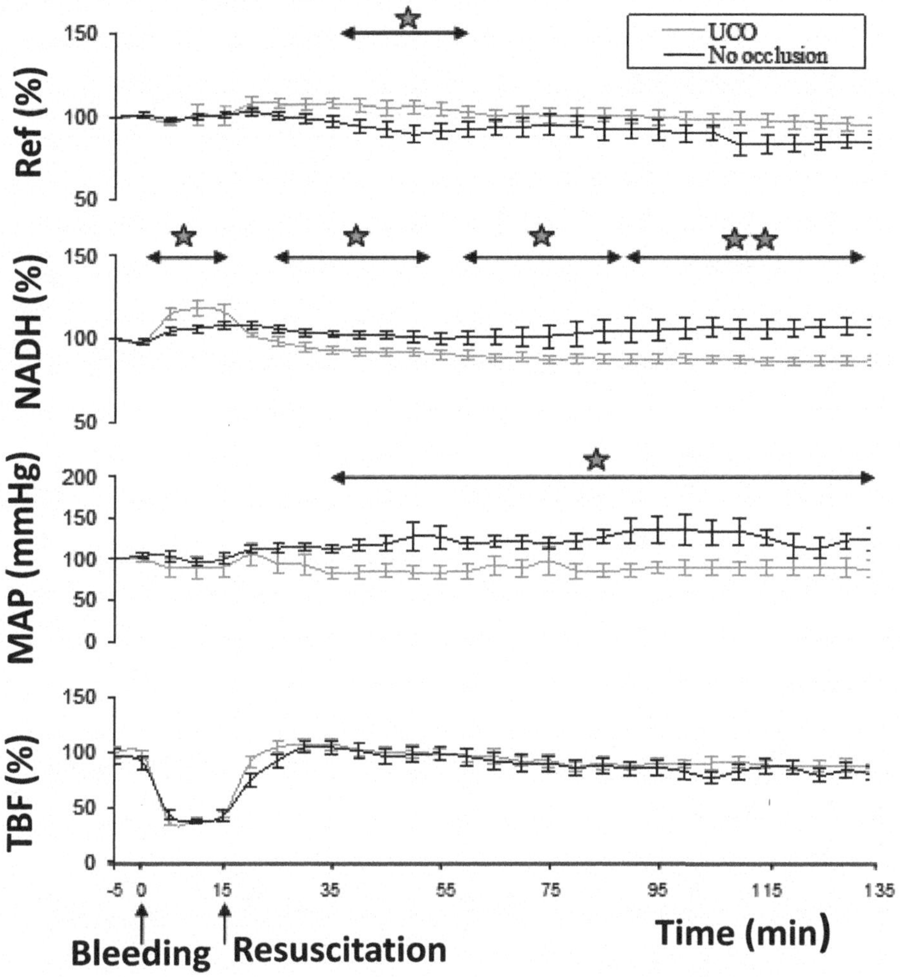

Fig. 5.48 Comparison of the responses of MAP and cerebral parameters to controlled hypotension for 15 min in rats with normal perfusion (No occlusion, black, $n = 9$) and rats that underwent unilateral carotid artery occlusion for 24 h (UCO, pale gray, $n = 9$, mean ± S.E.). The asterisks present levels of significance—$*p < 0.05$ and $**p < 0.01$. Ref reflectance, NADH mitochondrial NADH redox state, TBF tissue blood flow, MAP mean arterial pressure

Intestinal Responses

As can be observed in Fig. 5.49, only during the hypotension period significant differences were observed. In the UCO group, the IBF decreased to a greater extent ($p < 0.01$), and in accordance, both the reflectance and NADH increased to a greater extent ($p < 0.001$ and $p < 0.01$, respectively).

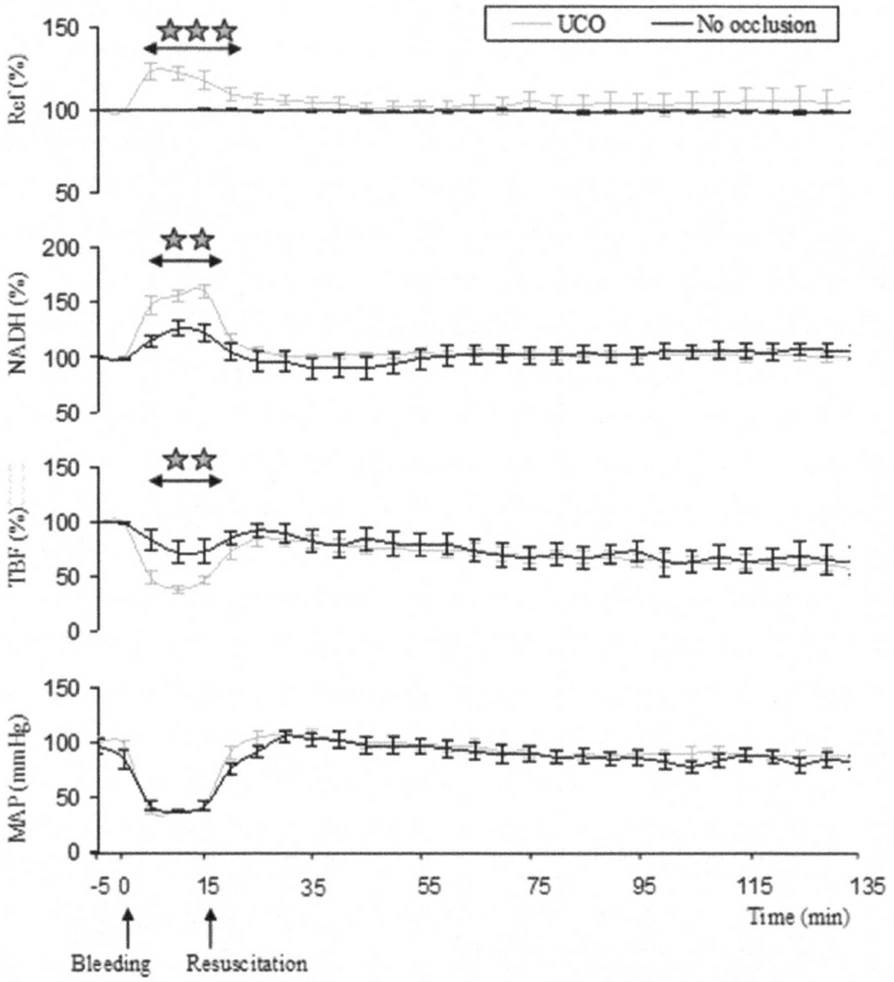

Fig. 5.49 Comparison of the responses of MAP and intestinal parameters to controlled hypotension for 15 min in rats with normal perfusion (No occlusion, black, $n = 9$) and rats that underwent unilateral carotid artery occlusion for 24 h (UCO, pale gray, $n = 9$, mean ± S.E.). The asterisks present levels of significance—$^{**}p < 0.01$ and $^{***}p < 0.001$. Ref reflectance, NADH mitochondrial NADH redox state, TBF tissue blood flow, MAP mean arterial pressure

5.3.7.2 Hemorrhage Following Bilateral Carotid Occlusion (BCO)

Bilateral Carotid Occlusion Control Group

Figure 5.50 presents a nontreatment monitoring of rats that underwent permanent bilateral carotid occlusion (BCO) 24 h prior to the monitoring ($n = 4$). Analysis of these results reveals no significant differences between the two organs in any of the parameters.

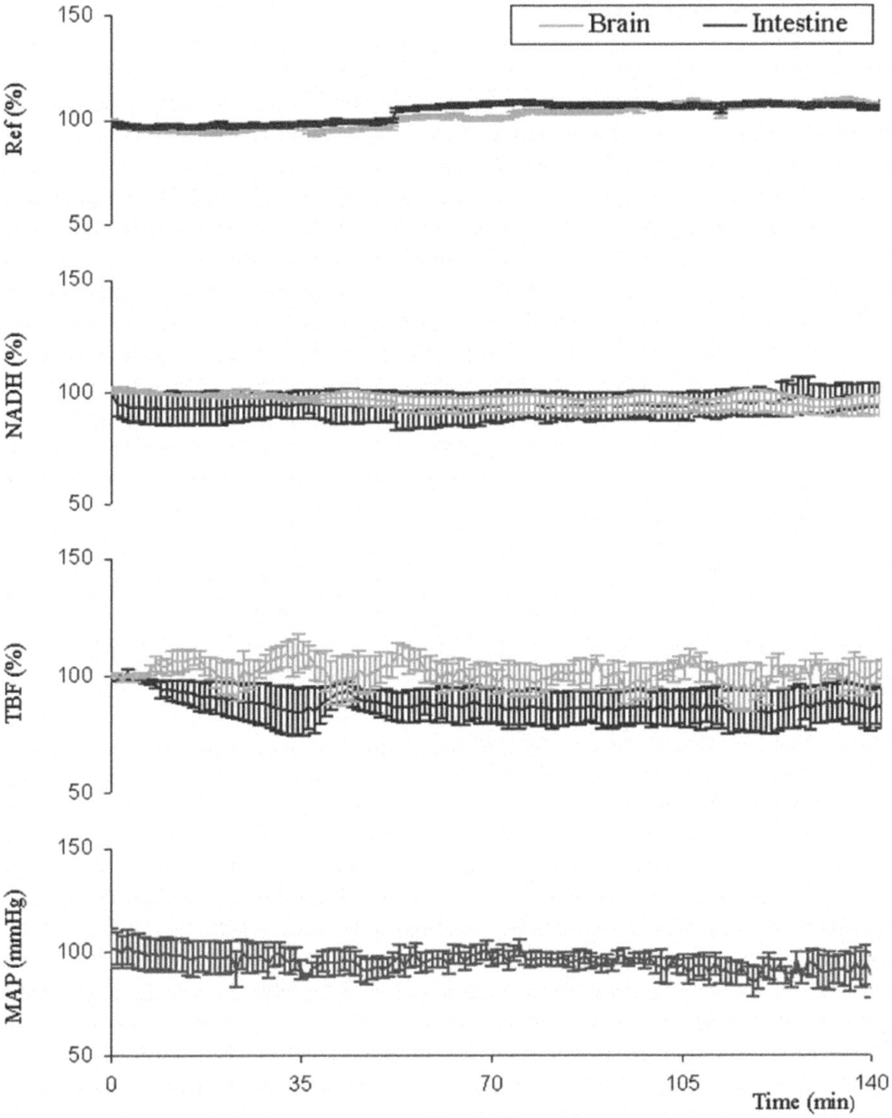

Fig. 5.50 Nontreatment monitoring of brain (gray), small intestine (black), and MAP in rats which underwent BCO 24 h earlier ($n = 4$). Ref reflectance, NADH mitochondrial NADH redox state, TBF tissue blood flow, MAP mean arterial pressure

Comparison Between Brain and Intestine During Hemorrhage Following BCO

Typical results in the permanently bilateral carotid occluded rats to controlled hypotension maintained for 15 min (figure not shown). The hypotension was induced by bleeding out 24% of the rat's total blood volume.

Following bleeding, MAP decreased sharply. Intestinal TBF immediately dropped and remained low for the entire period. The reflectance and NADH increased. TBF in the brain responded in phases, first it rapidly decreased, followed by a short increase for a few minutes, and then it decreased again and remained low for the entire period. The reflectance increased accordingly in two phases; however, a few minutes after the beginning of hemorrhage, the reflectance partially recovered and decreased. Cerebral NADH also increased in two phases.

During resuscitation, MAP increased above its basal level and remained high for the entire experimental period. Intestinal TBF increased only for about 30 min, and then it kept decreasing for the entire experimental period. The reflectance and NADH fully recovered. Hyperemia was observed in the brain, followed by recovery to the basal levels. During resuscitation, the reflectance rapidly decreased toward the basal level. NADH sharply decreased below the basal level, and about 30 min after resuscitation, it slowly increased back toward the basal level.

In the group of ischemic rats with controlled 40 mmHg hypotension maintained for 15 min (24 h following bilateral chronic carotid occlusion), only 7 rats out of 9 survived (Fig. 5.51). This level of hypotension was induced by bleeding out 22% $\pm$ 3 of the rats' total blood volume.

As a result of the bleeding, MAP significantly decreased by 53 $\pm$ 3.5 mmHg ($p < 0.001$). Both organs responded in a similar trend, namely, TBF decreased rapidly and remained low for the entire hemorrhage period. Although the intestine showed a sharp decrease of 56 $\pm$ 10% ($p < 0.001$), the brain responded more severely by a decrease of 79.5 $\pm$ 8% ($p < 0.001$). Accordingly, cerebral reflectance also showed a larger increase than the intestinal reflectance (66.554 $\pm$ 19%, $p < 0.05$ and 30.8 $\pm$ 9%, $p < 0.001$, respectively). Intestinal NADH increased by 45 $\pm$ 12% ($p < 0.01$), whereas cerebral NADH increased by 37 $\pm$ 9% ($p < 0.01$). Both peaks occurred at the end of the hypotension period.

After resuscitation, MAP increased, reaching the level of 112 $\pm$ 9 mmHg (hyperemia), which was followed by a full recovery to the basal levels. IBF increased until reaching a level of 86 $\pm$ 10.5% and then decreased to 46 $\pm$ 8% ($p < 0.001$). In contrast, the CBF increased until reaching a level of 133 $\pm$ 23%, then it decreased to 79 $\pm$ 10.8% ($p < 0.05$). Cerebral reflectance gradually decreased to the basal level. The intestinal reflectance showed a transient recovery, with a decrease and stabilization for 30 min followed by an elevation back to the same levels as were observed during the hypotension (133 $\pm$ 22%). NADH in both organs fully recovered, although in a different pattern. While intestinal NADH rapidly decreased and stabilized around the basal level, cerebral NADH at first rapidly decreased below the basal level and then fully recovered.

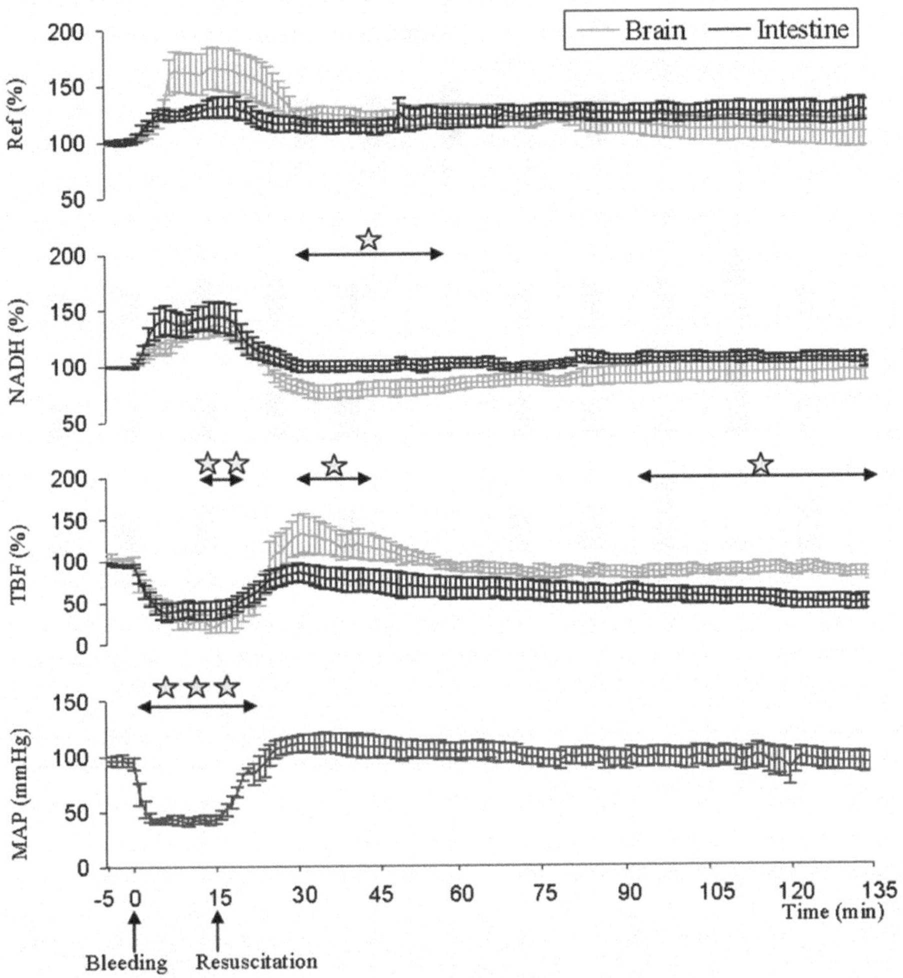

Fig. 5.51 The responses of the brain (gray), small intestine (black), and MAP to controlled hypotension for 15 min following 24 h of permanent BCO in a group ($n = 7$, mean ± S.E.). The arrows represent the period in which significant differences were found between the two organs and the asterisks represent significance levels—$*p < 0.05$, $**p < 0.01$, and $***p < 0.001$. Asterisks in Ref, NADH, and TBF mark significance between brain and intestine; asterisks in MAP mark significance compared to baseline. Ref reflectance, NADH mitochondrial NADH redox state, TBF tissue blood flow, MAP mean arterial pressure

When comparing both organs, during hypotension, a significant difference was observed only for the TBF ($p < 0.01$). Following resuscitation, there were two episodes of significant differences in TBF, immediately after resuscitation itself and an hour later. A significant difference was also observed in NADH during the first 30 min after the resuscitation.

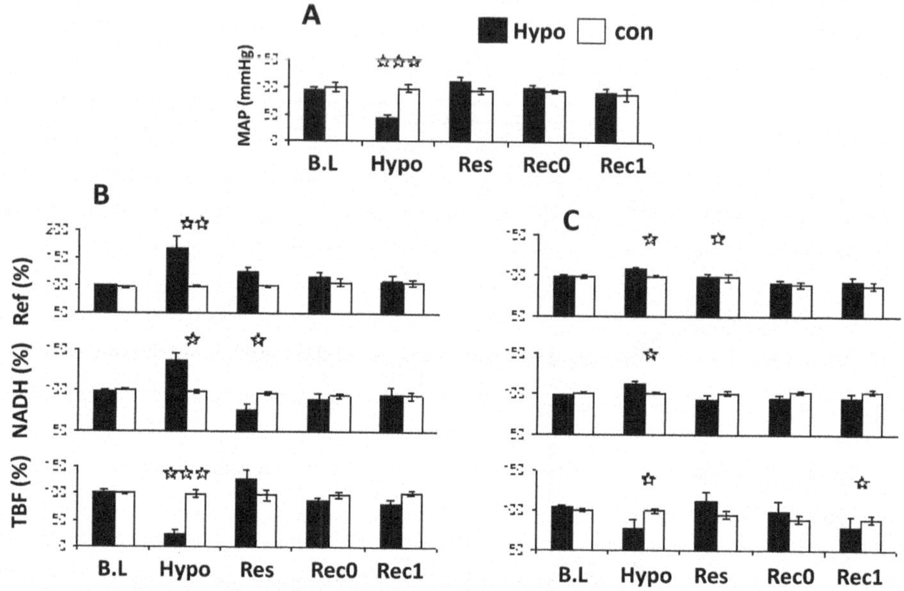

Fig. 5.52 Comparison of the MAP (**a**), cerebral (**b**), and intestinal (**c**) responses in the bilateral carotid occluded rats that underwent controlled hypotension (hypo-black, $n = 7$) and the bilateral carotid occluded rats alone (con-white, $n = 4$, mean ± S.E.). The asterisks represent the levels of significance: $*p < 0.05$, $**p < 0.01$, and $***p < 0.001$. B.L. basal level, Hypo end of the hypotension period, Res end of resuscitation, Rec_0 1 h following resuscitation, Rec_1 2 h following resuscitation, Ref reflectance, NADH mitochondrial NADH redox state, TBF tissue blood flow, MAP mean arterial pressure

BCO With or Without Hemorrhage

As Fig. 5.52a demonstrates, during Hypotension, MAP levels were significantly lower in the ischemic controlled hypotension group as compared to the MAP level in the ischemic control group ($p < 0.001$). Cerebral TBF (Fig. 5.52b) was significantly lower in the ischemic controlled hypotension group during the hypotension phase ($p < 0.001$), while reflectance and NADH were significantly higher at this point ($p < 0.01$ and $p < 0.05$, respectively). During resuscitation, NADH significantly decreased in the ischemic controlled hypotension group ($p < 0.05$).

Intestinal TBF (Fig. 5.52c) was significantly lower during the hypotension ($p < 0.05$), yet, during resuscitation, it greatly increased, followed by deterioration, which became significantly worse toward the end of the experiment. During both hypotension and resuscitation, the intestinal reflectance was significantly higher in the ischemic controlled hypotension group ($p < 0.05$). Intestinal NADH was also significantly higher during hypotension in the ischemic-controlled hypotension group.

Hemorrhage With or Without BCO

MAP Responses

In both the ischemic and the normal rats, MAP responded similarly and no significant differences were observed between the two groups (Fig. 5.51). Comparison of the responses of the normal and ischemic brain showed major differences.

Cerebral Responses

As represented in Fig. 5.53, in the ischemic brain followed by hemorrhage, TBF dramatically decreased and the reflectance and NADH increased, while in the normal brain only minor changes were observed in these parameters, therefore there is a $p < 0.001$ significant difference for CBF and $p < 0.01$ for both reflectance and NADH. Following resuscitation, TBF in the ischemic brain increased to hyperemic levels. Consequently, NADH and the reflectance decreased. However, in the normal brain, the parameters remained steady. These responses were significantly different for the two groups during the major parts of the experimental period ($p < 0.05$–$p < 0.01$).

Intestinal Responses

As can be seen in Fig. 5.54, in both models, the small intestine responded with a similar pattern, except for the reflectance which was significantly higher in the ischemic rats versus the normal rats through the entire experimental period, with various significance levels ($p < 0.01$ and $p < 0.05$).

Survival Rates

In this group, only the hypotension and resuscitation periods were normalized, since the two non-surviving rats died right after the resuscitation phase. ANCOVA analysis revealed no effect of the group (either survivors or non-survivors) on the time interval until reaching 40 mmHg, only an effect of time ($F(1,28) = 22.624$, $p < 0.001$).

Hypotension Phase

MAP Responses
Following hemorrhage (Fig. 5.55a), MAP in the non-surviving group decreased more rapidly than MAP in the surviving group. After that, MAP in both groups remained at a relatively stable level of 43 mmHg.

Cerebral Responses
CBF (Fig. 5.55b) in the non-surviving group sharply decreased to 7 ± 4% followed by stabilization around those levels, while the blood flow in the surviving group

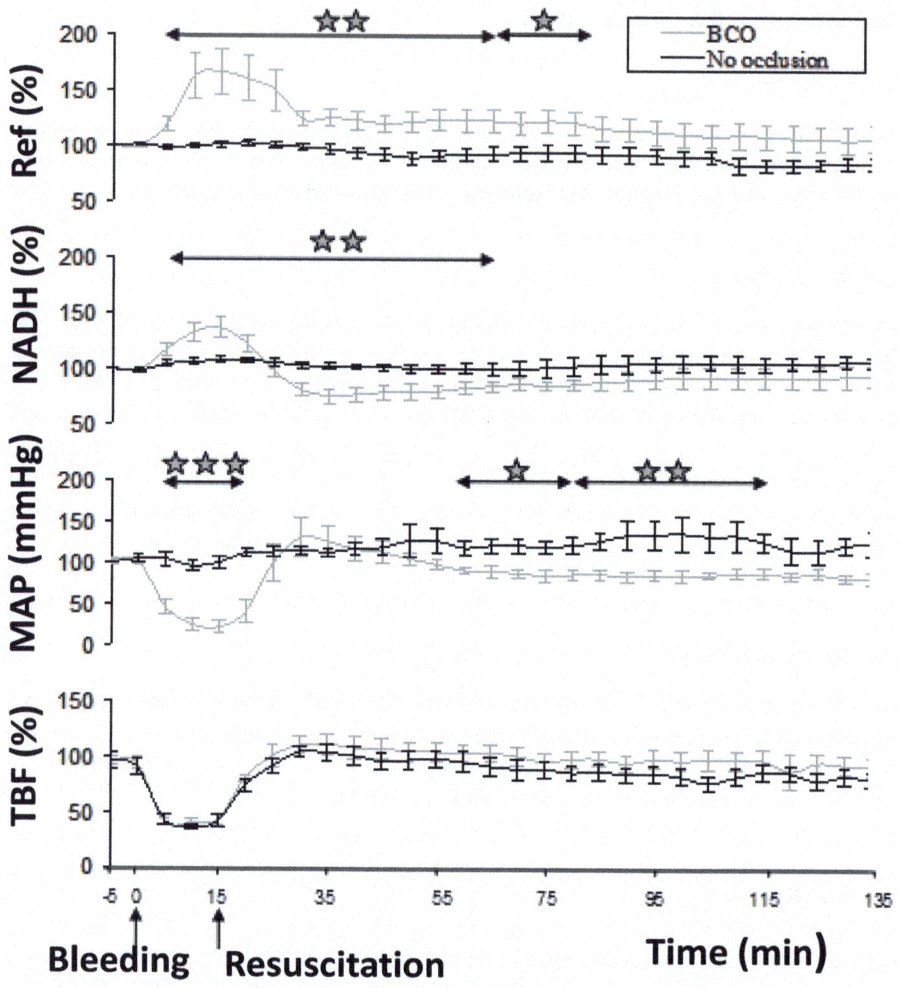

Fig. 5.53 Comparison of the responses of MAP and brain parameters to controlled hypotension for 15 min in rats with normal perfused rats (No occlusion, black, $n = 7$) and rats that underwent permanent BCO (gray, $n = 9$, mean ± S.E.). The asterisks present levels of significance—$^*p < 0.05$, $^{**}p < 0.01$, and $^{***}p < 0.001$. Ref reflectance, NADH mitochondrial NADH redox state, TBF tissue blood flow, MAP mean arterial pressure

gradually decreased to 22 ± 8%. Following a similar increase in both groups, the reflectance in the surviving group increased further to a level of 164 ± 23% and stabilized around this level, whereas the reflectance in the non-surviving rats increased only until a level of 141 ± 16% and then remained stable. NADH, in the non-surviving rats, showed a more moderate increase than in the non-surviving group (30 ± 8% and 60 ± 16%, respectively).

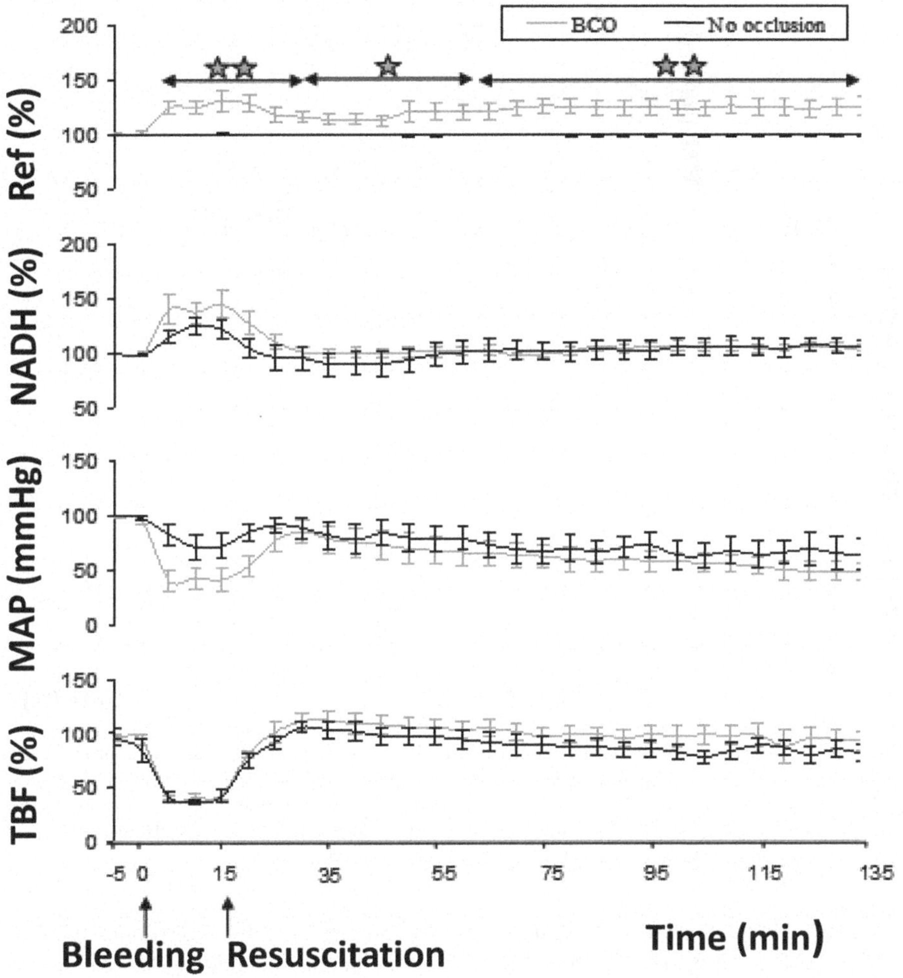

Fig. 5.54 Comparison of the responses of MAP and the intestine to controlled hypotension for 15 min in rats with normal perfusion (No occlusion, black, $n = 9$) and rats that underwent permanent BCO (gray, $n = 7$, mean ± S.E.). The asterisks present levels of significance—$^*p < 0.05$, $^{**}p < 0.01$, and $^{***}p < 0.001$. Ref reflectance, NADH mitochondrial NADH redox state, TBF tissue blood flow, MAP mean arterial pressure

Intestinal Responses

Both groups maintained relatively similar patterns of decrease of the IBF levels down to 40% and stabilizing around this level (Fig. 5.55c). However, toward the end of the period, TBF of the non-surviving rats further decreased to a level of 28 ± 3%. The reflectance of the surviving rats demonstrated a steady pattern of increase until a level of 131 ± 9%, whereas the reflectance in the non-surviving group demonstrated

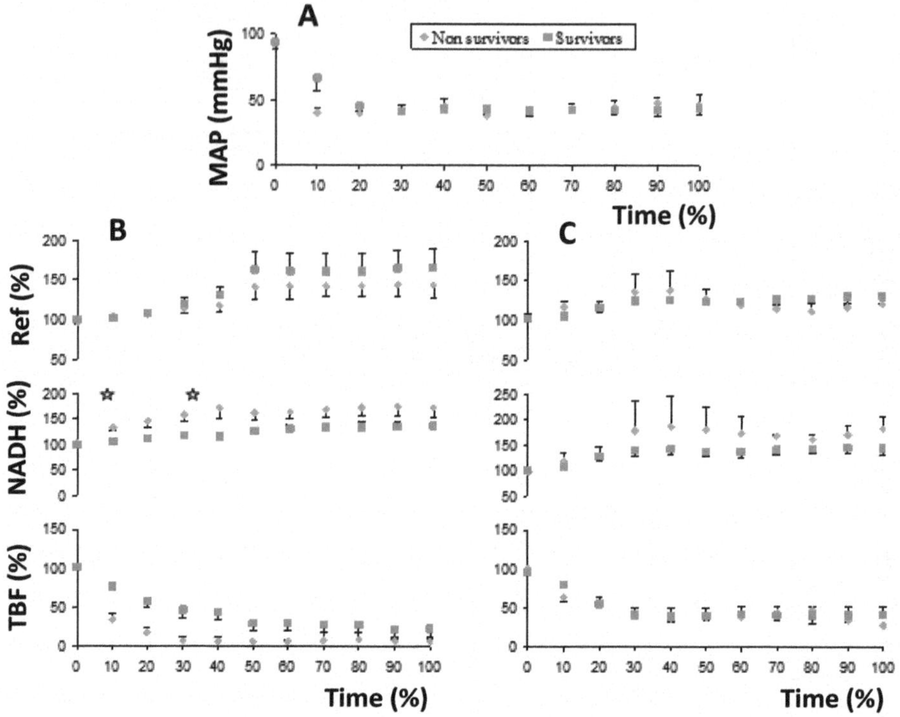

Fig. 5.55 Comparison between the responses of the MAP (**a**), the brain (**b**), and intestine (**c**) in the surviving rats ($n = 7$, squares) and the non-surviving rats ($n = 2$, circles) during hypotension following 24 h of permanent BCO (time was normalized with 10% intervals, mean ± S.E.). The asterisks represent the level of significance: $*p < 0.05$, Ref reflectance, NADH mitochondrial NADH redox state, TBF tissue blood flow, MAP mean arterial pressure

an unstable behavior with an initial increase to 138% ± 24, followed by a decrease to 112 ± 16% and toward the end of the period it increased again up to 121 ± 2%. NADH level, in the surviving rats, increased gradually up to 142 ± 12% and stabilized, while in the non-surviving rats, it rapidly increased to a level of 186 ± 60% and then decreased down to a level of 162 ± 8%, followed by another increase up to 182 ± 25%.

Resuscitation Phase

MAP Responses

During resuscitation (Fig. 5.56a), MAP of the surviving rats increased until a level of 109 ± 7 mmHg and stabilized, whereas in the non-surviving animals, MAP presented a more moderate increase up to 102 ± 22 mmHg, followed by a period of stabilization and then decreased to 87 ± 6%.

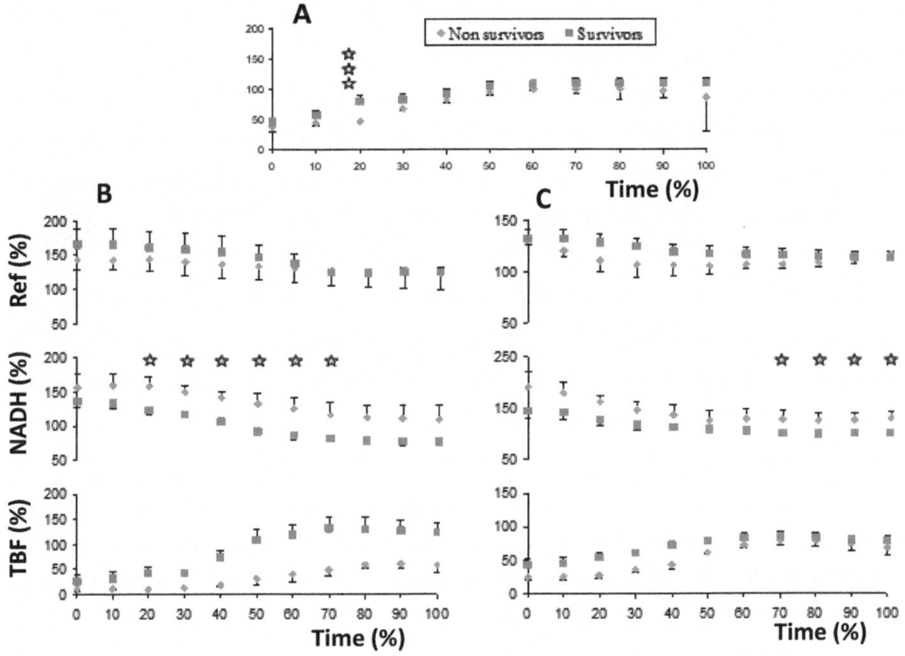

Fig. 5.56 Comparison between the responses of the MAP (**a**), the brain (**b**), and intestine (**c**) in the surviving BCO rats ($n = 7$, squares) and the non-surviving BCO rats ($n = 2$, circles) during resuscitation (time was normalized with 10% intervals, mean ± S.E). The asterisks represent the levels of significance: $*p < 0.05$, $***p < 0.001$, Ref reflectance, NADH mitochondrial NADH redox state, TBF tissue blood flow, MAP mean arterial pressure

Cerebral Responses

TBF in the brain (Fig. 5.56b) of the surviving rats immediately increased and fully recovered to above the basal level (131 ± 23%), yet, during the last 30% of this period, it decreased down to 124 ± 17%. In the non-surviving rats, TBF only partially recovered by reaching a level of 57 ± 15%. In both groups, the reflectance partially recovered and decreased down to 124%; however, in the non-surviving rats, the reflectance decreased more moderately, since its initial level was lower. NADH in both groups presented similar trends of decrease, although with different levels. Cerebral NADH of the non-surviving rats decreased from 157 ± 18% to 109 ± 2%, and in the surviving group, it decreased from 134 ± 8% to 76 ± 6%.

Intestinal Responses

IBF (Fig. 5.56c) in both groups partially recovered around the same level (79 ± 8% in the non-surviving rats and 85 ± 10% in the surviving rats); however, the initial level in the non-surviving group was lower than in the surviving group (24 ± 5% and 43 ± 10%, respectively). The recovery in both groups was followed by a slight decrease. The reflectance in both groups partially recovered; however, in the surviving animals, it showed a moderate decrease down to 113 ± 5%, whereas in the

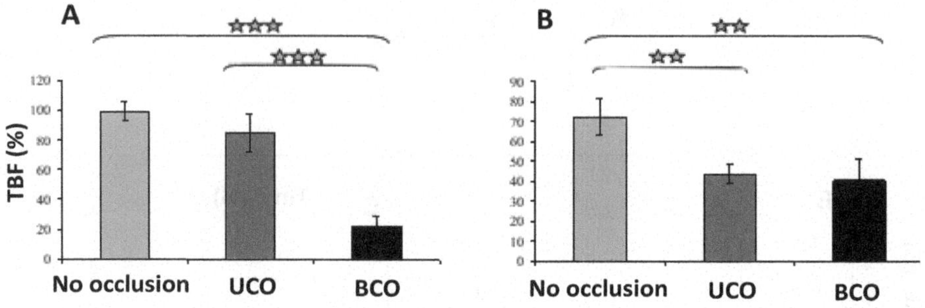

Fig. 5.57 The effect of BCO for 24 h on TBF levels during hypotension. (**a**) the effect on cerebral blood flow, (**b**) the effect on intestinal blood flow (mean ± S.E.). Pale gray—no occlusion, dark gray, UCO Unilateral Carotid Occlusion, and black—BCO Bilateral Carotid Occlusion. The asterisks present levels of significance: $^{**}p < 0.01$ and $^{***}p < 0.001$

non-surviving rats, the reflectance decreased rapidly down to a level of 106 ± 9% and was followed by an elevation to a level of 116 ± 4%. NADH in the surviving group fully recovered reaching the basal level, whereas NADH in the non-surviving group only partially recovered reaching a level of 130 ± 13%. The decrease of NADH in the non-surviving group was more rapid than that of the surviving group.

5.3.7.3 Comparison Between UCO and BCO

Tissue Blood Flow

As Fig. 5.57a shows, the averaged CBF level in the BCO group was significantly lower from the CBF level in both groups of normal perfused brain (no occlusion) ($p < 0.001$) and brains that underwent UCO ($p < 0.001$), while no significant level was observed between the group with no occlusion and the group with UCO.

Whereas in the intestine (Fig. 5.57b), the IBF level in the group with normal perfused brain was significantly higher than in the two other groups; the UCO group ($p < 0.01$) and the BCO group ($p < 0.01$). And again, no difference between the two other groups was observed.

NADH

As can be seen from Fig. 5.58a, the cerebral NADH level in the BCO group was significantly higher than in the two other groups: the group with the normal perfused brain ($p < 0.001$) and the UCO group ($p < 0.01$). No significant difference was found between the two other groups.

In the intestine (Fig. 5.58b), the NADH level in the group with the normal perfused brain was significantly lower than the groups with ischemic brain; the UCO group ($p < 0.05$) and the BCO group ($p < 0.01$).

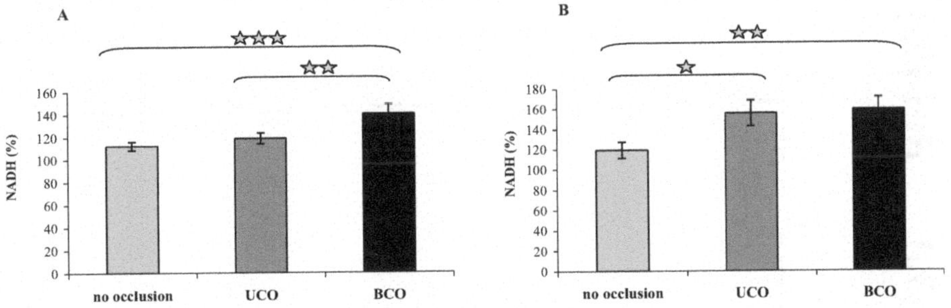

Fig. 5.58 The effect of BCO for 24 h on NADH levels during hypotension. (**a**) the effect on cerebral blood flow, (**b**) the effect on intestinal blood flow (mean ± S.E.). No occlusion—rats with normal perfused brain (pale gray), UCO—rats that underwent unilateral carotid artery occlusion (dark gray), and BCO—rats that underwent bilateral carotid artery occlusion (black). The asterisks present levels of significance: $*p < 0.05$, $**p < 0.01$, and $***p < 0.001$

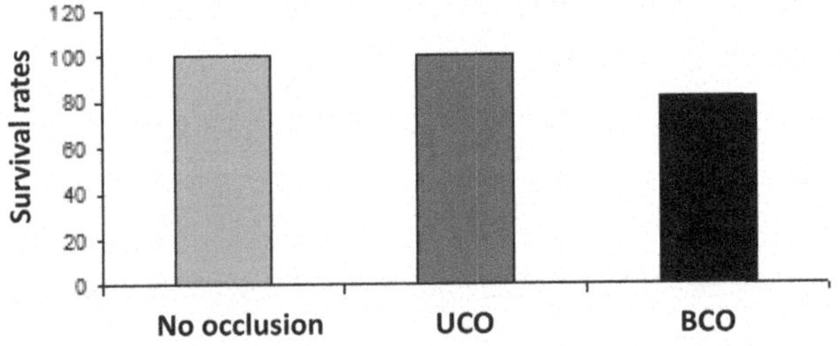

Fig. 5.59 Survival rates under controlled hypotension. No occlusion—rats with normal perfused brain (pale gray), UCO—rats that underwent unilateral carotid artery occlusion (dark gray), and BCO—rats that underwent bilateral carotid artery occlusion (black)

Survival Rates

During controlled hypotension, 100% of the rats with normal perfused brain or UCO brains survived, while only 82% of the BCO rats survived (Fig. 5.59).

5.3.7.4 Hemorrhage Following Middle Cerebral Artery Occlusion (MCAO)

Middle Cerebral Artery Occlusion Control Group

Figure 5.60 shows rats that underwent transient MCAO for 1 h, perfused for 24 h, and then monitored for 2 h ($n = 4$). During the monitoring, no perturbation was conducted. As can be seen, no significant differences in any of the parameters were found between the two organs and within the organs.

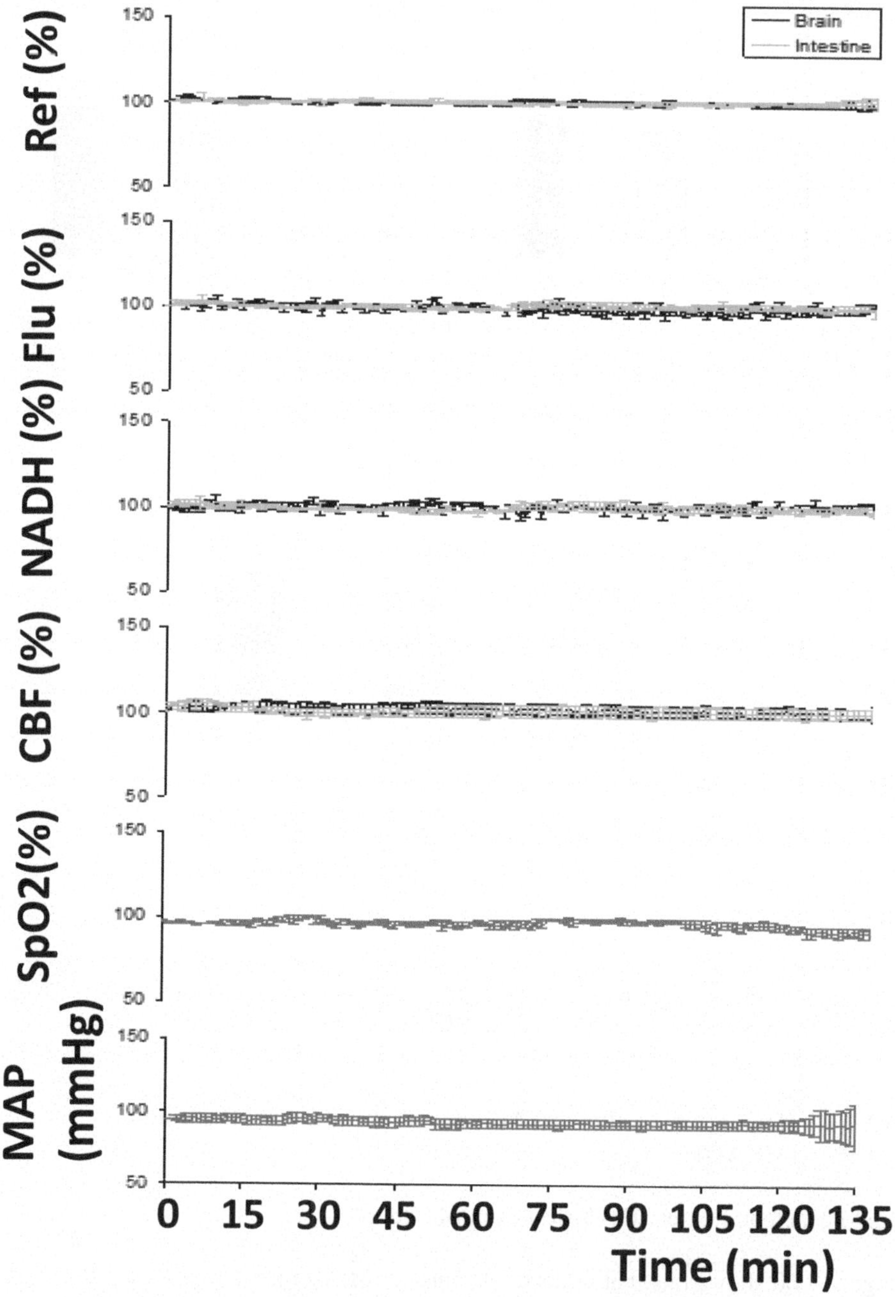

Fig. 5.60 Nontreatment monitoring of the brain (black), the small intestine (gray) SpO_2, and MAP in rats that underwent 1 h of transient MCAO 24 h before the monitoring ($n = 4$). Ref reflectance, NADH mitochondrial NADH redox state, TBF tissue blood flow, SpO_2 Saturation of Peripheral Oxygen, and MAP mean arterial pressure

Comparison Between Brain and Intestine During Hemorrhage Following MCAO

During the bleeding session, MAP immediately decreased and remained low for the entire session (figure not shown), while SpO_2 remained stable. Both the Cerebral and IBF decreased, reaching a low level. The cerebral reflectance remained stable while the intestinal reflectance increased sharply. The cerebral NADH only slightly increased and maintained a relatively stable level, while the intestinal NADH sharply and greatly increased.

Following resuscitation, MAP fully recovered and remained stable for the entire trial. SpO_2 was not altered by the resuscitation and remained stable. The CBF showed a minor increase, which was followed by additional deterioration later on. About 15 min later, CBF recovered and remained stable for the entire trial. The IBF recovered in two steps; at first, it only partially recovered, but then after it fully recovered and stayed so for about an hour. Then after, it deteriorated again. The cerebral reflectance was slightly decreased but fully recovered later on, while the intestinal reflectance fully recovered, followed by a slight increase further on. Both the cerebral and intestinal NADH fully recovered. Yet, while the NADH in the brain remained stable, the NADH in the intestine greatly increased later on.

As can be seen from Fig. 5.61, during the hemorrhage session, MAP decreased, reaching a low level of 40 ± 4 mmHg ($p < 0.005$), while SpO_2 exhibited no significant change. The blood flow in both organs immediately decreased, reaching a low level of $63 \pm 10\%$ in the brain and $51 \pm 8\%$ ($p < 0.005$) in the intestine ($p < 0.001$). In both organs, the reflectance remained relatively stable, while the NADH increased ($113 \pm 4\%$ ($p < 0.05$) in the brain and $137 \pm 11\%$ in the intestine ($p < 0.05$).

The resuscitation responses, in opposition to the bleeding responses, showed two steps. The MAP, at first, only partially recovered maintaining a stable level for around 20 min (79 ± 17 mmHg, $p < 0.05$). Then after, it fully recovered (95 ± 8 mmHg) and remained stable until the end of the trial. On the other hand, the SpO_2 was not affected by the resuscitation. The CBF was affected by the resuscitation. However, the changes were not significant due to a high standard error. At first, it fully recovered reaching a level of $100 \pm 8\%$, but immediately after it was deteriorated again by $24 \pm 11\%$ for about 15 min and then partially recovered reaching a level of $92 \pm 17\%$ and remained stable of the entire trial. The IBF, on the other hand, at first only slightly recovered reaching a level of $66 \pm 7\%$ ($p < 0.05$) and then after it partially recovered reaching a level of $90 \pm 6\%$. But, from that point on, the IBF slowly decreased until reaching a low level of $53 \pm 10\%$ ($p < 0.005$).

The cerebral reflectance did not respond to the resuscitation and remained stable for most of the trial. Toward the end of the trial, it decreased but did not present a significant change due to a high standard error ($79 \pm 10\%$). The intestinal reflectance remained relatively stable. Cerebral NADH fully recovered ($101 \pm 3\%$), but then after started slowly deteriorating, reaching the low level of $79 \pm 6\%$ ($p < 0.05$) at the end of the trial. In opposition to that, the intestinal NADH recovered in two steps: at first it only partially recovered ($125 \pm 9\%$, $p < 0.05$) and then after it fully recovered

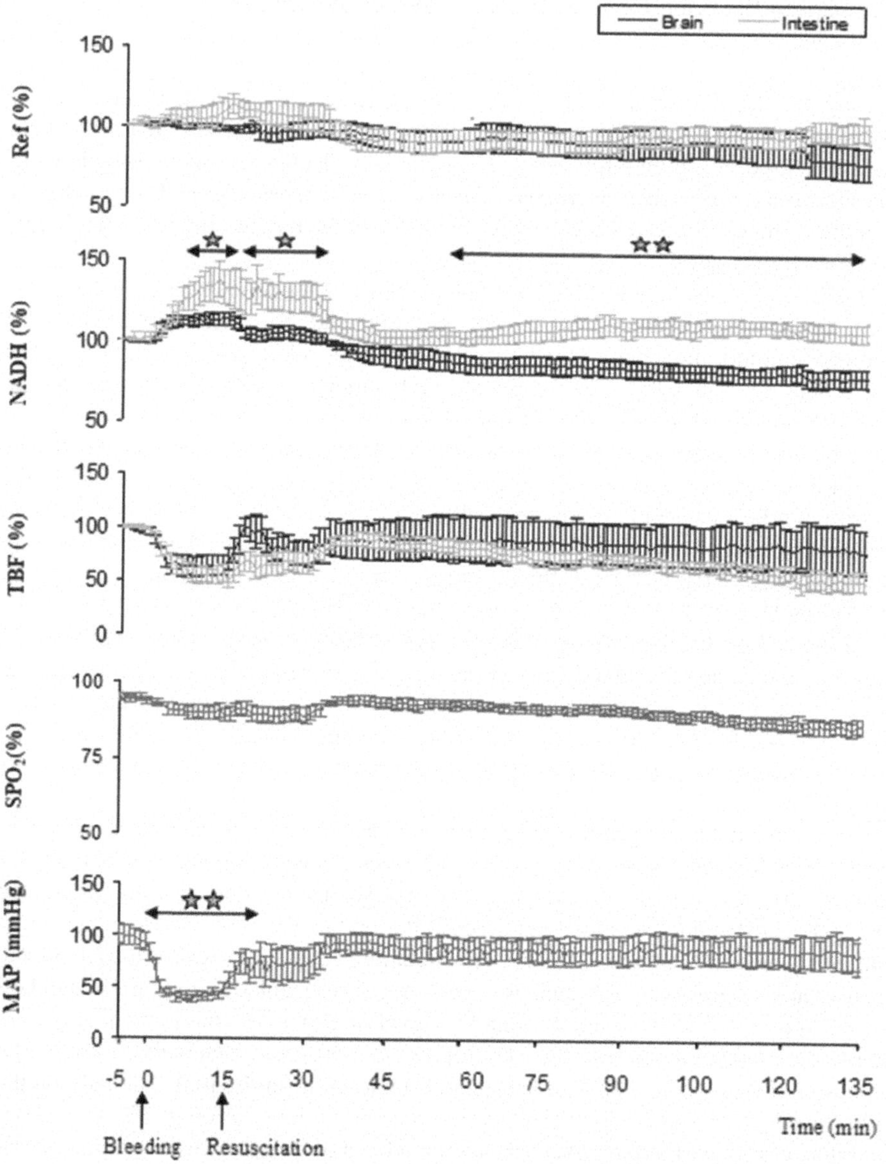

Fig. 5.61 The changes in all parameters in both organs: the brain (black) and the small intestine (gray) to controlled hemorrhage 24 h following transient MCAO ($n = 8$, mean ± S.E). The arrows represent the period in which significant differences were found between the two organs for each minute, and the asterisks represent levels of significance: $*p < 0.05$, $**\ p < 0.01$, and $***\ p < 0.001$. Asterisks in Ref, NADH, and TBF mark significance between brain and intestine; asterisks in MAP mark significance compared to baseline. Ref reflectance, NADH mitochondrial NADH redox state, TBF tissue blood flow, MAP mean arterial pressure (gray), and SpO_2 Saturation of peripheral oxygen (gray)

(102 ± 4%). Comparing the organs' response to the bleeding and resuscitation sessions reveals significant differences only in NADH ($p < 0.05$).

MCAO With or Without Hemorrhage

As can be seen from Fig. 5.62, a comparison between the focal ischemic controlled hypotension group and the focal ischemic control group was carried out. In MAP during both hypotension and following resuscitation, significant differences were observed ($p < 0.005$). In the brain, only the NADH and CBF presented significant

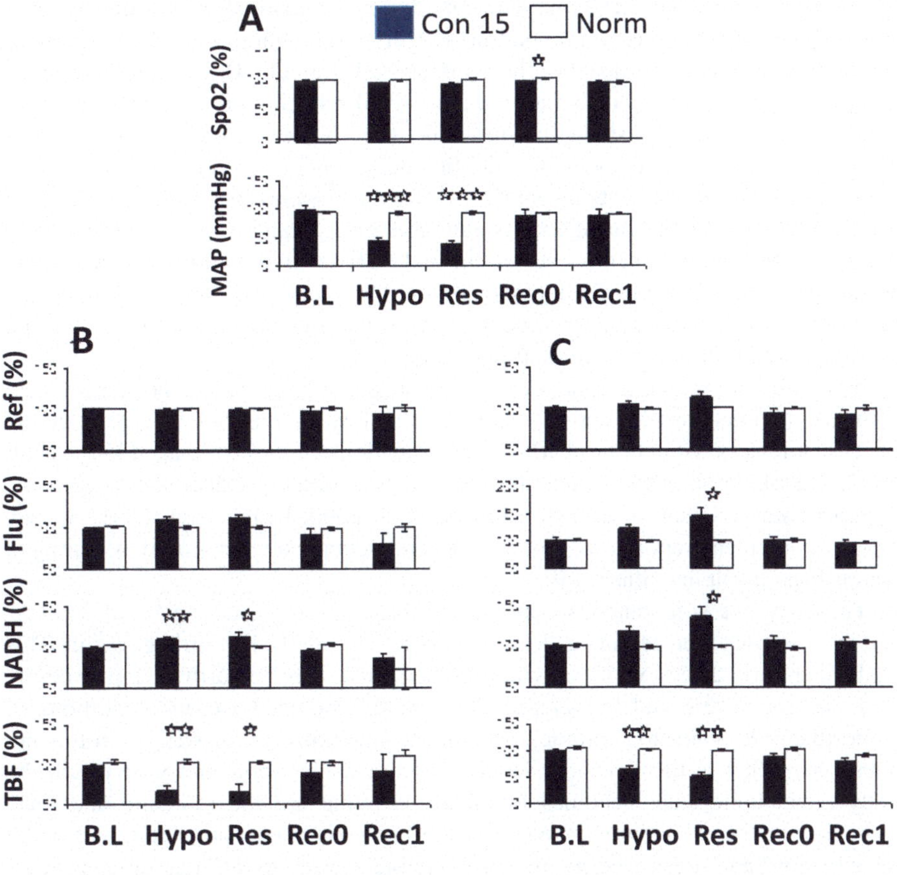

Fig. 5.62 Comparison of the SpO_2, MAP (**a**), cerebral (**b**), and intestinal (**c**) parameter responses in animals that underwent MCAO with (hypo-black, $n = 7$) or without (con-white, $n = 4$, mean ± S. E.) controlled hypotension. The asterisks represent the levels of significance: *$p < 0.05$, **$p < 0.01$, and ***$p < 0.001$. B.L. basal level, Hypo end of the hypotension period, Res end of resuscitation, Rec0 1 h following resuscitation, Rec1 2 h following resuscitation, Ref reflectance, NADH mitochondrial NADH redox state, TBF tissue blood flow, SpO_2 Saturation of Peripheral Oxygen, MAP mean arterial pressure

differences. In both parameters, it was found during hypotension ($p < 0.01$) and following resuscitation ($p < 0.05$). In the intestine, both fluorescence and NADH exhibited a difference following resuscitation, while in IBF, it was exhibited during both hypotension and following resuscitation ($p < 0.01$).

5.4 Discussion

Mitochondria are the juncture of homeostasis and dysfunction, allowing an early response to pathophysiological conditions such as stress and shock, since these conditions initiate high metabolic demands during the disorder of various systems and organs. Furthermore, mitochondrial NADH redox state is the most sensitive component of oxidative phosphorylation (Hubbard et al. 2004). Despite these facts, only less than ten studies of hemorrhage involved mitochondrial NADH measurements (usually in vitro studies). Moreover, the characteristics and outcome of hemorrhage are well documented in the literature, and proper regimens of resuscitation from hemorrhage are being sought. However, even less researches concern the early detection of hemorrhage and its outcome, even though it is agreed that early diagnosis of the hemodynamic catastrophe and early resuscitation from hemorrhagic shock are crucial for the final outcome. In this study, we conducted real-time monitoring of hemorrhagic insults, including the monitoring of mitochondrial NADH in addition to the traditional parameters.

It is well known that during severe body stress, autoregulation mechanisms are activated to maintain the activity and functionality of the "vital organs" at the expense of "less vital organs" (Pries et al. 1996; Vagnozzi et al. 1997; Jakob 2002; Zakaria et al. 2005). Since the intestine is extremely vulnerable to systemic hypoperfusion (Nelson et al. 1987; Morini et al. 2000; Dubin et al. 2001), it may serve as an ideal reporter organ of oxygen metabolism deterioration during a hemorrhage insult and afterward.

Therefore, we simultaneously monitored the brain, representing "vital organs," and the intestine, representing a "less vital organ" during the hemorrhage insult. This enabled an early detection of the symptoms accompanying hemorrhage injury before they become severe and irreversible. This work examined various conditions of hemorrhagic hypotension, mimicking clinical situations, to answer the following questions: What is the threshold level of hemorrhagic stress injury at which the body's metabolic state becomes imbalanced? What are the systemic and local responses in these conditions? When is the brain no longer protected, and under what hemorrhage stress severity, may irreversible damage to different organs occur?

5.4.1 The Monitoring Model Development

Severe hemorrhage causes a disproportionate compensatory reaction in the mesenteric circulation (Morini et al. 2000; Vajda et al. 2004a, b). A previous study reported that the mucosa and serosa in the jejunum respond differentially to isovolemic hemodilution. The mucosa preserved oxygen supply, while in the serosa, oxygen supply declined. This was presumably due to a redistribution of blood flow, and therefore oxygen supply, within the jejunal wall in favor of the mucosal layer (Haisjackl et al. 1997). Another study posited that increased sympathetic stimulation, during hemodilution, plays a role in the hemodynamic response and redistribution of blood flow, not only to organs but also within the organs, leading to vasoconstriction in the serosal and probably muscularis layers, while metabolic regulation within the mucosal-submucosal layers preserves the blood flow to this tissue compartment by vasodilation (Sjovall et al. 1983). In rats, the thickness of the intestinal wall is about 1 mm and the penetration depth of the excitation light is 1.5–2 mm for LDF and about 1 mm for the fluorometer; therefore, we obviously monitored both layers together. We tested this monitoring model, namely, placing the probe on the intestinal serosa and mucosa, using two perturbations with well-known results (Kozuka 1995; Arumugam et al. 2003), in order to find the best location for stable monitoring, rather than finding the differences in these layers' responses to the hemorrhagic insult. Since there were no significant differences between the layers in the two protocols, and the serosa location was easier to manipulate, provided better measurements, and, most importantly, was less invasive, we conducted the monitoring of the hemorrhage models from the serosa.

5.4.2 The Hemorrhage Models

Various models of hemorrhagic hypotension were tested to mimic different clinical situations. The first model was that of uncontrolled hypotension, a common situation in the clinic, especially in trauma victims. Nevertheless, it has been rarely used in experimental studies. Severe injury to a large vessel can cause rapid and severe hypotension. In many cases, the bleeding is stopped, but it may take considerable time until resuscitation or reinfusion of the shed blood is carried out. The second group of models, those of controlled hypotension, was tested to isolate the effects of hypotension duration on the various body responses. The third model was that of graded hemorrhages, in which we evaluated the effect of the duration of the intervals between two following bleeding sessions. The fourth group was of hemorrhagic hypotension during different cases of chronic brain ischemia, designed to mimic hemorrhagic insult in cases of low cerebral perfusion, as may be seen in older patients suffering from arterial calcification.

It has been extensively reported that the normal microcirculatory perfusion is characterized by temporal and spatial variations of capillary flow (Vagnozzi et al.

1997; Vajda et al. 2004a). Essentially, the pathogenesis of circulatory collapse, after blood loss, results from a disparity between central and peripheral regulation mechanisms, under progressive tissue hypoperfusion. Hemorrhagic shock sets off a series of physiologic compensatory adjustments, aimed at the preservation of the perfusion of vital organs, such as the heart and the brain (Vatner 1974; Vajda et al. 2004a; Zakaria et al. 2005). In hemorrhagic shock, the loss of blood volume causes a decrease in CO, leading to a vasoconstrictive response, mediated by the sympathetic nervous system, the renin-angiotensin system, and vasopressin. When adrenergic receptors on postcapillary mesenteric venules and veins are stimulated, these vessels will vasoconstrict; therefore, the CO will be improved by increasing the cardiac preload. Selective vasoconstriction of the afferent mesenteric arterioles serves to sustain systemic vascular resistance and therefore to maintain systemic blood pressure and sustain the perfusion of vital organs at the expense of mesenteric organ perfusion (the "fight or flight" response). This response is dependent, to a limited degree, on the sympathetic nervous system, but is mainly mediated by the renin-angiotensin axis and vasopressin (Jakob 2002; Vajda et al. 2004a; Zakaria et al. 2005).

This preferential redistribution of CO to vital organs at the expense of less vital organs was also demonstrated in our work. During hypotension in normal rats, significant differences in blood flow were observed between the small intestine and the brain in most cases. The intestine responded by an immediate decrease in IBF, while CBF remained intact or less affected. These effects can be explained by an increase in intestinal microcirculation resistance (Garrison and Zakaria 2005; Zakaria et al. 2005). The decrease in the IBF is coupled to a decrease in blood volume, resulting in the elevation of the intestinal reflectance in contrast to the cerebral reflectance, which remained relatively stable. The decrease in the IBF was also associated with a decrease in oxygen delivery; hence, mitochondrial NADH redox state increased in both organs. It is important to point out that the decrease in IBF was greater than the increase in mitochondrial NADH redox state, emphasizing the high ability of the intestine to enhance oxygen extraction, in order to postpone, if not prevent, metabolic tissue stress, as previously reported (Takala 1996). The fact that the cerebral mitochondrial NADH redox state was elevated, though the CBF was intact, points to the high sensitivity of the brain to a decrease in oxygen delivery and the importance of mitochondrial NADH redox state monitoring as an indicator of oxidative stress.

During resuscitation, a recovery in all of the parameters (either full or partial) was recorded in most of our models. MAP was elevated above the basal level. IBF increased toward the basal levels and CBF became hyperemic. In both organs, the reflectance decreased back toward the basal levels and mitochondrial NADH was oxidized back to the basal level. However, after a while, several parameters became deteriorated. MAP failed to maintain high levels, though, in most of the groups, it was preserved at the basal levels. IBF acted in accordance with the MAP and decreased to low levels, as was observed in other studies (Garrison and Zakaria 2005; Zakaria et al. 2003, 2005). In contrast, CBF became hyperemic, probably due to arterioles' vasodilation during the hypotension period, where the blood flow increased during blood reinfusion. Although the IBF decreased, NADH remained stable. This finding further indicates the ability of the intestine to increase oxygen extraction. In the literature, this deterioration is termed IRI. An interruption of blood supply results in ischemic injury, which rapidly damages metabolically active tissues. Paradoxically, the restoration of blood flow to the ischemic tissue initiates a cascade of events that may lead to additional cell injury, known as reperfusion injury. This reperfusion damage frequently exceeds the original ischemic insult with further necrosis, and it can also be manifested in distant organs (Carden and Granger 2000; Mallick et al. 2004; Cerqueira et al. 2005). It has been emphasized that the initial site affected by ischemia is the mitochondria (Jassem et al. 2002; Mallick et al. 2004; Cerqueira et al. 2005). In the mitochondria, oxidative phosphorylation becomes compromised as a result of an inadequate supply of oxygen, leading to ATP depletion and to the accumulation of metabolites, which can directly or indirectly cause cellular injury. The damage is dramatically magnified by a large number of events, such as oxygen free radical formation, release of iron storage, damage to the microvasculature, inflammatory cytokines release, complement activation, and neutrophil infiltration at the site of injury. The enterocytes, in particular, have been shown to produce increased amounts of cytokines during hemorrhagic shock (Chance et al. 1973; Lu et al. 2005). These hazardous effects intensify and can cause a circulatory shock due to an increase in the permeability of the vascular and intestinal epithelial barrier (Wattanasirichaigoon et al. 1999; Mallick et al. 2004; Cerqueira et al. 2005). These events are likely to terminate in the multiple organ dysfunction syndrome, also known as multiple organ failure (MOF), and death (Reilly et al. 2001; Arumugam et al. 2003). Among the internal organs, the intestine is probably the most sensitive to IRI (Boldt 2002; Boulpaep 2003). The decrease in tissue ATP content was shown in our work by the increase in NADH redox level, as this is the first component in the oxidative phosphorylation chain.

5.4.2.1 Two Models of Hypotension

In our study, we examined two models of hypotension; acute hemorrhage and graded hemorrhage. The former group is comprised of another two models; uncontrolled hemorrhage and controlled hemorrhage.

The Acute Hemorrhage Models

The uncontrolled hypotension model was used to mimic clinical situations associated with hemorrhagic shock in trauma patients, such as severe injury of a large vessel. In many clinical cases, the bleeding is mostly stopped after a while, but it often involves a delay until resuscitation or reinfusion of the shed blood. The controlled hypotension model is widely used in research, although there are no identical conditions in patients. Nevertheless, this model is widely employed since it enables the isolation of the hypotension factor and hence improves the understanding of its involvement in the pathogenesis of hemorrhagic shock (Lomas-Niera et al. 2005).

The immediate response to hypotension was the same for both the uncontrolled and controlled hypotension groups. When the bleeding session was stopped, the intestinal responses of these two groups were significantly different. In the uncontrolled hypotension group, there was a partial or full recovery of the intestinal parameters within the hypotension period. The decrease in IBF was due to the intense vasoconstriction of the intestinal arterioles, as was observed in a similar study (Cryer et al. 2005). In both models, the brain responded similarly by an elevation of blood flow, yet, only in the controlled hypotension group, the reflectance and NADH were elevated. However, no significant differences were observed between TBF and NADH in both models during the entire hypotension period, as well as when comparing the maximal responses. This indicates that although the brain in each model was exposed to hemorrhagic stress of different severity, due to the different periods of hypotension, the brain remained intact during the hypotension period. Thus, in order to maintain CBF, vasodilatation must be ensured, as was previously demonstrated by a decrease in cerebrovascular resistance (Yu et al. 2002; Eike and Palmer 2004).

Usually, when blood flow levels increase, the reflectance decreases, since changes in blood volume are the major parameter influencing the reflectance. However, in the controlled hypotension group, though the CBF increased, the reflectance increased as well, probably due to a movement of ions and water through the different cellular compartments (Mayevsky and Rogatsky 2007).

The MAP in both models responded similarly to resuscitation. The initial response of the brain and small intestine to resuscitation was similar, namely, both organs fully recovered. However, as time passed, differences between mitochondrial functions were observed: after a long hypotension period, as in the controlled hypotension group, NADH increased (despite a high level of CBF), whereas after a short hypotension period, as in the uncontrolled hypotension group, mitochondrial function remained intact. This specifies the importance of mitochondrial NADH measurement as an indicator of oxidative stress. In addition, it appeared that the post-hemorrhage injury is a type of IRI, as was previously reported (Cryer et al. 2005; Zakaria et al. 2005), insofar as the deterioration of the cerebral tissue was observed only in the resuscitation and recovery phases and not earlier during the bleeding insult.

It should be noted that, in our study, intestinal deterioration was observed even though MAP remained stable at the basal levels. Other studies also demonstrated that, despite the normalization of blood pressure, heart rate, and urine output, tissue hypoperfusion persists in 80–85% of the patients (Zakaria et al. 2005), indicating the importance of monitoring tissue perfusion for the evaluation of recovery rather than MAP (Garrison and Zakaria 2005; Zakaria et al. 2005).

These findings further emphasize the utility of our unique monitoring device, which elucidated the tissue condition instead of systemic parameters. The systemic parameters proved to be less accurate for a proper evaluation of the tissue state.

Moreover, the monitoring of a peripheral organ, such as the small intestine, may serve as a surrogate organ in similar clinical conditions.

The Graded Hemorrhage Models

The importance of the graded hemorrhage models is in their ability to discriminate the influence of each session of blood removal on the physiologic and metabolic response of the body. In our study, we were able to reveal the turning point of the body's autoregulation mechanisms. Both models evolved with large variability since the response to a specific amount of blood loss differs among individuals (Rixen and Neugebauer 2004; Lomas-Niera et al. 2005). We found in both models, graded hemorrhage with 10- or 5-minute intervals, that the bleeding session of the 20% hemorrhage was the point at which the systemic and intestinal parameters were less capable of maintaining homeostasis. In addition, in the group with 10-minute intervals, the endpoint of blood removal was the 35% hemorrhage. During this withdrawal of blood and for further 10 min, the MAP was no longer able to recover. These results are confirmed by the scientific literature, which describes the body's response to hemorrhage as "biphasic hemodynamic response" (Vagg et al. 2008). This biphasic response is divided into two ranges: 0–15% hemorrhage, which is considered to be a moderate hemorrhage, and 20–35% hemorrhage, which is referred to as severe hemorrhage (Vagg et al. 2008).

Furthermore, both models evolved with prominent differences between the brain and the intestine. While the intestinal parameters were altered very quickly after the first hemorrhage, the brain remained unaltered at least until a more progressive bleeding session. Additionally, the vast response of the intestinal NADH, in opposition to the cerebral NADH, highlights the importance of mitochondrial NADH monitoring in addition to the microcirculatory blood flow.

5.4.3 The Effect of the Duration of Hypotension

We compared the responses of three controlled hypotension models differing in the durations of hypotension; 15, 30, and 60 min, to study the effect of hemorrhage stress severity on the responses of the brain and the intestine.

These three groups were carried out twice, due to differences in the rat batches yielding different responses to the hemorrhagic protocol.

We found that during the hemorrhagic hypotension, the systemic parameters, MAP and SpO_2, were not affected by the duration of hypotension, as opposed to the local parameters and especially the TBF and NADH level. The change in both the cerebral and IBF at the 15-minute model was significantly moderate than in the two other models. In accordance with that, in both organs, the NADH level in both the 30- and 60-minute model was significantly higher than in the 15-minute model.

Furthermore, the 60-minute model also presented a significant difference in opposition to the 30-minute model, pointing toward an effect of duration.

As for the recovery phase, it can be seen that once again, no effect of duration was found for the systemic parameters. As for the local parameters, the effect of duration was found for TBF and NADH level, and it was found for the 60-minute model. This model presented vast significant differences from the other two models. For CBF, it was significantly different throughout the entire phase from the 15-minute model, and for the IBF, it was significantly different for about a third of the phase. In comparison to the 30-minute model in both organs, it was significantly different only during the resuscitation phase. For NADH too, the 60-minute model was significantly different from the 15-minute in both organs during the entire period. Regarding the 30-minute model, it was totally different from the intestinal NADH, and for almost half of the period, from the cerebral NADH.

The effect of the duration of hypotension on the mitochondrial NADH redox state was relatively high, emphasizing the high sensitivity of this parameter to hemorrhage stress and the importance of its monitoring as an indicator of the hemorrhage stress severity. These results also further highlight the importance of monitoring of the local tissues and not only the systemic parameters.

The effect of the duration of hypotension was also demonstrated in the responses of the non-surviving rats under the 60-minute model. As the duration of hypotension lengthened (60 min), the rats died throughout all of the experimental phases. However, when the duration of hypotension was shorter (15 and 30 min), all rats survived.

5.4.4 *The Influence of the Different Methods of Resuscitation on the Recovery from Hemorrhagic Hypotension*

Successful treatment of a massive hemorrhage and hemorrhagic shock must include surgical control of the hemorrhage and the restoration of proper blood pressure and tissue perfusion (Lu et al. 2005).

There is no doubt about the need for fluid resuscitation in order to solve the latter. Yet, the questions "with what" and "how much" to resuscitate are left open (Craig and Poole 1994; Lu et al. 2005).

What Fluid to Infuse?

The necessity of blood banks and blood unit is in a continuous increase worldwide. Hence, there is an urgent need for the development of universal blood substitutes. The Mortality from hemorrhagic shock in rats was reduced to zero by resuscitation with a hemoglobin-based oxygen carrier, whereas conventional fluid therapy did not achieve similar results, since the fluid infusion dilutes endogenous hemoglobin and therefore reduces the oxygen transport capacity of blood (Chang and Varma 1992; Kleen et al. 2000).

One such blood substitute is glutaraldehyde-polymerized hemoglobin (Eike and Palmer 2004).

At the first stage, we tested the effects of the blood substitute on normoxic brain and intestine compared to infusion of saline, as seen during short anoxia with N_2. This protocol served to test the effect of the blood substitute on the vitality of the measured tissues, as well as to check the ability of the MSMP system to detect the effects of the blood substitute on the monitored parameters.

The temporary decrease in MAP, which is observed during short anoxia, was less severe following blood substitute infusion as compared to saline injection. This demonstrates the ability of the glutaraldehyde-polymerized hemoglobin to serve as an effective blood substitute, probably due to its high ability to carry hemoglobin and therefore cause a lower degree of vasodilatation in the major arteries. This effect was also demonstrated in cerebral and intestinal blood supply. The increase and decrease in CBF during short anoxia following blood substitute infusion were lower compared to saline infusion. This, together with a lower increase in NADH following blood substitute, shows that due to the higher levels of hemoglobin, less modifications in blood supply are needed in order to maintain sufficient oxygen supply to the brain. Following blood substitute infusion, elevated IBF and lower reflectance levels were observed, indicating that even the intestine, which usually serves as a less vital organ and therefore suffers from reduced blood supply during stressful events, gains more blood supply following an anoxic episode after the infusion of blood substitute.

After testing the efficacy of glutaraldehyde-polymerized hemoglobin in normotensive rats, we tested its effects as a blood substitute following hemorrhage. Infusion of blood substitute following 15 min of hemorrhage leads to an immediate recovery in systemic, cerebral, and intestinal measured parameters. When comparing the ability of the tissue to withstand a metabolic challenge in the form of short anoxia, better responses were observed in all measured parameters 2 h following resuscitation, compared to the same responses immediately after resuscitation. These results demonstrate that glutaraldehyde-polymerized hemoglobin serves as an efficient blood substitute when infused following hemorrhage.

When comparing the metabolic effects of the blood substitute infusion compared to other means of resuscitation, which were used in the current work (infusion of saline, blood, or blood + saline), all groups reacted the same at the stage of hemorrhage. This unified response was expected, since in this stage all animals underwent the same study procedures.

Resuscitation with blood substitute led to an immediate recovery in MAP, to equivalent or elevated levels compared to the other resuscitation means. This demonstrates the ability of glutaraldehyde-polymerized hemoglobin to serve as an efficient blood substitute at the systemic level. Blood supply to the brain was lower, and intestinal blood supply was equivalent or higher compared to the other groups. In both organs, mitochondrial NADH levels, immediately following resuscitation with the substitute, were mostly equivalent to their levels following other resuscitation forms. These results demonstrate the ability of glutaraldehyde-polymerized hemoglobin to recover cerebral and intestinal blood supply and mitochondrial functioning following hemorrhage.

When observing the metabolic responses up to 2 h following resuscitation with the blood substitute, it is visible that most parameters recuperated at least as well as in other forms of resuscitation, except for blood supply to the intestine, which was lower compared to all other groups. This can result from one of the following: either the intestinal blood supply deteriorates in time, or a lower supply of blood is required by the intestine. Since intestinal mitochondrial activity following resuscitation with the blood substitute is maintained, it seems that intestinal blood supply is not damaged; rather, lower blood supply is required in order to maintain sufficient functioning of the intestine.

In conclusion, the results of the current study demonstrate that glutaraldehyde-polymerized hemoglobin may serve as an effective infused blood substitute following systemic hemorrhage, with hemodynamic and metabolic beneficial influences which are at least equal to those of infused saline and/or blood. In that, the current work corresponds with other studies which demonstrate the use of glutaraldehyde-polymerized hemoglobin as an effective blood substitute (Eike and Palmer 2004; Buehler et al. 2005).

Further research is required in order to determine the safety and efficacy of glutaraldehyde-polymerized hemoglobin (Alayash and Cashon 1995; Zhu et al. 2007). Nevertheless, we showed no beneficial effects of this material over blood.

How Much Fluid to Infuse?

In order to try and answer what is the volume of fluid that is needed for a proper recovery, we carried out two models. In the first one we infused back the shed blood during resuscitation and in the second one we infused the shed blood plus saline in a volume of two times the shed blood. The latter model was chosen according to previously reported clinical and laboratory studies, which show that when resuscitation was conducted using shed blood with two to three times that volume of crystalloid solution managed to reverse the outcomes of hemorrhagic shock (Revell et al. 2003).

In our study, no prominent differences were found between both groups; however, the differences that were found were in the small intestine. Following resuscitation with shed blood and saline, the IBF was significantly lower and in accordance the intestinal mitochondrial NADH level was significantly greater. Furthermore, until the end of the trial, the NADH level remained high. This means the small intestine did not recover from the hemorrhage despite the fluid infusion. Moreover, apparently, the resuscitation only caused a further deterioration of the intestinal condition via IRI (Carden and Granger 2000; Mallick et al. 2004; Cerqueira et al. 2005). These findings point toward too aggressive fluid resuscitation (Smail et al. 1998; Revell et al. 2003; Cotton et al. 2006).

The study of Lu (2005) also found that the apoptosis in the visceral organs, namely, the small intestine, liver, and kidney, was statistically higher following an aggressive resuscitation (Revell et al. 2003). These findings can be explained by: (1) Aggressive fluid infusion seriously disturbs the internal environment and exacerbates cellular metabolism, which can induce abnormal apoptosis of visceral organs. (2) High perfusion pressure of tissue and organ, followed by aggressive fluid resuscitation, can cause severe IRI by inducing a large amount of oxygen-derived free radicals (Revell et al. 2003; Cotton et al. 2006). Our results also highlight the importance of monitoring the less vital organs and especially the monitoring of mitochondrial NADH redox state in the less vital organ.

5.4.5 Hemorrhage Under Cerebral Ischemia

5.4.5.1 The Interaction Between Global Cerebral Ischemia and Hemorrhagic Hypotension

So far, we have tested the effects of different hemorrhage models in normal rats with normal cerebral perfusion. This paragraph will present the responses of the ischemic brain to these hemorrhage models. For this purpose, we employed the following three models of chronic (24 h) global ischemia; UCO, bilateral carotid artery occlusion (BCO), and MCAO, after which controlled hypotension was induced (for 15 min).

The reason for the induction of cerebral ischemia 24 h before the hemorrhage was to create a model in which the cerebral tissue is already in a stabilized new steady state characterized by low perfusion (Lipton 1999). This condition may mimic clinical situations, such as stroke and calcification of blood vessels in older people (Gorelick 1986). Numerous experimental and clinical studies report the effects of either hemorrhagic hypotension or cerebral ischemia on cerebral variables. Yet, only a few experimental studies have demonstrated the combined effect of both hemorrhagic hypotension and chronic cerebral ischemia (Rise and Kirkeby 1998).

Hemorrhage Under Unilateral Carotid Occlusion

A study of the effect of cerebral ischemia on the cerebrovascular responses to hemorrhage revealed, during hemorrhagic hypotension alone, an insignificant increase in CBF and a decrease in cerebral microcirculation resistance. When hemorrhage was induced after carotid artery occlusion, CBF decreased significantly, with a partial recovery, and cerebral microcirculation resistance increased (Rise and Kirkeby 1998). It should be noted that, in that study, acute ischemia was used and bleeding was not controlled.

In our study, we found out that hemorrhagic hypotension under UCO caused a slight decrease in CBF which was associated with a slight increase in the cerebral mitochondrial NADH. The intestinal responses in the UCO group had the same pattern as in the normally perfused brains group, but to a greater extent. It can be summarized that, surprisingly, the dominant effect of the UCO was on the intestinal response. We assume it was probably due to an interruption to the blood supply to the brain, which caused a further referring of blood from the intestine to the brain. Nevertheless, another study of acute unilateral common carotid artery occlusion with hypotension reported a decrease of oxygen availability to peripheral tissues and organs and to the CNS (O'Sullivan et al. 2007), which can explain the increase in both the cerebral and intestinal NADH.

Hemorrhage under Bilateral Carotid Occlusion

During the controlled hypotension period and afterward, under normal conditions, the brain remained protected, while in the ischemic model of BCO (Figs. 5.50, 5.51, and 5.53), the deterioration of the cerebral parameters was demonstrated by an immediate decrease in CBF to its minimal level, along with an increase in the reflectance and NADH redox state. These changes were significantly different from the response of the cerebral parameters in the normal brain and are compatible with the literature, which showed a significant decrease in CBF, with a partial recovery following resuscitation (Rise and Kirkeby 1998). These results indicate the impairment of autoregulation mechanisms, which was further demonstrated by resemblance between the responses of the ischemic brain and the intestine. All these findings indicate that, during a vast cerebral ischemia (like in the case of BCO), the compensatory mechanisms responsible for the preservation of systemic circulation during hemorrhagic hypotension are impaired, mainly the sympathetic discharge. As a result, cerebral microcirculation resistance increases instead of decreasing (Rise and Kirkeby 1998). Such an oxygen deficit can explain our finding of a significant increase in NADH redox state in both the brain and small intestine.

Briefly, during controlled hypotension under conditions of BCO, both the brain and the small intestine are no longer protected by autoregulatory mechanisms.

The responses of the cerebral parameters to hypotension exhibited biphasic changes, where the immediate response was followed by a further response. Such a biphasic change is referred to in the literature as "Ischemic-Depolarization" (ID) or "Hypoxic Spreading Depression-Like Depolarization" (Strandgaard 1976; Lubbers 1995; Mayevsky et al. 1995). ID is characterized by nearly complete neuronal depolarization, decrease in membrane resistance, and redistribution of ions across cell membranes. During ID, the neurons release K^+ and organic anions, including glutamate, to the interstitial fluid. At the same time, Na^+, Ca^{2+}, and Cl^- flow into the neurons. As a result, water enters the cells, causing their swelling. Before the appearance of ID, the mitochondrial oxidative enzymes become redox, as there is insufficient oxygen for an oxidation response, hence maximum reduction of the mitochondrial enzymes occurs (Somjen 2001). If the depolarization is prolonged, beyond a threshold time, irreversible damage is induced. Yet, if the depolarization does not induce an influx of Ca^{2+} into neurons, the neurons can regain their function (Somjen 2001). In our work, in most animals (namely the survivors), the hypotension probably did not continue beyond this threshold time, since no irreversible damage was observed, but rather a partial or even full recovery after resuscitation in all the cerebral parameters was seen.

The secondary vast increase in reflectance is called "secondary reflectance increase" (SRI), known to occur under massive constriction of blood vessels in the cerebral microcirculation, leading to a rapid decrease in blood volume. The SRI usually appears in severe ischemia, such as global ischemia, and death (Tomlinson et al. 1993; Mayevsky and Chance 2007). This further indicates the extensive reduction of cerebral perfusion, during both global ischemia and hemorrhagic hypotension, resulting from an increase in cerebral microcirculation resistance.

Similar models of a combination of hemorrhagic hypotension and BCO are usually used to induce complete cerebral ischemia; however, in such cases, the occlusion of both carotid arteries is done immediately before the onset of hemorrhage and not 24 h before bleeding (Lipton 1999).

5.4.5.2 Hemorrhage Under MCAO

Studies indicate that transient MCAO leads to interrupted blood supply and mitochondrial damage, which is visible even 24 h after reperfusion (Solenski et al. 2002). In this study, we demonstrated for the first time the effects of preliminary transient MCAO on hemorrhagic hypotension and resuscitation 24 h following ischemic reperfusion.

MCAO induced for duration of 1 h, occurring 24 h before the hemorrhage session (15 min bleeding) lead to a biphasic response during resuscitation: after an initial recovery, the MAP, CBF, and NADH suffered from a secondary decrease, and recuperated only at a later stage.

Following resuscitation, MAP recovered only partially, and reached full recovery only 20 min later. This partial recovery demonstrates that even though MCAO affects mainly cerebral blood supply, when combined with hemorrhage 24 h later may have systemic influences as well.

Cerebral blood supply suffered from an additional decrease after an initial recovery, while the IBF remained low after resuscitation, recovering only in a later stage. Mitochondrial activity in the brain increased following resuscitation, but returned to normal activity only shortly after, as demonstrated by NADH levels (Mayevsky et al. 1995). Intestinal mitochondrial activity decreased throughout the bleeding period and shortly after resuscitation, although it improved 20 min after resuscitation suffered from a moderate deterioration for up to 2 h following resuscitation.

These phenomena were absent in the other hemorrhage models shown in this work. The results of the study show that the transient ischemic penumbra suffers from difficulties in recovering from hemorrhage in the resuscitation period, compared to normoxic brain tissue.

The results of the study also demonstrate that transient MCAO 24 h prior to hemorrhage and resuscitation has an impact on intestinal response as well, as demonstrated by a mild recovery in blood supply and mitochondrial dysfunction during the hemorrhagic period and following resuscitation. Even following transient ischemia, the autoregulatory mechanism leads to an increased cerebral blood supply during hemorrhage and resuscitation, accompanied by recuperation in mitochondrial function.

Even though the cerebral penumbra underwent additional damage in the form of transient MCAO in addition to hemorrhage, the penumbra—as part of the brain—maintain the position of a vital organ, compared to the less vital intestine.

5.4.6 *The Surviving and Non-surviving Rats*

In three of our models, some of the rats did not survive the experimental period. We compared the responses of the surviving and non-surviving rats to try and identify a discriminating parameter that determines the final outcome. Such a parameter can be used as a warning sign for a possible deterioration that may lead to the rat's death.

5.4.6.1 Surviving Versus Non-surviving Rats Under Normal Conditions

Investigators disagree on the cause of death after hemorrhage; however, several possible explanations have been suggested, including total oxygen deficit, the volume of bleeding, the duration of hypotension, and the volume of reinfusion (Crowell and Smith 1964; Vivaldi et al. 1983; Schoenberg et al. 1985). In our study, we found that the longer the duration of hypotension, the higher the rate of death (Fig. 5.16), and the earlier its stage (as noted above). During the experiment, an increasing volume of blood was drawn; however, this was due to the need to maintain low MAP for longer durations of hypotension, and therefore, it cannot be considered a reason for the increasing mortality rate in the different models.

In our work, the main differences between the surviving and non-surviving rats in the controlled hypotension models maintained for 30 and 60 min, during the entire experiment, were shown by the levels of MAP and TBF. The differences between MAP of the two groups appeared during resuscitation. In the 30-minute model, MAP was significantly higher in the non-surviving group, whereas in the 60-minute model it was significantly lower in the non-surviving group. Concerning TBF, during the hypotension in the 30-minutes model, the decrease of IBF in the non-surviving rats was more intense than in the surviving cohort. CBF in the non-surviving group also decreased, though only at the end of the period after the decrease in the IBF. The massive decrease in the IBF was due to an increase in the resistance, likely causing an ischemic injury. Following resuscitation, in the 30-minute model, the intestinal decrease in perfusion was greater in the non-surviving group, and, at the end of the experiment, the cerebral perfusion in this group also deteriorated. The deterioration of the intestinal perfusion in both groups was probably due to IRI leading to MOF, as was earlier reported (Carden and Granger 2000; Cerqueira et al. 2005). In the 60-minute model, the CBF in the non-surviving rats was initially fully recovered, but then it decreased. On the other hand, in the surviving rats, even though it was partially recovered, the attained level remained stable until the end of the experiment. In the surviving rats, the IBF recovered, whereas in the non-surviving rats, the blood flow further deteriorated. Toward the end of the experiment, the IBF also deteriorated in the surviving group. The recovery of the intestinal perfusion in the surviving rats was probably due to the vasoreactivity of the intestinal microcirculation, leading to the return of the blood vessels to normal tonus, while the deterioration was due to ischemic injury. This deterioration in the non-surviving group can be explained by a reduction in the vascular responsiveness to endogenous vasoconstrictors during hemorrhagic shock, which leads to a disability to maintain blood pressure, therefore causing organ dysfunction and even irreversible injury and death due to vasodilatory shock (Mayevsky and Ziv 1991; Ellis et al. 2005; Peng et al. 2006). A previous study reported that the pressor and mesenteric constrictor responses to angiotensin II (one of the major parts in the hormonal autoregulation mechanisms) become significantly blunted as early as 60 min post hemorrhage (Pieber et al. 1999). This can explain why, only in the 60-minute model, but not in the 15- and 30-minute models, the intestine was unable to recover from the hypotension.

This sequence of events—a primary injury in the intestine followed by a deterioration of MAP and, at last, brain injury—suggests that the non-surviving rats died from a collapse of the microcirculation autoregulatory mechanisms at the level of the organ (intestine) or at the cardiovascular level, and not as a result of brain injury.

The finding that the earliest decrease in perfusion took place in the intestine emphasizes the importance of monitoring the intestine as a surrogate organ for the early detection of a forthcoming catastrophe and its prevention.

5.4.6.2 Surviving Versus Non-Surviving Rats Under Incomplete Global Cerebral Ischemia

During the hypotension, the MAP in the non-surviving rats decreased more rapidly than in the surviving rats; whereas during the recovery, MAP in both groups was fully restored. Nevertheless, the non-surviving rats failed to maintain this MAP level and it deteriorated within the resuscitation phase.

The decrease in the CBF after the hemorrhage insult was more rapid in the non-surviving group and reached a lower level of perfusion. Accordingly, the mitochondrial NADH redox state was higher in the non-surviving group. On the other hand, the decrease in the IBF was similar for the two groups, even though the increase in mitochondrial NADH was higher in the non-surviving group. We suggest that this regulation was due to a better O_2 extraction in the intestinal tissue of the surviving rats. Moreover, the fact that the intestinal hypoperfusion reached the same level in both group may indicate that, during hemorrhage, the intestinal perfusion reached its minimum level, and thus there was no further decrease in the non-surviving rats. During resuscitation, the non-surviving rats were unable to fully recover the cerebral perfusion, as did the surviving rats, and, accordingly, the level of mitochondrial NADH was higher in the non-surviving group. With reference to the IBF, both groups reached the same level of perfusion, though the mitochondrial NADH redox state was higher in the non-surviving group (as also occurred during the hypotension period). This indicates that these rats suffered from a higher hemorrhagic stress. This finding also highlights the importance of mitochondrial NADH redox state monitoring. In the ischemic rats, the deterioration of the non-surviving group was initiated in the brain, as opposed to the surviving group, where deterioration initially appeared in the intestine. The cause of death was presumably a vasodilatory shock, as a result of the extensive vasodilatation in the ischemic brain that was further injured by IRI.

5.4.7 Conclusion

The main conclusions of this study are as follows:

1. The intestine may serve as a surrogate organ for hemorrhagic insults, due to its capability for early detection of whole-body deterioration.
2. The monitoring of mitochondrial NADH redox state can be used as an indicator of the severity of the hemorrhagic stress.
3. The survival rate is affected by the duration of hypotension.
4. The duration of hypotension plays a crucial role in the severity of injury after hemorrhage.
5. An aggressive fluid resuscitation leads to further deterioration of the small intestine.
6. Resuscitation using a blood substitute leads to better outcomes by better maintaining the small intestinal vitality.
7. The combination of global ischemia as a result of bilateral carotid occlusion and hemorrhagic hypotension blurs the differences between the brain and the small intestine, due to the impairment of cerebral autoregulation mechanisms.
8. During hemorrhage, the cerebral state under global ischemia as a result of bilateral carotid occlusion is similar to the cerebral state under complete cerebral ischemia.
9. The death of the non-surviving rats following hemorrhagic hypotension was due to intestinal irreversible damage leading to systemic failure.

In view of all the above, the application of the Multisite-Multiparametric monitoring system has clearly some advantages under hemorrhagic hypotension. Further research will shed additional light on such conditions.

References

Aaslid R, Lindegaard KF, Sorteberg W, Nornes H (1989) Cerebral autoregulation dynamics in humans. Stroke 20(1):45–52

Acker T, Acker H (2004) Cellular oxygen sensing need in CNS function: physiological and pathological implications. J Exp Biol 207(Pt 18):3171–3188

Alayash AI, Cashon RE (1995) Hemoglobin and free radicals: implications for the development of a safe blood substitute. Mol Med Today 1(3):122–127

Arumugam TV, Arnold N, Proctor LM, Newman M, Reid RC, Hansford KA, Fairlie DP, Shiels IA, Taylor SM (2003) Comparative protection against rat intestinal reperfusion injury by a new inhibitor of sPLA2, COX-1 and COX-2 selective inhibitors, and an LTC4 receptor antagonist. Br J Pharmacol 140(1):71–80

Barbiro-Michaely E, Tolmasov M, Rinkevich-Shop S, Sonn J, Mayevsky A (2007) Can the "brain-sparing effect" be detected in a small-animal model? Med Sci Monit 13(10):Br211–219

Boldt J (2002) Clinical review: hemodynamic monitoring in the intensive care unit. Crit Care 6(1): 52–59

Boquist L (1972) Obesity and pancreatic islet hyperplasia in the Mongolian gerbil. Diabetologia 8(4):274–282

Boulpaep EL (2003) Organization of the cardiovascular system. In: Boron WF, Boulpaep EL (eds) Medical Physiology. Saunders, New York

Boulpaep EL, Boron WF (2003a) Foundation of Physiology. In: Boulpaep EL, Boron WF (eds) Medical Physiology. Saunders, Philadelphia

Boulpaep EL, Boron WF (2003b) Regulation of arterial pressure and cardiac output. In: Boulpaep EL, Boron WF (eds) Medical physiology. Saunders, Philadelphia

Brun-Buisson C, Doyon F, Carlet J (1996) Bacteremia and severe sepsis in adults: a multicenter prospective survey in ICUs and wards of 24 hospitals. French Bacteremia-Sepsis Study Group. Am J Respir Crit Care Med 154(3 Pt 1):617–624

Buehler PW, Boykins RA, Jia Y, Norris S, Freedberg DI, Alayash AI (2005) Structural and functional characterization of glutaraldehyde-polymerized bovine hemoglobin and its isolated fractions. Anal Chem 77(11):3466–3478

Carden DL, Granger DN (2000) Pathophysiology of ischaemia-reperfusion injury. J Pathol 190(3): 255–266

Cerqueira NF, Hussni CA, Yoshida WB (2005) Pathophysiology of mesenteric ischemia/reperfusion: a review. Acta Cir Bras 20(4):336–343

Chance B, Oshino N, Sugano T, Mayevsky A (1973) Basic principles of tissue oxygen determination from mitochondrial signals, oxygen transport to tissue. Adv Exp Med Biol 37A:277–292

Chang TM, Varma R (1992) Effect of a single replacement of one of Ringer lactate, hypertonic saline/dextran, 7g% albumin, stroma-free hemoglobin, o-raffinose polyhemoglobin or whole blood on the long term survival of unanesthetized rats with lethal hemorrhagic shock after 67% acute blood loss. Biomater Artif Cell Immobil Biotechnol 20(2–4):503–510

Chen RY, Fan FC, Schuessler GB, Simchon S, Kim S, Chien S (1984) Regional cerebral blood flow and oxygen consumption of the canine brain during hemorrhagic hypotension. Stroke 15(2): 343–350

Corbett EJ, Barry BN, Pollard SG, Lodge JP, Bellamy MC (2000) Laser Doppler flowmetry is useful in the clinical management of small bowel transplantation. The Liver Transplant Group. Gut 47(4):580–583

Cotton BA, Guy JS, Morris JA Jr, Abumrad NN (2006) The cellular, metabolic, and systemic consequences of aggressive fluid resuscitation strategies. Shock 26(2):115–121

Coyle P, Panzenbeck MJ (1990) Collateral development after carotid artery occlusion in Fischer 344 rats. Stroke 21(2):316–321

Craig RL, Poole GV (1994) Resuscitation in uncontrolled hemorrhage. Am Surg 60(1):59–62

Crowell JW, Smith EE (1964) Oxygen deficit and irreversible hemorrhagic shock. Am J Phys 206: 313–316

Cryer HM, Gosche J, Harbrecht J, Anigian G, Garrison N (2005) The effect of hypertonic saline resuscitation on responses to severe hemorrhagic shock by the skeletal muscle, intestinal, and renal microcirculation systems: seeing is believing. Am J Surg 190(2):305–313

Cuzzocrea S, Chatterjee PK, Mazzon E, Dugo L, De Sarro A, Van de Loo FA, Caputi AP, Thiemermann C (2002) Role of induced nitric oxide in the initiation of the inflammatory response after postischemic injury. Shock 18(2):169–176

den Uil CA, Klijn E, Lagrand WK, Brugts JJ, Ince C, Spronk PE, Simoons ML (2008) The microcirculation in health and critical disease. Prog Cardiovasc Dis 51(2):161–170

Dieter RS, Laird JR (2005) Carotid artery stenting: update. Int J Cardiovasc Interv 7(3):126–133
Dubin A, Estenssoro E, Murias G, Canales H, Sottile P, Badie J, Baran M, Palizas F, Laporte M, Rivas Diaz M (2001) Effects of hemorrhage on gastrointestinal oxygenation. Intensive Care Med 27(12):1931–1936
Duling BR (1972) Microvascular responses to alterations in oxygen tension. Circ Res 31(4): 481–489
Eike JH, Palmer AF (2004) Effect of $NaBH_4$ concentration and reaction time on physical properties of glutaraldehyde-polymerized hemoglobin. Biotechnol Prog 20(3):946–952
Ellis CG, Jagger J, Sharpe M (2005) The microcirculation as a functional system. Crit Care 9(Suppl 4):S3–S8
Friesenecker BE, Tsai AG, Martini J, Ulmer H, Wenzel V, Hasibeder WR, Intaglietta M, Dunser MW (2006) Arteriolar vasoconstrictive response: comparing the effects of arginine vasopressin and norepinephrine. Crit Care 10(3):R75
Ganong WF (1991) Circulation through special regions. In: Appleton and Lange medical book. Review of medical physiology, pp 562–582
Garrison RN, Zakaria ER (2005) Peritoneal resuscitation. Am J Surg 190(2):181–185
Gjedde A (2005) The pathways of oxygen in brain. I. Delivery and metabolism of oxygen. Adv Exp Med Biol 566:269–275
Gorelick PB (1986) Cerebrovascular disease. Pathophysiology and diagnosis. Nurs Clin North Am 21(2):275–288
Gruhn N, Larsen FS, Boesgaard S, Knudsen GM, Mortensen SA, Thomsen G, Aldershvile J (2001) Cerebral blood flow in patients with chronic heart failure before and after heart transplantation. Stroke 32(11):2530–2533
Grum CM (1993) Tissue oxygenation in low flow states and during hypoxemia. Crit Care Med 21(2 Suppl):S44–S49
Gutierrez G, Reines HD, Wulf-Gutierrez ME (2004) Clinical review: hemorrhagic shock. Crit Care 8(5):373–381
Guyton AC (1999a) Functional organization of the human body and control of the "Internal Environment". In: Wonsiewicz MJ (ed) Textbook of Physiology. W.B. Saunders Company, Philadelphia
Guyton AC (1999b) Nervous regulation of the circulation and rapid control of arterial pressure. In: Wonsiewicz MJ (ed) Textbook of medical physiology. W.B. Saunders Company, Philadelphia
Haisjackl M, Luz G, Sparr H, Germann R, Salak N, Friesenecker B, Deusch E, Meusburger S, Hasibeder W (1997) The effects of progressive anemia on jejunal mucosal and serosal tissue oxygenation in pigs. Anesth Analg 84(3):538–544
Harbrecht BG, Alarcon LH, Peitzman AB (2004) Management of shock. In: Moore EE, Feliciano DV, Mattox KL (eds) Trauma, 5th edn. McGraw-Hill, New York, pp 201–226
Hendrikse J, Hartkamp MJ, Hillen B, Mali WP, van der Grond J (2001) Collateral ability of the circle of Willis in patients with unilateral internal carotid artery occlusion: border zone infarcts and clinical symptoms. Stroke 32(12):2768–2773
Hossmann KA, Traystman RJ (2009) Cerebral blood flow and the ischemic penumbra. Handb Clin Neurol 92:67–92
Huang YC (2005) Monitoring oxygen delivery in the critically ill. Chest 128(5 Suppl 2):554S–560S
Hubbard WJ, Bland KI, Chaudry IH (2004) The role of the mitochondrion in trauma and shock. Shock 22(5):395–402
Jakob SM (2002) Clinical review: splanchnic ischaemia. Crit Care 6(4):306–312
Jakob SM, Tenhunen JJ, Laitinen S, Heino A, Alhava E, Takala J (2001) Effects of systemic arterial hypoperfusion on splanchnic hemodynamics and hepatic arterial buffer response in pigs. Am J Physiol Gastrointest Liver Physiol 280(5):G819–G827
Jassem W, Fuggle SV, Rela M, Koo DD, Heaton ND (2002) The role of mitochondria in ischemia/reperfusion injury. Transplantation 73(4):493–499
Johnson PC (1986) Autoregulation of blood flow. Circ Res 59(5):483–495

Johnson PC (1988) Microvascular regulation: normal function and disturbance in disease. In: Manabe H, Zweifach BW, Messmer K (eds) Microcirculation in circulatory disorders. Springer-Verlag, New York

Josko J, Gwozdz B (2003) Regulation of cerebral blood flow in ischemia. Wiad Lek 56(11–12): 564–568

Klabunde RE (2004) Cardiovascular physiology concepts. Lippincott Williams & Wilkins

Kleen M, Habler O, Meisner F, Kemming G, Pape A, Messmer K (2000) Effects of primary resuscitation from shock on distribution of myocardial blood flow. J Appl Physiol (1985) 88(2): 373–385

Kozuka M (1995) Changes in brain energy metabolism, neurotransmitters, and choline during and after incomplete cerebral ischemia in spontaneously hypertensive rats. Neurochem Res 20(1): 23–30

Kraut A, Barbiro-Michaely E, Mayevsky A (2004) Differential effects of norepinephrine on brain and other less vital organs detected by a multisite multiparametric monitoring system. Med Sci Monit 10(7):BR215-BR220

Krejci V, Hiltebrand L, Banic A, Erni D, Wheatley AM, Sigurdsson GH (2000) Continuous measurements of microcirculatory blood flow in gastrointestinal organs during acute haemorrhage. Br J Anaesth 84(4):468–475

Lavi S, Egbarya R, Lavi R, Jacob G (2003) Role of nitric oxide in the regulation of cerebral blood flow in humans: chemoregulation versus mechanoregulation. Circulation 107(14):1901–1905

Leach RM, Treacher DF (1998) Oxygen transport-2. Tissue hypoxia. BMJ 317(7169):1370–1373

Lemasters JJ, Nieminen AL, Qian T, Trost LC, Herman B (1997) The mitochondrial permeability transition in toxic, hypoxic and reperfusion injury. Mol Cell Biochem 174(1–2):159–165

Levin BE (2006) Metabolic sensing neurons and the control of energy homeostasis. Physiol Behav 89(4):486–489

Levin BE, Dunn-Meynell AA, Routh VH (1999) Brain glucose sensing and body energy homeostasis: role in obesity and diabetes. Am J Phys 276(5):R1223–R1231

Lipton P (1999) Ischemic cell death in brain neurons. Physiol Rev 79(4):1431–1568

Lomas-Niera JL, Perl M, Chung CS, Ayala A (2005) Shock and hemorrhage: an overview of animal models. Shock 24(Suppl 1):33–39

Longa EZ, Weinstein PR, Carlson S, Cummins R (1989) Reversible middle cerebral artery occlusion without craniectomy in rats. Stroke 20(1):84–91

Lu YQ, Cai XJ, Gu LH, Fan YJ, Wang Q, Bao DG (2005) Effects of three fluid resuscitation methods on apoptosis of visceral organs in rats with hemorrhagic shock. J Zhejiang Univ Sci B 6(9):907–912

Lubbers DW (1995) Optical sensors for clinical monitoring. Acta Anaesth Scand Suppl 39(104): 37–54

MacKenzie ET, Farrar JK, Fitch W, Graham DI, Gregory PC, Harper AM (1979) Effects of hemorrhagic hypotension on the cerebral circulation. I. Cerebral blood flow and pial arteriolar caliber. Stroke 10(6):711–718

Mallick IH, Yang W, Winslet MC, Seifalian AM (2004) Ischemia-reperfusion injury of the intestine and protective strategies against injury. Dig Dis Sci 49(9):1359–1377

Mayevsky A (1976) Brain energy metabolism of the conscious rat exposed to various physiological and pathological situations. Brain Res 113:327–338

Mayevsky A (1984) Brain NADH redox state monitored *in vivo* by fiber optic surface fluorometry. Brain Res Rev 7:49–68

Mayevsky A (1993) Biochemical and physiological activities of the brain as *in vivo* markers of brain pathology. In: Bernstein EF, Callow AD, Nicolaides AN, Shifrin EG (eds) Cerebral Revascularization. Med-Orion Pub, pp 51–69

Mayevsky A, Barbiro-Michaely E (2013) Shedding light on mitochondrial function by real time monitoring of NADH fluorescence: II: human studies. J Clin Monit Comp 27:125–145

Mayevsky A, Chance B (2007) Oxidation-reduction states of NADH *in vivo*: from animals to clinical use. Mitochondrion 7:330–339

Mayevsky A, Rogatsky G (2007) Mitochondrial function *in vivo* evaluated by NADH fluorescence: from animal models to human studies. Am J Physiol Cell Physiol 292:C615–C640
Mayevsky A, Ziv I (1991) Oscillations of cortical oxidative metabolism and microcirculation in the ischaemic brain. Neurol Res 13(1):39–47
Mayevsky A, Duckrow RB, Yoles E, Zarchin N, Kanshansky D (1990) Brain mitochondrial redox state, tissue hemodynamic and extracellular ion responses to four-vessel occlusion and spreading depression in the rat. Neurol Res 12:243–248
Mayevsky A, Meilin S, Rogatsky GG, Zarchin N, Sonn J (1995) Multiparametric monitoring of the awake brain exposed to carbon monoxide. J Appl Physiol 78:1188–1196
Mayevsky A, Meilin S, Manor T, Zarchin N, Sonn J (1999) Optical monitoring of NADH redox state and blood flow as indicators of brain energy balance. Adv Exp Med Biol 471:133–140
Mayevsky A, Manor T, Meilin S, Razon N, Ouaknine GE (2001) Multiparametric monitoring of tissue vitality in clinical situations. SPIE 4255:33–39
Md S, Moochhala SM, Siew-Yang KL (2003) The role of inducible nitric oxide synthase inhibitor on the arteriolar hyporesponsiveness in hemorrhagic-shocked rats. Life Sci 73(14):1825–1834
Morini S, Yacoub W, Rastellini C, Gaudio E, Watkins SC, Cicalese L (2000) Intestinal microvascular patterns during hemorrhagic shock. Dig Dis Sci 45(4):710–722
Murphy SJ, Kirsch JR, Zhang W, Grafe MR, West GA, del Zoppo GJ, Traystman RJ, Hum PD (2008) Can gender differences be evaluated in a rhesus macaque (Macaca mulatta) model of focal cerebral ischemia? Comp Med 58(6):588–596
Nelson DP, King CE, Dodd SL, Schumacker PT, Cain SM (1987) Systemic and intestinal limits of O2 extraction in the dog. J Appl Physiol (1985) 63(1):387–394
Novak V, Novak P, Spies JM, Low PA (1998) Autoregulation of cerebral blood flow in orthostatic hypotension. Stroke 29(1):104–111
O'Sullivan JC, Yao XL, Alam H, McCabe JT (2007) Diazoxide, as a postconditioning and delayed preconditioning trigger, increases HSP25 and HSP70 in the central nervous system following combined cerebral stroke and hemorrhagic shock. J Neurotrauma 24(3):532–546
Payne SJ, Tarassenko L (2006) Combined transfer function analysis and modelling of cerebral autoregulation. Ann Biomed Eng 34(5):847–858
Peitzman AB, Billiar TR, Harbrecht BG, Kelly E, Udekwu AO, Simmons RL (1995) Hemorrhagic shock. Curr Probl Surg 32(11):925–1002
Peng TC, Liao KW, Lai HL, Chao YF, Chang FM, Harn HJ, Lee RP (2006) The physiological changes of cumulative hemorrhagic shock in conscious rats. J Biomed Sci 13(3):385–394
Pieber D, Horina G, Sandner-Kiesling A, Pieber TR, Heinemann A (1999) Pressor and mesenteric arterial hyporesponsiveness to angiotensin II is an early event in haemorrhagic hypotension in anaesthetised rats. Cardiovasc Res 44(1):166–175
Powers WJ, Grubb RL Jr, Darriet D, Raichle ME (1985) Cerebral blood flow and cerebral metabolic rate of oxygen requirements for cerebral function and viability in humans. J Cereb Blood Flow Metab 5(4):600–608
Pries AR, Secomb TW, Gaehtgens P (1996) Biophysical aspects of blood flow in the microvasculature. Cardiovasc Res 32(4):654–667
Reilly PM, Wilkins KB, Fuh KC, Haglund U, Bulkley GB (2001) The mesenteric hemodynamic response to circulatory shock: an overview. Shock 15(5):329–343
Revell M, Greaves I, Porter K (2003) Endpoints for fluid resuscitation in hemorrhagic shock. J Trauma 54(5 Suppl):S63–S67
Rise IR, Kirkeby OJ (1998) Effect of cerebral ischaemia on the cerebrovascular and cardiovascular response to haemorrhage. Acta Neurochir 140(7):699–705
Rixen D, Neugebauer EAM (2004) Changing paradigms in animal models of traumatic/hemorrhagic shock. Eur J Trauma 30:279–288
Sander-Jensen K, Secher NH, Bie P, Warberg J, Schwartz TW (1986) Vagal slowing of the heart during haemorrhage: observations from 20 consecutive hypotensive patients. Br Med J (Clin Res Ed) 292(6517):364–366
Saraste M (1999) Oxidative phosphorylation at the fin de siecle. Science 283(5407):1488–1493

Schertel ER, Valentine AK, Schmall LM, Allen DA, Muir WW (1991) Vagotomy alters the hemodynamic response of dogs in hemorrhagic shock. Circ Shock 34(4):393–397
Schoenberg MH, Smedegard G, Gerdin B, Messmer K, Arfors KE (1985) Hemorrhagic shock in the dog. I. Correlation between survival and severity of shock. Res Exp Med (Berl) 185(1):21–33
Sherwood L (1997a) The blood vessels and blood pressure. In: Sherwood L (ed) Human physiology from cells to systems. Wadsworth Publising Company, Boston
Sherwood L (1997b) Homeostasis: the foundation of physiology. In: Sherwood L (ed) Human physiology from cells to systems. Wadsworth Publishing Company, Boston
Sims NR, Muyderman H (2010) Mitochondria, oxidative metabolism and cell death in stroke. Biochim Biophys Acta 1802(1):80–91
Singel DJ, Stamler JS (2005) Chemical physiology of blood flow regulation by red blood cells: the role of nitric oxide and S-nitrosohemoglobin. Annu Rev Physiol 67:99–145
Sjovall H, Redfors S, Hallback DA, Eklund S, Jodal M, Lundgren O (1983) The effect of splanchnic nerve stimulation on blood flow distribution, villous tissue osmolality and fluid and electrolyte transport in the small intestine of the cat. Acta Physiol Scand 117(3):359–365
Smail N, Wang P, Cioffi WG, Bland KI, Chaudry IH (1998) Resuscitation after uncontrolled venous hemorrhage: does increased resuscitation volume improve regional perfusion? J Trauma 44(4):701–708
Solenski NJ, diPierro CG, Trimmer PA, Kwan AL, Helm GA (2002) Ultrastructural changes of neuronal mitochondria after transient and permanent cerebral ischemia. Stroke 33(3):816–824
Somjen GG (2001) Mechanisms of spreading depression and hypoxic spreading depression-like depolarization. Physiol Rev 81(3):1065–1096
Strandgaard S (1976) Autoregulation of cerebral blood flow in hypertensive patients. The modifying influence of prolonged antihypertensive treatment on the tolerance to acute, drug-induced hypotension. Circulation 53(4):720–727
Strandgaard S, Paulson OB (1984) Cerebral autoregulation. Stroke 15(3):413–416
Takala J (1996) Determinants of splanchnic blood flow. Br J Anaesth 77(1):50–58
Tomlinson FH, Anderson RE, Meyer FB (1993) Brain pH_i, cerebral blood flow, and NADH fluorescence during severe incomplete global ischemia in rabbits. Stroke 24:435–443
Torres Filho IP, Spiess BD, Pittman RN, Barbee RW, Ward KR (2005) Experimental analysis of critical oxygen delivery. Am J Physiol Heart Circ Physiol 288(3):H1071–H1079
Torres LN, Torres Filho IP, Barbee RW, Tiba MH, Ward KR, Pittman RN (2004a) Systemic responses to prolonged hemorrhagic hypotension. Am J Physiol Heart Circ Physiol 286:H1811–H1820
Torres LN, Torres Filho IP, Barbee RW, Tiba MH, Ward KR, Pittman RN (2004b) Continuous peripheral resistance measurement during hemorrhagic hypotension. Am J Physiol Heart Circ Physiol 287(5):H2341–H2345
Treacher DF, Leach RM (1998) Oxygen transport-1. Basic principles. BMJ 317(7168):1302–1306
Vagg DJ, Bandler R, Keay KA (2008) Hypovolemic shock: critical involvement of a projection from the ventrolateral periaqueductal gray to the caudal midline medulla. Neuroscience 152(4): 1099–1109
Vagnozzi R, Tavazzi B, Di Pierro D, Giardina B, Fraioli B, Signoretti S, Distefano S, Galvano M, Lazzarino G (1997) Effects of increasing times of incomplete cerebral ischemia upon the energy state and lipid peroxidation in the rat. Exp Brain Res 117(3):411–418
Vajda K, Szabo A, Boros M (2004a) Heterogeneous microcirculation in the rat small intestine during hemorrhagic shock: quantification of the effects of hypertonic-hyperoncotic resuscitation. Eur Surg Res 36(6):338–344
Vajda K, Szabo A, Kucsa K, Suki B, Boros M (2004b) Microcirculatory heterogeneity in the rat small intestine during compromised flow conditions. Microcirculation 11(4):307–315
Vatner SF (1974) Effects of hemorrhage on regional blood flow distribution in dogs and primates. J Clin Invest 54(2):225–235
Vivaldi E, Macinelli S, Gunther B (1983) Experimental hemorrhagic shock in dogs: standardization. Res Exp Med (Berl) 182(2):127–137

Wattanasirichaigoon S, Menconi MJ, Delude RL, Fink MP (1999) Effect of mesenteric ischemia and reperfusion or hemorrhagic shock on intestinal mucosal permeability and ATP content in rats. Shock 12(2):127–133

Woolf PD (1986) Endocrinology of shock. Ann Emerg Med 15(12):1401–1405

Yu ZY, Ono S, Spatz M, McCarron RM (2002) Effect of hemorrhagic shock on apoptosis and energy-dependent efflux system in the brain. Neurochem Res 27(12):1625–1632

Zakaria ER, Garrison RN, Spain DA, Matheson PJ, Harris PD, Richardson JD (2003) Intraperitoneal resuscitation improves intestinal blood flow following hemorrhagic shock. Ann Surg 237(5):704-711; discussion 711-703

Zakaria ER, Ehringer WD, Tsakadze N, Li N, Garrison RN (2005) Direct energy delivery improves tissue perfusion after resuscitated shock. Surgery 138(2):195–203

Zhan C, Yang J (2006) Protective effects of isoliquiritigenin in transient middle cerebral artery occlusion-induced focal cerebral ischemia in rats. Pharmacol Res 53(3):303–309

Zhu X, Chu W, Wang T, Wang F, Fan D, Dan N, Chen C (2007) Variations in dominant antigen determinants of glutaraldehyde polymerized human, bovine and porcine hemoglobin. Artif Cells Blood Substit Immobil Biotechnol 35(5):518–532

Part III
Clinical Monitoring

Chapter 6
Technology Transfer: From Experimental Animals to Clinical Use

6.1 Introduction

One of the main goals of performing experiments in animal models as well as developing new monitoring tools is to enable the clinicians the understanding of pathophysiological situations in patients, as well as to develop new tools for daily use in patients. As of today, monitoring of mitochondrial function simultaneously with microcirculatory blood flow and hemoglobin oxygenation is very rare in clinical practice. One of the main factors enabling normal tissue and organs function is the appropriate level of oxygen supply to the mitochondria needed for ATP production. In this chapter, we are summarizing the published material regarding the monitoring of NADH redox state in various organs in patients in vivo.

6.2 History of NADH Monitoring in Patients

Most of the published material was investigated in the intact human brain but other body organs were monitored as well under in vivo conditions. Monitoring of mitochondrial function alone is not enough, and therefore over the years, the multiparametric monitoring approach has been adopted in various clinical studies. Chance et al. (2005) concluded that "the redox state of mitochondrial NADH will complement blood gas analysis for measuring the health and welfare of human tissues." Nevertheless, as of today, this idea has not been implemented in daily clinical practice.

© The Author(s), under exclusive license to Springer Nature Switzerland AG 2025
A. Mayevsky et al., *Body Oxygen Homeostasis and Mitochondrial Function*,
https://doi.org/10.1007/978-3-031-94113-9_6

6.2.1 Monitoring the Human Brain

Although NADH was monitored in isolated cells from human tissues (Kohen et al. 1966), the first attempt to apply NADH fluorometry to human tissues in vivo was made by Jobsis et al. in 1971. The fluorometric technique was used in patients (Fig. 6.1) in order to evaluate the relative rate of oxygen consumption and the bioenergetics activity of the intact cerebral cortex of patients. Using NADH fluorescence microfluorometry (6.1A+B), they monitored the exposed brain of neurosurgical patients undergoing treatments for focal cerebral seizures. They correlated the electrocorticographic data to the NADH redox state under direct cortical stimulation of the monitored area. The recorded changes (C), namely, the oxidation of NADH (decrease of fluorescence), were very similar to those obtained in analogous procedures in the cat brain (Rosenthal and Jobsis 1971).

A few years later, the collaboration between Austin et al. (1978a, b) led to the recording of NADH in the brain of patients subjected to microanastomosis of the superficial temporal artery (STA) to the middle cerebral artery (MCA). They found

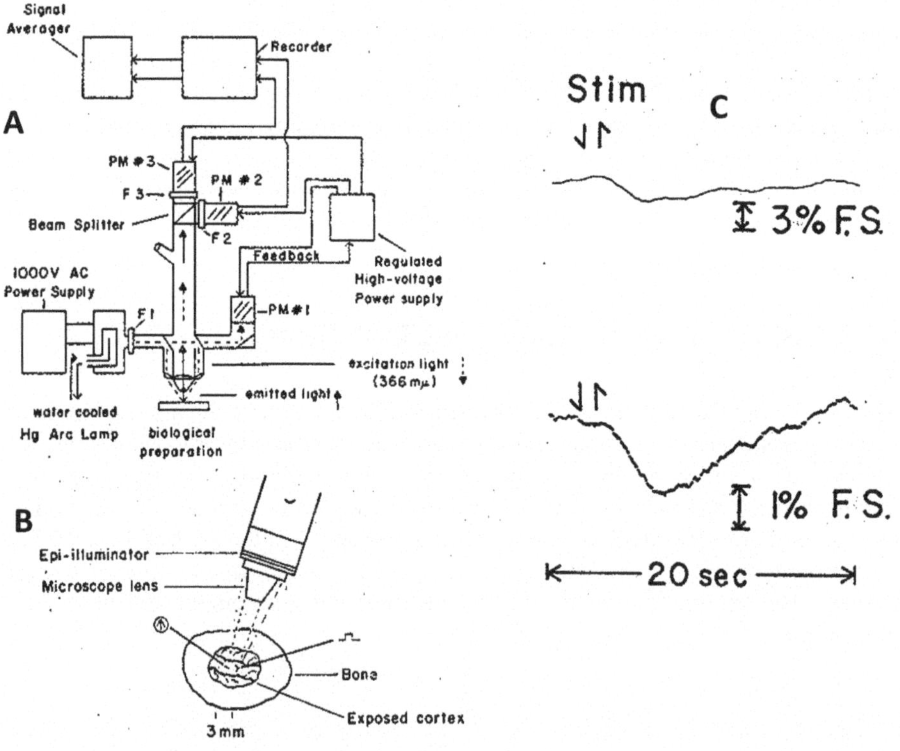

Fig. 6.1 The technology and typical results measured by Jobsis et al. (1971) from human cerebral cortex. The microfluorometer (**a**) and its tip (**b**) placed above the surface of the brain is shown on the left side. The responses to direct cortical stimulation are shown on the right side (**c**). (Reprinted with permission from Elsevier)

an improvement of cerebral oxidative metabolism after the anastomosis, which was correlated to the elevated blood flow and increased tissue pO_2. Using a different type of fluorometer, Van Buren et al. showed a decrease in NADH (oxidation) due to cortical stimulation in epileptic patients (Van Buren et al. 1978). Cortical electrical activity was compared with NADH fluorescence in patients undergoing operation following focal epilepsy. NADH was oxidized by 3–15% in response to cortical stimulation only in 60% of the patients. The topography of spread of NADH changes after cortical stimulation was usually asymmetrical in human epileptigenic cortex.

Barlow et al. (1979) used the fluorometric technique to measure NADH and flavoprotein fluorescence. In patients, they monitored the fluorescence of flavoprotein in the human brain after anastomosis of the superficial temporal artery to the middle cerebral artery. They "demonstrated increased rate of cortical oxidative metabolism after the surgical procedures."

During the same year Fein and Jobsis studied the changes in brain energetic in patients undergoing STA-MCA microanastomosis (Fein and Jobsis 1979). They tested six patients and concluded that "the mechanism of energy production may be more sensitive to ischemic anoxia than the mechanism of energy utilization." Later on Fein and Olinger monitored patients after transient ischemic attacks (TIAs) (Fein and Olinger 1982; Fein 1983). The brain of these patients, who had undergone an extracranial-intracranial bypass, was stimulated, and changes in NADH were recorded. They concluded that "persistent reversible abnormalities of cortical mitochondrial metabolism exist in a significant number of patients with history of TIAs who are suitable candidates for bypass surgery." Chance in a review article in 1983 compares three types of brain imaging techniques; optical—the surface fluorescence, NMR, and radio isotopes, emphasizing the advantages and disadvantages of each one.

6.2.2 Monitoring the Heart and Skeletal Muscle

The laser-based fluorimeter developed by Renault et al. (1984) was used to monitor NADH redox state in the endomyocardium of patients during left heart catheterization (Duboc et al. 1986). They injected contrast medium into the right coronary artery and recorded a large transient increase in NADH fluorescence. It was hard to interpret the response since a controlled injection of saline solution was not reported. Later on, the same group monitored the human heart during induced variations in left ventricular end-diastolic pressure (LVEDP) and during coronary arteriography. NADH laser fluorometry therefore seems to be promising as a means of exploring myocardial metabolism during cardiac catheterization (Toussaint et al. 1987, 1988). During coronary angiography, injection of the contrast medium induces a rise in NADH in patients with a significant degree of coronary stenosis. Figure 6.2 shows the effect of catheterization on NADH fluorescence measured in patients. In another study, NADH was monitored in patients with McArdle's disease (muscular phosphorylase deficiency). This supports the hypothesis that in this disease when the flux

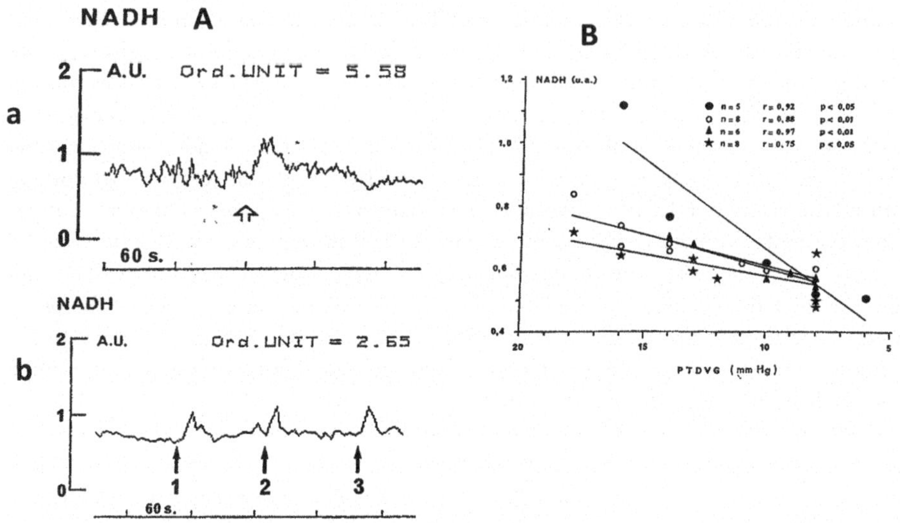

Fig. 6.2 (**Aa**) Direct effect of the coronary angiography on the concentration of NADH. The arrow indicates the injection of the contrast agent. The NADH concentration, expressed in arbitrary units (AU). (**Ab**) Direct effect of the coronary angiography on the concentration of NADH. Arrow no.1 indicates the time of injection of radioselectan*. Arrow no. 2 and 3 indicate the time of injection of hexabrix* (Toussaint et al. 1987). (**b**) The correlation between the telediastolic pressure of the left ventricle (PTDVG) and the NADH Concentration. N number of measurements for each patient. AU arbitrary units (Toussaint et al. 1987)

through the Embden-Meyerhof pathway is reduced the mitochondrial ability to produce NADH during ischemia decreases (Duboc et al. 1987). Later on, NADH was monitored in the knee extensor muscle in 30 healthy males. NADH increased maximally within the first seconds of contraction (Guezennec et al. 1991).

Under amyotrophic sclerosis, there is impairment of mitochondrial function as presented by the decrease in the activity of NADH:CoQ oxidoreductase as compared to the control group. In this study, NADH was not monitored in vivo but rather an isolated muscle fiber was used (Wiedemann et al. 1998).

The same group of investigators monitored muscle fiber in vitro isolated from patients suffering from Sporadic Amyotrophic Lateral Sclerosis (SALS). They monitored flavoprotein/NAD(P)H ratio using autofluorescence imaging and found heterogeneous distribution of mitochondrial respiratory chain defect in 11/17 patients with SALS (Vielhaber et al. 2000).

Intracellular fluorophores were monitored in isolated muscle cells having the advantage of avoiding the problems of photoxicity and photobleaching (Kunz et al. 2002).

It is important to note that other studies in which spectroscopy was used to detect cancer cells or tumors were published. In most of them, the monitoring was not directly related to the changes to NADH redox state itself but rather the wide spectrum was used and not only the NADH emission spectrum (Galeotti et al. 1970, 1972; Paschen et al. 1984; Kaplan 1985; Chandra et al. 2007) (Kohen et al. 1968, 1971, 1975a, b).

In the last 10–20 years, the low temperature scanner of NADH and Fp was applied to the identification of cancer and its malignancy (Zhang et al. 2004a, b; Li et al. 2009; Sivabalan et al. 2010; Xu et al. 2010, 2011).

6.2.3 Monitoring of Visceral Organs

Mayevsky and his group attempted to apply the NADH monitoring system in human kidney during a transplantation procedure. We measured the transplanted kidney cortex immediately after the surgical procedures (Mayevsky et al. 2000, 2003).

As of today, in vivo monitoring of the liver in human was not performed as yet. Nevertheless, two papers described the changes in liver redox state after transplantation. A correlation between flavoproteins/pyridine nucleotides in human liver frozen tissue and the ratio between arterial blood acetoacetate and beta-hydroxybutirate was found. They concluded that hepatic mitochondrial redox state of pyridine nucleotide can be evaluated by arterial blood (Ozawa et al. 1992). Intramitochondrial oxido-reduction state was evaluated in frozen samples of liver grafts obtained from living donors. There was a delay in the oxidation of the intramitochondrial oxidation-reduction state as compared to blood oxygenation, probably due to the cold preservation condition. Those findings indicate the importance of microcirculation in the normalization of oxidation-reduction state (Tanaka et al. 1993).

6.2.4 Monitoring of Cancer Cells and Tissues

Although the current review is dealing with in vivo monitoring of NADH in human organs, we are including a short review of the monitoring in cells and tissues. Spectroscopy of cells and tissues in vitro is used in order to differentiate between normal and cancer cells of human organs.

One of the problems of determining melanoma in situ is due to the lack of suitable techniques. This group has developed a fluorescence technique for a qualitative and quantitative determination of nuclear cataract in human lenses. This technique is based on the fluorescence of a natural chromophore in the tissue that is excited at 366 nm and excites maximally at 475 nm (probably NADH) and is the first to be changed following metabolic changes (Lohmann and Paul 1988). Lohmann et al. (1990) reported the monitoring of lung tissue. The fluorescence tomography (with a peak at 460 nm), together with other established methods, might be used for the determination of the histology of a biopsy sample. Schneckenburger et al. (1992) found that skin fibroblasts from patients with mitochondrial disease showed high levels of fluorescence. This technique was also used on patients with moles of melanoma, since there is an increase in the metabolic rate in cancerous cells, and this is mainly observed at the borders between healthy tissue and cancerous tissue. The fluorescence in the melanoma tissue was sixfold higher than that of the normal tissue.

Autofluorescence was used in colonic tissue to distinguish between neoplastic (abnormal tissue growing) and hyperplasic tissue (normal tissue growing). This technique is significantly ideal for tissue diagnosis since it does not require the administration of exogenous fluorescent dyes to the patients (Schomacker et al. 1992).

Konig et al. (1994) claim that early diagnosis of a diseased tissue especially in the field of cancer is not available yet. Characterization of the tissue metabolic state can successfully distinguish between healthy and malignant tissue. This study monitored bladder mucosa carcinoma and prostatic tissue. Laser-induced fluorescence is a technique that could perform "optical biopsy" in the gastrointestinal mucosa. Monitoring of esophageal specimens showed good accuracy in distinguishing between normal versus malignant tissue (Wang et al. 1994).

Since cancer cells usually show higher glycolytic activity, this activity effects their level of NADH/NAD^+. Frozen slices of bronchial tissue were monitored. NADH fluorescence was higher in the normal tissue versus the malignant tissue (Betz et al. 1994).

Sterenborg et al. (1995/9) found that the detection of skin cancer using the fluorescence of natural chromophores, especially NADH was possible. Pfeifer et al. (1995) used the fluorescence method to distinguish between healthy, cancerous, and necrotic tissue in lung tumor diagnostics. It was demonstrated that NADH fluorescence can be used for differentiation between tumorous and necrotic tissue, while flavine measurements are suitable for the differentiation between healthy and nonhealthy tissue.

A compact probe of laser spectrometer was used in various models: hippocampal slices, to differentiate between necrotic and non-necrotic lung tissue, for monitoring cell growth, bladder carcinoma diagnosis (the last one showed lower fluorescence in malignant cells than that of normal cells (Pfeifer et al. 1996).

6.3 Clinical Monitoring Using Noncommercial Devices

6.3.1 The First Study

6.3.1.1 Introduction

The new era in monitoring of mitochondrial function in patients started in 1990, when our team developed a unique multiparametric monitoring system that included the measurement of NADH fluorescence, using a light-guide-based device (Mayevsky et al. 1991). This system was initially applied to monitor neurosurgical patients undergoing brain surgery. In the second half of the 1990s (Mayevsky et al. 1996, 1998b; Meilin et al. 1999), we were able to monitor the brain of comatose patients in the ICU and described, for the first time, the development of cortical spreading depression (CSD)-like responses, including marked NADH responses that were correlated to the microcirculatory blood flow. A modified multiparametric monitoring system was applied for NADH measurement during neurosurgical procedures in the operating room (Mayevsky et al. 1998a, 1999, 2001, 2002).

Monitoring of patient in real time could be divided into two possible categories as shown in Chap. 2, Fig. 2.2. In part A, when a specific organ is monitored, the information accumulated represents the events occurring in the organ itself. For example, monitoring the intra cranial pressure (ICP) or the Electroencephalography (EEG) in neurosurgical patients will provide information on the events developed in the brain itself. The information is dependent initially upon local pathological events occurred in the brain. Monitoring the Electrocardiography (ECG) represents the electrical activity of the heart and most of the changes are correlated to local and not systemic events initially.

The second approach in monitoring patients is shown in Chap. 2, Fig. 2.2b. This approach is divided into two subgroups labeled as B1 and B2. The parameters shown in B1 are measured routinely from the cardiovascular and respiratory system. It provides information on systemic events such as changes in gas exchange in the respiratory system or cardiac output affecting the perfusion of various organs in the body. The second group of parameters, shown in part B2 is measured at the tissue level and the accumulated results represent the oxygen balance of the entire body. The physiological mechanism behind the monitoring approach shown in part B2 of Fig. 2.2 is illustrated in Fig. 3.1.

The pattern of pathophysiological cascade of events that may occur in many emergency clinical situations in adult patients, which may lead to morbidity and mortality, is shown in Fig. 2.2. As shown, various pathological states may lead to metabolic disturbances and may end up in cellular energy derangement (Hotchkiss and Karl 1992; Meier-Hellmann and Reinhart 1995; Marik and Varon 1998; Ince and Sinaasappel 1999).

The six pathological states, shown in the upper part of Fig. 3.1 (Mayevsky and Barbiro-Michaely 2013), are the most common events that may develop in clinical practice. It may develop due to a specific clear event, such as a major operation, or during slow process of body deterioration, such as in sepsis or shock. The definition of each of those six states is not so well established and some overlapping may exist. Under all those situations, the metabolic state of the body will be deteriorated and energy failure will develop.

The situations shown in Fig. 3.1 (Mayevsky and Barbiro-Michaely 2013) could develop in patients hospitalized in various operating rooms or ICU. It includes the respiratory, neurosurgical, cardiac, and neonatal ICU. Also, all patients undergoing severe operations such as cardiac bypass, neurosurgical, or organ transplantation may develop the Body Emergency Metabolic State (BEMS). Other patients who may develop the BEMS are newborns during delivery or elderly patients treated in the internal medicine departments.

As a central protection mechanism, blood flow redistribution will occur and the three protected organs (brain, heart, and adrenal gland) will receive more blood and O_2, while the peripheral organs or areas (skin and muscles), as well as others less vital visceral organs, will undergo vasoconstriction and a decrease in blood flow and O_2 supply will occur. Monitoring at the cellular function is the most significant indicator of the metabolic state of patients in critical care medicine (Fullerton and Singer 2011; Ekbal et al. 2013).

The energy balance in the most vital organs will remain positive due to higher blood flow while the less vital organs will be hypo perfused and a negative energy balance will have developed. As presented in Fig. 3.1 (Mayevsky and Barbiro-Michaely 2013), the blood flow redistribution mechanism will affect the energy production by the mitochondria in the most vital and less vital organs in the body. This change in mitochondrial function will affect the production of ATP.

6.3.1.2 Methods

Over the years, we had developed few types of multiparametric monitoring systems which were applied to patients at various clinical locations. The concept of our development was to provide the clinicians new physiological information that will improve the ability to diagnose the stage of pathological condition in real time.

The Multiparametric Monitoring System

In the 1980s, we developed the first multiparametric monitoring system that was adapted to experimental animals exposed to various pathological conditions (Friedli et al. 1982; Mayevsky 1983b; Mayevsky and Sclarsky 1983; Zhu et al. 2007). This system is presented in Fig. 6.3. Later on, we improved the system in order to introduce it to the neurosurgical operating room or ICU for patient monitoring. The concept behind the development was that the more parameters that are monitored, the better the diagnosis of the situation will be as presented in Fig. 6.3. The ideal system was to monitor eight parameters from the brain (right side) in addition to the various systemic parameters monitored in every patient hospitalized (left side). The long-term vision was that all parameters monitored by the two monitoring systems will be integrated in the same data bank and an expert system will be developed for better diagnosis of the patients. The translation of this concept into a practical tool and monitoring device is presented in Fig. 6.4. Based on the well-developed MultiProbe Assembly (MPA) shown in Fig. 6.4 and used for animal experiments (Mayevsky et al. 1995), a new MPA was developed and applied to patients monitored in the neurosurgical operation rooms and ICU. This device allowed us to monitor, in real time, the hemodynamic, metabolic, ionic, and electrical activities in the brain of comatose patients. All details regarding the technology and the clinical setup appear in our published paper (Mayevsky et al. 1996).

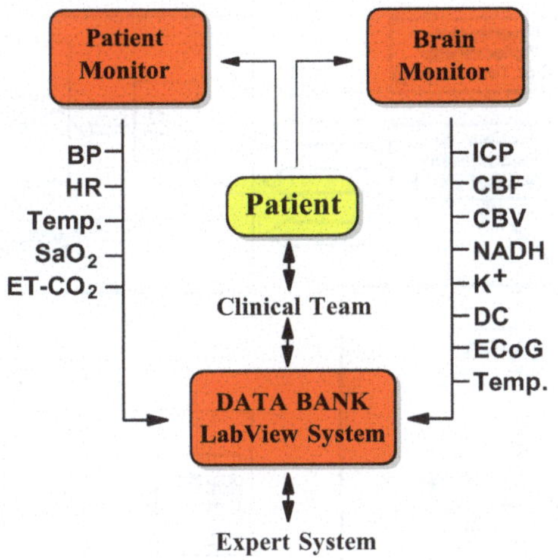

Fig. 6.3 Neurological patient monitoring usually includes systemic parameters (Left); however, we suggest the monitoring of specific brain parameters (Right). One Data bank (LabView system) will concentrate this data, which will be analyzed by the clinical team. BP blood pressure; HR heart rate; Temp. body temperature; SaO_2 oxygen saturation; ET CO_2 End Tidal CO_2; ICP intracranial pressure; CBF or CBV cerebral blood flow or volume; NADH mitochondrial NADH redox state; K^+ extracellular potassium level; DC steady DC potential; ECoG electrocorticography; Temp. Brain Temperature (Mayevsky and Barbiro-Michaely 2013)

Figure 6.4a shows the longitudinal section of the MPA (measuring 8 parameters) used in the neurosurgical operating room and ICU.

This development of the MPA was a response to the challenge made by the statement of Wiebers et al. (1990) as follows: "Ultimately, however, the answers to many of our questions regarding the underlying pathophysiology and treatment of stroke do not lie with continued attempts to model the human situation perfectly in animals, but rather with development of techniques to enable the study of more basic metabolism, pathophysiology and anatomical imaging detail in living humans."

All details regarding the preparation of the various sensors used as well as the construction of the MPA were described in our paper (Mayevsky et al. 1991) and are presented in the next paragraphs.

NADH Redox State Fluorometry The principle of NADH monitoring from the surface of the brain is that excitation light (366 nm) is passed from the fluorometer to the brain via a bundle of optical fibers made of quartz. The emitted light (450 nm), together with the reflected light at the excitation wavelength, is transferred to the fluorometer via another bundle (Mayevsky 1984). The changes in the reflected light are correlated to changes in tissue blood volume and also serve for correction of hemodynamic artifacts appearing in the NADH measurement. Calibration of the system is done as follows: the reflectance and fluorescence signals obtained from the photomultipliers (RCA 931B) are calibrated to a standard signal (0.5 V) by variation of photomultiplier dynode voltage obtained from the high-voltage power supply. The standard 0.5 V signal is set to give a half-scale deflection on the recorder

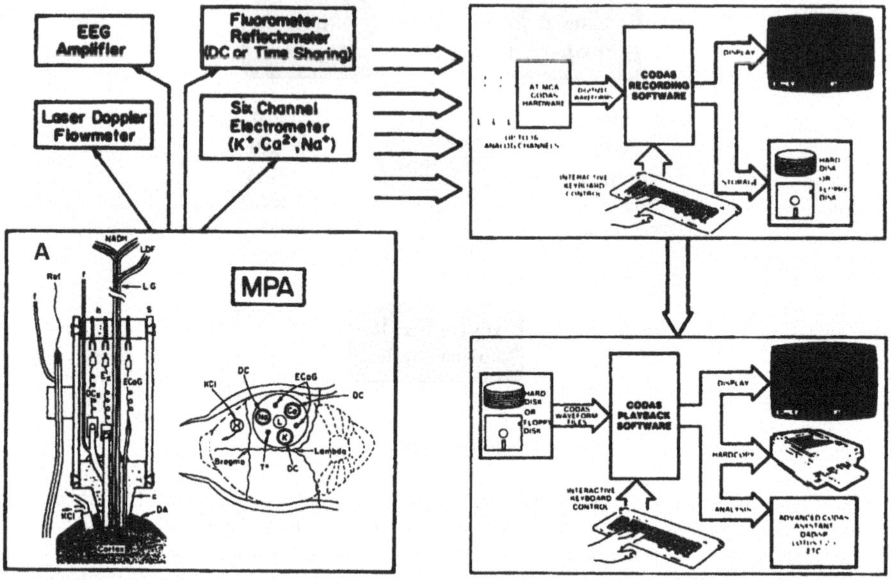

Fig. 6.4 Experimental setup used in this study to monitor various brain functions in real time. The multiprobe assembly (MPA) is connected to the various monitoring instruments. The analog signals are transferred to the A/D input of the acquisition setup. C Plexiglass probe holder; DA dental acrylic cement; KCl push–pull cannula for KCl; ECoG Electrocortical electrodes; K, Ca, Na ion-specific electrodes; DC area of DC steady potential monitoring; NADH 2 arms of NADH monitoring light guide; h connectors holder; s Plexiglass sleeve; Ref reference electrode; f feeling tube of reference of an X electrode (Mayevsky et al. 1991)

(2.5 cm) with pen resting at mid-scale. The gain is increased as required by a factor of two or four to give 50% or 25% of the full scale, respectively. The changes in fluorescence and reflectance signals are calculated relative to the calibrated signals under normoxic conditions. This type of calibration is not absolute, but provides reliable and reproducible results from different animals and between different fluorometers. More details on this technique have been published by our group (Mayevsky and Chance 1982; Mayevsky et al. 1982).

Laser Doppler Flowmetry In order to measure in real time, the cerebral blood flow (CBF) from the same cortical area of the MPA location, we applied the Laser Doppler flowmetry (LDF) technique (Stern et al. 1977). During the last year, this technique was calibrated against the H_2 clearance probe (Haberl et al. 1989) or the (^{14}C) iodoantipyrine autoradiography (Dirnagl et al. 1989), two well-established ways to monitor quantitatively CBF. It seems that this LDF provides only relative changes, which were correlated very well with relative changes in CBF measured with the two other approaches.

Multiprobe Assembly This approach was developed by us several years ago (Friedli et al. 1982; Mayevsky 1983b) and had been used successfully in rats and gerbils. The MPA used in this study has the following features:

(a) All parameters have the same type of noninvasive surface contact with the brain cortex.
(b) The MPA can be removed without any damage from the brain at the end of the measurements, so that repetitive applications can be performed in a short period of time with minimal technical support.
(c) The same MPA could be used in animal models as well as in the operation room and therefore may contribute to the better understanding of the results measured in the human brain.

All details regarding the preparation of electrodes were described previously (Friedli et al. 1982, Mayevsky 1983b). Figure 6.4 shows the MPA used by us. For this study, the modified assembly developed had the following differences:

1. Ca^{2+} and Na^{+} electrodes were added to the MPA.
2. The DC steady potential is measured by a highly stable Ag/AgC1 electrode located around the K^{+}, Na^{+}, and Ca^{2+} electrodes.
3. The reference electrode is an Ag/AgCl electrode (via a salt bridge), located below the skin of the animal's neck or the exposed skull of the patient.
4. Ca^{2+} and Na^{+} electrodes have the same construction as the K^{+} electrode and is based on the calcium and sodium ionophore embedded in a PVC membrane (WPI Inc. New Haven, Conn.).
5. The MPA is constructed in a new way (using epoxy glue) to hold the electrodes to the cannula and therefore is suitable for usage in the awake state or to avoid artifacts in the OR environment.

The multiprobe holder (made of Delrin) having a rectangular shape (4.5 × 7 mm) contained a bundle of fibers transmitting light to and from the brain as well as three ion selective electrodes (K^{+}, Ca^{2+}, and Na^{+}) combined with DC steady potential electrode (Ag/AgCl). The common part of the light guide (1.8 mm in diameter) contained two groups of fibers. For the LDF one 50-micron input fiber and two 100-micron output fibers were glued in a triangular shape (0.7 mm separation) and connected to the standard commercial plug of the Laser Doppler flowmeter (TSI Inc., St. Paul, MN or Perimed Inc. Piscataway, NJ.). For the monitoring of NADH redox state, 10 excitation and 10 emission 200 micron fibers (General Fiber Inc. Cedar Grove, N.Y.) were randomly mixed between and around the fibers used for the LDF (Fig. 6.9a, b). This configuration of the fibers enabled us to monitor CBF and NADH redox state from about the same tissue volume. The ion-selective electrodes were connected to an Ag/AgCl electrode holder (WPI, New Haven, CT) and protected by a Plexiglass sleeve. The entire MPA was sterilized by the standard gas sterilization routine and was checked for electrode integrity and calibration inside the operating room 24 h later. The MPA was located on the exposed cortex using a micromanipulator, and data collection started immediately after, using a microcomputer-based data acquisition system as shown in Fig. 6.4 (386 Dell Processor and DATAQ ADC and software). After recording of baseline levels of CBF, NADH redox state, and extracellular ion levels, the responses to CBF decrease (occlusion of a blood vessel) were recorded followed by the recovery period.

Animal Experiments In order to test the integrity and function of the MPA, we used the Mongolian gerbil as an animal model. The animal was anesthetized by Equithesin (each ml contains: pentobarbital 9.72 mg, chloral hydrate 42.51 mg, magnesium sulfate 21.25 mg, propylene glycol 44.34% w/v, alcohol 11.5%, and water) injected IP (0.3 ml/100 g). We have been using this anesthetic for about 40 years, and it never showed a significant effect on mitochondrial activity unless we gave an overdose amount. The skull was exposed by midline incision of the skin. An appropriate hole (about 6 mm) was drilled in the parietal bone area. The dura mater was removed gently, and the MPA was located on the brain so that extra pressure on the brain was avoided. In order to elicit spreading depression waves (Leao 1944), a push–pull cannula was implanted epidurally 2 mm anteriorly to the MPA (Fig. 6.4a). The MPA, the cannula, and three stainless steel screws were cemented to the skull by dental acrylic cement.

6.3.1.3 Results

Most of the results to be presented were collected in animal experiments in which the new system was tested. Figure 6.5 shows the responses of the gerbil brain to ischemia induced by occlusion of the two carotid arteries (the MPA was implanted on the right hemisphere). After occlusion of the right artery, ischemia was induced (50% of the maximum) as seen by the decrease of the LDF signal and an increase in the NADH redox state (corrected fluorescence [CF] signal). The addition of the left artery occlusion led to maximum level of ischemia.

This two step of ischemia is seen also in the responses of the ECoG as well as a different slope of the potassium leakage to the extracellular space. After the occlusion of the right artery, the ischemia level did not reach a steady state (partial recovery of flow and CF) due to compensation of blood flow through the anterior cerebral artery. After reopening of the two carotid arteries, a fast reperfusion was recorded as seen in the LDF trace. A wave of spreading depression was developed around the calcium electrode simultaneously with the reopening of the blood vessels. In this animal, we have used two potassium electrodes instead of one K^+ and one Na^+ electrode. The wave of Spreading Depression (SD) propagated to the rest of the hemisphere as seen by the rest of the parameters measured. Due to the extra ion pumping activity during the SD wave, a large increase in flow followed the wave of oxidation of NADH (CF decrease). All the signals returned to the baseline level 5 min after the wave of SD. Figure 6.6 shows the responses of the same gerbil brain to SD wave initiated by application of 1 M KC1. The wave started in the K^+ and Ca^{2+} electrodes which were located in the same line in front of the wave (A). A large increase in CBF was recorded (LDF) together with the oxidation of NADH (decrease of CF) seen in line B. The reflectance trace (R) shows an initial increase followed by

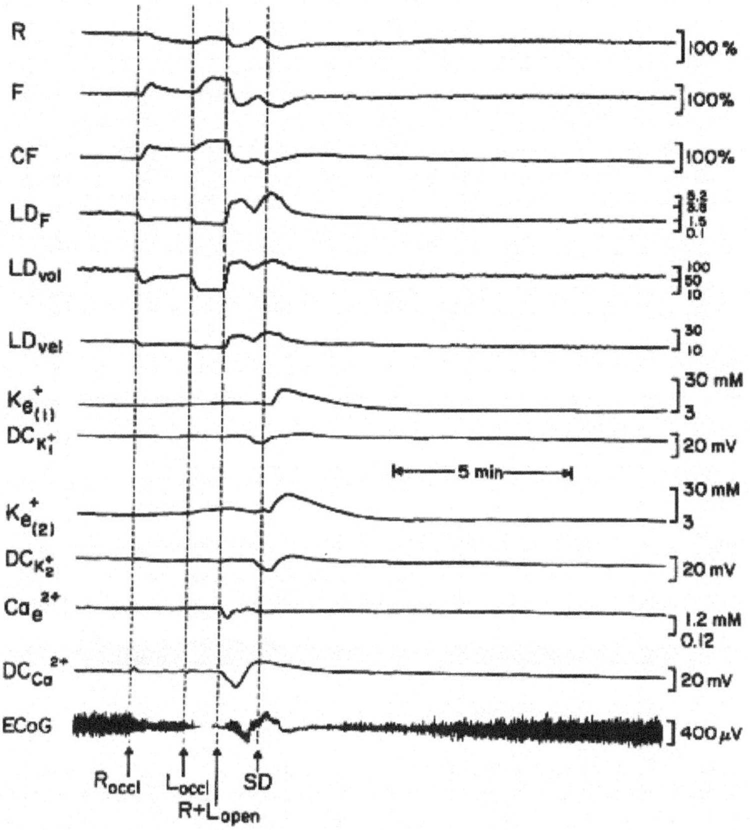

Fig. 6.5 The effects of unilateral and bilateral carotid artery occlusion on the metabolic, ionic, and electrical activities in the Mongolian gerbil. R, F, and CF—366 nm Reflectance, NADH fluorescence, and NADH corrected fluorescence; LDF, LD01, and LDvei Laser Doppler flow, Laser Doppler volume, and Laser Doppler velocity; K^+(1 K^+(2), Ca^{2+}—extracellular level of 2 potassium and 1 calcium electrodes; DCK^+(I) DCK^+(2), DCa^{2+}—DC steady potential around the 2 potassium and calcium electrodes; ECOG Electrocorticogram; loccl, Loccl right and left carotid artery occlusion; SD Spreading Depression (Mayevsky et al. 1991)

a longer decrease which correspond to the large increase in blood volume (LD vol). Occlusion of the right carotid artery for 5 min led to partial ischemia as shown in Fig. 6.7. After about 2 min an ischemic depolarization (ID) event was developed and led to an increase in K^+ as well as a decrease in Ca^{2+} concomitantly with a negative shift of the DC steady potential. The optical signals show a secondary decrease in blood flow and volume during the ID event, suggesting a vasoconstriction response.

This change was correlated to the increase in the reflectance signal. The recovery from this event was slower as compared to those shown in Figs. 6.5 and 6.6. Adaptations in the MPA, necessary for operating room conditions, enabled us to monitor and record the optical signals during patient operations. The ion measurements were very noisy, and some modifications are needed in order to improve the signal-to-noise ratio.

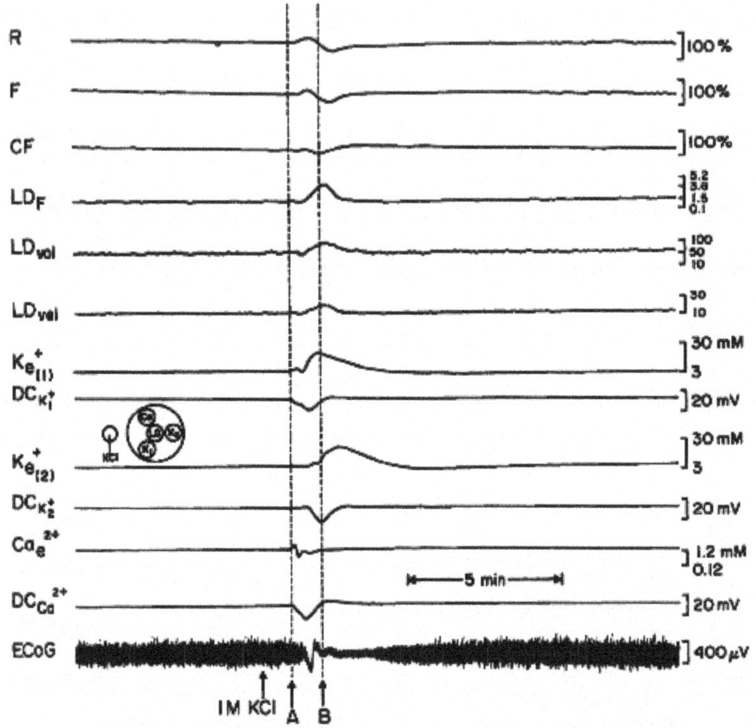

Fig. 6.6 Effects of cortical spreading depression (induced by 1 M KCl) on the various activities in the gerbil brain. The wave entered the area of K_1 and Ca electrodes initially (line A) and propagated to the rest of the hemisphere. Line B indicates the peak of oxidation in the NADH levels, which precedes the maximum increase in CBF (LDF). All abbreviations are as in Fig. 6.5 (Mayevsky et al. 1991)

Figure 6.10 shows the response of the human cerebral cortex to a short ischemia episode induced by occlusion of the carotid artery prepared for the aneurysm procedure. Immediately after the occlusion, a drop in CBF was recorded (decrease in LDF signal) simultaneously with an increase in NADH redox state (CF signal). Very small changes in the reflected light was recorded (R) during the occlusion. The recovery of the flow and the oxidation of NADH were very quick, as seen in both traces. A large overshoot in CBF was noted without a parallel undershoot in the NADH redox state during the reperfusion period as found also in animal models exposed to ischemic conditions (Mayevsky 1990).

6.3.1.4 Discussion

The aim of this study was to test the possibility of using the MPA developed a few years ago (Friedli et al. 1982; Mayevsky 1983b) in human brain monitoring intraoperatively. As concluded by Rosenørn (Rosenorn 1989) "Self-retaining brain

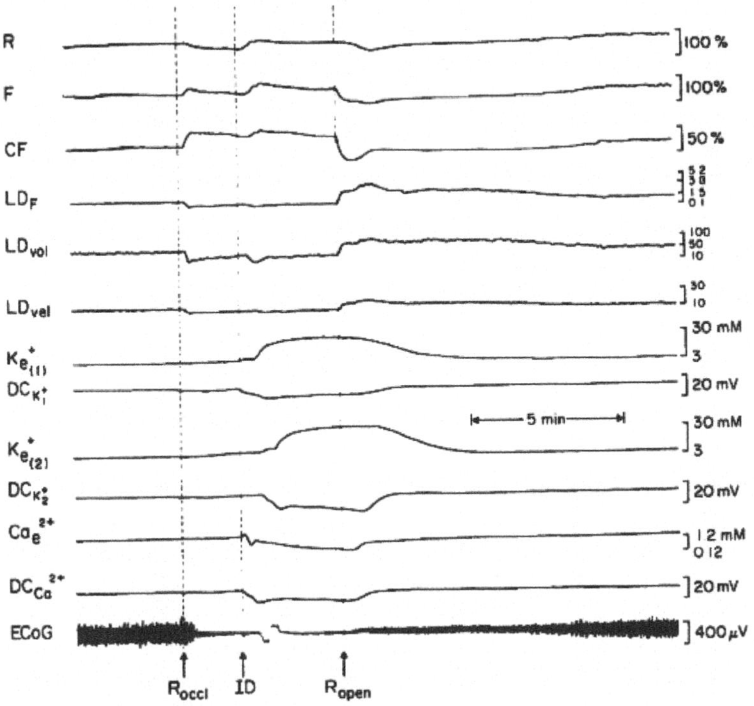

Fig. 6.7 The development of ischemic depolarization (ID) during 5 min of partial ischemia (Roccl) in the gerbil brain. All abbreviations are as in Fig. 6.5 (Mayevsky et al. 1991)

retractors are indispensable in intracranial surgery, but may in some circumstances provide ischemic brain damage." He suggested that development of recording instrument for intraoperative conditions is very desirable. The MPA developed by us for animal experiments is probably very complex for the operation room environment and the short time available for monitoring under those conditions requires a modified version of the MPA. According to the preliminary data accumulated in animal experiments, we believe that only three probes are required in order to monitor the functional state of the human brain. The usage of the Laser Doppler flowmeter (relative CBF), surface fluorometry reflectometry (intramitochondrial NADH redox state), and extracellular level of K+ will provide the necessary information. Monitoring CBF or NADH redox state alone will not provide reliable information due to various unclear responses to pathological events such as hypoxia, ischemia, or brain stimulation (epileptic activity or spreading depression). To prove our argument, we present the correlation between the relative CBF and the NADH redox state measured in a group of eight gerbils exposed to graded hypoxia, ischemia, and spreading depression.

As shown in Fig. 6.8, a clear and significant correlation was found under ischemia and hypoxia. In ischemia, the decrease in flow, induced by occlusion of one or two carotid arteries, led to an increase in NADH. Under hypoxia, due to the

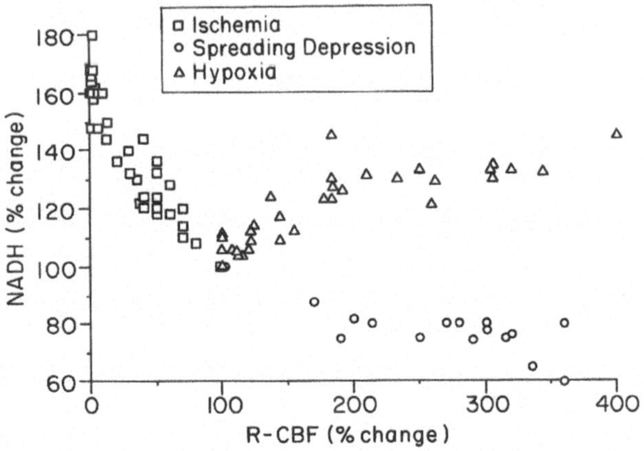

Fig. 6.8 Correlation between relative CBF and NADH redox state in the gerbil brain exposed to hypoxia, ischemia, and spreading depression. The results are expressed as percent change related to the 100% level at the control period (Mayevsky et al. 1991)

autoregulation response, an increase in CBF was recorded simultaneously with the increase in NADH (Mayevsky et al. 1986). When spreading depression was induced, the increase in energy requirement led to an activation of the mitochondrial respiration (Chance and Williams 1955; Chance et al. 1962) and oxidation of NADH was recorded (decrease in CF). This increase in O_2 consumption (Lukyanov and Bures 1967) led to an increase in CBF in the range of up to 200–350% as compared to the normoxic values. If CBF is the only parameter to be monitored, the differentiation between hypoxia and spreading depression is impossible. The same is true for the differentiation between hypoxia and ischemia if one is monitoring the NADH redox state by itself. By monitoring the CBF and NADH redox state together and using the appropriate algorithm, it will be possible to predict and describe more accurately the pathological state. The outcome of any pathological state in the brain is projected in the ionic homeostasis situation (Hansen 1985). Therefore, monitoring of this parameter is necessary for the evaluation of the functional state of the brain. Due to the large energy consumption by the Na^+-K^+-ATPase (Erecinska and Silver 1989) any change in the energy supply will be correlated to the extracellular level of K^+ (Mayevsky 1983a; Hansen 1985; Mayevsky 1990). Since changes in extracellular levels of Ca^{2+} and Na^+ are expected mainly during massive depolarization event, we do not expect it to happen during surgical situations. Therefore, the monitoring of extracellular K^+ will represent the ionic state of the brain. Also, if massive depolarization will occur it will be detected by the potassium levels (Rosenthal and Sick 1988). The sage of an animal model is the only way by which one can develop the appropriate algorithm of the clinical situation. We have used the Mongolian gerbil due to the simplicity of induction of ischemia (Levine and Payan 1966) as well as hypoxia and spreading depression. The same kind of measurements were done by other groups but not in real time and not with a multiprobe approach (Lauritzen et al. 1982; Lauritzen 1987; Hertelendy et al. 2017; Richter et al. 2017). Very recently we were able to correlate in vivo data of NADH redox state and extracellular K+ with CBF and O_2 consumption data measured by microspectrophotometry (Mayevsky and Weiss 1991). Significant correlation was found between those parameters in rats exposed to spreading depression. In the MPA, we are using surface probes which do not penetrate the tissue. This kind of geometry may affect the results as compared to those obtained by microelectrode studies (Hansen 1985). On the other hand, this approach is applicable to the human brain and also allows the comparison with other surface probes such as the NADH redox state and Laser Doppler flowmeter. The difference between ID and SD shown in Figs. 6.5 and 6.7 are very significant in prediction of the functional state of human brain under observation. Since potassium levels may be comparable in the two conditions, the NADH and CBF will be in opposite directions and may determine the actual state (Figs. 6.9 and 6.10).

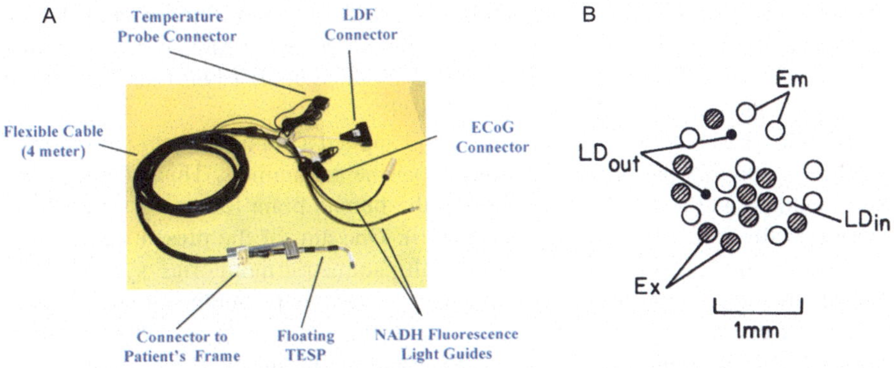

Fig. 6.9 (**a**) Floating tissue energy state probe (TESP) used during various surgical procedures, including optical fibers for NADH fluorometry and for Laser Doppler Flowmetry as well as a temperature probe. ECoG electrocorticography probe. (reprinted with permission from Elsevier (Mayevsky et al. 2003). (**b**) Cross-section of the combined light guide for monitoring of CBF and NADH redox state. A single fiber is used for LDF light input and 2 for the output. Ex and Em represent the two groups of fibers used for NADH excitation (Ex) and Emission (Em) (Mayevsky et al. 1991)

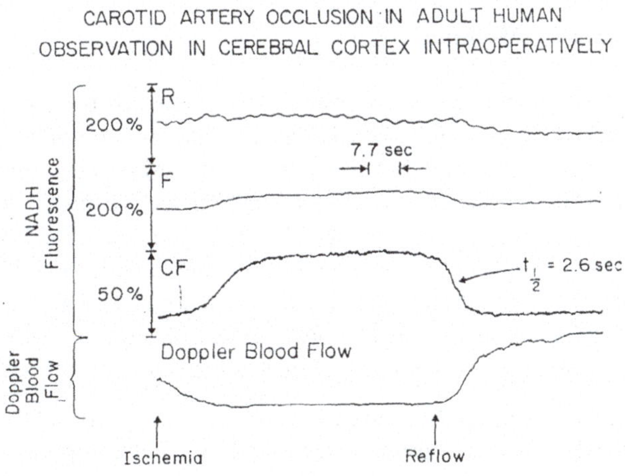

Fig. 6.10 Responses of the human cerebral cortex to a decrease in blood flow induced by common carotid artery occlusion. R, F, and CF Reflectance, Fluorescence, and corrected fluorescence. (SPIE, reprinted with permission (Mayevsky et al. 1991))

6.3.2 The Second Study: The Discovery of Cortical Spreading Depression in Comatose Patient

6.3.2.1 Introduction

Leao first discovered CSD of the EEG in the rabbit brain about 80 years ago (Leao 1944). Since then, this event has been identified and described in various animals including primates (Bures et al. 1974). The depolarization caused by the electrical stimulation used by Leao to induce SD was characterized by EEG depression (Bures et al. 1974), pial vasodilation (Lauritzen 1994), and a slow negative potential shift (Leao 1947). Only one study in human stereotactic neurosurgery reported SD in the caudate nucleus or the hippocampus following KC1 injection (Sramka et al. 1977). No direct evidence exists for spontaneous or elicited SD in the human cerebral cortex in vitro or in vivo (Pearce 1985; Bures 1993; Somjen et al. 1993; Lauritzen 1994; McLachlan and Girvin 1994; Clough et al. 1998).

Over the past years, we have developed and applied a multiparametric monitoring system for studying the brain of various experimental animals. This technique has been modified and adapted for monitoring the human brain following severe head injury (Friedli et al. 1982; Mayevsky 1983b). One aim of the present report is to demonstrate the feasibility of using this multiparametric monitoring system under clinical situations. CSD has been implicated in various pathophysiological events including ischemia or stroke (Somjen et al. 1990; Dietrich et al. 1994), seizures (Marshall 1959), migraine (Pearce 1985), and brain injury (Oka et al. 1977; Rogatsky et al. 1996). SD waves cause transient changes in hemodynamic, metabolic, ionic, and electrical activities (Vyskocil et al. 1972; Rosenthal and Somjen 1973; Mayevsky et al. 1974; Hansen and Zeuthen 1981; Lauritzen 1988; Mayevsky et al. 1992; Lauritzen 1994) and thus it is preferable to monitor all these parameters simultaneously. A second aim of this study was to present results indicating, for the first time, that CSD occurs also in the human brain.

6.3.2.2 Methods

The multiparametric assembly (MPA) enables the simultaneous, continuous online measurement of relative CBF and cerebral blood volume (CBV) (Haberl et al. 1989); intramitochondrial NADH redox state (Mayevsky and Chance 1982; Mayevsky 1984); extracellular K^+ concentrations (Crowe et al. 1977); spontaneous electrocorticography (ECoG); DC steady potential and ICP (Crutchfield et al. 1990). The principles of the various sensors and parameters included in the MPA have been described in previous publications (Crowe et al. 1977; Friedli et al. 1982; Mayevsky 1983b; Mayevsky et al. 1991, 1992). Therefore, only a short description of the principles of the various parameters measured by the MPA is presented in Fig. 6.11.

Laser Doppler Flowmetry Over the past years, LDF has been calibrated against the H_2 clearance probe or against [^{14}C] iodoantipyrine autoradiography which are both well-established methods for quantitative monitoring of CBF. LDF apparently

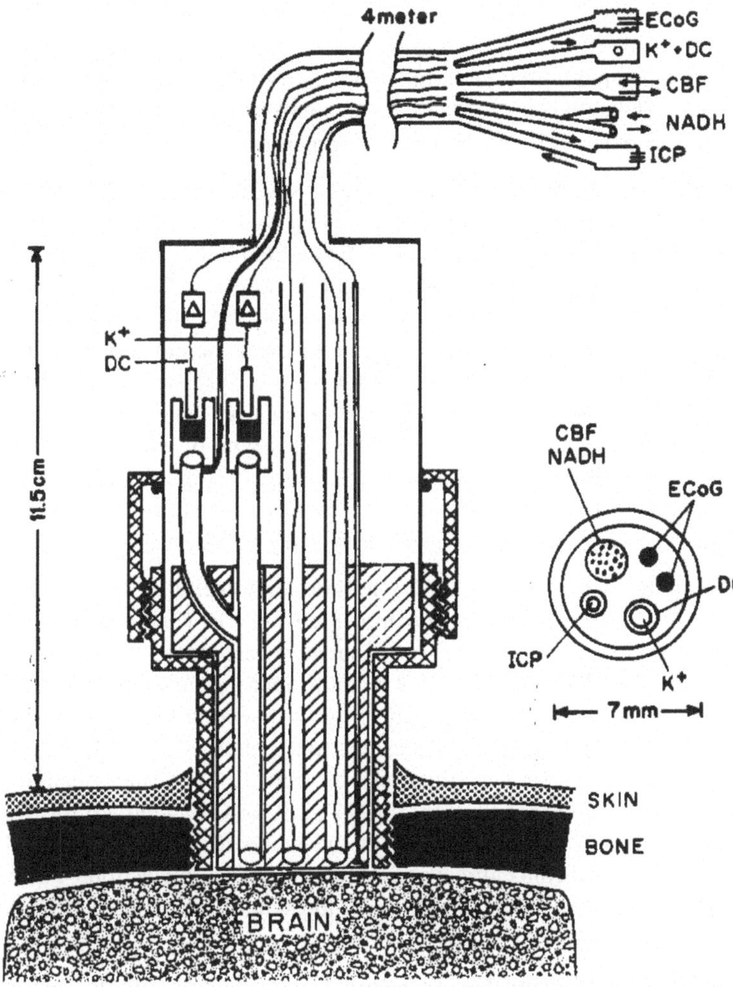

Fig. 6.11 Multiprobe assembly used for monitoring the brain of head-injured patients. The MPA is connected to the brain via a special holder screwed into the skull. ICP, intracranial pressure probe; CBF, NADH, fiber optic light-guide probe to measure local cerebral blood flow (CBF) and mitochondrial redox state (NADH); K^+, DC, extracellular K^+ surface mini-electrode surrounded by a DC steady potential monitoring space; ECoG, bipolar electrocortical electrodes located inside the fiber optic bundle (Mayevsky et al. 1996)

measures relative changes demonstrating a significant correlation to the relative changes in CBF measured by the two other approaches (for review, see Ref. (Wadhwani et al. 1990). The principle of LDF is to utilize the Doppler shift (frequency change) that light undergoes when reflected by moving red blood cells. A beam of low power laser light is transmitted to the brain by an optical fiber. After multiple scattering of the light, two other optical fibers pick up the reflected light, which enters the photodetector. The run signal is analyzed by a complicated algorithm developed by the manufacturer, and the results are presented as a percentage of the full scale (0–100%), providing relative blood flow values. The change in the total back-scattered light is an indirect indicator of the blood volume in the tested tissue volume. The calibration of the CBV is more complicated as compared to the CBF due to the construction of the Laser Doppler Flowmeter by the manufacturer.

NADH Surface Fluorometry The principle of NADH monitoring from the surface of the brain is that excitation light (366 nm) is passed from the fluorometer to the brain via a bundle of quartz optical fibers. The emitted light (450 nm), together with the reflected light at the excitation wavelength, is transferred to the fluorometer via another bundle of fibers. The changes in the reflected light are correlated to changes in tissue blood volume and also serve to correct hemodynamic artifacts appearing in the NADH measurement (Mayevsky and Chance 1982; Mayevsky 1984). The changes in the fluorescence and reflectance signals are calculated relative to the calibrated signals under normoxic conditions. This type of calibration is not absolute, but provides reliable and reproducible results from different animals and between different laboratories. Details on this technique have been published by our group (Mayevsky and Chance 1982, Mayevsky 1984).

Extracellular K^+ Levels In order to monitor extracellular K^+ concentrations, specially designed mini-electrodes manufactured by WPI (Sarasota, Florida, USA) were used. A flexible PVC tube (1.2 mm diameter) was sealed at one end with a K^+ sensitive membrane. The tube was filled with the appropriate solution and connected to an electrode holder (WPI) with a salt bridge connecting between the membrane and the Ag/AgCl pellet located inside the electrode holder. The interface between the PVC tubing and the holder was glued using 5-min Epoxy. The selectivity coefficients (-log) of the K^+ electrode is 4.0 for Na^+, 3.9 for Ca^{2+}, and 3.0 for Mg^{2+}. The impedance of the new K^+ electrode is in the range of 10^9–10^{10} ohms. The DC potential was measured concentrically around the potassium electrode. An Ag/AgCl electrode connected via a saline bridge under the skin of the patient's head was used as a reference electrode.

Construction of the Multiprobe Assembly The solution used for the sensor holder (Fig. 6.11) was basically a modification of the Lucite cannula described by Crowe et al. (1977) for the light guide and the potassium-sensitive electrode. The hole accommodating the K^+ electrode was divergent at the top to facilitate handling and sealing of the probes (Friedli et al. 1982). An additional hole was drilled obliquely from the upper surface to merge with the K^+ electrode channel at approximately mid-distance to the lower surface. This hole was used to accommodate a polyethylene tubing for recording the local DC potential concentrically to the K^+ electrode. The long and rigid steel stem of the light guide used in this study occupies a straight vertical hole in the cannula and serves as an axis to hold the cannula and the cable holder at a convenient distance from each other. Spontaneous electrical activity (ECoG) was measured by two polished platinum rods inserted into the MPA holder. Parenchymal ICP monitoring was performed using a fiber optic ICP probe (Camino Lab., USA) which was inserted into the MPA so as to have the same contact area with the brain surface as the other probes. The values measured by the optical probe correlated well with the standard ICP monitoring device (Crutchfield et al. 1990). The complete assembly was protected and shielded by an aluminum sleeve sliding over the cable holder. For stronger construction, the sleeve was permanently screwed into the cannula and cable holder. The MPA was prepared for clinical use by gas sterilization which was found to be harmless to all the sensors.

Patient Preparation Informed consent was obtained from 14 patients in accordance with institutional review board procedures (Tel Aviv Medical Center). The patients monitored by this MPA were hospitalized due to a comatose state induced by severe head injury (GCS <8). Preparation of the patient for brain monitoring was performed in the neurosurgical operating theater under general anesthesia. The patients were prepared for monitoring as follows.

1. A 15 mm diameter hole was carefully drilled in the frontal bone area selected according to the CT scan. The hole was drilled in a "normal" looking section of the brain.
2. After cleaning the exposed area of bone residues, the dura mater was cut to enable direct contact between the MPA probes and the brain.
3. The MPA holder was screwed into the skull to a predetermined depth to avoid any undue pressure on the brain.
4. The operated area was closed by suturing the skin around the MPA holder (covered until the beginning of the monitoring).
5. The patient was transferred to the ICU (either general or neurosurgical) and after stabilization and connection to the routine patient monitoring system the sterilized MPA was introduced into the holder (for contact with the brain) and fastened to it in order to ensure measurements for up to 48 h (sometimes even longer). Figure 6.12 shows the monitoring system connected to the brain of a comatose patient treated in the Neurosurgical ICU. The average monitoring time in the 14 patients was 40 h. Three out of the 14 have died during the monitoring period due to the severity of the trauma upon admission. The other 11 recovered and were discharged from the hospital.

6.3.2.3 Results

In this study, the brain function of 14 patients suffering from severe head injury was monitored using a multiparametric analyzer. Due to the circumstances of the traumas, it was not possible to measure all the parameters in all patients. In general, it was possible to monitor ECoG, ICP and CBF in most patients. NADH fluorescence was less easily measured, and the most problematic parameter was the extracellular K^+ concentration. This was probably due to the fact that the K^+ electrode is a high impedance electrode, which is very sensitive to ground noise. The typical response of most patients to treatments aimed at reducing the ICP, such as mannitol infusion and hyperventilation, was a measurable decrease in ICP and a resultant increase in CBF and oxidation of NADH (Fig. 6.13).

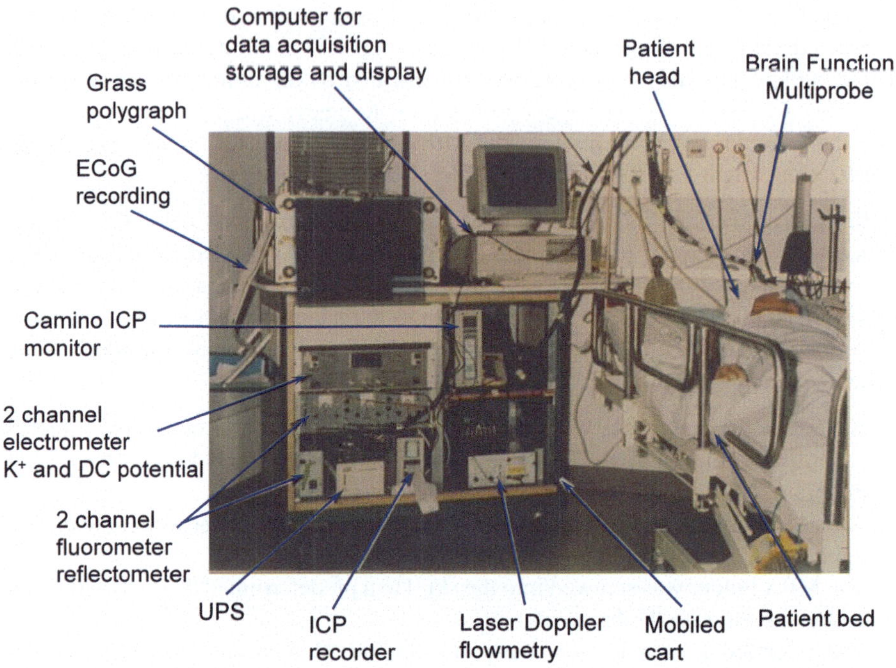

Fig. 6.12 The first clinical monitoring setup ("Brain Monitor") used in the neurosurgical intensive care unit. The multiparametric monitoring system consists of various devices installed in the same cart. The MPA shown in Fig. 6.9a connects the brain of the patient to the brain monitor (Mayevsky and Barbiro-Michaely 2013)

Spreading Depression Cycles Only one of the 14 monitored patients had developed spontaneous changes in all parameters similar to the typical responses to CSD recorded in animals (Mayevsky et al. 1991, Mayevsky and Weiss 1991). These changes were recorded 4.5 h after the beginning of monitoring, which was 7 h after admittance to hospital. During the measuring period, this patient was bilaterally irresponsive to pain, his pupils were dilated and nonreactive to light (6 mm), he was mechanically ventilated, and his brain CT scan showed evidence of severe brain edema in the left hemisphere and right parietal hemorrhagic contusion. The measurements were taken from the right frontal lobe.

The first cycle of CSD was developed in parallel to the standard procedure in comatose mechanically ventilated patient, namely, suction of the respiratory system tubing. This response is presented in Fig. 6.14. The ICP was elevated slightly and recovered to baseline after few minutes. The integrated ECoG shows a significant decrease as also very clear from the raw ECoG response seen in Fig. 6.14. As seen in Fig. 6.14, an elevated extracellular K^+ and a negative shift in the DC potential were recorded. We assumed that due to the suction procedure the patient moved and therefore those changes in potassium and DC signal were not recovered to the base line and recalibration of the two signal was necessary. Hemodynamic and metabolic

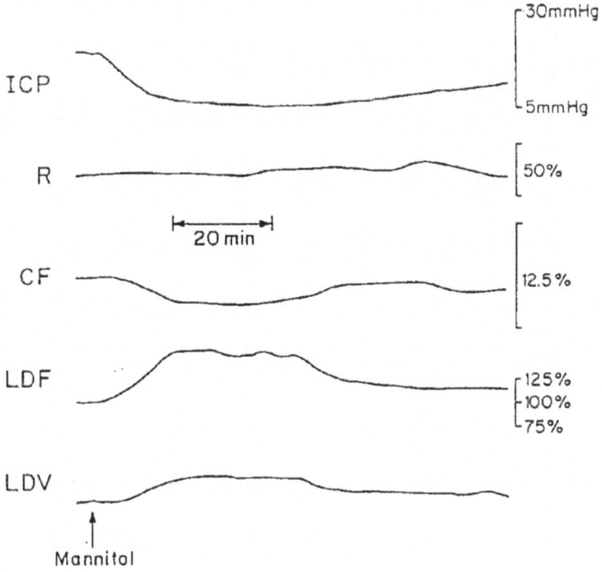

Fig. 6.13 Metabolic and hemodynamic responses of a severely head-injured patient to an IV infusion of mannitol. ICP, intracranial pressure; R, CF, 366 nm reflectance and 450 nm corrected NADH fluorescence: CBF and CBV, cerebral blood flow and volume (Mayevsky and Barbiro-Michaely 2013)

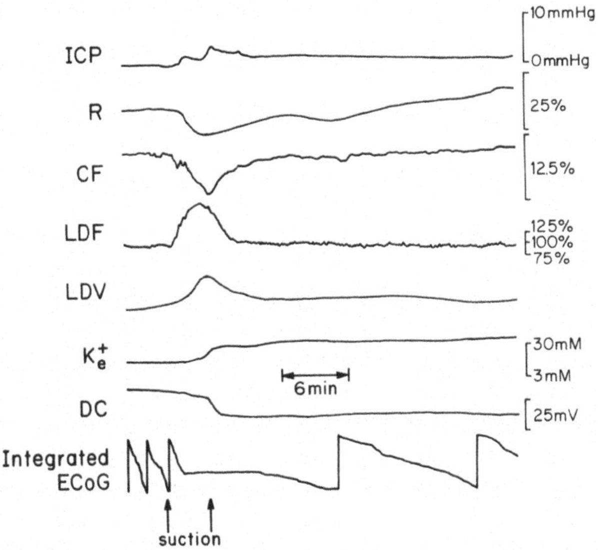

Fig. 6.14 The effect of "suction procedure" on the development of Cortical Spreading Depression (CSD) in a head-injured patient treated in the Neurosurgical ICU. All abbreviations are as in Fig. 6.15 (Mayevsky and Barbiro-Michaely 2013)

compensation for the depolarization was observed in the other parameters. An oxidation cycle was observed in the NADH redox state (decreased CF), while blood flow and volume increased. The calibration of the CBV trace shown in Fig. 6.14 is missing due to an error recorded during this interval of time.

As seen in Fig. 6.15, the ECoG became depressed for 10–15 min and, at the same time, a cycle of elevated extracellular K^+ and a small negative shift (in this cycle) in the DC potential were recorded. These three changes are typical of transient

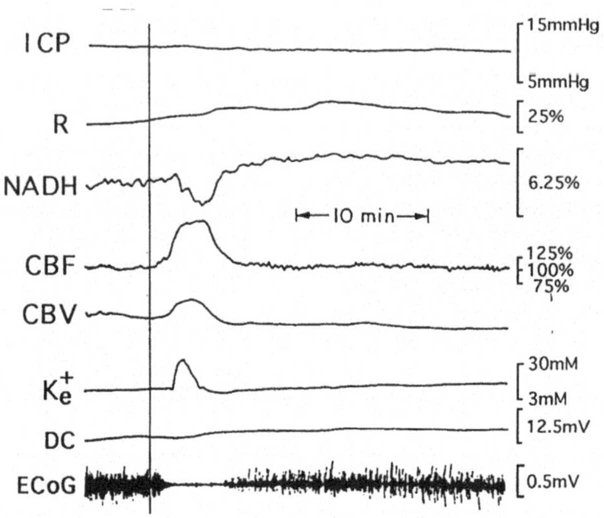

Fig. 6.15 Spontaneously developed cortical spreading depression cycle in a head-injured patient. ICP, intracranial pressure measured by the parenchymal probe; R, CF, 366 nm reflectance and 450 nm corrected NADH fluorescence; CBF and CBV, cerebral blood flow and volume; K^+_e, DC, extracellular potassium and DC steady potential; Integrated ECoG, amplitude-integrated electro cortical activity (Mayevsky and Barbiro-Michaely 2013)

depolarization, which is a dominant part of spreading depression. Hemodynamic and metabolic compensation for the depolarization was observed in the other parameters. An oxidation cycle was observed in the NADH redox state (decreased CF), while blood flow and volume increased. The calibration of the CBV trace shown in Figs. 6.13, 6.14, and 6.15 is missing due to an error recorded during this interval of time.

This patient exhibited repetitive CSD cycles every 20–30 min, although at various times during the monitoring period, CSD cycles were recorded at shorter or longer intervals. *The first five cycles* recorded in this patient were very similar, i.e., NADH was oxidized while blood flow increased significantly during the CSD wave (as in Figs. 6.14 and 6.15). However, in all the other cycles recorded later in the same patient (starting 4.5 h after the SD cycle shown in Fig. 6.14), the metabolic and hemodynamic responses were different (Fig. 6.16) from the initial cycles. The leakage of potassium and negativation of the DC potential were very similar in all SD cycles, whereas the NADH oxidation cycle was replaced by a biphasic cycle comprised mainly of a phase of increased NADH followed by a small oxidation phase. The compensation of blood flow and volume was also reversed at this time. The monophasic increase in CBF and CBV (Figs. 6.14 and 6.15) was replaced by an initial decrease followed by a smaller increase in flow and volume. Significant correlations were found between the amplitude of the changes in the monitored parameters shown in Fig. 6.17—NADH and CBF: $p < 0.05$ ($r = 0.903$, df $= 4$); NADH and K^+: $p < 0.001$ ($r = 0.994$, df $= 4$); CBF and K^+: $p < 0.01$ ($r = 0.940$, df $= 4$).

Five fragments of 1 h analog and integrated ECoG recordings are presented in Fig. 6.18. It can be seen that the deep depression lasted between 5 and 10 min while complete recovery was not achieved even after 30–45 min. In part E of the figure,

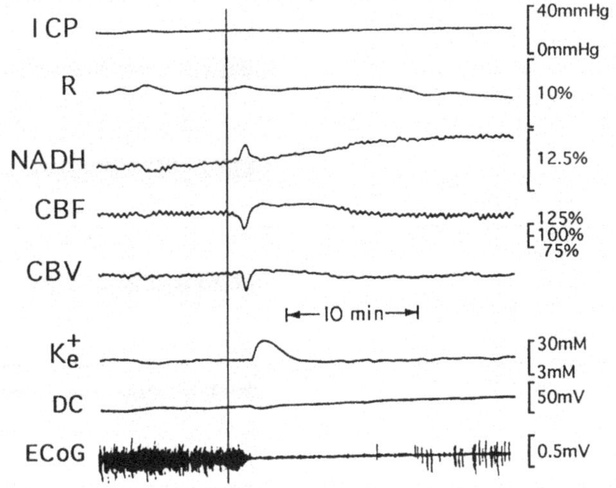

Fig. 6.16 Spontaneously developed cortical spreading depression cycle in a head-injured patient. ICP, intracranial pressure measured by the parenchymal probe; R, CF, 366 nm reflectance and 450 nm corrected NADH fluorescence; CBF and CBV, cerebral blood flow and volume; K^+_e, DC, extracellular potassium and DC steady potential; Integrated ECoG, amplitude-integrated electro cortical activity (Mayevsky and Barbiro-Michaely 2013)

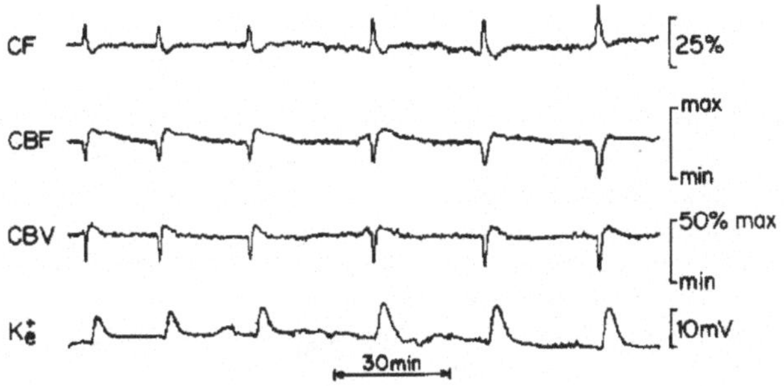

Fig. 6.17 C—Repetitive spreading depression cycles monitored in a head-injured patient 29.5 h after the beginning of monitoring. All abbreviations are as in Fig. 6.15 (Mayevsky and Barbiro-Michaely 2013)

which corresponds to the CSD cycle shown in Fig. 6.15, the ECoG demonstrated very slow recovery as compared to the other parameters. The DC potential negative shifts of those five depolarization cycles were 4 ± 0.35 mV (S.E.M).

Figure 6.19 present another set of CSD cycles developed in the same patient during a period of significant increase in the ICP (above 20 mmHg). It can be seen that in parallel to the increase in the ICP, the CBF decrease and the NADH was significantly elevated.

Figure 6.20 shows eight consecutive SD cycles recorded from the same patient. Downward deflections in the DC potential represents a negative shift and the average value calculated is 5.3 ± 0.3 mV (S.E.M). The comparable values in gerbils were 10 mV (Mayevsky et al. 1985) and in rats 14–16 mV (Sonn and Mayevsky, unpublished data) using the same monitoring approach.

The patient who developed SD cycles (presented in Figs. 6.14, 6.15 and 6.16) showed signs of brain deterioration leading to brain death. The tracing presented in Fig. 6.21 demonstrates that metabolic and ionic changes preceded brain death by several hours. The ECoG became isoelectric 3 h prior to the beginning of the recording in Fig. 6.21 (68 h from beginning of monitoring). A gradual increase in

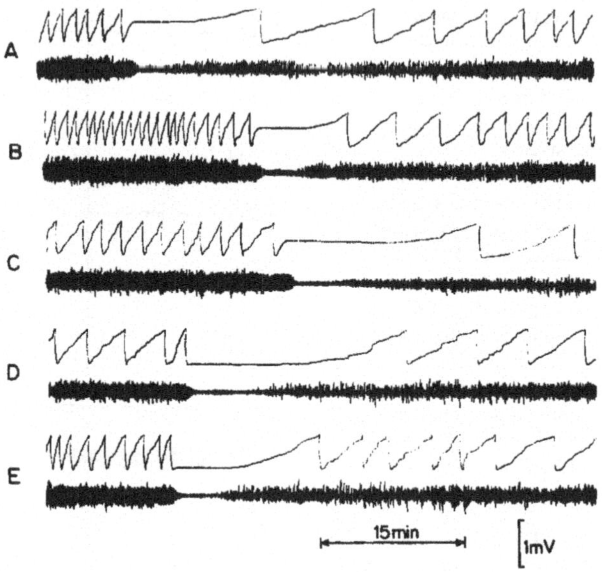

Fig. 6.18 Raw analog and integrated ECoG signals monitored during five spreading depression cycles. Records A–B were taken 4.4, 6.6, 7.6, 9.4, and 13.5 h after the beginning of the monitoring. During the initial phase of Part A, a suction procedure was used on the patient while the other four SD cycles occurred spontaneously (Mayevsky and Barbiro-Michaely 2013)

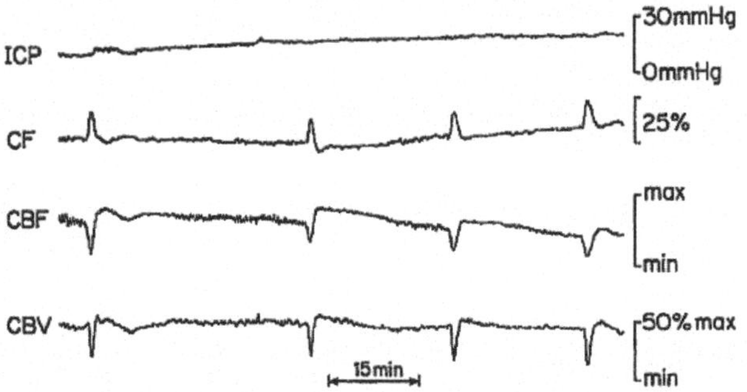

Fig. 6.19 Cycles of "reversed" spreading depression waves recorded from a head-injured patient. ICP Intercranial Pressure; CF NADH, CBF and CBV Cerebral Blood Flow and Volume, measured by Laser Doppler Flowmetry (Mayevsky and Barbiro-Michaely 2013)

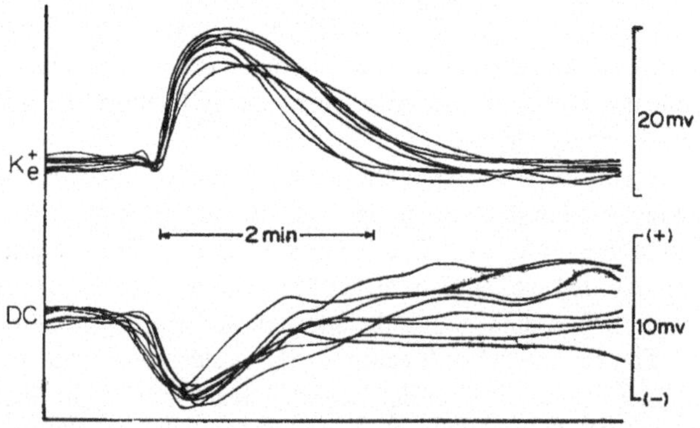

Fig. 6.20 Responses of extracellular K^+ (K^+_e) and DC steady potential to eight repetitive SD cycles. The absolute values of K^+ were estimated to be 3–4 mM, but due to measurement artifacts occurring periodically, absolute values are missing (Mayevsky and Barbiro-Michaely 2013)

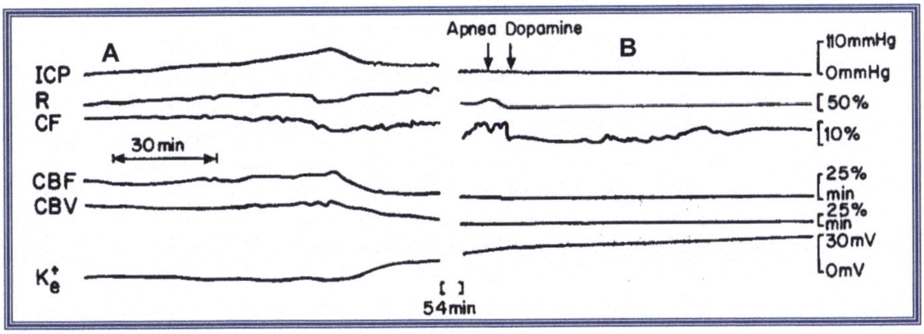

Fig. 6.21 Multiparametric recording of a patient 5 h before brain death. All abbreviations are as in Fig. 6.3 and Fig. 6.5. The absolute values of extracellular K^+ were calculated by the final calibration of the electrode done after its removal from the brain. All abbreviations are as in Fig. 6.15 (Mayevsky and Barbiro-Michaely 2013)

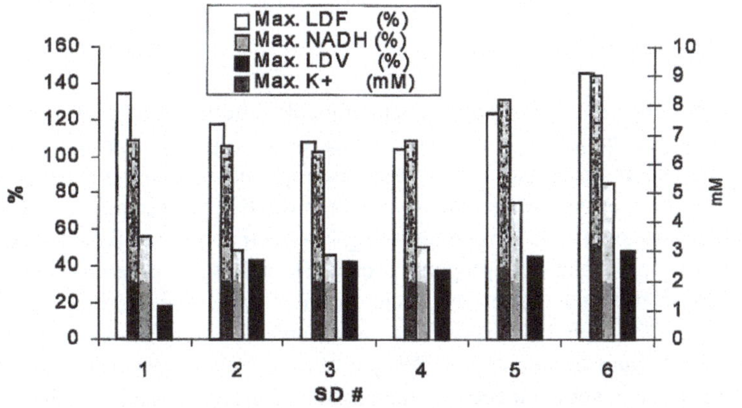

Fig. 6.22 The amplitudes of changes in a series of repetitive SD waves. NADH, LDF, and LDV are presented as percent of change, when 100% is defined as baseline at the beginning of measurement (Mayevsky et al. 2001)

ICP up to the level of 100 mmHg was measured, followed by a sharp decrease in CBF and CBV (Fig. 6.21a). During the terminal ischemic phase, a large increase in the extracellular K^+ level was recorded. Part B is a continuation of Part A after a 54 min interval due to a technical problem in the computer recording system.

The amplitudes of changes in NADH, LDF, LDV, and extracellular potassium, in ANOTHER series of six CSD waves, are presented in Fig. 6.22, with correlation coefficients of $r = 0.903$ (df = 4, $p < 0.05$) between NADH and CBF, $r = 0.994$ (df = 4, $p < 0.001$) between NADH and K^+, and $r = 0.940$ (df = 4, p < 0.01) between CBF and K^+.

Other clear effects of mannitol in the same patient are presented in Fig. 6.23. Due to the decrease in the ICP, CBF was elevated and NADH became more oxidized (Decreased).

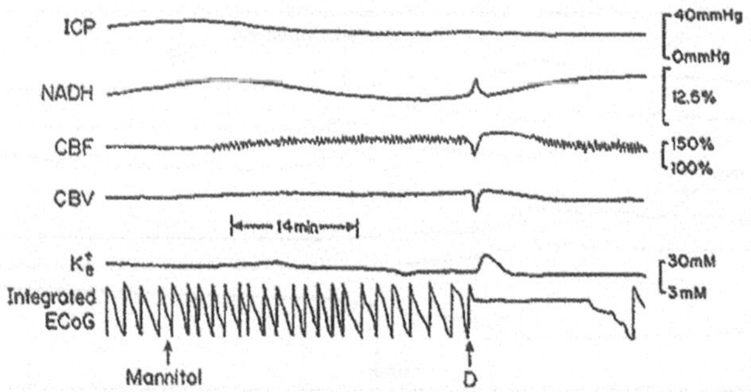

Fig. 6.23 Effects of IV infusion of mannitol on brain hemodynamic, metabolic, and ionic activities in a head-injured patient. ICP intracranial pressure; R 366 nm reflectance; CBF and CBV cerebral blood flow and volume; K^+_e extracellular potassium; D spontaneous depolarization developed repeatedly every 30 min. (Mayevsky et al. 1998a)

6.3.2.4 Discussion

Monitoring of the human brain exposed to clinical pathophysiological conditions has long presented a challenge. However, the lack of appropriate measuring techniques has so far prevented progress in this field. Attempts have been made to monitor the human brain in the operating room as well as in the ICU. For example, the standard technique of intraoperative EEG monitoring has been applied to patients exposed to surgical excision of their epileptogenic focus (Gloor 1986). Spreading depression of the cerebral cortex was never recorded. Sramka et al. (1977), using DC steady potential together with EEG recording, were able to elicit and record SD in the caudate nucleus and hippocampus. Somjen et al. (1993) stated that "The electrical signs of SD have never been actually demonstrated in any condition of intact human brain," but they were able to do so in hippocampal tissue slices. CBF has been monitored in the operating room (OR) using LDF (Rosenblum et al. 1987). Other groups have applied spectrophotometric techniques to the monitoring of mitochondrial redox state in the neurosurgical OR (Jobsis et al. 1971). Nevertheless, until now a standard practical approach for monitoring the brain in the OR or in the ICU has not been adopted. Even in less severe neurological disorders such as stroke, the need exists for a reliable monitoring technique. Wiebers et al. (1990) made the following very lucid statement on the issue of human brain monitoring: "Ultimately, however, the answers to many of our questions regarding the underlying pathophysiology and treatment of stroke do not lie with continued attempts to model the human situation perfectly in animals, but rather with development of techniques to enable the study of more basic metabolism, pathophysiology and anatomical imaging detail in living humans." The results presented in this study suggest that the human brain can be monitored under various pathophysiological conditions. The development and implementation of the surface probes organized in the MPA for animal studies (Crowe et al. 1977; Friedli et al. 1982; Mayevsky and Chance 1982), has made it relatively simple to adapt this assembly to clinical situations. Indeed, in a preliminary study (Mayevsky et al. 1991), we demonstrated a correlation between CBF and NADH redox state measured under partial ischemia induced in the human brain during neurosurgical procedures.

The recording of spreading depression in the human cerebral cortex in one of the patients, which was found to be qualitatively similar to that observed in rats, gerbils, or cats (Rosenthal and Somjen 1973; Mayevsky et al. 1991; Mayevsky and Weiss 1991), suggests that the basic mechanisms of pathophysiological processes may be similar in the animal and human brain. It is well documented that during the SD wave, which is basically the propagation of depolarization through the entire hemisphere, there is a concomitant large increase in O_2 consumptions due to stimulation of the ion pumps (for review, see Ref. (Mayevsky and Weiss 1991). A large increase in CBF (measured by various techniques) compensates for this increased O_2 demand (Van Harreveld and Stamm 1952; Hansen et al. 1980; Lauritzen 1994). When SD is induced in animals under normal conditions, complete coupling occurs between the increased O_2 consumption and the increase in CBF. If the O_2 supply is limited (ischemia and hypoxia), the responses of the brain to SD are different. We have found that under various pathophysiological conditions NADH demonstrates a cycle of reduction (increased NADH) instead of the typical normal oxidation cycle (decreased NADH) (Mayevsky et al. 1982). The combined measurement of NADH and CBF used in this study indicated that the reversed NADH response (an increase instead of a decrease) was due to the limited compensation of CBF as indicated by its initial decrease (Fig. 6.17). The results presented in this report demonstrate that CSD can develop in the human brain. We suspect that with this patient, one of the suction procedures (used to clear the airway system) was the stimulus for initiating the first SD wave. We do not know what caused the repetitive SD waves. The fact that the nature of the hemodynamic and metabolic responses changed in this patient as a function of time may be due to the deterioration of CBF autoregulation in the head-injured patient. Indeed. Oka et al. (1977) claim that in head-injured patients, a unique traumatic spreading depression syndrome could be identified, although they were unable to correlate behavior and physiological parameters.

In conclusion, this study demonstrates that the multiparametric analyzer can be used for monitoring the human brain in the neurosurgical as well as the general ICU. This assembly could be used for monitoring other pathophysiological conditions, such as ischemia, epilepsy, and migraine, which are considered inducive of spreading depression, especially since it has been shown in this study that spreading depression may occur in vivo in the human cerebral cortex.

6.3.3 The Third Study: Monitoring of Patients in the Neurosurgical OR

6.3.3.1 Introduction

In our previous publication (Mayevsky et al. 1999), we showed that multiparametric monitoring of neurotrauma patients is feasible and has a significant contribution to the diagnosis of brain functional state. In neurosurgical patients, it is necessary to monitor brain function either in the operating room or in the neurosurgical ICU. The time available in the OR is much shorter and as a result, one should apply technology that will not disturb or significantly delay the surgical procedure. At the same time. A single parameter monitoring device such as laser flowmeter (Dirnagl et al. 1989) will not be enough to provide valuable information in real time. Therefore, we have decided to apply the Tissue Spectroscope to be used in real time during various neurosurgical procedures.

Most of the neurochemical monitoring techniques applied to head-injured patients are not practical for the short time available in the OR. There are three components that may affect brain viablity status, namely, supply, demand, and energy balance.

The intracellular organelle responsible for the major synthesis of ATP is the mitochondrion dependent upon O_2 availability (Chance and Williams 1955).

Real-time evaluation of brain viability in situ could be done by monitoring different parameters, related to energy metabolism, which are complementary to each other. Over the past 40 years, the following four techniques were developed and applied to monitor, in real time, the energy supply and balance in the brain in situ:

1. CBF using LDF (Haberl et al. 1989).
2. Hemoglobin oxygenation or saturation (HbO_2) by dual wavelength reflectometry (Rampil et al. 1992) or spectral analysis.
3. Tissue average oxygenation using oxygen electrodes (Mayevsky et al. 1980).
4. Mitochondrial redox state by monitoring of NADH fluorescence using surface fluorometry (Chance et al. 1962).

All those techniques are based on the monitoring of a single parameter which make it very limited in terms of viability diagnosis in clinical situations. To monitor the second part of the equation, namely, the energy demand of the brain, it is possible to evaluate the ionic homeostasis by monitoring the extracellular K^+ levels using ion-specific electrodes in the brain (Mayevsky 1993).

In order to be able to evaluate the oxygen balance of the tissue in situ, it is important to monitor more than one parameter simultaneously in real time as shown in Fig. 6.24. Various combinations of the abovementioned parameters were made in the past and have been reported (Rampil et al. 1992; Mayevsky et al. 1996). We have developed and used a combination of two optical techniques, namely, CBF and NADH redox state, in a single bundle of optical fibers thus enabling the correlation of the two optical parameters that could be monitored together.

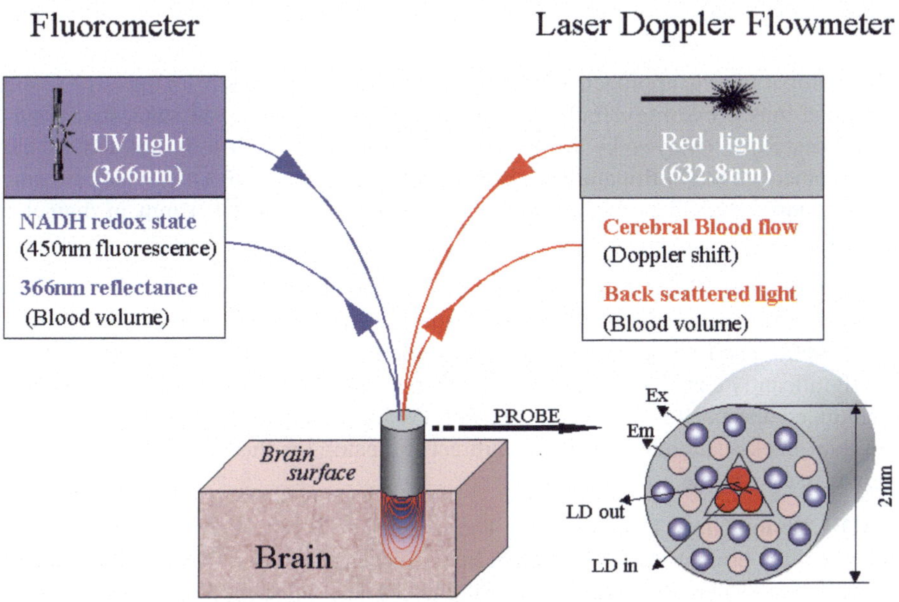

Fig. 6.24 The tissue spectroscope used in this study enables to monitor simultaneously cerebral blood flow (Laser Doppler Flowmeter) and mitochondrial NADH redox state (by surface fluorometry). E_x, E_m Excitation and emission transmitting fibers of the fluorometer; LD_{in}, LD_{out} Fibers connecting the brain to the Laser Doppler Flowmeter (Mayevsky et al. 2001)

The pioneering work published by Chance and Williams in Chance and Williams 1955 and later in 1962 by Chance et al. established the foundation for the close to 800 publications that have appeared since then on NADH fluorescence as an optical probe for mitochondrial function evaluation. In our previous work, we combined the monitoring of the oxygenation of hemoglobin and NADH (Rampil et al. 1992) or tissue pO_2 and NADH (Mayevsky et al. 1980), but it seems to us that the best combination is to monitor CBF and NADH redox state simultaneously. This approach provides information about the two most important parameters that represent the regulated processes in brain's energy state, namely, CBF and mitochondrial function or redox state.

Since CBF is a regulated process, it was hypothesized that simultaneous monitoring of CBF and NADH redox state will provide significant knowledge as to the energy state of the brain. In order to evaluate in real time both mitochondrial NADH redox state and brain flow, we combined the optical fibers attached to the two instruments into one common bundle as shown also in Fig. 6.24 (Mayevsky et al. 1996). Using this approach were able to correlate changes in CBF and mitochondrial redox state in the brain exposed to various pathological states in experimental animals (Mayevsky et al. 1995) and in *clinical* research (Mayevsky et al. 1996).

6.3.3.2 Methods

The multiparametric approach, of measuring several parameters from the same tissue area, and thus providing detailed information about its functional state, was previously described (Mayevsky et al. 1995; Meilin et al. 1996). It has led to the development of the Multiparametric Monitoring Assembly (MPA), in which various sensors and optic fibers are assembled together to form a flat 6 mm in diameter surface:

Local Cerebral Blood Flow In order to measure the CBF in real time from the same cortical area as the MPA location, the LDF technique was used (Wadhwani et al. 1990). The LDF apparently measures relative changes (0–100% range) which are significantly correlated to the relative changes in CBF measured by the two other quantitated approaches (for review see Ref. (Wadhwani et al. 1990)). The change in the total back-scattered light is an indirect indicator for the blood volume in the tested tissue volume.

NADH Redox State Fluorometry The principle of NADH monitoring from the surface of the brain is that excitation of light (366 nm) is passed from the fluorometer to the brain via a bundle of quartz optical fibers. The emitted light (450 nm), together with the reflected light at the excitation wavelength, is transferred to the fluorometer via another bundle of fibers. The changes in the reflected light are correlated to changes in tissue blood volume and also serve to correct hemodynamic artifacts appearing in the NADH measurement.

The changes in the fluorescence and reflectance signals are calculated relative to the calibrated signals under normoxic conditions. More details on this technique and the calculation of the results have been published by our group (Mayevsky 1976; Mayevsky and Chance 1982; Mayevsky 1984, 1990).

In this study, the two optical techniques were combined into a single fiber optic probe (Fig. 6.25). The probe connects the brain to the two instruments, namely, the fluorometer and the Laser Doppler Flowmeter as seen in Fig. 6.24.

6.3.3.3 Results

The optical multiprobe was applied to clinical situation, and a significant correlation was found between CBF and NADH redox state. Figure 6.26 shows the effects of STA-MCA anastomosis operation on the CBF and NADH recorded. Gradual occlusion of the internal carotid artery (Occl) while the aneurysmal area was temporarily occluded led to a large decrease in CBF while NADH increased significantly.

In a group of patients that underwent various neurosurgical procedures (aneurysm correction or tumor removal), we monitored CBF and mitochondrial NADH redox state. In each patient, we induced a short hypercapnic episode and the response to the change in CO_2 was recorded (Fig. 6.27). We analyzed the responses and found that the patients could be divided into a nonresponsive and a responsive group. In the

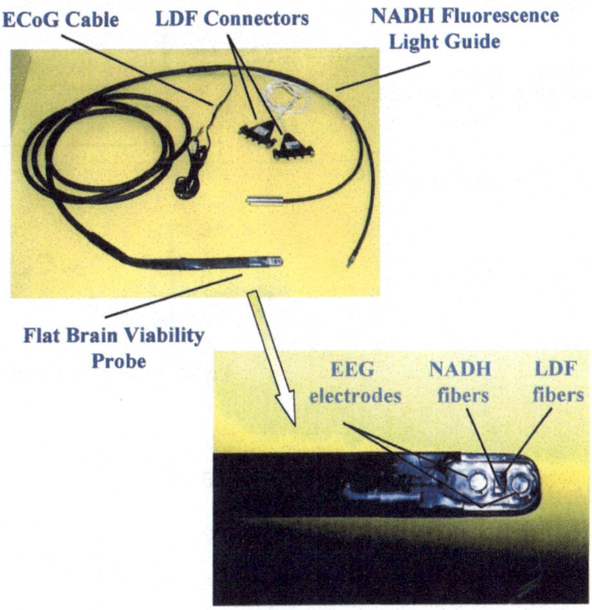

Fig. 6.25 A flat tissue energy state probe used during various surgical procedures including optical fibers for NADH fluorometry and for Laser Doppler Flowmetry probe as well as ECoG electrocorticography probe

nonresponsive group, elevation of the end tidal CO_2 did not change the CBF or the NADH significantly (Fig. 6.27a). As seen, in a group of seven patients, elevation of the end tidal CO_2 was not effective in the autoregulation process. In the responsive group, two types of responses to changes in the end tidal CO_2 were recorded.

The averages of the responses measured in six patients are shown in Fig. 6.27b. In this group of patients, the CBF increased significantly without any significant response in the NADH trace. In two patients, the change in end-tidal CO_2 led to the phenomenon of blood stealing as shown in Fig. 6.27c. The example shown was recorded in brain region near a removed tumor. As seen, the high CO_2 led to a decrease in CBF while the NADH was elevated. When the data recorded in the responsive group of patients was analyzed, the correlation between the two signals was found to be statistically significant ($R = -0.5$, $p < 0.05$, $n = 15$).

In order to demonstrate the potential use of our technology in clinical neurosurgery, the results obtained are presented in Fig. 6.28a. When papaverine was injected locally, a clear elevation of CBF was recorded together with small oxidation of NADH (decreased signal).

The effects of changes in CPP induced by blood pressure elevation (by epidirine injection) on brain CBF and NADH are shown in Fig. 6.28b. As seen, the increase in blood pressure led to an increase in CBF as well as a clear oxidation of NADH (decreased signal). The same type of correlation between CBF and NADH was found, where by decreasing the ICP (removal of CSF) a clear increase in CBF was recorded together with the decrease in NADH. The reflectance trace remained stable during the changes in CPP and blood pressure. During transplantation procedure, the organ, such as the kidney, preserved under low temperature is shifted toward the

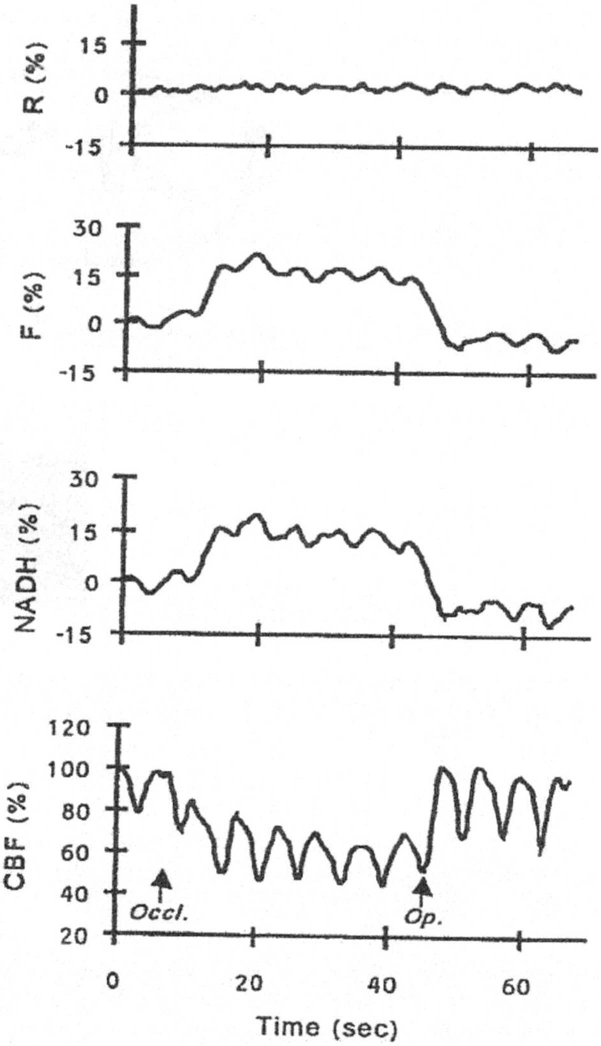

Fig. 6.26 The metabolic and hemodynamic responses to a short occlusion of the internal carotid artery during an STA-MCA bypass operation (Mayevsky et al. 1999)

normothermic range. This temperature change shifts the mitochondria toward a more oxidized state and will depend upon adequate tissue oxygenation. The successful rate of organ transplantation is dependent upon adequate microcirculatory blood flow (O_2 supply) as well as recovery of mitochondrial function to the normal range.

In two patients, the changes in end-tidal CO_2 led to the phenomenon of blood stealing. When the data recorded in the responsive group of patients was analyzed as seen in Fig. 6.29a, the correlation between the two signals was found to be statistically significant ($R = -0.5$, $p < 0.05$, $n = 15$).

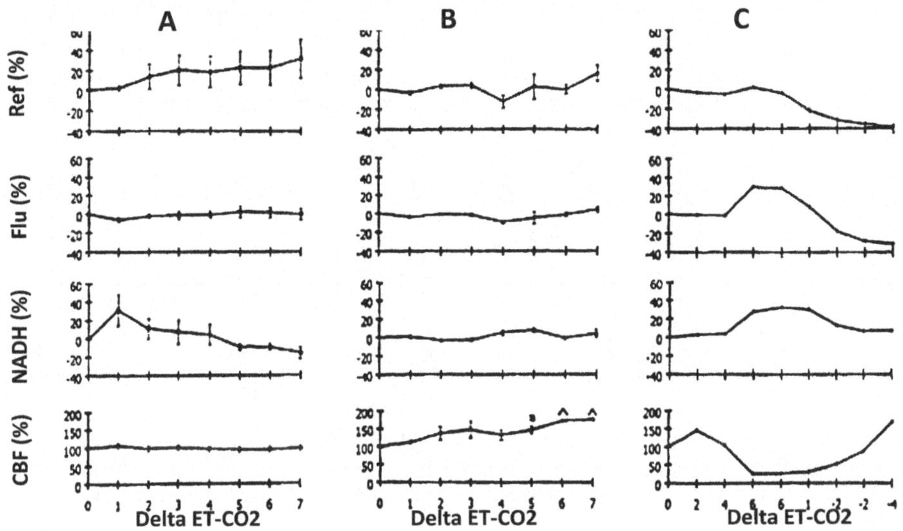

Fig. 6.27 Responses of the human brain to changes in end tidal CO_2 induced in 14 patients. In group A, the averages of the A patients monitored (both CBF and NADH) were not significantly affected. In Group B, the CBF was significantly elevated. *$p < 0.05$, ^$p < 0.01$. In Part C, the stealing of blood was recorded in one patient (Mayevsky et al. 2001)

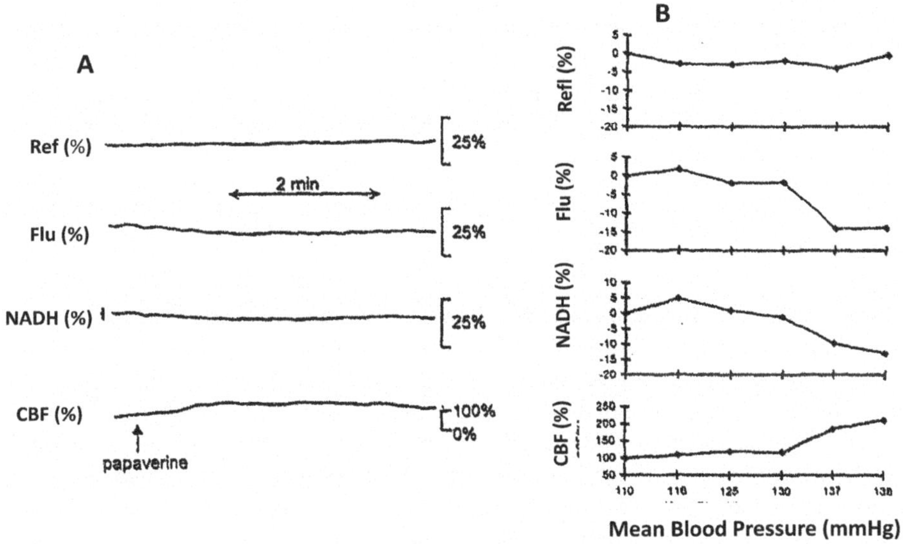

Fig. 6.28 (**a**) Metabolic and hemodynamic responses to the injection of papaverine (Mayevsky et al. 2001). (**b**) Effects of mean arterial blood pressure on brain CBF and NADH (Mayevsky et al. 2001)

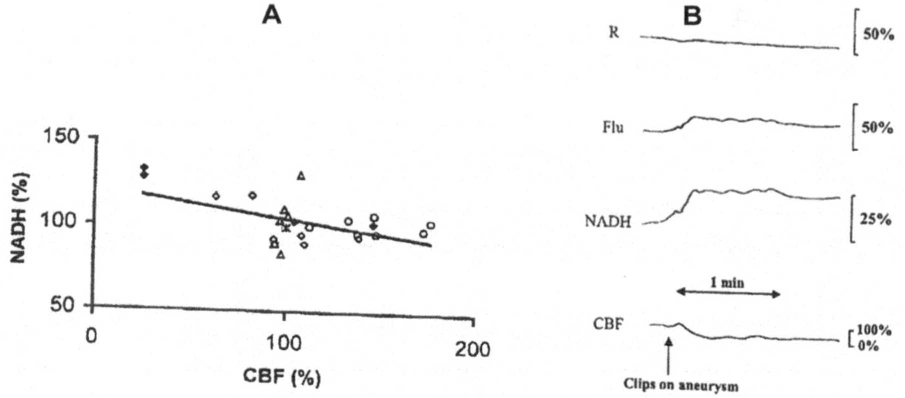

Fig. 6.29 (**a**) The correlation between CBF and NADH redox state measured in neurosurgical patients exposed to hypercapnia during the operation (Mayevsky et al. 2002). (**b**) Metabolic and hemodynamic responses to occlusion of an aneurysm located in one of the brain arteries (Mayevsky et al. 2002)

In order to demonstrate the potential use of our technology in clinical neurosurgery, the results obtained are presented. The patient underwent a carotid artery aneurysm correction procedure (Fig. 6.29b). As seen, occlusion of aneurysm area led to a clear decrease in CBF while mitochondrial NADH was increased significantly.

The effects of changes in CPP induced by a decrease in ICP by removal of 40 ml CSF on brain CBF and NADH are shown in Fig. 6.30. As seen, a clear correlation between CBF and NADH is shown. By decreasing the ICP, a clear increase in CBF was recorded together with the decrease in NADH. The reflectance trace remained stable during the changes in CPP.

6.3.3.4 Discussion

Tissue vitality and function could be expressed by various organs in the body. The most important parameters, common to all tissues in the body, are tissue microcirculatory blood flow and mitochondrial NADH redox state. In normal healthy tissue the ratio, or balance, between oxygen supply and oxygen demand is positive. That is, the supply mechanism (oxygenated blood circulation) is able to provide the amount of O_2 needed for the demand mechanisms (ATP production for various types of work). The energy balance reflects the cell's functional capacity to do work. In this way, energy balance reflects the metabolic state of the tissue.

Regarding the vitality of the brain, we have found that it is necessary to monitor various parameters, which represent the hemodynamic, metabolic, ionic, and electrical activities simultaneously. The advantage of the multiparametric approach is clear when one looks at the interpretation of monitoring a single parameter when the patient undergoes different clinical situations. In both states of hypoxia and

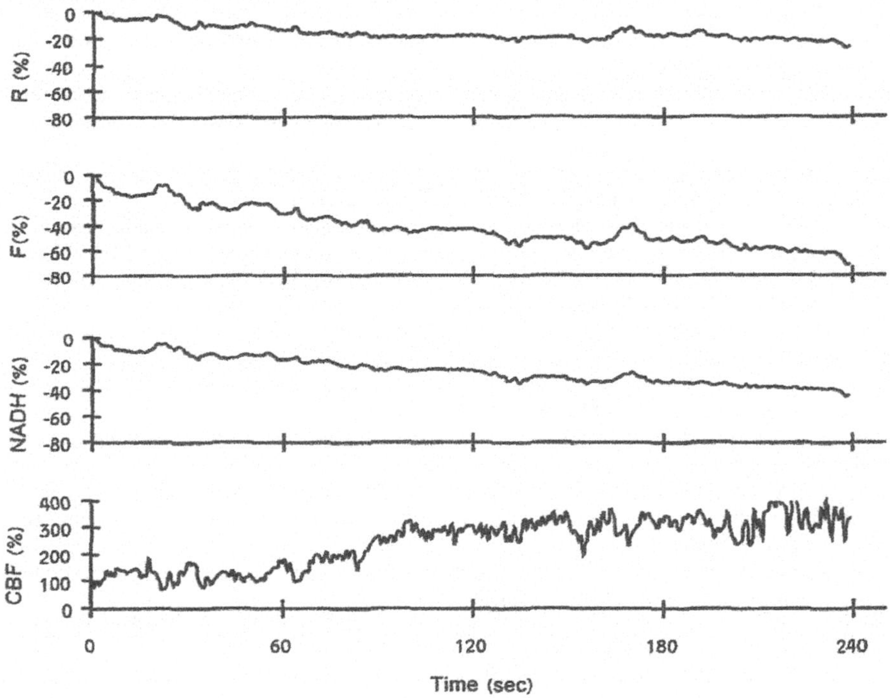

Fig. 6.30 Effects of a change in CPP induced by withdrawal of CSF on brain CBF and NADH

hypercapnia, CBF increases, and discrimination between the two cannot be made. However, if NADH was measured together with CBF, Hypoxia will be reflected by an increase in NADH (reduction), whereas Hypercapnia will be reflected by a decrease (oxidation). Furthermore, adding other parameters such as EEG temperature, extracellular K^+ levels, DC potential, and ICP increases the resolution of the diagnosis of an unclear pathological state of the brain.

Situations that develop under clinical treatment, such as elevated ICP, spreading depression waves, increased blood pressure, and responses to pharmacological treatment, have a direct effect on the prognosis of the patient.

References

Austin G, Jutzy R, Chance B, Barlow C (1978a) Noninvasive monitoring of human brain oxidative metabolism. Adv Exp Med Biol III:1445–1455

Austin G, Jutzy R, Hill W, Chance B, Barlow C (1978b) Cortical oxidative metabolism: a measure of collateral flow in microanastomoses. In: Proceedings of the world microsurgery Congress

Barlow CH, Harden WR III, Harken AH, Simson MB, Haselgrove JC, Chance B, O'Connor M, Austin G (1979) Fluorescence mapping of mitochondrial redox changes in heart and brain. Crit Care Med 7(9):402–406

Betz V, Schneckenburger H, Alleroder HP, Sybrecht GW, Meyer JU (1994) Evaluation of changes in the NADH level between carcinogenic and normal tissue samples by use of fluorescence spectroscopy. SPIE 2324:284–291
Bures J (1993) History of experimental spreading depression. In: Lehmenkuhler A, Grotemeyer K-H, Tegtmeier F (eds) Migraine: basic mechanism and treatment. Urhan and Schwarzenberg, pp 279–292
Bures J, Buresova O, Krivanek J (1974) The mechanism and applications of Leao's spreading depression of electroencephalographic activity. Academic, New York
Chance B (1983) A noninvasive biochemical assay and imaging of animal and human tissues by optical and nuclear magnetic resonance techniques. Proc Am Philos Soc 127:1–25
Chance B, Williams GR (1955) Respiratory enzymes in oxidative phosphorylation (I- Kinetics of oxygen utilization). J Biol Chem 217:383–393
Chance B, Cohen P, Jobsis F, Schoener B (1962) Intracellular oxidation-reduction states in vivo. Science 137:499–508
Chance B, Nioka S, Warren W, Yurtsever G (2005) Mitochondrial NADH as the bellwether of tissue O_2 delivery. Adv Exp Med Biol 566:231–242
Chandra M, Scheiman J, Heidt D, Simeone D, McKenna B, Mycek M-A (2007) Probing pancreatic disease using tissue optical spectroscopy. J Biomed Opt 12(6):060501
Clough GF, Bennett AR, Church MK (1998) Effects of H1 antagonists on the cutaneous vascular response to histamine and bradykinin: a study using scanning laser Doppler imaging. Br J Dermatol 138(5):806–814
Crowe W, Mayevsky A, Mela L (1977) Application of a solid membrane ion selective electrode to in vivo measurements. Am J Phys 233:C56–C60
Crutchfield JS, Narayan RK, Robertson CS, Michael LH (1990) Evaluation of a fiberoptic intracranial pressure monitor. J Neurosurg 72(3):482–487
Dietrich WD, Feng ZC, Leistra H, Watson BD, Rosenthal M (1994) Photothrombotic infarction triggers multiple episodes of cortical spreading depression in distant brain regions. J Cereb Blood Flow Metab 14(1):20–28
Dirnagl U, Kaplan B, Jacewicz M, Pulsinelli W (1989) Continuous measurement of cerebral cortical blood flow by laser-Doppler flowmetry in a rat stroke model. J CBF Metab 9:589–596
Duboc D, Toussaint M, Donsez D, Weber S, Guerin F, Degeorges M, Renault G, Polianski J, Pocidalo JJ (1986) Detection of regional myocardial ischaemia by NADH laser fluorimetry during human left heart catheterization. Lancet 2:522–522
Duboc D, Renault G, Polianski J, Muffat-Joly M, Toussaint M, Guerin F, Pocidalo J-J, Fardeau M (1987) NADH measured by laser fluorimetry in skeletal muscle in McArdle's disease. New Engl J Med 316(26):1664–1665
Ekbal NJ, Dyson A, Black C, Singer M (2013) Monitoring tissue perfusion, oxygenation, and metabolism in critically ill patients. Chest 143(6):1799–1808
Erecinska M, Silver IA (1989) ATP and brain function. J Cereb Blood Flow Metab 9(1):2–19
Fein JM (1983) NADH kinetics in patients with unruptured aneurysms versus cerebrovascular occlusive disease. J CBF Metab 3:S29–S30
Fein J, Jobsis F (1979) Brain energetics in patients undergoing STA-MCA microanastomosis. Acta Neurol Scand Suppl 72:504–505
Fein JM, Olinger R (1982) Cortical nicotinamide adenine dinucleotide (NADH) kinetics in patients undergoing extracranial-intracranial bypass. Neurosurgery 10(4):428–436
Friedli CM, Sclarsky DS, Mayevsky A (1982) Multiprobe monitoring of ionic, metabolic, and electrical activities in the awake brain. Am J Phys 243(3):R462–R469
Fullerton JN, Singer M (2011) Organ failure in the ICU: cellular alterations. Semin Respir Crit Care Med 32(5):581–586
Galeotti T, van Rossum GDV, Mayer D, Chance B (1970) Fluorescence studies of NAD(P)H binding in intact cells. Hoppe Seylers Z Physiol Chem 351:274–274
Galeotti T, Cittadini A, Chance B (1972) $NAD(P)^+$ redox compartmentation in Ehrlich ascites tumor cells. Biochim Biophys Acta 267:1–6

Gloor P (1986) Migraine and regional cerebral blood flow. Trends Neurosci 9:21
Guezennec CY, Lienhard F, Louisy F, Renault G, Tusseau MH, Portero P (1991) In situ NADH laser fluorimetry during muscle contraction in humans. Eur J Appl Physiol 63:36–42
Haberl RL, Heizer ML, Marmarou A, Ellis EF (1989) Laser-Doppler assessment of brain microcirculation: effect of systemic alterations. Am J Physiol Heart Circ Physiol 256:H1247–H1254
Hansen AJ (1985) Effect of anoxia on ion distribution in the brain. Physiol Rev 65:101–148
Hansen AJ, Zeuthen T (1981) Extracellular ion concentrations during spreading depression and ischemia in the rat brain cortex. Acta Physiol Scand 113(4):437–445
Hansen AJ, Quistorff B, Gjedde A (1980) Relationship between local changes in cortical blood flow and extracellular K^+ during spreading depression. Acta Physiol Scand 109(1):1–6
Hertelendy P, Menyhart A, Makra P, Sule Z, Kiss T, Toth G, Ivankovits-Kiss O, Bari F, Farkas E (2017) Advancing age and ischemia elevate the electric threshold to elicit spreading depolarization in the cerebral cortex of young adult rats. J Cereb Blood Flow Metab 37(5):1763–1775
Hotchkiss RS, Karl IE (1992) Reevaluation of the role of cellular hypoxia and bioenergetic failure in sepsis. JAMA 267:1503–1510
Ince C, Sinaasappel M (1999) Microcirculatory oxygenation and shunting in sepsis and shock. Crit Care Med 27:1369–1377
Jobsis FF, O'Connor MJ, Rosenthal M, Van Buren JM (1971) Fluorometric monitoring of metabolic activity in the intact cerebral cortex, Excerpta medica international congress series, NO. 253. G. G. Somjen, Paris, pp 18–26
Kaplan NO (1985) The role of pyridine-nucleotides in regulating cellular-metabolism. Curr Top Cell Regul 26:371–381
Kohen E, Kohen C, Jenkins W (1966) The influence of microelectrophoretically introduced metabolites on pyridine nucleotide reduction in giant tissue culture ascites cells. Exp Cell Res 44:175–194
Kohen E, Kohen C, Thorell B (1968) The effect of cytostatic drugs on the metabolic response of ascites tumor cells. Acta Pharmacol Toxicol 26:556–570
Kohen E, Kohen C, Thorell B (1971) Rapid microfluorimetry of enzyme reactions in single living cells. Biochim Biophys Acta 234:531–536
Kohen E, Hirschberg JG, Kohen C, Wouters A, Pearson A, Salmon J-M, Thorell B (1975a) Multichannel microspectrofluorometry for topographic and spectral analysis of NAD(P)H fluorescence in single living cells. Biochim Biophys Acta 396:149–154
Kohen E, Kohen C, Thorell B, Schachtschabel D (1975b) Multisite analysis of metabolic transients in single living cells by multichannel microfluorometry. Mikrochim Acta 1:223–236
Konig F, Schnorr D, Loening SA, Paul R, Pfeifer L, Lehmenkuhler A, Scheer A, Fink F (1994) Laser-induced autofluorescence of prostate- and bladder tissue. Laser – tissue interaction V photochemical mechanisms. SPIE 2134:284
Kunz D, Winkler K, Elger CE, Kunz WS (2002) Functional imaging of mitochondrial redox state. In: Abelson JN, Simon MI (eds) Methods in enzymology. Elsevier Science/Academic, pp 135–150
Lauritzen M (1987) Cerebral blood flow in migraine and cortical spreading depression. Acta Neurol Scand Suppl 113:1–40
Lauritzen M (1988) Blood flow and metabolism in cortical spreading depression. In: Somjen G (ed) Mechanisms of cerebral hypoxia and stroke. Advances in behavioral biology. Springer, Boston, p 35
Lauritzen M (1994) Pathophysiology of the migraine aura. The spreading depression theory. Brain 117:199–210
Lauritzen M, Jorgensen MB, Diemer NH, Gjedde A, Hansen AJ (1982) Persistent oligemia of rat cerebral cortex in the wake of spreading depression. Ann Neurol 12(5):469–474
Leao AAP (1944) Spreading depression of activity in cerebral cortex. J Neurophysiol 7:359–390
Leao AAP (1947) Further observations on the spreading depression of activity in the cerebral cortex. J Neurophysiol 10:409–414

Levine S, Payan H (1966) Effects of ischemia and other procedures on the brain and retina of the gerbil (Meriones unguiculatus). Exp Neurol 16(3):255–262

Li LZ, Zhou R, Xu HN, Moon L, Zhong T, Kim EJ, Qiao H, Reddy R, Leeper D, Chance B, Glickson JD (2009) Quantitative magnetic resonance and optical imaging biomarkers of melanoma metastatic potential. Proc Natl Acad Sci USA 106:6608–6613

Lohmann W, Paul E (1988) In situ detection of melanomas by fluorescence measurements. Naturwissenschaften 75:201–202

Lohmann W, Hirzinger B, Braun J, Schwemmle K, Muhrer K-H, Schulz A (1990) Fluorescence studies on lung tumors. Z Naturforsch 45c:1063–1066

Lukyanov LD, Bures J (1967) Changes in pO_2 due to spreading depression in cortex and nucleus caudatus of rat. Physiol Bohemoslov 16(5):449–455

Marik PE, Varon J (1998) The hemodynamic derangements in sepsis: implications for treatment strategies. Chest 114:854–860

Marshall WH (1959) Spreading cortical depression of Leão. Physiol Rev 39(2):239–279

Mayevsky A (1976) Brain energy metabolism of the conscious rat exposed to various physiological and pathological situations. Brain Res 113:327–338

Mayevsky A (1983a) Metabolic, ionic and electrical responses to experimental epilepsy in the awake rat. In: Baldy M, Moulinier DH, Ingvar DH, Meldrum BS (eds) Proceedings first international Congress cerebral blood flow, metabolism & epilepsy. John Libbey, pp 263–270

Mayevsky A (1983b) Multiparameter monitoring of the awake brain under hyperbaric oxygenation. J Appl Physiol 54(3):740–748

Mayevsky A (1984) Brain NADH redox state monitored in vivo by fiber optic surface fluorometry. Brain Res Rev 7:49–68

Mayevsky A (1990) Level of ischemia and brain functions in the Mongolian gerbil in vivo. Brain Res 524:1–9

Mayevsky A (1993) Biochemical and physiological activities of the brain as in vivo markers of brain pathology. In: Bernstein EF, Callow AD, Nicolaides AN, Shifrin EG (eds) Cerebral revascularization. Med-Orion Pub, pp 51–69

Mayevsky A, Barbiro-Michaely E (2013) Shedding light on mitochondrial function by real time monitoring of NADH fluorescence: II: human studies. J Clin Monit Comput 27:125–145

Mayevsky A, Chance B (1982) Intracellular oxidation-reduction state measured in situ by a multicannel fiber-optic surface fluorometer. Science 217:537–540

Mayevsky A, Sclarsky DS (1983) Correlation of brain NADH redox state, K^+, PO_2 and electrical activity during hypoxia, ischemia and spreading depression. Adv Exp Med Biol 159:129–141

Mayevsky A, Weiss HR (1991) Cerebral blood flow and oxygen consumption in cortical spreading depression. J CBF Metab 11:829–836

Mayevsky A, Zeuthen T, Chance B (1974) Measurements of extracellular potassium, ECoG and pyridine nucleotide levels during cortical spreading depression in rats. Brain Res 76:347–349

Mayevsky A, Lebourdais S, Chance B (1980) The interrelation between brain PO_2 and NADH oxidation- reduction state in the gerbil. J Neurosci Res 5:173–182

Mayevsky A, Zarchin N, Friedli CM (1982) Factors affecting the oxygen balance in the awake cerebral cortex exposed to spreading depression. Brain Res 236:93–105

Mayevsky A, Friedli CM, Reivich M (1985) Metabolic, ionic and electrical responses of the gerbil brain to ischemia. Am J Phys 248:R99–R107

Mayevsky A, Zarchin N, Yoles E, Tannenbaum B (1986) Oxygen supply to the brain in hypoxic and hyperoxic conditions. In: Nicolau C (ed) Advances in the biosciences. "O_2-transport in red blood cells", vol 54. Pergamon Press, pp 119–132

Mayevsky A, Flamm ES, Pennie W, Chance B (1991) A fiber optic based multiprobes system for intraoperative monitoring of brain functions. SPIE 1431:303–313

Mayevsky A, Frank K, Muck M, Nioka S, Kessler M, Chance B (1992) Multiparametric evaluation of brain functions in the Mongolian gerbil in vivo. J Basic Clin Physiol Pharmacol 3:323–342

Mayevsky A, Meilin S, Rogatsky GG, Zarchin N, Sonn J (1995) Multiparametric monitoring of the awake brain exposed to carbon monoxide. J Appl Physiol 78:1188–1196

Mayevsky A, Doron A, Manor T, Meilin S, Zarchin N, Ouaknine GE (1996) Cortical spreading depression recorded from the human brain using a multiparametric monitoring system. Brain Res 740:268–274

Mayevsky A, Manor T, Meilin S, Doron A, Ouaknine GE (1998a) Real-time multiparametric monitoring of the injured human cerebral cortex – a new approach. Acta Neurochir Suppl 71: 78–81

Mayevsky A, Meilin S, Manor T, Ornstein E, Zarchin N, Sonn J (1998b) Multiparametric monitoring of brain oxygen balance under experimental and clinical conditions. Neurol Res 20:S76–S80

Mayevsky A, Doron A, Meilin S, Manor T, Ornstein E, Ouaknine GE (1999) Brain viability and function analyzer: multiparametric real-time monitoring in neurosurgical patients. Acta Neurochir Suppl (Wien) 75:63–66

Mayevsky A, Nakache R, Merhav H, Luger-Hamer M, Sonn J (2000) Real time monitoring of intraoperative allograft vitality. Transplant Proc 32:684–685

Mayevsky A, Manor T, Meilin S, Razon N, Ouaknine GE (2001) Multiparametric monitoring of tissue vitality in clinical situations. SPIE 4255:33–39

Mayevsky A, Ornstein E, Meilin S, Razon N, Ouaknine GE (2002) The evaluation of brain CBF and mitochondrial function by a fiber optic tissue spectroscope in neurosurgical patients. Acta Neurochir Suppl 81:367–371

Mayevsky A, Sonn J, Luger-Hamer M, Nakache R (2003) Real time assessment of organ vitality during the transplantation procedure. Transplant Rev 17:96–116

McLachlan RS, Girvin JP (1994) Spreading depression of Leao in rodent and human cortex. Brain Res 666(1):133–136

Meier-Hellmann A, Reinhart K (1995) Effects of catecholamines on regional perfusion and oxygenation in critically ill patients. Acta Anaesthesiol Scand Suppl 107:239–248

Meilin S, Rogatsky GG, Thom SR, Zarchin N, Guggenheimer-Furman E, Mayevsky A (1996) Effects of carbon monoxide exposure on the brain may be mediated by nitric oxide. J Appl Physiol 81:1078–1083

Meilin S, Mendelman A, Sonn J, Manor T, Zarchin N, Mayevsky A (1999) Metabolic and hemodynamic oscillations monitored optically in the brain exposed to various pathological states. Adv Exp Med Biol 471:141–146

Oka H, Kako M, Matsushima M, Ando K (1977) Traumatic spreading depression syndrome. Review of a particular type of head injury in 37 patients. Brain 100(2):287–298

Ozawa K, Chance B, Tanaka A, Iwata S, Kitai T, Ikai I (1992) Linear correlation between acetoacetate/b-hydroxybutyrate in arterial blood and oxidized flavoprotein/reduced pyridine nucleotide in freeze-trapped human liver tissue. Biochim Biophys Acta 1138:350–352

Paschen W, Shima T, Hossmann K-A (1984) Pial arterial pressure in cats following middle cerebral artery occlusion. II. Relationship to regional disturbance of energy metabolism. Stroke 15:686–690

Pearce JM (1985) Is migraine explained by Leao's spreading depression? Lancet 2(8458):763–766

Pfeifer L, Schmalzigaug K, Paul R, Lichey J, Kemnitz K, Fink F (1995) Time-resolved autofluorescence measurements for the differentiation of lung-tissue states. Barcelona, Spain

Pfeifer L, Paul R, Yalcin E, Marx U, Konig F, Fink F (1996) A time-gated laser spectrometer using optical fibres for detecting fluorescent biomolecules in cells and tissue. In: Gonzalez-Mora JL, Borges R, Mas M (eds) . Santa Cruz de Tenerife, University of La Laguna, pp 42–43

Rampil IJ, Litt L, Mayevsky A (1992) Correlated, simultaneous, multiple-wavelength optical monitoring in vivo of localized cerebrocortical NADH and brain microvessel hemoglobin oxygen saturation. J Clin Monit 8(3):216–225

Renault G, Raynal E, Sinet M, Muffat-Joly M, Berthier J-P, Cornillault J, Godard B, Pocidalo J-J (1984) In situ double-beam NADH laser fluorimetry: choice of a reference wavelength. Am J Phys 246:H491–H499

Richter F, Eitner A, Leuchtweis J, Bauer R, Lehmenkuhler A, Schaible HG (2017) Effects of interleukin-1ss on cortical spreading depolarization and cerebral vasculature. J Cereb Blood Flow Metab 37(5):1791–1802

Rogatsky GG, Mayevsky A, Zarchin N, Doron A (1996) Continuous multiparametric monitoring of brain activities following fluid-percussion injury in rats: preliminary results. J Basic Clin Physiol Pharmacol 7:23–43

Rosenblum BR, Bonner RF, Oldfield EH (1987) Intraoperative measurement of cortical blood flow adjacent to cerebral AVM using laser Doppler velocimetry. J Neurosurg 66(3):396–399

Rosenorn J (1989) The risk of ischaemic brain damage during the use of self-retaining brain retractors. Acta Neurol Scand Suppl 120:1–30

Rosenthal M, Jobsis FF (1971) Intracellular redox changes in functioning cerebral cortex. II. Effects of direct cortical stimulation. J Neurophysiol 34:750–762

Rosenthal M, Sick TJ (1988) Measurement of metabolic activity associated with ion shifts. In: Neuronal microenvironment, pp 187–246

Rosenthal M, Somjen G (1973) Spreading depression, sustained potential shifts, and metabolic activity of cerebral cortex of cats. J Neurophysiol 36:739–749

Schneckenburger H, Gessler P, Pavenstadt-Grupp I (1992) Measurements of mitochondrial deficiencies in living cells by microspectrofluorometry. J Histochem Cyrochem 40(10):1573–1578

Schomacker KT, Frisoli JK, Compton CC, Flotte TJ, Richter JM, Nishioka NS, Deutsch TF (1992) Ultraviolet laser-induced fluorescence of colonic tissue: basic biology and diagnostic potential. Lasers Surg Med 12:63–78

Sivabalan S, Vedeswari CP, Jayachandran S, Koteeswaran D, Pravda C, Aruna PR, Ganesan S (2010) In vivo native fluorescence spectroscopy and nicotinamide adinine dinucleotide/flavin adenine dinucleotide reduction and oxidation states of oral submucous fibrosis for chemopreventive drug monitoring. J Biomed Opt 15:017010

Somjen GG, Aitken PG, Balestrino M, Herreras O, Kawasaki K (1990) Spreading depression-like depolarization and selective vulnerability of neurons. A brief review. Stroke 21(11 Suppl): III179–III183

Somjen GG, Aitken PG, Czeh G, Herreras O, Jing J, Young JN (1993) Spreading depression in Hippocampus: membrane currents and ion mechanisms. In: Lehmenkuhler A, Grotemeyer K-H, Tegtmeier F (eds) Migraine: basic mechanisms and treatment. Urban and Schwarzenberg, pp 329–344

Sramka M, Brozek G, Bures J, Nadvornik P (1977) Functional ablation by spreading depression: possible use in human stereotactic neurosurgery. Appl Neurophysiol 40(1):48–61

Sterenborg NJ, Thomsen S, Jacques SL, Duvic M, Motamedi M, Wagner RF Jr (1995/9) In vivo fluorescence spectroscopy and imaging of human skin tumors [letter]. Dermatologic Surg 21(9): 821–822

Stern MD, Lappe DL, Bowen PD, Chimosky JE, Holloway GA Jr, Keiser HR, Bowman RL (1977) Continuous measurement of tissue blood flow by laser-Doppler spectroscopy. Am J Phys 232(4):H441–H448

Tanaka A, Kitai T, Iwata S, Hirao K, Tokuka A, Sato B, Yanabu N, Mori S, Inomoto T, Yamaoka Y, Tanaka K, Ozawa K, Chance B (1993) Delayed oxidation of intramitochondrial pyridine nucleotide oxidoreduction state as compared with tissue oxygenation in human liver transplantation. Biochim Biophys Acta 1182:250–256

Toussaint M, Duboc D, Renault G, Polianski J, Schved M, Donsez D, Weber S, Dessault O, Pocidalo JJ, Guerin F (1987) Study of myocardial metabolism of NADH by laser fluorimetry during cardiac catheterization. (In French). Arch Mal Coeur Vaiss 80(9):1341–1349

Toussaint M, Duboc D, Renault G, Guerin F, Degeorges M (1988) Laser fluorimetry of NADH. Arch Mal Coeur Vaiss 81 Spec No:47–51

Van Buren JM, Lewis MD, Schuette WH, Whitehouse WC, Marsan CA (1978) Fluorometric monitoring of NADH levels in cerebral cortex: preliminary observations in human epilepsy. Neurosurgery 2(2):114–121

Van Harreveld A, Stamm JS (1952) Vascular concomitants of spreading cortical depression. J Neurophysiol 15(6):487–496

Vielhaber S, Kunz D, Winkler K, Wiedemann FR, Kirches E, Feistner H, Heinze HJ, Elger CE, Schubert W, Kunz WS (2000) Mitochondrial DNA abnormalities in skeletal muscle of patients with sporadic amyotrophic lateral sclerosis. Brain 123:1339–1348

Vyskocil F, Kritz N, Bures J (1972) Potassium-selective microelectrodes used for measuring the extracellular brain potassium during spreading depression and anoxic depolarization in rats. Brain Res 39(1):255–259

Wadhwani KC, Rapoport SI, Shepherd AP, Oberg PA (1990) Blood flow in the central and peripheral nervous systems. In: Shephrd AP, Oberg PA (eds) Laser doppler blood flowmetry, vol 107. Kluwer Academic Pub, Boston, pp 265–304

Wang KK, Gutta K, Laukka M, Densmore J (1994) Laser induced fluorescence in the detection of esophageal carcinoma. SPIE 2324:14–18

Wiebers DO, Adams HP Jr, Whisnant JP (1990) Animal models of stroke: are they relevant to human disease? Stroke 21(1):1–3

Wiedemann FR, Winkler K, Kuznetsov AV, Bartels C, Vielhaber S, Feistner H, Kunz WS (1998) Impairment of mitochondrial function in skeletal muscle of patients with amyotrophic lateral sclerosis. J Neurol Sci 156(1):65–72

Xu H, Nioka S, Glickson JD, Chance B, Li LZ, Cachecho R (2010) Quantitative mitochondrial redox imaging of breast cancer metastatic potential. J Biomed Opt 15(3):036010

Xu HN, Nioka S, Chance B, Li LZ (2011) Heterogeneity of mitochondrial redox state in premalignant pancreas in a PTEN null transgenic mouse model. Adv Exp Med Biol 701:207–213

Zhang ZH, Li H, Liu Q, Zhou LL, Zhang M, Luo QM, Glickson J, Chance B, Zheng G (2004a) Metabolic imaging of tumors using intrinsic and extrinsic fluorescent markers. Biosens Bioelectron 20(3):643–650

Zhang Z, Blessington D, Li H, Busch TM, Glickson J, Luo Q, Chance B, Zheng G (2004b) Redox ratio of mitochondria as an indicator for the response of photodynamic therapy. J Biomed Opt 9(4):772–778

Zhu X, Zuo L, Cardounel AJ, Zweier JL, He G (2007) Characterization of in vivo tissue redox status, oxygenation, and formation of reactive oxygen species in postischemic myocardium. Antioxid Redox Signal 9:447–455

Chapter 7
Clinical Monitoring Using Commercial Devices

7.1 The Transition from Noncommercial to Commercial Monitoring System

7.1.1 Introduction

Real-time monitoring of local tissue or organ vitality, during various surgical as well as other clinical procedures, is a critical factor in the future development of modern medicine. As of today most of the techniques available, provide single time point, anatomical or physiological, information that can be used in the intensive care units (ICUs) or during the pre- or postoperative periods. The most significant technology that is missing is such that can simultaneously evaluate a number of parameters which represent tissue or organ vitality in real time. The physiological and biochemical parameters that are most needed in clinical practice are those related to oxygen balance (supply/demand) in tissues exposed to changes in O_2 supply (i.e., ischemia or reperfusion). In normal healthy tissue, the ratio or balance between O_2 supply and O_2 demand is positive. That is, the uniform supply mechanisms (blood flow and hemoglobin oxygenation) are able to provide the amount of O_2 needed for the demand mechanisms (typical to each tissue). Insufficient O_2 supply results in impaired function of the tissue and puts its survival at risk. When reviewing the various techniques available for real-time monitoring of tissues, it is possible to divide them according to different criteria. These include the site of monitoring (i.e., OR or ICU), duration of monitoring (acute vs. chronic), usefulness in animal studies or in clinical situations, or the level of invasiveness to the monitored tissue. In this chapter, we will use another criterion for technology classification, namely, the compartment of the tissue, analyzed by the various techniques.

As shown in Fig. 7.1, the different tissue compartments that could be analyzed in real time are labeled A to C, starting in the interior of blood vessels (A) through the extra- and intracellular spaces (B and C). The lowest point in the O_2 gradient is the mitochondria.

© The Author(s), under exclusive license to Springer Nature Switzerland AG 2025

A. Mayevsky et al., *Body Oxygen Homeostasis and Mitochondrial Function*,
https://doi.org/10.1007/978-3-031-94113-9_7

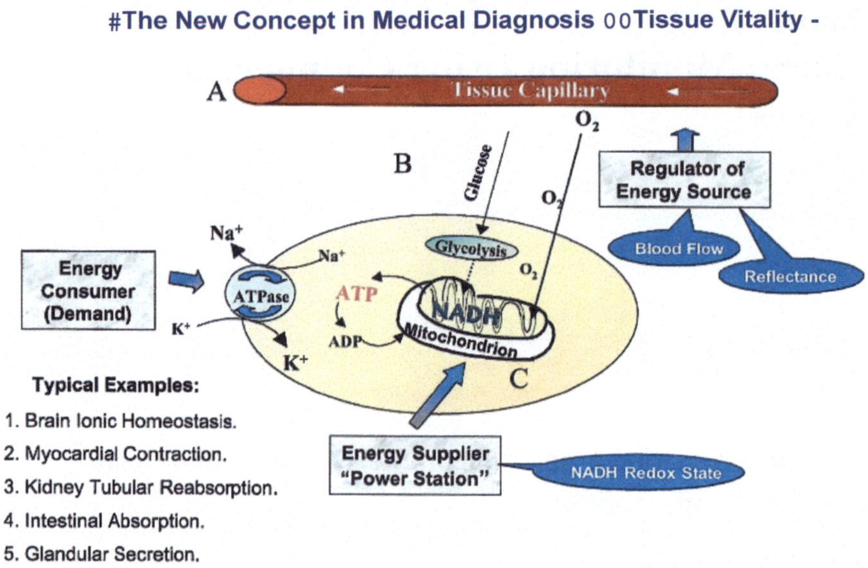

Fig. 7.1 Schematic presentation of the three major components of the tissue O_2 balance equation, namely, O_2 supply, demand, and balance. (Mayevsky et al. 2002 [Modified])

Two possible monitored parameters are related to the microvascular compartment (A) and represent two different physiological processes. The level of hemoglobin (Hb) oxygenation is a result of the binding capacity of O_2 to Hb as well as the delivery of O_2 in the small capillaries. The microcirculatory blood flow, on the contrary, is the main regulated process in adjusting the supply of O_2 to meet the changes in O_2 demand.

Monitoring of extracellular levels of ions represents very specific information on the ionic homeostasis integrity around the cellular membrane, a process which is sensitive to the status of energy or O_2 balance in the tissue partial oxygen pressure (pO_2) values, for example, are an average of the three compartments and any change in tissue blood flow (TBF) (A) or intracellular O_2 consumption (C) will affect the pO_2 without a clear diagnostic ability.

The mitochondrial signal of NADH redox state shown in C represents biochemical processes which occur in the respiratory chain components located in the inner membrane of the mitochondria. NADH is very sensitive to the O_2 balance in the mitochondria, and can be monitored optically using its fluorescence characteristics.

The main disadvantage of the many available techniques is that they provide information on a single physiological or biochemical parameter. This fact makes their diagnostic value inferior to a multiparametric monitoring approach. In conclusion, it seems that future breakthroughs in tissue monitoring during surgery or other clinical situations will depend upon the development of multiparametric monitoring approaches.

In this study, we describe the use of a new Tissue Spectroscope (TiSpec), a device which enables to simultaneously monitor microcirculatory blood flow, tissue reflectance, and mitochondrial NADH redox state.

7.1.2 Methods

7.1.2.1 Tissue Blood Flow—TBF

Today, real-time microcirculatory blood flow is mainly measured by Laser Doppler Flowmetry (LDF), which is a continuous monitoring. The principle of monitoring TBF by LDF is that moving red blood cells, exposed to laser light, reflect that light at a slightly different wavelength (Doppler shift). Tissue components, including nonmoving red blood cells, reflect the light at the same wavelength as the excitation light. Usually the changes in TBF are calculated as percent change as compared to a control level, defined as 100% of TBF. The LDF signal after death is defined as 0% TBF. These definitions allow for the calculation of relative changes between 0% and 100% TBF, and above. The main disadvantage according to most of the users is that relative values are obtained instead of absolute units (Dirnagl et al. 1989; Haberl et al. 1989; Fabricius and Lauritzen 1996). LDF has widespread use, which is especially growing with potential in the dermatology field as noninvasive monitoring (Schiller et al. 1997; Park et al. 1998; Yuen and Feng 1998). As a minimal invasive technique, it has already proved its usefulness in the clinics at various applications, namely, organ transplantation (Seifalian et al. 1997; Sundset et al. 1997), neurosurgery (Arbit et al. 1989; Bolognese et al. 1993; Kuchiwaki et al. 1996), and gastrointestinal procedures (Hallbook et al. 1996; Tateishi et al. 1997; Redaelli et al. 1998).

Continuous real-time monitoring of tissue microcirculatory blood flow was possible since the development of the Helium-Neon (He-Ne) laser in 1961 by Javan. After a few years of fluid flow measurements (Yeh and Cummins 1964), a laser-Doppler Velocimeter (LDV) for blood flow measurements was developed (Riva et al. 1972). In the early stage, the LDV was used for single vessel monitoring, and only in 1974, Stern was able to suggest the application of LDV to measure tissue perfusion (Stern 1975; Stern et al. 1977). The need for LDF measurement in experimental and clinical practice has led to a growing number of patents as well as companies that offer various models of monitors (Shepherd 1990).

7.1.2.2 Mitochondrial NADH Redox State

Various participants of the respiratory chain, located in the mitochondria, exhibit different optical properties in their reduced and oxidized forms thus enabling noninvasive monitoring of various organs *in vitro* as well as *in vivo*. The oxidation–reduction state of NADH can serve as an indicator of ATP formation rate in the tissue. The absorption and oxidation–reduction state of NADH can serve as an indicator of ATP formation rate in the tissue. The absorption and fluorescence spectra of NADH (reduced form) have been well characterized at different levels of organizational complexity, i.e., in solution, mitochondrial and cell suspensions, tissue slices, and *in vivo*. NADH has an optical absorption band at about 300–380 nm and a fluorescence emission band at 420–480 nm. The spectra are considered to be the same for different environments and measurement conditions, although there are small differences in the shape and maxima of the spectra. It is important to note that NAD^+ (the oxidized form), does not absorb light at the mentioned excitation wavelength and thus emits no fluorescence.

In 1955, Chance and Williams (1955b) defined various metabolic states of the mitochondria *in vitro*. These states depend upon the availability of oxygen, substrate, and phosphate acceptor (ADP) as shown in Fig. 7.2 (left side). Using the autofluorescence of the mitochondrial NADH, the excitation and emission spectra of NADH in a suspension of rat liver mitochondria were recorded. The near maximum intensity (99%) recorded in the resting state, State 4, decreased (53%) during the transition to the active state, State 3, in which the electron transport chain was the limiting factor. By depriving the mitochondria from substrate, State 2, a further decrease in NADH was observed. No significant increase in NADH was recorded by exposing the State 4 mitochondria to anoxic conditions (State 5). Other members of the respiratory chain, such as flavoproteins, cytochromes b, c, and aa3, also provide good signals for state 4–3-4 transition. The smallest response of the chain to the various states was found when cytochromes aa3 (cytaa3) were monitored (Chance and Williams 1955b; Chance et al. 1973).

Mitochondria In-Vitro*			
Respiration Rate	Limiting Substance	NADH %	State #
0	Oxygen	~100	5
Slow	ADP	99	4
Fast	Resp. Chain	53	3
Slow	Substrate	~0	2

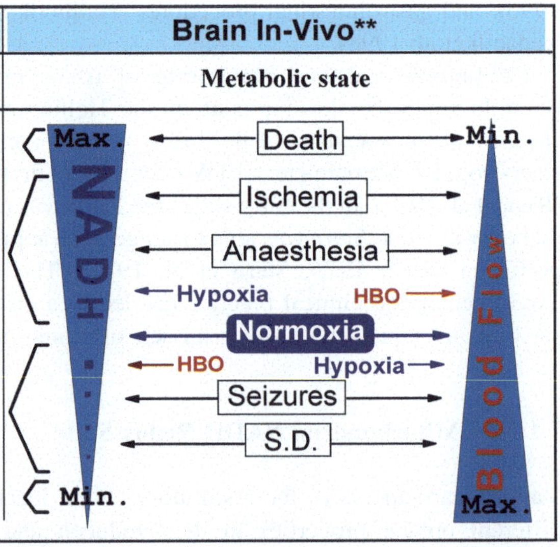

Fig. 7.2 Comparison between *in vitro* mitochondrial NADH redox state and the metabolic states taking place under *in vivo* pathophysiological situation. The correlation between NADH and tissue blood flow is shown on the right-hand side. (Mayevsky et al. 2002)

In 1962, Chance and coworkers (1962a, b; Chance and Schoener 1962) described for the first time the measurement of NADH in exposed brain cortex, kidney, and other organs of anesthetized rat, using surface fluorometry. The principle of the method was that the exposed tissue was illuminated by 355 nm to excite the NADH, and the emitted fluorescence light was measured in the range of 420–480 nm. They studied the effect of anoxia on the oxidation–reduction state of NADH and correlated it to the activation state of the brain. They showed that convulsive activity (O'Connor et al. 1973a), direct cortical stimulation (Rosenthal and Jobsis 1971), or spreading depression (SD) (Rosenthal and Somjen 1973) produced a transient oxidation of NADH which was interpreted as a state 4–3-4 transition. In 1975, Sundt and Anderson (1975) applied *in vivo* fluorometry to study the brain of squirrel monkey. At the same time, Harbig et al. (1976), using cat brain, presented a clear increase in the NADH spectrum between normoxic and anoxic conditions.

In 1972, we have introduced for the first time optical fibers into the surface fluorometry replacing the old 'rigid' optical system used until then and, as a result, measurement of NADH from the brain of an awake animal became possible (Chance et al. 1973, 1974; Mayevsky and Chance 1973).

During the last three decades, this technique has been developed, improved, and applied to many experimental setups in various organs *in vivo* and *in vitro* (Cummins and Elliott 1976; Balaban and Mandel 1980; Balaban et al. 1980; Franke et al. 1980; Ji et al. 1980; Kedem et al. 1981; Balaban and Blum 1982; Dora 1984; Renault et al. 1984; Koretsky et al. 1987; Bickler et al. 1988; Breuer and Mayevsky 1992; Kitai et al. 1992; Ince et al. 1993).

The principle of *in vitro* mitochondrial NADH monitoring is identical to the *in vivo* setup, although the interpretation of the results is different. As shown on the right side of Fig. 7.2, the NADH dynamic range is from minimal (state 2) to maximal (state 5), changing in a continuous manner rather than between defined states (Mayevsky 1984). The normoxic awake brain is scaled midway between state 3 and state 4. Various physiological and pathological perturbations affect the NADH redox state, as shown in Fig. 7.2.

7.1.2.3 Tissue Spectroscope

The main disadvantage of our previous approach was that two different instruments and technologies were used in combination (Barbiro-Michaely et al. 1998; Mayevsky et al. 1991, 1996, 1998). Furthermore, due to the usage of two different wavelengths in the two instruments, the penetration depth of the light was different. In order to improve the monitoring system, a device called TiSpec, was developed and the results of its preliminary testing in experimental animals are presented. This new device can provide continuous, real-time parameters correlated to tissue vitality.

In order to demonstrate the principles of the TiSpec, a short description of the light path in the monitored tissue will be given. The light source, on which the TiSpec function is based, is a low-power UV laser (325 nm or 390 nm) that guided the tissue by a flexible optical fiber. The emitted light from the tissue contains three main components:

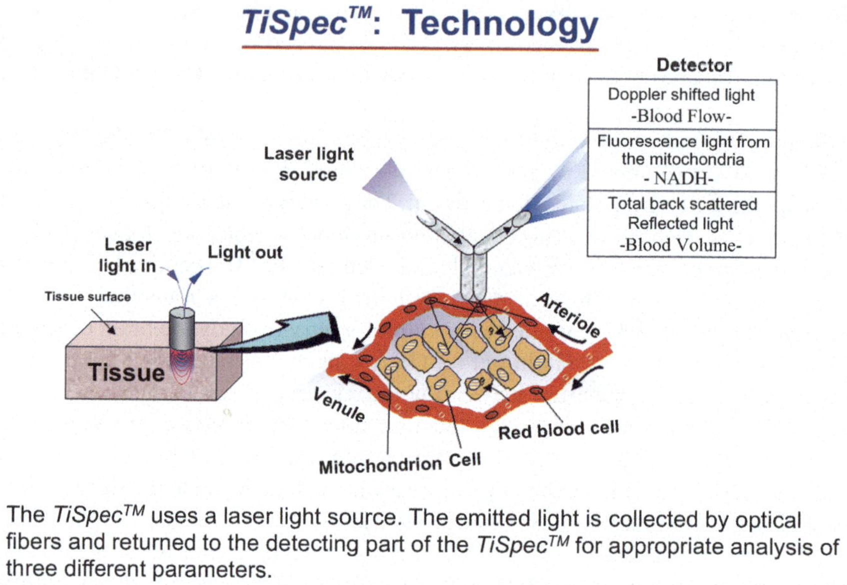

Fig. 7.3 Experimental principles of the simultaneous monitoring of tissue blood flow, tissue reflectance, and mitochondrial NADH redox state using the same light source and the same optical fibers. (Mayevsky et al. 2002 [Modified])

A. Reflected light at exactly the same wavelength as the excitation light.
B. Doppler-shifted light reflected from the red blood cells moving inside small blood vessels (small arterioles and capillaries), representing the microcirculation.
C. Fluorescence light (420–480 nm), originated and emitted from intracellular NADH molecules, excited by the 325 nm or 390 nm light.

The emitted light is collected by other optical fibers and returned to the detecting part of the TiSpec for appropriate analysis.

Using the UV light source (Fig. 7.3), the TiSpec can measure the following three different parameters:

1. Total backscattered or reflected light (R), which is correlated to blood volume and other optical changes in the tissue.
2. Microcirculatory TBF.
3. Intra-mitochondrial NADH redox state.

Figure 7.4 shows the first TiSpec that was constructed and used in our studies.

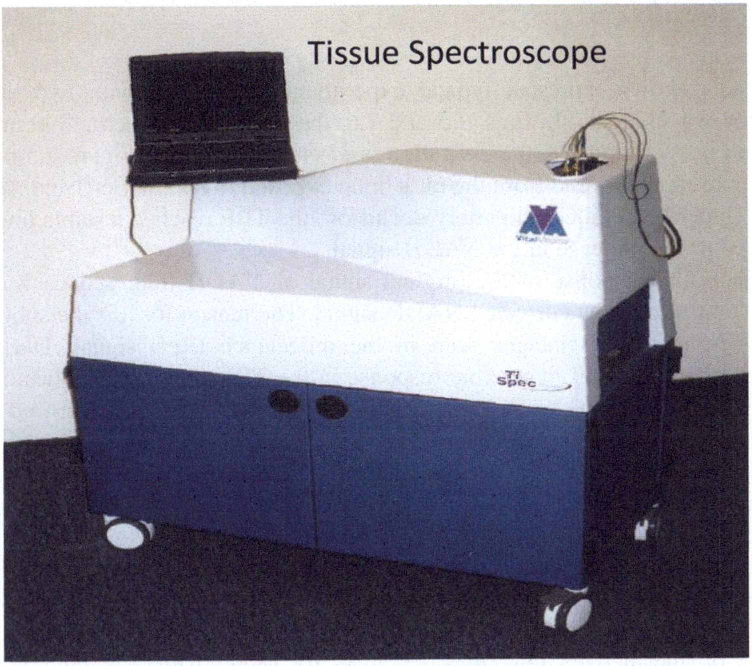

Fig. 7.4 The first Tissue Spectroscope (TiSpec1) developed commercially. ([Modified])

7.1.2.4 Surgical Preparation

The animals (rats, 180–220 g, or gerbils, 50–70 g) were anesthetized by an IP injection (0.3 ml/100 g) of Equithesin (each ml contains pentobarbital 9.72 mg, chloral hydrate 42.51 mg, magnesium sulfate 21.25 mg, propylene glycol 44.34% w/v, alcohol 11.5%, and water). We have been using this anesthetic agent for approximately 50 years, and it has never shown significant effects on mitochondrial activity. Furthermore, we have not been able to change the response of the rat brain to SD by using an increased dose of Equithesin, suggesting that this is a safe drug. The skull was exposed, and an appropriate hole was drilled in the parietal bone. The light guide holder was located on the brain so that undue pressure on the brain was avoided. The cannula and screws were fixed to the skull by dental acrylic cement in 2–4 different locations. In order to elicit a SD wave, a special push–pull cannula (2 mm diameter) was located epidurally, 2–3 mm from the monitoring site. If needed, additional injections of anesthetic were given to the rats or gerbils every 30 min during the operation (0.1 ml Equithesin/100 g body weight).

7.1.3 Results

The results shown represent typical experiments performed with two different excitation wavelengths. In Figs. 7.5 and 7.6, the TiSpec contained a 390 nm laser while in Figs. 7.7 and 7.8, the excitation laser emitted 325 nm. Figure 7.5 presents two responses, monitored from the rat's brain exposed to anoxia (left) and ischemia (right). Under bilateral carotid artery occlusion, the TBF reached a stable low level, correlated to the increase in the NADH signal.

As seen, the response of the original signal of NADH fluorescence was even higher compared to the corrected NADH signal. The reason for it is the subtraction of the hemodynamic changes seen in the reflectance (Ref) signal. During the recovery phase, a small hyperemic response in the TBF signal was noticed. When anoxia was induced (by exposing the rat to 100% N_2), the NADH showed a very similar response to that of the ischemia. The TBF presents an unstable decrease initially followed by large hyperemia. In Fig. 7.6, the gerbil was exposed to anoxia very similar to that of the rat brain (Fig. 7.5 left side). In this specific experiment, a spontaneous wave of SD was developed. As seen, the NADH signal shows a very large decrease (oxidation) followed by a significant increase in TBF. All signals returned to baseline levels within a few minutes.

The responses presented in Figs. 7.7 and 7.8 were monitored from the small intestine of anesthetized rats ($n = 13$, $n = 10$, respectively). In this study, we compared two different instruments for the two monitored parameters. The TBF measured by the TiSpec was compared to the LDF made by Perimed (Sweden). The NADH measured by the TiSpec was compared to the standard NADH fluorometer used in our laboratory for the last 25 years.

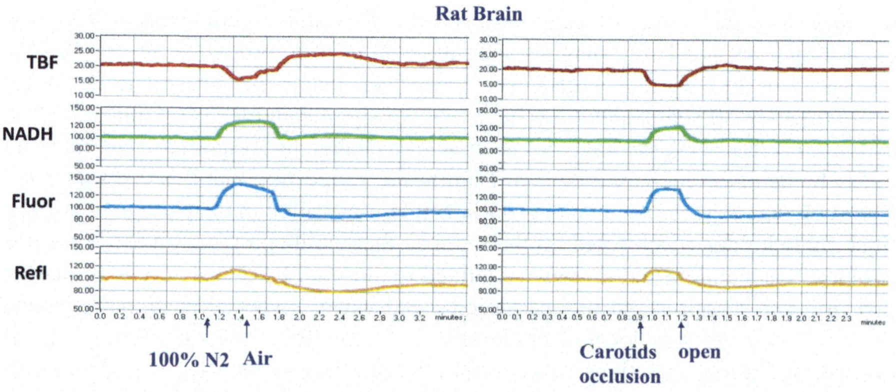

Fig. 7.5 Effect of anoxia (left) and ischemia (right) on the signals monitored from the rat brain by the Tissue Spectroscope. TBF Tissue Blood Flow, Ref Reflectance, F Fluorescence, NADH Corrected NADH redox state. (Mayevsky et al. 2002 [Modified])

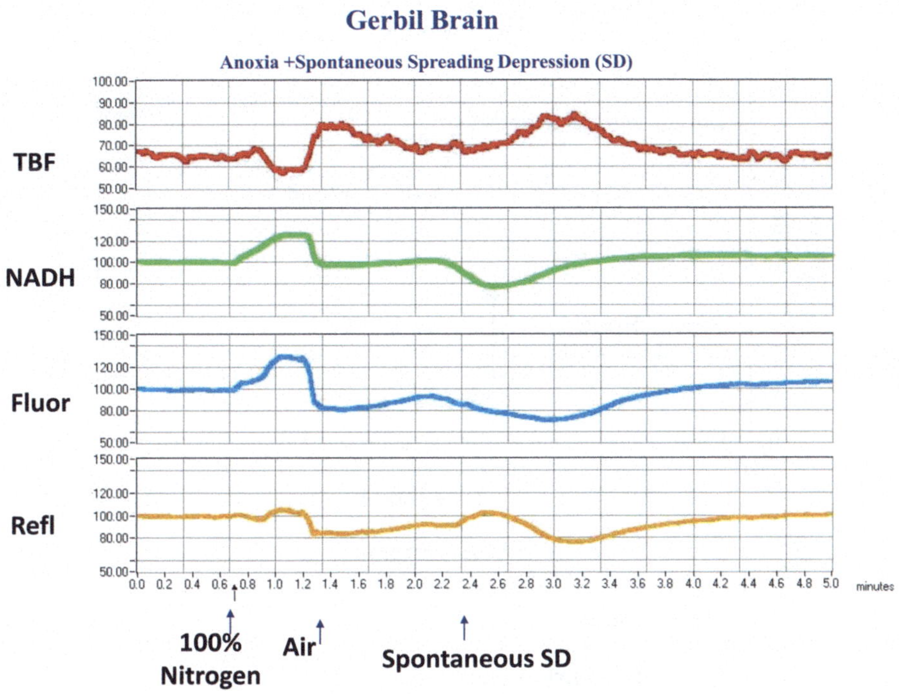

Fig. 7.6 Responses of the gerbil brain to anoxia. The spreading depression (SD) wave developed spontaneously due to the anoxia. Abbreviations are as in Fig. 7.5. (Mayevsky et al. 2002 [Modified])

As seen, a very good correlation was found between the three instruments used to monitor the four parameters. Under anoxia (Fig. 7.7), a clear increase was recorded together with the decrease in TBF. The effect of cardiac arrest shows an inverse decrease in TBF and a significant increase in the NADH redox state.

7.1.4 Discussion

The added value of combined monitoring of several parameters is significantly higher than that of single parameter devices. The ability to look at the processes of different tissue compartments simultaneously enhances diagnostic capabilities.

The TiSpec is indicated for *in vivo* monitoring of changes in mitochondrial activity (NADH redox state) and microvascular perfusion in tissue. Changes in these parameters provide information on tissue metabolic activity. The TiSpec can be used for quality control of various surgical procedures as well as for diagnostic assessment of tissues exposed to various pathological situations. Specific indications are as follows:

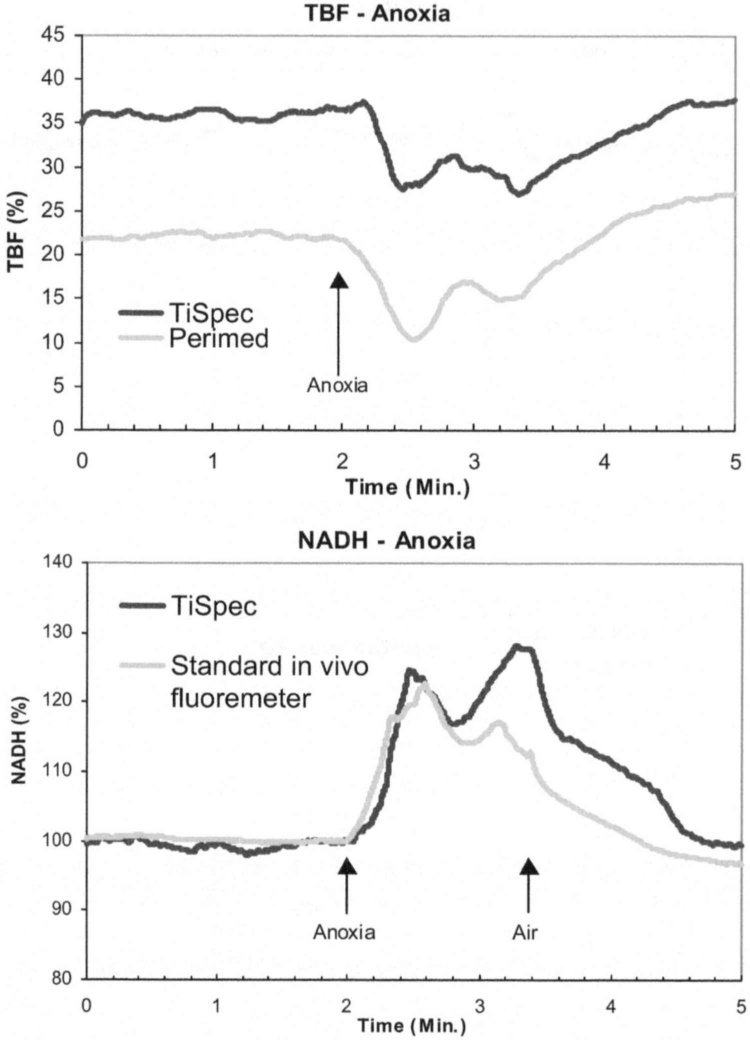

Fig. 7.7 Responses of the rat small intestine to anoxia. ($R_{TBF} = 0.756$, $R_{NADH} = 0.9$). (Mayevsky et al. 2002 [Modified])

- Assessment of organ metabolic state and activity before, during, and after the transplanation procedure.
- Evaluation of brain and spinal cord metabolic activity during and after the performance of various neurosurgical procedures.
- Monitoring of tissue viability in the cardiovascular system. This includes the evaluation of the Epicardium during bypass surgery and endocardial mapping before selection of the treatment.

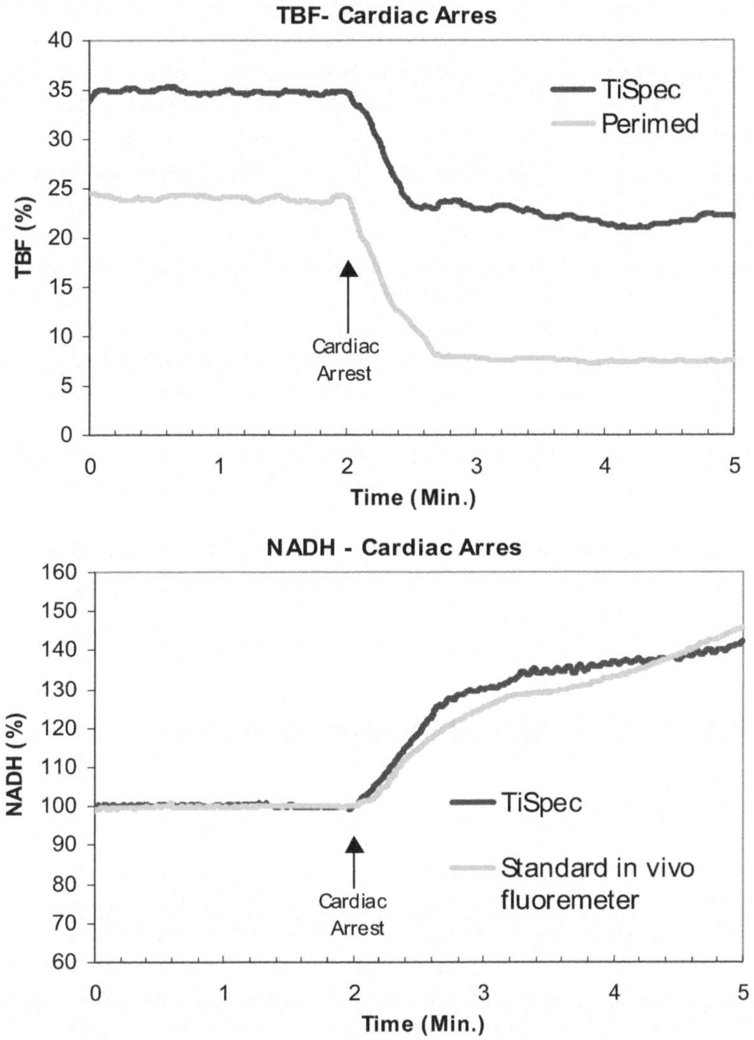

Fig. 7.8 Effect of cardiac arrest induced by IV injection of Equithesin solution on the small intestine responses. ($R_{TBF} = 0.892$, $R_{NADH} = 0.965$). (Mayevsky et al. 2002 [Modified])

- Assessment of metabolic activity in the wall of the Gastrointestinal (GI) anastomosis procedures.
- Continuous real-time assessment of brain metabolic state of patients hospitalized in Neuro-ICU after Brain Trauma.
- Monitoring of skin metabolic activity before and after various surgical procedures such as reconstructive surgery, burn injury treatment, or plastic surgery.
- Assessment of skin metabolic activity during delivery or in high-risk neonates treated in the neonate ICU.

As shown in Fig. 7.2, the qualitative correlation between NADH redox state monitored in the mitochondria *in vitro* and *in vivo* is clear. The NADH redox state in intact tissues under normoxia represents the so-called mitochondria at rest. In comparison to the *in vitro* "scale," the *in vivo* levels are midway between State 4 (resting) and State 3 (active). The reason for it is probably due to the difference in oxygen gradients toward mitochondrial activation under *in vitro* and *in vivo* conditions. In isolated mitochondria, a transition between State 4 and 5 induces a negligible increase in NADH (from 99% to 100%) while under *in vivo* conditions, the change is very large. Furthermore, the transition from State 4 to 3 *in vitro* is a very clear-cut step function while under *in vivo* conditions we believe that it is a continuous function dependent upon the turnover of the ATP utilization.

As mentioned before, it is necessary to combine the TBF monitoring to the NADH redox state in order to be able to analyze the metabolic state of the tissue *in vivo*.

Another reason is that the responses of various tissues or organs to the same perturbations are different due to central regulation and distribution of blood flow to various organs.

From the results obtained in this study, it is very clear that NADH could be excited by the "classical" Hg arc as a 366 nm light source or by 325 or 390 nm emitted from laser light source.

7.2 The First Commercial Monitoring System

7.2.1 Introduction

Monitoring of organ and tissue physiology in clinical practice include patients hospitalized in various types of operation rooms (ORs) as well as all types of ICUs. Figure 7.9 shows the two approaches toward tissue and body vitality monitoring of patients in clinical practice. In ORs and ICUs, a number of vital signs are monitored routinely as shown in part A of the figure. The monitored parameters represent the integrity and function of the cardiovascular and respiratory systems. The second approach of real-time monitoring is related to the vitality of a specific organ (B) in a patient exposed to severe operation or hospitalized in the neurosurgical ICU where intracranial pressure is monitored. It has to be noted that the number of devices available for specific organ monitoring is very small.

The purpose of this chapter is to describe the state of the art of the updated Tissue Spectroscope (TiSpec02) developed in our laboratory in order to provide a new medical device to be used in daily clinical practice. In our previous publication (Mayevsky et al. 2002), the principles of monitoring of tissue vitality were described in detail using the TiSpec. In this device, we had used a single light source—the He-Cd laser (325 nm) as an excitation light. This TiSpec01 has provided three different parameters that represent tissue vitality. The main disadvantage of the 325 nm-based TiSpec01 was the large dimensions of the device, which limits its

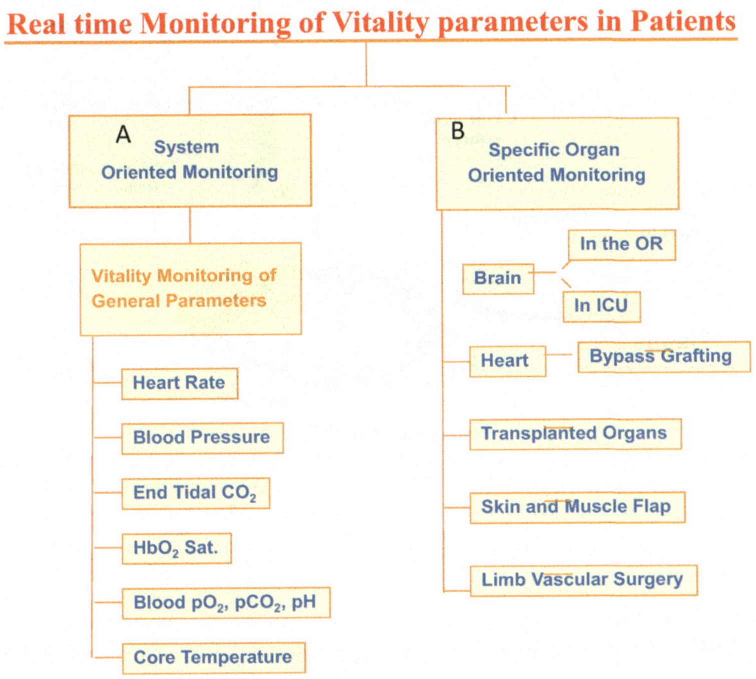

Fig. 7.9 General presentation of the two main approaches used in real-time monitoring of patients in the operating room or in the various intensive care units. (Pevzner et al. 2003 [Modified])

clinical practicability. The development of Ultraviolet (UV) laser diode by Nichia, Japan, enabled us to redesign the TiSpec into a significant smaller device. This study aims to describe the new TiSpec02 based on a laser diode emitting at 390 nm.

7.2.1.1 Monitoring of Tissue Vitality by the Tissue Spectroscope

In our previous publication (Mayevsky et al. 2002), the device called TiSpec was described. This TiSpec enables the *in vivo* monitoring of three physiological parameters: microcirculatory TBF by LDF, mitochondrial NADH redox state by measurement of NADH autofluorescence, and tissue blood volume by measurement of total back Ref. Those three parameters represent the main components of tissue O_2 balance under *in vivo* conditions. All three parameters were measured (in clinical studies) optically by single use sterile fiber optic probe. The system incorporates single excitation light source for measurement of all three parameters (Fig. 7.10). This light source for the TiSpec01 was built around a bulky He-Cd gas laser, which has several disadvantages as described below.

The physical dimensions of the laser, i.e., its length and the weight of the power supply emerge in a relatively bulky final product. The limited floor space of the operating room (OR) can hardly incorporate such bulky TiSpec; therefore, the smaller TiSpec device was highly desirable.

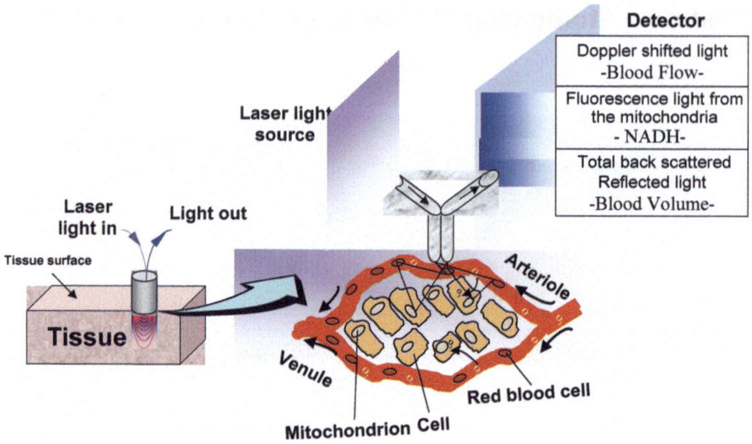

Fig. 7.10 The technical principles of the TiSpec. A single light source was used and the emitted light from the tissue is collected by an optical fiber to the detection system. (Pevzner et al. 2003 [Modified])

In order to perform LDF measurement, single transverse mode He-Cd laser was utilized. This He-Cd laser has 11 longitudinal modes spaced over the 3GHz gain profile. This multimode nature, along with non-negligible intensity noise (RIN ~0.5 to 1% RMS DC-20 KHz), deteriorates the LDF measurements. In order to enable the performance of reliable LDF measurements differential detection scheme was used (Nilsson et al. 1984). This differential detection imposes the use of two sets of collection fibers and two detection channels. This increases the complexity of the TiSpec device and complicates the construction of the single-use fiber optic probe. The preferred design will incorporate a single detection channel. The UV light intensity needed for reliable LDF measurements is close to the safety limit as defined by various laser radiation safety standards. The maximum permitted exposure (MPE) is 1 mW/cm^2 for UVA laser radiation from 315 nm to 400 nm (Ansi 1993). In order to reduce the mean irradiation intensity, the TiSpec utilizes a chopping excitation and synchronous detection scheme. It is generally known that the shorter UV wavelengths have more adverse influence on tissue than longer ones (Sliney 2000). At longer wavelengths, the responsively of the photodiode detector is higher; therefore, the excitation intensity for LDF measurements may be lower. Additionally, too high intensity of irradiation inside the absorption spectrum of NADH can be potentially problematic for fluorescence measurements as the photo-bleaching effect may be significant. Therefore, excitation at higher wavelengths and lower intensity is favorable.

Recently released to the market, UV-Violet laser diodes by Nichia (Nichia Chemical Industries Ltd., Anan, Japan) have emission wavelengths at UVA that coincide with the absorption spectrum of NADH. The light source based on laser diode may be much smaller in weight and dimensions than the He-Cd based system. Additionally, the noise characteristic of laser diodes is much better than He-Cd laser;

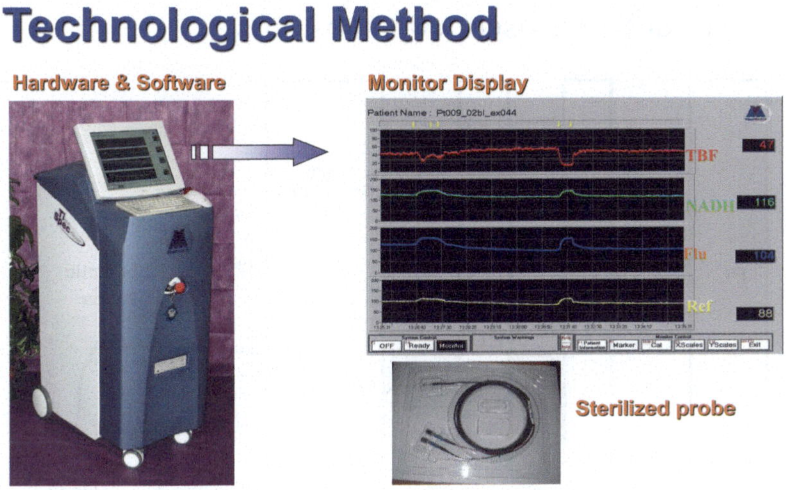

Fig. 7.11 The new TiSpec02. (Pevzner et al. 2003 [Modified])

therefore, single-channel detection system can be utilized. These considerations led us to design a new model of TiSpec device, called TiSpec02, shown in Fig. 7.11. This TiSpec02 device is based on the new laser diode light source made by Nichia. Here we describe the new TiSpec02 device and show some preliminary measurements performed in experimental animals.

7.2.2 *Methods*

7.2.2.1 Description of the TiSpec

Overview

The TiSpec02 device consists of several subunits as described in Fig. 7.12. The Light Source Unit (LSU) comprises the grating stabilized laser head DL100 from Toptica Photonics AG (Munich, Germany) with a Nichia laser diode. All necessary accessories for laser diode beam shaping and intensity modulation are included in the LSU. The light produced by the LSU is coupled to the excitation optical fiber of the fiber optic probe. The excitation and emission fibers are connected to the LSU and to the Detector Unit (DTU), respectively, by subminiature version (SMA) optical connectors. The excitation light is guided to the tissue through a multimode step-index silica fiber. The distal tip of the fiber optic probe is attached to the tissue by means of miscellaneous distal end attachments, as will be described later. A small fraction of the light that exits from inside the bulk of the tissue to its surface is collected by the collection fibers and guided to the DTU. Several detectors are used to measure the light intensity correlated to the three different monitored parameters

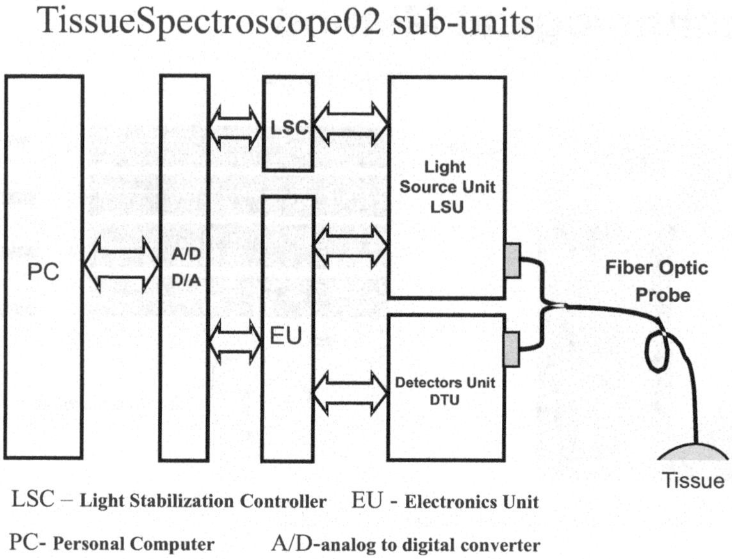

Fig. 7.12 TiSpec02 subunits. LSC Light Stabilization Controller, EU Electronics Unit, A/D Analog to Digital Converter, PC Personal Computer. (Pevzner et al. 2003 [Modified])

and convert the light intensity into electric signals. The Electronics Unit (EU) processes these electrical signals, and the output of the EU is relatively slowly changing voltages that represent the three measured parameters and several auxiliary parameters. These signals feed into the Analog-to-Digital (A/D) converter. The A/D card enables data acquisition and device control through the Personal Computer (PC). The collected data is stored and displayed on the screen by dedicated user-friendly software, which enables the user to control the entire operation of the TiSpec02 device. The Laser Stabilization Controller (LSC) provides the laser head, which is a part of the LSU, the correct laser diode current and temperature settings in order to operate in single longitudinal mode.

LSU and LSC

The schematic diagram of the LSU and LSC is shown in Fig. 7.13. The LDF measurements require a high coherence length or a narrow line-width and as low as possible amplitude noise. In laser diodes in general, there is a strong coupling between the amplitude noise and the frequency noise. The free-running Nichia UV violet laser diode emits several longitudinal modes that span over relatively broad linewidth of about 400 MHz. Mutual competition of these modes expresses relatively high-intensity noise. This problematic line-width may be dramatically reduced by utilizing the external cavity (Nakamura and Kaenders 1999). The laser head comprises the external cavity stabilized laser diode (ECLD) DL100 from Toptica

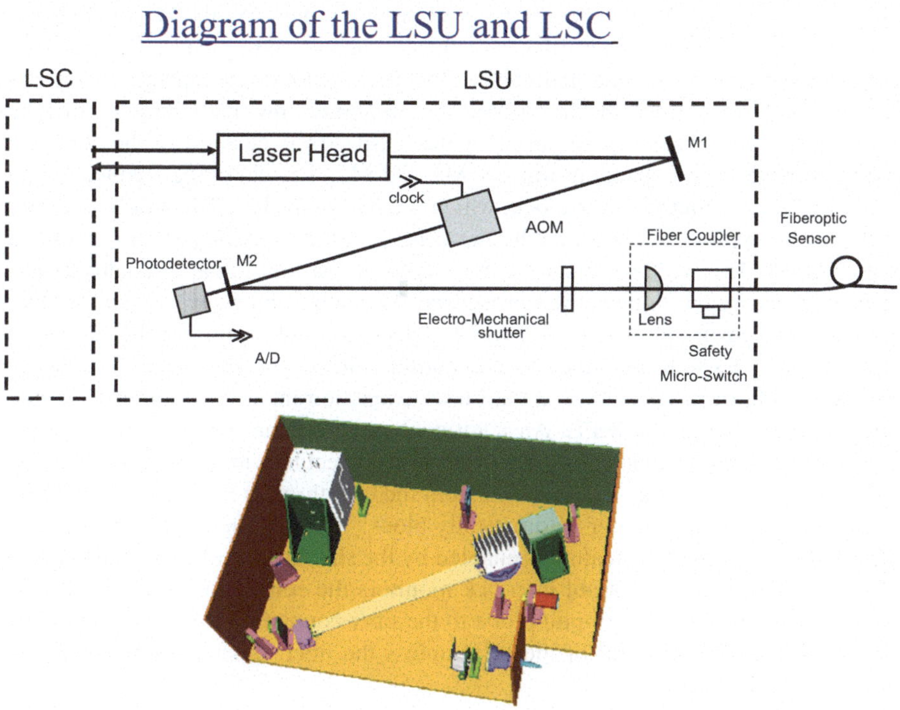

Fig. 7.13 Schematic diagram of the LSU and LSC. (Pevzner et al. 2003 [Modified])

Photonics AG (Munich, Germany). The light from the Nichia UV laser diode at a wavelength of about 390 nm is collimated by a lens toward the grating. The grating serves as the front mirror of the ECLD resonator, while the rear facet of the laser diode comprises the back mirror. The grating is used to feedback part of the light, after spectral filtration, back to the laser diode. The first order is back reflected to the laser while the zero order is coupled out. This kind of grating settings is called Littrow design (Arnold et al. 1998). The resulting output of DL100 with Nichia's UV-violet laser diode is the single longitudinal mode with a typical line-width of about 4 MHz. Such a narrow line-width is sufficient for LDF measurements. Also, when a laser diode operates in such a longitudinal single mode, the amplitude noise is also reduced. As a result of various instabilities, the system may drift after a while from the single longitudinal operation mode to some other adverse operation mode. The LSC system monitors the laser status and changes the laser current in order to ensure single longitudinal operation mode. The laser head also comprises the Peltier cooler and temperature stabilization controlling circuit (not shown on the scheme) that is operated by the temperature controller, which is a part of the LSC. The laser head operates in Continuous Wave (CW) mode in order to obtain the lowest possible noise characteristic. The chopping modulation is performed by means of an Acousto-Optic Modulator (AOM) (AA sa Opto-Electronic Division, St Remy les

Chevreuse, France). In order to get better chopping efficiency, the laser ray passes through lambda/2 polarization plane rotator and anamorphic prism pair prior (not shown on the scheme) to modulation by AOM. In order to ensure separation between the first and zero order of the AOM, the minimum ray path length must be established. The total small footprint of the TiSpec02 is ensured by bending the laser ray twice by two banding mirrors M1 and M2. The chopping frequency and duty cycle is determined by the clock, which is a part of the EU. The output intensity of the excitation light is controlled by the voltage of the clock pulse that is supplied to the AOM. This voltage is set by the software, via the digital-to-analog (D/A) converter, during the initialization procedure. The output intensity is monitored by the intensity sensing photodetector, which detects a small portion of the excitation intensity that passes through the second bending mirror. This photodetector enables the monitoring of momentary and total laser excitation intensity in order to ensure safe operation below the MPE. An additional safety feature incorporated into the LSU is the electro-mechanical shutter. The shutter precludes irradiation of the tissue unless an explicit command is received from the user through the software. In order to enhance safety, the shutter automatically blocks the excitation light when one of the interlocks opens. This feature is enabled by the shutter controller, which is a part of the EU. The most important interlock monitors the connection of the excitation SMA connector of the fiber optic probe to the fiber coupler of the LSU, precluding laser light from coming out of the LSU unless the fiber optic probe is connected firmly to the TiSpec02.

DTU and EU

The collecting optical fibers, at the DTU end of the fiber optic probe, are connected to the DTU by the SMA connector. The light from the fibers is collimated by a lens and passes through the dichroic beam-splitter, which separates backscattered and Doppler-shifted light at 390 nm from the NADH fluorescence light at 420–480 nm. Most of the NADH fluorescence light passes through the beam splitter while the backscattered and Doppler–shifted signals are reflected toward the photodiode detector, which converts it to a voltage. The output of the detector is fed into a synchronous detector, allowing for the detection of the chopped light as initialized by the light source unit. The output of the synchronous detector is a combination of DC and AC voltages that are similar to those that would be acquired from a regular Continuous Wave (CW) (non-chopped) excitation. The reflected light is a very slow-changing parameter represented by the DC voltage level at the amplifier output. The Doppler-shifted light signal produces rapid voltage level changes in the audio frequency range (from several Hz to a few kHz). Those changes appear as AC ripple on the relatively high DC level. Both signals appear at the output of the synchronous detector. The Doppler signals pass through an AC amplifier while the backscatter signals pass through a DC amplifier. After amplification, both signals are fed into the Doppler processor, and its output consists of two signals. The first signal is correlated to Doppler-shifted light that is related to microcirculatory TBF. The seconds signal is correlated to the backscattered light intensity that is related to Tissue Blood Volume. When leaving the Doppler processor both signal types are slow changing, with a time constant of 3 s. The AC and DC amplifiers and the Doppler processor are part of the EU. The light that passed through the dichroic beam splitter is focused on the active area of the photomultiplier (PMT) detector by a lens. Before reaching the PMT fluorescence detector, the light passes through a band-pass filter that transfers only NADH fluorescence wavelengths (420–480 nm). The fluorescence detector signal, after amplification, is transferred to the A/D. The amplifier also works as a synchronous detector by accepting the timing signal from the clock. The Fluorometer controller, which is a part of the EU, performs the set of PMT gain. The software starts the initialization of the PMT gain setting by DIO of the A/D card. When leaving the EU, all signal types are slow changing, with a time constant of 3 s. The signals at the output of the EU are forwarded to the TiSpec02 software through the A/D.

A/D Converter, PC, and Software

The analog signals undergo A/D conversion in order to be processed on the digital computer. TiSpec02 utilizes DAQPad-1200 from National Instruments (Austin, TX, USA). This standard multichannel A/D converter also enables several D/A outputs. The device also has several Digital Inputs–Outputs (DIO) that enable switching between ON/OFF states. This feature is useful for control of various TiSpec02 subunits, such as the detectors, amplifiers, and electro-mechanical shutter. Since all three TiSpec02 signals, i.e., Doppler, NADH fluorescence, and backscatter are slow-changing signals, the sampling rate for these channels is as slow as 10 Hz. The A/D is connected to the parallel port of the Pentium II computer running Win2000 operating system. The TiSpec02a device is controlled by dedicated software—the TiSpec Operating Software (TiSpecOS). All of the TiSpec02 functions, such as laser operation, measurement initialization, and system calibration, are controlled through software. The acquired data is automatically displayed on the screen and stored on the hard disk for further data analysis. The TiSpecOS program is based on LabView development environment (National Instruments, Austin, TX, USA).

The Probes

Fiber optic probe is required in order to enable excitation light and optical signal transmission between the TiSpec and the monitored patient. Since the TiSpec is used mainly in OR and ICUs, the probe cable should be long enough to enable the device to be located relatively remotely from the patient, minimizing the disturbance to the routine medical practice. The probe is based on a bundle of optical fibers, divided into two optical channels: the excitation channel and the collection channel. In the probe tip, all fibers are adjunct, while in the machine-end, the bundle is split into separate excitation channel and collection channel. The excitation channel is

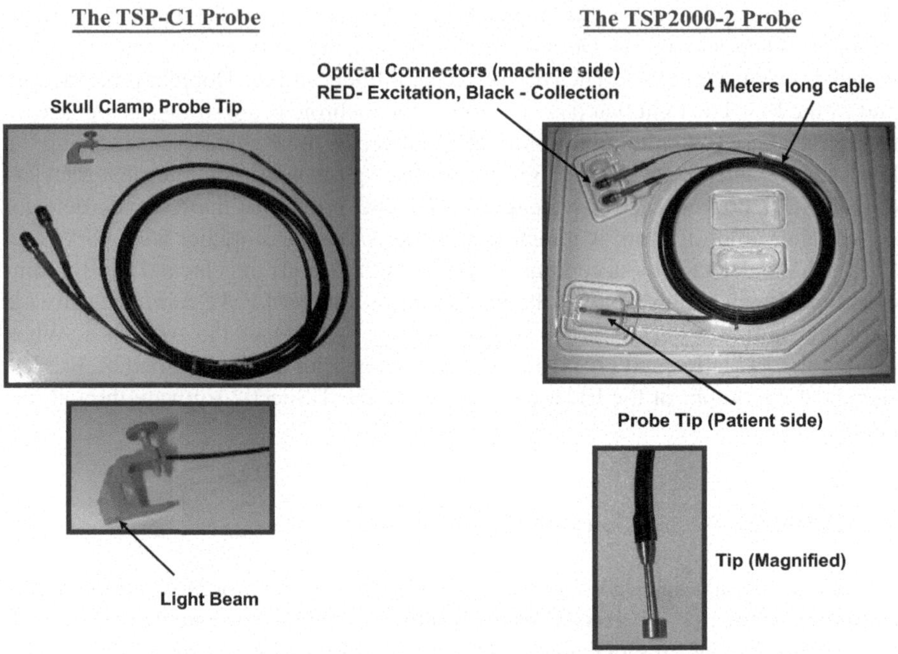

Fig. 7.14 The TSP2000-2 probe for brain cortex monitoring. (Pevzner et al. 2003 [Modified])

connected to the TiSpec light source by an optical SMA connector, and the collection channel is connected to the device's detection system (the DTU) by another SMA connector. The probe is equipped with a special tip, carefully designed to obtain continuous contact with the monitored tissue. A reliable contact with the tissue is crucial; however, no extensive force is allowed, since the physiological situation is very sensitive to applied pressure. Two probe types were developed: the TSP2000-2 and TSP-NB1. The TSP2000-2 probe was designed to monitor exposed brain cortex in various locations under the craniotomy during neurosurgical procedures. The tip is formed from a thin metal tube, which accommodates the bundled optical fibers. A 5 mm OD cylindrical element is attached to the end of the metal tube, in order to enlarge the footprint of the probe and enable the probe to float upon the cortex (Fig. 7.14 right side). Since the human brain is not static during the surgery, an adjustable floating arm was designed in order to track the restless brain tissue. The basis of the floating arm is firmly connected to a Yasargil arm (Fig. 7.15), which is a common spatial positioning device in neurosurgery. Yet, the floating arm exploits the inherent flexibility of the probe cable to produce spring-like movements of the tip, applying slight thus harmless pressure upon the brain cortex. This mechanism provides the required continuous contact of the probe and the tissue (Fig. 7.15).

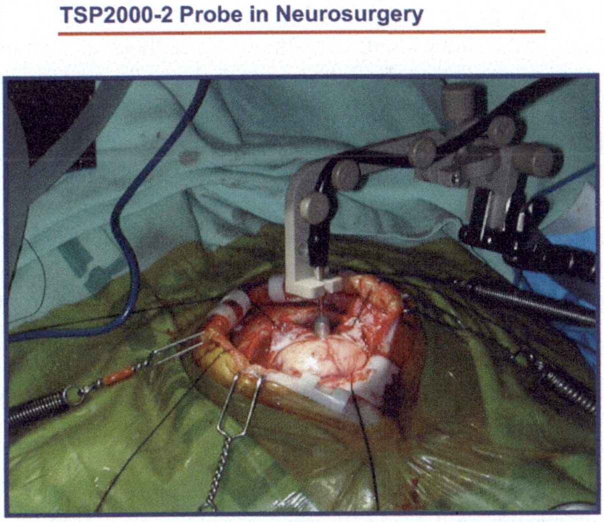

Fig. 7.15 The TSP2000-2 in a neurosurgical procedure. The probe is attached to a floating arm mounted on Yasargil positioning arm. (Pevzner et al. 2003 [Modified])

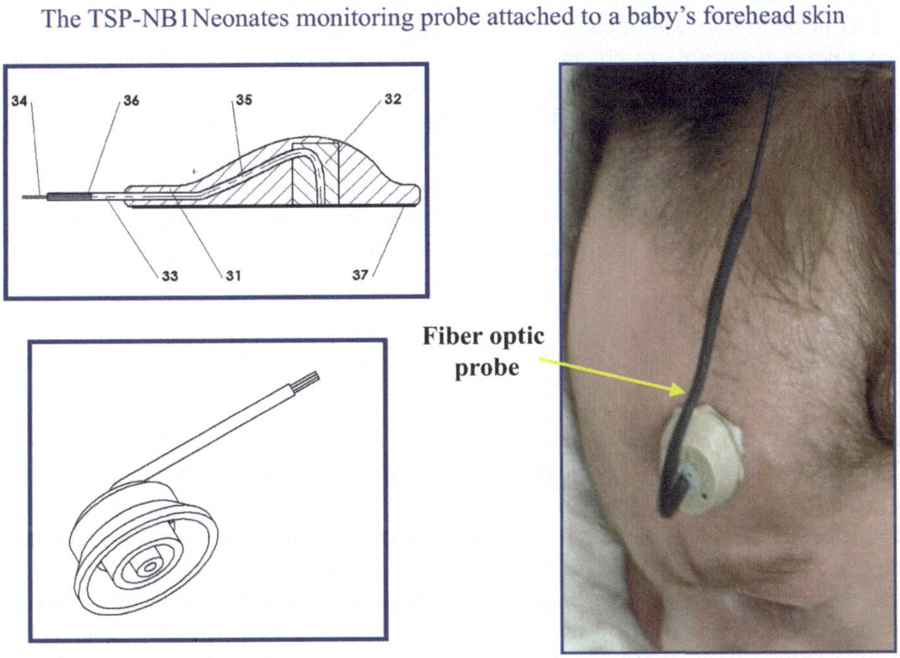

Fig. 7.16 The TSP-NB1 Neonates monitoring probe (Left) attached to a baby's forehead skin (right). (Pevzner et al. 2003 [Modified])

The TSP-NB1 probe (Fig. 7.16) was designed to monitor neonates hospitalized in ICU. The probe was designed to be attached to the neonates' skin, since their skin is very thin and lack epidermis. The extremely thin neonatal skin enables, on the one hand, the penetration of the highly absorbed UV light to labile tissues under the skin, but, on the other hand, bans the use of adhesives or medical adhesive tapes, due to aggressive peeling. In order to attach the probe to the neonatal skin, yet not exposing the delicate skin to the dangerous effect of peeling an adhesive, a button shaped probe was designed. The probe features a low-profile cylindrical tip, in which the fiber optic bundle is 90° bent within. The tip has two round grooves on its face, concentric with the fiber optic bundle. The outermost groove is to be overfilled with an EEG/ECG gel, then pressed upon the skin. The gel sticks the probe to the skin while excessive gel is forced out of the probe or inside into the innermost groove that serves as a moat. The moat keeps the optical fibers from being screened with gel.

7.2.2.2 Surgical Preparation for Animal Experiments

The animals (gerbils, 50–70 g) were anesthetized by an IP injection (0.3 ml/100 g) of Equithesin (each ml contains: pentobarbital 9.72 mg, chloral hydrate 42.1 mg, magnesium sulfate 21.25 mg, propylene glycol 44.34% w/v, alcohol 11.5%, and water). We have been using this anesthetizing agent for approximately 25 years, and it has never shown significant effects on mitochondrial activity. Furthermore, we have never been able to change the response of the rat brain to SD by using an increased dose of Equithesin, suggesting that this is a safe drug. The skull was exposed, and an appropriate hole was drilled in the parietal bone. The light guide holder was located on the brain so that undue pressure on the brain is avoided. The cannula and screws were fixed to the skull by dental acrylic cement in 2–4 different locations. If needed, additional injections of anesthetic were given to the gerbils every 30 min during the operation (0.1 ml Equithesin/100 g body weight).

7.2.3 Results

In order to test the performance of the new TiSpec, in which the excitation light is 390 nm, animal studies in rats and gerbils were performed. In gerbils we have tested the effect of anoxia and ischemia on brain energy metabolism. In rats we have tested the effect of anoxia, ischemia or adrenaline injection on the three monitored parameters from the small intestine, skeletal muscle or the internal side of the skin covering the head (result not shown). Figure 7.17 shows typical response of the brain to ischemia induced in a gerbil by occlusion of the two carotid arteries. As seen, the cerebral blood flow (CBF) was decreased initially to a very low level followed by a small spontaneous recovery during the occlusion period. The decrease in blood flow led to decrease in O_2 supply to the brain as seen in the large increase in the mitochondrial NADH. The net change in NADH is calculated by the correction of the fluorescence for the hemodynamic artifacts measured as seen in the Ref trace. During the ischemia, the Ref was elevated due to a decrease in blood volume, followed by slow recovery to the baseline toward the end of the occlusion. The

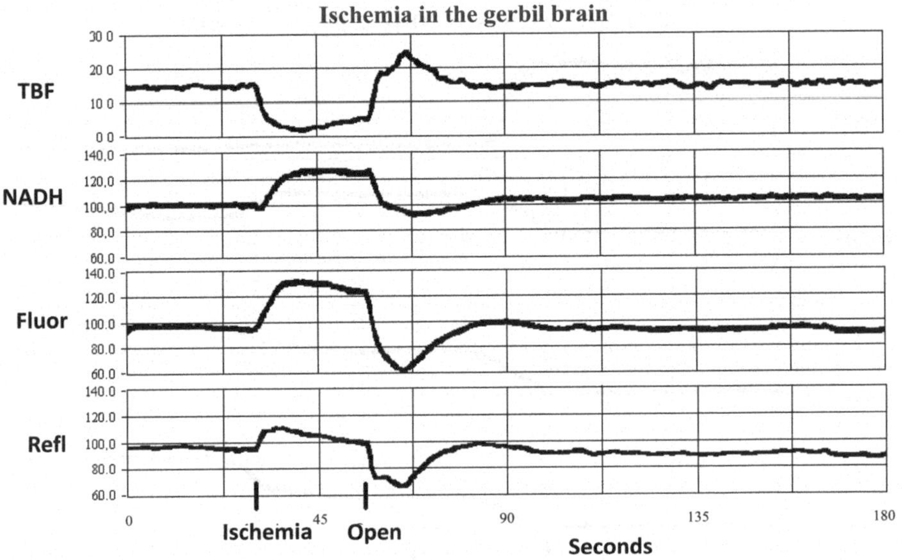

Fig. 7.17 Effects of ischemia of brain energy metabolism in the gerbil. TBF Tissue Blood Flow, F Fluorescence, Ref Reflectance. (Pevzner et al. 2003 [Modified])

recovery processes recorded after the reopening of the two carotid arteries were very fast, as seen by the increase in CBF, having also a hyperemic phase. The NADH was reoxidized, showing also a small undershoot immediately after the reperfusion, probably due to the large amount of blood releasing more O_2 to the tissue. The complete recovery of all traces to its baseline levels was recorded 2 min after the reperfusion. In Fig. 7.18, the responses to terminal anoxia, induced by breathing 100% N_2, are shown. The initial CBF change was small in parallel to the increase in NADH, followed by a fast decrease to the minimal level of flow. About 2 min after breathing N_2, a large increase in the Ref trace was noted at the stage where the blood flow was minimal and the NADH was stable at its maximal level. This phenomenon of spontaneous reflectance increase (SRI) was noted many years ago in rat studies and is probably due to a massive increase in the extracellular level of potassium, leading to a large vasoconstrictive effect (Mayevsky 1984, 1990). Due to this SRI event, a large decrease in NADH was recorded (in the absence of O_2) and it is due to the correction procedure that cannot take care of the large change in tissue absorption properties and not by blood volume change alone. This decrease in NADH was recovered back to the maximal level recorded previous to this artifact. In Fig. 7.19, the two episodes shown were recorded from the same gerbil brain. Here, we have compared the responses of the brain to ischemia to those induced by anoxia. On the left side, a typical response to carotid artery occlusion in the gerbil is shown. The large increase in the Ref is clear, but the fluorescence change is larger during the ischemic episode, as seen clearly in the CBF trace. During anoxia, shown on the

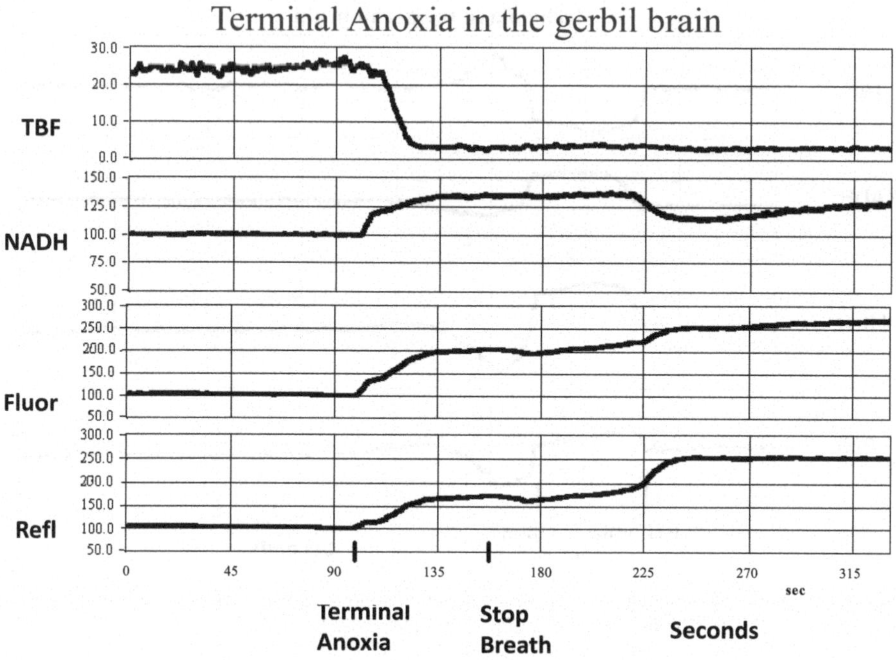

Fig. 7.18 Responses of the gerbil brain to terminal anoxia induced by exposing the animal to 100% N2. Abbreviations are as in Fig. 7.16. (Pevzner et al. 2003 [Modified])

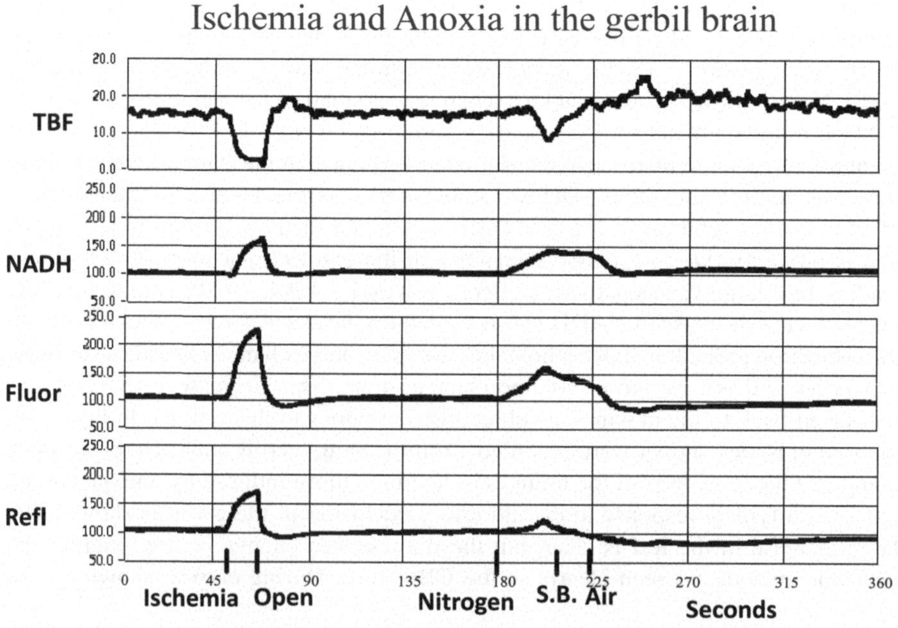

Fig. 7.19 Responses of the gerbil brain to ischemia (left side) and anoxia (right side). Abbreviations are as in Fig. 7.16. (Pevzner et al. 2003 [Modified])

right side, the NADH reached its maximal level while the CBF showed a clear decrease. Upon recovery, a clear hyperemic response is shown in the CBF trace although the NADH was stable in its baseline level.

7.2.4 Discussion

The TiSpec described in the present article has small dimensions that could be used in ORs and ICUs which are limited in free space for new medical devices. The search for reliable medical device that could be used in real time is an ongoing process and became more critical due to the high mortality rate recorded in the various ICUs around the world. In addition to the monitoring approaches shown in Fig. 7.20, a new principle of patient monitoring was developed in the last years. This new approach is seeking to identify an early warning signal that will be monitored in real time and will be detected significantly earlier than the existing monitoring devices. Under conditions where the BEMS is developed, the body responds by the activation of homeostasis mechanisms dominated by the sympathetic nervous system (Ganong 1991a). The immediate response to such a situation is blood flow

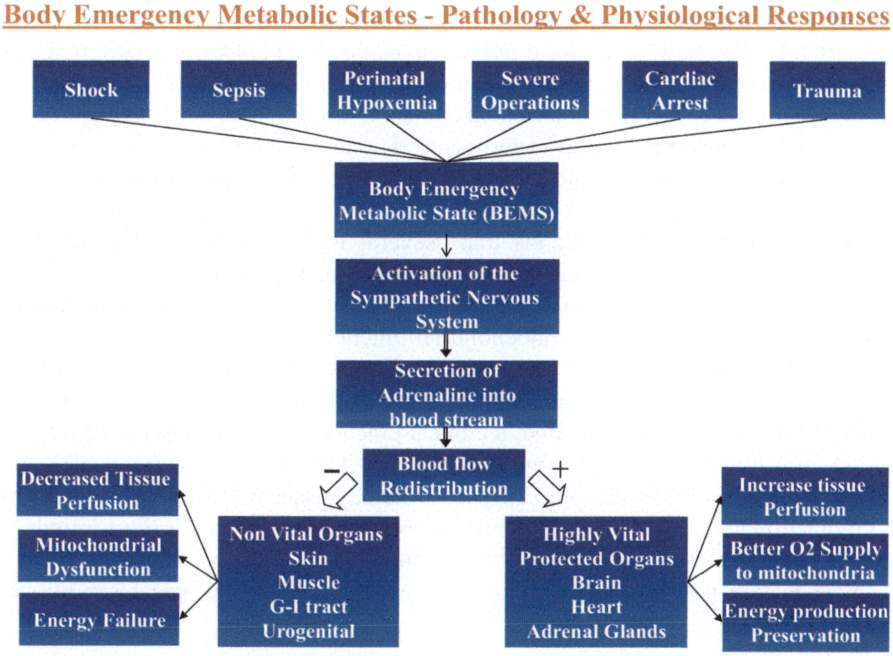

Fig. 7.20 Various clinical pathological conditions (e.g., shock, sepsis) that lead to the development of Body Emergency Metabolic State—BEMS in critically ill patients (see text for detailed explanation). (Pevzner et al. 2003 [Modified])

redistribution between more vital organs and less vital organs. The vital organ will receive more blood and the better oxygenation will ensure continuous supply of energy. The perfusion to the non-vital organs will decrease, and as a result, O_2 supply to the mitochondria will diminish significantly. Ischemia will be detected in the non-vital organs, and energy failure will occur. It is assumed that these changes in the non-vital organs could be used as an early warning signal for the development of BEMS. The assumption is that the standard monitored systemic parameters such as systemic blood pressure, heart rate, or even the pulse oximeter will respond to the BEMS in a late stage; therefore, monitoring of non-vital organ is the solution for such situation. At present, we were able to test the TiSpec02 in neurosurgical OR. The dimensions of the TiSpec02 were acceptable in the OR, and the routine activity of the medical team was not disturbed.

7.3 Tissue Spectroscope—A Novel in Vivo Approach to Monitor Tissue Vitality in Real Time

7.3.1 Introduction

The monitoring of tissue or organ vitality in real time is a critical factor in the process of surgery as well as in preoperative and ICU management.

Currently, the main monitoring devices available for clinical use are aiming to provide hemodynamic and respiratory parameters (Boldt 2002). Local monitoring of tissue parameters is very rare or absent in clinical routine, and most of the monitoring effort has been made in animal experiments. The multiple organ dysfunction syndrome accounts for most deaths in the ICU (Boldt 2002), and inadequate tissue perfusion and oxygenation are likely to contribute to the development of organ failures. Also, Fink (2001) claims that "several lines of evidence indicate that cellular energetics are deranged in sepsis, not by inadequate tissue perfusion but rather by impaired mitochondrial respiration." Therefore, monitoring of parameters indicating tissue perfusion and mitochondrial function is necessary for the improvement of patient outcome.

Presently, monitoring of mitochondrial function by monitoring NADH fluorescence *in vivo* has not been possible in clinical practice due to the absence of a Food and Drug Administration (FDA) approved device. The main aim of this book is the introduction of a new device—the TiSpec—whereby tissue perfusion and mitochondrial function are monitored in real time. Before going into detail, a short introduction to tissue energy metabolism will be given.

Note that in this chapter, we examine several pathophysiological conditions, defined as follows:

1. Ischemia is a decrease in blood supply to the tissue due to obstruction or constriction of a blood vessel.

2. Hypoxia is a decrease in the level of oxygen in the breathing mixture to which the animal is exposed.
3. Anoxia is a reduction in the level of oxygen to 0 by exposing the animal to 100% N_2.
4. Hypoxemia is a decrease in the amount of oxygen in the arterial blood.

7.3.1.1 Tissue Energy Metabolism

The functional capacity of a tissue is related to its ability to perform its work. It is possible to assess this ability through the knowledge of changes in the oxygen balance, i.e., the ratio of oxygen supply to oxygen demand in the tissue.

Let us first briefly describe how cells are able to do work (Fig. 7.21). Healthy tissue cells perform various types of work, such as muscle contraction, secretion, the nervous system activity, and the synthesis of cell constituents. The energy needed to perform these types of work is derived from the degradation of foodstuffs, primarily through biological oxidation reactions, mediated by a respiratory chain comprised of several complex enzyme systems, in which oxygen is the ultimate electron acceptor. The electron transfers (oxidations/reductions) down the respiratory chain result in the production of energy-rich pyrophosphate bonds of adenosine triphosphate

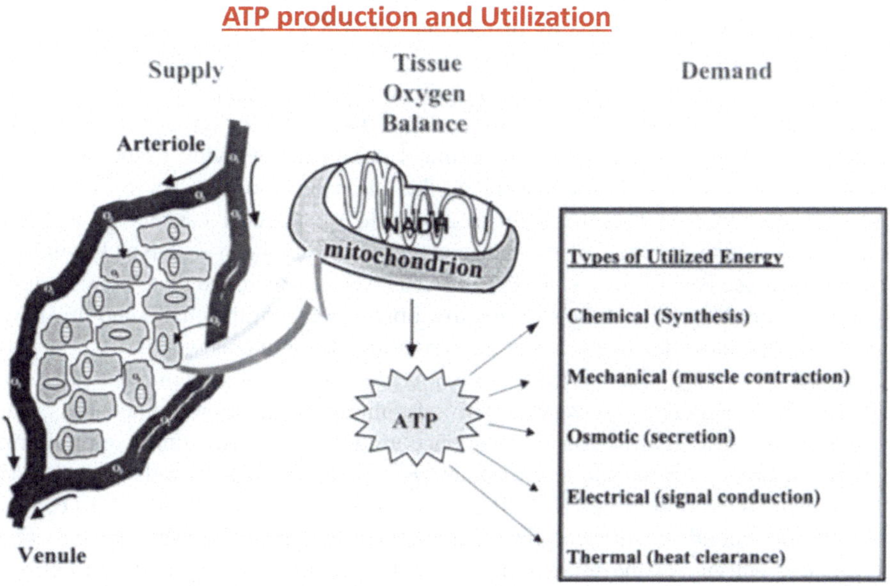

Fig. 7.21 Schematic representation of the way cells are able to do work. The blood supply is the source for the substrate and O_2, providing the energy needs of the tissue. There exists a tissue oxygen balance, which can be defined as a ratio between the oxygen supply and demand. Every tissue may utilize the energy for various types of activity (the demand). Most of the ATP is synthesized in the mitochondria in oxidative phosphorylation. (Mayevsky et al. 2004a [Modified])

(ATP). Concomitantly with the electron transport, the respiratory chain components switch between reduced and oxidized states, each of which has different spectroscopic properties. Hydrolysis of the pyrophosphate bonds provides the energy necessary for the cell's work.

The formation of the pyrophosphate bonds depends on the sufficiency of sugar and oxygen supply to the tissue by the bloodstream. Without a sufficient supply, cells cannot function properly and can, ultimately, die. Since most of the energy consumed by tissues is dependent on the availability of oxygen, the terms "energy" and "oxygen" are used here synonymously.

In a normal healthy tissue, the ratio—or balance—between oxygen supply and oxygen demand is positive and reflects the cell's functional capacity to do work. That is, the supply mechanism (oxygenated blood circulation) is able to provide the amount of O_2 needed for the demand mechanisms (ATP production for various types of work).

The TiSpec monitors the parameters related to oxygen (energy) supply and oxygen (energy) demand.

1. Oxygenated blood availability represents the supply aspect, and its measurement relies on the specific optical properties of static and moving blood cells.
2. On the part of demand, the spectroscopic properties of the respiratory chain components, unique to their redox status, are used as internal markers of the state of oxidative phosphorylation.

The intracellular concentration of mitochondrial NADH (the reduced form) is a parameter related to the tissue's state of energy metabolism. Energy exchange depends on pO_2. Information regarding pO_2 in the tissue, therefore, is helpful for the evaluation of the tissue metabolic activity. The need for an intracellular pO_2 indicator, as a physiological and biochemical parameter of living tissue, emerged more than 50 years ago. Mitochondria are the intracellular organelles that consume most of the oxygen. Therefore, the redox state of electron carriers in isolated mitochondria as a function of oxygen concentration has been extensively studied (for a review see Ref. Chance et al. 1973). Chance et al. concluded that "For a system at equilibrium, NADH is at the extreme low potential end of the chain, and this may be the oxygen indicator of choice in mitochondria and tissue as well."

In excitable tissues, such as the brain or muscle tissue, as well as in other cells, the activity of Na-K-ATPase is very sensitive to alterations in ionic homeostasis. An increase in extracellular potassium ion concentration, $[K^+]$, will stimulate pumping activity in order to bring the extracellular $[K^+]$ back to normal levels (i.e., 3 mM range). The activation of Na-K-ATPase will increase the hydrolysis of ATP, and thus the mitochondria will phosphorylate the ADP molecules that are released. The accelerated activity of the mitochondria will be accompanied by a more oxidized state of the respiratory chain components, as well as by an increase in oxygen consumption.

To deliver more oxygen to the cells, blood flow and blood volume will be increased, and thus the extra oxygen needed by the tissue will be provided. This coupling between energy consumption and energy production is maintained as long as the O_2 supply is well regulated.

Under conditions where the oxygen supply or delivery is limited, e.g., after a stroke or heart attack, the energy supplier, i.e., the mitochondria, will not be able to produce the amount of ATP needed. As a result, energy-demanding processes will be restricted. The net effect of the imbalance between energy demand and supply will be manifested by a decrease in the tissue's ability to do work. This can lead to the development of various pathological deviations.

7.3.1.2 Real-Time Monitoring of Tissue Vitality

The monitoring of various tissue parameters is among the tools employed by clinicians to improve diagnosis. The major techniques developed for real-time analysis of tissue vitality are shown in Fig. 7.22. Their classification is based on the extent of the technology/probe invasiveness in the tested tissue.

Noninvasive Techniques

To evaluate the tissue hemodynamic and metabolic activity noninvasively and in real time, LDF can be utilized. LDF provides information correlated to tissue microcirculatory blood flow (TBF) (Wardell et al. 1994; Clough et al. 1998). Another noninvasive approach is the use of a pulse oximeter that provides continuous data on hemoglobin saturation in the arteries of the monitored area (e.g., a fingertip). The development of the near-IR technology has opened up the possibilities of monitoring the saturation of hemoglobin and even the oxidation-reduction state of the mitochondria (Jobsis 1977; Villringer et al. 1993; Villringer and Chance 1997).

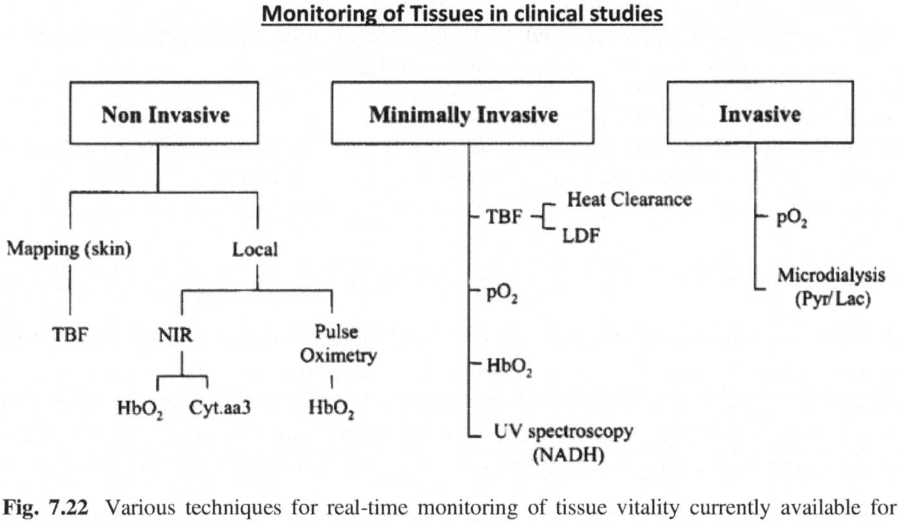

Fig. 7.22 Various techniques for real-time monitoring of tissue vitality currently available for clinical studies, classified according to their level of invasiveness (see text for details). (Mayevsky et al. 2004a [Modified])

Minimally Invasive Techniques

Our definition of this approach is that the monitoring probe has to be in contact with the examined tissue. This group of techniques includes the following four devices, as shown in Fig. 7.22.

Tissue Blood Flow (TBF)

To evaluate microcirculatory blood flow, at least two approaches have been developed and clinically applied. Heat clearance is the oldest one and requires heating of the sampled tissue to measure the heat clearance thereafter. The advantage of this technology is the ability to calibrate the flow in absolute units. However, the heating procedure is the main disadvantage of this approach. The predominant technology employed for monitoring the microcirculatory blood flow is LDF, which enables continuous monitoring without the need for tissue preheating. Its main drawback, according to most users, is that relative values are obtained instead of absolute values (Dirnagl et al. 1989; Haberl et al. 1989).

Oxygen Partial Pressure (pO_2)

Oxygen-sensitive electrodes have been used since the early 1960s and even earlier, in various *in vitro* as well as *in vivo* situations. Recently, continual monitoring of brain pO_2 has been implemented and a large number of clinical reports have been published (Dings et al. 1998). Monitoring of pO_2 from the tissue surface (for instance, during a surgical procedure) may also be affected by the surrounding atmosphere.

Saturation of Hemoglobin

In order to determine the level of hemoglobin oxygenation in the circulating blood, a methodology was developed based on the absorption spectra of oxy and deoxy hemoglobin (Frank et al. 1989).

Another approach, described by Rampil et al. (1992), is based on two narrow wavelength windows that are used in a time-sharing reflectometer device. By this technique, only relative oxygenation values are monitored.

Mitochondrial NADH Redox State

The conceptual foundations for mitochondrial NADH fluorometry were established (Chance and Williams 1955b) and published in the early 1950s.

Invasive Techniques

According to our definition, this group of techniques involves penetration of the monitoring probe into the tissue. The main disadvantage of this approach is the damage to the integrity of the monitored tissue. After the probe penetration, a new microenvironment is developed in the monitored area, and one must take this point into consideration (Dings et al. 1998).

7.3.1.3 Aims of this Study

The aims of this study can be summarized as follows.

1. We aim to describe the scientific and technological principles of the "TiSpec," whereby microcirculatory blood flow, tissue Ref, and mitochondrial NADH redox state are measured by means of a single light source.
2. We intend to review the typical published papers on the monitored parameters under *in vitro* and *in vivo* conditions.
3. Since the TiSpec provides information on three different physiological and biochemical parameters, it was necessary to compare it to various existing instruments; therefore, this aim was divided into three stages, as follows.

   A. We aim to demonstrate the ability to measure NADH fluorescence (F) signal comparable to a clinical fluorometer, by comparing measurements of aqueous NADH solutions at various concentrations by the TiSpec versus the Victor^2 1420 Fluorometer (Fink 2001).
   B. We intend to demonstrate the correlation between the TiSpec and the Victor^2 1420 Fluorometer with respect to NADH fluorescence measurements in solution.
   C. We compare, in a short pilot study, the signals monitored *in vivo* by the TiSpec to those of standard *in vivo* monitoring devices by which TBF, NADH fluorescence, and Ref are measured.

7.3.2 *TiSpec*

7.3.2.1 Principles of Monitoring

The TiSpec is an innovative monitoring device, developed by a commercial company, Vital-Medical Ltd., based on extensive long-term studies conducted by A. Mayevsky over the past 50 years. It monitors continuous, real-time parameters that provide information related to the tissue's ability to perform its work. The principles of the TiSpec measurements and their relation to the tissue's ability to perform its function are as follows:

The TiSpec uses a single light source to illuminate the tissue with UV excitation light, and detects light that is subsequently emitted from that tissue (Fig. 7.23). Light is conducted to and collected from the tissue by flexible optical fibers, combined in a single probe. The light emitted from the tissue contains three different components analyzed by the TiSpec:

1. Doppler-shifted light at wavelengths very close to the excitation wavelength. This light is scattered by the moving red blood cells inside small blood vessels (small arterioles, venules, and capillaries). Changes in this component are correlated with changes in the microcirculatory TBF.

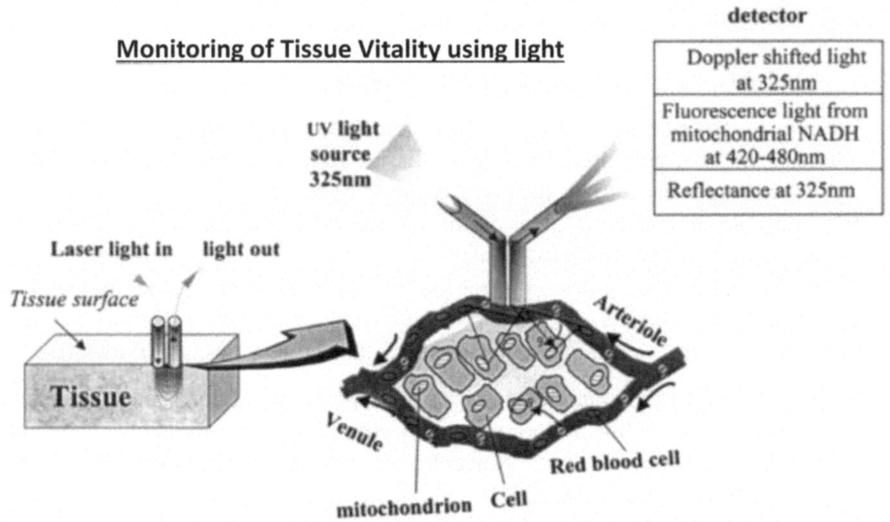

Fig. 7.23 Illustration of the experimental principles of the TiSpec simultaneous monitoring of TBF, tissue reflectance, and mitochondrial NADH redox status using the same light source and the same optic fibers. (Mayevsky et al. 2004a [Modified])

2. Fluorescence light (420–480 nm) emitted from excited intracellular NADH molecules. Changes in the F signal intensity are correlated with changes in the intramitochondrial NADH redox state (Flu).
3. Reflected light (Ref) at exactly the same wavelength as the excitation light. This is also known as total backscatter light. Changes in Ref are correlated mainly with blood volume changes in the tissue (Jobsis et al. 1971a; Rosenthal and Jobsis 1971; Mayevsky and Chance 1973, 1982; Dora 1984; Dora et al. 1984; Mayevsky 1984; Rampil et al. 1992).

Changes in TBF, blood volume, and NADH redox state bring about changes in the emitted light signals. The principles and significance of the three different monitored parameters are described in the following.

Tissue Blood Flow—TBF

The principle behind monitoring TBF by LDF is that coherent light from a laser source is delivered to the tissue under investigation by optical fibers. Photons that enter biological tissue undergo random scattering both by stationary tissue and by moving blood cells. Light that was scattered by moving blood cells undergoes a frequency shift due to the Doppler effect. A portion of scattered photons, both Doppler shifted and unshifted, is collected by the optic fibers and transferred to the detector. The Doppler-shifted light is superimposed with unshifted light on the photodetector. The photodetector produces DC levels related to the total backscattered light (see later in this chapter), while the AC ripple originates from the Doppler signal (Shepherd and Oberg 1990). To deduce the blood cell flux from the power spectra, Bonner and Nossal (1981) proposed an algorithm based on the theoretical model of Doppler scattering.

Continuous real-time monitoring of TBF became possible after the development of the helium-neon (He-Ne) laser in 1961 by Javan et al. (1961). Following a few years of fluid flow measurements (Yeh and Cummins 1964), the laser-Doppler velocimeter (LDV) was developed for blood flow measurements (Riva et al. 1972). At the early stage, the LDV was used for single vessel monitoring, and only in 1974 was Stern able to suggest the application of LDV for tissue measurement (Stern 1975; Stern et al. 1977). The main steps in the development of the Laser Doppler blood Flowmetry were described previously (Obeid et al. 1990; Oberg 1990; Shepherd and Oberg 1990).

A wide range of tissues were tested (Stern et al. 1977, 1979; Bonner and Nossal 1981; Dirnagl et al. 1989; Haberl et al. 1989; Shepherd and Oberg 1990; Vongsavan and Matthews 1993; Belcaro et al. 1994) by LDF. Usually, the changes in TBF are calculated as a percent change as compared to a control level, defined as 100% of TBF. The LDF signal after death in animal studies is defined as 0 TBF. In many animal studies (Mayevsky et al. 1990c, 1992; Mayevsky 1992), as well as in the human brain (Mayevsky et al. 1991, 1996), the output of the Doppler analyzer was very close to 0. These definitions enable the calculation of relative changes between 0 and 100 TBF, and above, where the 0-mV value of the TBF analyzer was referenced as the 0 TBF.

Mitochondrial NADH Fluorescence

Only the reduced form of NAD^+, i.e., NADH, absorbs UV light at almost the entire UVA spectrum area (315–400 nm), and fluoresces at 420–480 nm. A small part of the excitation light delivered to the tissue by the excitation optic fiber will be absorbed and re-emitted as fluorescence by NADH. The total number of NAD^+ and NADH molecules in a cell is constant, and any decrease in NADH will be accompanied by an increase in NAD^+. Therefore, any change in the redox status of NAD^+/NADH will be reflected in the intensity of the emitted fluorescence, while the more reduced state (NADH) shows more intense fluorescence than the oxidized state (NAD^+). The reduced states, i.e., having higher NADH concentrations, have greater fluorescence intensities.

The redox state of NADH, and hence the fluorescence, depends on the availability of oxygen as well as on the rate of ATP turnover (ADP availability). When ATP consumption is constant, a decrease in intracellular O_2 shifts the redox potential to a more reduced state thus increasing NADH and fluorescence intensity. When the O_2 level is 0, NADH accumulates to its maximal level. On the other hand, when $[O_2]$ is not a limiting factor, any increase in ATP turnover (ADP availability) induces a shift in the mitochondrial NADH redox potential to a more oxidized state thus decreasing the F signal intensity. Due to the dual dependence of the NADH redox state on O_2

availability and ATP turnover, the cause for a change in F signal is not unambiguous. However, combined with the information provided by TBF monitoring, a more accurate interpretation of the fluorescence changes can be accomplished.

There are two types of pyridine nucleotides (PNs), NAD^+ and $NADP^+$ (until 1964, these were called DPN and TPN, respectively). The spectroscopic as well as the fluoroscopic properties of the PN are being used in *in vitro* assays as well as for the determination of intact cell energy metabolism. Very recently, Lubbers, a world leader in tissue metabolism research concluded "The most important intrinsic luminescence indicator is NADH, an enzyme of which the reaction is connected with tissue respiration and energy metabolism" (Lubbers 1995).

There is a general agreement on the characteristics of the spectra and their biochemical significance when measured *in vivo* and *in vitro*. The absorption and fluorescence spectra of NADH (the reduced form) have been well characterized at different levels of organizational complexity: in solution (Chance and Baltscheffsky 1958; Galeotti et al. 1969; Chance et al. 1979; Koretsky and Balaban 1987), in cell suspension (Kohen et al. 1969; Eng et al. 1989), in tissue slices (Chance et al. 1962a; Pappajohn et al. 1972), in perfused organs (Koretsky et al. 1987), and under *in vivo* monitoring (Chance 1967; Sundy and Anderson 1975; Harbig et al. 1976; Mayevsky 1984).

The conceptual foundations for tissue mitochondrial NADH fluorometry were established (Chance and Williams 1955b) in the early 1950s. These defined various metabolic states of activity or rest for *in vitro* mitochondria. These metabolic states depend on the availability of oxygen, substrate, and phosphate acceptor (ADP), and are associated with different redox states of NAD^+/NADH.

During the past four decades, NADH fluorometry techniques have been developed, improved, and applied to *in vivo* measurements in various organs. The principles of *in vivo* NADH monitoring are identical to the *in vitro* monitoring, though the interpretation of the results is different. As shown in Fig. 7.24, the *in vivo* NADH dynamic range is from minimum (corresponding to the *in vitro* state 2) to maximum (corresponding to the *in vitro* state 5), and it changes in a continuous manner rather than discretely between defined states. The brain is given as an *in vivo* example due to the relatively large amount of published material, yet the same scaling could be presented for other organs.

The awake brain is situated in the middle of the NADH range, i.e., between states 4 and 3. Any increase in the level of NADH indicates that the mitochondria are proceeding toward state 4 or 5. Activation of the brain results in a decrease in NADH and a shift toward state 3. In Fig. 7.24, selected metabolic conditions of the brain are shown covering the entire range of NADH responses.

The responses of mitochondrial NADH to metabolic perturbations have been studied in the brain as well as in other body organs (Chance 1976). Jobsis and his group correlated the NADH fluorescence changes with the activation state of the brain and showed that convulsive activity (O'Connor et al. 1973b), direct cortical stimulation (Rosenthal and Jobsis 1971), or SD (Rosenthal and Somjen 1971, 1973) produced transient oxidation of NADH, which was interpreted as a state 4–3-4 transition.

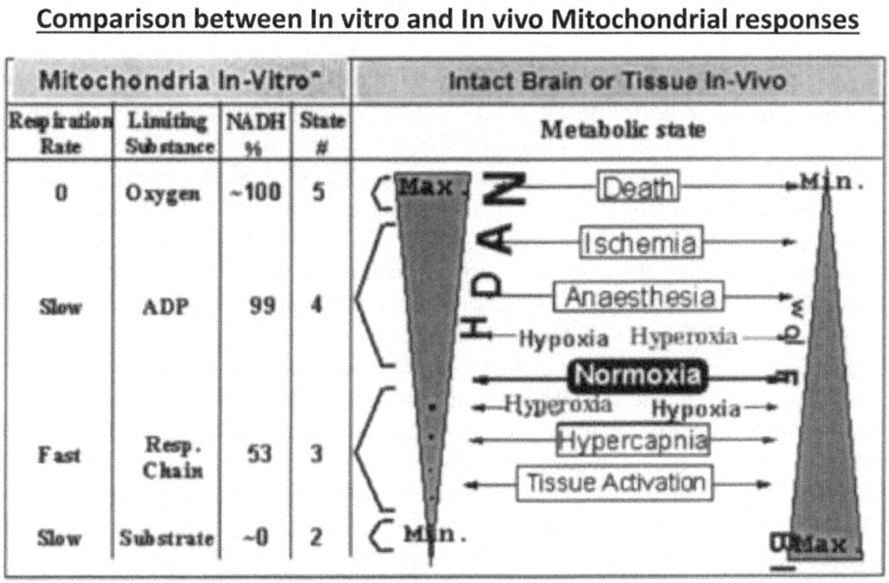

Fig. 7.24 Schematic representation of tissue blood flow and NADH redox state *in vivo* responses to various perturbations, and the metabolic states of isolated mitochondria as defined under controlled conditions. In most perturbations, there is a negative correlation between blood flow and NADH redox state. (Mayevsky et al. 2004a [Modified])

In 1975, the same group of investigators also introduced the measurement of cyt aa3 (by Ref spectrophotometry) to study the brain *in situ*, in various functional conditions (LaManna et al. 1975; Rosenthal et al. 1976; Jobsis 1977; Jobsis et al. 1977). Chance et al. (1973) have shown in detail that NADH is the most oxygen sensitive component of the respiratory chain, and therefore is best to serve as an indicator of intracellular O_2 concentration and mitochondrial activity.

In 1972, the use of optical fibers in surface fluorometry was introduced, replacing the old "rigid" optical systems. As a result, measurement of NADH from the awake animal brain (Chance et al. 1973; Mayevsky and Chance 1973) and from the heart became possible (Chance et al. 1974).

During the past decades, this technique has been developed, improved, and applied in many experimental setups in various organs *in vivo* and *in vitro* (Kohen et al. 1969; Cummins and Elliott 1976; Balaban and Mandel 1980; Balaban et al. 1980; Franke et al. 1980; Ji et al. 1980; Kedem et al. 1981; Balaban and Blum 1982; Dora 1984; Dora et al. 1984; Renault et al. 1984; Koretsky et al. 1987; Kitai et al. 1992; Mayevsky et al. 1992; Ince et al. 1993).

Total Backscattered Light-Reflectance Signal

Due to the scattering of excitation light by the tissue, some small part of the excitation light is reflected to the tissue surface, collected by the collecting fibers, and transmitted to the photodetector. This light intensity is referred to as a backscattered or Ref signal. A certain amount of light is absorbed by the tissue (primarily due to hemoglobin). The absorption of light and the Ref signals are inversely correlated.

Hemoglobin being the main absorption element, on blood elimination from the tissue, the reflected signal Ref increases dramatically. Conversely, when the tissue becomes more perfused with blood, Ref decreases. In this way, R provides information regarding blood volume in the tissue. This feature of the Ref signal also helps correct "artifacts" in the NADH F signal due to changes in tissue blood volume. When the blood volume changes, this affects both the accessibility of excitation light and the intensity of the fluorescent emission light. When the blood volume increases, the NADH signal decreases, and vice versa. Various algorithms, including F/Ref and F-Ref, have been used for correction. In the simplest case, the corrected fluorescence is expressed as a linear relationship between the measured fluorescence and the reflected excitation light (Jobsis et al. 1971a, b; Mayevsky and Chance 1974; Harbig et al. 1976; Ji et al. 1977; Mayevsky and Chance 1982; Dora et al. 1984). We have found that F-Ref provides good correction, therefore, this correction technique was used both in the TiSpec (TS) and in the predicate laboratory fluorometer used for the *in vivo* comparative studies. This correction technique is necessary when NADH is monitored in a blood-perfused organ *in vivo*.

A large increase in Ref is measured *in vivo* (e.g., in the brain) upon animal death: oxygen is no longer available and a massive vasoconstriction occurs. This event of a large increase in the Ref signal, termed secondary reflectance increase (SRI), was described by Mayevsky and Chance (1976), Mayevsky (1978a), Mayevsky and Zarchin (1981), and Zarchin and Mayevsky (1981). In 1990, Mayevsky showed that SRI occurred when a large change in extracellular K^+ was recorded (Mayevsky 1990). Later, we found that during the SRI, a decrease in blood volume is recorded (Mayevsky et al. 1990c). The same principle applies when tissue absorption is increased due to vasodilatation and increased tissue blood volume. Under brain hypoxia, for example, the diminished O_2 induces an elevation in the blood volume and the Ref signal drops as compared to the 100 control values. The same holds true when a high level of CO_2 (hypercapnia) induces a large increase in blood volume and a significant decline in the Ref signal.

Experimental results have clearly shown that changes in blood volume affect changes in the Ref signal. Those studies were summarized elsewhere (Mayevsky and Chance 1982; Mayevsky 1984).

Relationship Between Mitochondrial Function and Tissue Microcirculatory Blood Flow

The mitochondrial metabolic state, in the concept of Chance and Williams (1955b), can be related to TBF and NADH redox state, as shown in Fig. 7.24. The responses of NADH or TBF to various disturbances are well documented (Mayevsky 1992).

If monitoring only one of the parameters, the information would be insufficient to determine whether a change in tissue vitality occurred. For example, if examining TBF alone, it would be impossible to know whether its decrease indicates a normal physiological response (to a factor such as lowered energy demand) or a pathological change such as ischemia. Likewise, NADH alone does not provide sufficient information regarding changes in the metabolic state. For example, if NADH declines (becomes more oxidized), it indicates an increase in oxidative phosphorylation. Without other data, it cannot be determined whether this is a normal response to hypercapnia (increased CO_2).

Once we measure changes in the NADH signal and correlate them to the changes in TBF, we can better identify the metabolic state that effected these changes. In this chapter, we use the term hypoxia to describe a decrease in the partial pressure of O_2 in the inspired air, whereas ischemia is defined as a decrease in the blood flow to the tissue. Let us assume that a decrease in TBF is seen simultaneously with an increase in NADH. In other words, both the blood flow and oxidative phosphorylation have decreased. Depending on their magnitude, these changes could indicate a state of anesthesia, ischemia, or even death. If, on the other hand, we monitor simultaneously an increase in TBF, for example, in the brain, along with an increase in NADH (i.e., when the blood flow increases accompanied by a decrease in oxidative phosphorylation)—this would indicate a state of hypoxia. In other organs, such as the kidney under hypoxia, TBF is decreased and NADH is elevated. Further consideration of Fig. 7.24 clarifies the types of metabolic states that can be identified by changes in the measured variables.

7.3.2.2 TiSpec—Technological Aspects

The TS device contains all the necessary subunits for spectroscopic measurements of biological tissues. The TiSpec comprises a light source unit (LSU), a fiber optic probe, a detection unit (DTU), an A/D and D/A converter board, and a computer. The system is controlled by user-friendly software, which also enables processing, analyzing, and displaying the acquired data (see Fig. 7.25).

General Instrument Description

The TS can be better understood by examining its subunits. The light source is a He-Cd laser whose emitted light is directed to the tissue by a flexible optic fiber. The light interacts with the tissue and subsequently exits from the measured tissue

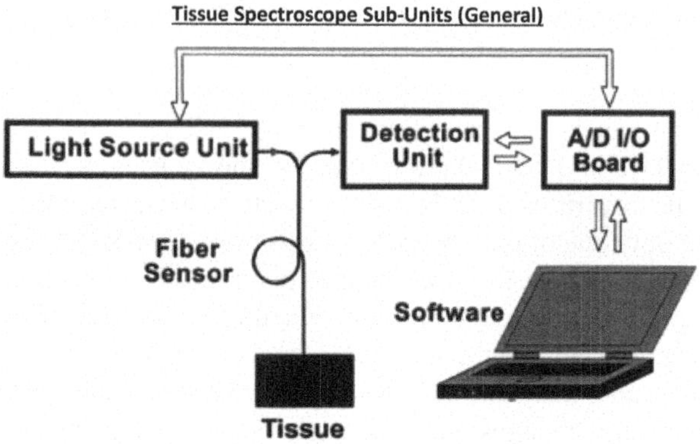

Fig. 7.25 TS system subunits. The LSU emits laser light into a fiber optic probe that delivers the light to the tissue and collects the returned light. The TDU converts light signals to electrical current, amplifies the signals, and performs Doppler detection. The A/D I/O board converts analog signals to a digital format for subsequent digital data processing by the PC. (Mayevsky et al. 2004a [Modified])

volume. The light emitted from the tissue contains three components as already described, namely:

1. Fluorescent emission at 420–480 nm
2. Doppler-shifted laser light (325 nm)
3. Total backscattered light (325 nm)

The receiving optical fibers collect both the fluorescent and reflected signals and direct them to the detection unit. While the total backscattered light and the Doppler-shifted light have nearly the same wavelength of 325 nm, the fluorescence is red shifted toward 450 ± 30 nm. At the detection unit, a dichroic mirror separates the various Ref signals from the fluorescence. The dichroic mirror reflects the 325 nm reflected light toward a photodiode that produces the total backscatter and Doppler signals. The longer wavelength F signal passes through the dichroic mirror. This light is again filtered by a band-pass (420–480 nm) filter and then enters a photomultiplier detector.

The photodetectors convert the light intensity to electrical signals. These electrical signals consist of two components: a DC component that is correlated to the total backscatter signal; and an AC component that is correlated to the Doppler signal. These signals are amplified and processed in the detection unit by the Doppler signal-processing subunit. The output of the Doppler processor is transmitted to an A/D converter. The A/D input–output (A/D I/O) board functions as an interface between the instrument control software and other subunits of the instrument. This board is responsible for the conversion of analog signals produced by the DTU into a digital form for further display and processing. It also supplies control signals to the LSU and to the DTU.

The system software is responsible for data acquisition, display, and storage, along with control over all the instrument units. Additionally, it enables the user to control the instrument functions and provides measured and calculated data.

Light Source Unit

The main part of the light source unit, shown in Fig. 7.26, is a CW Helium-Cadmium (He-Cd) laser that comprises a laser head generating 325 nm light at 6 mW maximum power from Melles Griot (Carlsbad, California). The power supply is controlled by the TS software through the A/D I/O board. The laser is protected from unauthorized use by a key switch. The same key switch also protects the entire system from unauthorized use by shutting off the power supply to the system.

A relatively high-intensity laser was chosen since it has better light stability. This feature is important for the TS measurements. The outgoing 6 mW light from the laser head is attenuated to a very low light intensity when it exits the distal tip of the probe. The TS conforms to the safety guidelines for laser output by the ANSIZ136.1 standard Safe Use of Lasers. The instrument was designed in such a way that the mean excitation intensity for clinical applications was 0.5 mW/cm^2, which is 0.5 of the threshold limiting values (TLVs). The TLVs also apply to the maximum permissible exposure (MPE) level (1 mW/cm^2). In the animal studies, the light intensity was in the range of 2.0 mW/cm^2, which is four times the TLV.

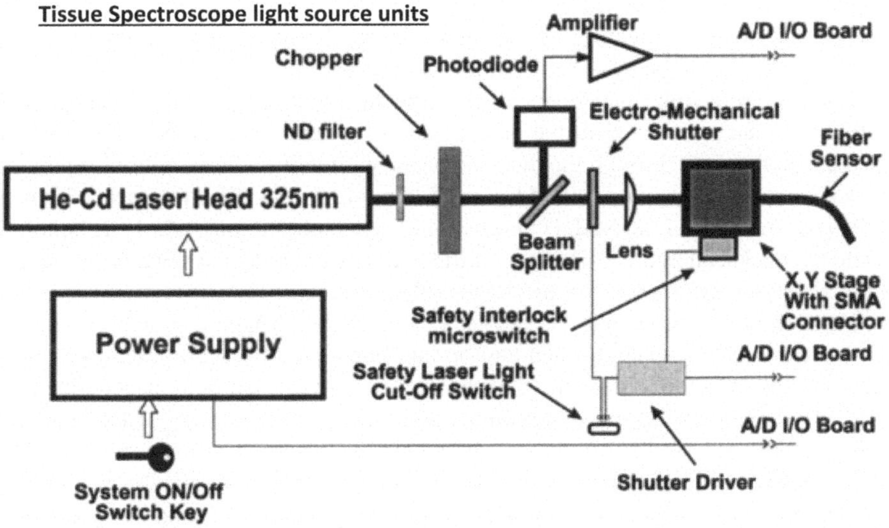

Fig. 7.26 LSU includes He-Cd laser head with laser power supply, neural density (ND) filter, chopper, electromechanical shutter, beam splitter, and photodiode for light intensity monitoring, and a lens for coupling the laser beam into the optic fiber which is connected by a SMA (subminiature version A) connector. Interlock and safety switches are installed for additional safety. (Mayevsky et al. 2004a [Modified])

After exiting the laser, the light passes through an acousto-optical modulator (AOM). This modulator chops the CW laser light with a duty cycle of 1/10. This chopping operation mode permits the usage of a synchronous detection scheme and enables additional reduction of the excitation intensity. The same AOM enables excitation intensity control by the software.

The chopped light then passes through a beamsplitter that splits about 5% of the total light intensity toward the intensity monitoring photodiode. The monitoring photodiode transmits the light intensity correlated voltage to the A/D I/O board after appropriate amplification. This voltage is used to monitor the output intensity for both normalization and safety purposes.

After exiting the beamsplitter, the light passes through an electromechanical shutter. This shutter is controlled by the software through the A/D I/O board and is closed when the instrument is not monitoring.

At the final stage, the light beam is coupled to the optical fiber by means of a focusing lens and an x-y stage, connected to the fiber by an SMA connector. A safety interlock micro-switch is placed near the SMA connector and precludes laser exposure unless the probe is connected.

Detection Unit

The detection unit is built according to the two-channel concept (Fig. 7.27), which enables, by subsequent subtraction of the two channels, the elimination of all synchronous noises (Nilsson et al. 1980). This arrangement enables efficient subtraction of noise originating in the laser light source, so that a longitudinal, multimode laser can be used for LDF measurements.

The two collecting optic fibers, at the detection unit end of the Y-shaped fiber optic probe, are connected to the detection unit by SMA connectors. The light of each channel passes through the dichroic beam splitter. This beam splitter separates backscattered and Doppler-shifted light at 325 nm from NADH fluorescence light at 420–480 nm. Most of the NADH fluorescence light passes through the beam splitter, while the backscattered and Doppler-shifted signals are reflected toward the photodiode detectors, which convert them to a voltage signal.

The output of the detector is fed to a synchronous detector, which comprises sample and hold (S/H) circuit and appropriate filters (not shown in the figure). The synchronous detector enables the detection of the duty-cycled light as generated by the LSU. The output of the synchronous detector is a combination of DC and AC voltages that are similar to those that would be acquired from a regular CW (non-duty-cycled) excitation.

The reflected light is a very slowly changing parameter and is, therefore, represented by the DC voltage level at the amplifier output. The Doppler-shifted light signal appears as AC ripple at the relatively high DC level. Both signals appear at the output of the synchronous detector. The Doppler signals pass through an AC amplifier, while the backscattered signals pass through a DC amplifier.

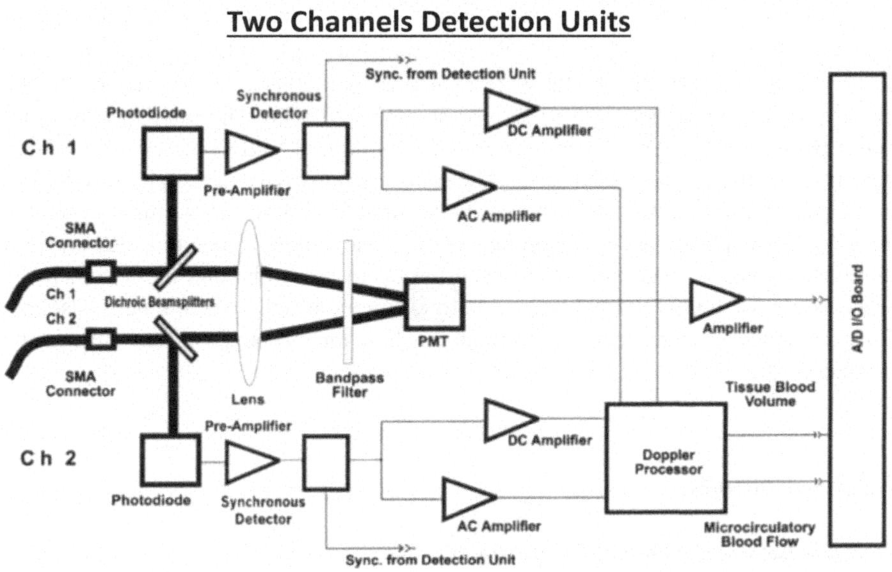

Fig. 7.27 DTU consists of two input channels (optic fibers) directed through beam splitters into two photodiodes (for Doppler analysis) and one photomultiplier tube (PMT) for NADH fluorescence analysis. (Mayevsky et al. 2004a [Modified])

After amplification, both signals are fed to the Doppler processor. The Doppler processor output consists of two signals. The first signal is correlated to Doppler-shifted light that is due to microcirculatory TBF. The second signal is correlated to the backscattered light intensity that is relevant to tissue blood volume. When exiting the Doppler processor, both signal types are slow changing, with a time constant of ~3 s.

The light that passes through the dichroic beam splitter of each channel is refracted toward a common detector by a lens. Before reaching the detector, the light passes through a band-pass filter that transmits only NADH fluorescence wavelengths (420–480 nm).

The fluorescence detector signals, after amplification, are transmitted to the A/D I/O board. The amplifier also operates as a synchronous detector by receiving the clock signal (not shown) from the clock unit. When exiting the fluorometer, the F signal is a slowly changing parameter, with a time constant of ~3 s. The output of the amplifier is connected to the A/D I/O board. The sampling rate for all three signals was 10 Hz.

TiSpec Performance

Since the excitation light of the TS is in the UV area (325 nm), light penetration is no greater than 1 mm. Most of the information is collected from a sphere having 0.6–0.7 mm in diameter. The stability of the three types of measurement was tested separately. The Doppler stability (tested in a solution of microspheres) was better than 1 per hour. The stability of the Ref parameter was tested by applying the tip of the probe to the surface of Delrin (a standard inert plastic white material). The minimal detection capacity of the TS was 50 nW at 325 nm. The stability was better than 1 per hour. To test the F signal, we used a solution of NADH dissolved in a large volume of saline to avoid bleaching. The minimal sensitivity of the photomultiplier was 1 pW at 460 nm. The stability was better than 1 per hour. Regarding a possible error due to the placement of the probe, note that the only way to avoid such errors is to have a constant good contact between the tip of the probe and the tissue. If the probe is moved relative to the tissue, the measurement will start at a new baseline. Only tendency changes can be compared when the probe is moved during the measurement.

7.3.3 Methods

7.3.3.1 Basic Experimental Approach

In Vitro Measurement

The ability to quantitatively measure NADH fluorescence, comparable to a standard clinical fluorometer, is demonstrated by direct fluorescence measurements in aqueous NADH solutions, tissue, and blood. The fluorometric analysis of aqueous NADH solution by the TS provides a suitable simulation of fluorometric tissue analysis, since the principles of *in vitro* and *in vivo* NADH monitoring are identical. There is an agreement on the characteristics of NADH spectra and their biochemical significance, when measured both *in vivo* and *in vitro*. The intensity of the fluorescence band is independent of the organizational level of the environment and is proportional to the concentration of NADH (the reduced form).

In Vivo Measurements

The ability to measure TBF comparable to another Laser Doppler Flowmeter is demonstrated by simultaneous *in vivo* measurements of TBF in different tissues of rats and gerbils using the TiSpec and the other commercial Laser Doppler Flowmeter. Metabolic perturbations were induced in the experimental animals, resulting in TBF alterations that were measured.

Additionally, to demonstrate the *in vivo* performance of the TS, we have compared the *in vivo* NADH signal measured by the TS with the measurements by a standard *in vivo* fluorometer currently used in academic research. Although the research fluorometers are not classified as medical devices, their effectiveness in carrying out *in vivo* fluorometric measurements is widely accepted. We felt that such a comparison might be useful to further demonstrate the effectiveness of the TS.

7.3.3.2 Instrumentation

Monitoring and Measurement Devices

The *in vitro* fluorometer—Victor^2 1420 (EG&G Wallace) is a multilabel, multitask plate counter, operating as a prompt fluorometer, time-resolved fluorometer, and a luminometer. For excitation of NADH, a 355 nm band-pass filter was used. The fluorescence emission was collected via a 460 nm band-pass filter. The excitation light was delivered to the sample from the up side. The emission light was also collected from the up side. This arrangement was necessary to ensure a short light path needed to eliminate sample auto absorption.

The Laser Doppler Flowmeter used was the Periflux model PF 2B Laser Doppler Flowmeter, Sweden. The Perimed PF 308 (standard, multipurpose) probe was utilized. The Laser Doppler Flowmeter was calibrated according to the manufacturer's instructions.

The TS is described in Sect. 7.2.2.

The standard *in vivo* laboratory fluorometer was described in detail by Mayevsky and Chance (1982) and Mayevsky (1984).

Fiber Optic Probes

The standard fiber optic probe (Fig. 7.28a) is composed of UV-enhanced fused silica fibers arranged in a bundle of excitation and emission fibers. The probe is designed for light transmission both from the laser toward the tissue (the excitation fiber) and from the tissue toward the detectors (the emission fibers). On the machine end, the probe is equipped with optical SMA connectors. On the tip end, the fibers are positioned and glued in a hypodermic stainless steel tube, where the emission fibers adjacently surround the excitation fiber (Fig. 7.28a). For clinical usage, the probes were packaged in a thermoformed tray, sealed and sterilized by exposure to ethylene oxide (Eto).

We developed three different types of disposable probes, designed for optical measurements of brain tissue vitality using the TS device. All these types have a similar fiber optic bundle, but feature different external design and accessories. The different types of probes were designed to enable the TS to monitor the brain tissue during different types of craniotomy and surgery.

TSP2000-2
This probe is designed for monitoring exposed brain cortex in various locations under craniotomy. The bundle end is fitted with a thin metal tube accommodating the optical fibers (Fig. 7.28b panels 1 and 2).

The spatial position of the brain is dynamic. The brain pulses as a result of breathing and the blood flow pulsation. Since the probe is sensitive to movement, and monitoring of continuous brain vitality trends is required, the probe must remain in continuous contact with a certain measurement point on the brain. We overcome

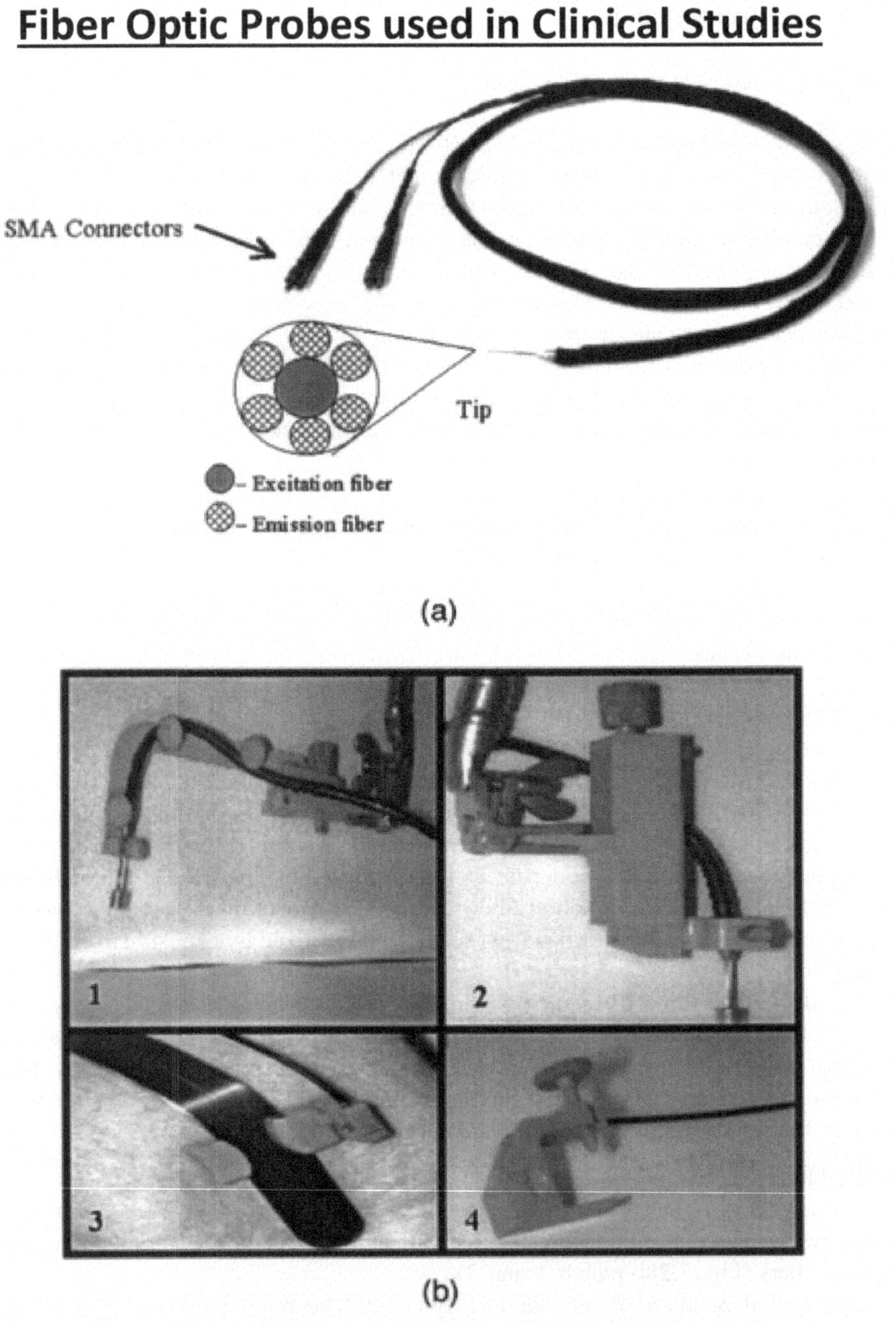

Fig. 7.28 Picture of a standard fiber optic probe (**a**) used to connect the *in vivo* monitored tissue to the TS, where the enlargement of the tip shows the arrangement of the excitation and emission fibers in the common part of the probe; (**b**) four clinical applications of the probe: 1, TSP2000–2 probe on floating arm; 2, TSP2000–2 probe on manipulator arm; 3, TSP-R1 probe and a slider attached to a neurosurgical retractor; and 4, TSP-C1 clamp-type probe. (Mayevsky et al. 2004a [Modified])

this difficulty by developing intermediate arms that, combined with the TSP2000-2 probe, enable the probe to track the changing height of the brain cortex. An adjustable floating arm (Fig. 7.28b, panel 1) and an adjustable manipulator (Fig. 7.28b, panel 2) were designed to handle the probe. Both attachments had to be applied to a standard flexible Yasargil arm, which is a common instrument in neurosurgery.

TSP-R1

This low-profile probe is designed for monitoring the metabolic stress of brain tissue pressed under a neurosurgical retractor, during profound surgery (Fig. 7.28b, panel 3). The probe is versatile and can be attached to various types of sliders to fit various widths of retractors commonly used in neurosurgery. The surgeon can easily click together the probe and the appropriate retractor attachment (slider). The elasticity of the slider enables a firm grasp of the retractor, along with an adjustment of the probe location along the retractor. The bending radius of the fibers was decreased to 5 mm, resulting in a total height of 7 mm of the plastic probe.

TSP-C1

This miniature probe (Fig. 7.28b, panel 4) is designed to monitor the exposed brain cortex, adjacent to the craniotomy area. The fibers are sheathed within a thin, flexible cable, equipped with a plastic clamp that is tightened upon the bone. In this probe, we succeeded, for the first time, in handling bending fibers without cracking them. The advantages of the TSP-C1 probe are its small dimensions and the direct attachment to the bone without any accessories.

7.3.3.3 Measurement Performance

In Vitro Measurements of NADH in Solution

A series of solutions with known concentrations of NADH (β-nicotinamide adenine dinucleotide, the reduced form, Lot 128H7001, Sigma Chemical, co. N-8129) in double-distilled water (DDW) were prepared. The samples were transferred into a microtiter plate with 96 wells for the measurements by the Victor^2 device. Every measurement was repeated twice for each duplicate, to obtain four reading values for each NADH concentration.

The same NADH solutions were placed in separate vials into which the “pencil type” probe (010) of the TS was placed. Fluorescence intensity was recorded by the TS in millivolt units.

Our preliminary studies revealed that the range of 0 to about 180 μM NADH in solution provides linear relationships with the fluorescence levels.

Statistical Analysis

The millivolt fluorescence values of the TS and the arbitrary units for the Victor^2 were plotted against concentration values. A linear plot is expected for the concentration range used. The regression line was calculated for each instrument separately.

Additionally, a correlation test was performed between the fluorescence values as measured by each instrument at identical NADH concentrations. This was done to evaluate the similarity of the two instruments. A linear plot with a high correlation coefficient was expected.

Since there is no gold standard for these measurements, we also used the Bland and Altman (1986, 1995) analysis to compare the two instruments.

Animal Preparation

Sixteen male Wistar rats (200–240 g) and 16 male Mongolian gerbils (*Meriones unguiculatus*) (45–55 g) were used. The rats were anesthetized by Equithesin (E-th = chloral hydrate 42.51 mg; magnesium sulfate 21.25 mg; alcohol 11.5; propylene glycol 44.34; and pentobarbital 9.72 mg) intraperitoneal (IP) injection 0.3 ml/100 g body weight. The left femoral vein was cannulated for intravenous (IV) injections. The small intestine was exposed by a minimal opening in the abdomen to minimize dehydration and loss of body heat. The exposed segment of the small intestine was placed on a tray and, except for the area covered by the probes, covered to maintain humidity. The probes (TS and PF308, Perimed) were held together (with the probe tips at the same height and 2 mm apart) by a flexible arm clamp and positioned on the surface of the intestine. The probe tips lightly touched the intestinal surface, and care was taken not to apply excess pressure. The animals were kept anesthetized during the operation as well as during the entire monitoring period, by IP injections of E-th 0.1 ml every 30 min. We have been using this anesthetic for approximately 50 years and it has never shown significant effects on mitochondrial activity. Furthermore, we have not been able to change the response of the rat brain to SD by using an increased dose of Equithesin, suggesting that this is a safe drug (Mayevsky et al. 1982). The addition of small volumes of E-th every 30 min kept the animal in a stable state. Body heat was measured by a rectal probe (YSI) and was regulated to be in the range of 35–37 °C using a heating blanket.

The gerbils were anesthetized by Equithesin (E-th) IP injection of 0.3 ml/100 g body weight and placed in a head holder. After a midline incision of the skin, a hole (5 mm in diameter) was drilled in the parietal bone of the left hemisphere. The dura mater remained intact. Two stainless steel screws in the right parietal bone were used, with dental acrylic cement, to fixate the probes, which were positioned by a micromanipulator on the cortex. The two common carotid arteries were isolated just before brain surgery, and ligatures of 4-0 silk threads were placed around them. Body heat was measured by a rectal probe (YSI) and was regulated to be in the range of 35–37 °C using a heating blanket. The animals were kept anesthetized during the operation as well as during the entire monitoring period, by IP injections of E-th 0.03–0.05 ml every 30 min.

Metabolic Perturbations

In the reported study, TBF and NADH in rat intestine (16 animals) and gerbil brain (16 animals) were monitored *in vivo*, simultaneously, by the TS and Perimed or standard *in vivo* fluorometer, while systemic perturbations were induced in the animals. The two probes were attached together (tips at the same height) and placed on the tissue, so to simultaneously measure the TBF and NADH of the tissue under the probes. In these animal studies, the light intensity at the probe tip did not exceed 1.3 mW. Under the present experimental conditions, we did not observe any obvious damage to the tissue. The intestine was not sensitive to the UV light, and the brain was protected by the dura mater.

There were three levels of comparison:

1. Different sites in the same tissue by the two instruments
2. Different tissues
3. Different species of animals

Due to the distance between the locations of the two probes in the same tissue, we did not compare two points of measurement by the same instrument because of the obvious difference between the optical characteristics of the tissue. If the same probe is located on different portions of the tissue or organ, the difference in baseline values will be larger than the response to several perturbations; therefore, only tendencies are compared at the same site.

The experimental animals were exposed to the following perturbations (the distribution of the various perturbations in the organs and animals is shown in Table 7.1):

Experimental Protocols

Monitoring commenced at 30 min post operation. Perturbations were then applied to the animals (a specific marker was entered into the computer), following the protocol of each individual experiment.

Table 7.1 The distribution of various perturbations in various organs and animals

Animal	Organ	Perturbation		
Rat	Intestine	Short anoxia $n = 16$	Anoxia $n = 16$	Cardiac arrest $n = 14$
Gerbil	Brain	Anoxia $n = 16$	Ischemia $n = 12$	Terminal anoxia $n = 10$

(1) *Anoxia*: The animals were exposed to oxygen deficient atmosphere by spontaneous breathing of N_2, for a short (10 s) or long (25 s) period. Terminal anoxia was induced by 100% N_2 until the animal stopped breathing

(2) *Cardiac arrest*: To induce cardiac arrest, an overdose of anesthetics (Equithesin) was injected IV

(3) *Ischemia*: Reversible occlusion (~1 min) of the two common carotid arteries (by constricting them with threads) led to brain ischemia in the gerbils

Data Collection, Analysis, and Statistics

Data Collection

The TS and Perimed display relative values of the blood cell flux and are not calibrated in absolute physiological units. The flux time constant for both instruments is ~3 s. The range of the recorder output is 0–10 V. The blood cell flux values are shown in percent units (percent of the blood flux corresponding to a full scale deflection, i.e., the adjustment yielding a full scale deflection at 10 V). Corrected fluorescence (CF) and Reflectance (Ref) are also measured by the TS and the standard *in vivo* fluorometer. The data were collected at a rate of 10 samples/s, stored, and saved in different channels of a computerized data acquisition program. The values were viewed on the screens and plotted in a chart during the experiment. All the calculations are presented as the average percent change from baseline levels (100). Each perturbation was initiated at the time when the monitored parameters were at a particular baseline value (in percents). The event was noted on the computer record by a marker entered at the moment of the perturbation commencement. For the TBF parameter, each animal started at its own, unique baseline level. For NADH, the initial baseline was always defined as 100.

Data Analysis

The TS TBF channel was passed through a running median filter for smoothing (±50 samples) to reach a noise level similar to that of the Periflux signal (the signal of the Periflux PF 2B is passed through various filters before being extracted). The signals from the TS NADH channel and the standard *in vivo* fluorometer NADH channel were both passed through a running median filter (± five samples) to reduce the noise level. The running median filter returns the median number of a specified set (in the case of TS TBF—100 data points) so that 50 data points are below it and 50 data points are above it. Starting from the first number of the set, the first median value is provided for data points 1–100, the second is for data points 2–101, the third for data points 3–102, and so on (running). The running median filter smoothes extreme single data points, usually defined as instrument noise, and displays the original recorded data points.

Statistical Analysis

The correlation coefficient (r) between the values measured by the TS and the PF 2B was calculated for each event, starting at the time of the marker entrance, for 2 min with a total of 1200 data points (anoxia, terminal anoxia, and ischemia) or for 80 s with a total of 800 data points (short anoxia).

The Fisher z transformation was applied to determine the confidence intervals for each event. Each r was transformed, and a confidence interval was computed on the *z* scale. Thereafter, the confidence limits were transformed back to obtain a confidence interval for *r*.

An average correlation coefficient ($r_{average}$) was calculated by averaging the *z* values to obtain a $z_{average}$ for all animals in each type of event in each species, and then transforming it back. The confidence intervals for $r_{average}$ were computed from the confidence intervals of the $z_{average}$.

7.3.4 *Results*

7.3.4.1 In Vitro Measurements of NADH Solution

Linear correlation coefficients were calculated, in which values close to 0 describe variables that are essentially uncorrelated and values close to 1 describe those that are highly correlated. The results of the readings of the TS and the Victor^2 plotted against NADH concentrations are presented in Fig. 7.29a. Linear regression was calculated for each instrument, and the linear equation and r^2 are presented in the graph.

The correlation coefficients between Victor^2, TS, and NADH concentration (C) were high:

The high degree of correlation implies the similarity of the two measurement systems. The coefficient of correlation between Victor^2 and TS was also calculated, $r = 0.9915$.

Figure 7.29b (the Bland-Altman plot) shows the difference against the average of the test and standard method measurements, with 95% limits of agreement (broken lines) and the regression line. As the two methods of NADH fluorescence measurement use different scale units, the measurements were normalized as a percentage of the maximal values observed. The mean difference (TS-Victor = new-old) is 3.8 and the standard deviation (SD) is 4.00. The correlation between difference and average is weak ($r^2 = 0.11$, $p = 0.17$ 95 CI from -4.04 to 11.64). The linear trend is very moderate and not statistically significant ($p = 0.167$). In the region of high absolute measurement values, the differences are small (3 or less). Thus, one can state that "the differences are small, and the new method can be applied."

7.3.4.2 In Vivo Measurements

All measurements under *in vivo* conditions were conducted in slightly anesthetized animals. The injection of additional amounts of the anesthetic material did not affect the monitored signals. Only large doses of E-th that may decrease systemic blood pressure will affect the monitored signals. The nature of the responses is clearly appreciated from the average responses. Figures 7.30 and 7.31 present the time plot graphs of the blood flow and NADH responses measured by both instruments, averaged over all the experiments, for each animal type for several perturbations.

Note that in Figs. 7.30 and 7.31, the curves of the TBF obtained by the TS and the commercial device are not overlapping due to the differences in the calibration of the two devices. The units presented in the TBF scale are relative and have meaning only when tendencies recorded in real time are compared. On the other hand, the NADH is calibrated in the same way in both the TS and the laboratory fluorometer. To compare the LDF values, it is possible to normalize the values to 100 at the perturbation onset. Since our comparative analysis was performed through correlation analysis, the use of this normalization technique was of no importance.

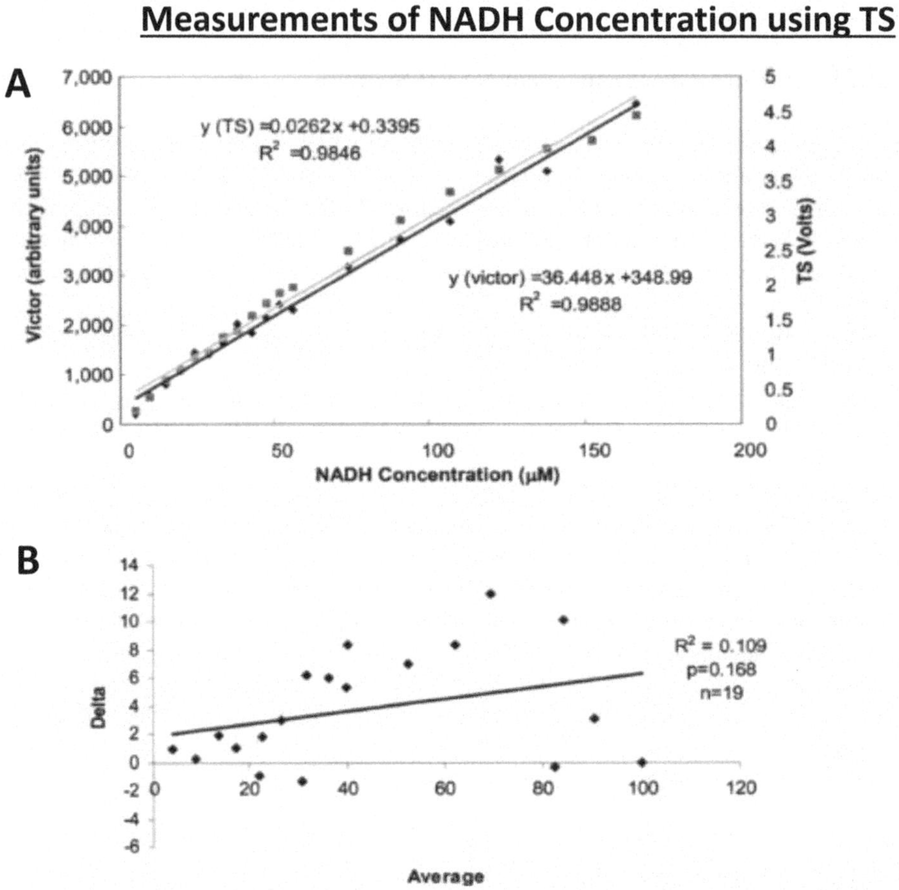

Fig. 7.29 (**a**) Readings of the TS (on the right Y axis) and readings of the Victor [2] (on the left Y axis) plotted against NADH concentrations. A linear regression was calculated for each instrument, and the equation and R^2 are presented. (**b**) The difference (delta) against the average for the tested instrument (TS) and the standard instrument (Victor[2]). (Mayevsky et al. 2004a [Modified])

Figures 7.30 and 7.31 (left) show the measurement results in the gerbil brain that reflect autoregulatory and metabolic responses to anoxia and terminal anoxia.

Figures 7.30 (left) shows the response of TBF and NADH to anoxia. As it is seen, the initial change due to anoxia is an increase in the NADH level. This is due to the decrease in the oxygen available to the brain mitochondria, which shifts the [NAD]/[NADH] ratio toward a more reduced state. Concomitantly, as we see, there is an initial autoregulatory decrease in the TBF due to the change in blood pressure. This is followed by an increase in TBF when the animal is allowed to breathe air again.

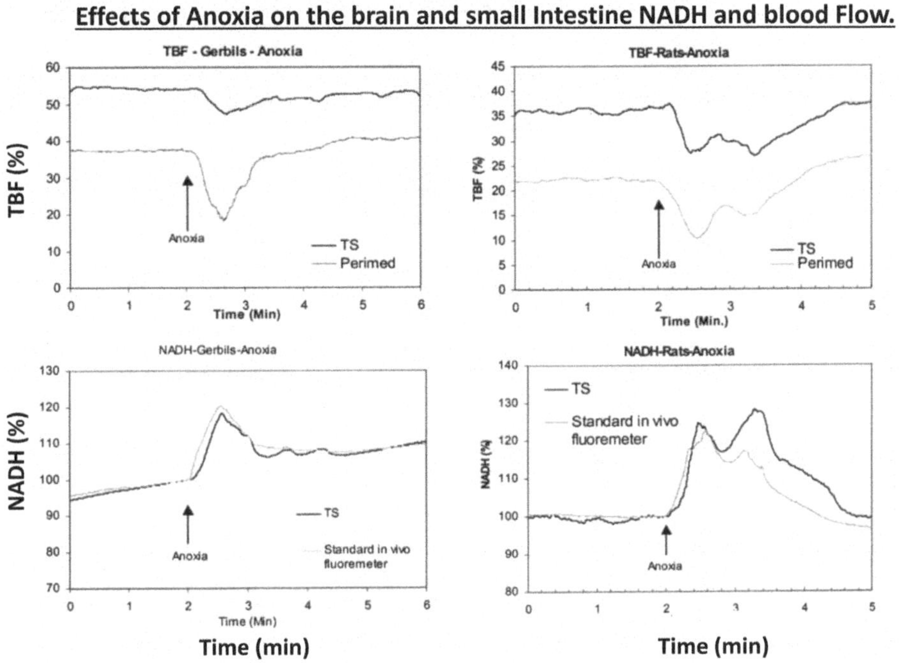

Fig. 7.30 Responses of the gerbil brain (on the left) and the rat small intestine (on the right) to reversible anoxia. Under anoxia, the NADH was elevated due to the lack of O_2 in the blood flowing to the organs. (Mayevsky et al. 2004a [Modified])

The initial change in the parameters in response to ischemia (not shown in the graphs) was a decrease in TBF. This resulted from the occlusion of the common carotid arteries. The decrease continued until the reopening of the arteries. The response of the mitochondrial NADH commenced after the initial drop in the blood flow and returned to the preischemic level after the reperfusion.

Figure 7.31 (left) shows the results regarding brain TBF and NADH due to terminal anoxia. As we see, there is an increase in the NADH level due to the decrease in the available oxygen. Insofar as the gerbil eventually ceases to breathe, the NADH signal remains high. As seen in this figure, there is an initial decrease in TBF, as in anoxia. The autoregulatory effect is seen as a "shoulder" in the TBF decrease. Insofar as the gerbil eventually ceases to breathe, the TBF signal drops and remains low.

Figures 7.30 and 7.31 (right) show the results of measurements in rat small intestine that reflect hemodynamic and metabolic responses to anoxia and cardiac arrest.

The response of the intestine to anoxia, shown in Fig. 7.30 (right), is a biphasic decrease in TBF and a biphasic transient increase in NADH. This is similar to the

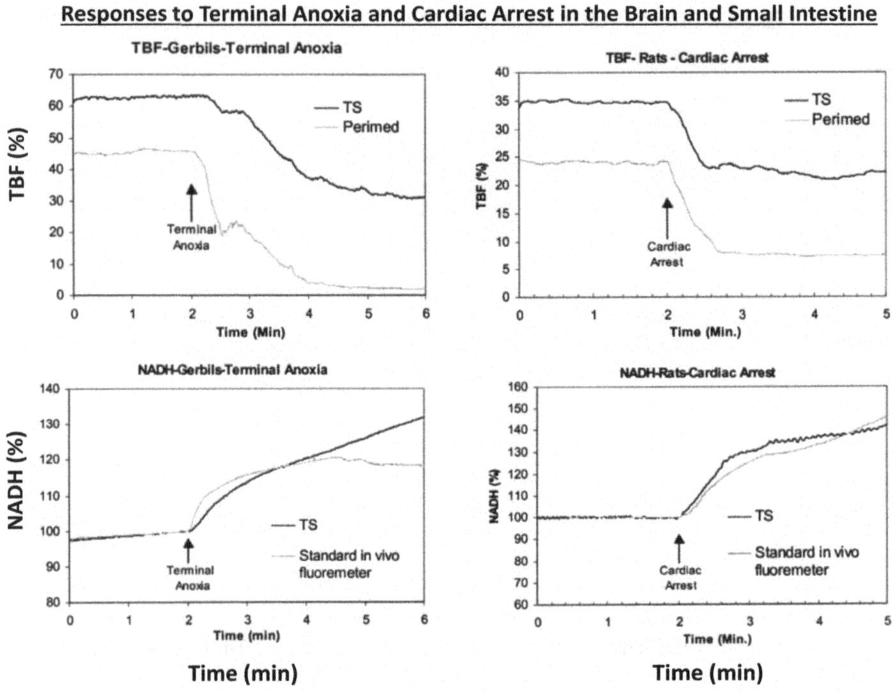

Fig. 7.31 Effect of terminal anoxia on the gerbil brain (on the left) and of cardiac arrest on the rat small intestine (on the right). Under both conditions, the blood flow reached its minimal level, while NADH accumulated to its maximal levels. (Mayevsky et al. 2004a [Modified])

responses to short anoxia, except that in short anoxia the effects are smaller and the TBF shows one phase response.

Figure 7.31 (right) shows the measurements of rat intestine TBF and NADH when exposed to cardiac arrest. As it is seen, the initial change due to cardiac arrest and the cessation of breathing is manifested by an increase in the NADH level. This is due to the decrease in the oxygen available to the cells mitochondria, which shifts the [NAD]/[NADH] ratio toward a more reduced state. Cardiac arrest causes the heart to cease pumping, which results in lower TBF. There is, understandably, no subsequent rise in TBF or decrease in NADH. Since the heart stops pumping blood, no autoregulatory effect is possible and no "shoulder" is seen.

The changes due to anoxia in rat small intestine, as seen in Fig. 7.30 (right), are similar to those in the gerbil brain because they conform to the same physiological mechanisms, although the response of the TBF and NADH in the intestine is biphasic.

Correlation

The TBF measurements by the TS and Periflux 2B correlated well, with an average correlation coefficient $R^2 = 0.70$. The NADH measurements by the TS and the standard research *in vivo* fluorometer had an average correlation coefficient $R^2 = 0.86$. The averaged correlation coefficients for each animal type and each perturbation type are shown in Fig. 7.32.

Variance

1. Using SAS software, two variance analyses were performed on the z-scores. First, to determine whether there is a statistically significant difference ($p < 0.05$) between rats and gerbils, a three-way analysis of variance (ANOVA) was performed with species (rats, gerbils), experiment number (within species), and perturbation (within species). Afterward, to determine whether there are statistically significant differences ($p < 0.05$) within animal species, between individual animals, as well as between different perturbations, a two-way ANOVA was performed within each species separately, with experiment number and perturbation. Each of these analyses was performed separately for TBF and NADH. The resulting p-values are presented in Table 7.2.

Gerbils

1. *TBF*: The correlation coefficient for ischemia (0.930) is significantly different from that of anoxia (0.750). The correlation coefficient for terminal anoxia (0.866) is in between and is not significantly different from either.

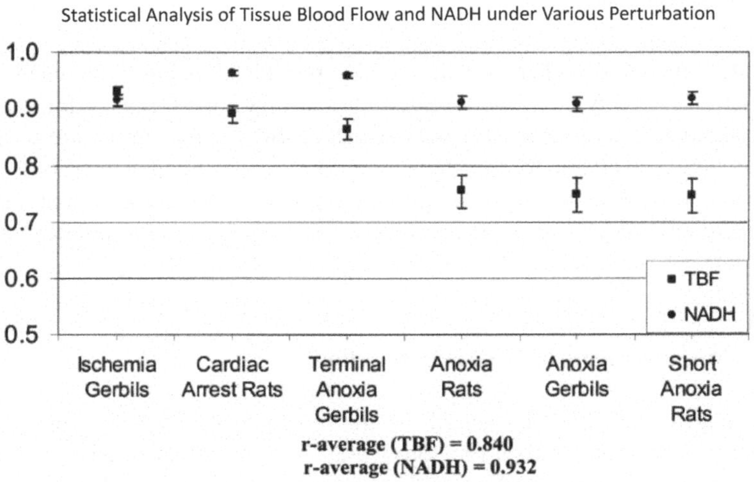

Fig. 7.32 Average correlation coefficients and confidence intervals of TBF and NADH for all six perturbations (rats and gerbils). (Mayevsky et al. 2004a [Modified])

Table 7.2 Two/three-way ANOVA and the resulting p values for TBF and NADH

Source	TBF p value	NADH p value
Between species	0.11	0.63
Gerbils—Between subjects	0.88	0.82
Gerbils—Between perturbations	0.002	0.007
Rats—Between subjects	0.37	0.20
Rats—Between perturbations	0.01	0.02

Using the Bonferroni method for multiple comparisons, significant differences of the transformed correlation coefficients were found between certain perturbations but not others (correlation coefficients are shown in Fig. 7.32, *p* values are not shown)

2. *NADH*: At the 5 level, there is no significant difference (p-value = 0.07). There is no difference between ischemia (0.916) and anoxia (0.909), but there seems to be a difference, although not statistically significant, between these and terminal anoxia (0.960).

Rats

1. *TBF*: There is no significant difference between the correlation coefficients for anoxia (0.756) and short anoxia (0.749), but there is a significant difference between these and cardiac arrest (0.892).
2. *NADH*: The correlation coefficient for cardiac arrest (0.965) is significantly different from that of anoxia (0.913). The correlation coefficient for short anoxia (0.920) is in-between and is not significantly different from either.

7.3.5 Discussion

The TS described in this chapter is designed to provide clinicians with a new device whereby tissue vitality could be assessed by real-time monitoring of three physiological parameters. To discuss the advantages of the TS, it is important to discuss other techniques available to clinicians today. In view of the general aim to utilize the TS as a medical device, it was important to prove its performance under *in vitro* as well as *in vivo* conditions. Note that there is as yet no device for monitoring tissue vitality at the tissue level available on the clinical market.

7.3.5.1 Measurements of NADH in Solution

Several investigators have measured the fluorescence of NADH in solution. Our findings are very similar to their previous work. Very recently, Cordeiro et al. (1995) plotted a calibration curve for NADH fluorescence in solutions (340 nm excitation). They found a high degree of correlation ($r^2 = 0.986$). Orr and Arthurs (1992) also showed a good correlation between NADH concentration (355 nm excitation) and fluorescence intensity. White and Wittenberg (1993) found a significant linear correlation between the percentage of NADH in a mixture with NAD and the fluorescence intensity.

In this study, both instruments, the TS and Victor^2, provide fluorescence outputs proportionate to NADH concentration, with correlations very close to 1.0. This shows that, in fact, the relationship to the NADH concentration is, quite expectedly, highly linear for each instrument.

The correlation between these two instruments is also close to unity thus unequivocally establishing similarity and even equivalence between them.

7.3.5.2 In Vivo Measurements

The results shown in Figs. 7.30 and 7.31 provide a good demonstration of how the TS measurements are correlated to standard, well-characterized physiological responses. All measurements by the TS correlate well with the Periflux 2B (TBF) and the standard research fluorometer (NADH), with an average correlation coefficient of $r = 0.84$ for TBF and $r = 0.93$ for NADH. By yielding high correlation coefficients for all groups, the results unequivocally demonstrate the qualitative similarity between the TS, the Periflux, and the accepted *in vivo* research fluorometer, when used in different physiological conditions as well as in different tissues and animal species. No difference was found between the various instruments. Note that the use of either of these devices diminishes the need for absolute values in the two LDF systems. One of the main reasons for the difference in the response amplitude lies in the depth of penetration by the two wavelengths: 632 nm of the Perimed versus 325 nm of the TS. Consequently, the sampled tissue volume is not the same. Because of this, only correlation analysis can be adequate. The difference between the sensitivity of the two instruments arises for the same reason. The sampled volume is very different, as also the size of the blood vessels analyzed by the two instruments. We were able to adjust the gain of the instruments so that they would resemble each other, but since only correlation analysis was planned, we left it as is.

As seen in Table 7.2, based on the variance analysis, there is no statistically significant difference between the correlation coefficients for animals of the same species or different species, or within a given type of perturbation. There is a statistically significant difference between the correlation coefficients for different types of perturbations, as should be expected since different metabolic perturbations induce different physiological responses.

Since all the responses to any of the perturbations are measured simultaneously by the same two devices, i.e., the TS and Periflux, or the TS and the research fluorometer, it might be expected that all the groups would have similar correlation coefficients. Nevertheless, we find differences between the correlations of different perturbations. This is understandable, however, since the physiological responses are slightly different for different perturbations. As a result of the difference in physiological responses, including autoregulatory effects on local blood flow, it can be expected that the correlation between the measured signal at two adjacent sites would be more similar for certain perturbations than for others. Similarly, when autoregulatory mechanisms are active, for example, in anoxia and short anoxia, variations in blood flow at different tissue sites can be expected.

We induced a variety of perturbations. Although all of them affect the vascular system, there are nevertheless differences between them. Some of the disturbances are systemic and affect the entire body (i.e., anoxia, when the animal breathes oxygen-deficient air, short anoxia, and terminal anoxia), while other perturbations are local (i.e., ischemia, when the arteries transporting blood to the tissue are occluded and the effect is localized to the tissue).

Additionally, the intensity of the perturbation effect depends on both its duration and nature. Duration is significant insofar as autoregulatory mechanisms are more active during a short perturbation and counteract the unwanted disturbance, whereas after a while (a few seconds) their ability to act decreases dramatically. The nature of the perturbation is significant insofar as transient or nonradical perturbations, such as anoxia or short anoxia, affect the vascular system in a lesser degree than an extreme or even terminal event, such as ischemia or terminal anoxia. The more drastic the effect on the body, the less able are the autoregulatory mechanisms to remedy the situation. For that reason, for example, the few cases of low correlation coefficients in gerbils were all seen in anoxia but not in ischemia. Anoxia is a less drastic intervention (less oxygen is supplied but continuous blood flow is maintained), when compared to ischemia, which is a complete arrest of CBF.

The results show that for NADH, all correlation coefficients are above 0.9, and for TBF, they are all about 0.75 for nonradical perturbations and about 0.9 for extreme events. These results are consistent among animals of the same and different species. The very few cases where a low correlation coefficient was found are doubtlessly due to differences in the local behavior of the vascular bed, that is, if one probe is located above a tiny blood vessel and the other is not.

The results of the comparison between different sites in the same tissue, different tissues, and different species emphasize the ability of the TS to measure all types of living tissue.

The effects of three types of systemic perturbations were applied on rats: "short anoxia," "anoxia," and "cardiac arrest." "Anoxia" is defined as "absence or almost complete absence of oxygen from inspired gases, arterial blood, or tissues" (Von Lubitz et al. 1995). Introduction of changes in the oxygen content of the inhaled air is an effective, easy, and reliable way to affect the blood flow of various organs, without further interference with the normal function of the organism, such as intravenous catheterization or introduction of chemicals to the body.

The perturbations applied on gerbils were "anoxia," "ischemia," and "terminal anoxia." "Ischemia" is defined as "local anemia due to mechanical obstruction (mainly arterial narrowing) of the blood supply" (Spraycar 1995) and was achieved by occlusion of the arteries transporting blood to the brain. The unique anatomy of the incomplete Willis circle, peculiar only to gerbils, permits a total cerebral blockade of the blood supply (Mayevsky et al. 1990c).

The effects of the abovementioned perturbations on the TBF and NADH in the brain and intestine are well documented (Anderson and Sundt 1975; Dora 1985a; Kaminogo 1989; Von Lubitz et al. 1995). Briefly, short anoxia, anoxia, and terminal anoxia, all cause a decrease in the blood flow, transient or terminal, in both the TBF of the brain and intestine. Ischemia causes the same TBF effect in the brain, and an increase in NADH level. The NADH signal increases in a mono- or biphasic manner when the brain or intestine is exposed to anoxia.

Variations in Blood Flow in Different Sites of the Same Tissue

Variations in the blood flow in different sites of the same tissue are very low, and are characterized by the coefficient of variation (Smits et al. 1986; Godden et al. 1991). The coefficient of variation relates to the extent of the homogeneity of the vascular bed anatomy and the capillary density of a certain tissue, that is to say how similar the blood flow is when measured at different sites of the same tissue. For example, the coefficients of variation are 25 for the skin, 15 for the small intestine, 33 for the stomach, and 34 for the gracilis muscle. In an earlier study, simultaneous measurements of TBF (by the Perimed) and NADH (by the standard *in vivo* research fluorometer) from four sites of the rat brain cortex were performed (Manor et al. 1996). The tendencies in the different sites are considerably similar with regard to the kinetics and amplitude when visually assessed (no statistical data available).

The magnitude of the TBF signal depends on a number of factors that probably differ between tissues, e.g., hematocrit, RBC (red blood cells) velocity, vascular geometry, tissue optical properties, and point-to-point variations of the blood flow within the tissue. These differences might prevent the calibration factors obtained in one vascular bed from being applied to another, meaning that the baseline levels of the blood flow in different tissues cannot be compared due to those factors. However, the methodology of inducing a change and measuring the net response minimizes the influence of those factors. For this reason, to compare the devices despite those variations, we applied perturbations and measured the changes simultaneously by both devices, rather than the baseline TBF values.

Different Tissues

Two types of organs were used to broaden the range of tissues measured. The brain is one of the primary organs whose blood flow and activity are of the highest importance for the survival of the organism. The intestine represents an internal organ that belongs to the supporting systems. Its blood flow and function are necessary but not crucial for the immediate survival of the organism (Mayevsky and Chance 1982).

Different Animals

Two kinds of animals were used (gerbils and rats). The measurements performed by the TS, such as LDF, tissue reflectance and NADH fluorescence, are based on the physical and physiological properties, common to most mammalian species and almost all types of tissue, like as autoregulated blood flow and oxygen dependency, the tissue composition of cells and blood capillaries, etc. The conception of the "identical principles of physiological function between different organisms" is the rationale behind the claim of the TS efficiency in any living tissue (Kraut et al. 1998).

Since both instruments conduct the measurements simultaneously, the variations resulting from different types of tissue, or different animal species, are detected by both instruments in a similar manner. Therefore, the comparison is indeed valid. Furthermore, the variations are minimized by the large sample size (15 repetitions in the average for each perturbation).

7.3.5.3 Conclusion

In the initial pilot study, the TS, when compared to the Periflux and the accredited *in vivo* research fluorometer, was found to be essentially equivalent. Further studies are required to prove its general clinical applicability. The results of comparison, obtained in the pilot study, between different sites in the same tissue, different tissues, and different species, confirm the ability of the TS to be broadly applied in the future. The TS performance is consistent in different species and different tissues thus attesting that the experimental model constitutes a valid representation of all mammalian tissue.

7.4 TiSpec03 Real-Time Monitoring of Microcirculatory Blood Flow, HbO_2, and Mitochondrial NADH Redox State

7.4.1 Introduction

The involvement of mitochondrial function in various pathophysiological conditions, developed in experimental and clinical situation, is widely documented. Nevertheless, real-time monitoring of mitochondrial function *in vivo* is very rare due to the lack of suitable instruments that could be operated by minimally trained researchers and clinical staff. The relationship between mitochondrial and neuronal survival was reviewed recently (Nicholls and Budd 2000) but the authors did not cite papers describing the *in vivo* monitoring of mitochondrial function at real-time mode. The question is, how do the mitochondria of intact tissues behave under *in vivo* conditions as compared to the isolated mitochondria? In our previous publication, we described the second generation of the TiSpec02 based on the illumination of the tissue with a stabilized 390 nm laser diode which enabled the monitoring of three physiological parameters *in vivo* (Pevzner et al. 2003). The TiSpec02 enabled us to monitor microcirculatory blood flow, tissue reflectance and mitochondrial NADH redox state using the same light source. Although the TiSpec02 was safer and relatively smaller as compared to the first generation TiSpec01, which was based on the He-Cd laser (325 nm) as an excitation light (Mayevsky et al. 2002), its price was quite high. The recent development of UV (i.e., 370 nm) LED enabled us to develop a very small TiSpec relative to the TiSpec02. Also, we have added another important physiological parameter, the level of blood oxygenation, to the three parameters measured by the TiSpec02. The aim of this study is to describe the new TiSpec03 which enables the monitoring of microcirculatory blood hemoglobin oxygenation in addition to the three parameters measured before, namely, tissue microcirculatory blood flow, Ref, and mitochondrial NADH redox state. Tissue vitality is correlated to oxygen or energy balance defined as the ratio between O_2 supply and demand. Each tissue in the body is using the generated ATP for specific activities taking place in that tissue as shown on the right side of Fig. 7.33. Therefore, in order to monitor the demand for energy of the tissue, one had to use a different technology for each tissue. For example, the brain consumes more than 50% of the total energy consumption for keeping normal ionic homeostasis (Erecinska and Silver 1989), so one can measure level of extracellular Potassium ions (K^+) as an indicator for brain energy demand (Sonn and Mayevsky 2000). The heart is using a major part of its energy consumption for muscle contraction. We have shown that it is possible to monitor the demand for energy by measuring the extracellular level of K^+ in the brain using ion specific electrode (Mayevsky 1990; Mayevsky et al. 1996). In contrary to the energy demand, energy or oxygen supply mechanisms are identical in all tissues and therefore could be monitored by the same techniques. As seen on the left side of Fig. 7.33, at least three parameters could be

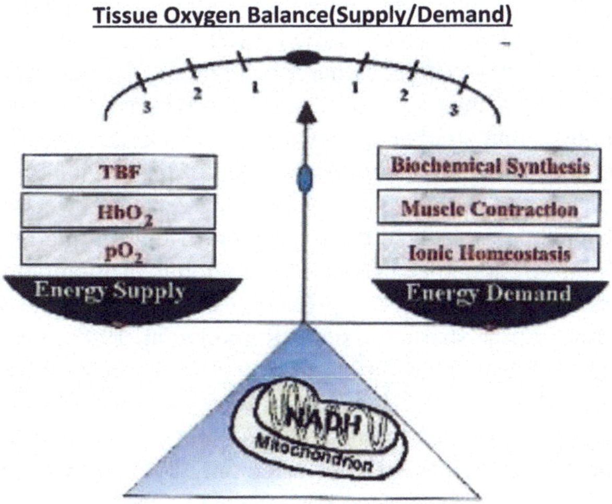

Fig. 7.33 Schematic presentation of the relationship between energy demand and supply in a typical tissue in the body. The demand processes are specific to each tissue, while the supply mechanisms are identical in all tissues. NADH Nicotine amid adenine dinucleotide; TBF Tissue Blood Flow; HbO_2 Hemoglobin Oxygenation; pO_2 Partial Pressure of Oxygen. (Deutsch et al. 2004 [Modified])

monitored from any tissue in order to identify the supply of O_2 to the tissue. Monitoring of the mitochondrial NADH redox state provides information regarding supply of energy as well as the tissue energy balance. In order to understand the functional state of a tissue, exposed to various pathophysiological conditions, the more parameters to be monitored the better the diagnosis of the tissue will be. The aim of this chapter is to describe a new integrated medical device—TiSpec03 which enabled, for the first time, to monitor four parameters from the tissue using a single optical probe (consists of two fiber optic bundles) attached to the tissue.

7.4.2 Methods

In our last communication (Pevzner et al. 2003), we described the TiSpec02 which employed a single UV laser source to optically measure three physiological parameters: TBF using LDF, NADH fluorescence excited by the UV light, and the Ref of the tissue by measuring the intensity of backscattered light. LDF requires an extremely stable laser source, usually within the red or near Infrared region. However, the excitation spectrum of the NADH lies in the near UV region. In the TiSpec02, we synthesized these apparently opposing spectral constrains by developing a unique, ultra-low noise UV light source. The light source of the TiSpec02 was based on an external cavity stabilized UV laser diode, chopped by means of an Acousto-Optic modulator (AOM). Beam chopping is necessary to neutralizing the detected signals from external room illumination. Moreover, the chopping reduces the UV irradiation upon the tissue. We preferred keeping the laser running on CW mode and modulate the AOM rather than modulating the driving current, in order to maintain highest stability in the laser cavity. It became clear that the complexity of system must be dramatically scaled down in order to produce an affordable and smaller device. We relied on the recent developments of UV LEDs, to design a completely new light source unit (LSU) comprising of a NIR laser diode for Doppler measurement and a UV LED for Ref measurement and NADH excitation. Additionally, the new TiSpec03 harbors a completely new module for the determination of the oxygenation level of hemoglobin (HbO_2), based on LEDs. The TiSpec03 hence utilizes four solid-state light sources to simultaneously monitor four types of optical signals, essential for the assessment of tissue vitality (Fig. 7.34).

For the monitoring of blood oxygenation, we utilized a two-wavelength Ref technique to analyze the absorption of light in the blood, exploiting the difference in absorption spectra of oxyhemoglobin (hemoglobin saturated with oxygen) and deoxyhemoglobin. The principles of the technique were demonstrated years ago using a time-sharing system (Rampil et al. 1992). Generally, absorption coefficient of oxyhemoglobin differs from that of deoxyhemoglobin, along most of the spectrum (Prahl 1999). However, in some certain isosbestic wavelengths, the absorption curves cross each other and represent the same value. The change in Ref in an isosbestic wavelength is affected only by blood volume, while a change in Ref in a non-isosbestic wavelength is affected by the oxy-deoxy ratio as well. By subtracting

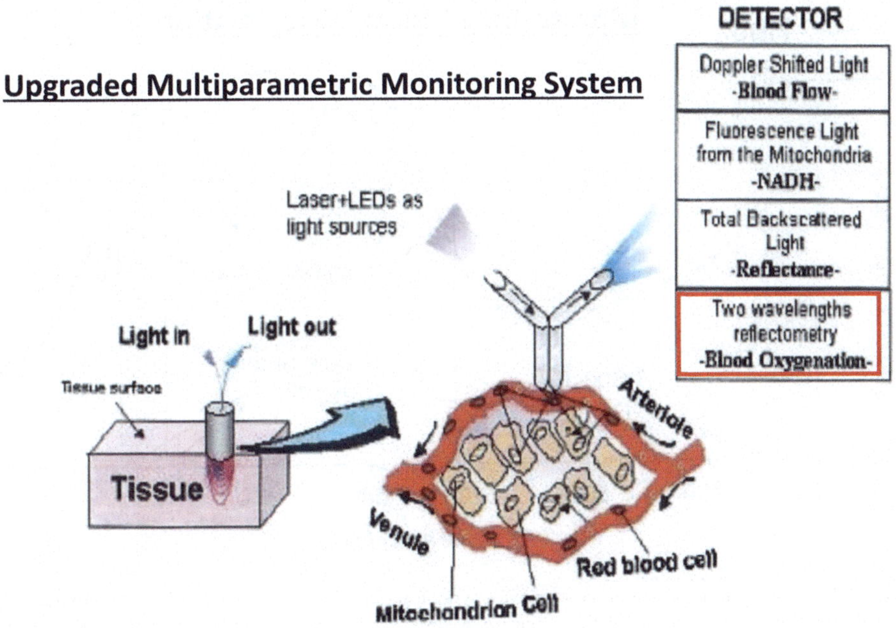

Fig. 7.34 Schematic presentation of the principles used for the construction of the TiSpec03 (for details, see text). (Deutsch et al. 2004 [Modified])

the back-reflectance in these two wavelengths, one can consider the difference between the measurements as qualitative representation of blood oxygenation level. As seen in Fig. 7.35, the former time-sharing system used two Hg-lamp emission bands to determine the oxydeoxy ratio *in vivo* by measuring the Ref in 585 nm (R585) and 577 nm (R577). Calculating—(R577–R585) yields the relative change in oxyhemoglobin state. We used the same principle with contemporary 525 nm (nearly isosbestic) and 470 nm (non-isosbestic) super bright LEDs. The relative change in oxyhemoglobin state is therefore derived by (R470–R525). It is important that the two wavelengths are close and roughly within the same range of absorption coefficient. Thus, the absorption of the tissue may be considered constant and get neutralized.

TiSpec03 uses the same A/D converter as TiSpec02, now connected to an external laptop PC. Software was rebuilt on LabVIEW 6.0 development environment (National Instruments, Austin, TX). The software employs two virtual digital oscilloscopes for signals synchronization and identification. Data is acquired, calibrated, and smoothed (using a median filter algorithm). The processed data is then displayed on the screen and stored on a local hard drive. Operator may add remarks or notes as required. Data storage format is compatible with Microsoft Excel electronic spreadsheet, and is readily available for offline review and analysis.

A schematic block diagram of the new TiSpec03 is shown in Fig. 7.36.

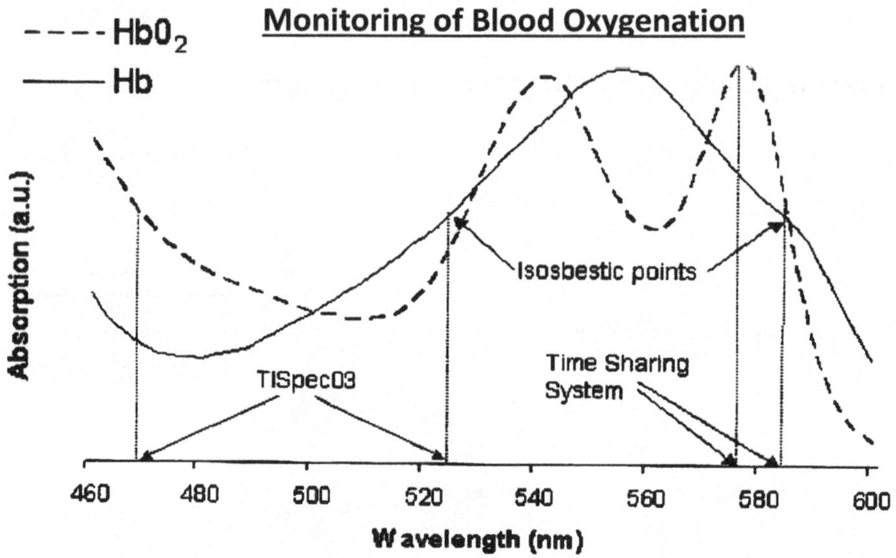

Fig. 7.35 Absorption spectra of oxy- and deoxyhemoglobin. Wavelengths involved in monitoring tissue blood oxygenation by the time-sharing system and by the TiSpec03 are marked. (Deutsch et al. 2004 [Modified])

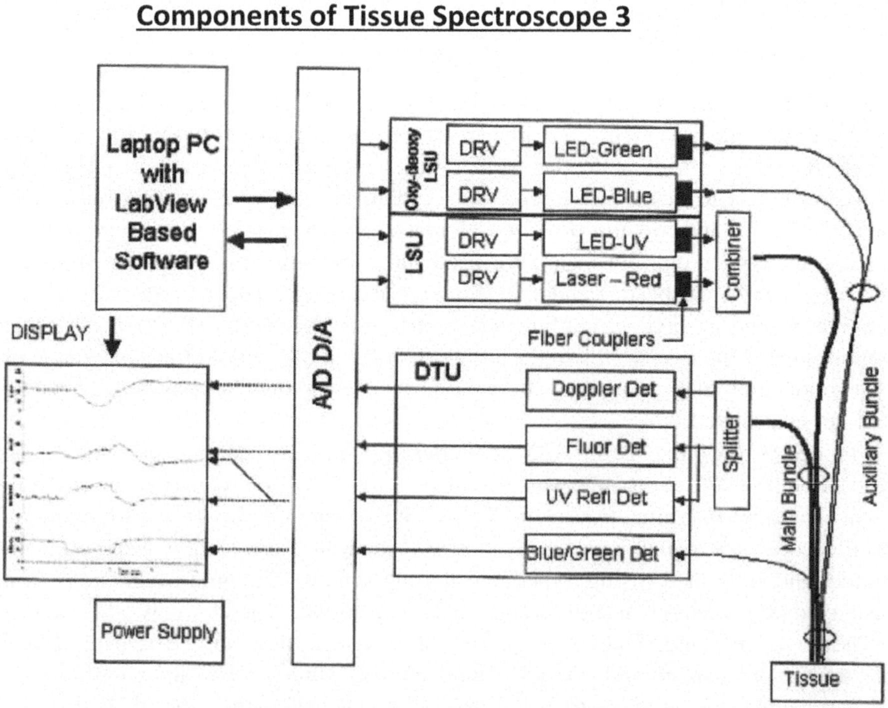

Fig. 7.36 Block diagram of the TiSpec03 (for details, see text). (Deutsch et al. 2004 [Modified])

Light Source Unit (LSU)
The light source unit of the TiSpec03 comprises a temperature stabilized and collimated 785 nm CW laser diode module (Power Technology Inc., Alexander, AR). An adjustable lens couples the laser to an optical fiber. Attenuation of the laser intensity and elimination of back-reflection are achieved by an APC fiber connector and non-paraxial coupling. A 370 nm LED by Nichia Chemical Industries (Anan, Japan) serves as a UV light source. The UV LED is mounted on a miniature XYZ stage, coupled to a 0.22NA fused silica fiber by means of an AR-coated aspheric lens from Thorlabs Inc. (Newton, NJ). Since the UV LED does not serve as a light source for the delicate Doppler measurement, a direct modulation of the LED is enabled. Consequently, the actual current to the LED in each pulse may be about three times higher than the nominal recommended CW current, yielding a similar amplification of the intensity during the pulse. Yet, no degradation or major life-time shortening were noticed so far.

Oxy-Deoxy LSU
Considering the abovementioned requirements for measuring hemoglobin saturation and LEDs availability, we chose super-bright LEDs as light sources. A 525 nm, 1.6 mW Super-Green LED, and an isosbestic wavelength and a 470 nm, 3.3 mW Super-blue LED for the non-isosbestic wavelength, both from Ledtronic Inc. (Torrance, CA). Having a narrow viewing angle of 15 degrees, these LEDs are ideal for coupling into 0.22NA fibers. LEDs are mounted in a two miniature XYZ stages, to optimize fiber coupling. The LEDs were sequentially modulated on and off using a 40 Hz pulse train.

Fiber Optic Bundles
Two separate external fiber bundles serve the TiSpec03: the main bundle is dedicated to the measurement of blood flow, Ref, and NADH, while an auxiliary bundle is dedicated to the oxy-deoxy hemoglobin measurement. The bundles have "excitation" channels that transmit light from the LSUs toward the tissue on the distal end of the bundle, and "collection" channels, transmitting the light signals from the tissue into the detection units (DTUs). The proximal ends (device side) of the fibers are terminated using SMA connectors, providing interchangeability of bundles using the connection ports located on the front panel of the TiSpec03. The distal ends (tissue side) of fibers in each bundle are compactly packed in a thin hypodermic stainless steel tube tip. The tips should be placed adjacently on the monitored tissue, enabling monitoring of roughly the same tissue area.

We put much effort in reducing the number of fibers running in the main bundles from the device toward the tissue. Among the reasons for fibers reduction are: lowering price and labor of the long bundle, cutting down the number of optical connectors, and minimizing fuss when connecting new bundles to the TiSpec. In order to achieve this goal, we multiplexed the fibers from the UV and NIR sources into a larger core fiber. Using a similar method inversely, we demultiplexed light signals from the detection channel into two detection channels. All multiplexing/demultiplexing stages were permanently and internally embedded in the TiSpec03. The intention of reducing the total number of external optical connectors is currently fulfilled only in the main bundle. The main bundle does have only one input and one output connectors (although it supports two light sources and three detectors of different wavelengths), whereas the auxiliary bundle has three connectors—each one faces a single LED or detector. However, we will shortly use the same multiplexing technique to reduce the number of connectors in the auxiliary bundle as well.

Detection Unit (DTU)
The modifications in light sources essentially resulted in appropriate modifications in the detection system. The detection system and primary signal processing of the three basic wavelengths (370, 460, and 785 nm) transmitted by the main bundle uses the same two-channels opto-mechanical design of the former TiSpec02. However, some essential adaptations were performed as follows. On one channel, the 785 nm light is collimated after emitting the fibers, then discriminated using high-pass mirror. The collimated beam is focused into the active area of a fast photodiode with a joined pre-amplifying circuit.

On the other channel, the light is collimated and directed through a dichroic mirror. The Back-reflected 370 nm light is reflected by the mirror into focusing lenses toward the active area of a second pre-amplified photodiode. NADH fluorescence, around 460 nm is merely affected by the mirror and directed toward a photomultiplier tube (PMT) via a band-pass filter. The band-pass filter attenuates residual 370 and 785 nm signals. PMT and photodiodes are the same as in TiSpec02 (Pevzner et al. 2003). The detection system of the auxiliary bundle is much simpler. It consists of a single pre-amplified photodiode that measures the back-reflected intensity of both green and blue-pulsed light. Due to the different on/off timing of the green and blue LEDs, the system can easily separate the two signals from the alternating output voltage of the single photodiode. The efforts for improving the hardware and the software resulted in the portable briefcase-size device, shown in Fig. 7.37.

7.4.3 *Results*

The TiSpec03 used in this study is different from the previous device, Tispec02, described in our previous publication (Pevzner et al. 2003). We have changed the light source for the excitation of the NADH and also added the monitoring of hemoglobin oxygenation (HbO_2). Therefore, we checked the performance of the TiSpec03 by using four rats and four gerbils exposed to various perturbations. The typical results to be presented were taken from one of the gerbils used. Figures 7.38 and 7.39 show the responses to two metabolic perturbations, namely, anoxia and ischemia. In the two events, a wave of cortical SD was developed (Sonn and Mayevsky 2000, 2001) and the responses were recorded as well. As seen in Fig. 7.38, exposing the gerbil to anoxia (100% N_2) led to a clear decrease in HbO_2 correlated very well to the increase in NADH (calculated by subtracting the reflectance [Ref] from the fluorescence [F] signal). CBF did not change dramatically

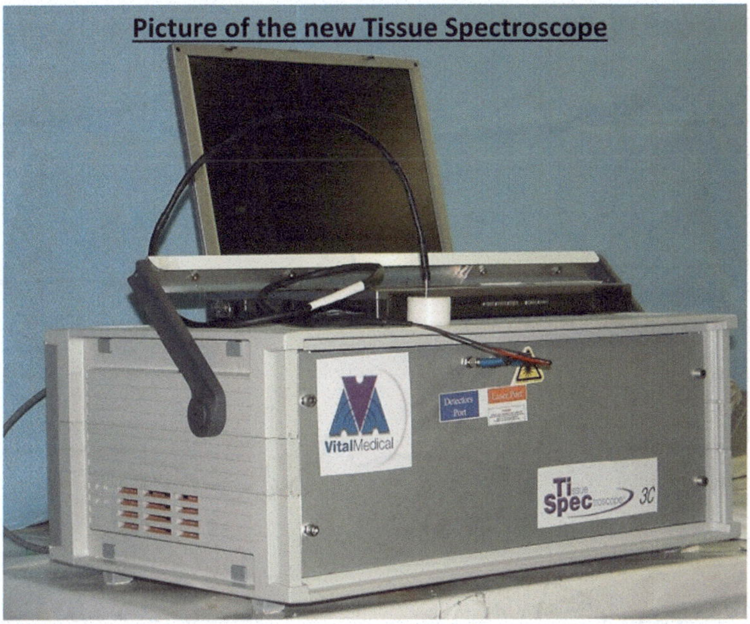

Fig. 7.37 The TiSpec03. (Deutsch et al. 2004 [Modified])

during the anoxia but a large hyperemic response was recorded during the recovery phase in parallel to the recovery of the NADH and HbO_2 to its baseline levels. About 1 min after the recovery, a wave of SD developed led to massive activation of the energy metabolism as expected (Mayevsky and Weiss 1991). Here we can see a clear correlation between the oxidation of the NADH and the increase in CBF. The HbO_2 response was biphasic starting with a small decrease followed by a relatively larger increase. Within 2–3 min, all signals returned to the baseline values. During the ischemic episode induced by the occlusion of the two carotid arteries (Fig. 7.39) very similar responses were recorded with the exception of the CBF response. As soon as the arteries were occluded the CBF decreased to its very low values and a recovery, including a short hyperemia, was recorded immediately after the opening of the carotid arteries. The other difference was the significant increase in the Ref signal during the ischemia. The responses to the SD developed after the recovery from the ischemia were similar to that recorded after the anoxia.

Figure 7.40 shows the responses of the brain to repetitive waves of SD induced by epidural application of KCl solution. The three responses were very similar to those recorded after the anoxia and ischemia (Figs. 7.38 and 7.39). Here again, activation of the energy metabolism due to the efflux of potassium from the intracellular to the extracellular space led to the oxidation of NADH (decrease in fluorescence) and a large increase in CBF. The SD waves led to a 100% increase in CBF in all three cycles recorded. It can be seen that there is a clear correlation between the oxidation of NADH and the increase in CBF. The relative change in the oxygenation of the

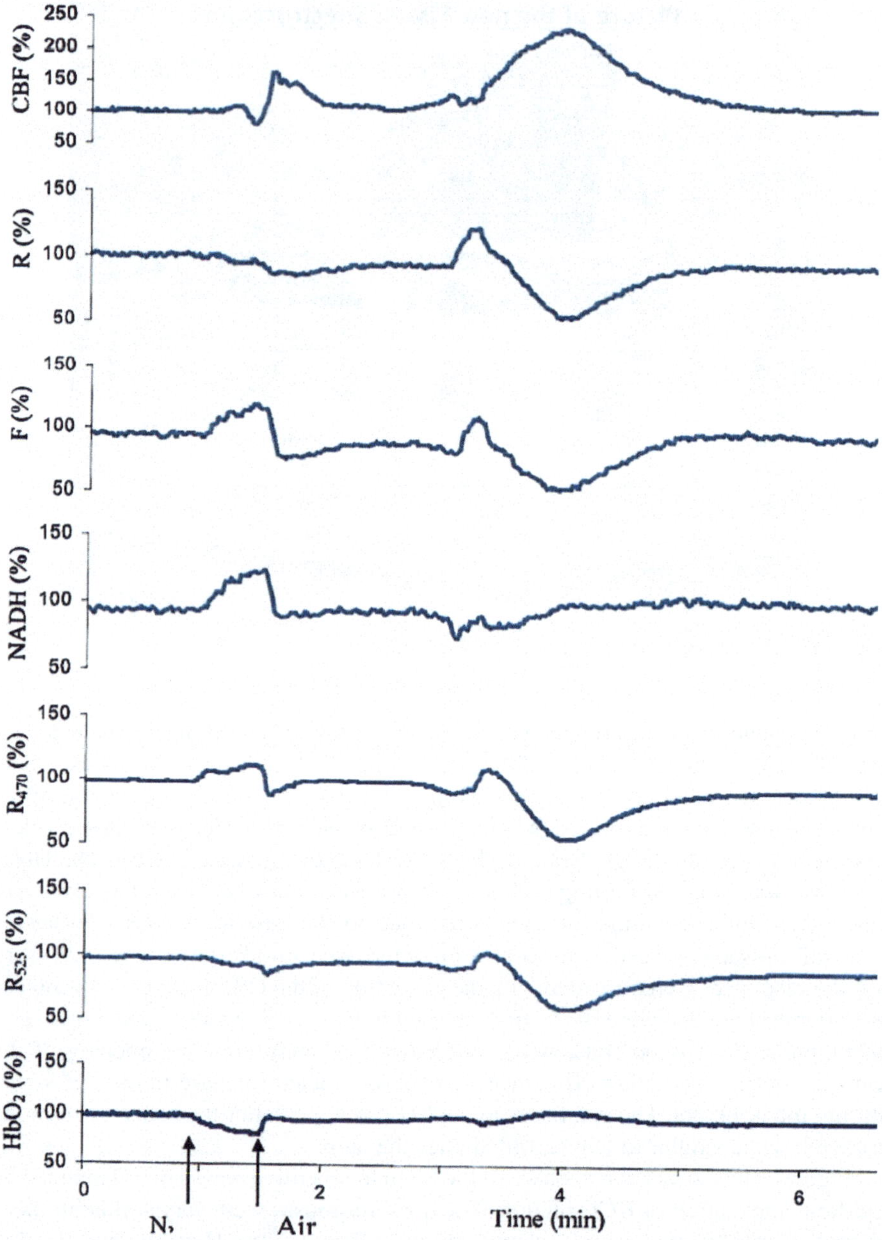

Fig. 7.38 Typical responses to anoxia (N_2) recorded from the cerebral cortex of a gerbil. CBF Cerebral blood flow; Ref Reflectance; F Fluorescence; R_{470}, R_{525}—Two wavelength reflectance signals used to calculate the oxygenation of hemoglobin—HbO_2. (Deutsch et al. 2004 [Modified])

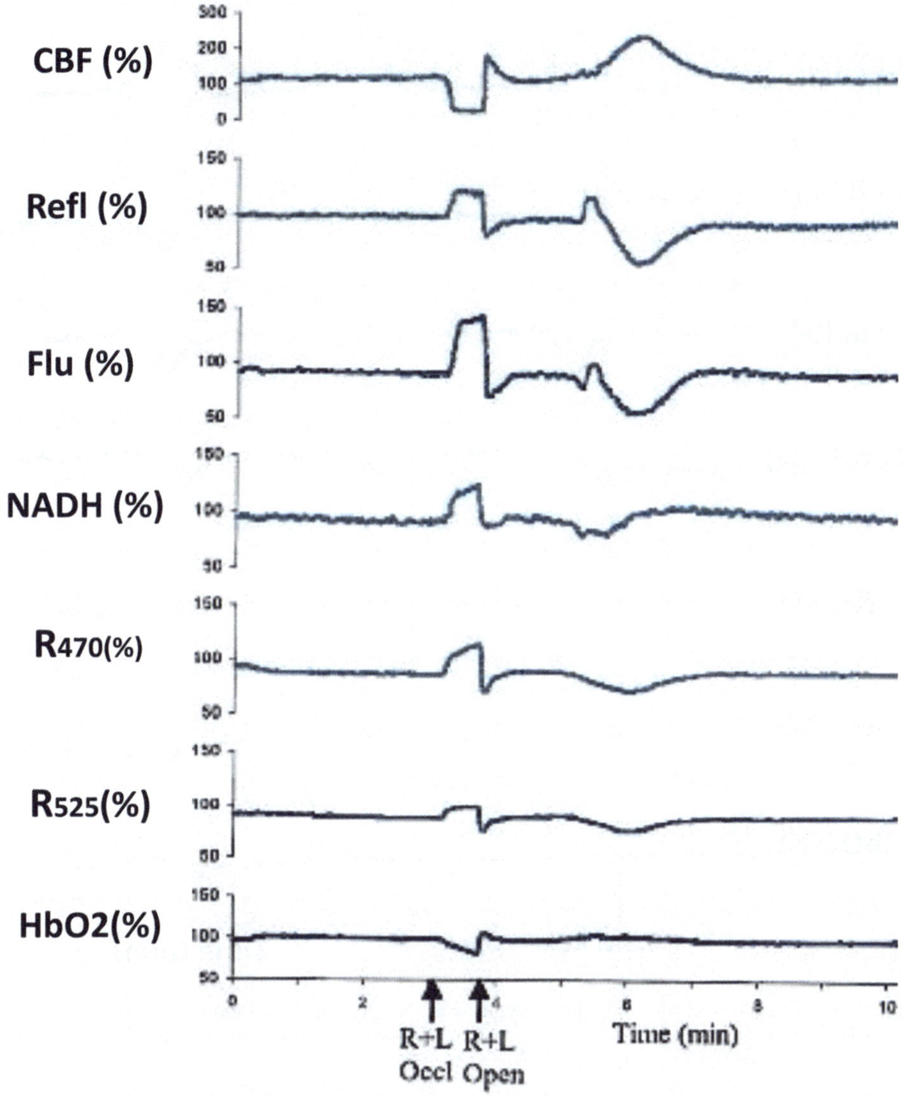

Fig. 7.39 Typical responses to a short complete ischemia induced by bilateral occlusion of the two carotid arteries. Abbreviations are as in Fig. 7.38. (Deutsch et al. 2004 [Modified])

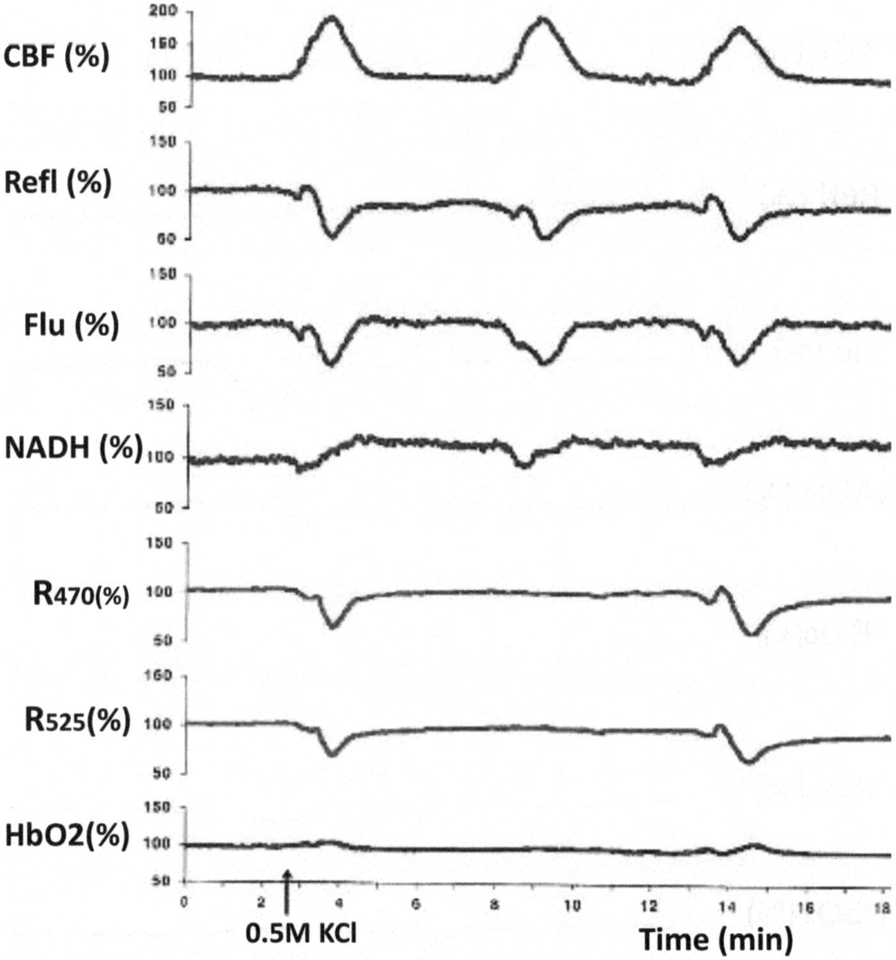

Fig. 7.40 Effects of cortical spreading depression (SD) on cerebral blood flow, mitochondrial NADH redox state, and the level of hemoglobin oxygenation. Abbreviations are as in Fig. 7.38. (Deutsch et al. 2004 [Modified])

hemoglobin was very small due to the normoxic state of the animal. The effect of 5-min ischemia is presented in Fig. 7.41. In this recording, the responses could be divided into two stages. In the initial step, due to the occlusion of the two carotid arteries, a dramatic drop in CBF was accompanied by a clear decrease in Oxyhemoglobin and a large increase in NADH. After about 1.5 min, an ischemic depolarization (ID) was developed and due to large increase in the Ref traces in 365 and

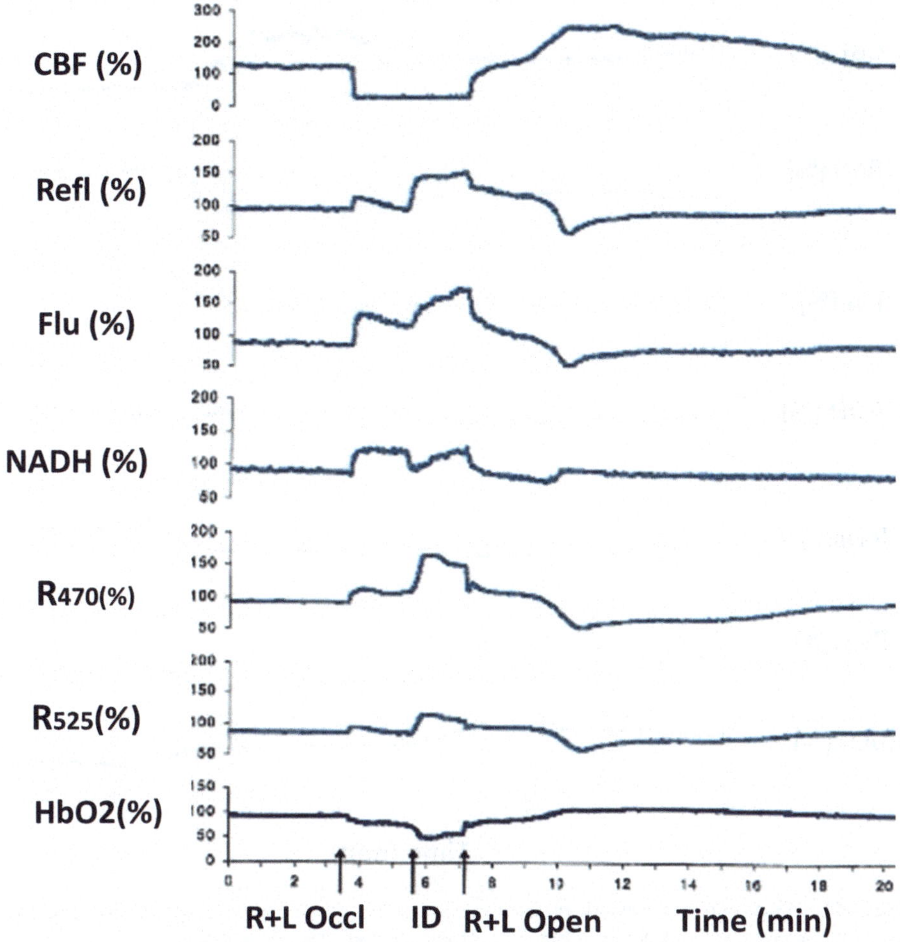

Fig. 7.41 Effects of 5 min complete ischemia, induced by bilateral occlusion of the carotid arteries in a gerbil model. ID Ischemic Depolarization. Other abbreviations as in Fig. 7.38. (Deutsch et al. 2004 [Modified])

525 nm, an artifactual oxidation of NADH and decrease in oxyhemoglobin were recorded. After the opening of the two carotid arteries the various parameters recovered to the pre-ischemic levels within 15 min. As can be seen the NADH signal recovered to base line much faster as compared to the other two parameters. The effects of long-term terminal anoxia are presented in Fig. 7.42. As seen, during the initial phase of the anoxia a decrease in oxyhemoglobin was accompanied by a large increase in NADH. Blood flow was elevated initially for about 2 min and then CBF fall to the minimal levels indicating the death of the gerbil. Here also, due to the

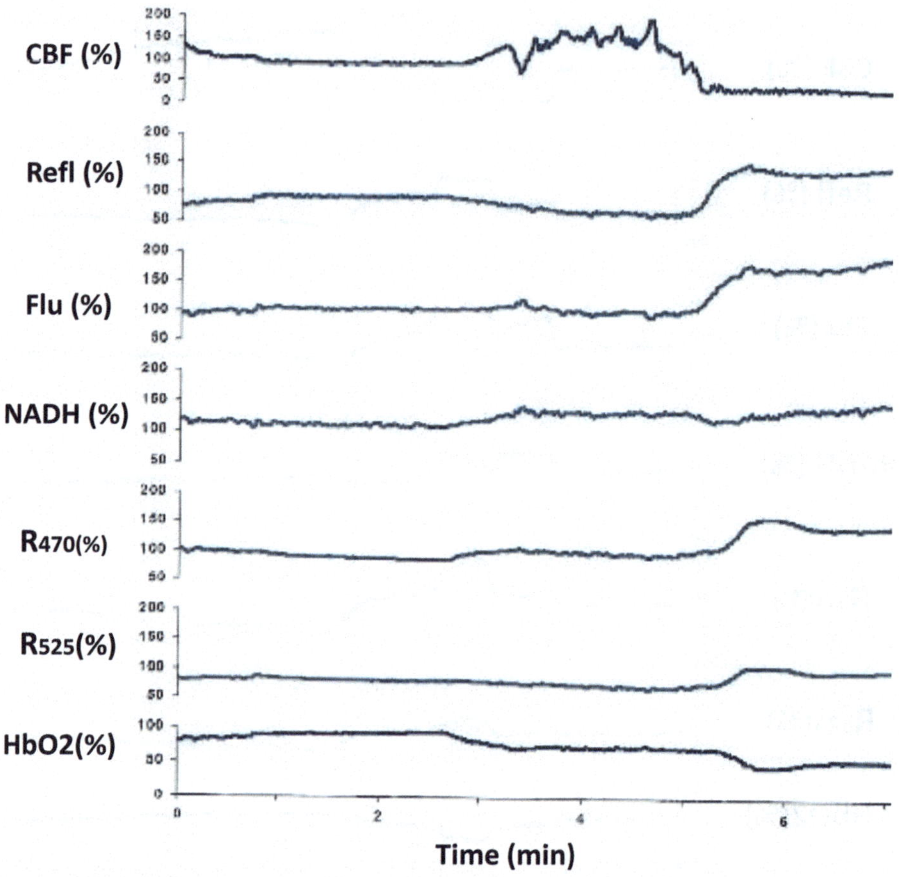

Fig. 7.42 Effects of terminal anoxia on the responses of the various parameters monitored from the gerbil brain. Abbreviations are as in Fig. 7.38. (Deutsch et al. 2004 [Modified])

anoxic depolarization, an apparent oxidation of NADH was recorded together with apparent large decrease in the oxyhemoglobin levels.

7.4.4 Discussion

Real-time optical monitoring of tissue vitality could be applied to many clinical situations in order to diagnose the metabolic state of the patient. In our previous presentation (Pevzner et al. 2003), we described the advantages of the TiSpec02 having smaller dimensions, and other advantages, as compared to the first TiSpec described in many details recently (Mayevsky et al. 2004b). This book presents the recent version of our TiSpec03, having two major new features that make this model very attractive to be used in clinical practice. The first issue is the addition of a new parameter to the three existing indicators of tissue vitality, namely, the oxygenation of hemoglobin. The second advantage is the technical improvement of the monitoring system and its smaller dimensions. Despite the fact that the presentation of the parameters is not in absolute values, it appears that the multiparametric nature of the system make it very practical for real-time evaluation as a function of time in the same tissue. The redox state of the NADH as measured *in vivo* is the most sensitive parameter of intracellular oxygen levels as compared to the other parameters monitored by the TiSpec. The changes in the NADH levels are also the best indicator of cellular energy balance in the tissue exposed to various pathophysiological conditions (for review see ref. (Chance et al. 1973)). The other three parameters monitored by the TiSpec are very important for the understanding of tissue vitality but having them measured without the NADH decreased the level of data interpretation. For example, in Fig. 7.38, under anoxia the changes in CBF are not significant at the early stage while the HbO_2 and NADH showed a better correlation to the availability of oxygen. Under Ischemia, all three parameters are very well correlated to the oxygen supply to the tissue. When the brain was exposed to SD the more significant responses were in NADH, CBF, and Ref while the HbO_2 changes were minimal due to the positive energy balance achieved by the large increase in CBF to compensate for the extra oxygen needed for the mitochondria. The decrease in dimensions of the TiSpec03 will enable to use it in the OR and in the various ICUs. In critically ill patients, the needs for better monitoring system are obvious and could be beneficial to the better outcome of the patients. The TiSpec could provide real-time information on various parameters representing changes in tissue vitality. These measurements may be performed on a so-called non-vital organ and will provide information that will serve as an early warning signal for the deterioration of the body (Ganong 1991a; Pevzner et al. 2003). This approach was tested in preliminary clinical studies where tissue vitality was monitored in the urethral wall using a three-way Foley catheter (Mayevsky et al., unpublished results).

7.5 The Critiview—A New Biomedical Device for Evaluation of Tissue Vitality in Critical Care Medicine

7.5.1 Introduction

Mitochondrial function is a prerequisite condition for the continuous supply of energy in order to perform all cellular functions in the body. Most of the Oxygen consumed in the body is utilized by the mitochondria (Waltemath 1970), which

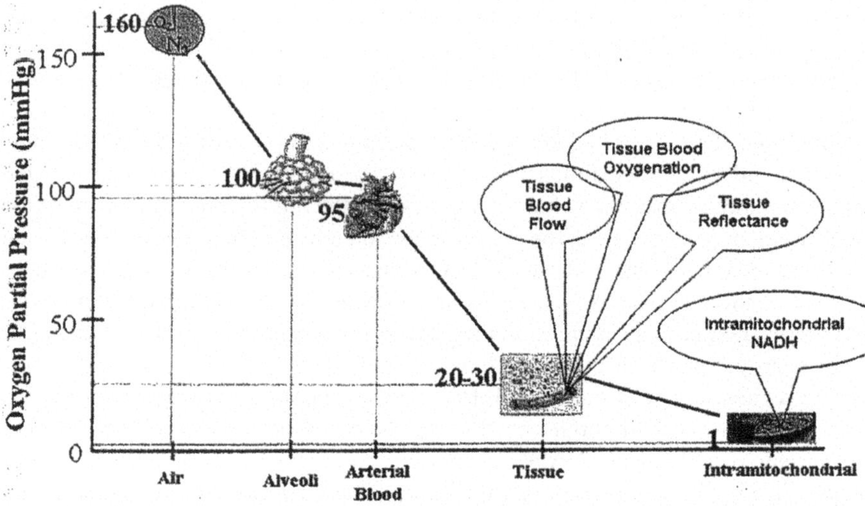

Fig. 7.43 Oxygen gradient in the body starting at the highest level in the lung (100 mmHg) and the lowest levels in the mitochondria (about 1 mmHg). The optical measurements of 4 parameters done by the CritiView represent events that occur in the microcirculation level as well as in the intracellular space. (Mayevsky et al. 2005 [Modified])

produces 95% of the ATP in normal tissues. As shown in Fig. 7.43, the levels of pO_2 needed for normal mitochondrial function is around 1 mmHg. The exact level of PO_2 exists in the mitochondria are not known due to the lack of direct monitoring technique for *in vivo* measurement. The PO_2 levels at the other body locations, shown in Fig. 7.43 are well known and documented as shown by the average values. The determination of pO_2 at the tissue levels is practical by using an appropriate oxygen electrode, representing the average levels of O_2 in the intravascular, extracellular, and intracellular compartments. The best approach to evaluate intracellular or intramitochondrial O_2 levels is to measure the spectroscopic nature of the various components of the respiratory chain (Chance et al. 1973). In this book, Chance concluded that "For a system in equilibrium, NADH is at the extreme low potential end of the chain, and this may be the oxygen indicator of choice in mitochondrial and tissue as well." The same conclusion was published by Lubbers stating "The most important intrinsic luminescence indicator is NADH, an enzyme of which the reaction is connected with tissue respiration and energy metabolism" (Lubbers 1995). The pioneering work by Chance and Williams opened up an era of 50 years of intensive research dealing with mitochondrial function *in vivo*. The evaluation, *in vitro* as well as *in vivo* conditions, was done by monitoring NADH redox state using the autofluorescence approach. Although the importance of normal mitochondrial function in various tissues is appreciated (Nicholls and Budd 2000; Scheffler 2000), the research effort done in this important field was not significant.

Most of the published material regarding *in vivo* monitoring of mitochondrial NADH was contributed by not more than 20 groups of scientists around the world. In his chapter, Scheffler ignored the contribution of *in vivo* monitoring of mitochondrial function (i.e., NADH fluorescence) to the achievements and perspectives. During the last 30 years, our group has published about 150 articles in this important subject. We found that monitoring of mitochondrial NADH fluorescence alone is very practical but it is not enough to understand the metabolic state of the tissue. We therefore developed the multiparametric monitoring approach to evaluate the functional state of various tissues or organs in the body (Mayevsky et al. 1995; Mayevsky et al. 2004b). The other parameters that could be monitored provide information that originates mainly from the vascular compartment. For this reason, the correlation between those three parameters may be significant under certain conditions, such as ischemia (decrease in blood supply to the tissue). The decrease in blood to the tissue (i.e., ischemia) will be accompanied by a decrease in HbO_2 as was in blood volume (increase Ref). The intramitochondrial NADH will show a significant increase as published earlier (Mayevsky 1984; Mayevsky and Weiss 1991). The contribution of HbO_2 monitoring to the understanding of tissue energy metabolism is not clear since it was not monitored intensively with the other three parameters namely, blood flow, tissue reflectance, and mitochondrial NADH (Mayevsky et al. 1990b; Rampil et al. 1992). In our previous communication (Deutsch et al. 2004), we presented preliminary results using the TiSpec-3. The aims of this study were to upgrade the monitoring system and technology published previously into the CritiView that could be applied in critical care medicine.

7.5.2 Methods

7.5.2.1 Animal Preparation and Experimental Protocols

All experimental protocols were approved by the Institutional Animal Care Committee under the instruction of the National Institute of Health. Male Wistar rats (200–240 g) were used. The rats were anesthetized by Equithesin (E-th = Chloral hydrate 42.51 mg; Magnesium sulfate 21.25 mg; Alcohol I I.5%; Propylene glycol 44.34%; and Pentobarbital 9.72 mg) IP injection 0.3 ml/100 g body weight. The animals were kept anesthetized during the operation as well as during the entire monitoring period, by IP injections of E-th 0.1 ml every 30 min. We have been using this anesthetic for approximately 50 years and it has never shown significant effects on mitochondrial activity. Furthermore, we have not been able to change the response of the rat brain to SD by using an increased dose of equithesin, suggesting that this is a safe drug. The addition of small volumes of E-th every 30 min kept the animal in a stable state. Body heat was measured by a rectal probe (YSI) and was regulated to be at the range of 35–37 °C using a heating blanket. The rat was anesthetized and placed in a head holder. After a midline incision of the skin, a hole (5 mm in diameter) was drilled in the parietal bone of the right hemisphere. The dura mater remained intact and an appropriate light guide holder was placed in the hole. Two stainless steel screws in the right parietal bone were used, with dental acrylic cement, to fixate the probes, which were positioned by a micromanipulator on the cortex. In order to induce cortical SD, a small push–pull cannula was placed epidurally 2 mm anterior to the light guide holder. The two common carotid arteries were isolated just before brain surgery, and ligatures of 4–0 silk threads were placed around them. The experimental animals were exposed to the following perturbations:

Anoxia: The animals were exposed to oxygen deficient atmosphere by spontaneous breathing of N_2, for a short (25 s) period. Terminal anoxia was induced by 100% N_2 until the animal stopped breathing.

Cortical Spreading Depression (SD): The surface of the dura was washed by 0.5–1.0 M KCl solution.

Ischemia: Reversible occlusion (about 1 min) of one or two common carotid arteries (by constricting them with threads) led to brain ischemia in the rats.

7.5.2.2 The Monitoring System—"CritiView"

The new device named CritiView is based on the principles of monitoring the four parameters as described below:

Mitochondrial Function

Mitochondrial function is evaluated by monitoring of NADH fluorescence. The measurement is done by excitation of the tissue by UV light (375 nm) and measuring the emitted F signal (420–480 nm). The principles of this method was established 50 years ago (Chance et al. 1973) and the main changes that were introduced since then were the type of light source used as well as the usage of bundle of optical fibers to connect the tissue to the fluorometer. The main light source that was used was Hg arc lamp having a strong emitted line at 366 nm absorbed by the reduced form, NADH and not by the oxidized form (NAD). The introduction of 375 nm LED by Nichia as a new light source in the present instrument (CritiView) significantly affects the size and price of this component in this medical device. The use of optical fibers led to a more flexible connection between the monitored tissue and the fluorometer. This enabled us to monitor the unanesthetized brain (Mayevsky 1984), or at different locations in the same animal (Mayevsky and Chance 1982). The fluorometer of NADH is a good indicator of the decrease in oxygen availability in the cellular compartment. As shown in Fig. 7.44 (upper left side), exposing the brain to anoxia (100% N_2) led to a significant increase in the NADH fluorescence spectrum.

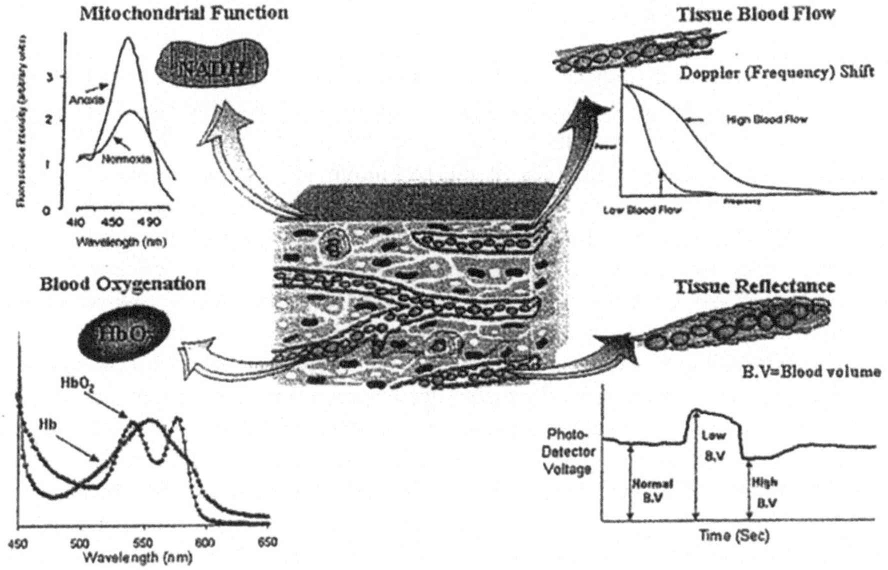

Fig. 7.44 The principles of tissue vitality monitoring performed by the CritiView. Three of the parameters were monitored from the intravascular compartment while the NADH parameter represents the function of the intramitochondrial space. (Mayevsky et al. 2005 [Modified])

Tissue Reflectance

The continuous changes in tissue absorption properties affect the intensity of tissue reflectance. It was found that this parameter could be used as a correction tool for the artifacts measured in the F signal (Jobsis et al. 1971a; Chance et al. 1973; Mayevsky et al. 2004b). As seen in Fig. 7.44, the main factor affecting the reflectance signal (Ref) changes by blood volume (Mayevsky and Chance 1982; Mayevsky 1984). When ischemia is induced, the decrease in blood volume will lead to an increase in the Ref signal. During the recovery phase from ischemia, the hyperemic response (increase in blood volume) will be recorded as a decrease in the Ref signal.

Tissue Blood Flow

Monitoring of microcirculatory TBF using the laser Doppler approach started more than 45 years ago (Stern 1975; Bonner and Nossal 1981) and is widely used in various animal model as well as in clinical applications (Shepherd 1990). The main advantage of the LDF approach is the real-time data provided by the optical nature of the signal. As of today, the large number of publications (more than 5000) in experimental animals and patients indicate its applicability (for more details see Ref. (Mayevsky et al. 2004b)).

Microcirculatory Blood Oxygenation

For the monitoring of blood oxygenation, we utilized a two-wavelength Ref technique to analyze the absorption of light in the blood, exploiting the difference in absorption spectra of oxyhemoglobin (hemoglobin saturated with oxygen) and deoxyhemoglobin. The principles of the technique were demonstrated years ago using a time-sharing system. Generally, absorption coefficient of oxyhemoglobin differs from that of deoxyhemoglobin, along most of the spectrum (Prahl 1999). However, in some certain isosbestic wavelength, the absorption curves cross each other and represent the same value. The change in Ref in an isosbestic wavelength is affected only by blood volume, while a change in Ref in a non-isosbestic wavelength is affected by the oxy-deoxy ratio as well. By subtracting the back-Ref in these two wavelengths, one can consider the difference between the measurements as qualitative representation of blood oxygenation at the microcirculatory level. As seen in Fig. 7.44, the former time-sharing system (Rampil et al. 1992; Mayevsky et al. 2004a) used two Hg-lamp emission bands to determine the oxy-deoxy ratio *in vivo* by measuring the Ref in 585 nm (R585) and 577 nm (R577). Calculating (R577-R535) yields the relative change in oxyhemoglobin state. In this study, we used the same principle with contemporary 530 nm (nearly isosbestic) and 470 nm (non-isosbestic) super bright LEDs. The relative change in oxyhemoglobin state is therefore derived by R70-R530. It is important that the two wavelengths are closed and roughly within the same range of absorption coefficient. Thus, the absorption of the tissue may be considered constant and get neutralized.

7.5.2.3 Device Description

In our last communication (Scheffler 2000), we described the TiSpecO3 which employed a light source unit (LSU) based on a NIR laser diode for Doppler measurement and a UV LED for Ref measurement and NADH excitation. Additionally, this device harbored a module for the determination of the oxygenation level of hemoglobin (HbO_7) by two wavelength reflectometry, based on Blue and Green LEDs. The NADH fluorescence emission spectrum peaks at about 450 nm while the blue LED reflection peaks at nearby 470 nm. Therefore, in order to eliminate the signal interference from the strong Ref signal to the weak F signal, the excitation fibers measuring the NADH and oxygenation parameters were separated by a few millimeters. Separate detectors were used for the detection of each parameter. In this new CritiView device new detection concept is utilized. This new concept enables detection of NADH fluorescence and the both blue and green reflection by a single detector. Additionally, it eliminates the need for the separation of the excitation fibers of NADH and oxygenation.

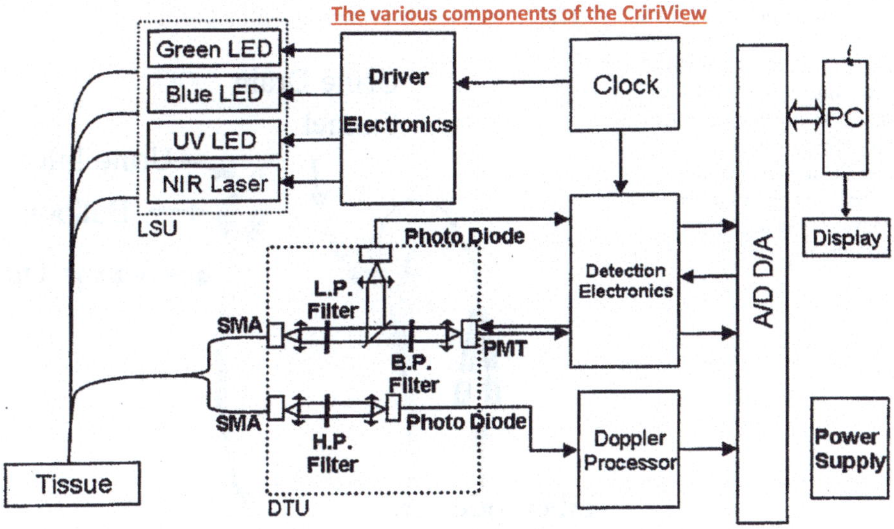

Fig. 7.45 Block diagram of the CritiView used in this study (for details see text). (Mayevsky et al. 2005 [Modified])

Light Source Unit (LSU)

The light source unit of the CritiView comprises a 785 nm CW laser diode which serves for laser Doppler measurement, a 375 nm LED for NADH fluorescence excitation and for total backscatter (or reflection) measurement, a 470 nm LED and 530 nm LED for oxygenation measurement. Each of the discrete light sources is fed by the appropriate electronics drivers as seen in Fig. 7.45.

Detection Unit (DTU)

The detection system is based on two channel detection. On one channel, the 785 nm light is collimated after emitting the fibers, then filtered by high-pass filter and focused into the active area of a fast photodiode with a joined pre-amplifying circuit. The detected signal is processed by an analog Doppler processor and the result values are digitized into the PC by an AID converter. On the other channel, the light is collimated and filtered by low pass filter in order to eliminate the 785 nm reflection. Then the light is split according to wavelengths by the dichroic mirror. The Back-reflected 375 nm light is reflected by the dichroic mirror toward focusing lenses and on the active area of a second pre-amplified photodiode. The NADH fluorescence around 450 nm and the reflections at 470 nm and 530 nm pass through

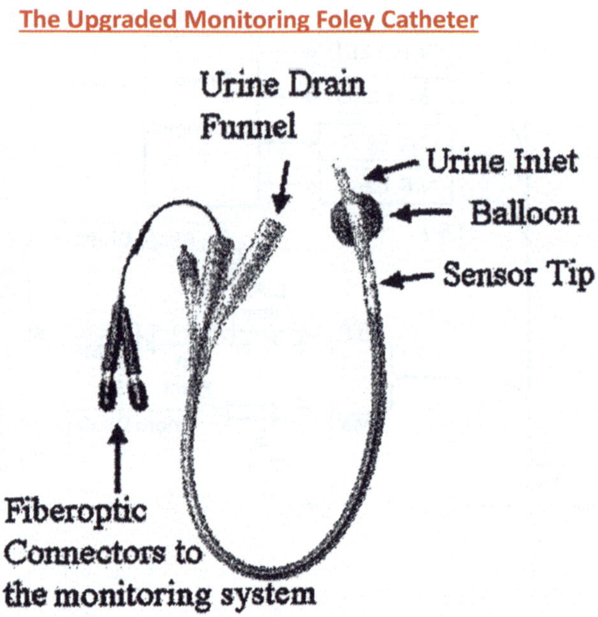

Fig. 7.46 A picture of a three-way Foley catheter used in patients hospitalized in the Intensive Care Unit (ICU). The balloon is filled with physiological saline after the insertion of the catheter into the bladder. (Mayevsky et al. 2005 [Modified])

the dichroic filter. These wavelengths pass through interference filter specified for NADH emission and are detected by photomultiplier (PMT). Since the reflection intensities of 470 nm and 530 nm signals are much stronger than the NADH fluorescence, the actual irradiation intensity of these light sources is reduced significantly in order to adjust the reflection intensity levels to be similar to the fluorescence level. This enables to utilize single PMT detector to detect all three wavelengths. This arrangement also enables to eliminate the need in two separate excitation fibers. The detected signals are processed by detection electronics that consists of sample and hold circuits that synchronized with the respective excitation pulses. The output of the detection electronics is acquired into PC by AID converter. In order to connect the experimental animal or the patients to the monitoring system, we used various types of probes. In our previous communication (Pevzner et al. 2003; Mayevsky et al. 2004b), we described in many details the basic structure of the probes used for brain studies in experimental animals as well as in clinical environment. In addition, a probe used in the neonate ICU was shown. Figure 7.46 shows the most advanced probe used in monitoring patients in the adult ICU.

7.5.3 Results

The new device—CritiView was tested in a group of rats exposed to changes in O_2 supply, (anoxia and ischemia) as well as during increase in O_2 demand (cortical SD).

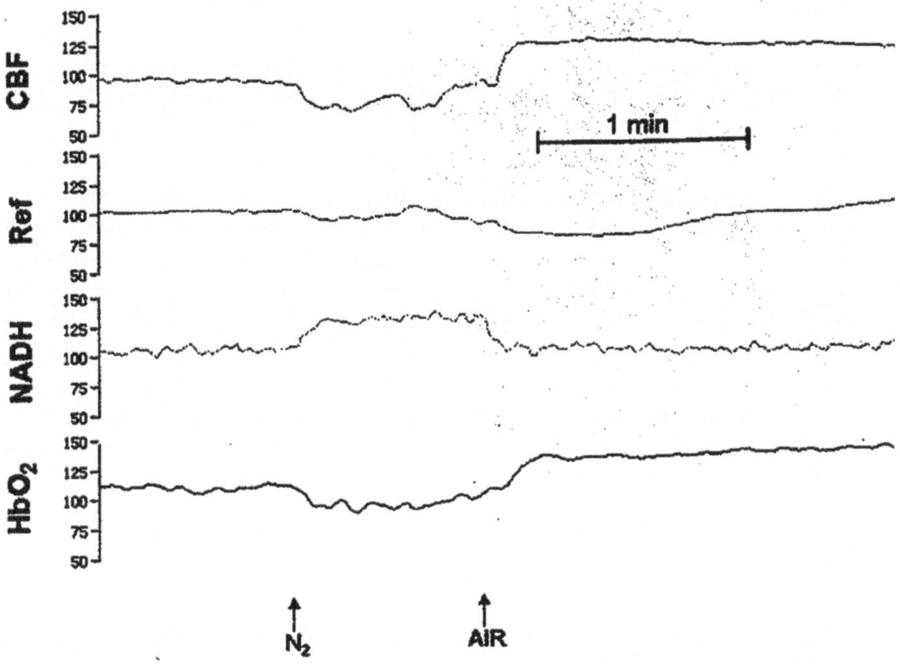

Fig. 7.47 Responses to complete deprivation of O_2 by breathing the rat with 100% O_2. CBF cerebral blood flow, Ref reflectance or total backscattered light at the 375 nm excitation wavelength, NADH net change in the NADH, HbO_2 microcirculatory blood oxygenation. The Y-scale of all parameters is expressed as percent change of the signal calibrated to 100% in the initial stage of the monitoring. (Mayevsky et al. 2005 [Modified])

In five figures presented in this communication, four signals are presented. The upper trace CBF is measured by the LDF and the trace represent the flux of red blood cells in the microcirculation. The 2nd and the 3rd traces are measured by the mitochondrial fluorometry/reflectometry. The NADH signal is calculated by subtraction of the Ref signal from the fluorescence as used by various investigators (for details see (Mayevsky and Chance 1982; Mayevsky 1984)).

The Y-scale of all parameters is expressed as percent change of the signal calibrated to 100% in the initial stage of the monitoring. In Fig. 7.47, the responses to an anoxic episode are shown. As expected the decrease in HbO_2 was accompanied by an increase in the NADH as well as a decrease in CBF. During the recovery phase, a large hyperemic response was noted in the HbO_2 and CBF. The decrease in the Ref signal also indicates an increase in blood volume. The changes in the NADH were quite symmetrical in terms of the percent change during the anoxia and the recovery to the same baseline. In Fig. 7.48, a typical response to 2 wave of SD are shown. The oxidation cycles (decrease in NADH) are typical of this event correlated to the large increase in CBF. In this rat, the HbO_2 signal also shows an increase in brain oxygenation. As seen, the mitochondrial response was much shorter as compared to the response recorded in the CBF. Figure 7.49 shows repetitive cycles of SD recorded from another rat that was presented in Fig. 7.48. The five oxidation cycles seen in the NADH signal are very symmetrical and similar to each other. The changes in CBF and HbO_2 were less stable and were correlated with each other. The baseline was shifted as passing from the 1st to the 5th cycle. When the response to the 6th SD cycle was started (vertical line), the right carotid artery was occluded (R_{occl}). As seen, the response to the SD was completed under the small change in blood supply and an oxidation cycle was recorded, but with a slower rate of recovery. During the occlusion, the HbO_2 shows that the decrease was followed by an increase in the signal correlated to the increase in CBF. After recovery from the SD response, the effect of anoxia was typical of the brain. After the recovery to normoxia, the last response to the SD stimulus was recorded. Figures 7.50 and 7.51 present responses to SD in normoxic brain (Fig. 7.50) and in partial ischemic brain (Fig. 7.51).

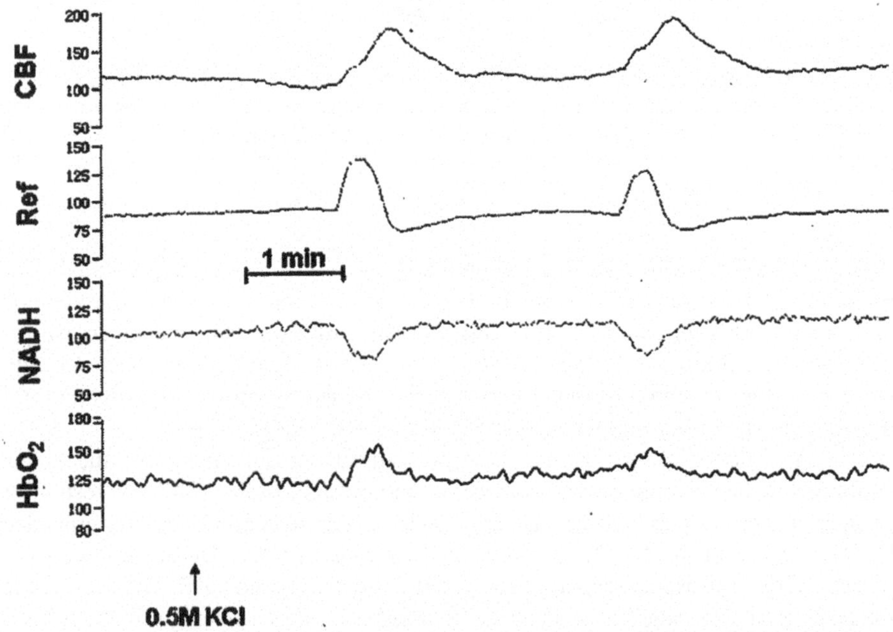

Fig. 7.48 Two responses to cortical spreading depression (SD) are shown. Abbreviations are as in Fig. 7.47. (Mayevsky et al. 2005 [Modified])

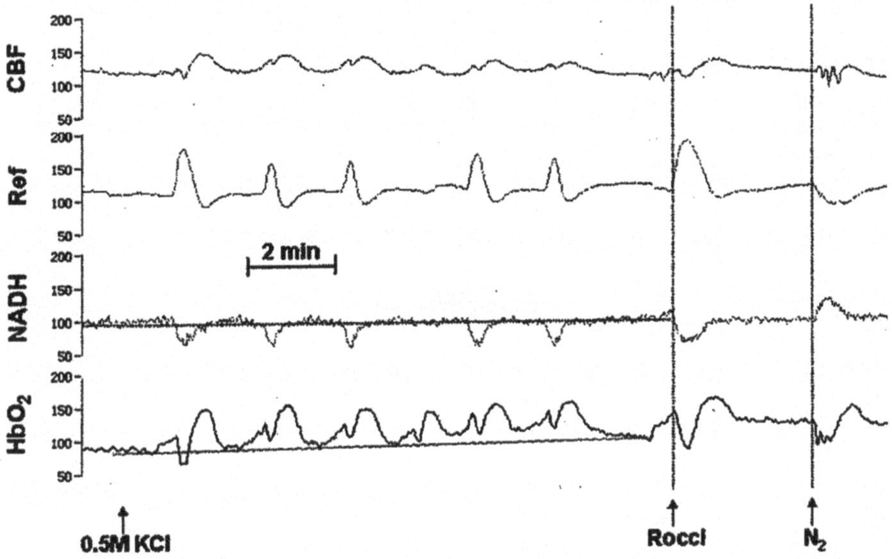

Fig. 7.49 Repetitive cycles of cortical spreading depression (SD) measured in the normoxic brain and the effect of partial ischemia on the response to SD. Abbreviations are as in Fig. 7.47. (Mayevsky et al. 2005 [Modified])

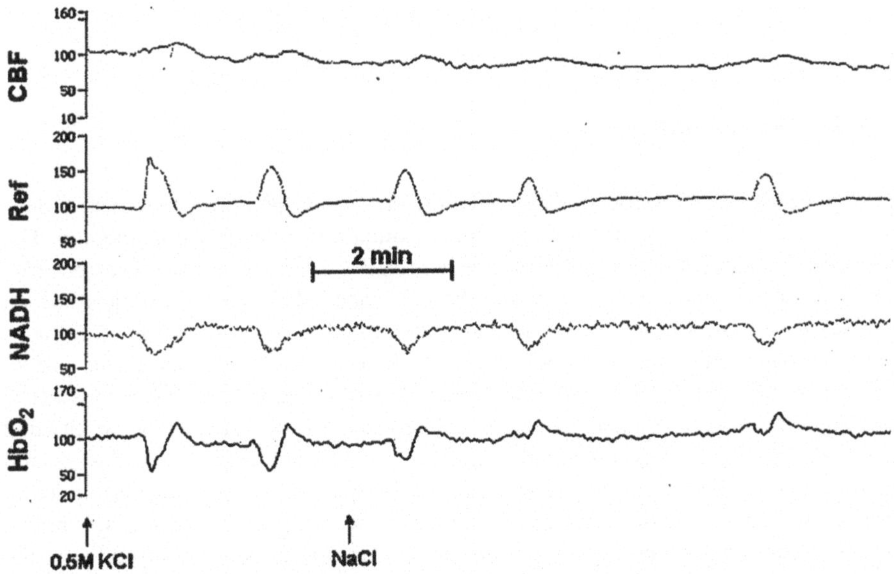

Fig. 7.50 Repetitive responses to cortical SD in the normoxic brain. Abbreviations are as in Fig. 7.47. (Mayevsky et al. 2005 [Modified])

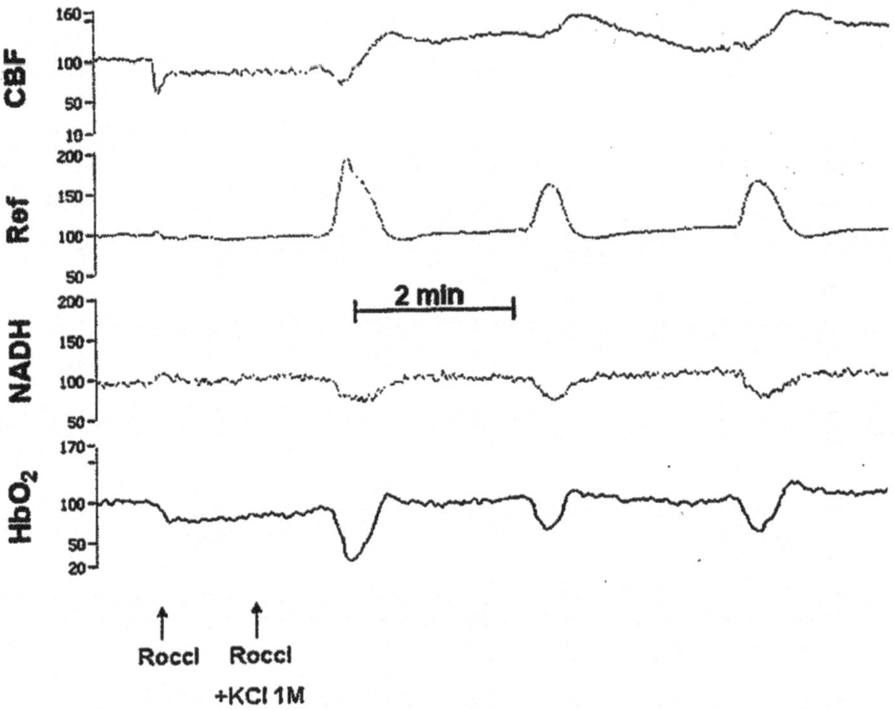

Fig. 7.51 The effect of partial ischemia (R_{occl}) on the responses to repetitive cycles of SD. Abbreviations are as in Fig. 7.47. (Mayevsky et al. 2005 [Modified])

7.5.4 Discussion

The involvement of mitochondrial function in various pathophysiological conditions, developed in experimental and clinical situations, is widely documented. The relationship between mitochondrial and neuronal survival was reviewed recently. The aim of this study was to describe the new medical device—CritiView, which enables the monitoring of microcirculatory hemoglobin oxygenation in addition to the three parameters measured before (Deutsch et al. 2004), namely, tissue microcirculatory blood flow, Ref, and mitochondrial NADH redox state. Tissue vitality is correlated to oxygen or energy balance defined as the ratio between O_2 supply and demand. Energy or oxygen supply mechanisms are identical in all tissues and therefore could be monitored by the same technique—CritiView, in various body organs. In order to understand the functional state of a tissue exposed to various pathophysiological conditions, the more parameters to be monitored, the better the diagnosis of the tissue will be. Until recently, most of the monitoring devices used in experimental animals or in the clinical environment were based on a single parameter. Out of the suggested four parameters monitored, LDF is the most widely distributed technology (Batista et al. 1996; Okunieff et al. 1998; Krejci et al. 2000;

Sailer et al. 2000; Ubbink et al. 2000), in experimental and clinical studies. Most of the 5000 studies were done by monitoring of TBF in a single organ (skin, brain, heart, etc.). Few medical devices were developed in order to monitor microcirculatory blood oxygenation (HbO_2). This kind of measurement is different from the widely used pulse oximeter that can measure oxygenation of blood that flows in large arteries. The changes in local O_2 consumption will affect local oxygenation but the pulse oximeter will not show any change. Monitoring of intramitochondrial NADH redox state *in vivo* was started close to 50 years ago (Chance 1957). The multiparametric monitoring approach was started close to 20 years ago when we were able to combine the measurement of NADH and PO_2 (Mayevsky et al. 1980), NADH and CBF or NADH, and CBF and HbO_2 (Mayevsky et al. 1992; Deutsch et al. 2004). Only recently we were able to upgrade the laboratory type of monitoring system into a medical device that was also approved by the FDA. In the present version of our device, we added to the three monitored parameters (TiSpec) the microcirculatory HbO_2, which will improve the diagnostic value of the CritiView. The possibility to correlate the systemic blood oxygenation (pulse oximetry) to the microcirculatory HbO_2 will add a new dimension to medical diagnosis made by the CritiView. This new device was also upgraded in the level of laser safety. The previous model (TiSpec) was classified as class 3B according to IEC standards (60825-1), while by improving the detection sensitivity (CritiView) the device is classified as class I. Also, the smaller dimensions of the CritiView will enable clinical team to have a monitoring unit near each bed in the ORs or in the ICUs. As seen in the results presented in this chapter, the most stable and representative parameter of tissue energy metabolism is the mitochondrial NADH redox state. The main factors affecting mitochondrial NADH are substrate and O_2 availabilities, levels of turnover of ADP, which are determined by metabolic activity of the tissue. The CBF and HbO_2, which are interconnected, are more sensitive to instability of regulation of blood flow, which is affected by many factors (NO, CO_2, and pH), and not only by oxygen consumption. We can conclude that monitoring mitochondrial NADH is the most appropriate parameter in the evaluation of tissue energy metabolism. The addition of the other three parameters is also significant to the diagnosis of tissue vitality.

7.6 Multiparametric Spectroscopy In Vivo: Toward a New Quantitative Tissue Vitality Index

7.6.1 Introduction

Monitoring of tissue physiological parameters, in real time, is a continuous challenge to scientists performing experimental animal studies as well as in clinical practice. Since the regulation of tissue physiology is dependent on many factors, monitoring of a single physiological parameter provides a very small fraction of the needed information for decision-making processes in clinical practice. The clinical diagnostic value of a single parameter monitoring device is a limiting factor preventing its implementation in daily clinical practice. In order to overcome this limitation, we developed a multiparametric monitoring system that can provide real-time information on changes in various physiological and biochemical parameters taking place in the brain (Friedli et al. 1982; Mayevsky 1983; Mayevsky et al. 1985), heart (Osbakken and Mayevsky 1996), kidney (Mayevsky et al. 2003) and liver (Barbiro-Michaely et al. 1998) exposed to various pathophysiological states. This approach was also used in preliminary clinical studies (Mayevsky et al. 1991, 1996, 1999a, 2001, 2003), but was never translated into a multiparametric monitoring device to be used in the daily clinical environment. The need for new approaches in evaluating the clinical status of patients is very wide. For example, the interpretation of the physiological status of the anesthetized patients is very subjective and therefore may be evaluated differently by various anesthesiologists. The same problem is demonstrated in patients hospitalized in the various ICUs that need additional monitoring approaches to evaluate objectively their condition. As of today, there is a discrepancy between the large amount of data collected by various techniques and the ability to integrate it into a practical objective tool to be used in clinical practice. Our effort to apply our developed multiparametric system to clinical daily use required the development of a new indexing technique for data interpretation. After the accumulation of a huge amount of experimental results, we decided to develop a preliminary tool named Tissue Vitality Index—TVI. In this study, we are presenting the preliminary developed vitality index that could be used in evaluating the vitality of the cerebral cortex in real time. We used the brain as a typical organ for the calculation of the vitality index, but the same approach could be used for other organs or tissues. The ultimate goal of our study is to develop the vitality index for tissues that could be monitored in the ICUs or in the ORs. In those medical environments, the need for a decision-making supporting system is very significant, as discussed by Hanson and Marshall (2001) and by Bates and Young (2003).

Figure 7.52 presents schematically the balance between O_2 supply and demand in a typical tissue. In order to calculate the vitality index of the tissue, it is necessary to use parameters representing the O_2 balance in the tissue. The supply of O_2 is evaluated by monitoring the microcirculatory blood flow using LDF as well as the level of hemoglobin oxygenation at the microvasculature. The evaluation of O_2 demand was not done directly in this study but O_2 balance was evaluated by monitoring of mitochondrial function using the NADH fluorometry/reflectometry approach. All three monitored parameters could be measured in all tissues or organs in the body, and therefore could be used for the calculation of the vitality index in all tissues using the same principles. Since we developed the formula of the vitality index on data accumulated in brain studies, a short introduction on this specific tissue is presented in Fig. 7.53. As seen, a typical brain tissue contains neurons, glial cells, and blood vessels. Our approach is to measure the physiological parameters from a large population of the tissue components which provide an average value of the parameter. By using the multiparametric monitoring approach, we are able to show the interrelation between various physiological and biochemical processes under

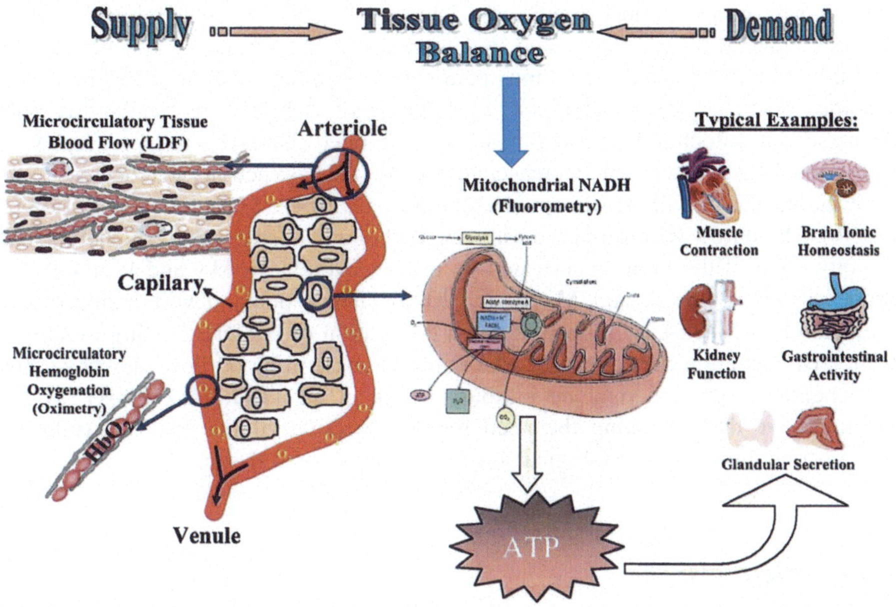

Fig. 7.52 The relationship between oxygen supply and demand. Tissue oxygen balance is evaluated by the integrity of mitochondrial function measured by the fluorescence of NADH *in vivo*. (Kutai-Asis et al. 2006 [Modified])

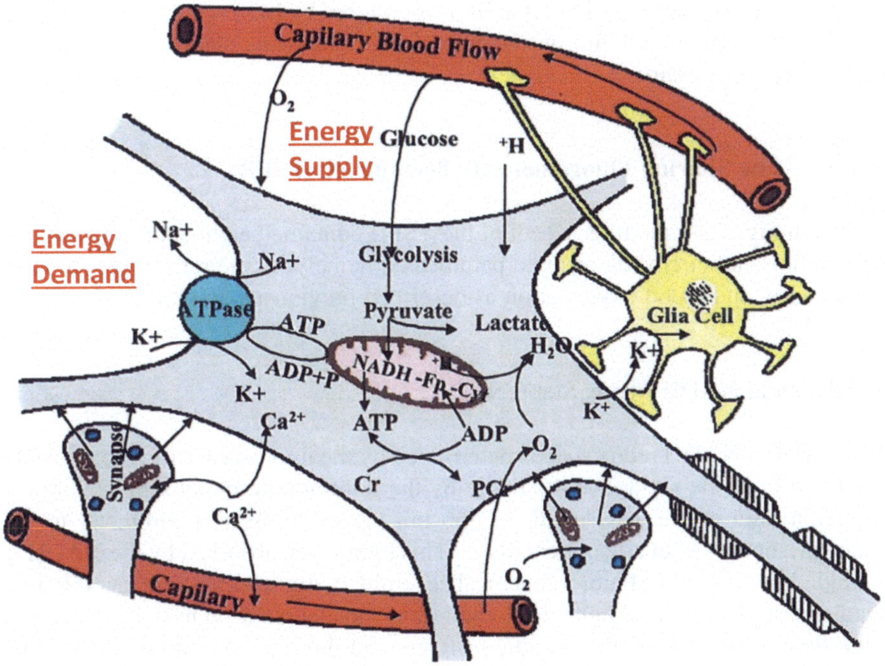

Fig. 7.53 Schematic presentation of the components of the brain tissue including neurons, Glia cells, and blood vessels and their involvement with the brain tissue metabolism. (Kutai-Asis et al. 2006 [Modified])

various pathophysiological conditions. The coupling or the uncoupling between microcirculatory blood supply and the intracellular mitochondrial function could be studied by monitoring the various parameters simultaneously in the same tissue volume. When oxygen is deprived, by hypoxia or ischemia, its availability will decrease and mitochondrial function will be inhibited (NADH will increase). As a result, the ATP turnover will decrease, the Na^+ K^+ ATPase activity will decrease and ionic homeostasis will be disrupted (Meilin et al. 1999; Mayevsky et al. 1999b, 2004b). Blood flow will increase during anoxia but will decrease under ischemia. On the other end, if the brain is activated by a SD wave (Mayevsky and Weiss 1991; Mayevsky 1992; Sonn and Mayevsky 2000) energy demand will increase and mitochondrial function will be stimulated (Oxidation of NADH) leading to significant CBF elevation. Under sepsis, a possible increase in blood flow may arise, but mitochondrial function may be inhibited (Fink et al. 1984; Fink 2001). Those examples are demonstrating the need for an objective vitality index in order to demonstrate the pathophysiological state of the monitored tissue.

7.6.2 Methods

The Multiparametric Monitoring System

The multiparametric monitoring system used in this study is based on the combination of several independent techniques. Simultaneous monitoring of mitochondrial NADH and HbO_2 from the brain surface was accomplished using the Time Sharing Fluorometer Reflectometer (TSFR). CBF is monitored by LDF (Fig. 7.54). All of the optical fibers were joined into one bundle of fibers that was placed on the parietal cortex of the rat's brain.

7.6.2.1 Time Sharing Fluorometer Reflectometer (TSFR)

As seen in Fig. 7.54, the filter wheel of the TSFR contained eight appropriate filters enabling the measurements of three parameters, namely, tissue reflectance, NADH fluorescence, and blood oxygenation as described previously (Rampil et al. 1992).

Mitochondrial NADH Redox State (NADH)

Mitochondrial NADH redox state is determined by the fluorescence of redox NADH monitored from the surface of the tissue by the fluorometric monitoring technique. Excitation light at the wavelength of 366 nm passes from the fluorometer to the tissue via a bundle of quartz optical fibers. This light when absorbed by the NADH is re-emitted at 450 nm (Fluorescence). The emitted fluorescent light at 450 nm, together with the reflected light at the excitation wavelength (366 nm), is transferred to the fluorometer via another bundle of fibers and through appropriate filters. The

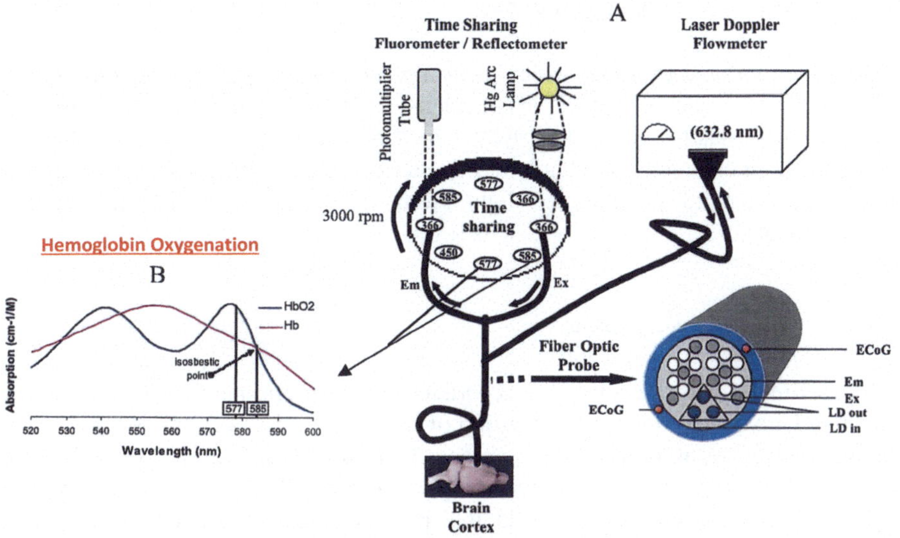

Fig. 7.54 (**a**) Schematic presentation of the Time Sharing Fluorometer Reflectometer (TSFR) combined with the Laser Doppler Flowmeter for blood flow monitoring. The time-sharing system includes a wheel that rotates at a speed of 3000 rpm with 8 filters: 4 for the measurements of mitochondrial NADH (366 nm and 450 nm) and 4 for oxyhemoglobin measurements (585 nm and 577 nm). The source of light is a mercury lamp. The probe includes optical fibers for NADH excitation (Ex) and emission (Em), laser Doppler excitation (LD in), laser Doppler emission (LD out) and needle electrodes for ECoG measurement (ECoG). (**b**) The absorption spectrum of Oxy and Deoxy-Hemoglobin. (Kutai-Asis et al. 2006 [Modified])

changes in the reflected light are correlated to changes in tissue blood volume and are used to correct the fluorescent signal for hemodynamic artifacts, which also affect the F signal at 450 nm. The changes in the corrected F signal are normalized relative to calibrated signals for nontreated animals under normoxic conditions (Mayevsky 1976, 1984; Mayevsky and Chance 1982; Mayevsky et al. 1990a).

Microcirculatory Blood Oxygenation

Tissue blood oxygenation was monitored based on the differences in the absorption properties of HbO_2 and deoxyhemoglobin. The brain surface is illuminated with light at two wavelengths (585 nm and 577 nm); 585 nm is an isosbestic point with the same absorption coefficient for HbO_2 and deoxyhemoglobin. Therefore, the absorption at this wavelength reflects changes in cerebral blood volume. At 577 nm, HbO_2 has a higher absorption coefficient than deoxyhemoglobin and by subtracting the 585 nm Ref from the 577 nm Ref a parameter correlated to blood oxygenation is obtained (Frank et al. 1989; Rampil et al. 1992).

7.6.2.2 Microcirculatory Blood Flow

TBF was monitored by LDF, which measures relative changes in microcirculatory blood flow where the basal level is considered as 100% and the level monitored on death is 0% (Wadhwani et al. 1990). The Doppler probe, which uses light of 632.8 nm, is included in the bundle of optical fibers used for the TSFR, and special filters are used to prevent red light from interfering with the fluorometric measurements.

7.6.2.3 Animal Preparation

All experiments were performed in accordance with the Animal Care Committee of Bar-Ilan University Guidelines. A group of male Wistar rats of 250–300 g (2.5–3-months old) were used. The rats were anesthetized with Equithesin (0.3 ml/100 g body weight, IP) (each ml contains: pentobarbital 9.72 mg, chloral hydrate 42.51 mg, magnesium sulfate 21.25 mg, propylene glycol 44.34% w/v, alcohol 11.5%, and distilled water). During the duration of the experiment steady anesthesia was maintained by 0.1 ml Equithesin injection every 30 min. Additionally, the rats were placed on a warming tray and body temperature was maintained at 37 °C. The rat was placed on an operating table with a special mouth holder. Two holes were drilled in the parietal bone, one for the monitoring probe and one for the pull–push cannula (4 mm and 2 mm, respectively) in order to induce SD. Additionally, two screws were attached to the parietal bone, for better fixation of the monitoring probe to the brain surface. The cannula designed for the probe, as well as the KCl cannula, for SD, were attached to the brain surface and fixed by acrylic dental cement. Then, the monitoring probe was placed in its cannula, avoiding extra pressure on the cerebral tissue (Mayevsky and Chance 1982, Mayevsky 1984).

In a group of rats, before craniotomy was induced, the rats were placed on the back and both carotid arteries were exposed and prepared for later occlusion using a surgical suture 4-0.

After a period of stabilization, all rats were exposed to a short anoxia, by pure nitrogen inhalation, until respiratory arrest. This short anoxia is a routine protocol conducted for the evaluation of proper placement of the probe on the brain surface and to ensure reliable monitoring. SD was induced by washing the brain surface with KCl solution (1–2 M), and stopped with NaCl 0.9%. Bilateral carotid artery occlusion was induced by pulling the sutures that surround the arteries for 30 s followed by releasing the sutures allowing blood flow through both carotid arteries to recover.

7.6.2.4 Real-Time Data Acquisition

All parameters were continuously monitored through the entire experimental period, and data was transferred to a multichannel computerized acquisition and recording system (Labview A/D software, National Instruments Co. USA). The data from the Labview was calculated and transferred to Excel for further analysis.

7.6.2.5 Tissue Vitality Index

The first step in the development of the Vitality Index was the preparation of special software using MATLAB, enabling various calculations of the data collected from the experiments. Several combinations with various partial importance (in percentage) were tested for each parameter (namely, TBF, NADH, and HbO_2). The ratios used included: 1. Equal values for each parameter, namely, 33% for TBF, NADH, and HbO_2. 2. 25% CBF: 50% NADH: 25% HbO_2. 3. 10% CBF: 80% NADH: 10% HbO_2. Each of these ratios was tested in all three models and was correlated to an expert opinion index.

7.6.3 Results

The responses of the three monitored parameters (excluding tissue reflectance) to the three types of perturbations are shown in Figs. 7.55, 7.56 and 7.57. The results of a typical anoxic episode and three Vitality Indexes in different ratios are presented in Fig. 7.55. As seen, inhalation of N_2 caused a decrease of 10% in CBF and HbO_2. Concomitantly, NADH increased by 47%. Those changes caused a decrease of 21–27% in indexes A and B, respectively (A—33:33:33, B—25:50:25) whereas, index C showed a decrease of 38% (C—10:80:10). Upon recovery, when the rat breathed air, a massive hyperemia in CBF to the level of 185% (as compared to the basal level) was observed and HbO_2 increased to the level of 116%. This massive hyperemia increased the level of index A to 130% and B to 120%; however, vitality index C was not affected by this hyperemia and returned to the basal level. When CBF returned to its basal level HbO_2, NADH and all vitality indexes fully recovered. Figure 7.56 presents the responses of the rat brain and three different vitality indexes, to ischemia induced by bilateral occlusion of the carotid arteries. As observed, the occlusion of both arteries caused a decrease in CBF and HbO_2 by 16%, while mitochondrial NADH increased by 23%. At this point, the vitality indexes decreased by 20%, 24%, and 27%, respectively, for indexes A, B, and C. When the occlusion was removed, CBF rapidly increased, reaching hyperemic levels of 116%, and HbO_2 increased to the level of 113%. Consequently, NADH was oxidized and decreased by 5% from its basal level. This oxidation was associated with elevation of all three vitality indexes by 11%, 9%, and 5% for indexes A, B, and C, respectively. Twelve seconds after the oxidation phase, all parameters, including the vitality indexes, showed full recovery. When SD was induced (Fig. 7.57), three waves developed. The first wave was characterized by a biphasic response of HbO_2, starting by HbO_2 decrease of 16% followed by a large increase of CBF (138%) and a 22% increase in HbO_2. As for the NADH, here also a two-step reaction was recorded. NADH was oxidized rapidly by 21% followed by a further oxidation of 35%. At this stage all the Vitality indexes (A, B, and C) showed a biphasic increase of 2%, 7%, and 15% followed by a further increase of 20%, 24%, and 26%, respectively. At the stage of

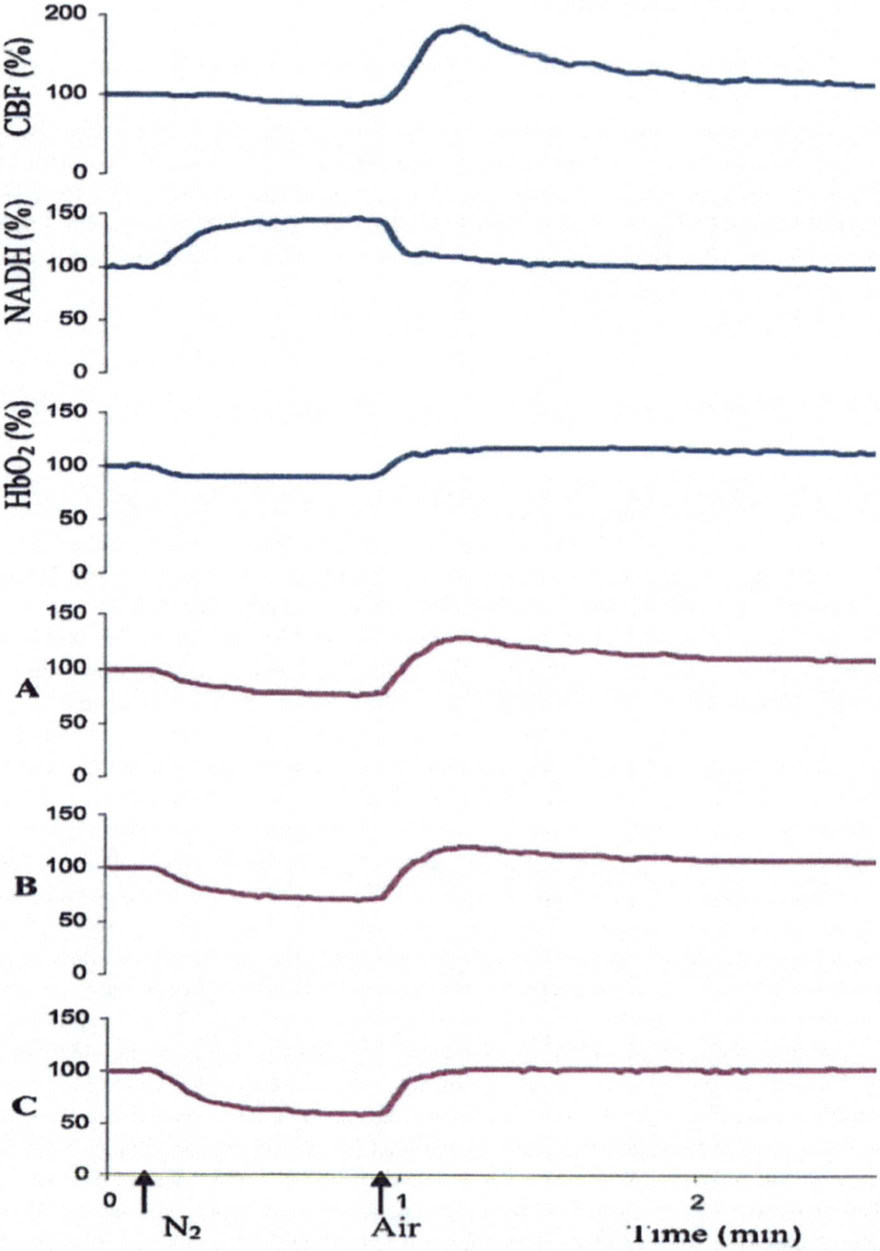

Fig. 7.55 The responses of Cerebral Blood Flow (CBF), mitochondrial NADH level (NADH), and hemoglobin oxygenation level (HbO_2) and three vitality indexes (**a**, **b**, and **c**) to short anoxia (N_2) in rat model. (**a**) an index with equal values for each parameter; (**b**) an index with 25% CBF: 50% NADH: 25% HbO_2; (**c**) an index with 10% CBF: 80% NADH: 10% HbO_2. (Kutai-Asis et al. 2006 [Modified])

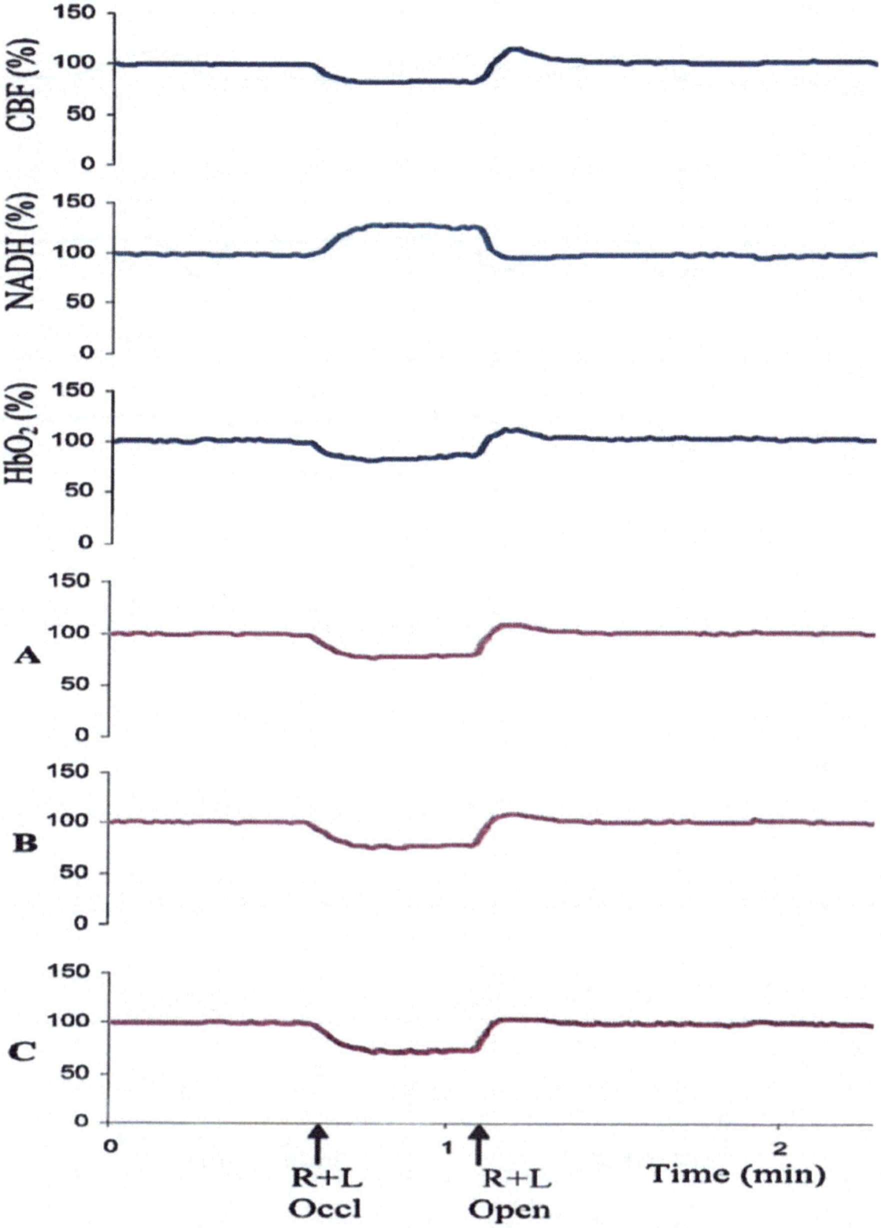

Fig. 7.56 The responses of Cerebral Blood Flow (CBF), mitochondrial NADH level (NADH), and hemoglobin oxygenation level (HbO_2) and three vitality indexes (**a**, **b**, **c**) to bilateral carotid arteries occlusion in rat model. (**a**) an index with equal values for each parameter; (**b**) an index with 25% CBF: 50% NADH: 25% HbO_2; (**c**) an index with 10% CBF: 80% NADH: 10% HbO_2. (Kutai-Asis et al. 2006 [Modified])

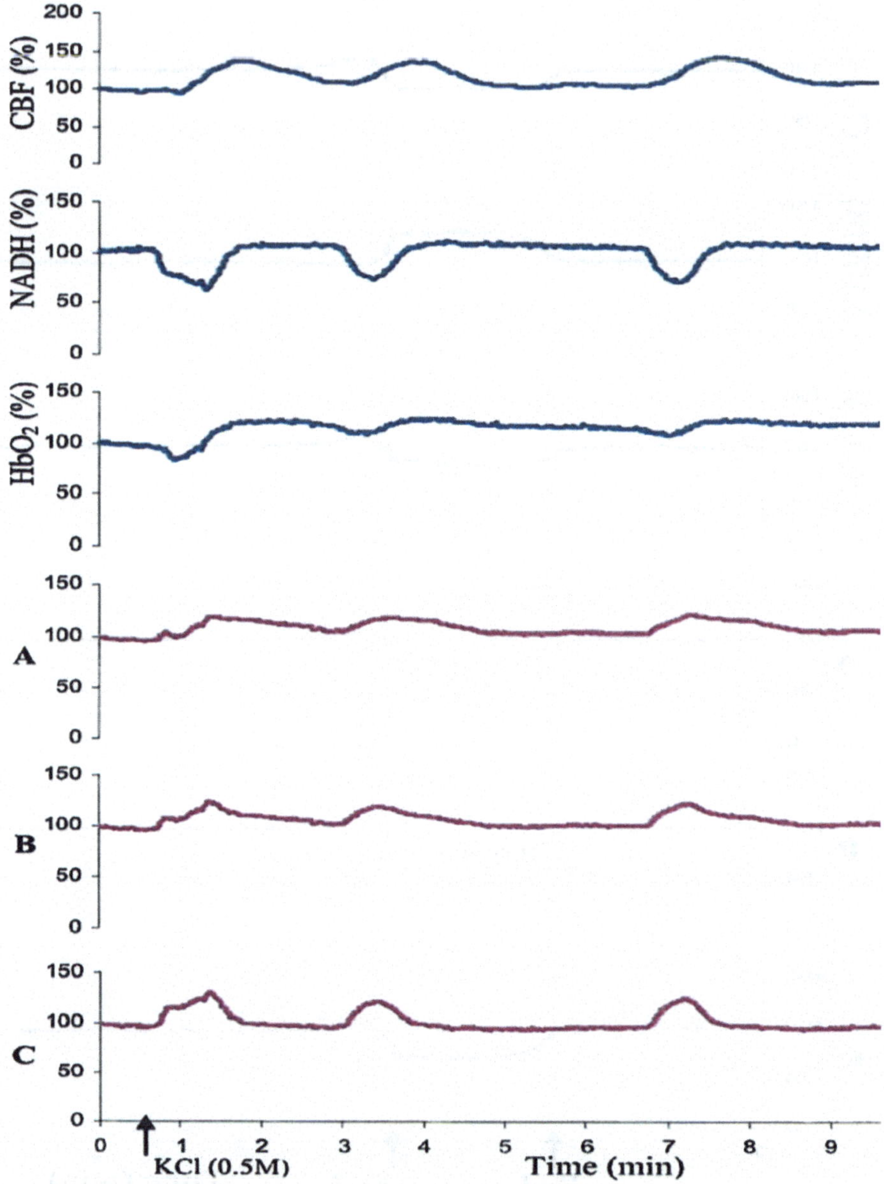

Fig. 7.57 The responses of Cerebral Blood Flow (CBF), mitochondrial NADH level (NADH), and hemoglobin oxygenation level (HbO_2) and three vitality indexes (**a**, **b**, **c**) to spreading depression (SD) induced by 0.5 KCl solution applied epidurally in rat model. (**a**) an index with equal values for each parameter; (**b**) an index with 25% CBF: 50% NADH: 25% HbO_2; (**c**) an index with 10% CBF: 80% NADH: 10% HbO_2. (Kutai-Asis et al. 2006 [Modified])

Table 7.3 The correlations (R^2) between calculated vitality indexes and an expert opinion under three pathophysiological conditions: SD spreading depression; Ischemia bilateral carotid artery occlusion; Anoxia pure N_2 inhalation

R^2 (SD)	R^2 (Ischemia)	R^2 (Anoxia)	Vitality index	
0.6468	0.9384	0.6686	A	CBF 33% NADH 33% HbO_2 33%
0.8035	0.9133	0.7960	B	CBF 25% NADH 50% HbO_2 25%
0.7755	0.9406	0.9187	C	CBF 10% NADH 80% HbO_2 10%

recovery from the first wave, none of the parameters fully recovered. CBF dropped to the level of 110%, HbO_2 decreased to the level of 118%, and NADH was reduced to the level of 107%. As for the vitality indexes, index C returned to its basal level immediately, while vitality indexes A and B recovered gradually. The second and third waves, showed an increase of 34–36% in CBF which was associated with HbO_2 oxygenation to the level of approximately 110%. As a result, NADH was oxidized by 24–31%. Regarding the changes observed in the vitality indexes, here also indexes A and B yielded similar pattern of smooth waves with moderate slopes whereas, index C waves were more instant and with steeper slopes.

Table 7.3 represents R^2 values in the three models used (SD, ischemia, and anoxia) when compared to an expert opinion. As seen, although vitality index A gave a high correlation coefficient in the ischemia model (0.9384), it showed low values in the two other models. The B vitality index showed better correlation in the SD model compared to index A; however, it still showed low correlation for the anoxia model. Vitality index C showed very high correlation coefficients in both ischemia and anoxia models, with a borderline value for the SD model.

7.6.4 Discussion

The aim of this study was to develop a new tissue vitality index based on the monitoring of three physiological parameters. The multiparametric monitoring system provides, simultaneously, information on microcirculatory blood flow and hemoglobin oxygenation as well as mitochondrial function. Under various pathophysiological conditions these three parameters are changing continuously and the expert is able to evaluate the vitality level of the tissue. In order to provide an objective evaluation of the state of the tissue, it was important to develop the vitality index. Few techniques are available in order to calculate and evaluate the changes in various physiological and biochemical parameters related to the metabolic activity of the tissue. In most cases, the monitoring device provides information on a single parameter. Monitoring of TBF, tissue PO_2, or blood oxygenation (HbO_2) provides information indicating the integrity of microcirculatory function. Nevertheless, these parameters do not represent the activity of the mitochondria, which is the most important functional unit in cellular energy metabolism. Therefore, we decided to calculate the tissue vitality index based on various parameters covering the entire range of oxygen gradient, namely, between the microcirculation and the intracellular energy transformation site—the mitochondria. Our results demonstrated that the most accurate vitality index to be used is vitality index C in which NADH is represented with the highest percentage—80%, and CBF and HbO_2 are presented with 10% each. As seen under anoxia, the hyperemia, which probably developed in order to compensate for the deficit in oxygen, yielded an increase in vitality indexes A and B, although this increase in CBF does not imply any improvement in the brain vitality state. However, index C was not influenced by this hyperemia and returned to basal levels concomitantly with the rat breathing air. Moreover, mitochondrial NADH levels fully recovered as the rat started to breathe air, a few seconds before CBF reached its maximum hyperemia. In other words, the hyperemia seen following anoxia is actually not essential for adequate O_2 supply to the mitochondria in the brain cortex. A similar phenomenon was observed under the ischemic model where following the decrease in CBF all three indexes were also decreased to the same level. However, the hyperemia observed immediately following both carotid arteries reopening caused a higher increase in vitality indexes A and B, compared with vitality index C. It should be clarified that this hyperemia does not improve tissue vitality state but rather is a result of microvessels contractility state. During ischemia, cortical arteries are vasodilated in order to improve blood supply to the impact cerebral tissue. In order to restore vasculature tone to its pre-occlusion state, several seconds are needed; meanwhile, vessels are relatively vasodilated compared to their normal state. Moreover, here also the hyperemia is needless, since full recovery of tissue energetic state (NADH) was accomplished immediately as arterial obstructions were removed, namely, before full hyperemia was achieved. The use of the vitality index under a condition in which the brain is activated (the SD model), also indicated that the NADH is a crucial parameter to be taken under consideration for vitality index calculations. In other words, although all three indexes yielded similar changes following SD, the waves observed in index C were the clearest, with sharp borders between increased vitality versus decreased vitality.

Furthermore, the advantage of index C is pronounced in the high correlation coefficient values presented in Table 7.1 under all three models used, where the calculated values were compared with an expert opinion. Since the main purpose of a vitality index is to give indication for the state in which a tissue or an organ is in, and since oxygen is a major factor in tissue viability, it is obvious that this index should include the combination between tissue oxygen supply and demand. The mitochondrion is the cell energy factory (Waltemath 1970), and mitochondrial NADH is the most sensitive indicator for oxygen balance (Mayevsky et al. 1980, 1992, 1998); thus, we believe NADH should get the highest importance in a formula for tissue vitality index.

7.7 From the TiSpec to the CritiView Used in Clinical Studies

7.7.1 Introduction

The need for a practical and simple medical device for *in vivo* monitoring of tissue physiology in real time had stimulated many scientists as well as various medical device companies. As of today, most of the developed medical devices provide information on one or two parameters representing tissue physiology and biochemistry. Few companies and researchers have developed Laser Doppler Flowmeters for TBF monitoring (Stern 1975; Bonner and Nossal 1981; Shepherd 1990; Batista et al. 1996), tissue hemoglobin saturation levels (HbO_2) (Rampil et al. 1992), tissue oxygen level (pO_2) (Mayevsky et al. 1980), tissue CO_2 and pH (Lubbers 1995) levels, and mitochondrial cytochrome oxidase redox state (Scheffler 2000). Nevertheless, not one of the mentioned techniques is used in daily clinical practice in any medical field. The most important parameter to be evaluated is the balance between O_2 supply and demand defined as oxygen balance and represented by mitochondrial redox state. As of today, a number of measured parameters, in daily clinical practice, represent the integrity of the cardiovascular as well as respiratory system. In our previous publications (Mayevsky et al. 2004b, 2005), we described a few devices aiming to monitor tissue vitality using a single-point probe that measured up to 4 optical parameters from a small tissue area. These devices had never been applied to routine clinical use due to factors such as size, price, or the ability to use and interpret the results as a clinically significant tool. The new device described in this chapter was developed in order to overcome all the disadvantages of the early developed devices. The prototype of the CritiView that was described last year (Mayevsky et al. 2005) was adapted to a single-point measurement in critical care medicine. The new model of the CritiView enables to integrate the measurements of three different locations of the tissue thus improving the statistical basis of the measurement. Also, the CRV3 is significantly smaller thus enabling its incorporation into the small space available in ORs or in the ICUs.

7.7.2 Methods

7.7.2.1 The Monitoring System—"CritiView"

The new device named CritiView is based on the principles of monitoring the four parameters as described below:

Mitochondrial Function

Mitochondrial function is evaluated by monitoring of NADH fluorescence. The measurement is done by excitation of the tissue by UV light (375 nm) and measuring the emitted F signal (420–480 nm). The principles of this method were established 50 years ago (Chance et al. 1955, 1962a), and the main changes that were introduced since then were the type of light source used as well as the usage of a bundle of optical fibers to connect the tissue to the fluorometer. The main light source that was used was Hg arc lamp having a strong emitted light at 366 nm absorbed by the reduced form, NADH and not by the oxidized form (NAD^+). The introduction of 375 nm LED by Nichia as a new light source in the present instrument (CritiView) significantly affects the size and the price of this component in this medical device. The use of optical fibers led to a more flexible connection between the monitored tissue and the fluorometer. This enabled us to monitor the unanesthetized brain (Chance et al. 1973; Mayevsky 1984), or at different locations in the same animal (Mayevsky and Chance 1982; Kraut et al. 2004). The fluorescence of NADH is a good indicator for the decrease in oxygen availability in the cellular compartment. As shown in Fig. 7.63, exposing the brain to anoxia (100% N_2) led to a significant increase in the NADH fluorescence.

Tissue Reflectance

The continuous changes in tissue absorption properties affect the intensity of tissue reflectance and fluorescence. It was found that Ref could be used as a correction tool for the artifacts measured in the NADH F signal (Chance et al. 1973; Mayevsky and Chance 1982; Mayevsky 1984). The main factor affecting the Ref signal changes by blood volume (Mayevsky and Chance 1982; Mayevsky 1984; Rampil et al. 1992; Mayevsky et al. 2004b). When ischemia is induced, the decrease in blood volume will lead to an increase in the Ref signal. During the recovery phase from ischemia, the hyperemic response (increase in blood volume) will be recorded as a decrease in the Ref signal.

Tissue Blood Flow

Monitoring of microcirculatory TBF using the laser Doppler approach started more than 35 years ago (Stern 1975; Bonner and Nossal 1981), and is widely used in various animal models as well as in clinical applications (Shepherd 1990). The main advantage of the LDF approach is the real-time data provided by the optical nature of the signal. As of today, the large number of publications (more than 5000) in experimental animals and patients indicate its applicability (Mayevsky et al. 2004b).

Microcirculatory Blood Oxygenation

For the monitoring of blood oxygenation we utilized a two-wavelength Ref technique to analyze the absorption of light in the blood, exploiting the difference in absorption spectra of oxyhemoglobin (hemoglobin saturated with oxygen) and deoxyhemoglobin. The principles of the technique were demonstrated years ago using a time-sharing system (Rampil et al. 1992). Generally, the absorption coefficient of oxyhemoglobin differs from that of deoxyhemoglobin, along most of the spectrum (Prahl 1999). However, in certain isosbestic wavelengths, the absorption curves cross each other and represent the same value. The change in Ref in an isosbestic wavelength is affected only by blood volume, while a change in Ref in a non-isosbestic wavelength is affected by the oxy-deoxy ratio as well. By subtracting the back-reflectance in these two wavelengths, one can consider the difference between the measurements as qualitative representation of blood oxygenation at the microcirculatory level. The former time-sharing system (Rampil et al. 1992; Mayevsky et al. 2004a) used two Hg lamp emission bands to determine the oxy-deoxy ratio *in vivo* by measuring the Ref in 585 nm (R585) and 577 nm (R577). Calculating—(R577-R585) yields the relative change in oxyhemoglobin state. In this study, we used the same principle with contemporary 530 nm (nearly isosbestic) and 470 nm (non-isosbestic) super bright LEDs. The relative change in oxyhemoglobin state is therefore derived by—(R470-R530). It is important that the two wavelengths are closed and roughly within the same range of absorption coefficient. Thus, the absorption of the tissue may be considered constant and neutralized.

7.7.2.2 Device Description

The CRV device described in this study is a platform that could be used in various experiments as well as clinical situations. The standalone recent model was developed during 2004–2005 as shown in Fig. 7.58. This very small unit evolved after two generations of the TiSpec (1999, 2001) and one generation of the CritiView (2003).

The general scheme of the CRV3 shown in Fig. 7.59 consists of a few units enabling the measurement of the four monitored parameters from three different points or areas in the tissue.

Light Source Unit

The light source unit consists of three identical floors. Each floor is responsible for supplying excitation light for a single excitation point and it comprises all excitation light sources required for CritiView measurements. In Fig. 7.60, one of the three light source floors and the detection unit are shown. It consists of near IR laser diode at 785 nm for laser Doppler measurements, LEDs at 530 nm and 470 nm for tissue oxygenation measurements and UV LED at 375 nm for excitation of NADH

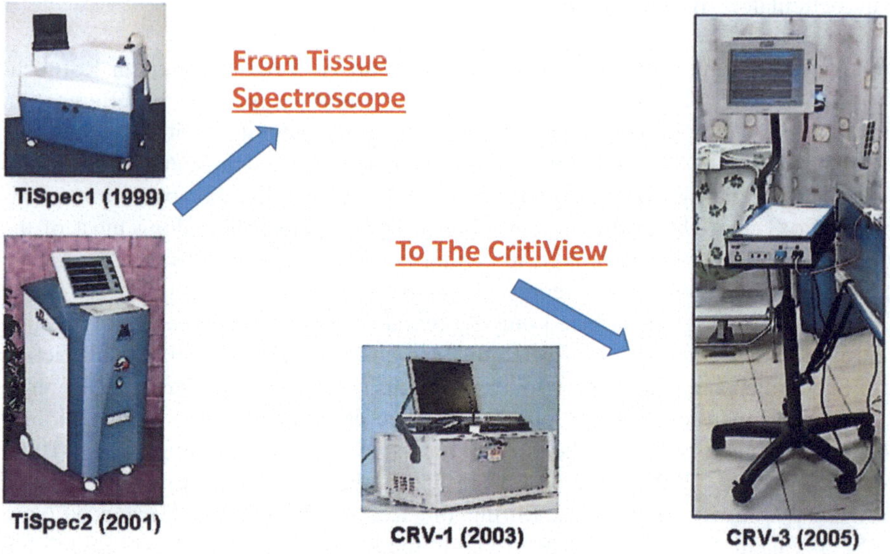

Fig. 7.58 Four generations of multiparametric monitoring system enabling the monitoring of tissue vitality of various tissues in real time. (Mayevsky et al. 2006 [Modified])

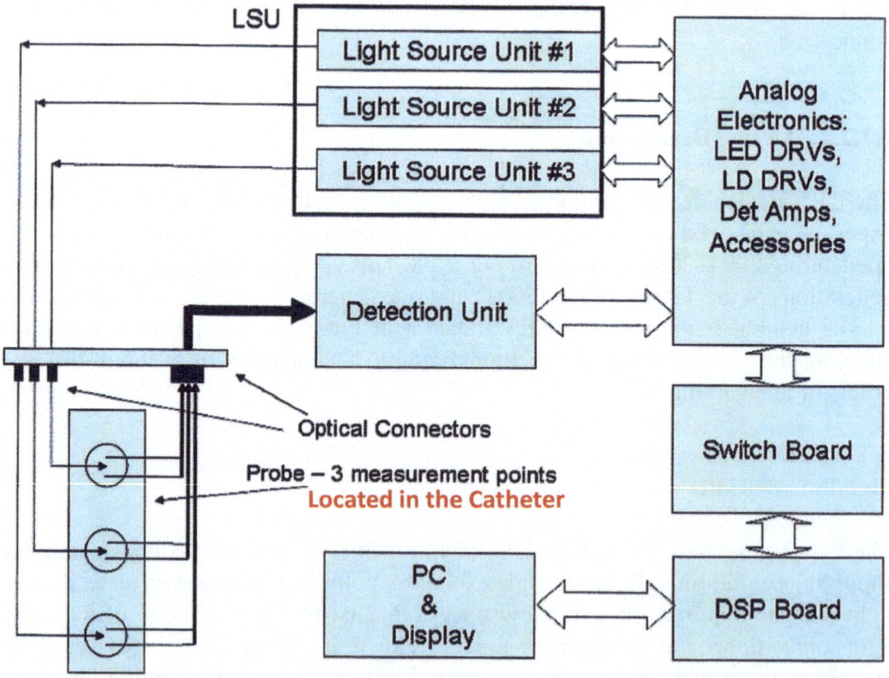

Fig. 7.59 Schematic presentation of the CRV3 used in this study (see details in text). (Mayevsky et al. 2006 [Modified])

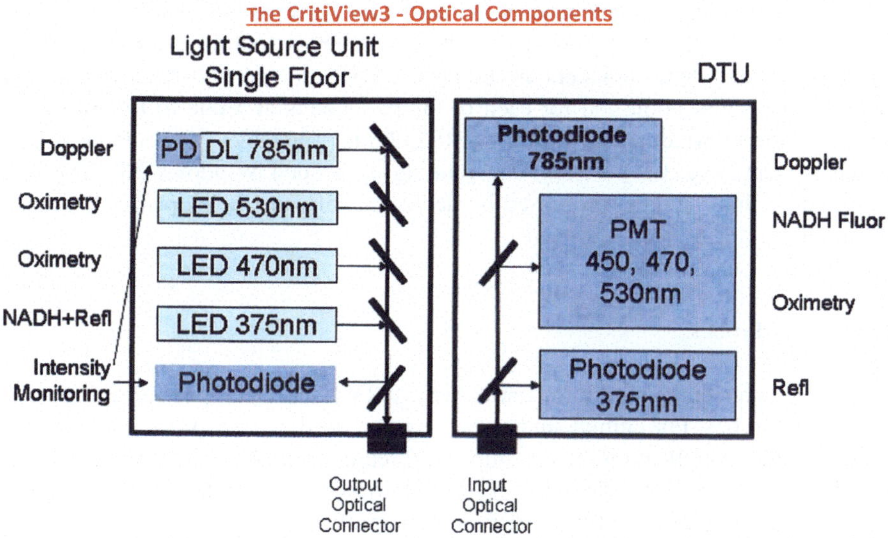

Fig. 7.60 Detailed schematic description of one of the floors of the CRV3 containing the light source unit (LSU) and the detection unit (DTU). Mayevsky et al. 2006 [Modified]

fluorescence and reflection measurements. All discrete light sources are collimated (not shown on the figure) and the different wavelengths are combined and coupled into a single excitation fiber by means of dichroic beam splitters. The light intensity of the LEDs is monitored by a discrete photodiode and the laser diode intensity is monitored by its internal photodiode.

Detection Unit

The detection unit consists of photodiode for Reflection measurements at 375 nm; photomultiplier (PMT) for NADH fluorescence at 450 nm and tissue oxygenation measurements by reflection measurements at two wavelengths of 470 nm and 530 nm; photodiode for laser Doppler measurements at 785 nm. The light incident into the detection unit is distributed according to the different wavelengths to appropriate detectors by means of dichroic beam splitters.

Electronics and DSP

The electronics unit includes all necessary drivers for LED and laser light sources and amplifiers for the detectors. The digital signal processor (DSP) P5 from Tern inc. (Davis, CA) is responsible for all the timing of the data acquisition. The system utilizes time-sharing operation mode in order to acquire signals from various light sources for each measurement point of the probe. In order to enable a sufficient number of A/D and D/A inputs for all light sources and detectors, additional multiplexers and demultiplexers are used. The DSP transmits all the acquired data for each measured point and for each of the parameters at a rate of five times per second to the external tablet PC unit via RS232 link. The tablet PC is a Hitachi Data Systems (San Jose, CA) VisionPlate running embedded Windows XP. The data acquired by the system is processed by dedicated software developed in Microsoft Visual Basic.

Fiber Optic Probes

In order to connect the monitored tissue to the CRV3, we developed various types of fiber optic probes. For animal studies, we used a "needle" type probe while for our present clinical application we developed a three-point probe embedded in a three-way Foley catheter as shown in Fig. 7.61. The cross-section of the catheter is shown in Fig. 7.62, including the tip of the fiber optic probe.

The CRV probe is built around three measurement point concept (see Fig. 7.62). The main advantage of the usage of multiple measurement points for assessment of tissue vitality parameters is in spatial averaging of the measured parameters. Indeed, microcirculatory TBF, tissue oxygenation, Ref, and NADH fluorescence may exhibit various levels as function of measurement site. The averaging of these parameters over three measurement points enables us to retrieve single value for each parameter that best represents the value of the parameter at the tissue under test. Each measurement point consists of one excitation fiber that delivers the excitation light from the CritiView device and two collection fibers that collect light from the tissue surface and direct it back to the CritiView device for detection. All fibers are embedded into a single microplastic fixture which holds each of the fiber in predefined place. While each measurement point receives the excitation light through single excitation fiber from separate Light Source Unit Floor, the collection fibers from all measurement points are combined at the collection optical connector into single bundle. Therefore, all collected light from all measurement points is detected by a common detection unit (Stern 1975; Bonner and Nossal 1981).

Animal Preparation and Experimental Protocols

All experimental protocols were approved by the Institutional Animal Care Committee under the instruction of the National Institute of Health. Male Mongolian gerbils (50–80 g) were used. The animal was anesthetized by Equithesin (E-th = Chloral hydrate 42.51 mg; Magnesium sulfate 21.25 mg; Alcohol 11.5%; Propylene glycol 44.34%; and Pentobarbital 9.72 mg) IP injection 0.3 ml/100 g body weight. The animals were kept anesthetized during the operation as well as during the entire monitoring period, by IP injections of E-th 0.03 ml every 30 min. We have been using this anesthetic for approximately 50 years and it has never shown significant effects on mitochondrial activity. Furthermore, we have not been able to change the response of the rat brain to SD by using an increased dose of

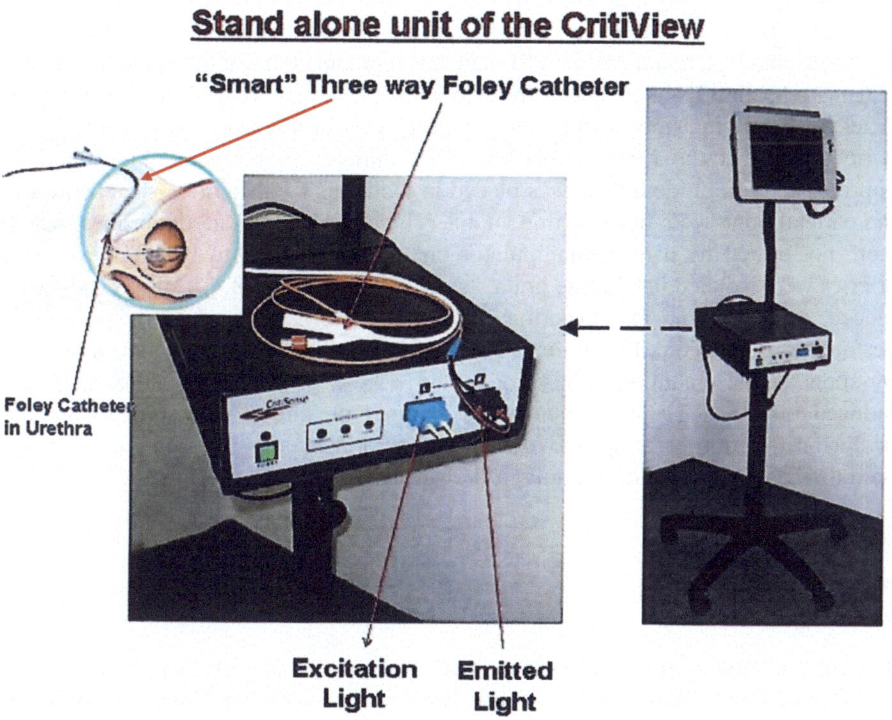

Fig. 7.61 The CRV3 model used for monitoring of four different parameters from the wall of the urethra. (Mayevsky et al. 2006 [Modified])

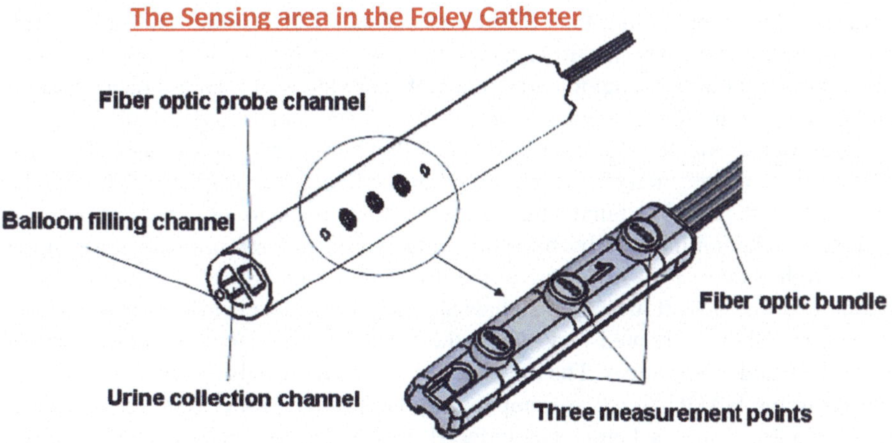

Fig. 7.62 Enlargement of the Foley catheter section where the fiber optic probe is embedded. (Mayevsky et al. 2006 [Modified])

Equithesin, suggesting that this is a safe drug (Mayevsky et al. 1982). The addition of small volumes of E-th every 30 min kept the animal in a stable state. Body heat was measured by a rectal probe (YSI) and was regulated to be at the range of 35–37 ° C using a heating blanket. The gerbil was anesthetized and placed in a head holder. After a midline incision of the skin, a hole (5 mm in diameter) was drilled in the parietal bone of the right hemisphere. The dura mater remained intact and an appropriate light guide holder was placed in the hole. Two stainless steel screws in the parietal bone were used, with dental acrylic cement, to fixate the probes, which were positioned by a micromanipulator on the cortex. The two common carotid arteries were isolated just before brain surgery, and ligatures of 4-0 silk threads were placed around them. The experimental animals were exposed to the following perturbations: Anoxia: The animals were exposed to oxygen deficient atmosphere by spontaneous breathing of N_2, for a short (25 s) period. Terminal anoxia was induced by 100% N_2 until the animal stopped breathing. Ischemia: Reversible occlusion (~1 min) of one or two common carotid arteries (by constricting them with threads) led to brain ischemia in the gerbils.

7.7.3 Results

The current version of the CritiView—CRV3 was tested in gerbil's brain as well as in rat's brain and small intestine. In this book, we are showing typical responses of the gerbils' brain to anoxia and ischemia. In Figs. 7.63 and 7.64, we present the five measured signals that are translated into the four physiological parameters, namely, TBF, R375, NADH, and HbO_2. The raw signal of fluorescence is translated into the net NADH signal by subtracting the R from the F signal. The HbO_2 signal is obtained by subtracting the R530 (isosbestic Ref) from the R470 (the non-isosbestic Ref). The net Ref change between the R470 and R530 is reversed in its polarity in order to demonstrate the HbO_2 in values of increase or decrease in HbO_2 and not the Ref change. As seen in Fig. 7.63, under anoxia (100% N_2) the initial response was the decrease in HbO_2 followed by an increase in NADH. The TBF showed a later increase which turned into a hyperemic response that was also recorded in the HbO_2 signal after recovery from the anoxia. The responses to complete ischemia induced by bilateral occlusion of the carotid arteries are shown on the right side of Fig. 7.63. Due to the decrease in blood flow and volume to the brain (TBF signal), all measured signals are increasing. The net changes are a large decrease in HbO_2 correlated very well to the NADH. A small hyperemia was noticed in the TBF and HbO_2 traces. The time needed for recovery to base line value is much shorter in the NADH signal as compared to the TBF and other Ref signals (R375, R470, R530). When ischemia was induced in two steps, namely, occlusion of one carotid artery followed by the occlusion of the second artery (while the first are remained occluded), the expected responses were recorded. The two levels of changes were recorded in the TBF, NADH, Ref, and HbO_2 signals.

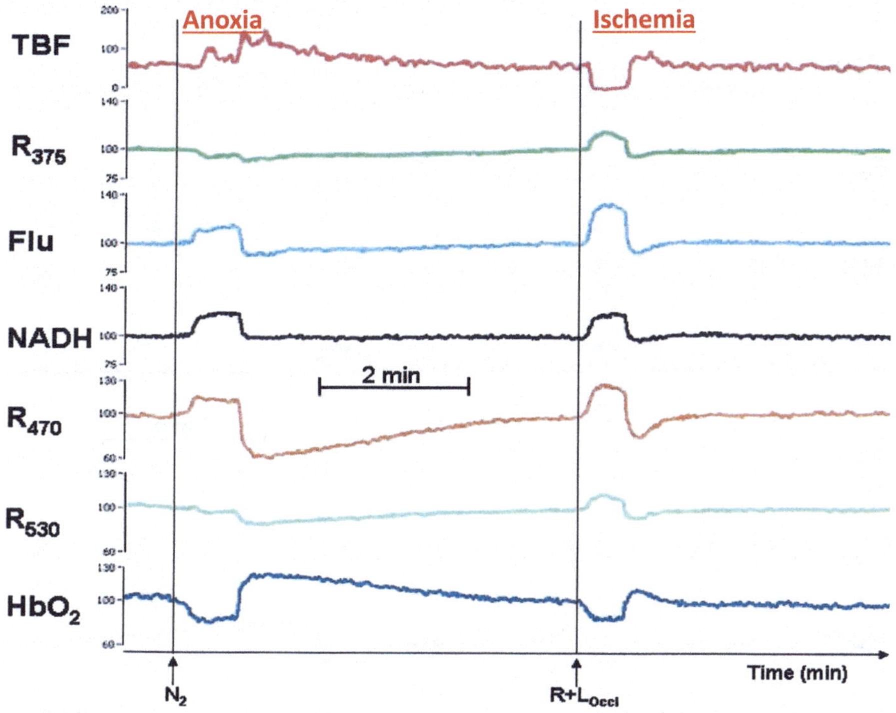

Fig. 7.63 Responses of the gerbil brain to anoxia (N_2) and complete ischemia ($R + L_{occl}$). TBF Tissue blood flow, F and NADH—uncorrected and corrected NADH fluorescence, R375, R470, R530—Reflected light at three wavelengths. HbO_2—Saturation level of hemoglobin. (Mayevsky et al. 2006 [Modified])

A small recovery was noted shortly after the unilateral occlusion due to blood flow compensation through the anterior part of the circle of Willis of the gerbil's brain. The responses to anoxia on the right side of Fig. 7.64 were very similar to those measured in the first anoxia shown in Fig. 7.63.

7.7.4 *Discussion*

The main progress, in our developmental effort to monitor tissue vitality, presented in the current report, is the significant increase in the better statistical-based monitoring approach. Most of the devices developed in the past measured parameters from a single small spot in the tissue. Due to variation in the vascular as well as in O_2 gradients across the tissue, the monitored parameter may not adequately represent the real behavior of the tissue. Only by increasing the number of monitored sites in the tissue, the variations will decrease and the results will better represent the physiological and biochemical processes taking place in the tissue. The monitoring

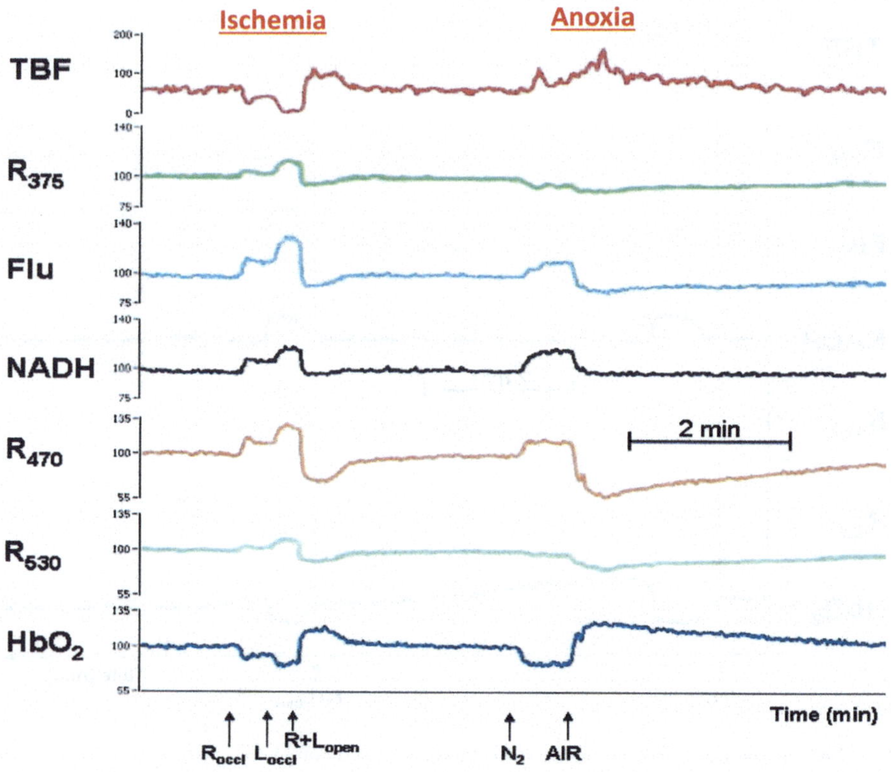

Fig. 7.64 Responses to partial (Roccl) and complete ischemia (L_{occl}) as well as anoxia (N_2) of the gerbil brain. Abbreviations are as in Fig. 7.63. (Mayevsky et al. 2006 [Modified])

of four different parameters enables the correlation between various processes taking place in the tissue. Three of the parameters are indicators of events related to the microcirculatory function. The fourth parameter—NADH redox state—originates from the intramitochondrial space. This parameter represents the physiological activity of the energy production system at the cellular level. The NADH redox state is sensitive to changes in O_2 availability in the mitochondria as well as to the consumption of ATP by various processes taking place in the cells. In our previous publication (Mayevsky and Chance 1982), we showed the simultaneous monitoring of NADH at four different location of the cerebral cortex. The responses to the lack of O_2 or to the stimulation of energy metabolism were very similar in the four monitoring sites. The volume of tissue that was monitored by the fiber optic probe was larger as compared to the size of tissue analyzed by our current single probe. Therefore, the integration of three monitored sites will increase the reliability of the results. The development of the sophisticated three-way Foley catheter including the three measuring points open up a new era in clinical monitoring of patients hospitalized in ICUs or in the ORs.

7.8 Mitochondrial Dysfunction: Bench-to-Bedside Optical Monitoring of Tissue Vitality

7.8.1 Introduction

The intracellular cytoplasmic organelles, the mitochondria (previously named bioblasts), were described by cytologists since the mid-nineteenth Century (Lehninger 1964). The name, mitochondrion was given by Benda in Benda 1898. Only in the mid-twentieth Century the role of the mitochondria in oxidative energy metabolism was established (Kennedy and Lehninger 1949). The invaluable pioneering work of Chance and collaborators since the early 1950s opened up a new era of studying mitochondrial function using spectroscopic techniques (Chance and Legallias 1951; Chance 1952, 1954). Monitoring of NADH redox state in isolated mitochondria (Chance and Baltscheffsky 1958) and later on under *in vivo* conditions (Chance et al. 1962a) established the foundations for the understanding of cellular bioenergetics developed under various pathophysiological conditions in experimental animals as well as in patients (Mayevsky and Chance 2007; Mayevsky and Rogatsky 2007). In parallel to the role of mitochondria in cellular bioenergetics, the role of mitochondrial dysfunction in various human diseases became very clear and obvious. For example, the involvement of the mitochondria in tumor cells was described initially by Warburg almost 100 years ago (Warburg 1930) followed by many studies later on (Warburg 1956; Weinhouse 1956). Since then, a large volume of papers showed the involvement of the mitochondria in many human diseases (Mizuno et al. 1995; Tatton and Olanow 1999; Preston et al. 2001; Crouser 2004; Ballinger 2005), as shown in Fig. 7.65. Nevertheless, the translation of the accumulated information into a practical clinical tool or medical device did not happen. The existing devices that are used during operations or in the ICUs provide real-time information on the activity of the respiratory and cardiovascular system. The information on the microenvironment activities, especially mitochondrial function, at the tissue level is missing. In the previous sections, we described the structure and function of a multiparameter monitoring system that provides information on the viability of the tested tissue (Mayevsky et al. 2005, 2007; Mayevsky and Rogatsky 2007).

The aims of this sections are as follows:

1. To present the upgraded technology of the CRV3 as compared to the previous model of the CritiView named CRV1 (Mayevsky et al. 2006).
2. To compare the performance of the CRV3 *in vitro* and *in vivo* to the CRV1 or other predicated devices cleared by the FDA.
3. To show preliminary results of a clinical study using the Foley catheter in patients.

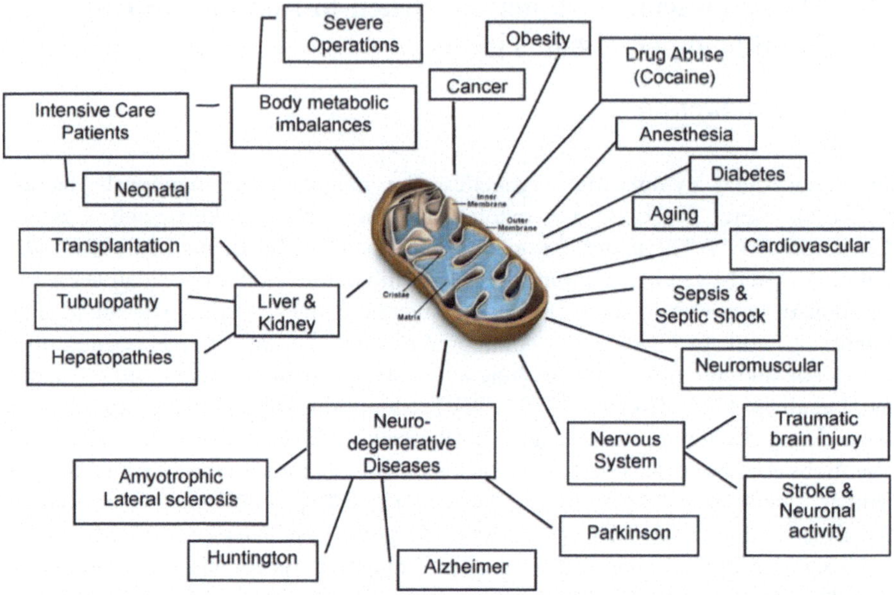

Fig. 7.65 The involvement of mitochondria in various pathological and pathophysiological conditions. (Mayevsky et al. 2008 [Modified])

7.8.2 Methods

In this section, a new updated version of the CritiView, named CRV3, will be described. This unit, cleared by the FDA in 2007, was tested in small animals (gerbils), large animals (pigs), and in more than 30 patients. The CRV3 device carries out certain *in vivo*, spectroscopic measurements as a multiparametric monitoring device intended for the measurement of tissue metabolic state. The CRV3 consists of the combination of a NADH Fluorometer, a Laser Doppler Flowmeter as well as a reflectometer for the monitoring of microcirculatory blood oxygenation. The CRV3 transmits radiation through tissue at known wavelengths and measures these parameters by detecting the intensities of light in the following wavelengths:

1. Mitochondrial NADH fluorescence emitted at 420–480 nm.
2. Total backscattered light (375 nm) reflected from the tissue. This parameter allows for correction of the NADH fluorescence measurement due to changes in tissue blood volume.
3. Doppler shifted laser light (785 nm) reflected from moving blood cells.
4. Two wavelength reflectometry (470 and 530 nm) enabling the monitoring of HbO_2.

The CRV3 does not carry out this assessment. It simply provides the measured information to the user. These parameters contain information pertaining to the redox state of mitochondrial NAD/NADH of the tissue, microvascular blood flow and oxygenation in the tissue, and total blood volume of the tissue. Changes in these values reflect changes in the balance between oxygen supply and oxygen demand. Since the CRV3 is intended as an adjunctive measurement device when used in conjunction with conventional patient monitors, this further supports the low safety risk of the CRV3.The basic scientific principles and technology of the CRV3 are identical to the predicate device (CRV1) described 2 years ago (Mayevsky et al. 2006).

7.8.2.1 Description of the Device—CRV3

The Light Source Unit (see Fig. 7.66) of the CRV3 comprises a 785 nm CW laser diode which serves for laser Doppler measurement, a UV LED (375 nm) for NADH fluorescence excitation and for total backscatter (or reflection) measurement, a Blue LED (470 nm) and a Green LED (530 nm) for HbO_2. In order to enable a very high measurement dynamic range of fluorescence and reflection parameters, the light source unit is designed to enable a very wide range of the excitation intensities. To enable this wide excitation range and linearity, the system is designed according to a three floors concept. Each floor comprises three LEDs one UV LED with emission peak at 375 nm, one Blue LED with emission peak at 470 nm and one Green LED with emission peak at 530 nm. The different wavelengths from all LEDs of the same floor are assembled together and coupled into a single fiber by a set of dichroic mirrors and appropriate collimation and focusing lenses. The current of each of the discrete LEDs is set by the appropriate electronics drivers directly controlled by a D/A of the DSP. The difference between the floors is the output intensity. There is a High, Medium, and Low intensity floors. The different excitation intensities are achieved by utilizing various pinholes while maintaining all other electro optical properties as the same for all three floors. The light from all three floors and a laser diode is combined into a single mixer fiber, therefore enabling precise setting of the excitation intensity within a very wide excitation range. The near-IR laser diode at 785 nm, for laser Doppler measurements, operates in CW operation mode. The UV LEDs, Blue LEDs, and Green LEDs operate in chopping mode. This enables the use of synchronous detection techniques in order to detect the NADH fluorescence and total backscatter light. Additionally, the chopping operation mode enables one to perform NADH measurements with very low excitation intensities well below the limits specified by the laser safety standards.

The Detection Unit (DTU) All six collection fibers of the fiber optic probe are assembled into a single male SC optical connector. The light from the probe passes through the panel connector into a single thick optical fiber that delivers the light to the DTU. At the DTU entrance, the collimation lens collimates the fiber output light. The collimated light is split according to the different wavelengths into the respective photo detectors by means of dichroic beam splitters. The first dichroic beam splitter reflects the total backscatter signal at 375 nm toward the photodiode detector.

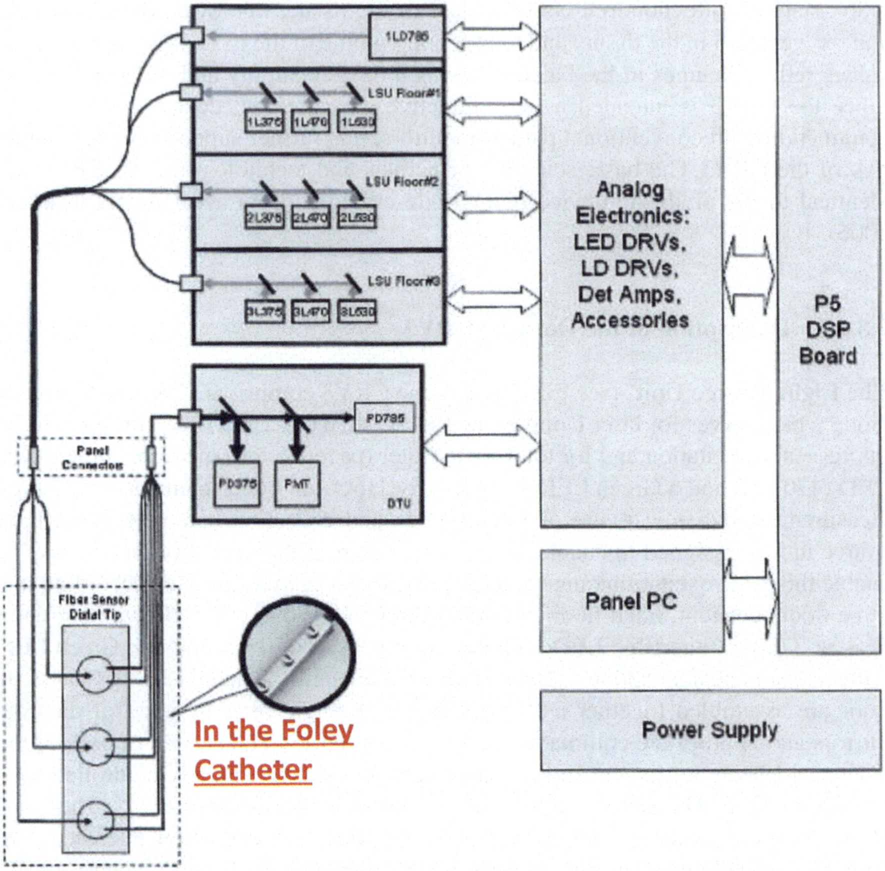

Fig. 7.66 The CRV3 block diagram and its Foley catheter probe design. (Mayevsky et al. 2008 [Modified])

The higher wavelengths pass through the first beam splitter toward the second dichroic beam splitter. The second dichroic beam splitter reflects the NADH F signal at 450 nm and total backscatter signals at 470 nm and 530 nm toward the photomultiplier detector. Due to the chopping operation of the LED's, the photomultiplier detector detects each of the abovementioned signals at different time, i.e., time-sharing operation detection mode. The second dichroic filter enables the laser Doppler signal at 785 nm to pass through it toward the photodiode detector. All acquired signals are digitized into the DSP by high-resolution 16-bit A/D.

The DSP is responsible for the whole system control, initial data processing, and calculation of the Doppler parameter. The DSP is built around Tern Inc. 586-Engine-P controller board with an AMD SC520 CPU. After initial data processing, the calculated values are transmitted to the panel computer for final data processing display through RS-232 serial interface. The CRV3 device utilizes medical-grade main power supply for all electronic circuits, including the panel computer.

7.8.2.2 The CRV3 Optical Probes

1. The pencil-style optical probe has one measurement point to be used on exposed skin/tissue. The excitation and collection fibers are held together in a stainless steel element. It is comprised of one excitation fiber connected to the respective excitation connector on the CRV3 panel, and 6 collection fibers connected to the single collection connector on the CRV3 front panel.
2. The Foley catheter optical probe for measuring the urethral wall. It is based upon a standard three-lumen Foley catheter (The Foley catheter is an inflatable balloon retention type catheter inserted through the urethra, and used to drain the bladder). This probe construction enables the measurement of tissue metabolism at the urethral wall while draining the bladder. The CRV3 Foley catheter optical probe is designed to perform optical measurements from three adjacent points on the tissue 3.5 mm apart. Each measurement point is comprised of one excitation fiber and two adjacent collection fibers. The three excitation fibers from the LSU are connected to the excitation connector on the CRV3 panel. All six collection fibers from the three measurement points are connected to the single collection connector on the CRV3 front panel. Figure 7.3 shows the finished probe (Fig. 7.67).

7.8.2.3 Testing the CRV3 In Vitro

The study aimed to demonstrate the efficiency of the CRV3 device for the measurement of NADH concentration in two steps:

1. Demonstrate the correlation between the NADH concentration in solution and the fluorescence intensity measured by the CRV3.
2. Compare the correlation between NADH fluorescence in solution measured in this study by the CRV3 to that measured by the CritiView-CRV1.

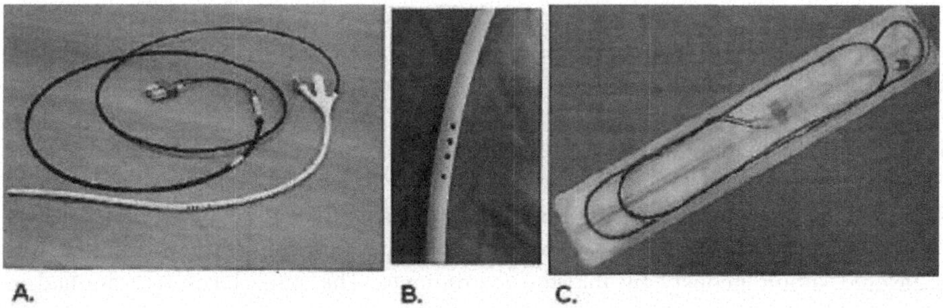

Fig. 7.67 Foley catheter probe assembly. (**a**) The fiber optic tip element embedded inside a standard three-way Foley catheter. (**b**) A close-up on the catheter measurement points. (**c**) The Foley catheter probe is packed in its blister package. (Mayevsky et al. 2008 [Modified])

The experimental methodology is straightforward. Various solutions of NADH were measured with the CRV3 as well as with the CritiView-CRV1. Insofar as the measurement with the CRV3 or the CritiView-CRV1 involves placing its probe tip into the solution, one might ask if this measurement is the same as placing the tip of the probe against a tissue surface. The equivalence of these two measurements can be understood if one considers that although immersion in aqueous solution and placement adjacent to tissue are, clearly, two different physical situations, nevertheless, the process and principle of measurement that takes place is the same for both. Aside from their contiguity, there is no interaction between the probe and the solution or any adjacent tissue. The probe merely collects emitted light from its immediate environment, i.e., a hemispherical volume adjacent to the probe tip, and transfers the collected light to detectors for measurement. Therefore, for the purposes of comparison of CRV3 spectroscopic characteristics to those of control NADH solution and to a clinical fluorometer, the CRV3 probe can be positioned on the surface of a tissue or a liquid, or it can be inserted into the tissue or liquid. Three sets of solutions with known concentrations of NADH + Double Distilled Water (DDW) were prepared in the following way: Stock solution of 1 mM NADH was prepared by dissolving 14.188 mg NADH in 20 ml DDW. The samples of NADH solutions (5 ml) were prepared in glass vials by the appropriate dilution of the stock solution with DDW, which is also used as a blank that was subtracted from the actual reading made by the CRV3. The "pencil-type" probe of the CRV3 was placed in the vials for 10 s and a reading was taken. The probe was taken out of the solution and a second reading was recorded. This was repeated again so that three reading were averaged. A second and a third set of diluted solution were prepared by the same procedure and were measured in the same way. The same procedure was used with the predicate CritiView-CRV1 with onset of NADH solutions.

7.8.2.4 Testing the CRV3 In Vivo

The aim of this study and report is to describe in detail the *in vivo* control study of the CRV3 device in order to demonstrate substantial equivalence to its predicate CritiView-CRV1. The performance of the CRV3 device in measurement of changes in TBF, NADH fluorescence, tissue reflectance, and HbO_2 resulting from standard metabolic perturbations in an animal model (gerbils) was done. All signals that are monitored by the CRV3 as well as the CRV1 are expressed as relative values. This brings all the calculations to be presented as "percent change of the signals." The TBF outputs of the CRV3 and the CRV1 are curves of signal measurements as functions of time. In order to assess the performance of CRV3 in measurement of TBF, we performed a series of experiments in which the TBF of animals was measured simultaneously by the two instruments. The same basis was applied to the other three parameters, namely, NADH, Ref, and HbO_2 in the same statistical analysis principles. The purpose of the statistical analysis is to demonstrate that the two curves of each experiment are equivalent. Since the scales of the instruments are not identical, it is meaningless to compute the difference between two curves. In fact,

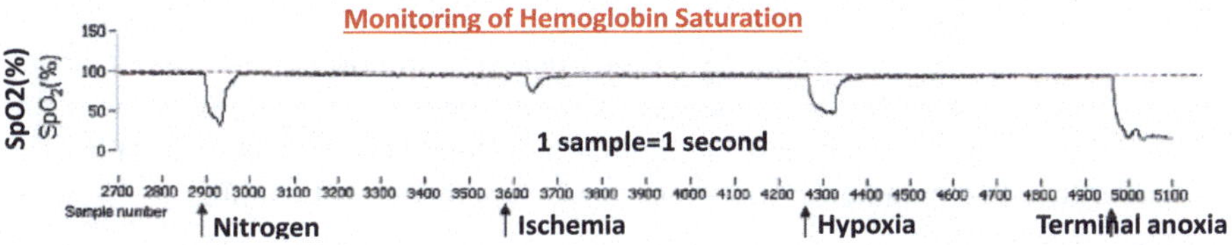

Fig. 7.68 Continuous record of the SpO_2 in one gerbil that was involved in the comparative study. N_2 = Anoxia. (Mayevsky et al. 2008 [Modified])

equivalence in this case means "having the same direction of change." That is, when one curve goes up, the other goes up; and when one goes down, the other goes down. The appropriate statistical measure for such behavior is the correlation coefficient. It is desirable to have a correlation coefficient as close to +1 as possible between every pair of corresponding curves. Thus, by computing correlation coefficients between two curves, it is possible to measure the statistical equivalence between them.

Animal Preparation In order to test the depth of anesthesia in each gerbil monitored, we used a surrogate systemic parameter that represents the function and the intactness of the cardiovascular and respiratory systems. The parameter is the level of the hemoglobin oxygenation in the systemic blood—SpO_2 measured by a standard veterinary pulse oximeter suitable for small animal experiments. In Fig. 7.68, a record of the SpO_2 during the entire experimental protocol of the gerbil is presented. As one can see, the SpO_2 was very high and stable during the entire period of monitoring. During the four perturbations imposed on the gerbil, a clear decrease of the SpO_2 was recorded. This decrease in SpO_2 during anoxia, hypoxia, and terminal anoxia is due to the availability of oxygen to the body. The decrease in SpO_2 during cerebral ischemia is probably due to stimulation of the sympathetic and parasympathetic nerves that are affected by the pulling of the carotid arteries during the ischemic episode.

Indeed, in each of the experiments included in the comparative study, we monitored the SpO_2 in addition to the parameters monitored from the brain. Operation procedure: Male Mongolian Gerbils (50–75 g) were used. We operated 25 gerbils but only 19 are reported in the results section. Six gerbils were excluded according to the predetermined criteria. The two common carotid arteries were isolated just before brain surgery, and ligatures of 4–0 silk thread were placed around them. The gerbil was placed in a head holder in the supine position. After a midline incision of the skin, an appropriate elliptic hole (about 3 × 8 mm) was drilled in the parietal bone of the right hemisphere. The dura mater remained intact, and a light guide holder cannula was placed in the drilled hole and extra pressure on the tissue was avoided. Two stainless steel screws in the left parietal bone were used to fix the cannula with dental acrylic cement. Body temperature was measured by a rectal probe (YSI) and was regulated to be within the range of 35–37 °C using a heating blanket. The animals were kept anesthetized during the operation as well as during the entire monitoring period, by IP injections of Eth 0.03–0.05 ml every 30 min, according to the existence of a pain response.

Metabolic Perturbations

Anoxia: Exposure of the animal to oxygen-deficient atmosphere by spontaneous breathing of 100% N_2, for a period of time (approx. 25 s) until changes are observed on the monitor, and the animal has stopped breathing. The animal is then allowed to breathe normal room air by spontaneous breathing or by short artificial respiration.

Ischemia: Reversible occlusion (30 s) of the two common carotid arteries by constricting them with threads.

Hypoxia: Exposure of the animal to spontaneous breathing of low oxygen concentration (6% oxygen in air) for a short period of time (60 s). The animal is then allowed to breathe normal room air by spontaneous breathing.

Terminal Anoxia: Exposure of the animal to complete irreversible anoxia by exposing the animal to 100% N_2 until the death of the gerbil.

The application of the various gases was done by placing a small tube, connected to the gas cylinder, around the nose of the gerbil. The flow of the gas was very slow (below 1 l per min) in order to avoid respiratory disturbances.

Experimental Protocol

Monitoring the various parameters was started immediately after the end of the operation. This was done by using two needle-type optical probes connected to the CRV3 and the Critiview-CRV1, and placed inside the cannula cemented to the skull. In order to check the intactness and the physiological status of the brain, a 20-s anoxia was induced. Gerbils that present abnormal response were discarded before running the experimental protocol. When body temperature reached the range of 435–37 °C, the exposure of the gerbils to the various perturbations started. The gerbils were exposed to the various conditions having interval of about 10 min between each test to allow recovery. The pulse oximeter value was used in order to ensure the complete recovery. The data was collected at a rate of 1 sample per second and stored in different channels of a computerized data acquisition program.

7.8.3 Results

7.8.3.1 In Vitro Testing

The data presented in Fig. 7.69 was collected from the three sets of NADH aqueous solutions. Final concentration of the 24 samples numbered as 1–24 prepared by diluting the stock solution with Double Distilled Water (DDW). The results of the three dilution sets are presented. Each sample was measured three times and the mean is shown. The fluorescence arbitrary unit values of the CRV3 were plotted against concentration values as seen in Fig. 7.69. A linear plot is expected for the concentration range used, and the three regression lines related to the three sets of measurements were calculated and plotted. From Fig. 7.69, one can see the linear

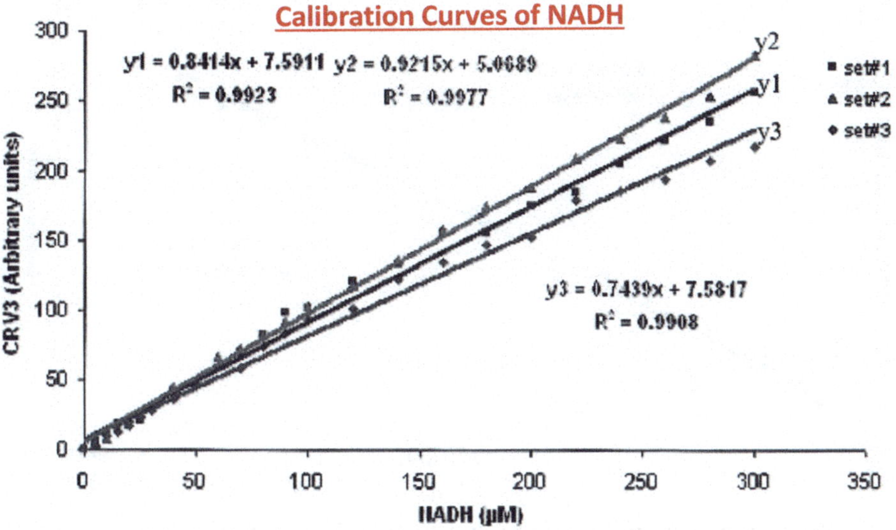

Fig. 7.69 Fluorescence readings (three sets) of the CRV3 plotted against NADH concentration after subtraction of the reading of the blank solution. A linear regression was calculated, and the equation and R2 are presented for the three sets of control solutions. (Mayevsky et al. 2008 [Modified])

correlation between NADH fluorescence to control NADH concentration in aqueous solution. There is no significant difference between the three sets of measurements.

Figure 7.70 presents readings of the predicated CritiView-CRV1 (on the right Y axis), and readings of the CRV3 (on the left Y axis) were plotted against NADH concentrations. A linear regression was calculated for each instrument, and the equation and R2 are presented.

It can be concluded that a linear relationship was established between the NADH concentrations and the F signals measured by the CRV3 with correlation coefficients that are very close to 1.0.

The correlation coefficients between CritiView-CRV1 or CRV3 and NADH concentration (C) were: r (CritiView-CRV1) = 0.9933; r(CRV3) = 0.9954.

This study proves the efficacy of the CRV3 in the measurement of NADH concentration in solutions and also that the CRV3 is substantially equivalent to its predicate, CritiView-CRV1, for the NADH measurement efficiency.

7.8.3.2 In Vivo Testing

The comparison between the two devices provided a large quantity of data that cannot be presented in the current chapter due to page limitations. Therefore, only the results of the NADH measured by the two instruments are presented. The responses of mitochondrial NADH to the lack of O_2 are unidirectional, namely, an

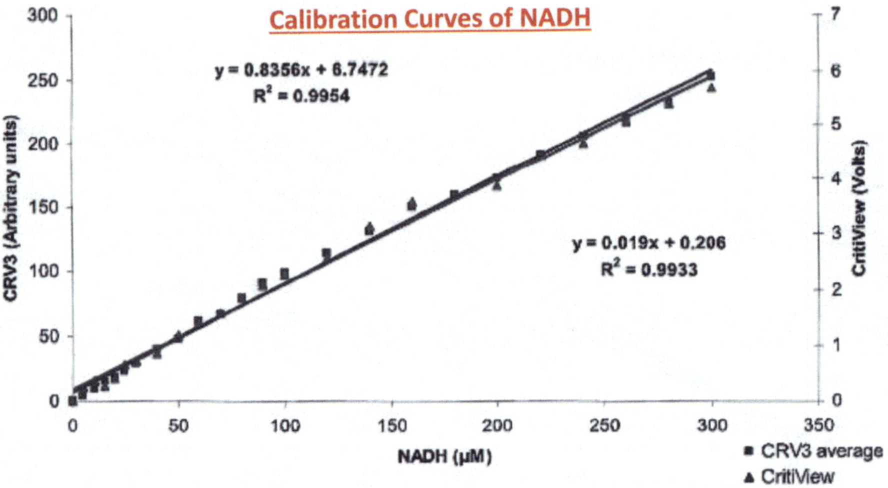

Fig. 7.70 Readings of the predicate CritiView-CRV1 (on the right Y axis), and readings of the CRV3 (on the left Y axis) plotted against NADH concentrations after subtraction of the reading of the blank solution. (Mayevsky et al. 2008 [Modified])

increase of signal in all perturbations used. As can be seen in Fig. 7.71a–d, under all perturbations, the CRV3 and the control instrument present very similar responses, and the correlations were statistically significant ($r > 0.87$). Under terminal anoxia, the duration of monitoring was 1 min before the perturbation and 1 min after the fast increase in NADH without a recovery phase.

7.8.3.3 Statistical Analysis

In almost all events induced in the various animals and type of perturbation that were recorded, the correlation coefficients were statistically significant. This indicates that mitochondrial responses and other parameters are very similar regardless of the instrument used or the perturbation induced.

7.8.4 Discussion

7.8.4.1 Responses to Metabolic Perturbations

In Fig. 7.71a–d, one sees the response of NADH to all four perturbations used. As seen in all four figures, the typical increase in the NADH level is very significant. This is due to the decrease in the available oxygen to the brain's mitochondria that shifts the [NAD]/[NADH] ratio to a more reduced state. This is followed by complete recovery when the animal is allowed to receive oxygen again. When

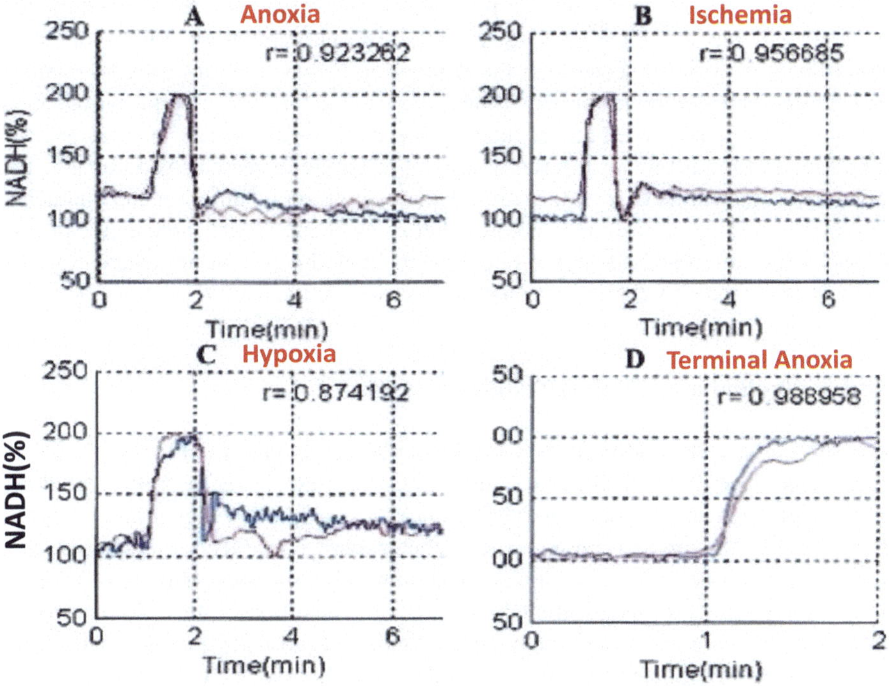

Fig. 7.71 Time plot of typical NADH response to (**a**) anoxia, (**b**) ischemia, (**c**) hypoxia, and (**d**) Terminal anoxia. The two plots in each perturbation were recorded by the CRV1 and CRV3. (Mayevsky et al. 2008 [Modified])

oxygen is deprived from the tissue, either by ischemia, hypoxia, or anoxia, the expected increase in NADH due to the inhibition of the respiratory chain located in the mitochondria was recorded. Under such conditions, the supply of energy is diminished while the demands for energy remains the same. Due to the negative oxygen balance, the brain will not be able to perform its various activities. TBF is a primary event under ischemia, while under anoxia or hypoxia it is secondary. The responses of TBF in the four perturbations were not identical. Under anoxia, the TBF response is showing a biphasic nature. The responses recorded by the Ref trace indicate that this parameter is correlated to blood volume in the tissue. The Ref was elevated under ischemia, indicating a decrease in blood volume.

7.8.4.2 Correlations

Since all responses to any of the perturbations are measured simultaneously by the same two devices, i.e., CRV3 and the CRV1, one might expect that all groups would have similar correlation coefficients. Nevertheless, we find differences between the correlations of different perturbations. This is understandable since the physiological responses are slightly different for the various perturbations. As a result of the difference in physiological responses, including autoregulatory effects on local blood flow, it can be expected that the correlation between the measured signals at two adjacent sites in the tissue would be more similar for certain perturbations than for others. Similarly, when autoregulatory mechanisms are active, such as anoxia, variations in blood flow at different tissue sites can be expected. Additionally, the intensity of the results of a perturbation depends both on its duration and character. Duration is significant insofar as autoregulatory mechanisms are more active during a short perturbation and attempt to reject the unwanted perturbation, whereas after a while (a few seconds), their ability to act decreases dramatically.

7.8.4.3 Preliminary Clinical Testing of the CRV3

In order to assess the ability of the CRV3 to measure the vitality of the urethral wall in patients, a preliminary clinical study was performed. In patients, we tested the viability of the urethral wall (a less vital tissue) by a three-way Foley urinary catheter that contains the optical probe. The catheter was introduced to patients who underwent open-heart bypass surgery or abdominal aorta aneurysm (AAA) operations. The monitoring started immediately after the insertion of the catheter into the patient and was stopped when the patient was discharged from the OR. Figure 7.72 presents the data collected from the Urethral wall during the occlusion period as well as after the reperfusion of the urethra in one of the six patients monitored. During the preparation for the occlusion of the large arteries, a small transient decrease in microcirculatory blood flow (TBF) together with an increase in NADH levels were recorded. This was an artefact of the machanical act. When the two arteries were occluded completely, the TBF decreased to near 0 levels while NADH reached its maximal levels. The HbO_2 decreased in parallel with the decrease in TBF. Under the

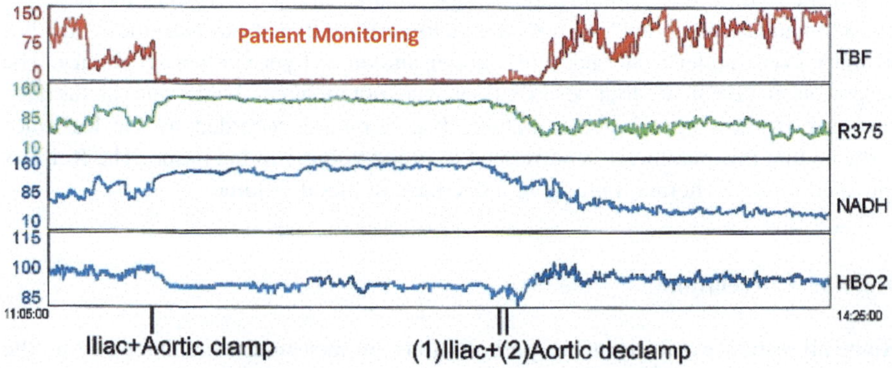

Fig. 7.72 Effects of occlusion of the Iliac artery and the abdominal aorta on urethral wall parameters represent the vitality at the tissue level in an AAA operation. The time scale represents a period of 3 h and 20 min. (Mayevsky et al. 2008 [Modified])

ischemic conditions developed the Ref signal increased due to the decrease in blood volume in the monitored area. Immediately after the reopening of the two blood vessels, all signals returned to the area of the baseline values. The results show that monitoring of the vitality of the Urethral wall provides information in correlation to the surgical procedure performed. In the AAA patients, the occlusion of the aorta led to severe ischemia developed in the urethral wall and recovery of signals was recorded only after the reperfusion of the urethra. In patients who underwent heart bypass surgery the urethra vitality was decreased dramatically during the operation and recovery was noted in most patients after the discharge of patient from the OR.

7.8.5 Conclusion

The CRV3 and the CRV1 are substantially similar when used in different physiological states. This is consistent with the concept of "identical principles of physiological function between different organisms," that forms the rationale for the ability to use the CRV3 in any living tissue. The results of the comparisons between different sites in the same tissue, different tissues, and different species were published recently (Mayevsky et al. 2004b), which confirms and emphasizes the ability of the CRV3 to measure all types of living tissues in animals as well as in patients.

7.9 Evaluation of the CritiView in a Pig Model of Abdominal Aortic Occlusion and Graded Hemorrhage

7.9.1 Introduction

During the last few years, we have shown that it is possible to measure various parameters *in vivo* that are indirect indicators of tissue viability by a multiparameter monitoring device (Mayevsky and Chance 2007; Mayevsky and Rogatsky 2007). We developed a compact device (Critiview) enabling real-time continuous monitoring of mitochondrial function, microcirculatory blood flow, tissue reflectance, and hemoglobin oxygenation levels. The technical aspects and their performance were described during the last few SPIE meetings (Kutai-Asis et al. 2006; Mayevsky et al. 2007; Mayevsky et al. 2008). The involvement of the mitochondria in many pathophysiological conditions was described by various investigators (Chance and Hollunger 1960; Jobsis 1964; Anderson and Sundt 1975; Mayevsky and Breuer 1992). After testing the CritiView *in vitro* as well as *in vivo* in small animals (rats and gerbils), we tested its performance in a large animal model (pigs) exposed to various pathophysiological conditions. By adding the abovementioned tissue-level parameters, measured by the CritiView, to the systemic hemodynamic parameters it will be possible to increase the diagnostic capabilities available in critical care medicine. Under various severe pathophysiological conditions, such as hemorrhage, the compensatory mechanisms of blood flow redistribution are taking place and as a result the sympathetic stimulation will induce a significant decrease in tissue oxygenation in less vital organs and tissues. Under these conditions the most vital organs of the body are protected by preserving increased blood flow and O_2 supply (Kraut et al. 2004; Barbiro-Michaely et al. 2007).

If the balance between O_2 supply and O_2 demand is negative, the function of the tissue or organ will be affected and a pathological state is developed. Maintaining tissue and cells oxygen supply during shock is the major objective of shock treatment. The effects of hemorrhagic shock on pig's urethral wall NADH redox state and TBF (measured by a laboratory experimental device) was described recently (Clavijo et al. 2008). Clinical diagnosis of shock is usually based on specific but very subjective clinical signs and systemic homodynamic parameters like hart rate, MAP, ECG, CVP, Hematocrit, urine output, and PaO_2. These parameters, which demonstrate circulatory insufficiency under the progressive stage of shock, are not well correlated with the pathophysiological condition and do not necessarily reflect tissue perfusion. The aims of this study are as follows: 1. To develop a pig model to be used with the CritiView. 2. To test the ability of the CritiView to monitor various tissue parameters representing tissue viability in less vital organs of a large animal model. 3. To test the effects of disturbances in systemic hemodynamics, such as abdominal aortic occlusion or graded hemorrhage on mitochondrial function, microcirculatory blood flow, and tissue reflectance in various less vital organs.

7.9.2 *Methods*

7.9.2.1 Monitoring Device

The CritiView device was developed by CritiSense LTD in order to monitor and provides parameters related to oxygen (energy) supply and oxygen demand at the tissue levels. The CritiView device is a tissue spectroscope that carries out *in vivo* spectroscopic measurements. It transmits light at known wavelengths through tissue and measures the properties and intensity of light collected from the tissue. The CritiView incorporates a light source unit, an optical probe (which transmit the light to the tissue and returns the collected light), a detector unit, analyzer of the specific wavelength as well as software. The CritiView measures four parameters: 1. Mitochondrial NADH fluorescence emitted at 420–480 nm. 2. Total backscattered light (370 nm) reflected from the tissue. This parameter allows correction of the NADH fluorescence measurement due to changes in tissue blood volume. 3. Doppler-shifted laser light (785 nm) reflected from moving blood cells. 4. HbO_2—Oxygen saturation of Hemoglobin (470 nm, 525 nm). More details on the technological aspects of the CritiView and its *in vitro* and small animal testing were published (Mayevsky et al. 2008).

7.9.2.2 Animal Preparation and Experimental Protocols

All protocols were approved by the Institutional Animal Care Committee of the Sheba Medical Center and operated under the instructions of the National Institute of Health. Female pigs (Sus scrofa domestica) were grown and supplied by the Lahav Research Institute—LRI in Israel. Pigs (45–55 Kg) were starved of food but allowed free access to water for 12 h before the study.

Anesthesia

After premedication by IM ketamine 5 mg/kg and xylazine 1 mg/kg, peripheral IV access was established. Induction of anesthesia was achieved by administration of 2 mg/kg of propofol. After endotracheal intubation, mechanical ventilation was commenced with FiO_2 of 0.5, tidal volume of 10–12 ml/kg, and respiratory rate of 8–12 per minute. This produced peak inspiratory pressures of 15–18 cmH_2O. Respiratory rate was adjusted so that $PetCO_2$ remained in the range of 35–40 mmHg. Muscle relaxation was produced by continuous infusion of vecuronium 0.1 mg/kg/h. Maintenance of anesthesia was achieved by the addition of 1.5 MAC of Isoflurane and continuous IV infusion of 2 mcg/kg/h of fentanyl. After the conclusion of the experiment, animals were sacrificed by IV injection of 40 meq of KCl.

Instrumentation

After the induction of anesthesia, a 4F thermistor, 50 cm-tipped catheter for thermodilution, and pulse contour analysis (PV 2014L50LGW) was inserted into the descending aorta via the femoral artery and connected to the PiCCO monitor (Pulsion, Munich, Germany) for the measurement of cardiac output and parameters of cardiac preload. A right internal jugular vein was cannulated with 8F cannula for blood extraction and retransfusion, and double lumen 14G cannula for central venous pressure monitoring, blood sampling and drug infusions.

CritiView Optical Probe

The optical probe (Foley catheter) was inserted into the pig's urethra to create contact with the mucosal wall and mechanically affixed by external means. Urine drainage was maintained by a suprapubic catheter. In few experiments, we used two sites of measurement in the same pig with the CritiView. As seen in Fig. 7.73, one catheter was introduced and placed in the urethra while the second probe was inserted and sutured to the bladder internal wall. We found that the incision of the abdominal wall as well as the bladder enabled us to insert the guiding device toward the vagina. The tip of the Foley catheter connected to the guide is inserted through

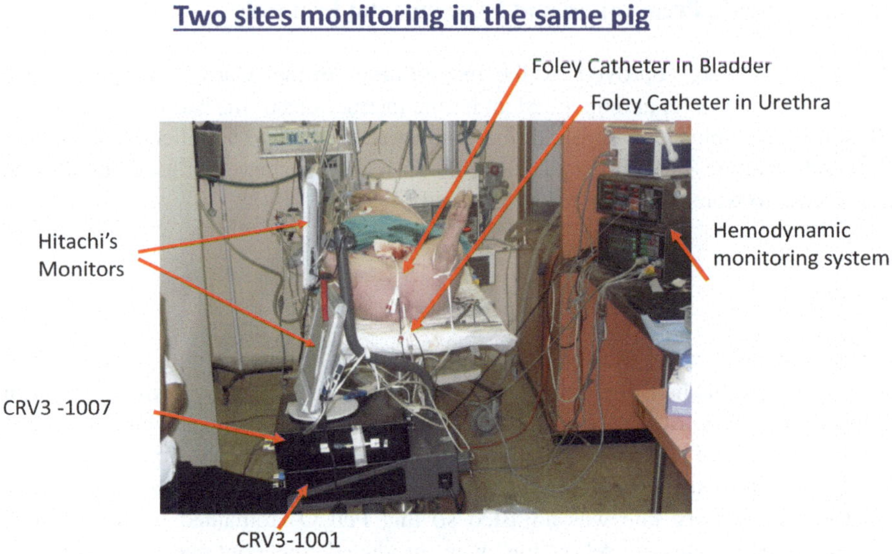

Fig. 7.73 The use of the CritiView in monitoring the female pig by two catheters located in the urethra and the bladder

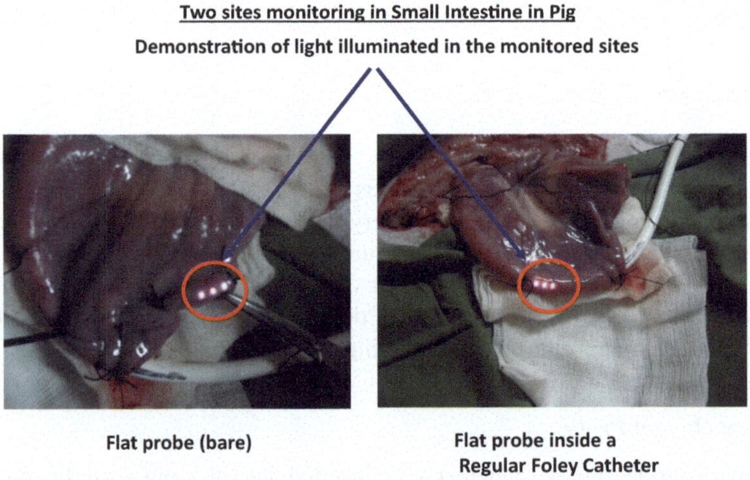

Fig. 7.74 Pictures of the small intestine monitored by two types of probes located inside the lumen and touching the mucosal side of the wall. (Mayevsky et al. 2009 [Modified])

the vagina toward the bladder. A second approach was to monitor two sites in the small intestine by placing two probes in the same pig as seen in Fig. 7.74. We have tested the responsiveness of the intestinal mucosa by occlusion of blood vessels providing blood to the monitored sites.

Experimental Procedure

After the induction of anesthesia, insertion of optical probe, cannulation and 30 min of a stabilization period, baseline measurements were performed. This set of measurements included basic hemodynamic data, PiCCO-derived parameters, CritiView parameters, and blood parameters. 1. Basic hemodynamic data by the CardioCap monitor **a**. BP (sys, dist, mean) **b**. HR (Heart Rate) **c**. CVP (Central Venous Pressure) 2. Cardiac output, intrathoracic blood volume and extravascular lung water were measured by means of transpulmonary thermodilution using PiCCO system 3. CritiView parameters are calculated in relative units and are not presented in absolute values therefore the device is used as a trend monitor. **a**. NADH redox state **b**. Ref at 375 nm **c**. TBF **d**. Microcirculatory Blood Oxygenation. 4. Arterial blood gases including lactate, hematocrit (Hct), and Central venous hemoglobin oxygen saturation (ScvO2). 5. $PetCO_2$ (end tidal partial pressure of CO_2). Blood sampling and recording of basic hemodynamic parameters were performed prior to the measurements of thermo dilution.

Experimental Protocols

Aortic Clamping

The animal was stabilized for 30 min, followed by aortic clamping for 40 min. The reperfusion period recording was for 60 min. Aortic occlusion was performed in three pigs only; therefore, statistical analysis was not practical.

Graded Hemorrhage

Following the initial measurements, pigs (total 7) were bled by increments of 5% of their blood volume (estimated to be 75 ml/kg). Each step of the hemorrhage lasted a total of 20 min, and included a graded bleeding over 6–7 min, 10-min stabilization, and 3–4 min for completion of all measurements. This procedure was repeated 8 times upto a blood loss of 40%. The bleeding steps were followed by a retransfusion step (the shed blood was retransfused over 35 min), and a fluid Load step of 20 ml/kg of hydroxyethyl starch (HAES-STERIL 6%). During the hemorrhage mean blood pressure was maintained at the level of 50 mmHg by means of infusion of escalating dose, 0.01–0.2 mcg/kg/min of noradrenalin.

0	1	2	3	4	5	6	7	8	7	8
BL	5%	10%	15%	20%	25%	30%	35%	40%	Retransfusion	Load

7.9.3 Results

As seen in Fig. 7.75, the clamping of the abdominal aorta led to a decrease in the microcirculatory blood flow. During the 50 min of the occlusion, the TBF was very low but was not at the zero level. Also the unstable laser led to an artifact in the TBF signal in the middle of the occlusion period. In parallel to the decrease in TBF, the HbO_2 showed a decrease. The response of NADH could be divided into two phases. During the first phase, 40–50% of the occlusion period, a clear increase in NADH was recorded. The NADH was reoxidized during the second half of the occlusion period. During the reperfusion period, TBF and HbO_2 recovered to the original values and even were above the control values during the post-ischemic period. A very small decrease in the NADH signal was recorded 30 min after the declamping of the aorta. The correlation between the main monitored parameters is shown in Fig. 7.76. As expected, the TBF and HbO_2 show an inverse correlation with the NADH. Also, when the TBF decreased, a significant increase in the Ref signal was noted due to the decrease in blood volume in the monitored site leading to a decrease in absorption—high reflectance.

Figure 7.77 presents responses to aortic clamping measured in a second pig. In this animal, the clamping led to a clear decrease in TBF, which recovered to normal after the reperfusion stage. The decrease in HbO_2 was not stable and showed partial recovery during the occlusion period. The response of the NADH was quite stable during the clamping period and showed a clear decrease after the declamping. In this figure, the effect of cessation of the heart activity (KCl injectia) is shown. The TBF and HbO_2 approached very low levels. The NADH was elevated to its maximal levels, followed by an artifact ("oxidation") 3 min after death.

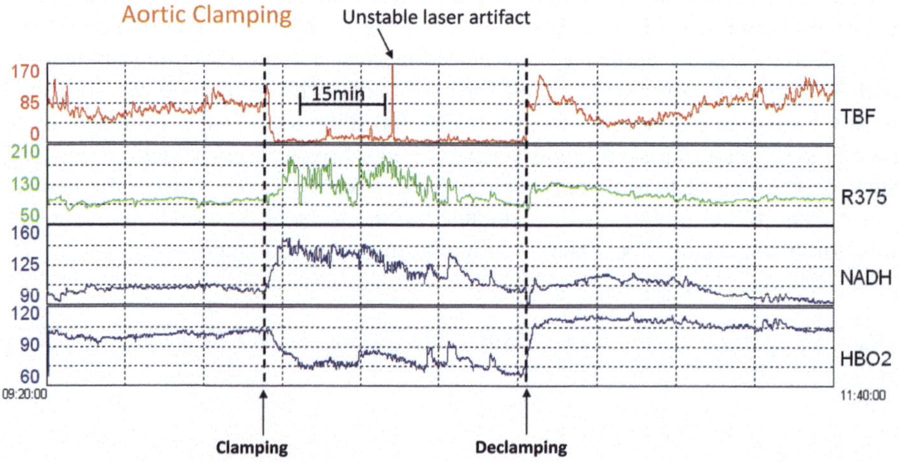

Fig. 7.75 Metabolic and hemodynamic responses to complete occlusion of the abdominal aorta in the urethra of a female pig. (Mayevsky et al. 2009 [Modified])

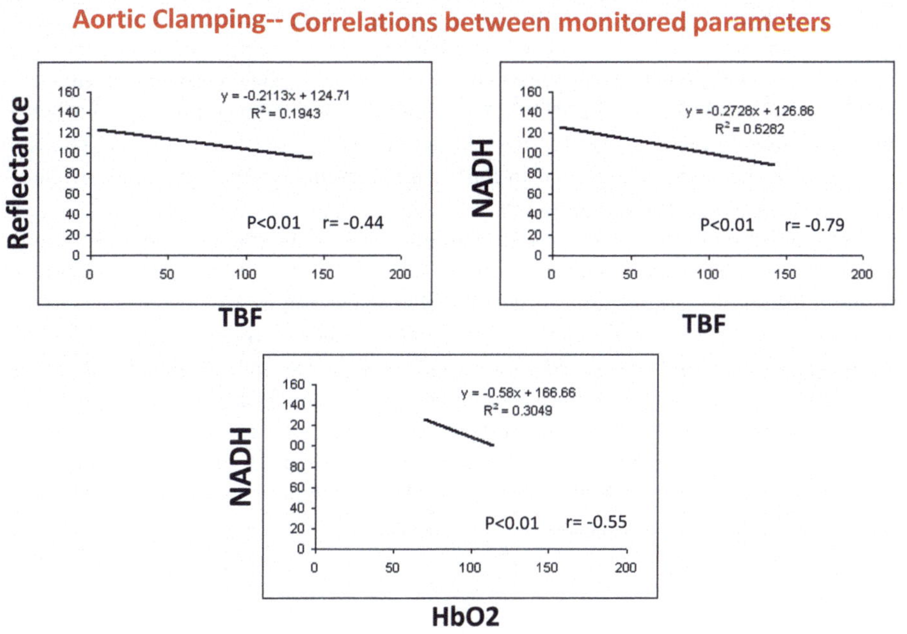

Fig. 7.76 The correlation between the main parameters monitored and presented in Fig. 7.74. (Mayevsky et al. 2009 [Modified])

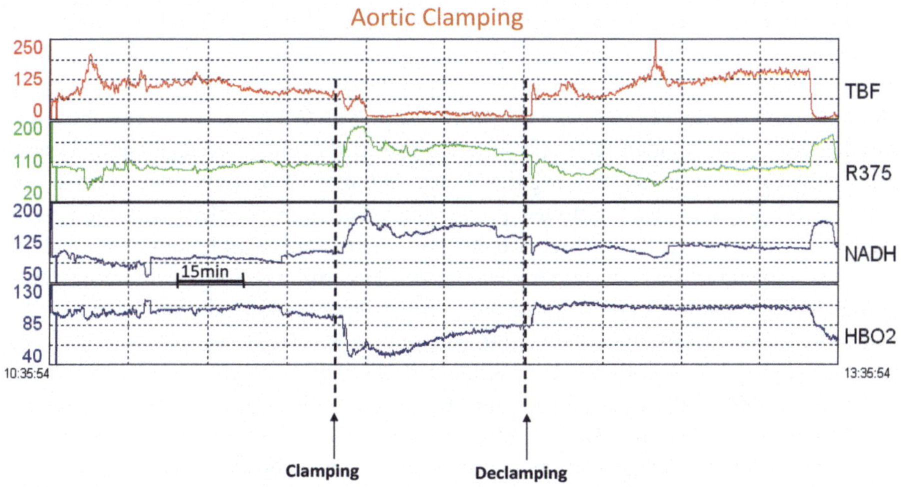

Fig. 7.77 The responses of the four monitored parameters to complete occlusion of clamping of the abdominal aorta. The death response was measured after IV injection of KCL solution. (Mayevsky et al. 2009 [Modified])

Nevertheless, the responses of the three pigs to AO are superimposed in the same figure as shown in Fig. 7.78. As shown in Part A, two of the pigs showed very clear responses of the TBF to AO, while the third one exhibited recovery during the occlusion period. In Part C, the NADH responses were similar in two pigs while the third one showed recovery during the second half of the occlusion period. The Ref responses of the three pigs shown in Part D are very noisy and were not correlated nicely with the other three parameters.

A second type of testing the monitoring model was the measurement at the two sites in the small intestine as shown in Fig. 7.74. We compared a regular flat probe. Figure 7.79 shows the responses of small flat probe, sutured to the wall to avoid movement artifacts, to reversible and irreversible ischemic inputs. In Fig. 7.79, the occlusion of the blood vessel led to a very clear decrease in TBF and HbO_2 while NADH was significantly elevated. During reperfusion the TBF showed a hyperemic response correlated to the recovery of the NADH reaching a more oxidized stage. In Fig. 7.80, the responses to death of the animal are shown. Here again the TBF and HbO_2 decreased while NADH was elevated. The change in the Ref was more gradual and smaller as compared to the response in Fig. 7.79.

7.9.4 Discussion

The evaluation of the hemodynamic and metabolic responses to systemic disturbances in oxygen delivery to less vital organs in an animal model has a very significant impact on the development of a clinical tool for critical care patients. Therefore, we tested the CritiView, previously successfully tested in small animal models (such as in the gerbil or rat brain), in a large animal model. The developed clinical probe inside a Foley catheter (Mayevsky et al. 2008) led us to use the same catheter in the pig model. The special anatomy of the pig male urogenital system prevented us from using the regular Foley catheter. The only way to test the developed Foley catheter was to use female pigs. The main problem with the female pig is the very short urethra, which affects the proper location and maintenance of the probe in stable contact with the mucosal tissue. The initial testing of the monitoring system was to use a relatively clear cut perturbation, namely, the occlusion of the abdominal aorta. In preliminary testing (Clavijo et al. 2008), we used a probe that was not a real Foley catheter but rather a flexible rod in which the fibers were arranged in 360° allowing a better sampling of the urethral tissue. Also the number of fibers was significantly larger leading to better signal-to-noise ratio. The results shown in this study indicate that using the Foley catheter in the female pig is not an ideal model. The same inconstancy of the responses will be found when the graded hemorrhage model was used.

The results obtained from the urethra were more reliable than those measured in the bladder due to the less optimal contact between the probe and the monitored tissue or due to the instability of the monitoring device. The effect of the Graded Hemorrhage (GH) were noticed in most animals only after the animal blood volume

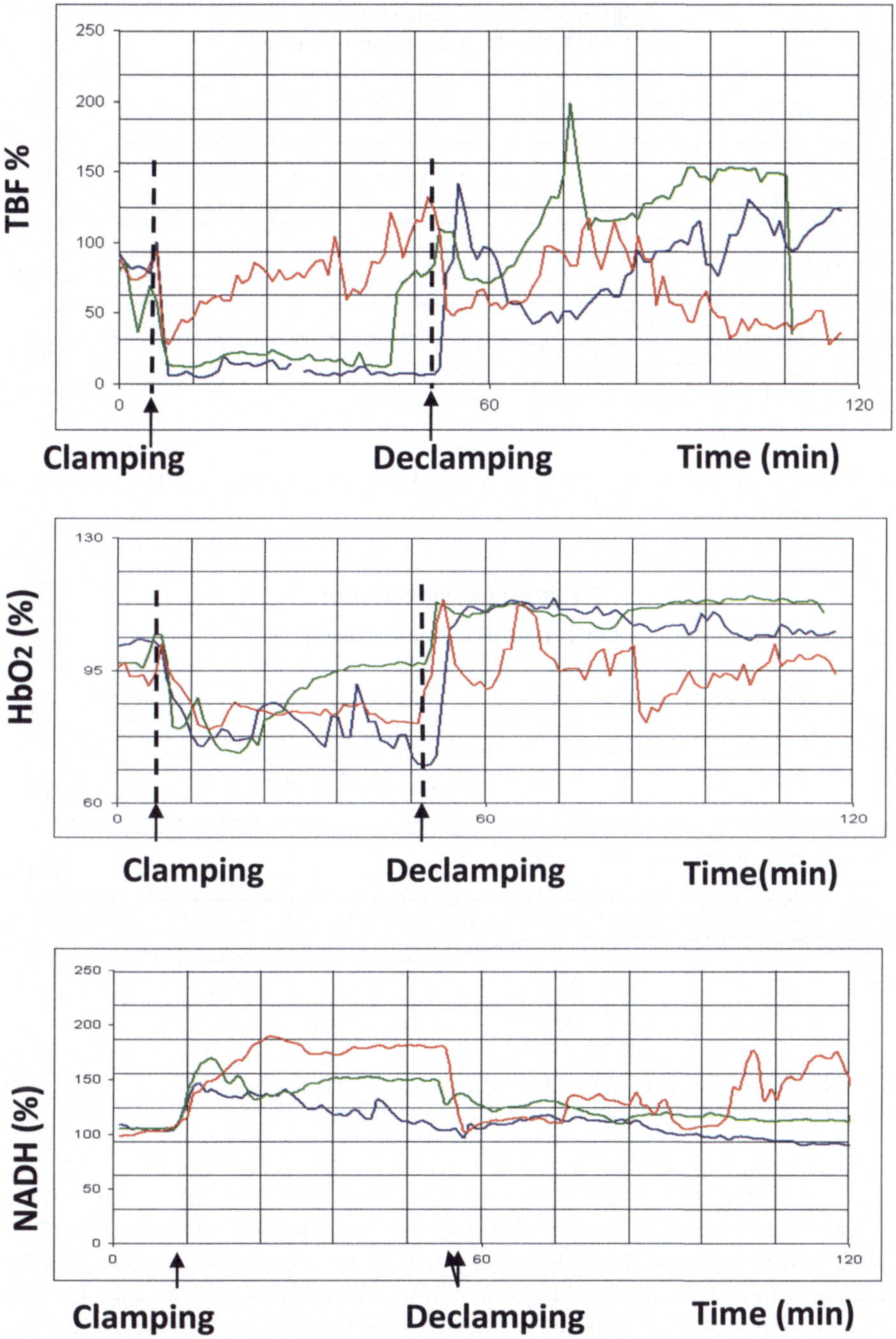

Fig. 7.78 The responses of three female pigs to aortic clamping as monitored by the human Foley catheter inserted to the urethra. (Mayevsky et al. 2009 [Modified])

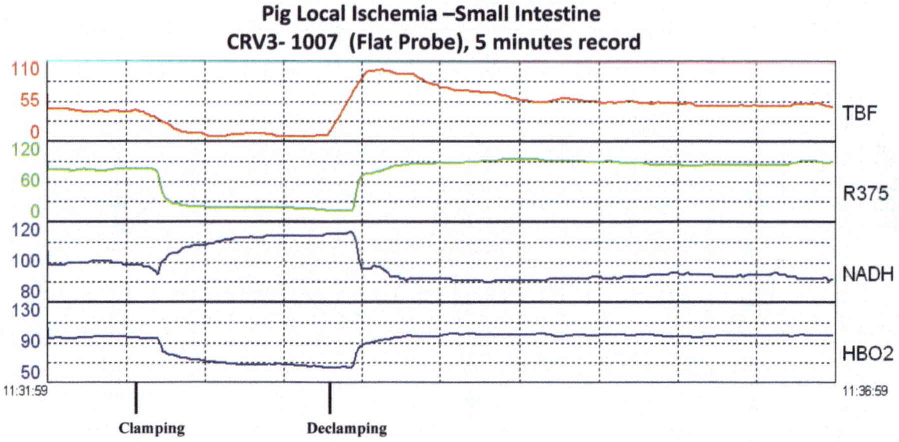

Fig. 7.79 The responses to reversible local ischemia measured in the mucosa of the small intestine in the pig model. (Mayevsky et al. 2009 [Modified])

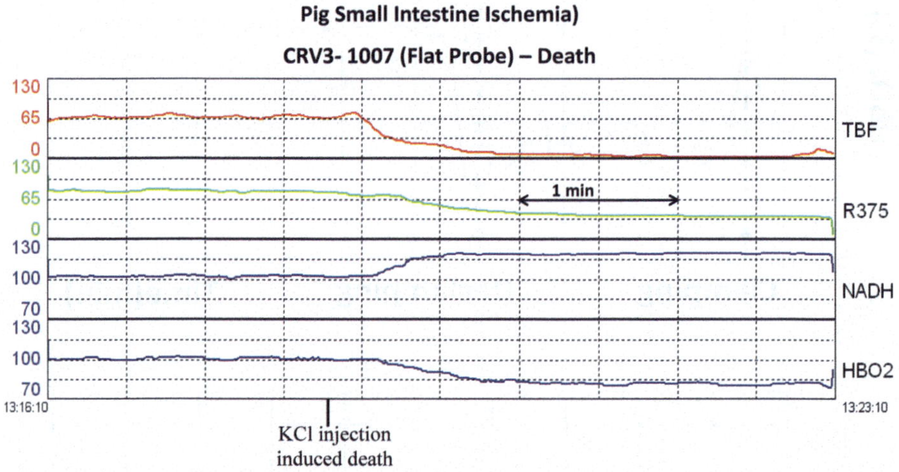

Fig. 7.80 The responses to irreversible local ischemia measured in the mucosa of the small intestine in the pig model. (Mayevsky et al. 2009 [Modified])

was decreased by 30% or more. Another monitoring site tested in this study was the mucosal side of the small intestine. In the specific measurements shown in Figs. 7.79 and 7.80, we located the flat probe in good contact with the mucosa after cleaning the internal surface with saline solution. Therefore, the responses were clear. In other pigs, we tried to avoid rinsing the tissue with saline in order to decrease the damage to the delicate tissue, and the results were less stable and significant. A possible approach is to use the small intestine and to develop a better monitoring preparation as done by Maier et al. (2009).

7.10 Mitochondrial Function and Tissue Vitality: Bench-To-Bedside Real-Time Optical Monitoring System

7.10.1 Introduction

The need for a practical and simple medical device for *in vivo* monitoring of tissue energy metabolism in real time has stimulated many scientists as well as various medical device companies. Few companies and researchers have developed Laser Doppler Flowmeters for TBF monitoring (Stern 1975; Bonner and Nossal 1981; Shepherd 1990; Batista et al. 1996), tissue hemoglobin saturation levels (HbO_2) (Rampil et al. 1992), tissue oxygen level (pO_2) (Mayevsky et al. 1980), tissue CO_2 and pH levels (Lubbers 1995), and mitochondrial cytochrome oxidase redox state (Scheffler 2000). Nevertheless, none of the mentioned techniques is used in daily clinical practice in any medical field. The most important parameter to be evaluated is the balance between O_2 supply and demand defined as oxygen balance and represented by mitochondrial redox state. As of today, a number of measured parameters, in daily clinical practice, represent the integrity of the cardiovascular as well as respiratory system. In our previous publications (Mayevsky et al. 2004b, 2005), we described a few devices aiming to monitor tissue vitality using a single-point probe that measured up to four optical parameters from a small tissue area. These devices never applied to routine clinical use due to factors such as size, price, or the ability to use and interpret the results as a clinical significant tool. The new device described in this section was developed in order to overcome all the disadvantages of the devices developed earlier. During the research and development period, two models of the new multiparametric monitoring device were tested and submitted to the FDA for approval. The prototype of the CritiView (CRV), which was described a few years ago (Mayevsky et al. 2005), was adapted to a single-point measurement in critical care medicine. The new model of the CRV enables one to integrate the measurements of three different locations of the tissue thus improving the statistical basis of the measurement. Also, the new CRV is significantly smaller thus enabling its incorporation into the small space available in ORs or in ICUs.

7.10.1.1 Patient Monitoring in Critical Care Medicine

Cardiovascular insufficiency commonly occurs in critically ill patients and those undergoing major surgical procedures. The consequences of delayed recognition and treatment of circulatory shock prior to ischemic tissue injury can be irreversible. Patients undergoing major surgery or hospitalized in the ICU frequently develop changes in systemic or peripheral hemodynamics. More than 20% of patients are expected to have acute cardiovascular dysfunction in the perioperative period of cardiac surgery (Mebazaa et al. 2010). Recently, Ospina-Tascon et al. (2008) concluded, based on a literature survey, that “there is no broad evidence that any form of monitoring improves outcomes in the ICU and most commonly used devices have not been evaluated by randomized controlled trials." Nevertheless, acute care practitioners are seeking new monitoring devices that would provide real-time data on the oxygen balance in patients at the tissue level (Clavijo-Alvarez et al. 2005; Creteur et al. 2007; Pinsky and Payen 2009). The pattern of the physio-pathological event cascade that usually occurs in clinical conditions associated with an imbalance between oxygen delivery and oxygen consumption, under hypoperfusion-induced tissue hypoxia, is the common end product of circulatory shock. Various unstable metabolic states, infection, trauma, and surgical stress lead to metabolic disturbances and often result in cellular energy deterioration. The most common example of such a pathological state is sepsis, which is a major cause of death in critically ill patients (Ince and Sinaasappel 1999; Angus et al. 2001). The autonomic nervous system, mainly its sympathetic branch, including the adrenal gland, plays a major role in enacting the body's compensatory mechanisms against a hemodynamic compromise. The rapid compensatory reaction to hypovolemia, for example, includes redistribution of the blood flow to various organs, with a preferential blood flow to the most vital organs of the body, namely, the brain, heart, and adrenal glands (Ganong 1991b; Barber 1994), while the peripheral organs or areas (the skin, GI tract, and muscles), as well as other less vital visceral organs, undergo vasoconstriction and subsequent decrease in blood flow and O_2 supply (Schlichtig et al. 1991). This decrease in tissue perfusion could be detected by monitoring mitochondrial NADH in real time in those organs that experienced an early diversion of the blood flow (Kraut et al. 2004; Clavijo et al. 2008).

Currently, the monitoring of mitochondrial function in real time *in vivo* is not employed in clinical monitoring. Gastric tonometry and sublingual capnography have been used to assess tissue perfusion but have not yet been adopted for daily clinical use (Maciel et al. 2004; Creteur 2006; Marik 2006). Tissue oxygen saturation measures have also been reported, but they can indicate decreasing values only at the end of a circulatory shock, not at its inception. The effects of cardiopulmonary bypass (CPB) on cerebral cortical oxygenation during open heart surgery are well documented (Talpahewa et al. 2004). Cerebral hypoperfusion may also occur during the early postoperative period after cardiovascular surgery (Millar et al. 2001). The monitoring of less vital organs (i.e., the skin or GI tract) during cardiovascular surgery has been extensively reported (Niinikoski and Kuttila 1993; Ohri et al. 1994; Haisjackl et al. 1998). In patients, the effects of hypothermic CPB on gut metabolism were an immediate decrease in blood flow followed by a recovery, while tissue oxygenation progressively declined after CPB (Ohri et al. 1997). Recently, Perner et al. (2005) reported that, during CPB, lactate concentration in the rectal lumen increases, indicating a metabolic dysfunction of the rectal mucosa. We tested the hypothesis that the urethral wall's mitochondrial function and microcirculatory blood flow in patients would be sensitive to macrocirculatory changes taking place during cardiovascular surgery. We chose this sample because of its existing invasive monitoring and the known alterations in the microcirculatory blood flow produced by surgery.

7.10.1.2 Energy Metabolism in Mammalian Cells

All cells in the body depend on a continuous supply of adenosine triphosphate (ATP) in order to perform their different physiological and biochemical activities. Figure 7.81 describes the basic processes occurring in a typical normal cell, using glucose as a major source of energy. The breakdown of glucose into water and CO_2 includes two steps, namely, glycolysis (the anaerobic phase) taking place in the cytoplasm and oxidative phosphorylation (the aerobic phase) occurring in the mitochondria. Of the total yield of 38 ATP per mole of glucose, two are produced in the glycolysis process and 36 during the oxidative phosphorylation. It is important to note that oxygen availability in the mitochondrion is a critical factor for the normal ATP production in the cell. The exact level of oxygen inside the mitochondria *in vivo* is not well defined due to the inaccuracy of the measurement tools. It is evaluated to be ~1 mm Hg as compared to the high pO_2, ~90 mmHg, existing in systemic circulation (Chance et al. 1973). Glycolysis depends on the entrance of glucose from the capillary into the cell via the glucose transporter. The end product of glycolysis, pyruvate, is transported into the mitochondria by a specific carrier protein. The pyruvate is transformed, in the matrix of the mitochondria, into acetyl

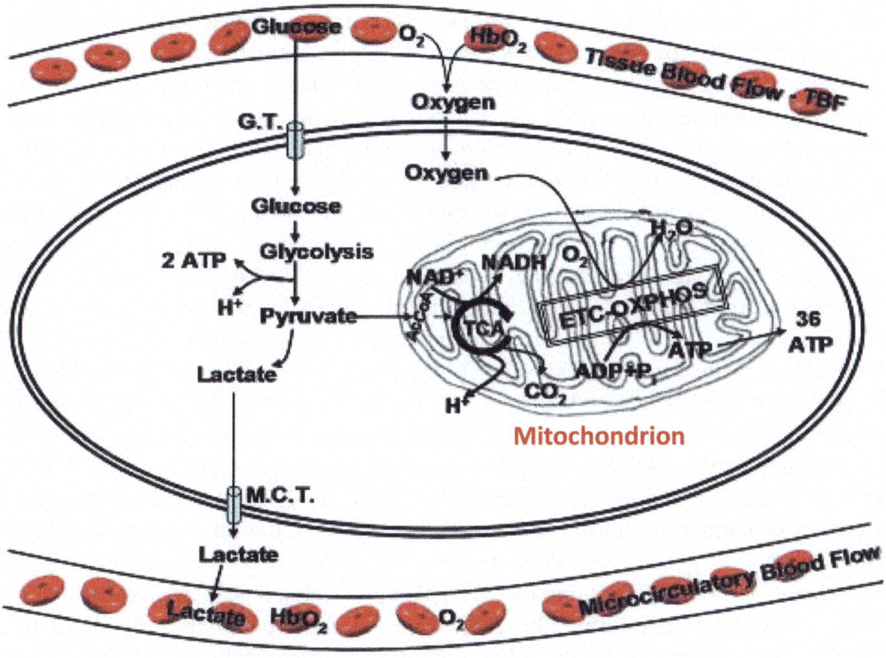

Fig. 7.81 Schematic presentation of tissue oxygen metabolism at the cellular and intramitochondrial level. GT glucose transporter; MCT monocarboxylate transporter; ETC electron transport chain; OXPHOS oxidative phosphorylation; TCA tricarboxylic acid cycle; AcCoA acetyl coenzyme A. (Mayevsky et al. 2011 [Modified])

coenzyme A, which activates the tricarboxylic acid (TCA) cycle. In the absence of oxygen, the end product of pyruvate is lactate, which may leave the cell and pass into the microcirculatory bloodstream via the monocarboxylate transporter located in the plasma membrane. In the mitochondria, the TCA cycle generates NADH (reduced nicotinamide adenine dinucleotide), which enters the electron transport chain (ETC) leading to the oxidative phosphorylation that generates ATP. The CO_2 released from the TCA cycle exits the cell as HCO_3^-, via the anion exchanger, into the bloodstream. The continuous supply of oxygen to the mitochondria depends on two microcirculatory parameters: TBF in the diffusible small vessels (small arterioles and capillaries) and the level of hemoglobin oxygenation or saturation (HbO_2) in the small vessels. Any change in the oxygen consumption by the mitochondria will be compensated either by downloading the extra oxygen needed from oxygenated hemoglobin or via an increase in the blood flow. Under a restriction of oxygen supply (hypoxia or ischemia), mitochondrial function will be inhibited and ATP production will decrease, while glycolysis will become stimulated. In most normal tissues and organs, this stimulation is not sufficient to provide the amount of ATP needed for the normal physiological and biochemical activities. The pioneering work of Chance and collaborators, started in the early 1950s, ushered in a new era of studying mitochondrial function, in health and disease, using spectroscopic techniques (Chance and Legallias 1951; Chance 1952, 1954). The monitoring of NADH redox state in isolated mitochondria *in vitro* (Chance and Baltscheffsky 1958), and later under *in vivo* conditions (Chance et al. 1962a), established the foundations for the understanding of cellular bioenergetics, under various pathophysiological conditions in experimental animals and patients (Mayevsky and Chance 2007, Mayevsky and Rogatsky 2007). The functional capacity of a tissue is related to its ability to perform its work. It is possible to assess this ability through the knowledge of changes in the oxygen balance [i.e., the ratio of oxygen supply to oxygen demand in the tissue (Fig. 7.82)]. The electron transfers (oxidations/reductions) down the respiratory chain resulting in the production of ATP. Concomitantly with the electron transport, the respiratory chain components switch between reduced and oxidized states, each of which has different spectroscopic properties. Hydrolysis of the pyrophosphate bonds provides the energy necessary for the cell's work. In order to assess the energy demand, it is necessary to measure different organ-specific parameters. In the brain, the energy demand can be evaluated by measuring the extracellular levels of K^+ that reflect the activity of the major ATP consumer Na^+/K^+ ATPase (Mayevsky and Chance 1982; Mayevsky 1984). In the heart, most of the energy is consumed by muscle contraction activity. On the other hand, the energy supply mechanism is the same in all tissues: oxygenated blood reaching the capillary bed releases O_2 that diffuses into the cells. Therefore, it is possible to evaluate tissue energy supply by monitoring the same four different parameters in all tissues, namely, mitochondrial function, TBF, volume, and oxygenation. The intracellular level of mitochondrial NADH (the reduced form) is a parameter related to the tissue's state of energy metabolism. Energy exchange depends on pO_2. Information regarding pO_2 in the tissue therefore is helpful for the evaluation of tissue metabolic activity. The sensitivity and accuracy of oxygen electrodes in the range of 1 mmHg

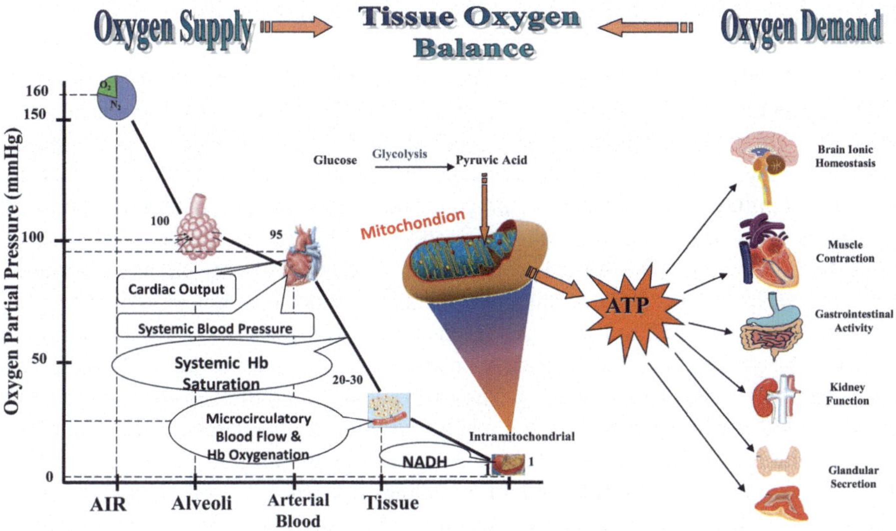

Fig. 7.82 Schematic presentation of the balance between oxygen supply and demand in a typical tissue. The supply is dependent on the availability of oxygen in the mitochondria, which is affected by the oxygen gradient between room air to the intracellular space (left side). The demand for energy is affected by the specific activities taking place in each organ. (Mayevsky et al. 2011 [Modified])

is not sufficient for the evaluation of mitochondrial function. The need for an intracellular pO_2 indicator, as a physiological and biochemical parameter of living tissue, emerged more than 50 years ago. Mitochondria are the intracellular organelles that consume most of the oxygen. Therefore, the redox state of electron carriers in isolated mitochondria *in vitro* as well as *in vivo* as a function of oxygen concentration has been extensively studied. Chance et al. concluded that "For a system at equilibrium, NADH is at the extreme low potential end of the chain, and this may be the oxygen indicator of choice in mitochondria and tissue as well" (Chance et al. 1973). Lubbers in 1995 concluded that "the most important intrinsic luminescence indicator is NADH, an enzyme of which the reaction is connected with tissue respiration and energy metabolism" (Lubbers 1995). Various parameters are measured in animal models or patients in order to evaluate the normality of the oxygen gradients, ranging from 160 mm Hg in room air to ~1 mm Hg in the mitochondria, as shown in Fig. 7.82. The respiratory function of the lungs is evaluated by measuring various parameters such as respiration rate and ventilation, as well as levels of blood gases (pO_2, pCO_2, and pH). The second monitoring point along the oxygen gradient is the hemodynamic parameters of the cardiovascular system, including cardiac output and systemic blood pressure. The next parameter is the saturation level of the arterial blood using the pulse oximetry. To date, the ability to measure tissue energy metabolism at the microcirculation and cellular level in clinical conditions is relatively limited.

Monitoring of NADH: History, Principles, and Basic Technology

Since the discovery of pyridine nucleotides by Harden and Young (1906), more than 1000 papers have been published on the use of NADH as a marker for mitochondrial function. This type of research was started in 1955 by Chance and Williams (1955a, c) by defining the mitochondrial metabolic state *in vitro*. In order to understand mitochondrial function *in vivo* and under various pathophysiological conditions, it is important to monitor the redox state of the respiratory chain in real time. The discovery of the optical properties of NADH led to very intensive research since the early 1950s. The establishment of the method to monitor NADH *in vivo* was done between 1952 and 1962. An intensive use of the *in vivo* NADH monitoring approach began in 1962. The "classical" paper on *in vivo* monitoring of NADH was published by Chance et al. (1962a). They were able to simultaneously monitor the brain and kidney of anesthetized rats using two microfluorometers. Chance et al. in (1962b) elaborated on this kind of *in vivo* monitoring and used it in other rat organs. The correction of the NADH F signal for changes in tissue absorption and scattering was initiated in the 1960s and 1970s by various investigators (Chance and Legallias 1963; Jobsis and Stainsby 1968; Jobsis et al. 1971a). Most of the published material in this field of *in vivo* NADH monitoring of tissues used the subtraction of the Ref signal (366 nm) from that of the NADH F signal measured to provide the corrected F signal (Harbig et al. 1976; Dora et al. 1984). A few attempts were made in order to improve the correction technique, but no significant improvement was found. In addition, the various correction techniques were listed and discussed in detail in Ince et al. (1992). Most of the published material was based on a 1:1 ratio, when subtracting the 366 nm Ref from the 450 nm F signal. We have found that subtracting the Ref from the F or dividing F by Ref provides similar net NADH changes. Bradley and Thorniley in Bradley and Thorniley 2005 reviewed the available techniques for correction of F signals. They concluded, "even though research has been conducted into correction techniques for over 30 years, the development of a successful and practical correction technique remains a considerable challenge."

7.10.1.3 Real-Time Monitoring of Tissue Vitality

Several real-time invasive and noninvasive techniques have been developed to determine tissue energy metabolism or tissue vitality *in vivo*. A short description of the available techniques for monitoring tissue viability described by various investigators, shown in Fig. 7.83, is presented here. These techniques could be adapted and used in experimental animals as well as in patients. In order to present the linkage between the systemic and tissue-level oxygenation, the pulse oximeter technique is also shown in Fig. 7.83; although the signal does not originate from the microcirculation.

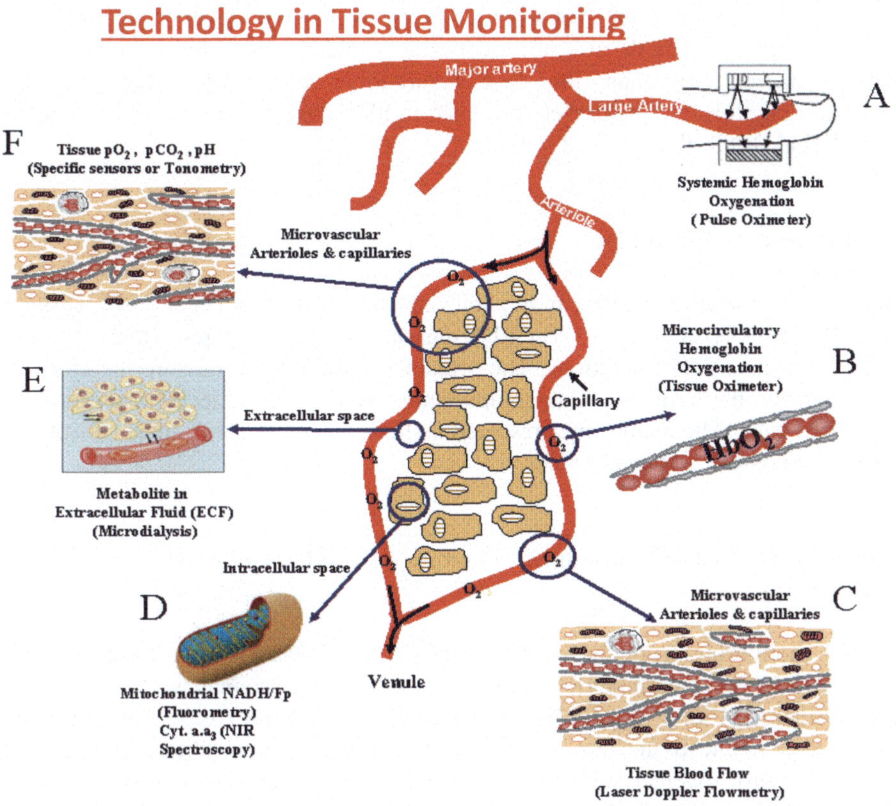

Fig. 7.83 The available techniques for real-time monitoring of energy metabolism at the tissue level. (**a**) The coupling between the macrocirculation monitored by the pulse oximeter and the microcirculation. (**b–f**) Monitoring of cellular and intracellular compartments (see text for details). (Mayevsky et al. 2011 [Modified])

Systemic Pulse Oximetry

This standard of care device is measuring the oxygenation level of the hemoglobin dependent on the pulse passing in the large arteries. In small vessels, the pulse disappears, and, therefore, this technique cannot provide information on hemoglobin saturation in the microcirculation. Because of local changes in oxygen consumption, the systemic pulse may be relatively stable but the local Hb saturation may differ.

Microcirculatory Hb Oxygenation

The level of oxygenated hemoglobin can be monitored at the microcirculatory level using the absorption spectrum of hemoglobin, which is different in its oxygenated or deoxygenated state. One possibility is to illuminate the tissue with light at the isosbestic wavelength point (not sensitive to the oxygenation level of hemoglobin), and a second wavelength, at a non-isosbestic point, in which the oxyhemoglobin absorbs more light than the deoxyhemoglobin form. By subtracting the two-wavelength Ref, a parameter correlated to blood oxygenation level is obtained (Frank et al. 1989; Rampil et al. 1992).

Microcirculatory Blood Flow

Real-time monitoring of TBF can be achieved using LDF (Stern et al. 1977; Dirnagl et al. 1989; Haberl et al. 1989). The LDF measures relative changes, which correlate with the relative TBF alterations. The principle of LDF is to measure the Doppler shift, namely, the frequency change undergone by the light reflected from moving red blood cells. The results are presented as percentages of a full scale (0–100%), providing relative perfusion values.

Mitochondrial NADH/Flavoproteins Fluorometry and Near-Infrared Spectroscopy

NADH monitoring from the organ surface (brain, kidney, liver, testis, etc.) is performed by the fluorometric technique based on the original work by Chance and Williams (1955a, b, c). The excitation light (366 nm) is passed from the fluorometer to the tissue through a bundle of quartz optical fibers. The emitted light (450 nm), together with the reflected light at the excitation wavelength, is transferred to the fluorometer through another bundle of fibers. The measured changes in F and Ref signals are calculated as percent values relative to the calibrated signals under normoxic conditions. This type of calibration is not absolute but provides reliable and reproducible results for different animals and different laboratories (Mayevsky and Chance 1982, Mayevsky 1984). Monitoring of flavoprotein (Fp) fluorescence in blood-perfused organs *in vivo* is very problematic as compared to blood-free organs *in vitro*. Two factors affect the very weak F signal. Blood oxygenation and blood volume changes in the microcirculation-induced artifacts in the measurements, which are very hard or impossible to be corrected.

Near-infrared spectroscopy (NIRS) is a noninvasive optical technique that evaluates the relationship between oxygen availability and consumption by monitoring of the oxidation-reduction state of Cytaa3, although the artifacts of the hemoglobin signal in the Cytaa3 signal remain problematic. Because tissue oxygenation is indicated by the state of cytochrome oxidase, the terminal enzyme of the respiratory chain, dynamic measurements of oxygen utilization, and perfusion can be achieved by NIRS (Jobsis 1977; Thorniley et al. 1997).

Microdialysis

The technique of tissue microdialysis allows continuous online monitoring of changes in tissue energy-related metabolites (e.g., glucose, lactate, pyruvate, adenosine, xanthine) (Johnston and Gupta 2002). It involves the insertion of a catheter lined with a polyamide dialysis membrane into brain parenchyma that is perfused with a physiological solution (e.g., Ringer's lactate) at ultralow flow rates using a precision pump. Small molecules diffuse from the extracellular space into the perfusion fluid, which is then collected into vials that are changed every 10–60 min, allowing for up to 70% equilibration across the dialysis membrane (Hutchinson et al. 2000; Hutchinson 2005). Cerebral microdialysis has great potential for exploring the pathophysiology of acute brain injury, drug pharmacokinetics within the central nervous system, and responses to therapeutic interventions.

Tissue pO_2, pCO_2, and pH

Measurements of tissue pO_2, pCO_2, and pH are used as a monitoring modality in the neurosciences and critical care units (Maas et al. 1993; Zauner et al. 1997; Doppenberg et al. 1998), and as a marker of tissue oxygenation in research protocols (Lacombe et al. 1992; van Rossem et al. 1992; van Hulst et al. 2002). The technique involves the insertion of a microsensor into the brain parenchyma, either through a bolt inserted into the skull or directly through the craniotomy site and tunneled under the skin. At least two commercial microsensors allow direct, continuous measurement of brain tissue gases (Chance and Schoener 1962).

The aims of this study are as follows:

1. To describe the principles and the technological aspects of tissue vitality optical monitoring.
2. To present the upgraded technology of the CRV developed from the previous model of the CRV (Mayevsky et al. 2006). Also, the laboratory device, the tissue vitality analyzer (TVA), is presented.
3. To compare the performance of the CRV and the TVA in animal models, the TVA was used as a predicate device, namely, a device which the functions of the CRV were compared to the FDA regulation process. This comparison includes the *in vitro* and *in vivo* small animal experiments.
4. To present preliminary clinical results obtained in critical care patients using the CRV.

7.10.2 Methods

In this study, the updated version of the CRV will be described and compared to the old laboratory monitoring system used in animal experiments named TVA. Two models of the CRV were cleared by the FDA. The first model was tested in small animal models (gerbils). The second model of the CRV was tested in small animals (gerbils), large animals (pigs), and in more than 30 patients. The TVA and CRV devices carries out certain *in vivo* spectroscopic measurements as a multiparametric monitoring device intended for the measurement of tissue in the metabolic state. The TVA consists of the combination of an NADH fluorometer, a Laser Doppler Flowmeter, as well as a reflectometer for the monitoring of microcirculatory blood oxygenation. The CRV is an integrated commercial unit that measures the same parameters. The two instruments are monitoring the following parameters (Fig. 7.84):

1. Mitochondrial NADH fluorescence emitted at 420–480 nm.
2. Total backscattered light at the excitation wavelength reflected from the tissue. This parameter allows for correction of the NADH fluorescence measurement due to changes in tissue blood volume.
3. Doppler-shifted laser light reflected from moving blood cells.
4. Two-wavelength reflectometry enables the monitoring of the saturation of HbO_2. The TVA and CRV do not carry out the assessment of the monitored parameters. They simply provide the measured information to the user. These parameters contain information pertaining to the redox state of mitochondrial NADH of the tissue, microvascular blood flow and oxygenation in the tissue, and total blood

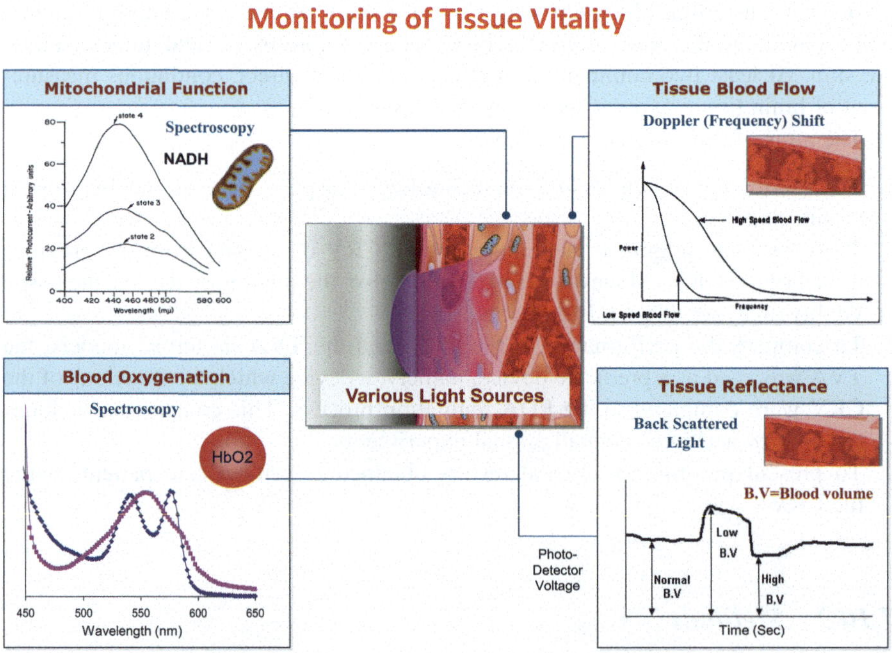

Fig. 7.84 Principles of tissue spectroscopy in order to evaluate tissue viability by the TVA and the CRV. Three of the parameters are monitored from the intravascular compartment, while the NADH parameter represents the function of the intramitochondrial space. (Mayevsky et al. 2011 [Modified])

volume of the tissue. Changes in these values reflect changes in the balance between oxygen supply and oxygen demand. Because the CRV is intended as an adjunctive measurement device when used in conjunction with conventional patient monitors, this further supports the low safety risk of the CRV. This means that the intensity of light in the various excitation wavelengths was below the level that may damage the tissue. The basic scientific principles of the TVA and CRV are identical, but the technology used is completely different.

7.10.2.1 Tissue Vitality Analyzer

In order to assess the hemodynamic and metabolic functions of the tissue, we used the TVA, which includes two devices: a time-sharing fluorometer-reflectometer for the mitochondrial NADH redox state and microcirculatory HbO_2 measurement (Rampil et al. 1992) combined with a LDF for TBF monitoring (Fig. 7.85a). The connection between the brain and the TVA was done by a flexible light guide. The

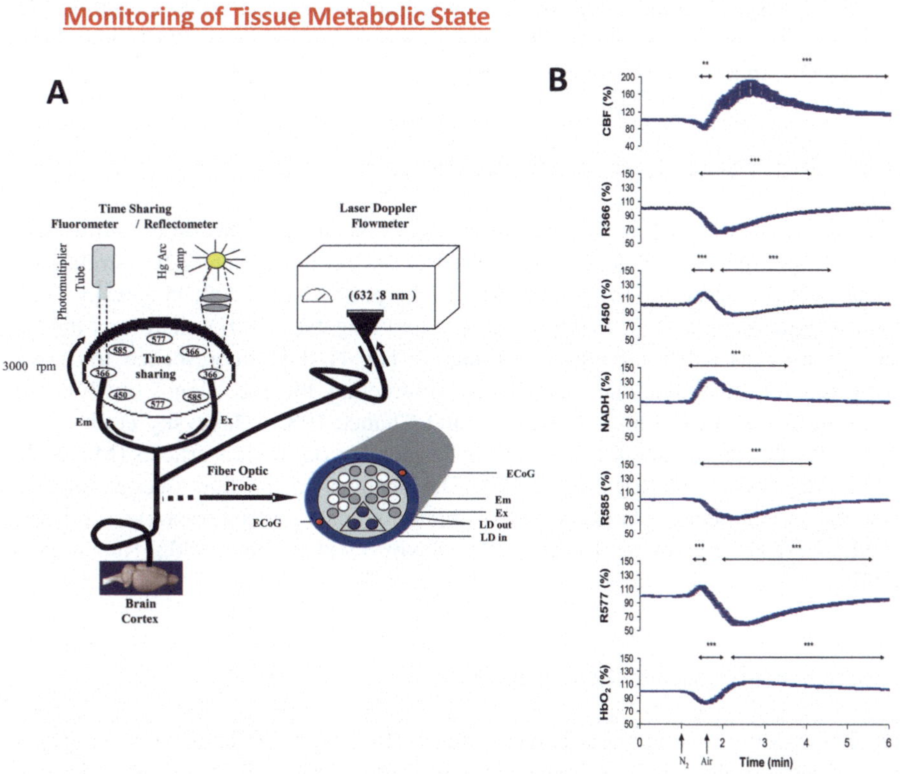

Fig. 7.85 (**a**) Schematic presentation of the TVA and (**b**) typical response to the lack of oxygen (anoxia). The details of the TVA components were described previously. TBF tissue microcirculatory blood flow; R366—reflected light at the excitation wavelength of NADH (366 nm); HbO_2—relative oxygenation level of hemoglobin. (Mayevsky et al. 2011 [Modified])

fiber optic probe includes fibers that were connected to the two instruments, as shown in the enlargement of the bundle tip. The measurement of HbO_2, Ref at 366 nm, and NADH was performed by the same excitation and emission fibers. The CBF was measured by three optical fibers located in the center of the time-sharing bundle of fibers. This system includes a rotating wheel with eight specific filters at the appropriate wavelengths (366 and 450 nm for NADH and 585 and 577 nm for oxyhemoglobin measurement) to four filters for excitation light and four filters for emitted light. The wheel rotates at ~2400 rpm, which is a very high speed with respect to the kinetics of physiological changes; thus, NADH and oxyhemoglobin are simultaneously monitored (Rampil et al. 1992; Mayevsky and Rogatsky 2007). Figure 7.85b presents typical responses to anoxia measured by the time-sharing device shown in Fig. 7.85a.

Microcirculatory Blood Flow

The optical fibers of the Laser Doppler Flowmeter were used for monitoring TBF. The LDF measures relative changes (in the 0–100% range), which are significantly correlated to the absolute values of TBF measured by other quantitative methods: H_2 clearance (Dirnagl et al. 1989), ^{14}C iodoantipyrine (Haberl et al. 1989), and radioactive microspheres (Kirkeby et al. 1995).

NADH Redox State Fluorometer/Reflectometer

The principle of NADH monitoring from the tissue surface was described in Sect. 7.10.1.2 and in 7.10.1.3. It is known that Ref at 366 nm is inversely correlated to changes in the tissue blood volume (Mayevsky et al. 1974, 1992; Mayevsky 1975; Dora 1985b; Rampil et al. 1992; Mayevsky and Rogatsky 2007); therefore, the Ref can be used as a hemodynamic parameter. The NADH fluorescence trace was corrected for hemodynamic artifacts by subtracting the Ref signal (at 366 nm) from the F signal (450 nm) (Mayevsky and Chance 1982; Mayevsky et al. 1992). A detailed discussion about this issue can be found in our review articles (Mayevsky 1984; Mayevsky and Rogatsky 2007). The changes in the fluorescence and Ref signals are calculated relative to the calibrated signals under normoxic conditions (100% NADH) (Mayevsky 1978b, 1984; Mayevsky et al. 1998, 2002; Kanner et al. 2002).

Hemoglobin Oxygenation Reflectometer

Oxyhemoglobin measurement is based on the differences in the absorption properties of hemoglobin in its oxygenated versus deoxygenated states. The tissue is illuminated at two wavelengths: 585 and 577 nm. At 585 nm, oxyhemoglobin and deoxyhemoglobin have the same absorption characteristics (the isosbestic point);

thus, the light emitted from the tissue reflects changes in tissue blood volume. At 577 nm, oxyhemoglobin has higher absorption ability (lower Ref) than deoxyhemoglobin; thus, at this wavelength, the emitted light intensities are affected by the oxygenation levels as well as by the changes in blood volume at the measurement site. Therefore, subtracting changes in 585 nm Ref from those at 577 nm provides a parameter correlated to net changes in blood oxygenation (Rampil et al. 1992; Deutsch et al. 2004).

7.10.2.2 CritiView

This new device, the CRV, measures the same four parameters as the TVA using different technology. The FDA approved the new device.

Light Source Unit

The Light Source Unit (LSU) [see Fig. 7.86a] of the CRV comprises a 785 nm CW laser diode, which serves for laser Doppler measurement; a UV light-emitting diode ([LED], 375 nm) for NADH fluorescence excitation and for total backscatter (or reflection) measurement; and a blue LED (470 nm) and green LED (530 nm) for HbO_2. In order to enable a very high measurement dynamic range of fluorescence

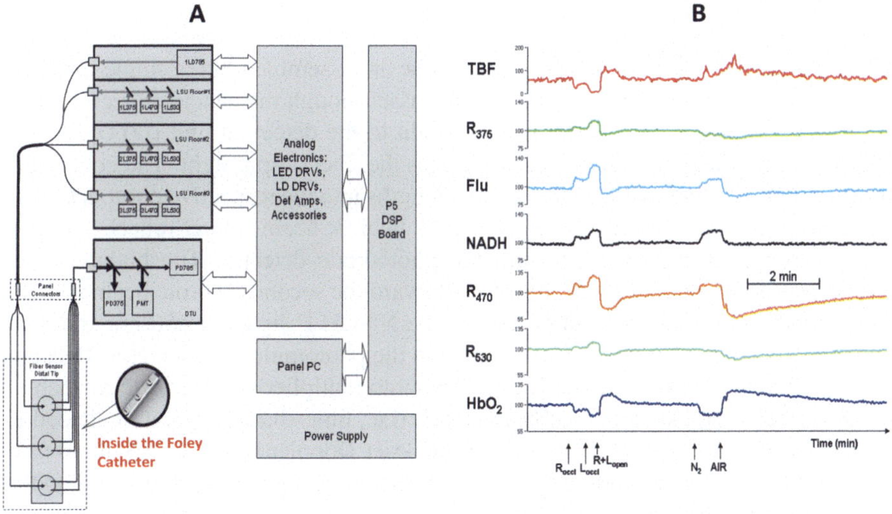

Fig. 7.86 (**a**) Schematic presentation of the CRV components. (**b**) Responses to ischemia and anoxia are presented. The effects of (left) ischemia and (right) anoxia on the four parameters measured from the brain of an anesthetized Gerbil. N_2—100% nitrogen; R_{occl}, L_{occl}—occlusion of the right or left common carotid artery; R + L_{open}—reopening of the two carotid arteries. (Mayevsky et al. 2011 [Modified])

and reflection parameters, the light source unit is designed to enable a very wide range of excitation intensities. To enable this wide excitation, range, and linearity, the system is designed according to a three-floor concept. Each floor comprises three LEDs: one UV LED with emission peak at 375 nm, one blue LED with emission peak at 470 nm, and one green LED with emission peak at 530 nm. The different wavelengths from all LEDs of the same floor are assembled together and coupled into a single fiber by a set of dichroic mirrors and appropriate collimation and focusing lenses. The current of each of the discrete LEDs is set by the appropriate electronics drivers directly controlled by a digital-to-analog (D/A) converter of the DSP. The difference between the floors is the output intensity. There are high-, medium-, and low-intensity floors. The different excitation intensities are achieved by utilizing various pinholes while maintaining all other electro-optical properties the same for all three floors. The light from all three floors and a laser diode is combined into a single mixer fiber, therefore, enabling precise setting of the excitation intensity within a very wide excitation range. The NIR laser diode at 785 nm, for laser Doppler measurements, operates in CW operation mode. The UV LEDs, blue LEDs, and green LEDs operate in chopping mode. This enables the use of synchronous detection techniques in order to detect the NADH fluorescence and total backscatter light. Additionally, the chopping operation mode enables one to perform NADH measurements with very low excitation intensities well below the limits specified by the laser safety standards.

Detection Unit

All six collection fibers of the fiber optic probe are assembled into a single male SC optical connector. The light from the probe passes though the panel connector into a single thick optical fiber that delivers the light to the detection unit (DTU). At the DTU entrance, the collimation lens collimates the fiber output light. The collimated light is split according to the different wavelengths into the respective photodetectors by means of dichroic beam splitters. The first dichroic beam splitter reflects the total backscatter signal at 375 nm toward the photodiode detector. The higher wavelengths pass through the first beam splitter toward the second dichroic beam splitter. The second dichroic beam splitter reflects the NADH F signal at 450 nm and total backscatter signals at 470 and 530 nm toward the photomultiplier detector. Because of the chopping operation of the LEDs, the photomultiplier detector detects each of the abovementioned signals at different times (i.e., time-sharing operation detection mode). The second dichroic filter enables the laser Doppler signal at 785 nm to pass through it toward the photodiode detector. All acquired signals are digitized into the DSP by high-resolution 16-bit A/D.

Digital Signal Processor

The DSP is responsible for whole system control, initial data processing, and calculation of the Doppler parameter. The DSP is built around a Tern Inc.

586-Engine-P controller board with AMD SC520 CPU. After initial data processing, the calculated values are transmitted to the panel computer for final data processing display through RS-232 serial interface. The CRV device utilizes medical-grade main power supply for all electronic circuits, including the panel computer.

CritiView Optical Probes

Two types of fiber optic probes were developed for experimental animal and for clinical applications. (i) The pencil-style optical probe has one measurement point to be used on exposed tissue. The excitation and collection fibers are held together in a stainless steel element. It is comprised of one excitation fiber connected to the respective excitation connector on the CRV panel and six collection fibers connected to the single collection connector on the CRV front panel. (ii) The Foley catheter optical probe for measuring the urethral wall vitality is based on a standard three-lumen Foley catheter, which is an inflatable balloon retention type catheter inserted through the urethra and used to drain the bladder. This probe construction enables the measurement of tissue metabolism at the urethral wall while draining urine from the bladder. The CRV Foley catheter optical probe is designed to perform optical measurements from three adjacent points on the tissue, 3.5-mm apart (Fig. 7.86a). Each measurement point is comprised of one excitation fiber and two adjacent collection fibers. The three excitation fibers from the LSU are connected to the excitation connector on the CRV panel. All six collection fibers from the three measurement points are connected to the single collection connector on the CRV front panel. Figure 7.87c shows the finished probe inserted inside the three-way Foley catheter.

7.10.2.3 Testing the CritiView In Vitro

The aim of this study was to demonstrate the efficiency of the CRV device for the measurement of NADH concentration. The experimental methodology is straightforward. Various solutions of NADH were measured with the CRV. Insofar as the measurement with the CRV involves placing its probe tip into the solution, one might ask if this measurement is the same as placing the tip of the probe against a tissue surface. The equivalence of these two measurements can be understood if one considers that, although immersion in aqueous solution and placement adjacent to tissue are clearly two different physical situations, nevertheless, the process and principle of measurement that takes place is the same for both. Aside from their contiguity, there is no interaction between the probe and the solution or any adjacent tissue. The probe merely collects emitted light from its immediate environment (i.e., a hemispherical volume adjacent to the probe tip) and transfers the collected light to detectors for measurement Therefore, for the purposes of comparison of CRV spectroscopic characteristics to those of control NADH solution and to a clinical fluorometer, the CRV probe can be positioned on the surface of a tissue or a liquid,

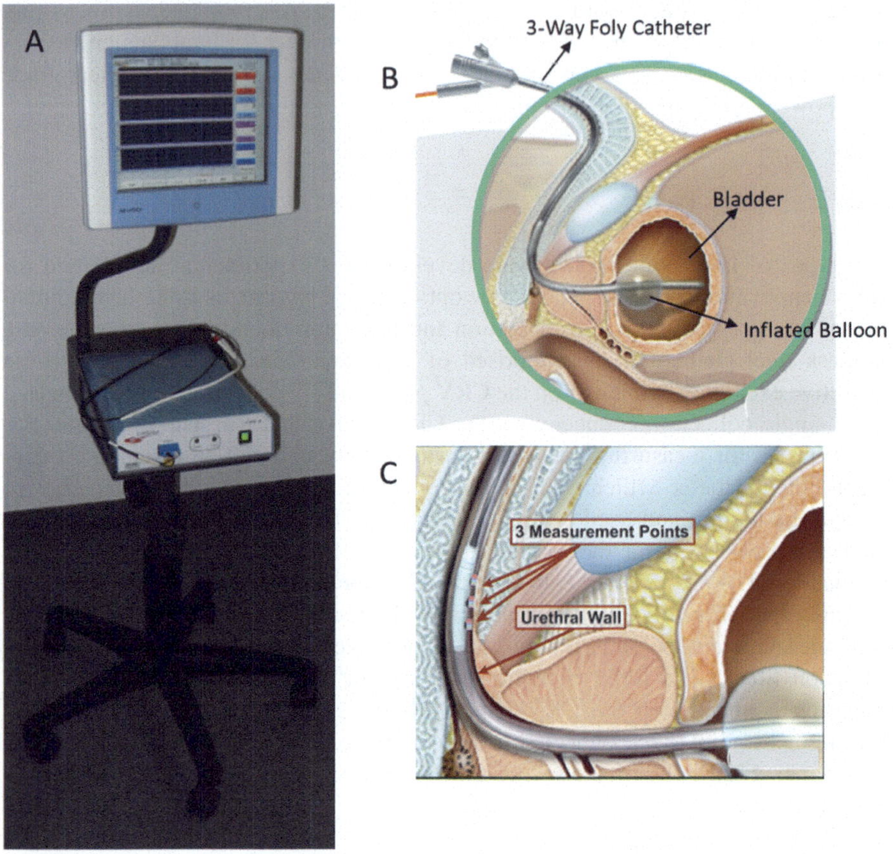

Fig. 7.87 (**a**) The CritiView device used in this study and (**b**, **c**)—location of the three-way Foley catheter and the monitoring sites in the urethral wall. (Mayevsky et al. 2011 [Modified])

or it can be inserted into the tissue or liquid. Three sets of solutions with known concentrations of NADH and DDW were prepared in the following way: Stock solution of 1 mM NADH was prepared by dissolving 14.188 mg NADH in 20 ml DDW. The samples of NADH solutions (5 ml) were prepared in glass vials by the appropriate dilution of the stock solution with DDW, which also used as a blank that was subtracted from the actual reading made by the CRV. The "pencil-type" probe of the CRV was placed in the vials for 10 s, and a reading was taken. The probe was taken out of the solution, and a second reading was recorded. This was repeated again so that three readings were averaged. A second and third set of diluted solution were prepared by the same procedure and were measured in the same way. The same procedure was used with the predicate CritiView-CRV1 with onset of NADH solutions.

7.10.2.4 Testing the CritiView in Animals

The aim of this study and report is to describe, in detail, the *in vivo* control study of the CRV device in order to demonstrate substantial equivalence to its predicate device. As mentioned before, the original animal laboratory device was the TVA. This device was used in our lab for a number of years before the development of the commercial devices: the CRV or the two models of the CritiView. Each of the three devices was tested in animal experiments in order to verify the performance of the device by itself. After this testing, the TVA was used as a predicate device for the newly developed CRV. The first CRV was compared to the TVA, and the second CRV was compared to the first CRV. In this report, we are presenting the comparison between the TVA and the first CRV. Basically, the comparison between the two CRV models was submitted to the FDA but is not presented here. The performance of the CRV device in measurement of changes in TBF NADH fluorescence, tissue reflectance, and HbO_2 resulting from standard metabolic perturbations in an animal model (gerbils) was done. All signals are monitored by the CRV as relative values. This brings all the calculations to be presented as "percent change of the signals." The TBF outputs of the CRV are curves of signal measurements as functions of time. In order to assess the performance of CRV and the TVA in the measurement of TBF, we performed a series of experiments in which the TBF of animals was measured simultaneously by the two instruments. The same basis was applied to the other three parameters, namely, NADH, Ref, and HbO_2, in the same statistical analysis principles. The response time of the different devices was the same (3 s). The purpose of the statistical analysis is to demonstrate that the two curves of each experiment are equivalent. Because the scales of the instruments are not identical, it is meaningless to compute the difference between the two curves. In fact, equivalence in this case means "having the same direction of change." That is, when one curve goes up, the other goes up, and when one goes down, the other goes down. The appropriate statistical measure for such behavior is the correlation coefficient that uses a large number of data points collected by each instrument in every episode. It is desirable to have a correlation coefficient as close to +1 as possible between every pair of corresponding curves. Thus, by computing correlation coefficients between two curves, it is possible to measure the statistical equivalence between them.

Animal Preparation

In order to test the depth of anesthesia in each gerbil monitored, we used a surrogate systemic parameter that represents the function and the intactness of the cardiovascular and respiratory systems. The parameter is the level of the hemoglobin oxygenation in the systemic blood, SpO_2 measured by a standard veterinary pulse oximeter suitable for small animal experiments (model 8600 V, Nonin Medical Inc., Minneapolis, Minnesota). The SpO_2 was very high and stable during the entire period of monitoring. During the four perturbations imposed on the gerbil, a clear decrease in the SpO_2 was recorded. This decrease in SpO_2 during anoxia, hypoxia, and terminal anoxia is due to the availability of oxygen to the body. The decrease in SpO_2 during cerebral ischemia is probably due to stimulation of the sympathetic and parasympathetic nerves that are affected by the pulling of the carotid arteries during the ischemic episode.

Operation Procedure

Male rats (250–300 g) and Mongolian gerbils (50–75 g) were used. We had operated 35 rats and 25 gerbils that were used for the various *in vivo* studies. This includes the testing of the various devices as well as the comparison studies between the TVA and the CRV. The animals were anesthetized by Equithesin (E-th = chloral hydrate 42.51 mg; magnesium sulfate 21.25 mg; alcohol 11.5%; propylene glycol 44.34%; and pentobarbital 9.72 mg) IP injection of 0.3 ml/100 g body weight. The animals were kept anesthetized during the operation, and during the entire monitoring period, by additional IP injections of E-th 0.03 ml in gerbils and 0.1 ml in rats, every 30 min. The addition of small volumes of E-th every 30 min kept the animals in a stable state. In the gerbils, the two common carotid arteries were isolated just before brain surgery, and ligatures of 4–0 silk thread were placed around them. The animal was placed in a head holder in the supine position. After a midline incision of the skin, an appropriate elliptic hole (about 3 × 8 mm) was drilled in the parietal bone of the right hemisphere. The dura mater remained intact, and a light guide holder—cannula—was placed in the drilled hole, and extra pressure on the tissue was avoided. Two stainless steel screws in the left parietal bone were used to fix the cannula, with dental acrylic cement. Body temperature was measured by a rectal probe (Yellow Springs Instrument, Yellow Springs, Ohio) and was regulated to be at the range of 35–37 °C using a heating blanket.

Metabolic Perturbations

Anoxia
Exposure of the animal to oxygen deficient atmosphere by spontaneous breathing of 100% N_2, for a period of time (~25 s), until changes are observed on the monitor and the animal has stopped breathing. The animal is then allowed to breathe normal room air by spontaneous breathing or by short artificial respiration.

Ischemia
Reversible occlusion (30 s) of the two common carotid arteries by constricting them with threads.

Hypoxia
Exposure of the animal to spontaneous breathing of low oxygen concentration (6% oxygen in air) for a short period of time (60 s). The animal is then allowed to breathe normal room air by spontaneous breathing.

Terminal Anoxia

Exposure of the animal to complete irreversible anoxia by exposing the animal to 100% N_2 until the death of the gerbil. The application of the various gases was done by placing a small tube, connected to the gas cylinder, around the nose of the animals. The flow of the gas was very slow (<1 l/min) in order to avoid respiratory disturbances.

Experimental Protocol

Monitoring the various parameters was started immediately after the end of the operation. This was done by using two needle-type optical probes connected to the CRV and the CritiView-CRV1 and placed inside the cannula cemented to the skull. In order to check the intactness and the physiological status of the brain, a 20 s anoxia was induced. Rats and gerbils that present abnormal response were discarded before running the experimental protocol. In each monitoring system, the variation in the response to short anoxia was in the range of ~10%. For example, if the increase in NADH was in the range of 50–60%, any animal that showed 20–30% or 70–80% was not involved in the study and discarded at this stage of the study. Normally, only 5–10% of the operated animals were discarded at this stage. When body temperature reached the range of 35–37 °C, the exposure of the animals to various perturbations started. The animals were exposed to various conditions, having an interval of ~10 min between each test to allow recovery. The pulse oximeter value was used in order to ensure the complete recovery. The data were collected at a rate of one sample per second and stored in different channels of a computerized data acquisition program.

7.10.2.5 Testing the CritiView in Patients

All patients included in the clinical testing signed an informed consent form after the approval of the protocols by the Helsinki committee of Sheba Medical Center, Israel. Using the CRV (Fig. 7.87a), we studied three groups of human subjects. The first group included general ICU patients ($N = 15$) monitored consecutively (Sheba Medical Center). During this period, we defined and improved the dynamic range of the monitored parameters and adjusted the tissue optical characteristics for the Human urethra, which were assumed to be different from those of the gerbil brain. The data obtained from these patients proved that the Foley catheter, shown in Fig. 7.87b, c, provided appropriate monitoring during at least the first 5 days after its insertion in the patient. The second group included five patients undergoing abdominal aortic aneurysm repair, where a defined period of aortic cross-clamping was needed for the surgical repair. The third group included 20 patients undergoing cardiac surgery, 18 of whom required cardiopulmonary bypass for 1–2 h. In the last group of patients, the hypothesis was as follows. The chest opening and connection of a patient to the heart and lung machine may develop body metabolic instability

Table 7.4 List of surgical procedures in cardiovascular patients

1.	Anesthesia
2.	Inserting of Foley catheter and beginning of monitoring
3.	Skin cut and chest opening 4
4.	Hypothermia (optional)
5.	Isolation of a. Mammary artery(ies). b. Vein(s)
6.	Preparation of large blood vessels (aorta and vena cava)
7.	Connecting the patient to HLM or pump off procedures
8.	Surgical procedures: a. Aortic valve or aneurysm repair. b. Heart valve's treatment or replacement. c. CABG
9.	Gradual recovery to the beating heart
10.	Closing of chest and skin
11.	Disconnection of Critisense's CRV unit
12.	Duration of monitoring period, 2–8 h

and negative oxygen balance even though the machine should provide the needs for oxygen. The objective was to test the CRV in an extreme, controlled event and to evaluate the responses during the entire operation period.

The protocol included the monitoring of CritiView parameters starting immediately after insertion of the three-way Foley catheter until the end of the operation procedure in the OR. The monitoring period included the following steps: opening of chest, vascular and heart procedures, body perfusion by heart and lung machine (HLM), and recovery to the beating heart. The details of the operation procedure are presented in Table 7.4.

7.10.2.6 Statistical Analysis

In Sect. 7.3, we present the data as mean and standard error of the mean. We preferred this presentation because most of the physiological journals prefer this approach. In any case, this presentation mode is not affecting the calculation of the significance level. Various statistical tests were used in this study. For the *in vitro* testing, regression analysis was performed. For the *in vivo* monitoring data analysis, the analysis of variance (ANOVA) test was used. The effects of surgery stages on the tissue microcirculatory blood flow level were analyzed using repeated measures ANOVA (GLM procedure of SPSS 15.0). For the analysis of patients in the third group, we used repeated measures ANOVA, sphericity assumed ($F_{2,20} = 11.170$, $p = 0.001$). The pairwise comparison (Bonferroni comparison) was utilized.

7.10.3 Results

7.10.3.1 Bench Tests and Stability

Bench test data are presented in Fig. 7.88a. Three regression lines, related to the three sets of measurements, were calculated and found to be highly statistically significant. As seen, the CRV is measuring the NADH accurately. In Fig. 7.88b, the stability of the microcirculatory blood flow and NADH for the brain and small intestine is shown. In terms of calculated stability, the values found for the microcirculatory blood flow were better than 2% per hour and for the other parameters were < 1%.

7.10.3.2 Small Animals (Rats and Gerbil Brain)

The two monitoring systems were tested *in vivo* in the rat and gerbil brain, and typical results are shown in Figs. 7.85a (TVA) and 7.86A (CRV). Figure 7.85b shows the responses of the brain to anoxia (100% nitrogen) measured by the TVA in a rat. As can be seen, the negative correlation between HbO_2 and NADH is very clear due to the decrease in the lack of oxygen. The microcirculatory blood flow, measured by the LDF, shows an initial decrease followed by long-lasting overshoot (hyperemia). Also, the changes in the two Ref signals (R366 and R585) are very similar under anoxia. Figure 7.86B presents the responses of tissue microcirculatory blood flow, Ref (R375), NADH, and HbO_2 measured by the CRV. Responses to

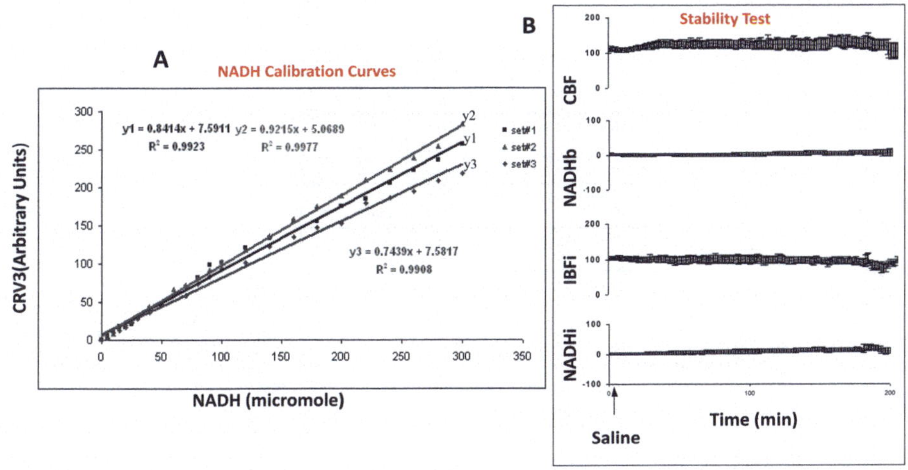

Fig. 7.88 (**a**) Fluorescence readings (three sets) of the CritiView plotted against NADH concentration after subtracting the reading of the blank solution. A linear regression was calculated, and the equation and R_2 are presented for the three sets of control solutions. (**b**) The stability for 200 min of the monitored parameters in two organs. The upper two traces were measured from the brain, while the other two are from the small intestine. (Mayevsky et al. 2011 [Modified])

ischemia (left) were induced in two steps: the occlusion of one carotid artery (R_{occl}) followed by the occlusion of the second artery (L_{occl}). A pronounced decrease in HbO_2 strongly correlated with an increase in NADH. Insignificant recovery was noted shortly after the unilateral occlusion, presumably due to microcirculatory blood flow compensation through the anterior part of the circle of Willis. Hyperemia was noted in the TBF and HbO_2 but not in the NADH traces after the reopening of the carotid arteries. The time needed for recovery to the baseline value is much shorter for the NADH signal as compared to the TBF and HbO_2 signals. As seen in Fig. 7.86b, under anoxia (100% N_2), the decrease in HbO_2 was followed by an increase in NADH. The TBF showed a later increase that turned into a hyperemic response that was also recorded in the HbO_2 signal after recovery from the anoxia. In order to compare the performance of the two devices, namely, the TVA and CRV, an appropriate number of animal experiments were performed. We used four perturbations in each animal and compared the responses of the four parameters. Typical results of those experiments are shown in Fig. 7.89. The two instruments were connected to the same brain of a gerbil, and the responses to various perturbations described in Sect. 7.10.2.4.3 were recorded. In the center of Fig. 7.89, the

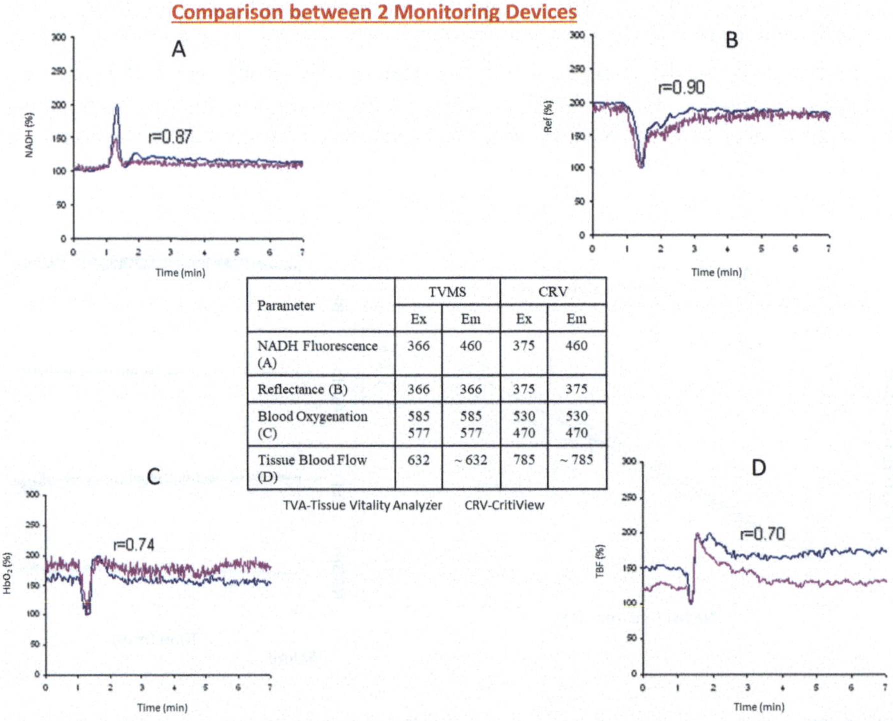

Parameter	TVMS		CRV	
	Ex	Em	Ex	Em
NADH Fluorescence (A)	366	460	375	460
Reflectance (B)	366	366	375	375
Blood Oxygenation (C)	585 577	585 577	530 470	530 470
Tissue Blood Flow (D)	632	~632	785	~785

Fig. 7.89 Comparison between the responses to anoxia measured by the TVA and CRV in the gerbil brain. In each of the four pictures, the two lines represent the results obtained by the two devices. In the center area, the various wavelengths used in the two devices are shown. (Mayevsky et al. 2011 [Modified])

wavelengths used in the two instruments to measure the four parameters are shown. In our previous studies, we were able to compare four monitored sites in the brain by using the multisite monitoring system (Mayevsky and Chance 1982). The responses to anoxia in all sites were very similar, although a different location was monitored. The main reason for this similarity is that the relatively large monitored site (>1 mm diameter) integrates a large number of cells and vascular components that provide similar responses. Therefore, we used this approach to test the equivalency of the two instruments. As can be seen, the responses to anoxia recorded by the two instruments were very similar in this typical animal. Statistical analysis of the group of at least 10 animals showed a significant correlation between the two responses of all four parameters measured. The total number of episodes used for the statistical analysis was 12 for TBF, 22 for NADH, 16 for Ref, and 17 for HbO_2. The statistical analysis of entire episodes was done and found to be significant. Because of space limitations, the details are not presented here.

7.10.3.3 Human Studies

Patients in the Intensive Care Unit

Three responses measured in ICU patients are shown in Fig. 7.90. We assumed that, under stressful situations, in a less vital organ (such as the urethral wall) the microcirculatory blood flow will be reduced and oxygen delivery to the mitochondria will diminish. In a female patient (Fig. 7.90a), the responses to a cessation of spontaneous respiration were recorded. As seen, a clear hypoperfusion was recorded in the urethral wall together with a significant increase in NADH levels. As soon as artificial ventilation was restarted, the signals returned to baseline levels. In another patient, a dramatic decrease in the urethral wall energy state was recorded during a standard airway-suction procedure. As Fig. 7.90b shows, a large decrease in the microcirculatory blood flow was measured simultaneously with a marked increase in NADH. These changes suggest that the cardiorespiratory stressful situation induced by suction could be detected by the decrease in the urethral wall viability. The third patient was monitored during the early postoperative period after coronary artery bypass graft (CABG) operation (Fig. 7.90c). The left side of Fig. 7.90c shows that, during the initial 3 h of monitoring, the NADH levels were relatively stable (±20% change). During the next 40 min of monitoring (right side of Figure 7.90c), a marked mitochondrial dysfunction was recorded (100% increase in NADH) in parallel to a decrease in TBF. The NADH response included two types of change, namely, a pronounced continuous increase in the NADH levels as well as a number of transient elevations lasting 1–5 min each. It was found that, during this period, a cardiac tamponade was developed and the patient was returned to the OR for resternotomy. The hemodynamic parameters (heart rate and blood pressure), shown in the boxes below Fig. 7.90, were not clearly correlated to the development of the tamponade.

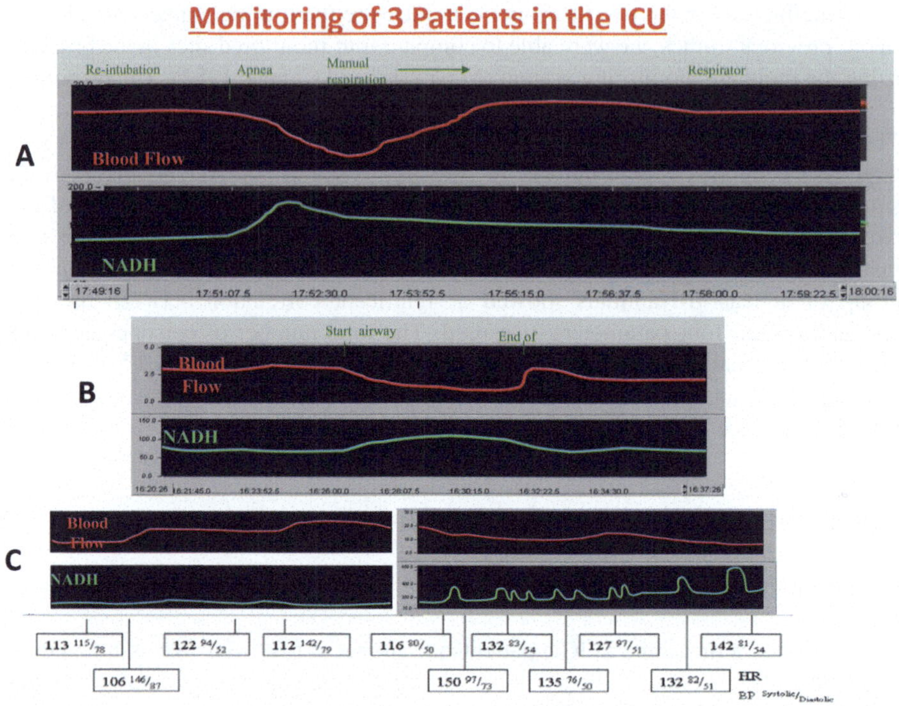

Fig. 7.90 The results of three patients monitored in the ICU using an early version of the device named the tissue spectroscope: (**a**) Responses to a cessation of spontaneous respiration in a female patient, (**b**) effects of the standard airway suction procedure on the viability of the urethral wall, and (**c**) monitoring of a patient in the cardiac ICU during the initial 4 h after the bypass surgery, showing the development of cardiac tamponade. (Mayevsky et al. 2011 [Modified])

Patients Undergoing Open Repair of Abdominal Aortic Aneurysm

In five patients, clear but not entirely consistent responses to the clamping of the abdominal aorta were observed and recorded. During this phase of the operation, the microcirculatory blood flow in the urethra decreased in all patients, while the results of NADH measurements showed some variability. Typical responses to aortic occlusion in one of the abdominal aortic aneurysm (AAA) patients are shown in Fig. 7.91a. The preparation of the aorta for the clamping procedure (marked as A) led to a clear transient decrease in TBF, as well as an increase in the NADH level. The clamping of the aorta led to a maximal decrease in TBF in parallel with the increase in NADH. In this patient, the monitored HbO_2 showed a clear decrease during the clamping interval. Because of a decrease in tissue blood volume, during the aortic occlusion, the Ref trace showed a large increase until the reopening of the occluded aorta. Immediately after declamping, the initial small increase in TBF led to a fast recovery of the HbO_2 and NADH redox state. All signals recovered to the same values measured during the short control period; although, the occlusion interval lasted for >80 min. In Fig. 7.91b, the typical correlation between the various

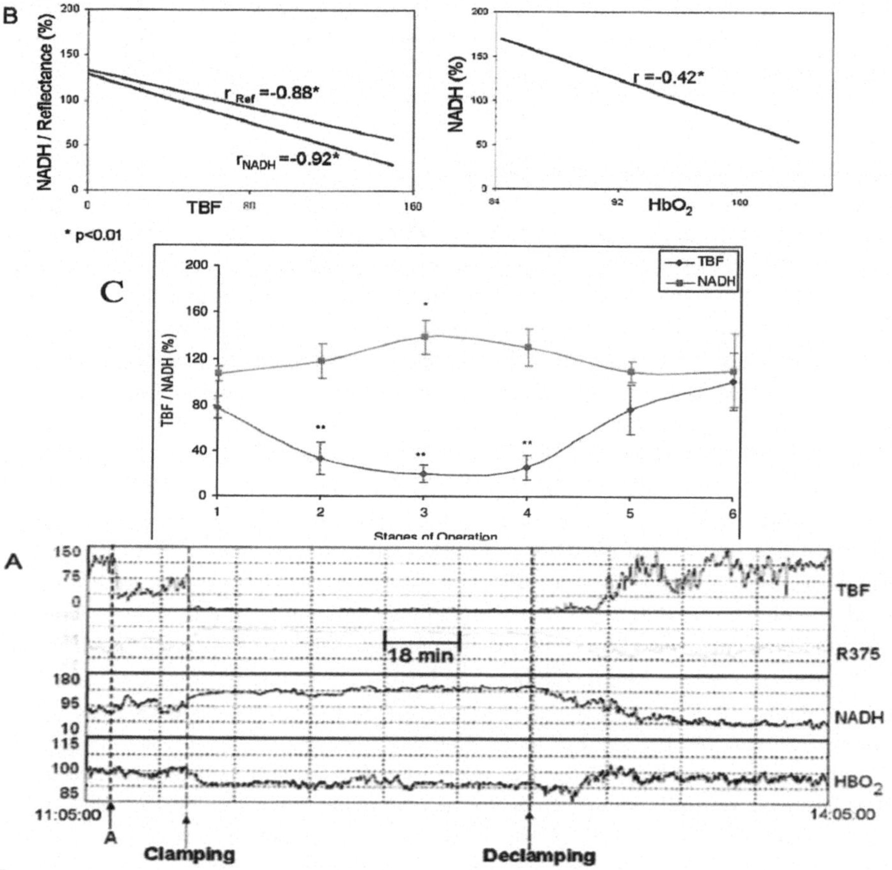

Fig. 7.91 (**a**) Effects of aortic occlusion in a patient who underwent an AAA repair operation, on the four parameters monitored by the CritiView. The technical preparation for aortic occlusion was performed at the marker A. (**b**) The correlations between TBF and NADH/reflectance or HbO_2 and NADH. (**c**) The Mean ± SE changes of TBF and NADH in five patients who underwent the AAA operation. (Mayevsky et al. 2011 [Modified])

parameters is presented. Data were extracted from the five patients of this group according to the six stages of the operation: Stage 1, 10 min before the clamp; Stage 2, 10 min after the clamp; Stage 3, midpoint of the clamping period; Stage 4, 10 min before the declamping; Stage 5, 10 min after the declamping; and Stage 6, the last 5 min of monitoring in the OR.

The results for the five patients (Fig. 7.91c) showed a pronounced decline in the mean TBF values, noted in our experiments from the second stage of the surgery up to the fourth stage. The dependence of TBF on time was statistically significant. The difference was significant for all stages of surgery (t1-t4, $p = 0.004$), but was nonsignificant ($p > 0.05$) for the partial intervals (t2-t3 and t3-t4). This analysis indicates that the only statistically significant change is between points 1 and 4. The changes in the NADH level were significant only when the basal level (t1) was compared to t3.

Patients Undergoing Open Heart Surgery

Thirteen patients underwent a coronary artery bypass using a cardiopulmonary bypass. Two other patients underwent an off pump coronary artery bypass. Another two patients had valve procedures, and two underwent aortic repair using a cardiopulmonary bypass. The last patient was exposed to aortic repair as well as the CABG procedure. Therefore, 18 out of 20 patients were operated on with the use of the cardiopulmonary bypass heart-lung machine and two were operated under beating heart conditions. Two out of the 20 patients were exposed to deep hypothermia. Three typical responses to the cardiovascular operation procedure are presented in Fig. 7.92. In the first patient [shown in Fig. 7.92a], the responses of TBF and NADH were recorded as soon as the preparation for operation started with the scrubbing of the chest area (marked as A). This very early response was found only in one patient monitored in the cardiovascular OR. The low level of TBF and the high level of NADH were recorded during the entire operation procedure, and the recovery began as soon as the chest was closed. When the patient left the OR for the cardiac ICU, the two parameters were close to the baseline levels. These fast responses are probably due to high sensitivity and a fast response of the autonomic nervous system. We could speculate that the level of adrenaline was elevated very early, and as a result, the microcirculatory blood flow was decreased dramatically and the mitochondrial NADH became highly reduced. In the second patient who underwent the routine CABG operation (Fig. 7.92b), the same type of response was recorded. TBF reduction and NADH elevation were observed during most of the surgical procedure. Spontaneous transient recovery of TBF and NADH was noted during the second half of the operative procedure. In the third patient, operated on for aortic repair (Fig. 7.92c), clear responses to the procedure were recorded. In this patient, at 16:49, the initiation of extracorporeal circulation led to a large decrease in TBF and a large increase in NADH. The signals reverted toward the initial values, although the baseline was not reached (the monitoring period ended at 18:14). In order to perform statistical analysis, we selected patients who underwent a similar operation procedure. Only 13 patients were exposed to the same surgical protocol, and data were extracted from 11 patients. In two patients from the CABG normothermic group, the data collection was interrupted in the middle of the operation, and we were unable to calculate the results to match the other 11 patients analyzed. The other seven patients were exposed to a different protocol and were not included in the statistical analysis. The Mean ± S.E. changes of TBF and NADH were obtained in 11 patients (Fig. 7.92D) following heart bypass operation during seven stages of the operation: Stage 1, 5 min baseline (after insertion of the catheter); Stage 2, 10 min after chest opening; Stage 3, 10 min after HLM or CPB was on; Stage 4, midpoint of HLM; Stage 5, 10 min before HLM was off; Stage 6, 10 min after HLM was off; and Stage

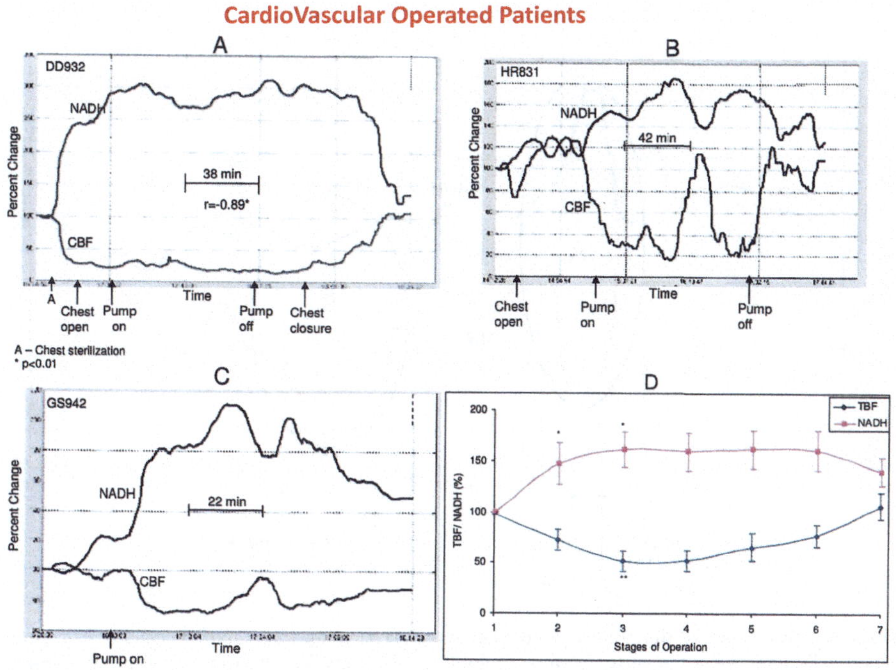

Fig. 7.92 Effects of the surgical procedure in (a–c) three cardiovascular operated patients on the urethral TBF and NADH redox state (see text for detailed explanations) and (d) the Mean ± SE changes of TBF and NADH in 11 patients during heart bypass operation. (Mayevsky et al. 2011 [Modified])

7, the last 5 min of monitoring. ANOVA multiple comparison tests were conducted on the first three stages of the operation. Significance was found between points 1 and 3 in the TBF and between points 1, 2, and 3 in NADH (*$p < 0.05$, **$p < 0.01$, $n = 11$). In one patient, it was possible to compare the responses of the urethral wall to ischemia induced under normothermic and hypothermic conditions. The results presented in Fig. 7.93 were measured in a patient who underwent aortic repair surgery. In the process of opening the chest, a major bleeding occurred and the patient was immediately put on the heart-lung machine and severe hypothermia was induced. As can be seen in Fig. 7.93a, while the patient was normothermic, large changes in TBF and NADH were noted during the major bleeding event. After cooling the patient and shifting him to the heart-lung machine, the signals recovered, although the TBF remained under the control values. While the patient was hypothermic (17 °C), the microcirculatory blood flow was directed mainly to the brain area (Fig. 7.93b), whereas the TBF to the urethra was decreasing significantly without any clear corresponding change in the NADH levels. This may suggest that the mitochondrial redox state behaves differently under hypothermia. It should be acknowledged that this response was recorded only in one patient.

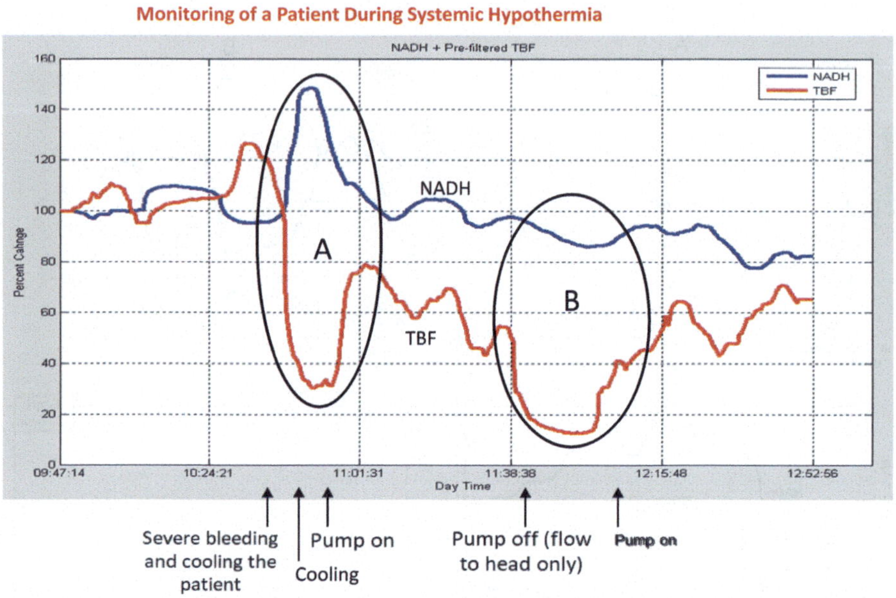

Fig. 7.93 Responses of the urethral wall TBF and NADH to systemic hypothermia (17 °C) in a patient who underwent the aortic repair operation: (**a**) Responses to bleeding under normothermia and (**b**) responses to a decrease of microcirculatory blood flow to the lower part of the body. (Mayevsky et al. 2011 [Modified])

7.10.4 Discussion

The main aim of this chapter is to describe the development of the FDA-approved multiparametric monitoring system based on a laboratory device used in animal experiments. This study demonstrates that the measures of local intracellular energy metabolism and microcirculatory blood flow assessed by an indwelling multiparameter sensor on a urethral catheter platform vary in a predictable fashion in response to transient severe changes in global and regional blood flow in patients exposed to cardiovascular surgery. Corresponding findings were obtained for the cerebral metabolism of gerbils. These semiquantitative data, presented as trends, suggest that changes in peripheral tissue perfusion may be detected rapidly and potentially before the occurrence of overt signs of end-organ hypoperfusion. Specifically, we were able to monitor the mitochondrial redox state, assessing the adequacy of oxygen balance in the urethral wall. We hypothesize that real-time monitoring of urethral mucosal NADH, measured concomitantly with urethral microcirculatory blood flow, may provide a rapid assessment of functional O_2 delivery. Insofar as elevated NADH is a sensitive measure of impaired visceral flow and tissue oxygen balance, the improvement of these parameters should indicate that other, metabolically more active organs have already replenished their oxygen debt. The multiparametric monitoring of the tissue in real time has several advantages over single parameter measures. To evaluate the factors related to oxygen balance (supply/demand) in situations involving changes in O_2 supply to the tissue (i.e., ischemia or reperfusion), it is necessary to simultaneously assess a mitochondrial energy state, tissue O_2 saturation, and microcirculatory blood flow. The monitoring of tissue microcirculatory blood flow, in addition to mitochondrial NADH, using the LDF has been previously shown to be useful for both experimental and clinical settings (Andrews and Muto 1992; Hoper and Gaab 1994). Because the monitoring of the blood oxygenation level does not always reflect oxygen supply to the tissue, this study monitored HbO_2 in the same location as the other parameters, namely, TBF and NADH. Nevertheless, the local monitoring of HbO_2 levels is not sufficient for an accurate evaluation of the balance between oxygen supply and demand, which is the key to tissue metabolism. In addition to TBF and HbO_2, we also monitored the most important parameter in energy metabolism, namely, the level of mitochondrial NADH. It is important to emphasize that *in vivo* monitoring of mitochondrial function could be done measuring NADH, Fp, or cytaa3. To date, in blood perfused tissues only the NADH could be monitored relatively accurate by using the subtraction technique. It was never convincing that Fp and Cytaa3 provide reliable data in blood perfused organs. The overlapping of the absorption spectra of hemoglobin, Fp, and Cytaa3, and the effects of blood volume or blood oxygenation could not be corrected properly. This issue was discussed in our review paper (Mayevsky and Rogatsky 2007). Because the mitochondria play a main role in energy metabolism at the cellular level, monitoring changes in the mitochondrial NADH redox state reflects the state of tissue viability (Coremans et al. 2000). Thus, the simultaneous monitoring of the four parameters provides a clearer picture of the tissue viability and of the primary causes of energy failure if it is present, as has been previously validated by our group (Wadhwani et al. 1990; Mayevsky et al. 1992; Rampil et al. 1992). In this study, we chose the urethral mucosal tissue as a sampling site for two reasons: convenience and physiological rationale. This is a commonly instrumented site, and it would therefore involve no further invasion for continuous, long-term monitoring. The urethral tissue has a low metabolic rate and is not a preferential tissue for blood flow redistribution; thus, its metabolic activity should accurately reflect the final vascular bed flow restoration, without irreversible components, during resuscitation. This contrasts to intestinal mucosa, which can easily demonstrate irreversible injury, making it a poor indicator of global resuscitation effectiveness.

The concept that a single parameter could provide information on the adequacy of resuscitation is rapidly losing validity among clinicians. It is clear that, despite the numerous publications on gastric tonometry, this technique has achieved neither a wide acceptance nor clinical applicability. Perhaps even more important is the fact that, presently, there is no state-of-the-art method to assess when an optimal level of cellular homeostasis has been achieved and, hence, to ascertain that there is no further metabolic stress or ongoing energetic failure at the cellular level after various clinical perturbations.

Ideally, a monitoring strategy capable of providing real-time information on cell viability should fulfill a number of requirements. First, it should be based on highly reliable specific markers. Second, the system must allow for an easy and minimally invasive or, preferably, noninvasive placement of the sensing element. It should be of low cost, low risk, and should provide fast feedback in order to improve the treatment of the patient. Our study documents that the urethral mucosa, an easily accessible region, can be used to assess effective tissue perfusion in a less vital organ. A primary limitation of this technique is its focus on a single monitoring site, the urethral mucosa. Because of variation in the vascular gradients, as well as O_2 gradients across the tissues, the monitored parameters may not adequately represent the real behavior of the tissue. Only by increasing the number of monitored sites in the tissue to three, we were able to collect better results during circulatory insufficiency states. The data accumulated in this preliminary study indicate that the CritiView device provides very clear results in performance tests under *in vitro* conditions as well as in small animal models. Testing the device in patients exposed to various cardiovascular surgical procedures produced fairly good preliminary results. We were aiming to correlate the real-time hemodynamic and respiratory parameters collected by the anesthesiology team to the results collected by the CritiView. We found that this aim could be fulfilled after investing time and money to build a special data-logging system. In the meantime, we manually tested the possible correlation but were unable to conclude statistically significant data. We are planning to do this in our next clinical study. Nevertheless, in one patient presented in Fig. 7.90c, we definitely found that our device provided an early warning signal to the deterioration of the patient due to the tamponade developed 3 h after the bypass operation. As seen on the right side of Fig. 7.90c, the dramatic elevation in urethral wall NADH level was recorded, while the systemic blood pressure and heart beat were not significantly different.

The fact that the responses in the AAA patients to the occlusion procedure were inconsistent could be due to the following reasons:

1. The urethra receives arterial flow directly from branches of the internal iliac and retrograde flow from branches that derivate from anastomoses between the superior and inferior mesenteric arteries. The origin of the superior mesenteric artery is situated in the abdominal aorta above the site of the clamping (The inferior mesenteric artery is distal to the clamping). In not all human subjects are these retrograde branches functional. This variation in the anatomy of blood vessels may explain why, in a few subjects, the decrease in microcirculatory blood flow occurred immediately after the occlusion (the retrograde branches are missing) and returned to normal after declamping. In other subjects, the retrograde flow may improve the level of TBF and saturation, and decrease NADH during the occlusion period. In these subjects, the declamping led to the more rapid recovery of all four parameters to the normal level.
2. Changes in blood volume due to bleeding (total blood loss that may be 500 ml up to 2–3 l) is a very important parameter with regard to changes in saturation, TBF, and NADH in subjects with a retrograde flow.

3. Also, a decrease in temperature may induce supplemental vasoconstriction (that can also be produced by a rapid and massive bleeding), which can change the real record of saturation. Although the number of patients monitored during open chest surgery was very small, some initial suppositions could be made. The main premise in selecting this group was the assumption that this severe operation would induce changes in the oxygen balance in the body. If this is the case, then the viability assay of the urethral wall could be utilized during the operation and, more importantly, during the postoperative period (up to 24 h). As shown in Fig. 7.90c, the monitoring of this patient postoperatively enabled us to detect early changes in body oxygen balance. Indeed, this patient developed a "tamponade" and NADH was elevated before the systemic parameters were clearly changed. This patient was moved to the OR, and the reopening of the chest and solving the problem of accumulated blood inside the pericardium saved his life. Another interesting case is the one presented in Fig. 7.92c. As shown, the patient was moved to the cardiac intensive care unit (CICU) while the urethral wall did not recover to the baseline levels of NADH and tissue microcirculatory blood flow. Our monitoring ended at the end of the operation; thus, it was impossible to evaluate the viability of the urethral wall during the postoperative period. It was reported to us that this patient died after 3 days in the CICU. These two cases provide preliminary indications suggesting that the monitoring of the urethral wall may have practical value in daily critical care medicine. It should be acknowledged, however, that the two patients constitute a very small number and, therefore, a large-scale clinical study is needed. We assume that most patients who will be monitored via the Foley catheter for the assessment of urethral wall viability will show a complete recovery at the end of the operation. Nevertheless, severely ill patients may deteriorate during the postoperative period, and the deterioration should be detected as early as possible.

7.10.5 Conclusion

The new approach to monitor patients in critical care by CritiView was tested successfully under *in vitro* and animal *in vivo* experiments as well as in patients. The conclusions from this study are as follows: (1) For the first time, NADH and microcirculatory blood flow and oxygenation were measured in the urethral wall of patients. (2) Patients exposed to vascular and open chest surgeries showed changes in the measured parameters during the stressful condition of the body. (3) The CritiView results may serve as early warning signals to the deterioration of the body or the endpoint of resuscitations during and after cardiovascular surgeries. A large-scale clinical study with positive results may open up a new era in the monitoring of critical care patients.

References

Anderson RE, Sundt TM (1975) Instrumentation for in vivo cerebral NADH studies in squirrel monkey. IEEE Trans Biomed Eng BME 22(3):220–224

Andrews RJ, Muto RP (1992) Retraction brain ischaemia: cerebral blood flow, evoked potentials, hypotension and hyperventilation in a new animal model. Neurol Res 14(1):12–18

Angus DC, Linde-Zwirble WT, Lidicker J, Clermont G, Carcillo J, Pinsky MR (2001) Epidemiology of severe sepsis in the United States: analysis of incidence, outcome, and associated costs of care. Crit Care Med 29(7):1303–1310

Ansi Z (1993) American National Standard for safe use of laser. American National Standards Institute, New York

Arbit E, DiResta GR, Bedford RF, Shah NK, Galicich JH (1989) Intraoperative measurement of cerebral and tumor blood flow with laser-Doppler flowmetry. Neurosurgery 24(2):166–170

Arnold AS, Wilson JS, Boshier MG (1998) A simple extended-cavity diode laser. Rev Sci Instrum 69:1236–1239

Balaban RS, Blum JJ (1982) Hormone-induced changes in NADH fluorescence and O_2 consumption of rat hepatocytes. Am J Phys 242:C172–C177

Balaban RS, Mandel LJ (1980) Coupling of aerobic metabolism to active ion transport in the kidney. J Physiol 304:331–348

Balaban RS, Mandel LJ, Soltoff SP, Storey JM (1980) Coupling of active ion transport and aerobic respiratory rate in isolated renal tubules. Proc Natl Acad Sci USA 77(1):447–451

Ballinger SW (2005) Mitochondrial dysfunction in cardiovascular disease. Free Rad Biol Med 38: 1278–1295

Barber A (1994) Shock. In: Schwartz SI (ed) Principles of surgery. McGraw-Hill, New York, pp 101–122

Barbiro-Michaely E, Zurovsky Y, Mayevsky A (1998) Real time monitoring of rat liver energy state during ischemia. Microvasc Res 56(3):253–260

Barbiro-Michaely E, Tolmasov M, Rinkevich-Shop S, Sonn J, Mayevsky A (2007) Can the "brain-sparing effect" be detected in a small-animal model? Med Sci Monit 13(10):BR211–BR219

Bates JH, Young MP (2003) Applying fuzzy logic to medical decision making in the intensive care unit. Am J Respir Crit Care Med 167(7):948–952

Batista JE, Wagner JR, Azadzoi KM, Krane RJ, Siroky MB (1996) Direct measurement of blood flow in the human bladder. J Urol 155(2):630–633

Belcaro G, Hoffmann U, Bollinger A, Nicolaides A (1994) Laser Doppler. Med-Orion Publishing, London/Los-Angeles/Nicosia

Benda C (1898) Ueber die Spermatogenese der Vertebraten und höherer Evertebraten, II. Theil: Die Histiogenese der Spermien. Arch Anat Physiol 73:393–398

Bickler PE, Litt L, Severinghaus JW (1988) Effects of acetazolamide on cerebrocortical NADH and blood volume. J Appl Physiol 65(1):428–433

Bland JM, Altman DG (1986) Statistical methods for assessing agreement between two methods of clinical measurement. Lancet 1(8476):307–310

Bland JM, Altman DG (1995) Comparing methods of measurement: why plotting difference against standard method is misleading. Lancet 346(8982):1085–1087

Boldt J (2002) Clinical review: hemodynamic monitoring in the intensive care unit. Crit Care 6(1): 52–59

Bolognese P, Miller JI, Heger IM, Milhorat TH (1993) Laser-Doppler flowmetry in neurosurgery. J Neurosurg Anesthesiol 5(3):151–158

Bonner R, Nossal R (1981) Model for laser Doppler measurements of blood flow in tissue. Appl Opt 20(12):2097–2107

Bradley RS, Thorniley MS (2005) A review of attenuation correction techniques for tissue fluorescence. J R Soc Interface 3:1–13

Breuer Z, Mayevsky A (1992) Brain vasculature and mitochondrial responses to ischemia in Gerbils: II. Strain differences and statistical evaluation. Brain Res 598:251–256

Chance B (1952) Spectra and reaction kinetics of respiratory pigments of homogenized and intact cells. Nature 169:215–221
Chance B (1954) Spectrophotometry of intracellular respiratory pigments. Science 120:767–775
Chance B (1957) Cellular oxygen requirements. Fed Proc 16:671–680
Chance B (1967) Biochemical studies of transitions from rest to activity. The Williams & Wilkins Company, Baltimore
Chance B (1976) Pyridine nucleotide as an indicator of the oxygen requirements for energy-linked functions of mitochondria. Circ Res 38(5):I-31–I-37
Chance B, Baltscheffsky H (1958) Respiratory enzymes in oxidative phosphorylation (VII – Binding of intramitochondrial reduced pyridine nucleotide). J Biol Chem 233(3):736–739
Chance B, Hollunger G (1960) Energy-linked reduction of mitochondrial pyridine nucleotide. Nature 185:666–672
Chance B, Legallias V (1951) Rapid and sensetive spectrophotometry. II. A stopped-flow attachment for a stabilized quartz spectrophotometer. Rev Sci Instrum 22:627–638
Chance B, Legallias V (1963) A spectrofluorometer for recording of intracellular oxidation-reduction states. IEEE Trans Biomed Eng BME-10:40–47
Chance B, Schoener B (1962) Correlation of oxidation-reduction changes of intracellular reduced pyridine nucleotide and changes in electro-encephalogram of the rat in anoxia. Nature 195:956–958
Chance B, Williams GR (1955a) A method for the localization of sites for oxidative phosphorylation. Nature 176(4475):250–254
Chance B, Williams GR (1955b) Respiratory enzymes in oxidative phosphorylation (I- Kinetics of oxygen utilization). J Biol Chem 217:383–393
Chance B, Williams GR (1955c) Respiratory enzymes in oxidative phosphorylation (III- The steady state). J Biol Chem 217:409–427
Chance B, Williams GR, Holmes WF, Higgins J (1955) Respiratory enzymes in oxidative phosphorylation (V- A mechanism for oxidative phosphorylation). J Biol Chem 217:439–451
Chance B, Cohen P, Jobsis F, Schoener B (1962a) Intracellular oxidation-reduction states in vivo. Science 137:499–508
Chance B, Legallias V, Schoener B (1962b) Metabolically linked changes in fluorescence emission spectra of cortex of rat brain, kidney and adrenal gland. Nature 195:1073–1075
Chance B, Oshino N, Sugano T, Mayevsky A (1973) "basic principles of tissue oxygen determination from mitochondrial signals." oxygen transport to tissue. Adv Exp Med Biol 37A:277–292
Chance B, Mayevsky A, Goodwin C, Mela L (1974) Factors in oxygen delivery to tissue. Microvasc Res 8:276–282
Chance B, Schoener B, Oshino R, Itshak F, Nakase Y (1979) Oxidation-reduction ratio studies of mitochondria in freeze- trapped samples. NADH and flavoprotein fluorescence signals. J Biol Chem 254(11):4764–4771
Clavijo JA, van Bastelaar J, Pinsky MR, Puyana JC, Mayevsky A (2008) Minimally invasive real time monitoring of mitochondrial NADH and tissue blood flow in the urethral wall during hemorrhage and resuscitation. Med Sci Monit 14:BR175-182
Clavijo-Alvarez JA, Sims CA, Pinsky MR, Puyana JC (2005) Monitoring skeletal muscle and subcutaneous tissue acid-base status and oxygenation during hemorrhagic shock and resuscitation. Shock 24:270–275
Clough GF, Bennett AR, Church MK (1998) Effects of H1 antagonists on the cutaneous vascular response to histamine and bradykinin: a study using scanning laser Doppler imaging. Br J Dermatol 138(5):806–814
Cordeiro PG, Kirschner RE, Hu Q-Y, Chiao JJC, Savage H, Alfano RR, Hoffman LA, Hidalgo DA (1995) Ultraviolet excitation fluorescence spectroscopy: a noninvasive method for the measurement of redox changes in ischemic myocutaneous flaps. Plast Reconstr Surg 96:673–680

Coremans JMCC, Van Aken M, Naus DC, Van Velthuysen ML, Bruining HA, Puppels GJ (2000) Pretransplantation assessment of renal viability with NADH fluorimetry. Kidney Int 57(2): 671–683

Creteur J (2006) Gastric and sublingual capnometry. Curr Opin Crit Care 12:272–277

Creteur J, Carollo T, Soldati G, Buchele G, De Backer D, Vincent JL (2007) The prognostic value of muscle StO2 in septic patients. Intensive Care Med 33(9):1549–1556

Crouser ED (2004) Mitochondrial dysfunction in septic shock and multiple organ dysfunction syndrome. Mitochondrion 4:729–741

Cummins JT, Elliott HW (1976) Effect of morphine on the potassium-induced change in the level of reduced pyridine nucleotides and cytochromes in brain slices. Biochem Pharmacol 25:893–896

Deutsch A, Pevzner E, Jaronkin A, Mayevsky A (2004) Real time evaluation of tissue vitality by monitoring of microcircultory blood flow, HbO2 and mitochondrial NADH redox state. SPIE Proc 5317:116–127

Dings J, Jager A, Meixensberger J, Roosen K (1998) Brain tissue pO_2 and outcome after severe head injury. Neurol Res 20(Suppl 1):S71–S75

Dirnagl U, Kaplan B, Jacewicz M, Pulsinelli W (1989) Continuous measurement of cerebral cortical blood flow by laser-Doppler flowmetry in a rat stroke model. J CBF Metab 9:589–596

Doppenberg EM, Zauner A, Bullock R, Ward JD, Fatouros PP, Young HF (1998) Correlations between brain tissue oxygen tension, carbon dioxide tension, pH, and cerebral blood flow–a better way of monitoring the severely injured brain? Surg Neurol 49(6):650–654

Dora E (1984) A simple cranial window technique for optical monitoring of cerebrocortical microcirculation and NAD/NADH redox state. Effect of mitochondrial electron transport inhibitors and anoxic anoxia. J Neurochem 42:101–108

Dora E (1985a) Effect of "flow anoxia" and "non flow anoxia" on the NAD/NADH redox state of the intact brain cortex of the cat. Pfluegers Arch Eur J Physiol 405:148–154

Dora E (1985b) Further studies on reflectometric monitoring of cerebrocortical microcirculation. Importance of lactate anions in coupling between cerebral blood flow and metabolism. Acta Physiol Hung 66(2):199–211

Dora E, Gyulai L, Kovach AGB (1984) Determinants of brain activation-induced cortical NAD/NADH responses in vivo. Brain Res 299:61–72

Eng J, Lynch RM, Balaban RS (1989) Nicotinamide adenine dinucleotide fluorescence spectroscopy and imaging of isolated cardiac myocytes. Biophys J 55:621–630

Erecinska M, Silver IA (1989) ATP and brain function. J Cereb Blood Flow Metab 9(1):2–19

Fabricius M, Lauritzen M (1996) Laser-Doppler evaluation of rat brain microcirculation: comparison with the [14C]-iodoantipyrine method suggests discordance during cerebral blood flow increases. J Cereb Blood Flow Metab 16(1):156–161

Fink MP (2001) Cytopathic hypoxia. Mitochondrial dysfunction as mechanism contributing to organ dysfunction in sepsis. Crit Care Clin 17(1):219–237

Fink S, Ray CW, McCartney S, Ehrlich H, Shoemaker WC (1984) Oxygen transport and utilization in hyperoxia and hypoxia: relation of conjunctival and transcutaneous oxygen tensions to hemodynamic and oxygen transport variables. Crit Care Med 12(11):943–948

Frank KH, Kessler M, Appelbaum K, Dummler W (1989) The Erlangen micro-lightguide spectrophotometer EMPHO I. Phys Med Biol 34:1883–1900

Franke H, Barlow CH, Chance B (1980) Surface fluorescence of reduced pyridine nucleotide of the perfused rat kidney: interrelation between metabolic and functional states. Contrib Nephrol 19: 240–247

Friedli CM, Sclarsky DS, Mayevsky A (1982) Multiprobe monitoring of ionic, metabolic, and electrical activities in the awake brain. Am J Phys 243(3):R462–R469

Galeotti T, van Rossum GDV, Mayer D, Chance B (1969) Spectrofluorimetric detection of "free" and "bound" forms of NAD(P)H in normal and tumoral cells. E. Quagliariello. Adriatica Editrice, Bari, pp 249–270

Ganong WF (1991a) Circulation through special regions. In: Appleton and lange medical book. Review of medical physiology, pp 562–582

Ganong WF (1991b) Review of medical physiology. Appelton & Lange, Norwalk

Godden DJ, Baile EM, Pare PD (1991) A comparison of laser Doppler flowmetry with the radiolabelled microsphere reference flow technique to measure tracheal blood flow in dogs. Acta Physiol Scand 142(1):49–57

Haberl RL, Heizer ML, Marmarou A, Ellis EF (1989) Laser-Doppler assessment of brain microcirculation: effect of systemic alterations. Am J Physiol Heart Circ Physiol 256:H1247–H1254

Haisjackl M, Birnbaum J, Redlin M, Schmutzler M, Waldenberger F, Lochs H, Konertz W, Kox W (1998) Splanchnic oxygen transport and lactate metabolism during normothermic cardiopulmonary bypass in humans. Anesth Analg 86(1):22–27

Hallbook O, Johansson K, Sjodahl R (1996) Laser Doppler blood flow measurement in rectal resection for carcinoma–comparison between the straight and colonic J pouch reconstruction. Br J Surg 83(3):389–392

Hanson CW 3rd, Marshall BE (2001) Artificial intelligence applications in the intensive care unit. Crit Care Med 29(2):427–435

Harbig K, Chance B, Kovach AGB, Reivich M (1976) In vivo measurement of pyridine nucleotide fluorescence from cat brain cortex. J Appl Physiol 41(4):480–488

Harden A, Young W (1906) Alcoholic ferment of yeast-juice part II co-ferment of yeast-juice. Proc Roy Soc B78:369–375

Hoper J, Gaab MR (1994) Effect of arterial PCO_2 on local HbO_2 and relative Hb concentration in the human brain–a study with the Erlangen micro-lightguide spectrophotometer (EMPHO). Physiol Meas 15(2):107–113

Hutchinson PJ (2005) Microdialysis in traumatic brain injury–methodology and pathophysiology. Acta Neurochir Suppl 95:441–445

Hutchinson PJ, O'Connell MT, Al-Rawi PG, Maskell LB, Kett-White R, Gupta AK, Richards HK, Hutchinson DB, Kirkpatrick PJ, Pickard JD (2000) Clinical cerebral microdialysis: a methodological study. J Neurosurg 93(1):37–43

Ince C, Sinaasappel M (1999) Microcirculatory oxygenation and shunting in sepsis and shock. Crit Care Med 27:1369–1377

Ince C, Coremans JMCC, Bruining HA (1992) In: Erdmann W, Bruley DF (eds) In vivo NADH fluorescence. Plenum Press, New York, pp 277–296

Ince C, Ashruf JF, Avontuur JAM, Wieringa PA, Spaan JAE, Bruining HA (1993) Heterogeneity of the hypoxic state in rat heart is determined at capillary level. Am J Phys 264:H294–H301

Javan A, Bennett WR, Herriott DR (1961) Population invertion and continuous optical laser oscillation in a gas discharge containing a he-ne mixture. Phys Rev Lett 6:106–110

Ji S, Chance B, Stuart BH, Nathan R (1977) Two-dimensional analysis of the redox state of the rat cerebral cortex in vivo by NADH fluorescence photography. Brain Res 119:357–373

Ji S, Lemasters JJ, Thurman RG (1980) A non-invasive method to study metabolic events within sublobular regions of hemoglobin-free perfused liver. FEBS Lett 113(1):37–41

Jobsis FF (1964) Basic processes in cellular respiration. In: Handbook of physiology – respiration Fenn and Rabn, pp 63–124

Jobsis FF (1977) Non-invasive, infra-red monitoring of cerebral O2 sufficiency, bloodvolume, HbO2-Hb shifts and bloodflow. Acta Neurol Scand Suppl 64:452–453

Jobsis FF, Stainsby WN (1968) Oxidation of NADH during contractions of circulated mammalian skeletal muscle. Respir Physiol 4:292–300

Jobsis FF, O'Connor M, Vitale A, Vreman H (1971a) Intracellular redox changes in functioning cerebral cortex. I. Metabolic effects of epileptiform activity. J Neurophysiol 3465:735–749

Jobsis FF, O'Connor MJ, Rosenthal M, Van Buren JM (1971b) Fluorometric monitoring of metabolic activity in the intact cerebral cortex, Excerpta medica international congress series, NO. 253, vol 20–22. G. G. Somjen, Paris, pp 18–26

Jobsis FF, Keizer JH, LaManna JC, Rosenthal M (1977) Reflectance spectrophotometry of cytochrome aa3 in vivo. J Appl Physiol Respir Environ Exerc Physiol 43(5):858–872

Johnston AJ, Gupta AK (2002) Advanced monitoring in the neurology intensive care unit: microdialysis. Curr Opin Crit Care 8(2):121–127

Kaminogo M (1989) Changes of cortical oxidative metabolism and cortical oxygen tension in hypoxia, anoxia and ischaemia. Neurol Res 11:139–114
Kanner AA, Rappaport ZH, Manor T, Mayevsky A (2002) Multiparametric monitoring of rat brain retraction. Proc SPIE 4623:206–213
Kedem J, Mayevsky A, Sonn J, Acad B (1981) An experimental approach for evaluation of the O_2 balance in local myocardial regions in vivo. Quart J Exp Physiol 66:501–514
Kennedy EP, Lehninger AL (1949) Oxidation of fatty acids and tricarboxylic acid cycle intermediates by isolated rat liver mitochondria. J Biol Chem 179(2):957–972
Kirkeby OJ, Rise IR, Nordsletten L, Skjeldal S, Hall C, Risoe C (1995) Cerebral blood flow measured with intracerebral laser-Dopplerflow probes and radioactive microspheres. J Appl Physiol 79(5):1479–1486
Kitai T, Tanaka A, Tokuka A, Ozawa K, Iwata S, Chance B (1992) Changes in the redox distribution of rat liver by ischemia. Anal Biochem 206:131–136
Kohen E, Kohen C, Thorell B (1969) Use of microfluorimetry to study the metabolism of intact cells. Biomed Eng 4:554–565
Koretsky AP, Balaban RS (1987) Changes in pyridine nucleotide levels alter oxygen consumption and extra-mitochondrial phosphates in isolated mitochondria: a ^{31}P-NMR and NAD(P)H fluorescence study. Biochim Biophys Acta 893:398–408
Koretsky AP, Katz LA, Balaban RS (1987) Determination of pyridine nucleotide fluorescence from the perfused heart using an internal standard. Am J Phys 253:H856–H862
Kraut A, Zurovsky Y, Mayevsky A (1998) Simultaneous hemodynamic and metabolic responses to oxygen deficiency in the brain and other organs of the rat. Neurosci Lett Suppl 51:S24
Kraut A, Barbiro-Michaely E, Mayevsky A (2004) Differential effects of norepinephrine on brain and other less vital organs detected by a multisite multiparametric monitoring system. Med Sci Monit 10(7):BR215-BR220
Krejci V, Hiltebrand L, Banic A, Erni D, Wheatley AM, Sigurdsson GH (2000) Continuous measurements of microcirculatory blood flow in gastrointestinal organs during acute haemorrhage. Br J Anaesth 84(4):468–475
Kuchiwaki H, Inao S, Kanaiwa H (1996) Intraoperative laser flowmetry reflects levels of cerebral blood flow during neurosurgical procedures. Neurol Res 18(5):457–461
Kutai-Asis H, Barbiro-Michaely E, Deutsch A, Mayevsky A (2006) Fiber optic based multiparametric spectroscopy in vivo: toward a new quantitative tissue vitality index. SPIE Proc 6083
Lacombe P, Sercombe R, Correze JL, Springhetti V, Seylaz J (1992) Spreading depression induces prolonged reduction of cortical blood flow reactivity in the rat. Exp Neurol 117(3):278–286
LaManna JC, Watkins W, Rosenthal M (1975) Relationship of inspired oxygen level of cytochrome a and ECoG in cats. Physiologist 18:284
Lehninger AL (1964) The mitochondrion. WA Benjamin Inc, New York
Lubbers DW (1995) Optical sensors for clinical monitoring. Acta Anaesth Scand Suppl 39(104): 37–54
Maas AI, Fleckenstein W, de Jong DA, van Santbrink H (1993) Monitoring cerebral oxygenation: experimental studies and preliminary clinical results of continuous monitoring of cerebrospinal fluid and brain tissue oxygen tension. Acta Neurochir Suppl (Wien) 59:50–57
Maciel AT, Creteur J, Vincent JL (2004) Tissue capnometry: does the answer lie under the tongue? Intensive Care Med 30(12):2157–2165
Maier S, Pajk W, Ulmer H, Hausdorfer H, Torgersen C, Klocker J, Hasibeder W, Knotzer H (2009) Epoprostenol improves mucosal tissue oxygen tension in an acute endotoxemic pig model. Shock 31(1):104–110
Manor T, Meilin A, Mayevsky A (1996) Monitoring different areas of the rat cortex in response to fluid percussion trauma. Isr J Med Sci 32:S39
Marik PE (2006) Sublingual capnometry: a non-invasive measure of microcirculatory dysfunction and tissue hypoxia. Physiol Meas 27(7):R37–R47

Mayevsky A (1975) The effect of trimethadione on brain energy metabolism and EEG activity of the conscious rat exposed to HPO. J Neurosci Res 1:131–142

Mayevsky A (1976) Brain energy metabolism of the conscious rat exposed to various physiological and pathological situations. Brain Res 113:327–338

Mayevsky A (1978a) Ischemia in the brain: the effects of carotid artery ligation and decapitation on the energy state of the awake and anesthetized rat. Brain Res 140:217–230

Mayevsky A (1978b) Shedding light on the awake brain. In: Dutton PL, Leigh J, Scarpa A (eds) Frontiers in bioenergetics: from electrons to tissues, vol II. Academic, New York, pp 1467–1476

Mayevsky A (1983) Metabolic, ionic and electrical responses to experimental epilepsy in the awake rat. In: Baldy M, Moulinier DH, Ingvar DH, Meldrum BS (eds) Proceedings of the first international Congress of the cerebral blood flow, metabolism & epilepsy. John Libbey, pp 263–270

Mayevsky A (1984) Brain NADH redox state monitored in vivo by fiber optic surface fluorometry. Brain Res Rev 7:49–68

Mayevsky A (1990) Level of ischemia and brain functions in the Mongolian gerbil in vivo. Brain Res 524:1–9

Mayevsky A (1992) Cerebral blood flow and brain mitochondrial redox state responses to various perturbations in gerbils. Adv Exp Med Biol 317:707–716

Mayevsky A, Breuer Z (1992) Brain vasculature and mitochondrial responses to ischemia in gerbils. I. Basic anatomical patterns and biochemical correlates. Brain Res 58:242–250

Mayevsky A, Chance B (1973) A new long-term method for the measurement of NADH fluorescence in intact rat brain with implanted cannula. Adv Exp Med Biol 37A:239–244

Mayevsky A, Chance B (1974) Repetitive patterns of metabolic changes during cortical spreading depression of the awake rat. Brain Res 65:529–533

Mayevsky A, Chance B (1976) The effect of decapitation on the oxidation-reduction state of NADH and ECoG in the brain of the awake rat. Adv Exp Med Biol 75:307–312

Mayevsky A, Chance B (1982) Intracellular oxidation-reduction state measured in situ by a multicannel fiber-optic surface fluorometer. Science 217:537–540

Mayevsky A, Chance B (2007) Oxidation-reduction states of NADH in vivo: from animals to clinical use. Mitochondrion 7:330–339

Mayevsky A, Rogatsky G (2007) Mitochondrial function in vivo evaluated by NADH fluorescence: from animal models to human studies. Am J Physiol Cell Physiol 292:C615–C640

Mayevsky A, Weiss HR (1991) Cerebral blood flow and oxygen consumption in cortical spreading depression. J CBF Metab 11:829–836

Mayevsky A, Zarchin N (1981) The effects of unilateral carotid occlusion on the responses to decapitation in the gerbil brain. Brain Res 206:155–160

Mayevsky A, Jamieson D, Chance B (1974) Oxygen poisoning in the unanesthetized brain: correlation of the oxidation-reduction state of pyridine nucleotide with electrical activity. Brain Res 76:481–491

Mayevsky A, Lebourdais S, Chance B (1980) The interrelation between brain PO_2 and NADH oxidation- reduction state in the gerbil. J Neurosci Res 5:173–182

Mayevsky A, Zarchin N, Friedli CM (1982) Factors affecting the oxygen balance in the awake cerebral cortex exposed to spreading depression. Brain Res 236:93–105

Mayevsky A, Friedli CM, Reivich M (1985) Metabolic, ionic and electrical responses of the gerbil brain to ischemia. Am J Phys 248:R99–R107

Mayevsky A, Duckrow RB, Yoles E, Zarchin N, Kanshansky D (1990a) Brain mitochondrial redox state, tissue hemodynamic and extracellular ion responses to four-vessel occlusion and spreading depression in the rat. Neurol Res 12:243–248

Mayevsky A, Frank KH, Nioka S, Kessler M, Chance B (1990b) Oxygen supply and brain function in vivo: a multiparametric monitoring approach in the Mongolian gerbil. In: Piper J, Goldstick TK, Meyer M (eds) Oxygen Transport to Tissue XII. Plenum Pub, pp 303–313

Mayevsky A, Yoles E, Zarchin N, Kaushansky D (1990c) Brain vascular ionic and metabolic responses to ischemia in the Mongolian gerbil. J Basic Clin Physiol Pharmacol 1:207–220

Mayevsky A, Flamm ES, Pennie W, Chance B (1991) A fiber optic based multiprobes system for intraoperative monitoring of brain functions. SPIE 1431:303–313

Mayevsky A, Frank K, Muck M, Nioka S, Kessler M, Chance B (1992) Multiparametric evaluation of brain functions in the Mongolian gerbil in vivo. J Basic Clin Physiol Pharmacol 3:323–342

Mayevsky A, Meilin S, Rogatsky GG, Zarchin N, Sonn J (1995) Multiparametric monitoring of the awake brain exposed to carbon monoxide. J Appl Physiol 78:1188–1196

Mayevsky A, Doron A, Manor T, Meilin S, Zarchin N, Ouaknine GE (1996) Cortical spreading depression recorded from the human brain using a multiparametric monitoring system. Brain Res 740:268–274

Mayevsky A, Meilin S, Manor T, Ornstein E, Zarchin N, Sonn J (1998) Multiparametric monitoring of brain oxygen balance under experimental and clinical conditions. Neurol Res 20:S76–S80

Mayevsky A, Doron A, Meilin S, Manor T, Ornstein E, Ouaknine GE (1999a) Brain viability and function analyzer: multiparametric real-time monitoring in neurosurgical patients. Acta Neurochir Suppl (Wien) 75:63–66

Mayevsky A, Meilin S, Manor T, Zarchin N, Sonn J (1999b) Optical monitoring of NADH redox state and blood flow as indicators of brain energy balance. Adv Exp Med Biol 471:133–140

Mayevsky A, Manor T, Meilin S, Razon N, Ouaknine GE (2001) Multiparametric monitoring of tissue vitality in clinical situations. SPIE 4255:33–39

Mayevsky A, Manor T, Pevzner E, Deutsch A, Etziony R, Dekel N (2002) Real-time optical monitoring of tissue vitality in vivo. Proc SPIE 4616:30–39

Mayevsky A, Sonn J, Luger-Hamer M, Nakache R (2003) Real time assessment of tissue vitality during the transplantation procedure. Transpl Rev 17:96–116

Mayevsky A, Barbiro-Micahely E, Kutai-Asis H, Deutsch A, Jaronkin A (2004a) Brain physiological state evaluated by real time multiparametric tissue spectroscopy in vivo. ProcSPIE 5326: 98–105

Mayevsky A, Manor T, Pevzner E, Deutsch A, Etziony R, Dekel N, Jaronkin A (2004b) Tissue spectroscope: a novel in vivo approach to real time monitoring of tissue vitality. J Biomed Opt 9(5):1028–1045

Mayevsky A, Deutsch A, Dekel N, Pevzner E, Jaronkin A (2005) A new biomedical device for in vivo multiparametric evaluation of tissue vitality in critical care medicine. In: Vo-Dinh T, Grundfest WS, Benaron DA, Cohn GG (eds) Advanced biomedical and clinical diagnosis system III, vol 5692. Proc SPIE, pp 60–70

Mayevsky A, Blum Y, Dekel N, Deutsch A, Halfon R, Kremer S, Pewzner E, Sherman E, Barnea O (2006) The CritiView – a new fiber optic based optical device for the assessment of tissue vitality. Proc SPIE 6083:0Z-1-0Z-9

Mayevsky A, Sherman E, Cohen-Kashi M, Dekel N, Pewzner E (2007) Optical assessment of intravascular and intracellular parameters related to tissue viability, Biomedical Optics (BiOS) 2007. International Society for Optics and Photonics

Mayevsky A, Dekel N, Oren L, Deutsch A, Pewzner E (2008) In: Mahadevan-Jansen A, Petrich W, Alfano RR, Katz A (eds) Mitochondrial dysfunction: bench-to-bedside optical monitoring of tissue vitality. Biomedical Optics (BiOS) 2008. International Society for Optics and Photonics, p 68531B-68511

Mebazaa A, Pitsis AA, Rudiger A, Toller W, Longrois D, Ricksten SE, Bobek I, De Hert S, Wieselthaler G, Schirmer U, von Segesser LK, Sander M, Poldermans D, Ranucci M, Karpati PC, Wouters P, Seeberger M, Schmid ER, Weder W, Follath F (2010) Clinical review: practical recommendations on the management of perioperative heart failure in cardiac surgery. Crit Care 14(2):201

Meilin S, Zarchin N, Mayevsky A (1999) Inter-relation between hemodynamic, metabolic, ionic and electrical activities during ischemia and reperfusion in the gerbil brain. Neurol Res 21(7): 699–704

Millar SM, Alston RP, Andrews PJ, Souter MJ (2001) Cerebral hypoperfusion in immediate postoperative period following coronary artery bypass grafting, heart valve, and abdominal aortic surgery. Br J Anaesth 87(2):229–236

Mizuno Y, Ikebe S, Hattori N, Nakagawa-Hattori Y, Mochizuki H, Tanaka M, Ozawa T (1995) Role of mitochondria in the etiology and pathogenesis of Parkinson's disease. Biochim Biophys Acta 1271(1):265–274
Nakamura S, Kaenders W (1999) Market-ready blue diodes excite spectroscopists. Laser Focus World
Nicholls DG, Budd SL (2000) Mitochondria and neuronal survival. Physiol Rev 80(1):315–360
Niinikoski J, Kuttila K (1993) Adequacy of tissue oxygenation in cardiac surgery: regional measurements. Crit Care Med 21(2 Suppl):S77–S83
Nilsson GE, Tenland T, Obert PA (1980) A new instrument for continuous measurement of tissue blood flow by light beating spectroscopy. IEEE Trans Biomed Eng 27(1):12–19
Nilsson GE, Tenland JT, Oberg PA (1984) Method and apparatus for measuring flow motions in a fluid. US Patent US4476875
Obeid AN, Dougherty G, Pettinger S (1990) In vivo comparison of a twin wavelength laser Doppler flowmeter using He-Ne and laser diode sources. J Med Eng Technol 14(3):102–110
Oberg PA (1990) Laser-Doppler flowmetry. Crit Rev Biomed Eng 18(2):125–163
O'Connor M, Lewis D, Herman C (1973a) Potassium effects on energy metabolism and epileptiform activity in the cat hippocampus. Stroke 4:375–376
O'Connor MJ, Lewis DV, Herman CJ (1973b) Effects of potassium on oxidative metabolism and seizures. Electroencephalogr Clin Neurophysiol 35:205–208
Ohri SK, Becket J, Brannan J, Keogh BE, Taylor KM (1994) Effects of cardiopulmonary bypass on gut blood flow, oxygen utilization, and intramucosal pH. Ann Thorac Surg 57(5):1193–1199
Ohri SK, Bowles CW, Mathie RT, Lawrence DR, Keogh BE, Taylor KM (1997) Effect of cardiopulmonary bypass perfusion protocols on gut tissue oxygenation and blood flow. Ann Thorac Surg 64(1):163–170
Okunieff P, Wang X, Rubin P, Finkelstein JN, Constine LS, Ding I (1998) Radiation-induced changes in bone perfusion and angiogenesis. Int J Radiat Oncol Biol Phys 42:885–889
Orr C-S, Arthurs SC (1992) Tissue viability measurement by in situ fluorometry. ASAIO Trans 38: M412–M415
Osbakken M, Mayevsky A (1996) Multiparameter monitoring and analysis of in vivo ischemic and hypoxic heart. J Basic Clin Physiol Pharmacol 7:97–113
Ospina-Tascon GA, Cordioli RL, Vincent JL (2008) What type of monitoring has been shown to improve outcomes in acutely ill patients? Intensive Care Med 34(5):800–820
Pappajohn DJ, Penneys R, Chance B (1972) NADH spectrofluorometry of rat skin. J Appl Physiol 33(5):684–687
Park DH, Hwang JW, Jang KS, Han DG, Ahn KY, Baik BS (1998) Use of laser Doppler flowmetry for estimation of the depth of burns. Plast Reconstr Surg 101(6):1516–1523
Perner A, Jorgensen VL, Poulsen TD, Steinbruchel D, Larsen B, Andersen LW (2005) Increased concentrations of L-lactate in the rectal lumen in patients undergoing cardiopulmonary bypass. Br J Anaesth 95(6):764–768
Pevzner E, Deutsch A, Manor T, Dekel N, Etziony R, Derzy I, Razon N, Mayevsky A (2003) Real-time multiparametric spectroscopy as a practical tool for evaluation of tissue vitality in vivo. Proc SPIE
Pinsky MR, Payen D (2009) Probing the limits of regional tissue oxygenation measures. Crit Care 13(Suppl 5):S1
Prahl S (1999) Optical absorption of hemoglobin. http://omlc.ogi.edu/spectra/hemoglobin/index.html
Preston TJ, Abadi A, Wilson L, Singh G (2001) Mitochondrial contributions to cancer cell physiology: potential for drug development. Adv Drug Deliv Rev 49(1–2):45–61
Rampil IJ, Litt L, Mayevsky A (1992) Correlated, simultaneous, multiple-wavelength optical monitoring in vivo of localized cerebrocortical NADH and brain microvessel hemoglobin oxygen saturation. J Clin Monit 8(3):216–225

Redaelli CA, Schilling MK, Carrel TP (1998) Intraoperative assessment of intestinal viability by laser Doppler flowmetry for surgery of ruptured abdominal aortic aneurysms. World J Surg 22(3):283–289
Renault G, Raynal E, Sinet M, Muffat-Joly M, Berthier J-P, Cornillault J, Godard B, Pocidalo J-J (1984) In situ double-beam NADH laser fluorimetry: choice of a reference wavelength. Am J Phys 246:H491–H499
Riva C, Ross B, Benedek GB (1972) Laser Doppler measurements of blood flow in capillary tubes and retinal arteries. Investig Ophthalmol 11(11):936–944
Rosenthal M, Jobsis FF (1971) Intracellular redox changes in functioning cerebral cortex. II. Effects of direct cortical stimulation. J Neurophysiol 34:750–762
Rosenthal M, Somjen G (1971) Sustained evoked potentials, spreading depression, and metabolic activity of the cerebral cortex measured in situ. In: International congress physical science in 25th Munich, p 479
Rosenthal M, Somjen G (1973) Spreading depression, sustained potential shifts, and metabolic activity of cerebral cortex of cats. J Neurophysiol 36:739–749
Rosenthal M, Lamanna JC, Jobsis FF, Levasseur JE, Kontos HA, Patterson JL (1976) Effects of respiratory gases on cytochrome A in intact cerebral cortex: is there a critical Po2? Brain Res 108(1):143–154
Sailer M, Debus ES, Fuchs KH, Beyerlein J, Thiede A (2000) Comparison of anastomotic microcirculatoin in coloanal J-pouches versus straight and side-to-end coloanal reconstruction: an experimental study in the pig. Int J Color Dis 15:114–117
Scheffler IE (2000) A century of mitochondrial research: achievements and perspectives. Mitochondrion 1:3–31
Schiller WR, Garren RL, Bay RC, Ruddell MH, Holloway GA Jr, Mohty A, Luekens CA (1997) Laser Doppler evaluation of burned hands predicts need for surgical grafting. J Trauma 43(1): 35–39; discussion 39-40
Schlichtig R, Kramer DJ, Pinsky MR (1991) Flow redistribution during progressive hemorrhage is a determinant of critical O_2 delivery. J Appl Physiol (1985) 70(1):169–178
Seifalian AM, Mallet SV, Rolles K, Davidson BR (1997) Hepatic microcirculation during human orthotopic liver transplantation. Br J Surg 84(10):1391–1395
Shepherd AP (1990) History of laser Doppler blood Flowmetry. In: Shepherd AP, Oberg PA (eds) Laser Doppler blood Flowmetry. Kluwer Academic Publishers, Boston/Dordrecht/London, pp 1–16
Shepherd P, Oberg PA (1990) Laser Doppler blood Flowmetry. Kluwer Academic, Nashwell
Sliney DH (2000) Ultraviolet radiation exposure criteria. Radiat Prot Dosim 91:213–222
Smits GJ, Roman RJ, Lombard JH (1986) Evaluation of laser-Doppler flowmetry as a measure of tissue blood flow. J Appl Physiol (1985) 61(2):666–672
Sonn J, Mayevsky A (2000) Effects of brain oxygenation on metabolic, hemodynamic, ionic and electrical responses to spreading depression in the rat. Brain Res 882(1–2):212–216
Sonn J, Mayevsky A (2001) The effect of ethanol on metabolic, hemodynamic and electrical responses to cortical spreading depression. Brain Res 908(2):174–186
Spraycar M (1995) Stedman's medical dictionary. Baltimore Williams & Wilkins
Stern MD (1975) In vivo evaluation of microcirculation by coherent light scattering. Nature 254(5495):56–58
Stern MD, Lappe DL, Bowen PD, Chimosky JE, Holloway GA Jr, Keiser HR, Bowman RL (1977) Continuous measurement of tissue blood flow by laser-Doppler spectroscopy. Am J Phys 232(4):H441–H448
Stern MD, Bowen PD, Parma R, Osgood RW, Bowman RL, Stein JH (1979) Measurement of renal cortical and medullary blood flow by laser-Doppler spectroscopy in the rat. Am J Phys 236(1): F80–F87
Sundset A, Tadjkarimi S, Khaghani A, Kvernebo K, Yacoub MH (1997) Human en bloc double-lung transplantation: bronchial artery revascularization improves airway perfusion. Ann Thorac Surg 63(3):790–795

Sundt TM, Anderson RE (1975) Reduced nicotinamide adenine dinucleotide fluorescence and cortical blood flow in ischemic and nonischemic squirrel monkey cortex. I. Animal preparation, instrumentation, and validity of model. Stroke 6:270–278

Sundy TM, Anderson RE (1975) Reduced nicotinamide adenine dinucleotide fluorescence and cortical blood flow in ischemic and nonischemic squirrel monkey cortex. I. Animal preparation, instrumentation, and validity of model. Stroke 6:270–278

Talpahewa SP, Lovell AT, Angelini GD, Ascione R (2004) Effect of cardiopulmonary bypass on cortical cerebral oxygenation during coronary artery bypass grafting. Eur J Cardiothorac Surg 26(4):676–681

Tateishi S, Arima S, Futami K (1997) Assessment of blood flow in the small intestine by laser Doppler flowmetry: comparison of healthy small intestine and small intestine in Crohn's disease. J Gastroenterol 32(4):457–463

Tatton WG, Olanow CW (1999) Apoptosis in neurodegenerative diseases: the role of mitochondria. Biochim Biophys Acta 1410:195–213

Thorniley MS, Simpkin S, Balogun E, Khaw K, Shurey C, Burton K, Green CJ (1997) Measurements of tissue viability in transplantation. Philos Trans R Soc Lond Ser B Biol Sci 352(1354): 685–696

Ubbink DT, Tulevski II, Jacobs MJ (2000) Can the green laser doppler measure skin-nutritive perfusion in patients with peripheral vascular disease? J Vasc Res 37(3):195–201

van Hulst RA, Hasan D, Lachmann B (2002) Intracranial pressure, brain PCO2, PO2, and pH during hypo- and hyperventilation at constant mean airway pressure in pigs. Intensive Care Med 28(1):68–73

van Rossem K, Vermarien H, Bourgain R (1992) Construction, calibration and evaluation of pO2 electrodes for chronical implantation in the rabbit brain cortex. Adv Exp Med Biol 316:85–101

Villringer A, Chance B (1997) Non-invasive optical spectroscopy and imaging of human brain function. Trends Neurosci 20(10):435–442

Villringer A, Planck J, Hock C, Schleinkofer L, Dirnagl U (1993) Near infrared spectroscopy (NIRS): a new tool to study hemodynamic changes during activation of brain function in human adults. Neurosci Lett 154(1–2):101–104

Von Lubitz DK, Lin RC, Jacobson KA (1995) Cerebral ischemia in gerbils: effects of acute and chronic treatment with adenosine A2A receptor agonist and antagonist. Eur J Pharmacol 287(3): 295–302

Vongsavan N, Matthews B (1993) Some aspects of the use of laser Doppler flow meters for recording tissue blood flow. Exp Physiol 78(1):1–14

Wadhwani KC, Rapoport SI, Shepherd AP, Oberg PA (1990) Blood flow in the central and peripheral nervous systems. In: Shephrd AP, Oberg PA (eds) Laser doppler blood flowmetry, vol 107. Kluwer Academic Pub, Boston, pp 265–304

Waltemath CL (1970) Oxygen, uptake, transport, and tissue utilization. Anesth Analg 49(1): 184–203

Warburg O (1930) Notiz uber den Stoffwechsel der Tumoren. Biochem J 228:257–258

Warburg O (1956) On the origin of cancer cells. Science 123(3191):309–314

Wardell K, Braverman IM, Silverman DG, Nilsson GE (1994) Spatial heterogeneity in normal skin perfusion recorded with laser Doppler imaging and flowmetry. Microvasc Res 48(1):26–38

Weinhouse S (1956) On respiratory impairment in cancer cells. Science (NY) 124:267–269

White RL, Wittenberg BA (1993) NADH fluorescence of isolated ventricular myocytes: effects of pacing, myoglobin, and oxygen supply. Biophys J 65:196–204

Yeh Y, Cummins HZ (1964) Localized fluid flow meansurements with an He-Ne laser spectometer. Appl Phys Lett 4:176–178

Yuen JC, Feng Z (1998) Reduced cost of extremity free flap monitoring. Ann Plast Surg 41(1): 36–40

Zarchin N, Mayevsky A (1981) The effects of age on the metabolic and electrical responses to decapitation in the awake and anesthetized rat brain. Mech Ageing Dev 16:285–294

Zauner A, Doppenberg E, Woodward JJ, Allen C, Jebraili S, Young HF, Bullock R (1997) Multiparametric continuous monitoring of brain metabolism and substrate delivery in neurosurgical patients. Neurol Res 19(3):265–273

Chapter 8
Discussion and Conclusions

8.1 Introduction

In the current new book, I tried to provide the reader with maximal information on the very challenging monitoring approach and system needed for patient monitoring. The central role of mitochondrial dysfunction in various human pathological processes was established and recognized by a large number of scientists active in the field of mitochondrial function. As of today, there is a large and strong body of knowledge claiming that the next step in patient monitoring is to develop monitoring devices that will evaluate in real-time tissue vitality of different organs (Chance et al. 2005; Singer 2007; Harrois et al. 2011).

Although the monitoring of NADH fluorescence started many decades ago, the implementation of the technology into a medical device started about 30 years ago. In this chapter, I will discuss a few issues that integrate a few of the main subjects that are critical in the future of this technology.

The first issue to be presented is the regulatory status of the devices developed for clinical use. Figure 8.1 presents the three developed devices that were cleared by the Food and Drug Administration (FDA) during the industrial activity years.

One of the main questions is, why monitoring of NADH fluorescence is the leading technology in monitoring mitochondrial function in vivo and in real time. It is clear that other electron carriers in the respiratory chain may have an advantage since they are localized only in the mitochondria. The NADH appears in addition to the mitochondria also in the cytosol. We have tried many options that will enable us to monitor flavoprotein (Fp), fluorescence together with NADH. It was found that Fp measurement is dramatically affected and sensitive to hemodynamic artefacts, and mainly to changes in hemoglobin oxygenation. Cytochrome oxidase monitoring suffers from the same problems due to the overlap with the hemoglobin absorption spectrum. Therefore, we concluded that the NADH is the best candidate to monitor mitochondrial function in patients as well as in experimental animals.

© The Author(s), under exclusive license to Springer Nature Switzerland AG 2025
A. Mayevsky et al., *Body Oxygen Homeostasis and Mitochondrial Function*,
https://doi.org/10.1007/978-3-031-94113-9_8

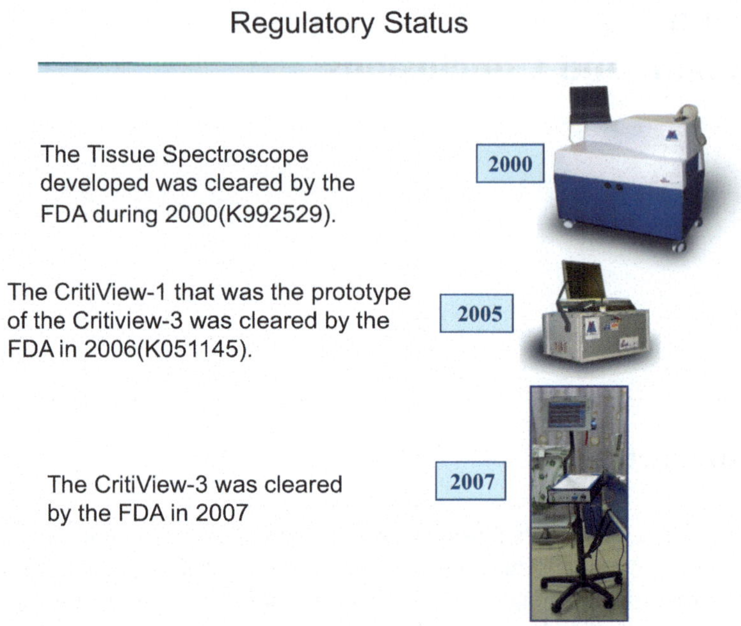

Fig. 8.1 The regulatory status of the devices developed in order to measure tissue energy state under in vivo conditions

8.2 The Isolated Mitochondria as a Basis to Measure NADH in Patients

The pioneer work of Chance, Williams, Connelly, and other collaborators, in the early 1950s, opened up new possibilities enabling the study of mitochondrial function in vitro, and later on in vivo monitoring became a reality (Chance and Williams 1954, Chance and Williams 1955a, b, 1956a; Connelly and Chance 1954). The shift from studying mitochondrial NADH redox state in isolated mitochondria to higher cellular and tissue organization level started in the late 1950s (Chance and Jobsis 1959). The main breakthrough occurred in 1962, a year that at least six papers described the in vivo monitoring of various organs in the anesthetized rat (Chance et al. 1962a, b, c, d; Chance and Schoener 1962, 1963).

After more than 70 years of monitoring NADH fluorescence in many biological systems, two issues need our attention:

Can we extrapolate from the results obtained in the monitoring of mitochondrial NADH in vitro to in vivo conditions?

Can we compare results obtained in animal models to events developed in patients?

Mitochondrial function and its metabolic states under in vivo condition are completely different from the definition made for isolated mitochondria by Chance and Williams in 1955. The main differences between the two situations are as follows:

1. The isolated mitochondria are not interconnected to other intracellular organelles and components.
2. The amount of oxygen available in the mitochondrial medium is not regulated by changes in blood flow or hemoglobin oxygenation under in vitro conditions.
3. One of the largest oxygen or energy consumers, i.e., various ions pumps, are missing from the isolated mitochondria preparation.
4. When ADP is added to State 4 mitochondria, ATP will be synthesized but will not be consumed in parallel as occurs under in vivo conditions.
5. The isolated mitochondria preparation is not exposed to systemic changes in hemodynamic and other physiological parameters occurring in the organism.

In the classical table published in 1955 by Chance and Williams (1955a, b), they show the relationship between the metabolic state and the level of NADH in the mitochondria (Fig. 1.13 in the current book). Out of the five metabolic states defined in the table, two of them are in the physiological range and the other three described pathophysiological conditions developed in the cell or tissue. The explanations given to the table in the original publication (Chance and Williams 1956b) are as follows:

States 2, 3, 4, and 5 are the ones relevant to the present discussion and will be described very briefly. State 2 is a "starved" state in which ADP has been added in order to exhaust the endogenous substrate. Alternatively, the mitochondria are freed of endogenous substrate by any process that will increase the endogenous ADP level sufficiently to promote rapid respiration and exhaust the substrate. In addition, uncoupling agents soon lead to State 2. State 5, like State 2, is characterized by zero respiration, but in this case, oxygen is lacking, and instead of the components being oxidized, all those associated with the respiratory pathway are reduced, since reducing substrate was added in order to produce State 5. The ADP level will be high in State 5.

State 4 is an aerobic state characterized by a low respiration rate—a "resting" state—even though substrate is present. Several components show large percentage reductions in this state, especially DPN. State 4 requires definite conditions for its establishment; mitochondria must be carefully prepared and supplied with substrate and without phosphate acceptor.

State 3 is the "active" state of rapid respiration and phosphorylation, with adequate supplies of substrate and phosphate acceptor.

Since 1962, a very large number of papers have described the use of NADH monitoring in intact animals exposed to a variety of physiological and pathological

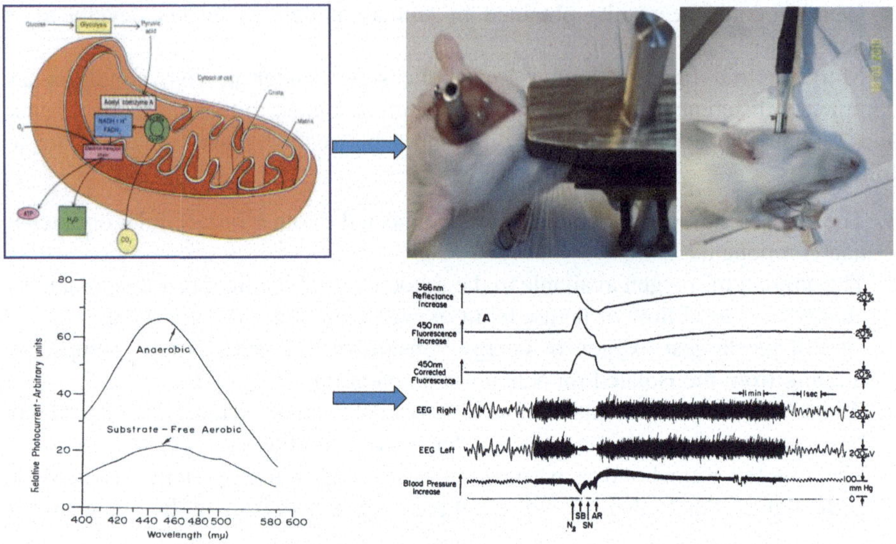

Fig. 8.2 The comparison between in vitro and in vivo monitoring of NADH fluorescence. The spectra in (**b**) were measured from isolated mitochondria (**a**) and the response of the intact brain to anoxia (**c** and **d**)

conditions. NADH was monitored in various animal models exposing most of the organs in the body to various perturbations (see Chap. 8 in (Mayevsky 2015)). The interpretation of the results obtained in vivo is based on the "Chance and Williams" concepts defined for in vitro conditions (Chance and Williams 1956b). The crucial question is could we interpret the in vivo data according to the concept developed for in vitro conditions.

Figure 8.2 illustrates the data interpretation dilemma. Parts A and B show the spectra of NADH fluorescence measured in the isolated mitochondria and the responses to shifts in the metabolic state of the mitochondria. Parts C and D show the response to anoxia measured from the intact rat brain in vivo under "State 4 to State 5" transition.

Can we compare and define the resting state (4) and active state (3) under in vivo situation to the in vitro definitions? As seen in Fig. 1.13 (Chap. 1), the maximal NADH level is achieved under complete O_2 deprivation that can be induced both under in vitro and in vivo conditions. This signifies that this definitive point can be used to determine State 5 in vivo as well. As can be seen in the table (Fig. 1.13 in the current book), the steady state NADH level at State 4 is 99%, namely that most of the NAD is in the reduced form NADH. When the mitochondria are shifted to State 3, by adding ADP, the NADH becomes more oxidized and the NADH level decreases to 53% of the reduced form. Moving from State 4 to State 5, eliminating the oxygen from the media will have a very small effect on the NADH level, namely, changing

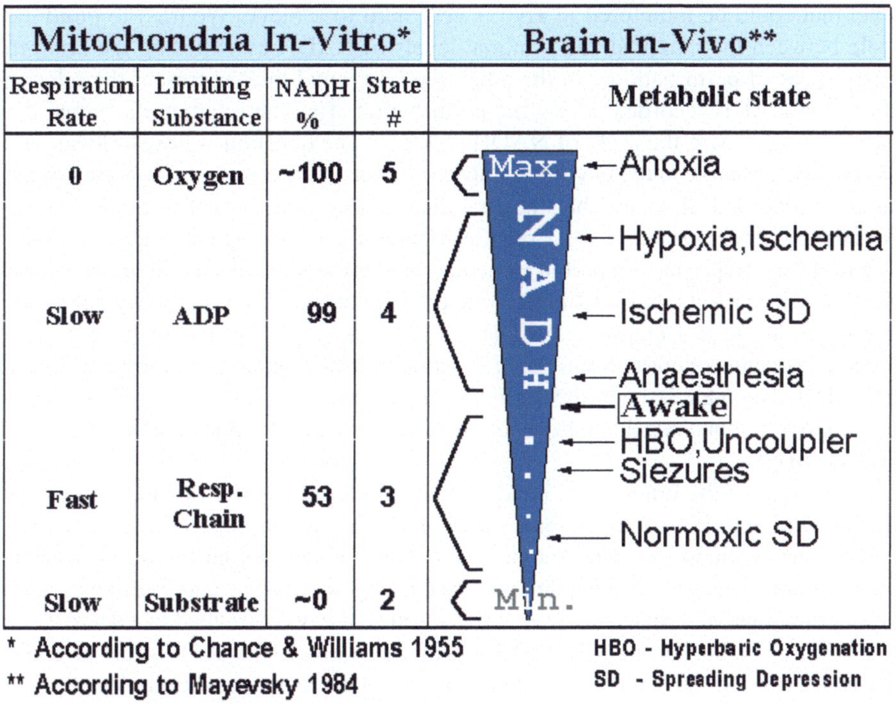

Fig. 8.3 Comparison between the mitochondrial metabolic state, defined by Chance and Williams and responses of the in vivo brain to changes in O_2 supply and brain activation. © American Physiological Society, reprinted with permission (Mayevsky and Rogatsky 2007)

from 99% to 100%. It is important to note that other members of the respiratory chain behave somewhat similarly to NADH. When the mitochondria are shifted to State 2, the NADH level is 0 since the NAD is the dominant state in the absence of substrate.

The problem is to determine the "resting" metabolic state of a tissue in an in vivo situation. If we adopt the in vitro value of a resting state (State 4), this would signify that the increase in NADH during State 5, induced by anoxia (0% O_2), would be only 1%, namely, from 99% to 100%. According to all in vivo studies, this is not the case, and during anoxia, the increase in NADH is much larger than the decrease under the State 4 to 3 transition.

When we started our monitoring of NADH redox state using fiber optic-based fluorometry/reflectometry, we used the brain in vivo exposed to various conditions in real time. We found that the NADH level could either increase or decrease from the baseline level calibrated in the normoxic animal. The initial results were plotted against the perturbation used as shown on the right side of Fig. 8.3.

The relative changes in brain NADH under various perturbations as compared to the Chance and Williams definitions. Although the maximal level of NADH could be achieved in vivo by exposing the animal to anoxia (100% nitrogen), the minimal level that could be monitored in vivo is very hard to achieve. We tried to build the scale between the maximal and minimal levels of NADH by exposing the brain to many types of perturbations. In the table, the horizontal lines represent the relative change in NADH recorded during the perturbation. The values are not accurate but rather trying to map the scale of NADH changes. The question is how to locate the in vivo scale relative to the in vitro definition shown on the left side. According to the changes recorded, it seems that the so-called resting brain is not at State 4 (99% NADH) but rather around 3.5. As seen, NADH will increase by various factors such as anesthesia, hypoxia, and partial ischemia. Under complete ischemia or anoxia the NADH will reach its maximal level—State 5. Exposing the animal to hyperoxia or uncoupler led to a decrease in NADH toward State 3. Activation of the brain by inducing seizures or cortical spreading depression (CSD) led to a very large decrease in NADH. The main issue in the scaling of the responses is how to "push" the brain in vivo to minimal NADH levels. As of now, it seems that this question remains open and unsolved.

Activation of the brain under conditions of restricted amount of oxygen, such as partial ischemia, led to an increase in NADH (Reduction) instead of a decrease (oxidation) as found in the normoxic brain. Those results found in various experimental conditions led to the dilemma. According to Chance and Williams, ATP production related to State 4 to State 3 transition is always correlated to a decrease or oxidation of NADH from 99% to 53%. The question is how ATP turnover could increase under NADH increase (reduction) and not decrease (oxidation). We want to make sure that the experimental results are not representing a technical artifact related to the monitoring itself. We assumed that by using appropriate modeling methods, the dilemma could be clarified or resolved. Figure 8.4 (Vatov et al. 2006) compared the response to spreading depression (SD) to the model calculated for normoxic (A) and the partial ischemic brain (B).

Now I will discuss the second issue mentioned earlier, namely, can we compare results obtained in animal models to events developed in patients? When a medical device for NADH monitoring is used as a diagnostic tool in patients, the user should compare the data to studies he performed in animal models. There are differences in the structure of the fluorometer to be used in patients.

Figure 8.5a shows the initial fluorometer developed for animal monitoring as compared to the clinical environment shown in Fig. 8.5b. The very dramatic shifts in the monitoring device between 1951 (Fig. 8.5c) and the one used in patients in the ICU (Fig. 8.5d) are presented.

As mentioned in the previous section, the responses to CSD in normoxic and ischemic brain are different. Figure 8.6 presents the responses to spreading depression (SD) in a rat model at the upper part and in patients at the lower part. When we induced SD in animal studies, we were sure about the oxygen supply to the animal. The blood supply to the brain decreased between part A and B. The shift in response to SD from the normoxic (A) to the ischemic (B) state is clear as discussed earlier (Mayevsky 2015). When we move to patients' monitoring, we want to use the NADH trace as an indicator of the physiological state of the patient or the brain in this example. As seen in the lower part, the two responses developed in the

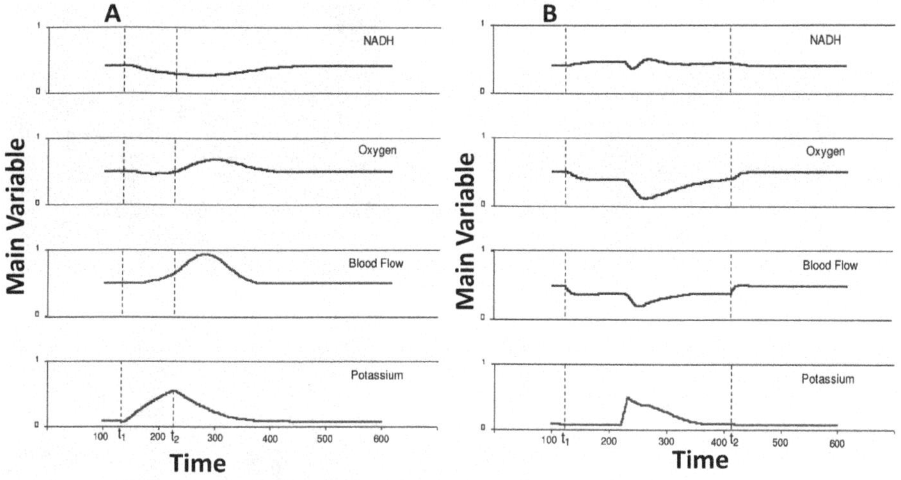

Fig. 8.4 (**a**) Simulated spreading depression, induced by an injection of potassium when it is at the resting level. The duration of the injection is from *t*1 to *t*2. (**b**) Simulated spreading depression induced after partial ischemia, resulting from a mild decrease in blood flow with a short injection of potassium. © Reprinted with kind permission of Springer Science + Business Media (Vatov et al. 2006)

neurosurgical patients (C and D) are not the same. There are similarities between A and C as well as between B and D. Therefore, we concluded that the brain of the patient in state C was probably normoxic, while in D the brain was partially ischemic. Indeed, this patient was deteriorating in time, and finally brain death was determined. This example demonstrates that the basic pathophysiological events developed in patients are very similar to those monitored in animal models.

8.3 From Single Parameter to Multiparametric Monitoring Approach

In order to understand the changes in NADH measured in vivo, it was necessary to move from a single parameter monitoring to a multiparametric approach. As described in Fig. 1.22, the NADH also represents the balance between oxygen consumption and supply. Therefore, the multiparametric approach provides a better understanding of the pathophysiological processes developed.

Figure 8.7 illustrates the improvement in the differentiation between various pathophysiological states developed in the brain. By using the two parameters, namely, cerebral blood flow (CBF) and NADH, the clinician will be able to diagnose the pathological state developed. Figures 2.1 and 2.4 demonstrate the options to

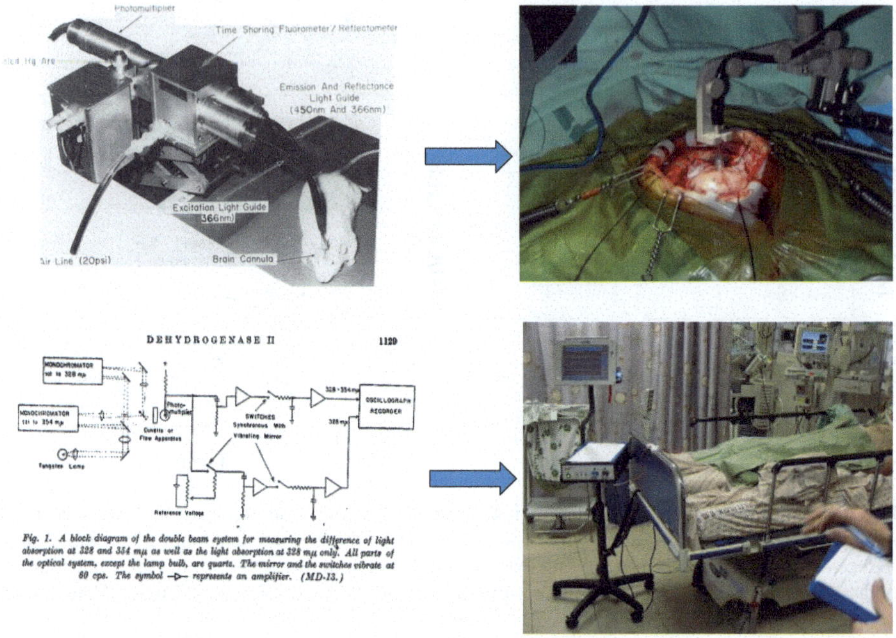

Fig. 8.5 Comparison between the monitoring of NADH in animal models (**a** and **c**) and the technological development in the clinical monitoring devices (**b** and **d**)

monitor the brain and other tissues as well. Figure 8.8 illustrates the responses of the two parameters monitored simultaneously from the brain in vivo under various perturbations. As can be seen, the inverse correlation between NADH and blood flow is not recorded under all conditions. For example, when the brain is exposed to hypoxia, the NADH increases together with the increase in CBF due to the compensation process in the brain.

The use of other physiological parameters in addition to NADH redox state enabled us to understand better the mechanism behind the development of pathophysiological states in various animal models used. This approach enables a clinician to diagnose better the development of pathological states in patients.

In Fig. 8.9, a few of the main events developed during CSD and ischemia were selected and shown in the center column. Under ischemia, the primary event is the decrease in CBF and O_2 delivery to the brain detected as an increase in the intramitochondrial NADH. The next step is the inhibition of the ion pumps, such as the Na^+K^+ ATPase, leading to the accumulation of K^+ in the extracellular space resulted in an ischemic depolarization phase. This situation will persist until the reperfusion starts and normal CBF was established.

In SD, the primary event is the depolarization, which propagates in the entire hemisphere. The Na^+K^+ ATPase is stimulated so that the net accumulation of K^+ in the extracellular space is smaller as compared to ischemia. The large increase in

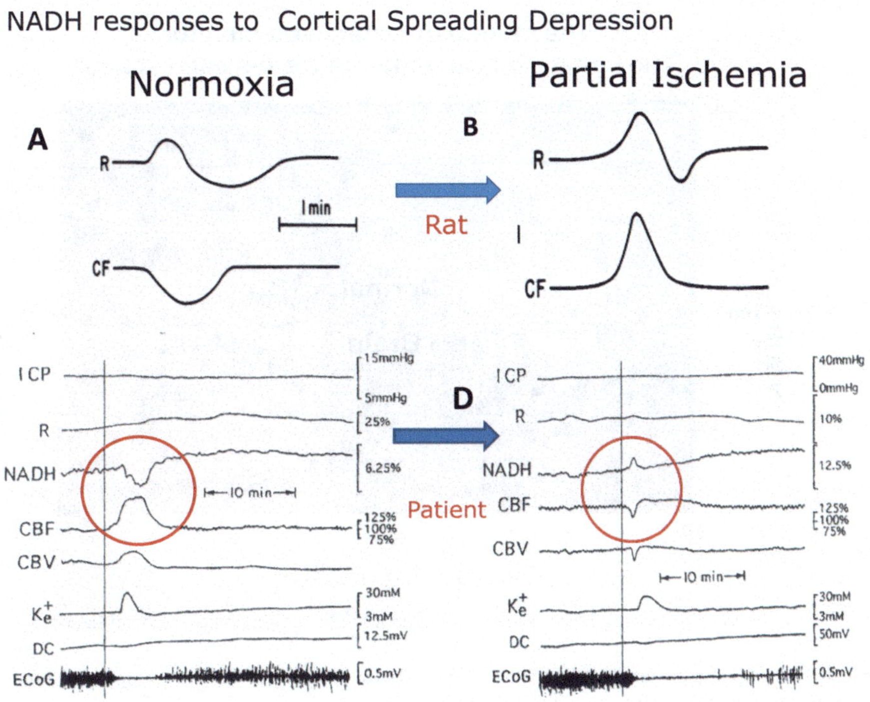

Fig. 8.6 The responses to cortical spreading depression measured in normoxic (**a**) and ischemic brain (**b**) in the rat. Parts **c** and **d** were measured in head-injured patient (see Chaps. 6 and 7)

energy utilization will be compensated by the increase in CBF seen in each of the SD waves. Although ischemia and SD have shared similar ionic disturbances, the differences between the two events are very large. The brain can tolerate SD waves without any damage created, while the damage created under ischemia is well documented.

8.4 Future Perspectives

It is now almost 60 years since the publication of Chance and Williams (1955a, b) summarizing various aspects of mitochondrial metabolic state in vitro. A few years later, in the early 1960s, the monitoring of Mitochondrial NADH in vivo was established in Philadelphia (Chance et al. 1962a) and many groups of scientists around the world adopted the approach and published 800–900 papers as presented and discussed in this book. My collaboration with the giant scientist, Professor Britton Chance, started in 1972 when I arrived to Philadelphia as a young post-doc fellow. The last time we met was during our stay in China (2009–2010) at the

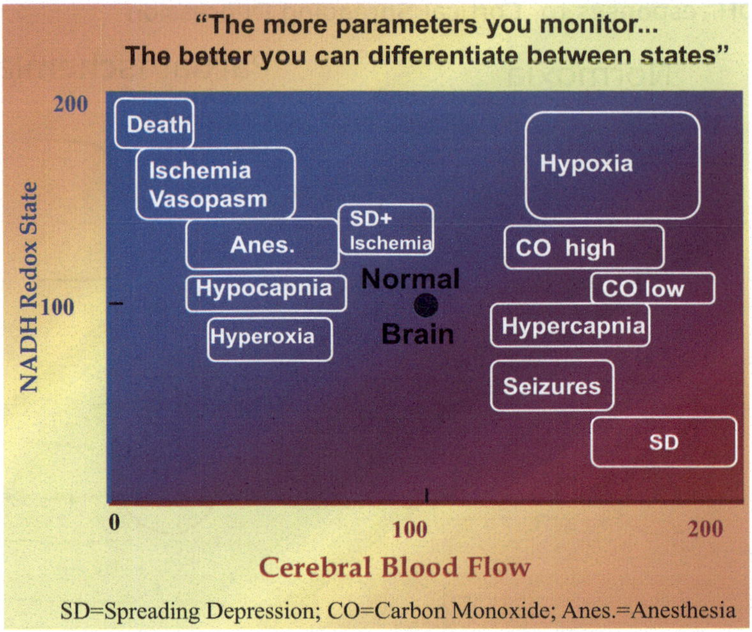

Fig. 8.7 Schematic presentation of the relationship between cerebral blood flow and mitochondrial NADH under various pathophysiological conditions

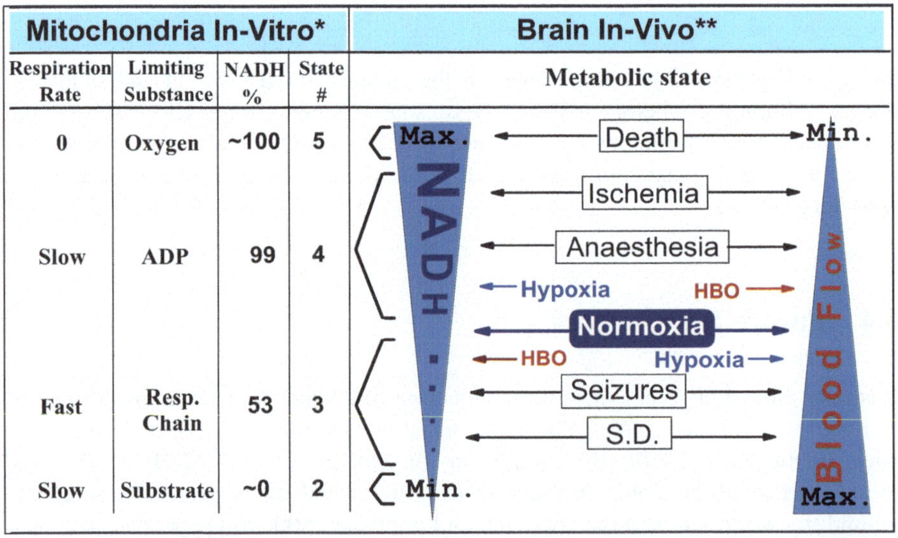

Fig. 8.8 Comparison between the classical definition of Chance and Williams regarding the metabolic state of isolated mitochondria and the metabolic state of the in vivo brain. *according to Chance and Williams (1955b). **according to Mayevsky (1984); HBO Hyperbaric oxygenation; SD Spreading depression. © Reprinted with kind permission of Springer Science+Business Media (Mayevsky et al. 1999)

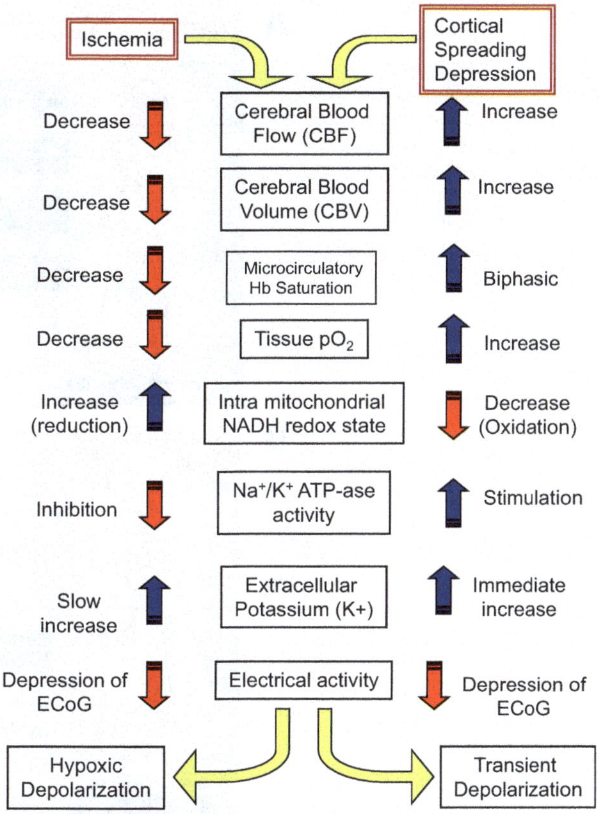

Fig. 8.9 Comparison between two pathophysiological situations, namely, ischemia and spreading depression (Mayevsky 1993)

Britton Chance Center for Biomedical Photonics established at the Wuhan National Laboratory for Optoelectronics (WNLO), Huazhong University of Science and Technology (HUST) Wuhan, China. During our many years of collaboration, he always mentioned the need to translate the technology and data collected in experimental animal studies into a practical clinical tool in order to help mankind. In the last decade, we were able to develop a practical tool that was approved by the FDA and was tested in preliminary clinical studies (Mayevsky et al. 2011). The concept of clinical monitoring of mitochondrial NADH as well as microcirculatory blood flow and oxygenation is well accepted. I have good reasons to believe that the multiparametric monitoring approach will be the standard tool in daily patient monitoring.

The next step in patient's monitoring device will be a very easy to use and practical tool as well as interpretation of the monitored parameters into a new Tissue Vitality Index. This new concept is presented in Fig. 8.10a, b.

Calibration of NADH fluorescence as well as the other monitored parameters in absolute units will be the next goal of the team dealing with patients monitoring.

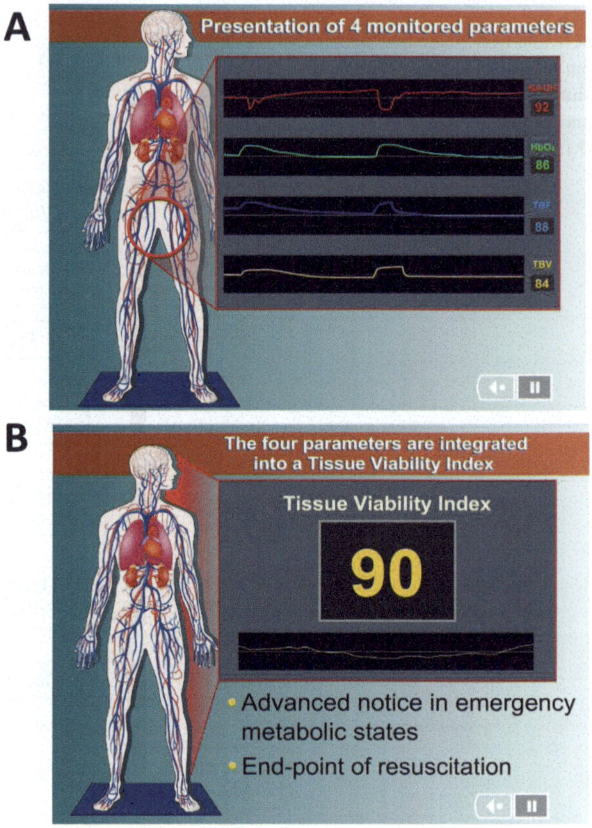

Fig. 8.10 (**a**) Schematic presentation of the four physiological parameters monitored in real time from the urethral wall in critical care patients. (**b**) Schematic demonstration of the Tissue Vitality Index (TVI) calculated from the parameters shown in Part A

References

Chance B, Jobsis F (1959) Changes in fluorescence in a frog sartorius muscle following a twitch. Nature 184:195–196

Chance B, Schoener B (1962) Correlation of oxidation-reduction changes of intracellular reduced pyridine nucleotide and changes in electro-encephalogram of the rat in anoxia. Nature 195:956–958

Chance B, Schoener B (1963) Control of oxidation-reduction state of NADH in the liver of anesthetized rats. In: Proceedings of the symposium on regulation of enzyme activity and synthesis in normal and neoplastic tissues, pp 169–185

Chance B, Williams GR (1954) Steady-state of reduced pyridine nucleotides in phosphorylating rat liver mitochondria. Am Soc Biol Chem 13:Abstract 633

Chance B, Williams GR (1955a) A method for the localization of sites for oxidative phosphorylation. Nature 176(4475):250–254

Chance B, Williams GR (1955b) Respiratory enzymes in oxidative phosphorylation (III- the steady state). J Biol Chem 217:409–427

Chance B, Williams GR (1956a) In: Nord FF (ed) The respiratory chain and oxidative phosphorylation. Interscience Publisher, Inc., New York, pp 65–134

Chance B, Williams GR (1956b) The respiratory chain and oxidative phosphorylation. In: Nord FF (ed) Advances in enzymology. Interscience Publisher, Inc, New York, pp 65–134

Chance B, Cohen P, Jobsis F, Schoener B (1962a) Intracellular oxidation-reduction states in vivo. Science 137:499–508

Chance B, Cohen P, Jobsis F, Schoener B (1962b) Localized fluorometry of oxidation-reduction states of intracellular pyridine nucleotide in brain and kidney cortex of the anesthetized rat. Science 136:325

Chance B, Legallias V, Schoener B (1962c) Metabolically linked changes in fluorescence emission spectra of cortex of rat brain, kidney and adrenal gland. Nature 195:1073–1075

Chance B, Schoener B, Fergusson JJ (1962d) In vivo induced oxidation by adrenocorticotrophic hormone of reduced pyridine nucleotide in the adrenal cortex of hypophysectomized rats. Nature 195:776–778

Chance B, Nioka S, Warren W, Yurtsever G (2005) Mitochondrial NADH as the bellwether of tissue O_2 delivery. Adv Exp Med Biol 566:231–242

Connelly CM, Chance B (1954) Kinetics of reduced pyridine nucleotides in stimulated frog muscle and nerve. Am Physiol Soc 13(1):29

Harrois A, Dupic L, Duranteau J (2011) Targeting the microcirculation in resuscitation of acutely unwell patients. Curr Opin Crit Care 17(3):303–307

Mayevsky A (1984) Brain NADH redox state monitored in vivo by fiber optic surface fluorometry. Brain Res Rev 7:49–68

Mayevsky A (1993) Biochemical and physiological activities of the brain as in vivo markers of brain pathology. In: Bernstein EF, Callow AD, Nicolaides AN, Shifrin EG (eds) Cerebral revascularization. Med-Orion Pub, pp 51–69

Mayevsky A (2015) Mitochondrial function in vivo evaluated by NADH fluorescence. Springer International

Mayevsky A, Rogatsky G (2007) Mitochondrial function in vivo evaluated by NADH fluorescence: from animal models to human studies. Am J Physiol Cell Physiol 292:C615–C640

Mayevsky A, Meilin S, Manor T, Zarchin N, Sonn J (1999) Optical monitoring of NADH redox state and blood flow as indicators of brain energy balance. Adv Exp Med Biol 471:133–140

Mayevsky A, Walden R, Pewzner E, Deutsch A, Heldenberg E, Lavee J, Tager S, Kachel E, Raanani E, Preisman S, Glauber V, Segal E (2011) Mitochondrial function and tissue vitality: bench-to-bedside real-time optical monitoring system. J Biomed Opt 16(067004): 067001–067021

Singer M (2007) Mitochondrial function in sepsis: acute phase versus multiple organ failure. Crit Care Med 35(9 Suppl):S441–S448

Vatov L, Kizner Z, Ruppin E, Meilin S, Manor T, Mayevsky A (2006) Modeling brain energy metabolism and function: a multiparametric monitoring approach. Bull Math Biol 68:275–291

Index

© The Editor(s) (if applicable) and The Author(s), under exclusive license to Springer Nature Switzerland AG 2025
A. Mayevsky et al., *Body Oxygen Homeostasis and Mitochondrial Function*,
https://doi.org/10.1007/978-3-031-94113-9